S0-CLZ-355

Methods in Enzymology

Volume 86

PROSTAGLANDINS AND ARACHIDONATE METABOLITES

METHODS IN ENZYMOLOGY

EDITORS-IN-CHIEF

Sidney P. Colowick   Nathan O. Kaplan

Methods in Enzymology

Volume 86

Prostaglandins and Arachidonate Metabolites

EDITED BY

William E. M. Lands

DEPARTMENT OF BIOLOGICAL CHEMISTRY
UNIVERSITY OF ILLINOIS MEDICAL CENTER
CHICAGO, ILLINOIS

William L. Smith

DEPARTMENT OF BIOCHEMISTRY
MICHIGAN STATE UNIVERSITY
EAST LANSING, MICHIGAN

1982

ACADEMIC PRESS

A Subsidiary of Harcourt Brace Jovanovich, Publishers

New York London
Paris San Diego San Francisco São Paulo Sydney Tokyo Toronto

ACADEMIC PRESS, INC.
111 Fifth Avenue, New York, New York 10003

United Kingdom Edition published by
ACADEMIC PRESS, INC. (LONDON) LTD.
24/28 Oval Road, London NW1 7DX

Library of Congress Cataloging in Publication Data
Main entry under title:

Prostaglandins and arachidonate metabolites.

(Methods in enzymology ; v. 86)
Includes bibliographical references and index.
1. Arachidonic acid--Research--Methodology.
2. Prostaglandins--Research--Methodology. I. Lands, William E. M., Date. II. Smith, William L.
III. Series.
QP601.M49 vol. 86 [QP752.A7] 574.19'25s 82-6791
ISBN 0-12-181986-8 [612'.405] AACR2

PRINTED IN THE UNITED STATES OF AMERICA

82 83 84 85 9 8 7 6 5 4 3 2 1

Table of Contents

Section I. Enzymes and Receptors: Purification and Assay

Section II. Immunochemical Assays of Enzymes and Metabolites

Section III. Substrates, Reagents, and Standards

Section IV. General Separation Procedures

Section V. Gas Chromatography–Mass Spectrometry of Prostaglandin Derivatives

Section VI. Biological Methods

Contributors to Volume 86

Article numbers are in parentheses following the names of contributors.
Affiliations listed are current.

ARIFF ALLY (61), *National Research Council, Laboratory of Ecotoxicology, Ottawa, Canada*

K. FRANK AUSTEN (37), *Department of Medicine, Harvard Medical School, and Department of Rheumatology and Immunology, Brigham and Women's Hospital, Boston, Massachusetts 02115*

MARTA I. AVELDAÑO (62), *Department of Physiological Chemistry, Ohio State University College of Medicine, Columbus, Ohio 43210*

KERSTIN BERNSTRÖM (5), *Department of Physiological Chemistry, Karolinska Institutet, S-10401 Stockholm, Sweden*

BISTRA BIZZINI (38), *INSERM* U. 207, *Institut Pasteur, Uria, 75015 Paris, France*

ALAN R. BRASH (68, 69), *Department of Pharmacology, Vanderbilt University, Nashville, Tennessee 37232*

M. BUYTENHEK (9, 13), *TNO–Gaubius Instituut, 2313 AD Leiden, The Netherlands*

DAVID GUEY-BIN CHANG (23), *Department of Immunotherapy, Wadley Institute of Molecular Medicine, Dallas, Texas 75235*

E. CHRIST-HAZELHOF (12), *Unilever Research Laboratorium, 3130 AC Vlaardingen, The Netherlands*

KEITH L. CLAY (64), *Department of Pharmacology, University of Colorado Medical Center, Denver, Colorado 80262*

CLINTON N. CORDER (22), *Department of Pharmacology, Oral Roberts University, Tulsa, Oklahoma 74171*

E. J. COREY (37), *Department of Chemistry, Harvard University, Cambridge, Massachusetts 02134*

JEFFREY S. DAY (34), *Department of Biochemistry, Michigan State University, East Lansing, Michigan 48824*

DAVID L. DEWITT (34, 35), *Department of Biochemistry, Michigan State University, East Lansing, Michigan 48824*

FERNAND DRAY (38, 42), *INSERM* U. 207, *Institut Pasteur, Uria, 75015 Paris, France*

N. S. EDWARDS (65, 71), *Research Institute, The Hospital for Sick Children, Toronto M5G 1X8, Canada*

THOMAS E. ELING (61), *National Institute of Environmental Health Sciences, Research Triangle Park, North Carolina 27713*

PIERRE FALARDEAU (69), *Institut de Recherche Cliniques de Montreal, Montreal, Quebec H2W 1RF, Canada*

SANDRA F. FALKENHEIN (54, 76), *Howard Hughes Medical Institute Laboratory, and Department of Internal Medicine, Division of Allergy and Immunology, Washington University School of Medicine, St. Louis, Missouri 63110*

F. A. FITZPATRICK (41, 43), *Pharmaceutical Research and Development, The Upjohn Company, Kalamazoo, Michigan 49008*

RODERICK J. FLOWER (14, 59), *Department of Prostaglandin Research, Wellcome Research Laboratories, Beckenham, Kent BR3 3BS, England*

CLAUDIO GALLI (74), *Institute of Pharmacology and Pharmacognosy, University of Milan, Milan 20129, Italy*

JOHN A. GAUGER (34), *Department of Biochemistry, Michigan State University, East Lansing, Michigan 48824*

JON M. GERRARD (75), *Department of Pediatrics, University of Manitoba, Winnipeg, Manitoba R3E 0V9, Canada*

GUSTAV GRAFF (49, 50), *Department of Biological Chemistry, University of Illinois Medical Center, Chicago, Illinois 60612*

ELISABETH GRANSTRÖM (43, 44, 45, 60), *Department of Physiological Chemistry, Karolinska Institutet, S-10401 Stockholm, Sweden*

SVEN HAMMARSTRÖM (5, 31), *Department of Physiological Chemistry, Karo-*

linska Institutet, S-10401 Stockholm, Sweden

HARALD S. HANSEN (26), *Biochemical Laboratory, Royal Danish School of Pharmacy, Copenhagen 2100, Denmark*

TIMOTHY HARPER (4), *Department of Pharmacology, University of Colorado Medical Center, Denver, Colorado 80262*

OSAMU HAYAISHI (11, 24), *Department of Medical Chemistry, Kyoto University Faculty of Medicine, Sakyo-ku, Kyoto 606, Japan*

YOKO HAYASHI (39), *Department of Biochemistry, Tokushima University School of Medicine, Kuramoto-cho, Tokushima 770, Japan*

LLOYD A. HORROCKS (62), *Department of Physiological Chemistry, Ohio State University College of Medicine, Columbus, Ohio 43210*

WALTER C. HUBBARD (47, 67, 72), *Department of Pharmacology, Vanderbilt University School of Medicine, Nashville, Tennessee 37232*

MARY M. HUBER (54, 76), *Howard Hughes Medical Institute Laboratory, and Department of Internal Medicine, Division of Allergy and Immunology, Washington University School of Medicine, St. Louis, Missouri 63110*

MAKOTO INOUE (48), *Department of Biological Chemistry, Faculty of Pharmaceutical Sciences, Nagoya City University, 3-1 Tanabedori, Mizuhoku, Nagoya 467, Japan*

BARBARA A. JAKSCHIK (4), *Department of Pharmacology, Washington University Medical School, St. Louis, Missouri 63110*

JOSEPH JARABAK 20, 25, 27), *Department of Medicine, University of Chicago, Chicago, Illinois 60637*

ROY A. JOHNSON (57), *Research Laboratories, The Upjohn Company, Kalamazoo, Michigan 49001*

HANS KINDAHL (43, 44, 45), *Department of Obstetrics and Gynecology, Swedish University of Agricultural Sciences, S-750 07 Uppsala, Sweden*

JEFFREY M. KORFF (25), *Hypertension–Endocrine Branch, National Heart, Lung, and Blood Institute, National Institutes of Health, Bethesda, Maryland 20205*

AKIRA KUMAGAI (3), *The Second Department of Internal Medicine, School of Medicine, Chiba University, Chiba City 280, Japan*

WILLIAM E. M. LANDS (10), *Department of Biological Chemistry, University of Illinois Medical Center, Chicago, Illinois 60612*

LAWRENCE LEVINE (36, 37), *Department of Biochemistry, Brandeis University, Waltham, Massachusetts 02254*

ROBERT A. LEWIS (37), *Department of Medicine, Harvard Medical School, and Department of Rheumatology and Immunology, Brigham and Women's Hospital, Boston, Massachusetts 02115*

RICHARD L. MAAS (70), *Department of Pharmacology, Vanderbilt University School of Medicine, Nashville, Tennessee 37232*

JACQUES MACLOUF (40, 73), *Unité INSERM 150, ERA 335 CNRS, Hôpital Lariboisière, 75475 Paris, France*

RONALD L. MAGOLDA (52), *Central Research Division, E. I. du Pont Experimental Station, Wilmington, Delaware 19899*

PHILIP W. MAJERUS (2), *Departments of Internal Medicine and Biochemistry, Division of Hematology–Oncology, Washington University School of Medicine, St. Louis, Missouri 63110*

SUZANNE MAMAS (38), *INSERM* U. 207, *Institut Pasteur, Uria, 75015 Paris, France*

W. RODNEY MATHEWS (53), *Department of Chemistry, Massachusetts Institute of Technology, Cambridge, Massachusetts 02139*

S. B. MITRA (30), *Department of Obstetrics/Gynecology, University of Louisville School of Medicine, Louisville, Kentucky 40292*

P. MOONEN (13), *Academisch Ziekenhuis, 2333 AA Leiden, The Netherlands*

RICHARD A. MORGAN (36), *Department of Biology, The Johns Hopkins University, Baltimore, Maryland 21218*

H. R. MORRIS (55), *Department of Biochemistry, Imperial College of Science and Technology, London SW7, England*

ROBERT C. MURPHY (4, 53, 64), *Department of Pharmacology, University of Colorado Medical Center, Denver, Colorado 80262*

SHUH NARUMIYA (6), *Department of Medical Chemistry, Kyoto University Faculty of Medicine, Sakyo-ku, Kyoto 606, Japan*

MARY NETTLEMAN (72), *Pulmonary Circulation Center, Departments of Medicine and Pharmacology, Vanderbilt University School of Medicine, Nashville, Tennessee 37232*

K. C. NICOLAOU (52), *Department of Chemistry, University of Pennsylvania, Philadelphia, Pennsylvania 19104*

D. H. NUGTEREN (7, 12, 13), *Unilever Research Laboratorium, 3130 AC Vlaardingen, The Netherlands*

MARTIN L. OGLETREE (72), *Pulmonary Circulation Center, Departments of Medicine and Pharmacology, Vanderbilt University School of Medicine, Nashville, Tennessee 37232*

HARUMI OKUYAMA (48), *Department of Biological Chemistry, Faculty of Pharmaceutical Sciences, Nagoya City University, 3-1 Tanabedori, Mizuhoku, Nagoya 467, Japan*

LARS ÖRNING (5), *Department of Physiological Chemistry, Karolinska Institutet, S-10401 Stockholm, Sweden*

C. R. PACE-ASCIAK (65, 71), *Research Institute, The Hospital for Sick Children, Toronto M5G 1X8, Canada*

CHARLES W. PARKER (54, 76), *Howard Hughes Medical Institute Laboratory, and Department of Internal Medicine, Division of Allergy and Immunology, Washington University School of Medicine, St. Louis, Missouri 63110*

MARK A. PHILLIPS (47), *Department of Pharmacology, Vanderbilt University School of Medicine, Nashville, Tennessee 37232*

PRISCILLA J. PIPER (55), *Department of Pharmacology, Institute of Basic Medical Sciences, Royal College of Surgeons of England, London WC2A 3PN, England*

WILLIAM S. POWELL (28, 58, 63), *Endocrine Laboratory, Royal Victoria Hospital, and Department of Medicine, McGill University, Montreal Quebec H3A 1A1, Canada*

STEPHEN M. PRESCOTT (2), *Department of Internal Medicine and Biochemistry, Division of Hematology–Oncology, Washington University School of Medicine, St. Louis, Missouri 63110*

GIORGIO RACAGNI (74), *Institute of Pharmacology and Pharmacognosy, University of Milan, Milan 20129, Italy*

CH. V. RAO (30), *Department of Obstetrics/Gynecology, University of Louisville School of Medicine, Louisville, Kentucky 40292*

MICHEL RIGAUD (73), *Service de Biochimie, CHU Dupuytren, 87031 Limoges, France*

SUSAN ERIKA RITTENHOUSE (1), *Departments of Medicine and Biochemistry, Brigham and Women's Hospital, Harvard Medical School, Boston, Massachusetts 02115*

L. JACKSON ROBERTS II (66, 70), *Departments of Pharmacology and Medicine, Vanderbilt University School of Medicine, Nashville, Tennessee 37232*

THOMAS E. ROLLINS (32), *Merck Institute for Therapeutic Research, Rahway, New Jersey 07065*

GERALD J. ROTH (33, 51), *Department of Medicine, University of Connecticut Health Center, Farmington, Connecticut 06032*

JOHN A. SALMON (6, 14, 59), *Department of Prostaglandin Research, Wellcome Research Laboratories, Beckenham, Kent BR3 3BS, England*

SHANKAR K. SANKARAPPA (46), *Department of Physiological Chemistry, Ohio State University, Columbus, Ohio 43210*

KENNETH SCHLESINGER (72), *Pulmonary*

Circulation Center, Departments of Medicine and Pharmacology, Vanderbilt University School of Medicine, Nashville, Tennessee 37232

TAKAO SHIMIZU (11, 24), *Department of Medical Chemistry, Kyoto University Faculty of Medicine, Sakyo-ku, Kyoto 606, Japan*

ADELAIDE M. SIEGL (29), *PRAT Fellow, Laboratory of Bioorganic Chemistry, National Institute of Arthritis, Diabetes, and Digestive and Kidney Diseases, National Institutes of Health, Bethesda, Maryland 20205*

CHARLES J. SIH (15), *School of Pharmacy, University of Wisconsin, Madison, Wisconsin 53706*

WILLIAM L. SMITH (32, 34, 35), *Department of Biochemistry, Michigan State University, East Lansing, Michigan 48824*

HOWARD SPRECHER (46, 62), *Department of Physiological Chemistry, Ohio State University College of Medicine, Columbus, Ohio 43210*

RANDALL G. STEHLE (56), *Pharmacy Research, The Upjohn Company, Kalamazoo, Michigan 49001*

DOUGLASS F. TABER (47, 70), *Departments of Pharmacology and Chemistry, Vanderbilt University School of Medicine, Nashville, Tennessee 37232*

HSIN-HSIUNG TAI (15, 17, 18, 21, 23), *College of Pharmacy, University of Kentucky, Lexington, Kentucky 40536*

BETH TAINER (61), *National Institute of Environmental Health Sciences, Research Triangle Park, North Carolina 27713*

G. W. TAYLOR (55), *Department of Biochemistry, Imperial College of Science and Technology, London SW7, England*

J. R. TIPPINS (55), *Department of Pharmacology, Institute of Basic Medical Sciences, Royal College of Surgeons of England, London WC2A 3PN, England*

BRADLEY G. TITUS (10), *Medical School, The University of Michigan, Ann Arbor, Michigan 48104*

HIDEKADO TOKUMOTO (24), *Department of Medical Chemistry, Kyoto University Faculty of Medicine, Sakyo-ku, Kyoto 606, Japan*

HISAO TOMIOKA (3), *The Second Department of Internal Medicine, School of Medicine, Chiba University, Chiba City 280, Japan*

F. J. G. VAN DER OUDERAA (9), *Unilever Research, Bebington, Wirral, Merseyside L63 3JW, England*

MIKE VAN ROLLINS (62), *Department of Pharmacology, University of Colorado Health Sciences Center, Denver, Colorado 80262*

ROBERT WARNOCK (61), *Department of Obstetrics and Gynecology, University of Arizona, Tucson, Arizona*

KIKUKO WATANABE (24), *Department of Medical Chemistry, Kyoto University Faculty of Medicine, Sakyo-ku, Kyoto 606, Japan*

PATRICK Y-K. WONG (19), *Department of Pharmacology, New York Medical College, Valhalla, New York 10595*

SHOZO YAMAMOTO (8, 11, 16, 39), *Department of Biochemistry, Tokushima University School of Medicine, Kuramoto-cho, Tokushima 770, Japan*

TANIHIRO YOSHIMOTO (16), *Department of Biochemistry, Tokushima University School of Medicine, Kuramoto-cho, Tokushima 770, Japan*

BARBARA YUAN (18), *Wyeth Laboratories, Philadelphia, Pennsylvania 19101*

Preface

Our intent in organizing this volume was to provide in one source a detailed description of the major techniques currently available for the study of significant arachidonic acid metabolites. In editing the contributions, we tried to make certain that adequate experimental detail was included. There are often different approaches that can be used to achieve a given result. One may be better suited to one laboratory than another; moreover, a seemingly minor detail may make the difference between success and failure in adapting a procedure. Therefore, in several instances, apparently redundant descriptions of methodologies (e.g., prostaglandin endoperoxide preparation) have been included.

While we accept responsibility for the final selection of topics, we would like to thank many of the contributors for their advice in this selection process. We also received important suggestions, comments, and advice from a number of other investigators whom we also wish to acknowledge, including Drs. Philip Needleman, John E. Pike, Robert R. Gorman, Lawrence J. Marnett, J. Bryan Smith, Aaron Marcus, and J. Throck Watson.

The organization, correspondence, and editing necessary to bring this effort to fruition would not have been possible without the skillful and cheerful assistance of Ms. Susan Uselton and Ms. Stephanie Lewis. We owe them, as well as the staff of Academic Press, a special debt of thanks for their help.

William E. M. Lands
William L. Smith

METHODS IN ENZYMOLOGY

EDITED BY

Sidney P. Colowick and Nathan O. Kaplan

VANDERBILT UNIVERSITY
SCHOOL OF MEDICINE
NASHVILLE, TENNESSEE

DEPARTMENT OF CHEMISTRY
UNIVERSITY OF CALIFORNIA
AT SAN DIEGO
LA JOLLA, CALIFORNIA

METHODS IN ENZYMOLOGY

EDITORS-IN-CHIEF

Sidney P. Colowick   Nathan O. Kaplan

VOLUME LX. Nucleic Acids and Protein Synthesis (Part H)
Edited by KIVIE MOLDAVE AND LAWRENCE GROSSMAN

VOLUME 61. Enzyme Structure (Part H)
Edited by C. H. W. HIRS AND SERGE N. TIMASHEFF

VOLUME 62. Vitamins and Coenzymes (Part D)
Edited by DONALD B. MCCORMICK AND LEMUEL D. WRIGHT

VOLUME 63. Enzyme Kinetics and Mechanism (Part A: Initial Rate and Inhibitor Methods)
Edited by DANIEL L. PURICH

VOLUME 64. Enzyme Kinetics and Mechanism (Part B: Isotopic Probes and Complex Enzyme Systems)
Edited by DANIEL L. PURICH

VOLUME 65. Nucleic Acids (Part I)
Edited by LAWRENCE GROSSMAN AND KIVIE MOLDAVE

VOLUME 66. Vitamins and Coenzymes (Part E)
Edited by DONALD B. MCCORMICK AND LEMUEL D. WRIGHT

VOLUME 67. Vitamins and Coenzymes (Part F)
Edited by DONALD B. MCCORMICK AND LEMUEL D. WRIGHT

VOLUME 68. Recombinant DNA
Edited by RAY WU

VOLUME 69. Photosynthesis and Nitrogen Fixation (Part C)
Edited by ANTHONY SAN PIETRO

VOLUME 70. Immunochemical Techniques (Part A)
Edited by HELEN VAN VUNAKIS AND JOHN J. LANGONE

VOLUME 71. Lipids (Part C)
Edited by JOHN M. LOWENSTEIN

VOLUME 72. Lipids (Part D)
Edited by JOHN M. LOWENSTEIN

VOLUME 73. Immunochemical Techniques (Part B)
Edited by JOHN J. LANGONE AND HELEN VAN VUNAKIS

Section I

Enzymes and Receptors: Purification and Assay

[1] Preparation of Selectively Labeled Phosphatidylinositol and Assay of Phosphatidylinositol-Specific Phospholipase C

By SUSAN ERIKA RITTENHOUSE[1]

Phospholipase C (PLC) is a phosphodiesterase that hydrolyzes glycerophospholipid, yielding diglyceride and the corresponding phosphorylated base. Phosphatidylinositol-specific phospholipase C (EC 3.1.4.10) displays a subset of this activity, acting only upon phosphatidylinositol and liberating, in many cases, myoinositol 1:2-cyclic phosphate in addition to myoinositol 1-phosphate and 1,2-diacylglycerol. The cyclic species is labile and can be converted completely to myoinositol phosphate in acidic media. In all cases described to date, the phosphatidylinositol-specific PLC, which is active at or near neutrality, is completely dependent upon Ca^{2+}; however, the pH at which optimal activity is observed can vary with the source of the enzyme. Phosphatidylinositol-specific PLC is ubiquitous in mammalian tissues. Evidence is accumulating rapidly that it is activated specifically in cells that have been stimulated by physiological agonists, and it is possible therefore that phosphatidylinositol-specific PLC plays a role in the promotion of special physiological events, such as secretion.

Human platelets are among the stimulatable cells that exhibit phosphatidylinositol-specific PLC activity.[2,3] Platelets hydrolyze phosphatidylinositol, produce diacylglycerol, and secrete stored substances in response to thrombin.[2] This chapter describes a procedure for the preparation of selectively labeled phosphatidylinositol substrate and the utilization of this substrate in assaying phosphatidylinositol-specific PLC obtained from human platelets. Platelets are a rich source of the soluble form of the enzyme, and homogeneous preparations of these cells can be obtained with minimal manipulation. The assay employs phosphatidylinositol labeled on myoinositol and is based upon the enzymic conversion of the labeled material from a lipid-soluble form (phosphatidylinositol) to a water-soluble form (myoinositol phosphate), which can be quantitated by liquid scintillation spectrophotometry.

[1] Also known as Susan Rittenhouse-Simmons. Dr. Rittenhouse is an Established Investigator of the American Heart Association.

[2] S. Rittenhouse-Simmons, *J. Clin. Invest.* **63,** 580 (1979).

[3] G. Mauco, H. Chap, and L. Douste-Blazy, *FEBS Lett.* **100,** 367 (1979).

ISBN 0-12-181986-8

Generation of [^{3}H]Myoinositol Phosphatidylinositol

Preparation of Microsomes

Fresh chicken liver microsomes are utilized as a source for myoinositol-incorporating enzyme; however, other sources of liver, if more convenient, may be used. Generally, mammalian phosphatidylinositol is rich in stearic acid and arachidonic acid and thus approximates the physiological substrate of the enzyme,[4,5] although a preferential hydrolysis of phosphatidylinositol containing these fatty acid substituents has not been demonstrated. Higher plants and yeasts contain inositol in sphingolipids and phosphatidylinositol mannosides,[6,7] and therefore we have avoided using yeast for generating labeled phosphatidylinositol.[3]

Fresh chicken livers (30 g, blotted, dry weight) are freed of bile sacs, major blood vessels, and fat and washed in cold NaCl (0.85%) to remove excess blood. The chicken livers are homogenized in a Waring blender in 120 ml of 0.25 *M* sucrose–1 m*M* EDTA–50 m*M* Trizma, pH 7.4, at 4°. Homogenization occurs during two 60-sec periods. Homogenate, minus foam, is spun at 11,200 rpm for 15 min at 4° in a fixed-angle 30-rotor (Beckman L2-65) ultracentrifuge to remove unbroken cells and nuclear and mitochondrial material. Supernatants are pooled and slowly titrated to pH 5 with 1 *M* HCl. The resulting precipitated microsomes are spun as above for 25 min, and the pellets are resuspended in an equal volume of 0.25 *M* sucrose and titrated to pH 7.6 with 1 *M* Trizma. The final volume of suspension is about 16 ml.

Labeling of Microsomal Phosphatidylinositol with [^{3}H]Myoinositol

Reagents

- [(*N*)-2-^{3}H]myoinositol, 12.5 Ci/mmol, 1 mCi/ml in ethanol–$H_2$0, 9:1 (New England Nuclear, Boston, Massachusetts)
- HEPES–NaOH, 100 m*M*, pH 7.4
- $MnCl_2$, 100 m*M*
- Myoinositol, 0.5 *M*
- Chloroform
- KCl, 2 *M*
- Chloroform–methanol–conc. HCl (10:10:0.06, v/v/v)
- HCl, 1 *M*

All solvents are of spectroquality.

[4] D. A. White, *in* "Form and Function of Phospholipids" (J. N. Hawthorne, G. B. Ansell, and R. M. C. Dawson, eds.), *BBA Libr.* **3,** 475 (1973).

[5] R. H. Michell, *Biochim. Biophys. Acta* **415,** 81 (1975).

[6] W. Tanner, *Ann. N. Y. Acad. Sci.* **165,** 726 (1969).

[7] H. E. Carter, D. R. Strobach, and J. N. Hawthorne, *Biochemistry* **8,** 383 (1969).

The labeling procedure is based upon a method described by Paulus and Kennedy.[8] To each of eight clean 50-ml conical centrifuge tubes are added 40 μl of [^{3}H]myoinositol. The solvent is evaporated under nitrogen flow, and 200 μl of HEPES buffer plus 20 μl of $MnCl_2$ are added. Finally, 1.8 ml of the microsomal suspension (see Preparation of Microsomes) is added to each conical tube. The mixtures are combined well by vortex agitation and incubated for 60 min at 37° in a shaking water bath. Incubations end with the addition of 7.7 ml of chloroform–methanol (1:2) and vortexing. After 15 min, 2.1 ml of chloroform, 2.0 ml of 2 *M* KCl, and 100 μl of 0.5 *M* myoinositol are added; the conical tubes are stoppered, and agitated to mix well. After a centrifugation at 900 rpm for 10 min at room temperature to facilitate separation of the phases, the upper aqueous phase is removed and discarded. The lower phases are filtered through Whatman No. 1 paper, which is rinsed with 2 ml of chloroform–methanol (2:1) for each sample. Filtrates are combined and divided among four clean 50-ml conical tubes and concentrated by rotoevaporation under vacuum.[9] To each of the residues is added 6.0 ml of chloroform–methanol–HCl (10:10:0.06, v/v/v), 1.7 ml of 1 *M* HCl, 100 μl of 0.5 *M* myoinositol, and 0.9 ml of H_20. Samples are mixed well, and the phases are defined by centrifugation. The lower (chloroform) phases are removed and concentrated by rotoevaporation in a clean 50-ml conical centrifuge tube. Each of the four residues is dissolved in 1 ml of chloroform.

Purification of [^{3}H]Phosphatidylinositol

Removal of Neutral Lipids. Four glass columns (3 cm in diameter) are packed with a slurry of silicic acid (Unisil, 100–200 mesh) in chloroform, 30 g of silicic acid per column. The silicic acid should have been activated by heating at 110° for 2 days to remove H_2O. The silicic acid columns are rinsed with five column-volumes of chloroform, after which the four [^{3}H]phosphatidylinositol-containing lipid mixtures are applied to the top of each, along with three 200-μl rinses of chloroform. Neutral lipids are removed by elution with 170 ml of chloroform. The phospholipid fraction, containing [^{3}H]phosphatidylinositol, is eluted with 175 ml of methanol and collected in 250-ml round-bottom flasks. The chloroform eluate contains less than 1% of the total radiolabel. The methanol eluates are concentrated by rotoevaporation and transferred, with chloroform rinses, to a clean 50-ml conical flask.

[8] H. Paulus and E. P. Kennedy, *J. Biol. Chem.* **235,** 1303 (1960).

[9] Concentration by rotoevaporation is achieved using apparatus similar to that described by M. Kates *in* "Techniques of Lipidology: Isolation, Analysis, and Identification of Lipids" (T. S. Work and E. Work, eds.), p. 343. Elsevier North-Holland, New York, 1972.

Precipitation of Cephalins.[10] The [^{3}H]phosphatidylinositol–phospholipid extract is brought to dryness by rotoevaporation in a 50-ml conical flask. To the reddish brown oily concentrate are added 700 μl of chloroform, which solubilizes the lipid. Then, slowly, 3.5 ml of ice-cold ethanol are added, with mixing. Some precipitate should form almost immediately. Nitrogen is run over the mixture for about 30 sec. The conical flask is then stoppered and refrigerated overnight. The precipitated [^{3}H]phosphatidylinositol-rich phospholipid is separated from the dark orange phosphatidylcholine-rich soluble material by centrifugation and removal of the supernatant with a Pasteur pipette. The residue is then dissolved in 12 ml of chloroform. This procedure removes about 70–80% of the phosphatidylcholine and sphingomyelin in the phospholipid extract.

In order to define the specific activity of [^{3}H]phosphatidylinositol and to quantitate the contaminating phospholipid remaining in the preparation of [^{3}H]phosphatidylinositol at this stage, 30 μl (approximately 40 nmol) of the chloroform solution are applied to each of four silicic acid-impregnated papers (Reeve-Angel; Whatman SG 81). The phospholipids are resolved by two-dimensional chromatography, as described by Wuthier,[11] and made visible with 0.05% Rhodamine 6G spray. Two of the [^{3}H]phosphatidylinositol spots are cut out and counted by scintillation spectrophotometry in 1 ml of H_2O + 10 ml of Hydrofluor (National Diagnostics), with suitable correction for any quenching. Additional radiolabel may also be present as lysophosphatidylinositol. Two [^{3}H]phosphatidylinositol spots and other resolved lipids are digested for phosphorus analysis as described by Wuthier.[11] At this stage, we have found 96% of the total radiolabel in phosphatidylinositol, 2% in lysophosphatidylinositol, and 2% at the origin (including myoinositol and polyphosphoinositides).[12] The specific activity of such preparations of phosphatidylinositol is about 5000 dpm/nmol.

[10] This method is based upon procedures described by J. Folch, *J. Biol. Chem.* **146,** 35 (1942).

[11] R. E. Wuthier, *J. Lip. Res.* **7,** 544 (1966). Of the various chromatographic solvent systems described, the most useful for us has been: direction 1, chloroform–methanol–diisobutylketone–acetic acid–H_2O (45:15:30:20:4, by volume); direction 2, chloroform–methanol–diisobutylketone–pyridine–H_2O (30:25:25:35:8, by volume).

[12] Contamination by other phospholipids at this stage is substantial. On a molar basis, other phospholipids are present in a ratio of phosphatidylcholine:phosphatidylethanolamine:phosphatidylserine:phosphatidylinositol = 1.0:1.8:0.5:1.0. Phosphatidic acid is also detectable in trace amounts. We have observed in preliminary studies that phosphatidylcholine is inhibitory for phosphatidylinositol-specific PLC. Other investigators [R. F. Irvine, N. Hemington, and R. M. C. Dawson, *Eur. J. Biochem.* **99,** 525 (1979)] have observed similar inhibitory effects on the phosphatidylinositol-specific PLC of rat brain, as well as stimulatory effects of acidic phospholipids and fatty acid on this enzyme.

Isolation of [^{3}H]Phosphatidylinositol by High-Performance Liquid Chromatography (HPLC). The final stage of purification of [^{3}H]phosphatidylinositol is undertaken using a modification by Patton *et al.* of a procedure described by Geurts van Kessel *et al.*[13] The use of HPLC eliminates the need for exposure of labeled phospholipid to the strongly basic or acidic modifying conditions that characterize preparative thin-layer chromatographic systems. A LiChrospher Si-100, 10 μm column (Hibar II, 250 mm × 4.6 mm, E. Merck) is employed in conjunction with a Waters high-performance liquid chromatograph Model ALC-GPC 244 (Waters Associates, Medford, Massachusetts) equipped with a Model 6000A pump, U6K injector, and a Model 450 variable-wavelength UV detector monitoring at 205 nm. The column is washed thoroughly, in succession, with HPLC (Fisher) grade chloroform (60 ml), acetone (60 ml), methanol (120 ml), and ethanol (120 ml) prior to equilibration with 120 ml of eluting solvent. The solvent consists of 90 ml of H_2O, 0.60 ml of 1 *M* potassium phosphate, pH 7.0, 735 ml of 2-propanol, 550 ml of *n*-hexane, and 150 ml of ethanol. After precipitation of salt overnight, the solvent is filtered through a 0.5 μm filter (Millipore) and deaerated briefly under vacuum. All solvents should be HPLC grade.

Approximately 0.6 mg of [^{3}H]phosphatidylinositol-containing lipid are dissolved in 60 μl of *n*-hexane–2-propanol–H_2O, 6:8:1 (v/v/v) and injected onto the column. Flow is controlled at 0.5 ml/min. After 15 min, flow rate is increased to 1.0 ml/min. [^{3}H]Phosphatidylinositol begins to appear in the eluate after about 25 min and continues to be eluted for another 9 min. Forty minutes after injection, flow is increased to 1.4 ml/min. The order of elution of phospholipid is the same as that described by Geurts van Kessel *et al.*[13] Fractions corresponding to material absorbing at 205 nm are monitored by scintillation spectrophotometry in H_2O:Hydrofluor (1:10 ml). The purity of the [^{3}H]phosphatidylinositol fraction obtained by this method, when checked by thin-layer chromatography, should be 100%. Successive applications to the column of the partially purified [^{3}H]phosphatidylinositol obtained after ethanol precipitation can be made once all lipid has been eluted from the column (approximately 80 min after sample injection). Solvent is evaporated from

[13] W. S. M. Geurts van Kessel, W. M. A. Hax, R. A. Demel, and J. deGier, *Biochim. Biophys. Acta* **486,** 524 (1977). The modification is described by G. M. Patton, J. M. Fasulo, and S. J. Robins in *J. Lip. Res.,* in press (1982).

[14] Should one wish to conserve radiolabeled material, the [^{3}H]phosphatidylinositol can be diluted at this stage with unlabeled substrate. Phosphatidylinositol obtained from commercial sources (e.g., bovine phosphatidylinositol from Avanti Polar Lipids, Birmingham, Alabama) can be used, provided that the investigator has determined that the preparation is free of lipid other than phosphatidylinositol.

the radiolabled fractions by nitrogen flow, and the residues are dissolved with several portions of chloroform–methanol (2:1).[14]

Using the procedures described above, one can obtain approximately 20 μmol of [^{3}H]phosphatidylinositol from 30 g of chicken liver.

Preparation of Platelet Phospholipase C

Blood is drawn from normal human donors, anticoagulated with acid–citrate–dextrose,[15] then spun at 150 *g* for 15 min at room temperature. Nalgene polypropylene or siliconized glassware is used throughout the processing of platelets. The platelet-rich plasma is removed carefully, leaving the lower red cell layer and buffy coat interface. The platelet-rich plasma is mixed with EGTA (pH 6.5) to a final concentration of 1 m*M* and spun at 2500 *g* for 5 min. Platelet pellets are gently resuspended in Tris–citrate–bicarbonate buffer, pH 6.5,[16] containing 2 m*M* EGTA, and spun again. Care is taken to separate platelets from any contaminating erythrocytes and neutrophils at the bottom of the pellet by sacrificing part of the platelet pellet. Twice-washed platelets are suspended to a final concentration of 3×10^9 cells/ml in the above buffer (minus EGTA) and sonicated in a volume of up to 2 ml for 15 sec using a microprobe attachment to a sonicator cell disrupter (Heat Systems Ultrasonics) at maximum output. A standard probe may be used for larger volumes. The lysates are spun under a layer of mineral oil at 105,000 *g* for 60 min at 4°. The mineral oil is useful both for filling the centrifuge tube when a small volume of sonicate is being spun and for absorbing any lipid liberated during sonication, which rises to the surface after centrifugation. The sonicate supernatant is removed from beneath the mineral oil layer. Care is taken not to disturb the pellet during this procedure. The supernatant is kept at 0° and used immediately or stored at −80° overnight. It should not be vortex-mixed. The supernatant should contain approximately 3 mg of protein per milliliter. If assays are performed on total sonicates, it is necessary to take into account both dilution of substrate specific activity by endogenous substrate and inhibitory or stimulatory effects of other phospholipids.

Assay of Phosphatidylinositol-Specific Phospholipase C

A simple and rapid assay is presented here for the detection of phosphatidylinositol-specific PLC. A control assay for nonspecific cleavage is also described in which [methyl-^{14}C]choline–phosphatidylcholine is substituted for [^{3}H]myoinositol–phosphatidylinositol. Those phospholipases

[15] R. H. Aster and J. H. Jandl, *J. Clin. Invest.* **43,** 843 (1964).

[16] S. Rittenhouse-Simmons and D. Deykin, *Biochim. Biophys. Acta* **426,** 588 (1976).

C that are active upon substrates other than phosphatidylinositol are usually active with phosphatidylcholine. Since phosphatidylcholine is one of the few phospholipids containing radiolabel on the polar base that is available commercially (New England Nuclear), we suggest the use of this substrate, diluted with purified egg phosphatidylcholine, as a convenient monitor of nonspecific activity.

Assay Mixture

The final concentrations of components of the mixture are given.

[^{3}H]Myoinositol phosphatidylinositol, 200 μM (0.1 μCi), or [methyl-^{14}C]Choline phosphatidylcholine, 200 μM (approximately 0.05 μCi)
$CaCl_2$, 5 mM
Tris acetate buffer, 60 mM, pH 4.5–7.5
Sodium deoxycholate 2 mg/ml (pH to be adjusted for assay conditions)
Supernatant protein, 1 mg/ml, from platelet sonicate or supernatant protein heated at 100° for 5 min
All in a total volume of 300 μl.

The chloroform solution of radiolabeled phosphatidylinositol or phosphatidylcholine, enough for several assays, is evaporated under nitrogen flow. The residue is dissolved in distilled H_2O with the aid of a sonicator bath (Heat Systems Ultrasonics). The final suspension should be opalescent, with no particulate material visible. Aliquots of this suspension are counted in H_2O–Hydrofluor (1:10 ml) to confirm the completeness of suspension. Generally, a substrate concentration six times that of the final concentration in the assay is convenient. The assay is initiated by the addition of active or heat-treated enzyme. The mixture should not be vortex-mixed at this stage, but swirled gently. Incubation continues at 37° for up to 30 min and is terminated by the addition of 1.5 ml of ice-cold chloroform–*n*-butanol–conc. HCl (10:10:0.06), v/v/v) and 0.45 ml of 1 M HCl. Mixtures are vortexed and then spun at room temperature at 1000 rpm for 10 min or until a sharp phase separation is achieved. The substitution of *n*-butanol for the more usual methanol[2] allows the efficient removal of any generated lysophosphatide from the aqueous phase. Duplicate 250-μl aliquots of the upper aqueous phase (total = 750 μl) are transferred to scintillation vials, and traces of quenching chloroform are removed under nitrogen flow for 30 sec. Then, H_2O–Hydrofluor (1:10 ml) is added, and vials are capped, vortexed, and counted. The conversion of substrate to water-soluble product is calculated on the basis of the specific activity of the substrate. The phosphatidylinositol-specific PLC of platelet sonicate supernatants exhibits an activity of approximately 10 nmol/min per milligram of protein (or 12 nmol/min per 10^9 platelets). Phospholipase C activity with respect to phosphatidylcholine is less than 1% of that with respect to phosphatidylinositol.

Descending Chromatography

A preliminary determination of the nature of the water-soluble radiolabeled myoinositol material can be made through the use of descending paper chromatography. When enzyme activity is indicated, assays are repeated, and the total aqueous upper phases are removed and evaporated under nitrogen flow. Residues are dissolved in 100 μl of water and applied as one spot 10 cm from the shorter edge of Whatman No. 1 chromatography paper, 20 cm × 55 cm. As standards, the following are also applied to the papers, 4 cm apart: an alkaline alcoholic digest of 5 nmol of [^{3}H]myoinositol phosphatidylinositol (approximately 25,000 dpm) as described by Dawson,[17] to be used as a monitor for glycerylphosphorylinositol; 0.01 μCi [^{3}H]myoinositol; and 2 μmol of myoinositol 2-phosphate (Sigma). Polar myoinositol-containing material is resolved by descending chromatography overnight (18 hr) in ethanol–13.5 *M* NH_3 (3:2) as described by Dawson and Clark.[18] The solvent front is marked, and papers are allowed to dry. Myoinositol 2-phosphate is detected by spraying the lane for this standard with 0.1% ethanolic $FeCl_3$ and 1% NH_4SCN in acetone, a negative stain for phosphate.[19] The rest of the chromatogram should be shielded during the spraying procedure to prevent contamination by the quenching brick-red stain. The other lanes are divided into 2-cm portions and counted by scintillation spectophotometry in H_2O–Hydrofluor (1:10 ml). Since the standards migrate to yield spots about 5 cm long, only approximate R_fs are presented here (mid-spot): myoinositol 2-phosphate, 0.29; myoinositol, 0.51; glycerylphosphorylinositol, 0.84. Inositol 1:2-cyclic phosphate migrates to an extent comparable to that of glycerylphosphorylinositol. However, this species should be present only in trace amounts after acidic extraction of the phosphatidylinositol-specific PLC incubation mixture.

Enzyme present in platelet sonicate supernatant produces water-soluble product, 85–90% of which migrates near inositol 2-phosphate, the remainder migrating with myoinositol, apparently as the result of some contaminating soluble phosphatase activity. Should most or more radioactivity migrate with myoinositol (an observation made for assays

[17] R. M. C. Dawson, *in* "Lipid Chromatographic Analysis" (G. W. Marinetti, ed.), p. 168. Dekker, New York, 1967.

[18] R. M. C. Dawson and N. Clarke, *Biochem. J.* **127,** 113 (1972).

[19] D. Oberleas, *Methods Biochem. Anal.* **20,** 87 (1971).

that utilize crude, unfractionated platelet sonicates), the presence of either phosphatidylinositol-specific PLC acting in conjunction with phosphatase, or phospholipase D, is implicated. Finally, should significant amounts of radioactivity be detected in the glycerylphosphorylinositol region for assay samples containing enzyme preparations other than those derived from platelets, phospholipase A plus lysophospholipase may be present. However, since phosphatidylinositol migrates near glycerylphosphorylinositol, it is possible that contaminating [^{3}H]myoinositol phosphatidylinositol is being detected. The presence of this species can be confirmed or ruled out after applying the upper aqueous phases to silicic acid-impregnated paper and running two-dimensional chromatography, as described earlier.[11] Glycerylphosphorylinositol should stay near the origin.

Acknowledgments

This work was completed during the tenure of an Established Investigatorship of the American Heart Association, with funds contributed in part by the Massachusetts affiliate of the American Heart Association and by Grant HL 27897 from the National Heart, Lung, and Blood Institute (Department of Health, Education, and Welfare). This work was also supported in part by a grant from the National Institutes of Health HL22502.

[2] Characterization and Assay of Diacylglycerol Lipase from Human Platelets

By PHILIP W. MAJERUS and STEPHEN M. PRESCOTT

1,2-Diacylglycerol →2-monoacylglycerol + fatty acid → fatty acid + glycerol

Diacylglycerol lipase was originally described in membranes from human platelets and is postulated to provide a major pathway for arachidonate release.[1] Very similar enzymes have since been described in pig thyroid plasma membranes[2] and human fetal membranes.[3] The enzyme catalyzes the hydrolysis of 1,2-diacylglycerol to 2-monoacylglycerol and then to glycerol and free fatty acid. Whether one or two enzymes are required to release arachidonate has not been determined.

[1] R. L. Bell, D. A. Kennerly, N. Stanford, and P. W. Majerus, *Proc. Natl. Acad. Sci. U.S.A.* **76,** 3238 (1979).

[2] Y. Igarashi and Y. Kondo, *Biochem. Biophys. Res. Commun.* **97,** 766 (1980).

[3] T. Okazaki, N. Sagawa, R. J. Okita, J. E. Bleasdale, P. C. MacDonald, and J. M. Johnston, *J. Biol. Chem.* **256,** 7316 (1981).

ISBN 0-12-181986-8

Assay Method: Release of Fatty Acid from 1,2-Diacylglycerol

Reagents

HEPES buffer, 0.05 *M*, containing 0.1 *M* NaCl, 5 m*M* $CaCl_2$, and with or without 0.005% Triton X-100

2-[1-^{14}C]Arachidonyl, or 2-[1-^{14}C]oleoyl, 1,2-diacylglycerol, 40 m*M* in petroleum ether, 0.5–5.0 μCi/μmol (prepared as described below)

Glutathione, reduced, 0.3 *M*, pH 5.5

The [^{14}C]diacylglycerol needed for a series of assays is placed in the bottom of a glass test tube, and petroleum ether is removed under nitrogen. HEPES buffer with Triton is then added to yield a final diacylglycerol concentration of 2 m*M*, and the suspension is sonicated at 100 W with a Biosonik IV sonifier equipped with a microprobe for 1 min until a stable, turbid suspension is achieved. Reactions are carried out in a total volume of 0.1 ml containing 5 μl of glutathione, 20–200 μg (protein) of platelet microsomes (see below), 25 μl of diacylglycerol substrate, and additional buffer without Triton X-100 to volume. Reactions are started with addition of substrate and immediate sonication for 5 sec. After incubation at 37° for 2–30 min in a shaking water bath, the reaction is stopped by addition of 1.5 ml of chloroform–methanol–heptane (1.25 : 1.4 : 1; v/v/v) and 0.5 ml of 50 m*M* K_2CO_3, pH 10.[4] After extraction, portions of the aqueous upper phase are counted in a scintillation spectrometer. The assay under the conditions described is not linear with either time or enzyme concentration, so that the activities determined are crude estimates of the total activity. Shorter times and lower enzyme concentration result in nearly linear conditions.[1]

Preparation of 1,2-Diacylglycerol

Method 1. Preparation from Lysophosphatidylcholine. Lysophosphatidylcholine is prepared from egg lecithin (or phosphatidylcholine with a defined fatty acid composition) obtained from Sigma as a 100 mg/ml solution in hexane. The hexane is evaporated from 800 mg of phosphatidylcholine (PC) under N_2 in a 250-ml glass-stoppered Erlenmeyer flask. Diethyl ether, 50 ml, is added followed by 1.2 ml of *Crotalus adamanteus* venom (Sigma), 4.5 mg/ml in 0.015 *M* Tris-chloride, pH 7.4 containing 0.14 *M* sodium chloride. This mixture is incubated at room temperature with occasional swirling for 90 min. The lysophosphatidylcholine precipitates during the reaction and is collected by centrifugation, washed with ethyl ether, and then dissolved in 5 ml of chloroform–methanol (2 : 1).

[4] P. Belfrage and M. Vaughn, *J. Lipid Res.* **10,** 341 (1969).

The yield of lysophosphatidylcholine is approximately 500 μmol, and it is >95% pure. 2-Arachidonyl PC is prepared using small amounts of [^{14}C]arachidonic acid for ease of analysis. The following are added to a 500-ml glass Erlenmeyer flask, and the solvent is evaporated under nitrogen: lysophosphatidylcholine, 100 μmol in chloroform–methanol; arachidonic acid, 0.2 ml of 0.5 *M* solution in ethanol; [1-^{14}C]arachidonic acid, 1 μCi. Sodium phosphate buffer, 150 ml, 0.1 *M*, pH 7.4, containing 50 μg of butylated hydroxytoluene (BHT) per milliliter, is added to the Erlenmeyer flask and sonicated to obtain a uniform, slightly cloudy suspension. A reaction mixture is prepared containing 600 μmol of $MgCl_2$, 25 μmol of coenzyme A (CoA), 600 μmol of ATP, and the microsomes obtained from 3 g of rat liver[5] in a final volume of 7 ml. One-third of this mixture is then added to the flask containing the substrates to initiate the reaction, and an additional one-third is added hourly while incubating in a shaking water bath at 37°. After 3 hr the reaction is stopped with 100 ml of 80% methanol in chloroform. After addition of 300 ml of 20% methanol in chloroform plus 100 ml of H_2O, the mixture is shaken, and the organic (lower) phase is collected and evaporated under N_2. The dried sample is dissolved in 10% methanol in chloroform and poured over a 20-g column of silicic acid (Mallinckrodt AR, 100 mesh) equilibrated in the same solvent. The column is washed with the same solvent to elute unreacted fatty acid. The PC is then eluted with 40% methanol in chloroform. The yield of PC based on phosphate assay is 15–20 μmol when the preparation is carried out on this scale. Highly labeled substrate is prepared on a 50-fold smaller scale by using 2 μmol of lyso-PC, 20 μCi of [1-^{14}C]arachidonic (or oleic) acid (56 μCi/μmol), and 1.6 μmol of unlabeled fatty acid dried under N_2 in an 18 × 150 mm glass test tube. Then 7.5 ml of 0.1 *M* sodium phosphate, pH 7.4, is added, and the sample is sonicated as described above. A reaction mixture is prepared containing 600 μmol of $MgCl_2$, 6 μmol of CoA, 600 μmol of ATP, and the microsomes from 60 mg of rat liver in 4 ml, and one-third is added to the flask hourly as described above. The reaction is stopped, and PC is extracted and purified on a 2-g silicic acid column as described above. Based on phosphate assay, approximately 1.5 μmol of PC are recovered with a specific activity of 8–9 μCi/μmol. The specific activity of the product is less than the starting arachidonic acid because of dilution by fatty acids and PC contained in rat liver microsomes. The [1-^{14}C]arachidonic acid is in the sn-2 position of PC (>97%) based on treatment of the product with *C. adamanteus* venom. The specific radioactivity of the [1-^{14}C]arachidonate-labeled PC is adjusted by mixing the highly labeled and lightly labeled products in varying proportions.

[5] A. F. Robertson and W. E. Lands, *Biochemistry* **1,** 804 (1962).

1,2-Diacylglycerol is prepared from PC using *Bacillus cereus* phospholipase C purified as described.[6] Phospholipase C (100 μg per micromole of substrate lecithin) is incubated in 30 mM sodium barbital, pH 7.4, containing 0.12 M NaCl and 1 mM $ZnSO_4$ for 20 min at 37° (0.5 ml of reaction volume per micromole of PC). The diacylglycerol is extracted into ethyl ether (2 ml × 3), and the ether is evaporated. The yield of diacylglycerol is 90–100% in this procedure, and it is 90–95% 1,2-diacylglycerol as determined by thin-layer chromatography (precoated silica gel 60 G, Merck-Darmstadt) in ethyl ether–hexane–acetic acid (70:30:1). In this system PC remains at the origin, 1,2-diacylglycerol has an R_f of 0.58, and 1,3-diacylglycerol has an R_f of 0.67. 1,2-Diacylglycerol is stored in small portions at −20° in petroleum ether to prevent isomerization to 1,3-diacylglycerol. The purity of the substrate is monitored periodically by thin-layer chromatography (TLC) and is discarded when >10% 1,3-diacylglycerol is formed.

Method 2. Preparation from LM Cells. Mouse LM cells (American Type Culture Collection) are grown in suspension culture in lipid-free, chemically defined medium as described.[7,8] [1-^{14}C]Arachidonic acid, 55 μCi/μmol, 50 μCi, and unlabeled arachidonic acid, 2 μmol, are added in 0.1 ml of ethanol to 75 ml of medium.[9] LM cells, 25 ml, growing logarithmically, are then added to yield a final concentration of 0.5×10^6 cells/ml. After 3 days at 37° in a shaking incubator, the cells (2.5 to 3×10^6/ml) are harvested, and the phospholipids are extracted and isolated by HPLC.[10] Under these conditions approximately 25% of the added radioactivity is recovered in phospholipids, and of this 40% is in PC. Based on the original specific activity of arachidonate this represents 50% arachidonate in the 2-position of PC.[11] This material can then be diluted with nonradioactive PC and converted to 1,2-diacylglycerol using *B. cereus* phospholipase C as described above.

Preparation of Platelet Microsomes

Human platelets are isolated as described[12] from fresh blood. The cells are suspended at 2 to 5×10^9 cells/ml in 50 mM HEPES, pH 7.0 containing 5 mM $CaCl_2$ and 10 mM 2-mercaptoethanol and sonicated twice for 15 sec at 100 W in a Biosonik IV sonifier equipped with a microprobe.

[6] C. Little, B. Aurebekk, and A. B. Otnaess, *FEBS Lett.* **52,** 175 (1975).

[7] K. A. Ferguson, M. Glaser, W. H. Bayer, and P. R. Vagelos, *Biochemistry* **14,** 146 (1975).

[8] K. Higuchi, *J. Cell. Physiol.* **75,** 65 (1970).

[9] [1-^{14}C)Oleic or linoleic acids have also been used.

[10] S. M. Prescott and P. W. Majerus, *J. Biol. Chem.* **256,** 579 (1981).

[11] A typical recovery of PC is 0.5 μmol with a specific activity of 10 μCi/μmol.

[12] N. L. Baenziger and P. W. Majerus, this series, Vol. 31, p. 149.

After centrifugation at 105,000 *g* for 45 min, the pellet is resuspended in the same buffer and sonicated briefly to make a uniform suspension. The microsomes can be stored at −70° indefinitely under these conditions.

Solubilization of Diacylglycerol Lipase from Platelets

Outdated frozen platelet (≅ 10^{12} platelets) concentrates are thawed at 37° and centrifuged (4°) at 2000 *g* for 10 min. The plasma supernatant is discarded, and the pellet is resuspended in 100 ml of phosphate buffer (33 m*M* pH 6.5) with 113 m*M* NaCl and 5.6 m*M* glucose.[12] The centrifugation is repeated, and the supernatant is discarded. The pellet is suspended in 100 ml of 50 m*M* Tris-HCl, pH 7.2, containing 200 m*M* NaCl, 10 m*M* 22-mercaptoethanol, and 0.05% deoxycholate. The cells are disrupted by sonication (Biosonik IV with the standard probe) at 200 W for 2 min (4 bursts of 30 sec) while on ice. The solution is centrifuged at 48,000 *g* for 30 min (4°), and the supernatant is removed and saved. The sonication and centrifugation steps are repeated twice, and all the supernatants are combined. Approximately 60% of the activity in the original sonicate is present in the supernatant at this stage (3 μmol of diacylglycerol hydrolyzed per minute per 10^{12} platelets in a 10-min assay at 250 μ*M* substrate).

The activity in this preparation can be purified 2- to 4-fold by ammonium sulfate precipitation. The protein concentration is adjusted to 3–4 mg/ml (volume of 400 ml), and the solution is adjusted to an ammonium sulfate concentration of 70% saturation by the slow addition of solid $(NH_4)_2SO_4$ (188.8 g) while stirring slowly. The solution is allowed to equilibrate for 1 hr at 4° and is centrifuged. The supernatant is discarded, and the pellet is extracted for 1 hr in the same buffer that contains 209 g of ammonium sulfate per liter (35% saturated). After centrifugation, the supernatant is discarded and the pellet is dissolved in the same buffer, without ammonium sulfate, and dialyzed overnight (4°) against the same buffer. The solution is centrifuged again at 48,000 *g* for 20 min. The supernatant contains 60–70% of the activity from the solubilized preparation.

Properties of Diacylglycerol Lipase

Very little is known about diacylglycerol lipase, since it has not been isolated from any tissue. It seems clear that the 1-position fatty acid is hydrolyzed before arachidonate (2 position) is released, but it is not clear whether one enzyme has both activities or whether two separate enzymes are required to release both fatty acids from 1,2-diacylglycerol. In our original studies we used [^{3}H]glycerol-labeled 2-[1-^{14}C]arachidonyl diglyceride and found that [^{3}H]glycerol was released mole for mole with [^{14}C]arachidonate. We found no monoglyceride product, thus indicating that the 1-position fatty acid was either hydrolyzed first or simultaneously with that in the 2 position. Studies using doubly labeled diacylglycerol indicate that the 1-position fatty acid is cleaved before the 2-position acid in thyroid[2] and placenta.[3] However, the activity is the same whether release of the 1-position or 2-position fatty acid is measured, implying that the first hydrolysis is the rate-limiting step. When lipase activity in platelets is measured using monoacylglycerol as substrate, much greater activity is found than with diacylglycerol (10- to 100-fold). In studies of 1,2-diacylglycerol of various fatty acid composition, there is a slight preference (2- to 3-fold) for hydrolysis of diacylglycerol with arachidonate in the 2 position compared to 2-oleyl diacylglycerol. Whether this degree of specificity can account for the fact that primarily arachidonate is released from thrombin-stimulated platelets remains to be determined.

Diacylglycerol lipase is stimulated 1.2- to 3-fold by reduced glutathione and is inhibited 85% by 0.7 m*M* *N*-ethylmaleimide and 75% by 0.07 m*M* *p*-chloromercuribenzoate, suggesting that the enzyme has a sulfhydryl group essential for catalytic activity. High concentrations (100 μg/ml) of indomethacin cause diacylglycerol to accumulate in stimulated platelets, implying that indomethacin can inhibit diacylglycerol lipase.[13] *p*-Bromophenacyl bromide (20–100 μM) also inhibits diglyceride lipase by 25–50%, and mepacrine at high concentrations (1 m*M*) inhibits diacylglycerol lipase very slightly (15%). The enzyme is inactivated by diisopropyl fluorophosphate (DFP). The $T_{1/2}$ for diacylglycerol lipase in 1 m*M* DFP is 4.5 min at 37°. Phenylmethylsulfonyl fluoride (PMSF) (0.12 m*M*) inhibits the enzyme by 50% in 1 hr at room temperature. EDTA only slightly inhibits enzyme activity, suggesting that calcium ions are not essential for activity although calcium ions stimulate activity up to 2-fold. The pH optimum is 7.0 with little activity below pH 6.0 or above pH 8.0. The activity of diacylglycerol lipase in platelet microsomes is sufficient to account for the arachidonate release that occurs in thrombin-stimulated platelets. Under optimal conditions (2-min assay, 750 μM substrate), we find 30 nmol of fatty acid release per minute per milligram of platelet microsomes. The actual potential activity of the enzyme is difficult to estimate, however, since the added exogenous substrate is extremely hydrophobic and tends to coalesce into aggregates that may not be available for catalysis in aqueous solutions. Inclusion of additional detergents inhibits diacylglycerol lipase activity, presumably by sequestering the added substrate. In the lipid environment of the cell membrane, it is possible that

[13] S. Rittenhouse-Simmons, *J. Biol. Chem.* **255**, 2259 (1980).

the enzyme is much more active than observed in cell extracts. In many experiments, in an effort to conserve substrate, incubations are carried out for times and at substrate concentrations at which the reaction is proportional to the amount of enzyme and substrate, but in a nonlinear manner. In this fashion apparent specific activities 10- to 20-fold lower than that stated above are often obtained as indicated in the data on solubilization of the enzyme reported above.

Acknowledgments

This research supported by Grants HLBI 14147 (Specialized Center in Thrombosis), HL 07088, and HL 16634 from the National Institutes of Health.

[3] Purified Human Placental Arylsulfatases: Their Actions on Slow-Reacting Substance of Anaphylaxis and Synthetic Leukotrienes

By AKIRA KUMAGAI and HISAO TOMIOKA

Slow-reacting substance of anaphylaxis (SRS-A) has been documented as one of major mediators for the pathogenesis of bronchial asthma. It is an acidic lipid-like mediator with a molecular weight of approximately 400 and is generated and released from a number of tissues or cells by immunological or nonimmunological stimuli. Its spasmogenic activity for smooth muscle preparation could be reversed by the specific antagonist FPL 55712, and it could be inactivated by arylsulfatases of limpet origin or eosinophil origin. The chemical structure of slow-reacting substance (SRS) from murine mastocytoma cells has been proposed by Hammarström *et al.*[1] to be a derivative of 5-hydroxy-7,9,11,14-eicosatetraenoic acid with glutathione in a thioether linkage at C-6. The proposed chemical structure without sulfate shown in Fig. 1A, however, cannot explain the fact that SRS-A can be inactivated by arylsulfatase. Therefore, the present study has been done in order to identify a possible structure of SRS-A that can be hydrolyzed and inactivated by purified human placental arylsulfatase B. Furthermore, it is examined whether synthetic leukotriene C_4 and leukotriene D_4 can be inactivated by limpet arylsulfatase (ASL) or purified human placental arylsulfatases.

[1] S. Hammarström, R. C. Murphy, B. Samuelsson, D. A. Clark, C. Mioskowski, and E. J. Corey, *Biochem. Biophys. Res. Commun.* **91,** 1266 (1979).

ISBN 0-12-181986-8

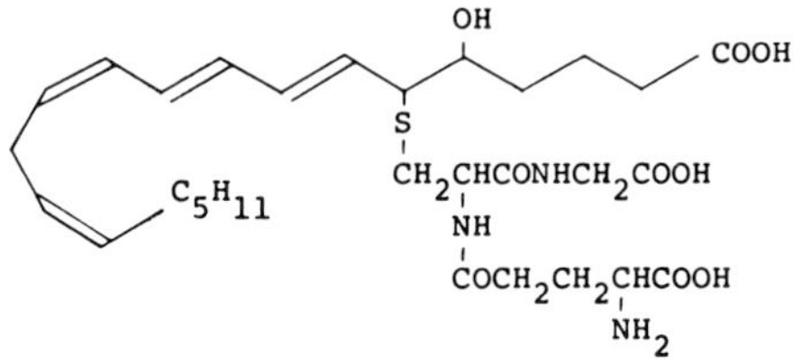

OH
COOH
S
C_5H_{11}
$CH_2CHCONHCH_2COOH$
NH
$COCH_2CH_2CHCOOH$
NH_2
(A)

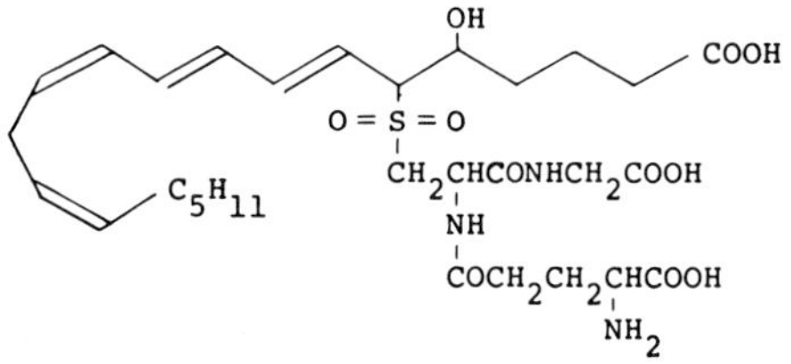

OH
COOH
O = S = O
C_5H_{11}
$CH_2CHCONHCH_2COOH$
NH
$COCH_2CH_2CHCOOH$
NH_2
(B)

FIG. 1. Proposed chemical structure of SRS(LTC_4) reported by Hammarström *et al.* (A) and SRS-A^{rat} determined in the present study (B).

Purification of Human Placental Arylsulfatase A (ASA) and Arylsulfatase B (ASB)

ASA and ASB were purified from fresh-frozen human placentas stored at −20° by the modified method of Gniot-Szulzycka[2] (Fig. 2). Arylsulfatase activity was quantified by measurement of the hydrolysis product (*p*-nitrocatechol) generated by the interaction of this enzyme and a substrate (*p*-nitrocatechol sulfate, Sigma). The enzyme preparation (0.5 ml) and of 0.01 *M* *p*-nitrocatechol sulfate (0.5 ml) were mixed and incubated at 37° for 10–90 min. The reaction was terminated with 1.5 ml of 1 *N* NaOH. The quantity of the *p*-nitrocatechol liberated was then measured by optical density at 515 nm, and 1 unit of enzyme activity was defined as the amount that liberated 1 nmol of *p*-nitrocatechol per minute. ASA and ASB were dissolved in 0.2 *M* acetate buffer containing 0.1% gelatin at pH 5.0 and pH 6.0, respectively. *p*-Nitrocatechol sulfate was dissolved in 0.5 *M* acetate buffer (pH 5.0)–0.01 *M* barium acetate to measure ASA, and in 0.5 *M* acetate buffer (pH 6.0)–1.66 *M* NaCl–0.05 *M* sodium pyrophosphate to measure ASB.[3]

The molecular weight determination was done on a Sepahdex G-100 column (2.6 × 95 cm) for ASB and on a Sephadex G-200 column (2.6 × 95 cm) for ASA with 0.9% NaCl as eluent. The standard proteins included human IgG (160,000 daltons), bovine serum albumin (68,000 dal-

[2] J. D. Gniot-Szulzycka, *Acta Biochim. Pol.* **19,** 181 (1972).

[3] H. Baum, K. S. Dodgson, and B. Spencer, *Clin. Chim. Acta* **4,** 453 (1959).

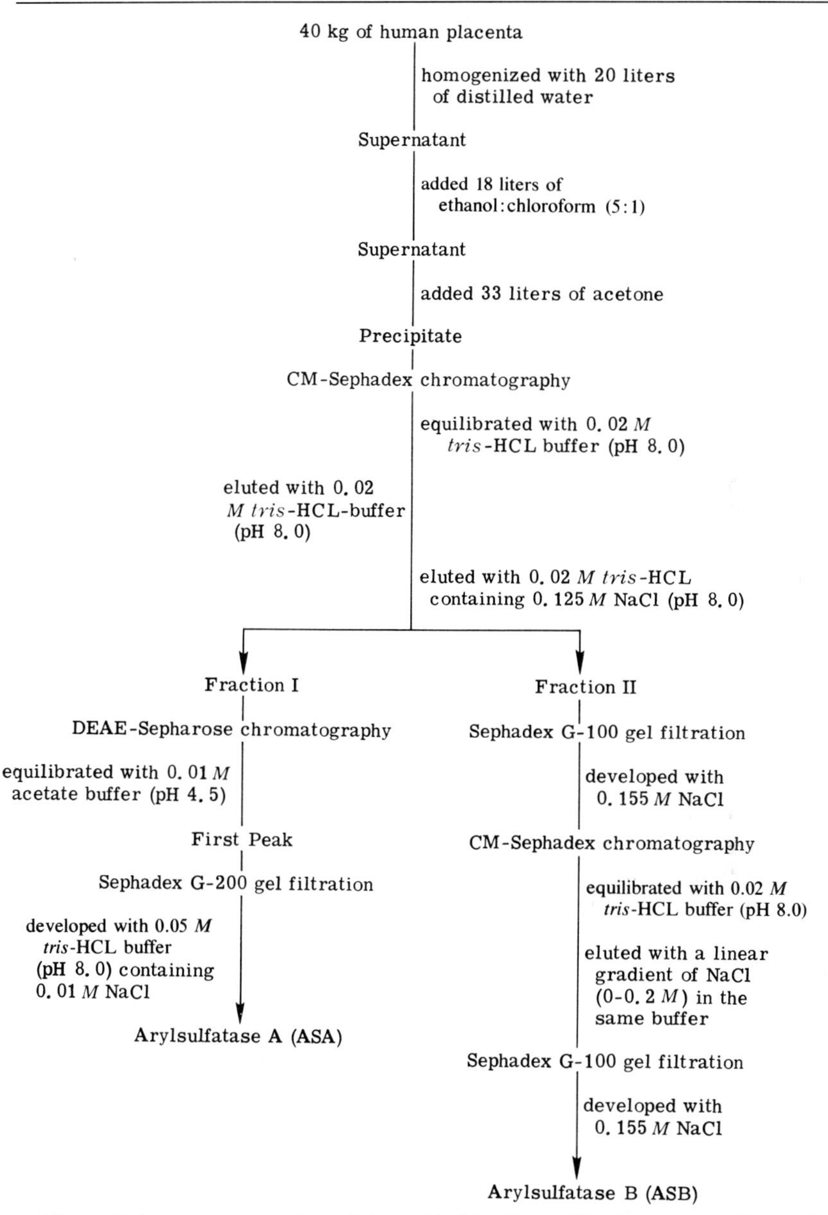

FIG. 2. Schematic presentation of the method for the purification of arylsulfatase A and arylsulfatase B from human placenta.

tons), ovalbumin (47,000 daltons), and chymotrypsinogen A (25,700 daltons). Protein concentration was determined by the method Lowry *et al.*[4]

The recovery of ASA with a molecular weight of 108,900 and ASB with a molecular weight of 51,500 from placenta were 24% and 27%, respectively. The specific activity of ASA was 45,500 units/mg and that of ASB was 38,800 units/mg. Table I shows the recovered activity and specific activity of arylsulfatases at each step of purification. Each arylsulfatase gave a single band on polyacrylamide gel electrophoresis. Furthermore, ASA and ASB showed a single precipitin line against a mixture of antisera to ASA and ASB. A thousand units of either arylsulfatase did not show any lipoxidase, peroxidase, carboxypeptidase A or B, or aminopeptidase activity.

Inactivation of SRS-A^{rat} and Other Chemical Mediators by Arylsulfatases

Preparation and Assay of SRS-A^{rat}. Wistar rats were immunized by intradermal injection of 0.5 mg of ovalbumin (OA) incorporated in Freund's complete adjuvant with simultaneous intraperitoneal injection of DPT-mixed vaccine (Takeda Pharmaceutical Co.). Twelve to fourteen days after immunization, SRS-A was generated by intraperitoneal challenge of 2 mg of OA with 6 mg of cysteine and 5 units of heparin in 5 ml of Tyrode's solution. Peritoneal washings from the challenged rat were subjected to 80% ethanol to extract crude SRS-A.

The spasmogenic activity of SRS-A was assayed on an isolated guinea pig ileum in the presence of 1×10^{-6} *M* mepyramine maleate and 5×10^{-7} *M* atropine sulfate. One unit of SRS-A was arbitrarily defined as the concentration required to produce a contraction with an amplitude equivalent to 5 ng of histamine base in that assay.[5]

In Vitro Inactivation of Crude SRS-A and Other Chemical Mediators by Arylsulfatases. Crude SRS-A (500 units) in 1 ml of 0.2 *M* acetate buffer at pH 5.7 was incubated at 37° with 750 units of ASA, 75 and 300 units of ASB, or 300 units of limpet arylsulfatase (ASL, Sigma, type V), respectively. The SRS-A activity remaining at 0, 30, and 90 min after incubation was assayed on the isolated guinea pig ileum as described above. The following mediators, 5×10^{-6} *M* histamine, 5×10^{-2} *M* serotonin, 5×10^{-6} *M* acetylcholine, 5×10^{-6} *M* prostaglandin E_1 (PGE_1), and 12.5×10^{-6} *M* prostaglandin $F_{2\alpha}$ ($PGF_{2\alpha}$) in 1 ml of 0.2 *M* acetate buffer, were

[4] O. H. Lowry, N. J. Rosebrough, A. L. Farr, and R. J. Randall, *J. Biol. Chem.* **193,** 265 (1951).

[5] D. J. Stechschulte, K. F. Austen, and K. J. Bloch, *J. Exp. Med.* **125,** 127 (1967).

TABLE I
RESULTS OF PURIFICATION OF PLACENTAL ARYLSULFATASE A (ASA) AND ARYLSULFATASE B (ASB)

Purification steps	Activity (units)	Protein (mg)	Specific activity (units/mg)	Purification factor
Arylsulfatase A				
Crude extract	1,008,000	1,409,600	0.72	1
Acetone ppt.	625,000	26,305	23.8	33
DEAE-Sepharose	349,800	112.5	3,109	4,320
Sephadex G-200	241,900	5.3	45,500	63,200
Arylsulfatase B				
Crude extract	1,187,000	1,409,600	0.84	1
Acetone ppt.	949,600	26,305	36.1	43
CM-Sephadex	713,400	1,118	638	760
Sephadex G-100	610,100	81.0	7,528	8,960
CM-Sephadex	477,200	24.6	19,400	23,100
Sephadex G-100	320,500	8.3	38,800	46,200

incubated with 1500 units of ASB at 37° for 20 min. The titration of those chemical mediators was carried out on the isolated guinea pig ileum by addition of 0.1 ml of each aliquot into a 10-ml bath.

ASB and ASL were capable of inactivating SRS-A in a time- and dose-dependent manner, but even a high dose (750 units) of ASA could not inactivate the spasmogenic activity of SRS-A (Table II). Although one half of the SRS-A was inactivated by 75 units of ASB, the spasmogenic activity of histamine, serotonin, acetylcholine, PGE_1, and $PGF_{2\alpha}$ was not reduced even by the incubation with high dose (1500 units) of ASB (Fig. 3).

TABLE II
In Vitro INACTIVATION OF SRS-A^{rat} BY ARYLSULFATASES

	SRS-A^{rat} activity[a] (units/ml)		Percent reduction of SRS-A^{rat} activity	
Arylsulfatases added[b]	30 min	90 min	30 min	90 min
Control	500	463	0.0	7.4
ASA, 750 units/ml	545	520	−9.0	−4.0
ASB, 75 units/ml	361	287	28.8	42.6
ASB, 300 units/ml	292	260	41.6	48.0
ASL, 300 units/ml	260	246	48.0	50.8

[a] Mean value of two experiments. Original activity of SRS-A^{rat} was 500 units/ml.
[b] Incubated in 0.2 M sodium acetate buffer (pH 5.7).

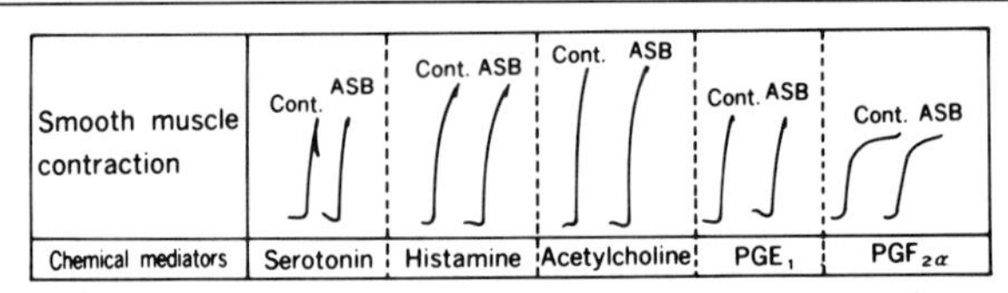

FIG. 3. Failure of inactivation of various spasmogenic mediators by arylsulfatase B *in vitro*. *In the presence of 1500 units of ASB per milliliter.

These findings clearly indicated that purified human placental ASB could inactivate SRS-A *in vitro*.

Purification of SRS-A^{rat} and Analysis of SRS-A Structure

Purification of SRS-A^{rat}. Crude SRS-A prepared from 400 sensitized Wistar rats was purified by the method of Orange *et al*[6] with addition of DEAE-Sephadex A-25 chromatography.[7] Final recovery of activity in 60% methanol eluted from DEAE-Sephadex A-25 chromatography was 16% of the starting crude SRS-A activity. Purification procedures and recovery are illustrated in Fig. 4. The ultraviolet spectrum of purified SRS-A had an absorbance maximum at 280 nm with shoulders at 270 and 292 nm, suggesting the presence of conjugated triene in the structure as suggested by Murphy *et al.*[8]

Analysis of Degradation Products from SRS-A by Arylsulfatase B. SRS-A (50,000 units) was incubated in acetate buffer at pH 5.7 with 30,000 units of ASB at 37° for 2 hr. Degradation products were submitted to a silica gel thin-layer chromatography (TLC) in ethanol–H_2O (7:3 v/v) as a solvent system and visualized with ninhydrin, sulfuric acid, or pinakryptol yellow. SRS-A gave a single spot at the origin, visualized with ninhydrin and sulfuric acid but not with pinakryptol yellow. ASB degradation products of SRS-A, however, gave three spots at R_f 0.93, R_f 0.25, and at the origin. The spot R_f 0.93 was visualized with only sulfuric acid. The spot at R_f 0.25 was visualized with ninhydrin and pinakryptol yellow. These results indicated that ASB resolved SRS-A into at least two components; the one with a low R_f value had both the amino group and the sulfonic group in it, but the component with a high R_f value had neither of them (Fig. 5).

A mass spectrum of the trimethylsilylated derivative of the product

[6] R. P. Orange, R. C. Murphy, M. L. Karnovsky, and K. F. Austen, *J. Immunol.* **110,** 760 (1973).

[7] S. Watanabe and A. Koda, *Immunology* **23,** 1009 (1979).

[8] R. C. Murphy, S. Hammarström, and B. Samuelsson, *Proc. Natl. Acad. Sci. U.S.A.* **76,** 4275 (1979).

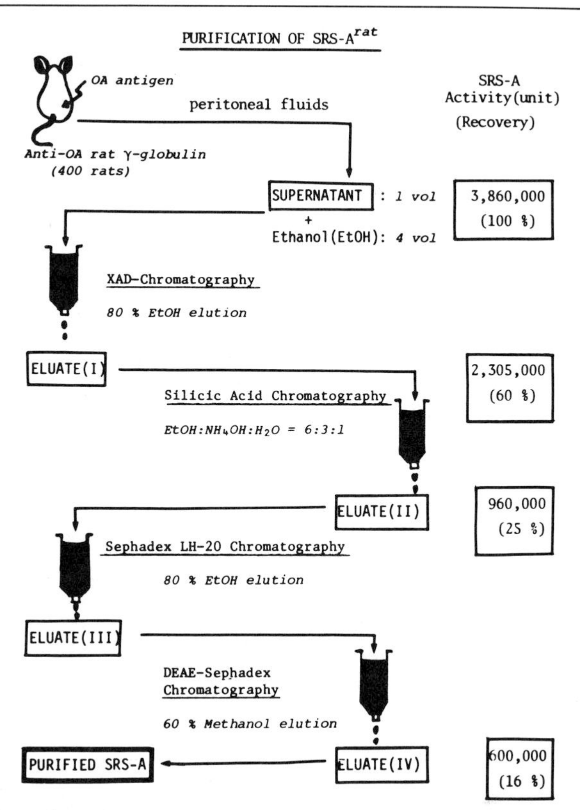

FIG. 4. Schematic presentation of the methods for the purification of SRS-A from passively sensitized rat peritoneal fluids. Four hundred Wistar rats were passively sensitized by anti-OA rat γ-globulin and then challenged with ovalbumin (OA).

having a high R_f value was measured by a gas chromatograph–mass spectrometer (GC–MS, JEOL-D 300) with 3% OV-1/Gas Chrom Q column and chemical ionization detector using isobutane as a reagent gas. Three major peaks at m/e 537, 463, and 391 were observed in its mass spectrum. as shown in Fig. 6. The peak at m/e 537 indicated a loss of methyl group from 5,6-dihydroxy-7,9,11,14-eicosatetraenoic acid containing three trimethylsilyl groups. Another peak at m/e 463 indicated an elimination of trimethylsilyl hydroxide from the quasi-molecular ion (M + 1). The base peak at m/e 391 was identified as a fragment ion at m/e 463. Each trimethylsilyl group was eliminated by a dehydration and a α-cleavage from trimethylsilylated 5,6-dihydroxy-7,9,11,14-eicosatetraenoic acid, suggesting the presence of a vicinal diol group at the allylic site of the double bond. Analysis for mass fragmentation and ultraviolet spectrum led to the con-

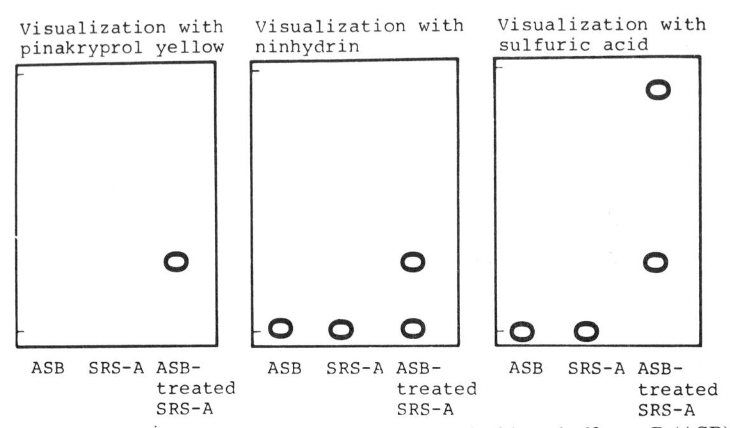

FIG. 5. Thin-layer chromatogram of SRS-A treated with arylsulfatase B (ASB).

clusion that the ASB degradation product having a high R_f value by TLC was 5,6,-dihydroxy-7,9,11,14-eicosatetraenoic acid.

Analysis of HCl Degradation Products from SRS-A. The amino acid composition of SRS-A was determined after HCl degradation. SRS-A (100,000 units) in 1 ml of 20% HCl was incubated at 110° for 24 hr. The incubation mixture was concentrated to dryness under vacuum, dissolved in 70% ethanol, subjected to TLC in ethanol–H_2O (7:3) or phenol–H_2O (3:1) as a solvent system, and visualized with ninhydrin or pinakryptol yellow. As shown in Fig. 7, analysis of the hydrolyzed products by TLC revealed the presence of glycine, glutamic acid, and cysteic acid. The component possessed the same R_f value as cysteic acid also reacted to pinakryptol yellow. Moreover, the mass spectrum of *N*,*N*-dimethylaminomethylene methyl-ester derivative of authentic cysteic acid was virtually

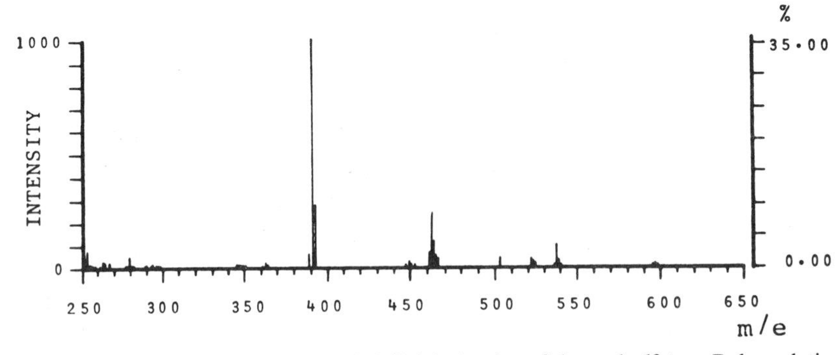

FIG. 6. Mass spectrum of the trimethylsilyl derivative of the arylsulfatase B degradation product having a high R_f.

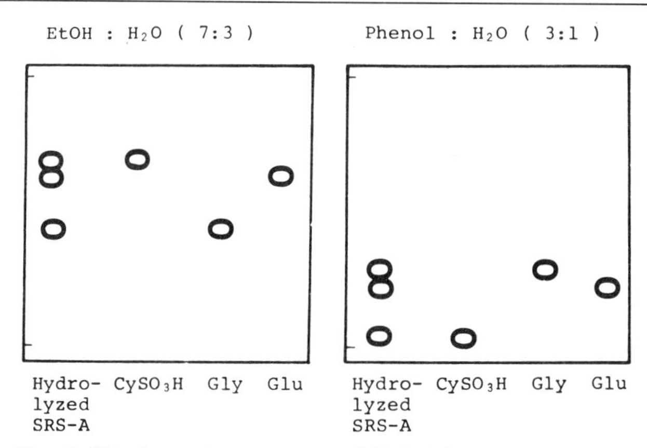

FIG. 7. Thin-layer chromatogram of SRS-A hydrolyzed by HCl.

identical to that of this component. These results indicate the presence of glycine and glutamic acid in SRS-A and the presence of cysteic acid or its derivative, which produced cysteic acid by HCl.

End Group Analysis of Peptide Moiety of SRS-A. To determine the C-terminal group of the peptide moiety, 50,000 units of SRS-A in 1 ml of anhydrous hydrazine were decomposed at 100° for 5 hr and analyzed for free amino acid by TLC in ethanol–H_2O(7:3) as a solvent system. For the N-terminal determination, 50,000 units of SRS-A in 0.5 ml of 66% ethanol containing 20 mg of sodium bicarbonate and 20 μl of dinitrofluorobenzene were incubated with agitation for 3 hr. The dinitrophenyl derivative of SRS-A was then hydrolyzed with HCl under the same conditions, and its degradation products were analyzed for dinitrophenylamino acid by TLC in chloroform–benzyl alcohol–acetic acid (70:30:3, v/v/v) as a developing solvent.

Analysis of the hydrazine degradation products by TLC revealed the presence of glycine and glutamic acid.

Dinitrophenylglutamic acid was identified in the analysis of HCl degradation products of dinitrophenylated SRS-A by TLC. These data demonstrated that the substituent at C-6 eicosatetraenoic acid consisted of three amino acids, viz. cysteic acid or its derivative, glutamic acid as the N-terminal residue, and glycine as the C-terminal residue.

Substrate Specificities of ASA and ASB. In order to examine the substrate specificity of purified human placental ASB, 500 units of SRS-A in 0.2 *M* acetate buffer at pH 5.7 were incubated with 750 units of ASA or 300 units of ASB at 37° for 90 min, and then the residual contractile activity of SRS-A was determined. Furthermore, 0.1 mg of compounds such as methionine, Boc-Cys(Bzl)-Val-Leu-Gly, *p*-nitrocatechol sulfate, *p*-to-

TABLE III
SUBSTRATE SPECIFICITIES OF ARYLSULFATASE A (ASA) AND ARYLSULFATASE B (ASB)[a]

No.	Substrates		Residual % of substrates	
			ASB	ASA
1	SRS-A		52	104
2	Methionine		98	99
3	Boc-Cys(Bzl)-Val-Leu-Gly		99	99
4	*p*-Nitrocatechol sulfate	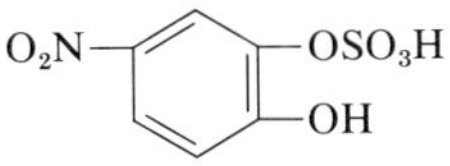 O_2N–(ring)–OSO_3H, –OH	13	11
5	*p*-Toluenesulfonic acid	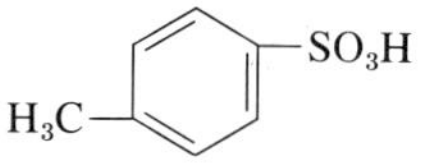 H_3C–(ring)–SO_3H	66	89
6	Benzenesulfinic acid	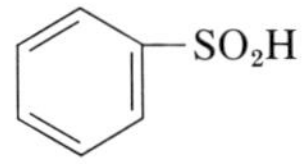 (ring)–SO_2H	34	86
7	2-(*S*-Methylsulfonyl)naphthalene	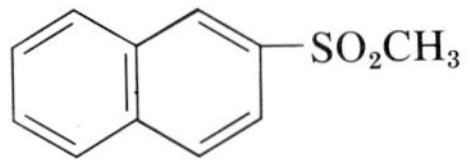 (naphthalene)–SO_2CH_3	85	97
8	Dermatan sulfate		22	83

[a] *p*-Cresol, phenol, or β-naphthol was detected as degradation product for substrates 5, 6, or 7, respectively, by means of GC–MS.

luenesulfonic acid, benzenesulfinic acid, 2-(*S*-methylsulfonyl)naphthalene, dermatan sulfate, and 20,000 units of ASA or ASB were incubated at 37° for 18 hr. After incubation, the residual material was determined by GC–MS or by reversed-phase high-performance liquid chromatography (HPLC; Hitachi 635) using a Hitachi gel No. 3010 column with H_2O–methanol (1:4). As shown in Table III, *p*-nitrocatechol sulfate was hydrolyzed by both ASA and ASB. SRS-A was inactivated only by ASB, but not by ASA. Compounds containing a sulfoxide or sulfone bond were also hydrolyzed by ASB, but not by ASA. The thioether bond was not hydrolyzed by either ASA or ASB.

These results suggested that SRS-A contained sulfone and are supported by the fact that degradation products from ASB and HCl treatment were visualized with pinakryptol yellow. On the basis that the Raney nickel degradation product of SRS was 5-hydroxyeicosaenoic acid, Murphy *et al.*[8] proposed that the glutathione would be attached by a thioether bond at C-6 of 5-hydroxy-7,9,11,14-eicosatetraenoic acid as shown in Fig. 1A. However, it is well known that sulfones are also

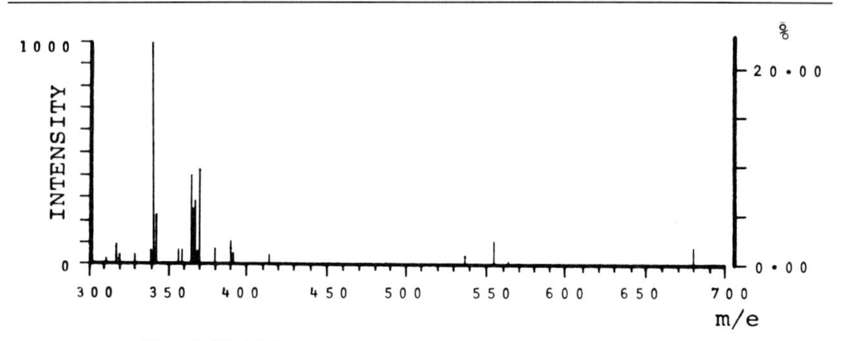

FIG. 8. Field desorption mass spectrum (FDMS) of SRS-A.

cleaved with Raney nickel during a sufficient degradation period.[9,10] In other words, Raney nickel degradation indicated not only the presence of a thioether bond, but also the presence of a sulfone bond in the structure.

Human placental ASB, giving a single band on polyacrylamide gel electrophoresis and a single precipitin line on immunoelectrophoresis, hydrolyzed SRS-A and sulfones, but did not hydrolyze the thioether bond of other compounds. Furthermore, the ASB degradation product had a sulfonic acid residue, supporting the idea that the peptide moiety was attached to 5-hydroxy-7,9,11,14-eicosatetraenoic acid at C-6 not in a thioether bond, but in a sulfone bond. All these results suggested that the chemical structure of SRS-A is expected to be [γ-glutamyl-4-(5-hydroxy-7,9,11,14-eicosatetraenoic acid-6-yl)-4,4-dioxocysteinyl]glycine, as shown in Fig. 1B.[11]

Field Desorption Mass Spectrum (FDMS) of SRS-A. To confirm the conclusion described above, FDMS of SRS-A was measured by GC–MS (JEOL-D 300) at 20 mA of emitter current. As shown in Fig. 8, distinguishable peaks were observed at *m/e* 680 and *m/e* 342 in FDMS of SRS-A. The peak at *m/e* 680 was identified as a parent peak of [γ-glutamyl-4-(5-hydroxy-7,9,11,14-eicosatetraenoic acid-6-yl)-4,4-dioxocysteinyl]glycine sodium salt. The base peak at *m/e* 342 was identified as the fragment peak with the peptide moiety eliminated from the parent peak. On the basis of these data, the structure of SRS-A^{rat} is determined as shown in Fig. 1B, which is different from the structure of SRS proposed by Hammarström *et al.*[1]

[9] R. Mozingo, U.S. Patent No. 2371642 (1945).

[10] J. S. Pizey, *in* "Synthetic Reagents (J. S. Pizey, ed.), Vol. 2, pp. 255–257. Wiley, New York, 1974.

[11] H. Ohnishi, H. Kosuzume, Y. Kitamura, K. Yamaguch, M. Nobuhara, Y. Suzuki, S. Yoshida, H. Tmoioka, and A. Kumagai, *Prostaglandins* **20,** 655 (1980).

Failure of the Inactivation of Synthetic Leukotrienes by Purified ASB

To confirm the result that the proposed structure of leukotriene C_4 (LTC_4) and leukotrinene D_4 (LTD_4) could not be inactivated by ASB, we have done experiments to inactivate synthetic LTC_4 and LTD_4 by purified ASB. Five hundred units of synthetic LTC_4 and LTD_4 (kindly supplied by Dr. Y. Terao, Takeda, Pharm. Co.) were incubated in 0.2 *M* acetate buffer at pH 5.7 with various amounts of ASB or limpet arylsulfatase (ASL, types IV and V, Sigma) for 60 min at 37°. The remaining contractile activity of SRS was compared with the activity at 0 min, and the results are shown in Tables IV and V. Neither LTC_4 nor LTD_4 could be inactivated by as much as 3000 units of ASB per milliliter. On the other hand, type IV ASL could inactivate both LTC_4 and LTD_4, and, more interestingly, type V ASL could not inactivate LTC_4, but could inactivate LTD_4.

The evidence that ASB cannot inactivate LTC_4 and LTD_4 was confirmed by the chemical method. LTC_4 or LTD_4 (2 μg) was incubated with 7500 units of ASA or ASB, respectively, in 0.2 *M* acetate buffer at pH 5.7, 37°. At 0, 30, 60, and 90 min after incubation, samples were taken and passed through Sep-Pak (Waters Associates, C_{18} cartridge) to adsorb remaining leukotrienes. The column was then washed with water and eluted by 80% ethanol. Recovered leukotrienes were concentrated and chromatographed by HPLC to separate and to quantify the remaining leukotrienes. The recovery of leukotrienes at different incubation times with or without ASA and ASB, respectively, is shown in Table VI. It is clear that no reduction of recovery of LTC_4 or LTD_4 was observed following ASA and ASB treatment compared to the recovery of untreated leukotrienes.

TABLE IV
INACTIVATION OF LTC_4 BY ARYLSULFATASES[a]

Arylsulfatases		Contractile response guinea pig ileum (cm)		Contractile response remaining (%)
		0	60 Min	
ASB	500 units	5.0	4.0	80.0
	500 units	4.5	6.2	137.0
	500 units	4.2	4.0	95.2
	3000 units	3.0	2.5	83.0
ASL	Type IV (140 units)	8.3	0.3	7.6
	Type V (100 units)	11.0	8.0	73.0
	Type V (500 units)	4.2	5.1	121.0

[a] LTC_4, 500 units/ml.

TABLE V
INACTIVATION OF LTD_4 BY ARYLSULFATASES[a]

Arylsulfatases		Contractile response guinea pig ileum (cm)		Contractile response remaining (%)
		0	60 Min	
ASB	300 units	5.4	5.1	94.4
	500 units	7.5	5.3	70.6
	3000 units	5.7	5.9	103.5
ASL	Type IV (140 units)	4.3	0.2	4.6
	Type V (500 units)	6.9	0.5	7.2
	Type V (500 units)	7.0	0.5	7.1

[a] LTD_4: 500 units/ml.

From these biological and chemical assays, we could definitely conclude that synthetic leukotrienes could not be inactivated by purified human placental arylsulfatases.

We should then answer the following questions: Why can ASL inactivate both SRS-A and synthetic leukotrienes? Why can type V ASL not inactivate LTC_4 only? At the moment there is no clear-cut answer to the latter question. For the former question, one can explain that contaminating enzymes that can hydrolyze peptides are present in the commercially available ASL preparation. Actually we confirmed that there is an aminopeptidase activity in type V ASL. Inactivation of synthetic leukotrienes by ASL through peptidase activity in the preparation was also reported by Houglum *et al.*[12] They performed the inactivation of purified SRS from cat paw and synthetic leukotrienes by type V ASL, and concluded that inactivation of SRS or synthetic leukotrienes by ASL was the result of the cleavage of the cysteinyl–glycine peptide bond, not the result of the hydrolysis of a sulfate ester. Therefore, we could conclude that the reported action of ASL on SRS-A inactivation was the result of the cleavage of peptide bond, but not the result of the hydrolysis of a sulfate ester.

However, we could not exclude the report that purified eosinophil-derived ASB and purified placental ASB, as we had reported, could inactivate SRS-A.[11,13] Therefore we should consider a possible structural difference between nonimmunologically generated SRS and immunologically generated SRS-A because the structures of LTC_4 and LTD_4, which could not be inactivated by purified human placental ASB, were deter-

[12] J. Houglum, J.-K. Pai, V. Atrache, D.-E. Sok, and C. J. Sih, *Proc. Natl. Acad. Sci. U.S.A.* **77,** 5688 (1980).

[13] S. I. Wasserman, E. J. Goetzl, and K. F. Austen, *J. Immunol.* **114,** 645 (1975).

TABLE VI
INACTIVATION OF LEUKOTRIENES BY HUMAN PLACENTAL ARYLSULFATASES[a]

	Recovery (%)					
	LTC_4			LTD_4		
Incubation time (min)	None	ASA	ASB	None	ASA	ASB
0	69	68	66	75	74	73
30	66	65	65	72	72	75
60	70	66	68	71	71	70
90	66	65	62	73	70	74

[a] 500 units/ml of LTC_4 and LTD_4 were incubated with 150 units/ml of arylsulfatase A and arylsulfatase B.

mined for SRS from calcium ionophore A 23187-stimulated murine mastocytoma cells. Therefore we did preliminary experiments to examine whether there were any differences in the susceptibility to ASB between nonimmunologically generated SRS and immunologically generated SRS-A.

Possible Difference between SRS and SRS-A. Crude human neutrophil-derived SRS generated by calcium ionophore A 23187 was subjected to inactivation by ASA, ASB, and ASL. Crude neutrophil-derived SRS, which has spasmogenic activity antagonized by FPL 55712, could not be inactivated by 500 units of ASA or ASB per milliliter. However, if the SRS-A from human lung fragments that were stimulated by anti-IgE was incubated with ASB, it was inactivated. Therefore, we are speculating that there are structural differences between nonimmunologically generated SRS and immunologically generated SRS-A.

[4] The 5-Lipoxygenase and Leukotriene Forming Enzymes

By BARBARA A. JAKSCHIK, TIMOTHY HARPER, and ROBERT C. MURPHY

The products of the 5-lipoxygenase pathway, including the leukotrienes (LT), have been described only recently. The pathway is outlined in Fig. 1. The first step, the addition of molecular oxygen to arachidonic acid to form 5-hydroperoxyeicosatetraenoic acid (5-HPETE) is catalyzed by a lipoxygenase. The conversion of LTA_4 to LTB_4 is catalyzed by a hy-

ISBN 0-12-181986-8

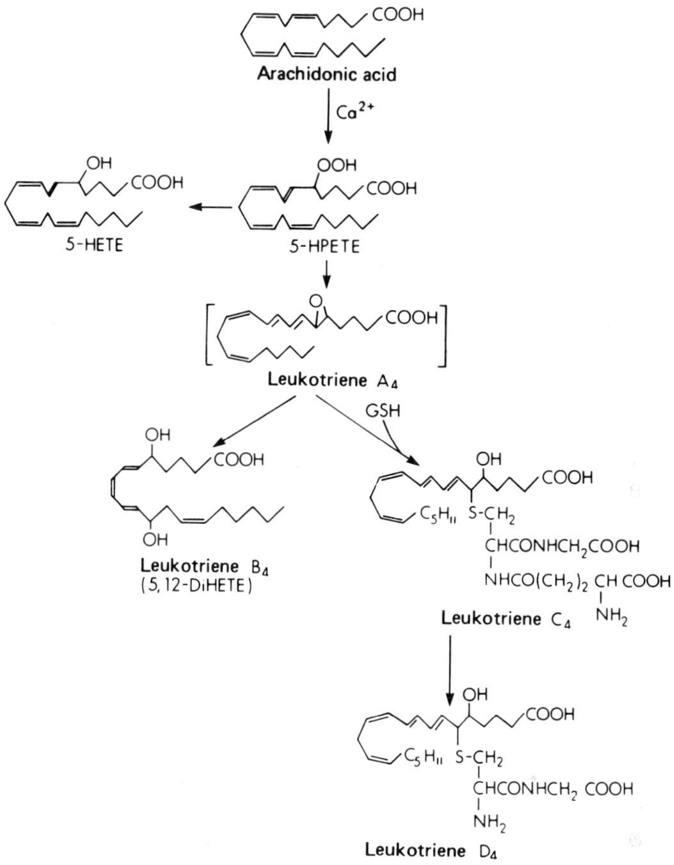

FIG. 1. 5-Lipoxygenase pathway leading to leukotriene formation.

drase with the addition of water.[1] In this chapter we describe the preparation of a single cell-free enzyme system that can be utilized for the study of the enzymic formation of LTB_4 and other dihydroxyeicosatetraenoic acids (di-HETE), LTC_4, LTD_4, and 5-HETE. The sum of these products reflects the total 5-lipoxygenase activity. We utilize for this preparation rat basophilic leukemia (RBL-1) cells.[2] However, this preparation is applicable to other cell types, such as mast cells or polymorphonuclear leukocytes.

[1] P. Borgeat and B. Samuelsson, *Proc. Natl. Acad. Sci. U.S.A.* **76,** 3213 (1979).

[2] B. A. Jakschik, A. Kulczycki, Jr., H. H. MacDonald, and C. W. Parker, *J. Immunol.* **119,** 618 (1977).

Preparation of Cell-Free Enzyme System

RBL-1 cells are grown in suspension cultures in Eagle's minimum essential medium supplemented with 5% heat-inactivated fetal calf serum, 2.5% heat-inactivated calf serum, 2 mM L-glutamine, 10 units of penicillin and 10 μg of streptomycin per milliliter in spinner bottles. Every second day, the cells are diluted to 0.3×10^6 cells/ml with the above medium. The cells are harvested by centrifugation at 400 g for 10 min. They are washed in the experimental buffer. In order to obtain a viable enzyme preparation, gentle conditions have to be used for breaking the cells. The cells are washed with 50 mM sodium phosphate buffer, pH 7, containing 1 mM EDTA, and 0.1% gelatin and resuspended at 5×10^7 cells/ml in ice-cold 35 mM sodium phosphate buffer, pH 7, containing 1 mM EDTA and 0.1% gelatin. The cells are homogenized with a Tekmar Tissuemizer, Model TR5T, setting 20, for twenty 25-sec periods.[3] It is important that the preparation be kept cold throughout the procedure. The Tissuemizer probe is precooled and cooled in between the 25-sec bursts of homogenization. We have obtained preparations with better activity when EDTA was present during the homogenization.

General Incubation Conditions

Calcium (1–2 mM) must be added to the incubation mixture to obtain optimum activity of the 5-lipoxygenase.[3] To synthesize LTC_4 or LTD_4 it is necessary to add glutathione (GSH) (1 mM) to the incubation mixture.[4] Therefore the incubation mixture contains: homogenate or 10,000 g supernatant, 1–2 mM Ca^{2+}, 1 mM reduced GSH (if LTC_4 or LTD_4 synthesis is desired), and arachidonic acid.

Thin-Layer Chromatography

One of the methods to evaluate 5-lipoxygenase activity is the separation of products and arachidonic acid by thin-layer chromatography. For these experiments, 0.5 ml of 10,000 g supernatant is incubated with 3–10 μM [^{14}C]arachidonic acid (specific activity 55 Ci/mol, New England Nuclear) and 1.5 mM calcium at 37° for 15 min. The volume of the total incubation mixture is 0.6 ml. The protein is precipitated with 2 volumes of acetone (1.2 ml), the pH is adjusted to 3.4 with 2 N formic acid, and extraction is performed twice with equal volumes (1.8 ml) of chloroform.

[3] B. A. Jakschik, F. F. Sun, L. H. Lee, and M. M. Steinhoff, *Biochem. Biophys. Res. Commun.* **95,** 103 (1980).

[4] B. A. Jakschik and L. H. Lee, *Nature (London)* **287,** 51 (1980).

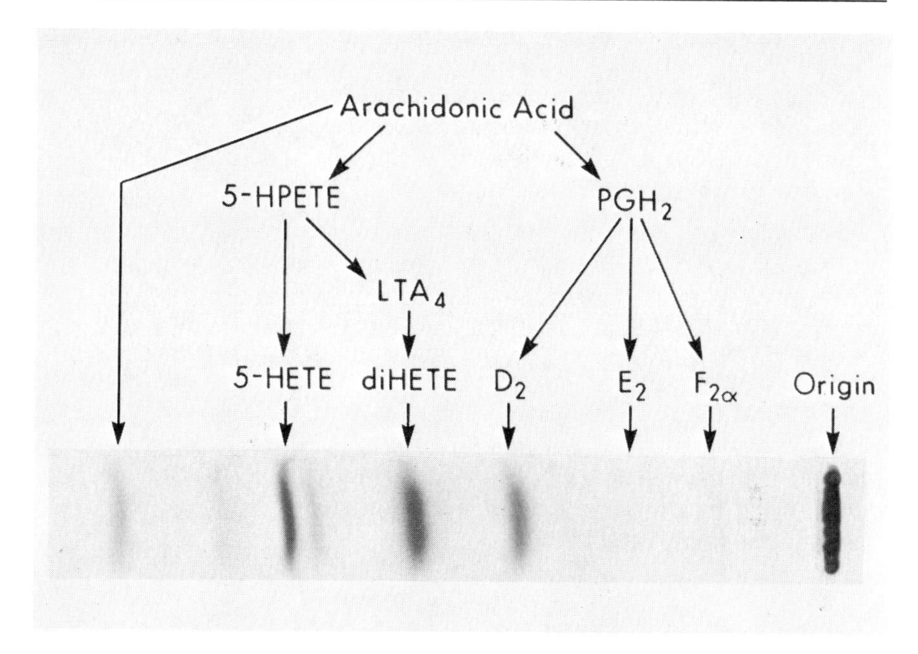

FIG. 2. Autoradiograph of thin-layer chromatography separation of 5-HETE, 5,12-di-HETE, prostaglandins, and arachidonic acid. Solvent system A9, the organic phase of ethyl acetate–2,2,4-trimethylpentane–acetic acid–water (110:50:20:100, v/v/v/v).

The extracts are combined and concentrated, standards are added, and the samples are applied to silica gel 60 plates (MCB Manufacturing Chemists, Inc.). The solvent system A9[5] [the organic phase of ethyl acetate–2,2,4-trimethylpentane–acetic acid–water (110:50:20:100, v/v/v/v)] separates 5-HETE and 5,12-diHETE effectively from prostaglandins (PG)[3,5] (Fig. 2). It is important that the solvent be prepared consistently in the same manner to obtain reproducible results. No paper curtains are used in the chromatography tank. Standards are visualized by iodine staining. Since 5-HETE and 5,12-di-HETE standards are not readily available, the interpretation of data can be difficult at times, especially if the inhibition of product formation is evaluated. Other compounds, such as PGB_2 or PGA_2, can be added to the samples prior to thin-layer chromatography and can be used as markers. Both PGB_2 and PGA_2 migrate between 5-HETE and 5,12-diHETE.[3]

This thin-layer chromatography system is quite adequate to evaluate 5-lipoxygenase activity. With no GSH added, the 5-lipoxygenase activity

[5] M. Hamberg and B. Samuelsson, *J. Biol. Chem.* **241,** 257 (1966).

is expressed in the formation of 5-HETE and 5,12-diHETE. Therefore, the sum of the 5-HETE and di-HETE bands will give the total 5-lipoxygenase activity. If radioactive arachidonic acid is used as a substrate, the bands can be visualized by radiochromatography (Kodak X-Omat RAX-5) and quantitation can be obtained by scraping and liquid scintillation counting. However, this system cannot be used to evaluate LTB_4 formation. The labile intermediate LTA_4 can readily break down to isomers of LTB_4[1], which are not biologically active, and to 5,6-di-HETE. All these di-HETEs comigrate in the A9 solvent system.[3] However, this method has the advantage that a large number of samples can be processed in a fairly short period of time.

No adequate thin-layer chromatography system is presently available for the separation of LTC_4 and LTD_4. Instead, see High-Performance Liquid Chromatography below and this volume [53]. The solvent systems published for slow-reacting substance[6] do not completely separate the compounds from phospholipids. In solvent system A9, LTC_4, and LTD_4 remain close to the origin.

Bioassay for LTC_4 and LTD_4

Bioassay for slow-reacting substance (LTC_4, LTD_4, and LTE_4) on the guinea pig ileum has been used for a long time. Either the superfusion[7] or the bath[8] method can be used. This assay is a quick, qualitative, and semiquantitative method (see this volume [76]). Even though the method has shortcomings, it has advantages: no expensive equipment is required, results are obtained quickly, and a large number of samples can readily be processed in a short period of time. We found Harvard smooth muscle transducers to be very sensitive for bioassay with smooth muscles, and any recorder that will couple to these myographs is adequate. Quantitation is difficult for a number of reasons. Standards are not readily available. However, the availability of standards would not solve the problem entirely since quite frequently a mixture of LTC_4 and LTD_4 and possibly LTE_4 is obtained. Each of these compounds has a different profile of contraction varying in height and width. We utilize the area under the curve of the contraction, which takes into account the shape of the contraction caused by LTC_4 as well as LTD_4.

For these experiments, the whole homogenate is incubated with 1.5 mM calcium, 1 mM GSH, and 3–17 μM arachidonic acid (NuChek

[6] K. Strandberg and B. Uvnas, *Acta Physiol. Scand.* **82,** 358 (1971).

[7] J. R. Vane, *Br. J. Pharmacol.* **35,** 209 (1969).

[8] Staff, Department of Pharmacology University of Edinburgh, "Pharmacological Experiments on Isolated Preparations." Livingstone, Edinburgh, 1970.

Prep) at 37° for 15–20 min, between 50 and 400 μl of the whole incubation mixture can be applied to the ileum with the superfusion technique. It is necessary that certain inhibitors be present in the bathing solution: 10^{-6} *M* pyrilamine maleate (an antihistamine); 5×10^{-7} *M* atroprine sulfate or hyoscine · HBr (to block acetylcholine and to reduce the spontaneous contraction); and 1 μg of indomethacin per milliliter (or another cyclooxygenase inhibitor). Exogenous arachidonic acid could be converted by the guinea pig ileum to lipoxygenase products, which could cause a contraction, and no adequate lipoxygenase inhibitors are available at present. Therefore, proper controls must be performed. The incubation mixture minus GSH (homogenate + CA^{2+} + arachidonic acid) should be tested. The presence of slow-reacting substance can be confirmed by the specific receptor antagonist FPL 55712.[9] It must also be kept in mind that guinea pig ileum can convert LTC_4 to LTD_4.

High-Performance Liquid Chromatography (HPLC)

To evaluate all the leukotrienes formed by this pathway, we developed an HPLC system that will separate all these compounds in one run. Owing to the high absorbance of the conjugated triene, the detection of leukotrienes is very sensitive at 280 nm. A number of systems have been published that separate the di-HETEs[10] or LTC_4 and LTD_4.[11,12] We have developed a system that will separate the diHETEs, LTC_4, and LTD_4 in one run[13] (Fig. 3).

The homogenate is incubated with 1.5 m*M* calcium, 1 m*M* GSH, and 10–35 μM arachidonic acid. The amount of homogenate used will depend on how much of the products one wishes to generate. The amount of products obtained from 0.5-1 ml of homogenate are sufficient to detect peaks readily at a setting of 0.02 absorbance unit equal to full scale on the recorder. We use routinely 5 ml of homogenate with a total incubation mixture of 6 ml.

The incubation mixture is processed in the following manner: the pH is adjusted to 6 with 2 *N* formic acid, and PGB_2 (0.2–0.6 μg per 1 ml of incubation mixture) is added as external standard. PGB_2 will absorb at

[9] J. Augstein, J. B. Farmer, T. B. Lee, P. Sheard, and M. L. Tattersall, *Nature (London), New Biol.* **245,** 215 (1973).

[10] P. Borgeat and B. Samuelsson, *J. Biol. Chem.* **254,** 7865 (1979).

[11] R. C. Murphy, S. Hammarström, and B. Samuelsson, *Proc. Natl. Acad. Sci. U.S.A.* **76,** 4275 (1979).

[12] L. Orning, S. Hammarström, and B. Samuelsson, *Proc. Natl. Acad. Sci. U.S.A.* **77,** 2014 (1980).

[13] B. A. Jakschik, T. Harper, and R. C. Murphy. *J. Biol. Chem.*, in press (1982).

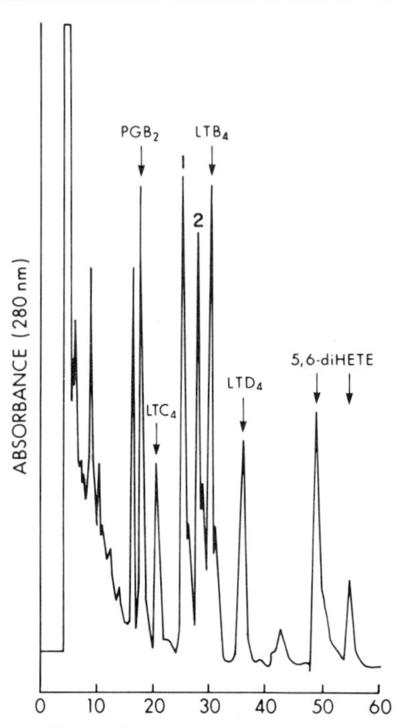

FIG. 3. High-performance liquid chromatography separation of leukotrienes. A nucleosil C_{18} column (5 μm particle size) was used with methanol–water (67:33), 0.02% acetic acid, pH adjusted to 4.7 with 1.5 *M* ammonium hydroxide, as mobile phase. LTC_4 and LTD_4 were identified by their ultraviolet (UV) spectra, comigration with standards, and activity on guinea pig ileum. The di-HETEs including LTB_4 were identified by their UV spectra and gas chromatography–mass spectrometry.[13] Peak 1: 5(*S*),12(*R*)-dihydroxy-6,8,10(all-*trans*)-14-*cis*-eicosatetraenoic acid; peak 2: 5(*S*),12(*S*),-6,8,10(all-*trans*)-14-*cis*-eicosatetraenoic acid.

280 nm, but the total UV spectrum of PGB_2 is quite different from that of the leukotrienes. After the addition of the external standards, the mixture (6 ml) is applied to an Amberlite XAD-7 column (20 ml). A rather large column is used, since the preparation is very crude. The column is washed with 125 ml of distilled water, and the sample is eluted with 20 ml of 80% ethanol. The eluate is dried under vacuum (40°) and reconstituted in methanol (0.2–0.4 ml) for storage. Any precipitate that will form upon standing is removed by centrifugation. It is not advisable to perform ethanol precipitation prior to application of the sample to Amberlite XAD-7 because ethoxy derivatives of LTA_4 will be formed that complicate the chromatographic pattern. Small volumes of sample (≤ 20 μl) are injected in methanol. If a larger volume of a sample is applied, the methanol is diluted to approximately 10% with water. The column used is Nucleosil

C_{18}, 5 μm particles (Machery Nagel), and the solvent system is methanol–water (67:33), 0.2% acetic acid adjusted to pH 4.7 with 1.5 *M* ammonium hydroxide. Small changes in pH may alter the elution time of LTC_4 and LTD_4, but not that of the di-HETEs or PGB_2. Changes in methanol concentrations will change the migration of all the compounds. Since the elution pattern can easily change with small variations in the solvent, it is advisable to test with standards every time new solvent is prepared. As can be seen in Fig. 3, LTC_4, LTD_4, the di-HETEs, and PGB_2 can be readily separated. If the 11-*trans* isomers of LTC_4 and LTD_4 are present, they will also separate and elute right after LTC_4 LTD_4, respectively. The peaks obtained can be characterized by a number of criteria.

1. Comigration with standards should be tested.
2. The UV spectrum should be determined. Leukotrienes have very characteristic spectra. LTB_4 has a UV maximum of 270 nm ($\epsilon \simeq$ 51,000) with shoulders at 260 ($\epsilon \simeq$ 38,000) and 281 nm ($\epsilon \simeq$ 39,500). For the other diHETEs the spectrum is shifted by approximately 2 nm.[10] LTC_4 and LTD_4 have a UV maximum of 280 ($\epsilon \simeq$ 40,000) with shoulders at 270 and 292 nm.[11,12] For the 11-*trans* isomers the spectrum is shifted to a UV maximum of 278 nm and shoulders at 268 and 290 nm.[14]
3. LTC_4 and LTD_4 can be identified by their activity on guinea pig ileum.
4. It is desirable to identify the peaks by gas chromatography–mass spectrometry.

Quantitation can be achieved by comparison with the external standard added. If an equivalent amount of PGB_2 is added to each sample prior to processing, the formation of LTC_4, LTD_4, and the diHETEs can be quantitated by comparing the areas of the respective peaks with that of PGB_2.

Acknowledgments

This work was supported by NIH grants 2R01 HL21874-04 and 5P30 CA16217-03 and HL25785 (R. C. M.).

[14] D. A. Clark, G. Goto, A. Marfat, E. J. Corey, S. Hammarström, and B. Samuelsson, *Biochem. Biophys. Res. Commun.* **94,** 1133 (1980).

[5] γ-Glutamyl Transpeptidase, a Leukotriene Metabolizing Enzyme

By Kerstin Bernström, Lars Örning, and Sven Hammarström

Leukotriene C + acceptor ⇌ leukotriene D + γ-glutamyl-acceptor
(acceptor: water, amino acid, or dipeptide)

Leukotrienes are a new group of biological mediators that probably have important roles in inflammatory and allergic responses.[1] They are formed from several polyunsaturated fatty acids that are converted to epoxides having three conjugated double bonds (leukotrienes A). Hydroperoxy derivatives, formed by lipoxygenase(s), are intermediates in these reactions. The epoxides are further transformed in two ways: by enzymic hydrolysis into dihydroxy acids (leukotrienes B) or by enzymic addition of glutathione to give leukotrienes C (LTC). The latter products, in addition to glutathione and glutathione disulfide, are endogenous substrates for γ-glutamyl transpeptidase (γ-glutamyltransferase, EC 2.3.2.2).

Preparation of Leukotrienes C

The enzyme can utilize the following LTC-type leukotrienes as substrates: C_3, C_4, C_5, 8,9-C_3 as well as 11-*trans* isomers of C_3, C_4, and C_5. No appreciable differences in the rate of conversion of the different C-type leukotrienes has been observed.

The substrates can be prepared biosynthetically from the appropriate precursor fatty acid by incubating the acid with murine mastocytoma cells stimulated with Ca^{2+} ionophore (A 23187).

Reagents

- Incubation buffer: 150 m*M* NaCl, 3.7 m*M* KCl, 3.0 m*M* Na_2HPO_4, 3.5 m*M* KH_2PO_4, 0.9 m*M* $CaCl_2$, 5.6 m*M* dextrose, pH 7
- CXBGABMCT-1 murine mastocytoma cells, 10^7 per milliliter
- L-cysteine, 0.2 *M*, dissolved in incubation buffer
- A 23187, 4 m*M*, and 10–20 m*M* precursor fatty acid, in 99% ethanol

The precursor fatty acids are eicosatrienoic acid (*n*-9), arachidonic acid, eicosapentaenoic acid (*n*-3), and eicosatrienoic acid (*n*-6) (dihomo-

[1] B. Samuelsson and S. Hammarström, *Vitam. Horm.* (*N.Y.*),in press (1982).

ISBN 0-12-181986-8

γ-linolenic acid) for leukotrienes C_3, C_4, C_5, and 8,9-C_3, respectively.[2–5]

Mastocytoma Cells. Murine mastocytoma cells (CXBGABMCT-1)[3] are kept as intramuscular tumors in hind legs of CXBG or CB_6F-1 mice. When the tumors reach a size of 0.5–1.5 cm^3 (17–20 days after injection), the animal is killed by excessive ether anesthesia. The tumors from both legs are rapidly excised and minced (e.g., using the blunt side of a Pasteur pipette) in 5–7 ml of sterile phosphate-buffered saline. Large tissue fragments are allowed to sediment for 5 min. The tumor cell suspension is then injected intraperitoneally (0.2 ml) or intramuscularly (0.1 ml per leg) into the same kinds of mice. A disposable 1-ml plastic syringe and a 25-gauge $\frac{5}{8}$-inch needle are used for the injections. Two weeks later, intraperitoneally injected animals are killed by decapitation under ether anesthesia. The abdominal skin is removed, and 5 ml of incubation buffer are injected through the peritoneum. After gentle massage of the abdomen, the fluid containing the ascites tumor cells is collected in a plastic beaker on ice after incision of the peritoneum with a pair of scissors. The peritoneal cavity is then rinsed several times with 1-ml portions of incubation buffer. Care is taken to avoid contamination by blood, bile, or urine. The mastocytoma cells are sedimented by a centrifugation in plastic tubes (250 *g*, 15 min, 4°). If contaminated by more than 1–2% erythrocytes, the sediments are resuspended in 5 ml of 0.155 *M* NH_4Cl–0.17 *M* Tris-HCl, pH 7.4 (9:1, v/v, freshly mixed) and incubated at 37° for 5 min. After centrifugation, the sediments are finally resuspended in 5 ml of incubation buffer (a 1-ml automatic pipette with a plastic tip is used for the resuspension). Cell concentration is determined with a hemacytometer and using a 100-fold dilution of the cell suspension. Up to this point the cells are kept at 0–4°.

Incubations.[3] Additional incubation buffer (37°) is added to make the cell concentration 10^7/ml. The cells are incubated in a plastic jar at 37° for 15–30 min. L-Cysteine solution is added (5% v/v), and incubation is continued for 2 min to allow synthesis of glutathione. The solution of ionophore and precursor fatty acid (0.5% v/v), is added dropwise and incubation is continued for another 20 min, with shaking. Four volumes of ethanol are added, and the white precipitate that forms after 3–16 hr at 4° is removed by filtration. The solvents are removed by rotary evaporation (water bath at 30°). When nearly dry, water (10–20% of the incubated volume) is added and the sample is applied to a column of Amberlite XAD-8

[2] S. Hammarström, *J. Biol. Chem.* **256,** 2275 (1981).

[3] R. C. Murphy, S. Hammarström, and B. Samuelsson, *Proc. Natl. Acad. Sci. U.S.A.* **76,** 4275 (1979).

[4] S. Hammarström, *J. Biol. Chem.* **255,** 7093 (1980).

[5] S. Hammarström, *J. Biol. Chem.* **256,** 7712 (1981).

TABLE I
ELUTION BEHAVIOR OF LEUKOTRIENES ON REVERSED-PHASE HIGH-PERFORMANCE LIQUID CHROMATOGRAPHY[a]

	Elution times (min)					
Isomers	LTC_3	LTD_3	LTC_4	LTD_4	LTC_5	LTD_5
11-*cis*	15.6	25.2	10.0	15.6	7.0	11.0
11-*trans*	17.4	27.2	12.2	17.6	9.0	13.2

8,9-LTC_3	8,9-LTD_3	Prostaglandin B_2
9.4	14.5	10.0

[a] Column, C_{18} Nucleosil (250 × 4.6 mm); solvent, methanol–water, 7:3 (v/v)–0.1% acetic acid, adjusted to pH 5.4 with NH_4OH; flow, 1 ml/min.

having the same volume. After application and elution with one more volume of water, two volumes, of 80% aqueous ethanol are used to elute leukotrienes. This eluate is concentrated to dryness on a rotary evaporator, and the residue is dissolved in methanol (10 ml). Silicic acid (SilicAr CC-7, 0.5 g) is added, and the methanol is evaporated to coat the residue on the silicic acid powder. This is transferred to the top of a 3.5-g column of SilicAr CC-7, packed in diethyl ether–hexane, 3:7 (v/v). The column is then eluted with 50-ml portions of the same solvent, ethyl acetate, 5, 10, and 50% (v/v) methanol in ethyl acetate. Leukotrienes are eluted with 50 ml of methanol. After evaporation of the methanol, the residue is dissolved in 2 ml of methanol–water, 7:3 (v/v), also containing 0.1% acetic acid and adjusted to pH 5.4 with NH_4OH. Nondissolved material is removed by centrifugation (500 *g*, 5 min), and the solution is injected to a C_{18} Polygosil (500 × 10 mm) reversed-phase high-performance liquid chromatography (HPLC) columns eluted with the same solvent (4.5 ml/min). Leukotrienes are detected from their absorption of ultraviolet light at 280 nm. Approximate elution volumes for leukotrienes, during subsequent analyses on a smaller column, are given in Table I.

Characterization of Leukotrienes C and D

Leukotrienes C_4 and C_5 have characteristic ultraviolet (UV) spectra with absorbance maximum at 280 nm and shoulders at 270 and 291 nm.[3,4] The spectrum is due to the presence of a conjugated triene chromophore with an allylic thioether substituent. The spectra of leukotriene C_3 and 8,9-leukotriene C_3 have absorbance maxima at 279 nm and shoulders at 269 and 290 nm.[2,5] The products obtained by treating these leukotrienes with γ-glutamyl transpeptidase (leukotrienes D_4, D_5, D_3, and 8,9-D_3) have the same UV spectra as the corresponding precursor leukotriene C. 11-*trans* Isomers of leukotrienes have UV spectra that are shifted 2 nm hypsochromatically in comparison to the 11-*cis* compounds (λ_{max} at 278 and 277 nm, respectively[2,6]). The conjugated triene in leukotrienes C_4, C_5, D_4, and D_5 is extended to a conjugated tetraene by treatment of the compounds with soybean lipoxygenase.[3] This enzyme acts on polyunsaturated fatty acids containing two methylene-interrupted cis double bonds at the ω-6 and ω-9 positions. One molecule of O_2 is introduced at ω-6, and the ω-6 double bond is isomerized to ω-7. The conversions of leukotrienes (2.5 nmol, dissolved in 1 ml of Tyrode's buffer) with soybean lipoxygenase (10 μg) is monitored by ultraviolet spectroscopy. A 28-nm bathochromic shift of the spectrum is observed if the leukotriene contains cis double bonds at the Δ^{11} and Δ^{14} positions and additional double bonds at Δ^7 and Δ^9.

Further characterization can be achieved by amino acid analyses after hydrolyses and by gas-liquid chromatography–mass spectrometry after catalytic desulfurization. For amino acid analysis, approximately 2 nmol of leukotriene (based on the absorbance at 280 nm; ϵ = 40,000) is introduced into a glass ampoule; the solvent is evaporated under a stream of nitrogen, and 0.5 ml of 6 *N* HCl–0.5% phenol, containing 10 nmol of norleucine as internal reference, is added. The ampoule is evacuated, using an oil pump, to a pressure of less than 0.05 mm of Hg and sealed. After 21 hr at 110°, the ampoule is allowed to cool to room temperature, then opened; the residual HCl is removed by evaporation under vacuum over KOH. The residue is dissolved in application buffer for the amino acid analyzer (Beckman 121 M or other instrument with similar sensitivity).

For mass spectrometric analyses, 5–10 nmol of leukotriene in 0.5 ml of ethanol are mixed with Raney nickel (W-2) suspension (10 mg in 0.5 ml of ethanol). The mixture is stirred for 30 min at 70°, cooled to room temperature, diluted with 1 ml of 1 *N* HCl, and extracted twice with diethyl ether. The washed extracts are evaporated to dryness, and the residue is treated in succession with ethereal diazomethane and pyridine–hexamethyldisilazane–trimethylchlorosilane, 5:2:1, v/v/v (50 μl) to convert the product into methyl ester, *O*-trimethylsilyl derivative. The derivative has a C value of 21.5 when analyzed on a column of 3% OV-17. The mass spectrum[3] has intense ions at *m/e* 399, 367, 313, and 203, formed by elimination of $\cdot CH_3$, $\cdot CH_3$ plus CH_3OH and by cleavages between C-4 and C-5 or between C-5 and C-6 with charge retention on the oxygen atom at C-5, respectively. This applies to C_{20} leukotrienes with a hydroxyl group at C-5

[6] D. A. Clark, G. Goto, A. Marfat, E. J. Corey, S. Hammarström, and B. Samuelsson, *Biochem. Biophys. Res. Commun.* **94,** 1133 (1980).

(leukotrienes C_3, C_4, C_5, D_3, D_4, D_5 and corresponding 11-*trans* isomers). 8,9-Substituted leukotrienes (8,9-leukotrienes C_3 and D_3) give a derivative that has a C value of 22.0 and ions at *m/e* 399, 367, 271, and 245 (the latter two ions formed by cleavage between C-7 and C-8 or between C-8 and C-9, respectively).[5] The amino acid analyses of C-type leukotrienes should indicate the presence of 1 mol/mol each of glutamic acid and glycine and ca. 0.4 mol/mol of half-cystine,[7] and for leukotriene D 1 mol/mol of glycine, ca. 0.4 mol/mol of half-cystine, but no glutamic acid.[8]

Assay Method

The assay is based on chromatographic (reversed-phase HPLC) separation of substrate and product with detection of UV light absorption at 280 nm.

Reagents

- Tris-HCl buffer, 0.1 *M*, pH 8
- Leukotriene C, 5–20 μM, in methanol–water 7:3 (v/v)–0.1% acetic acid adjusted to pH 5.4
- Enzyme, dissolved in Tris buffer
- Acetic acid, 0.5% (v/v) in methanol

Procedure. Leukotriene C solution (0.1–0.25 ml; 0.5–5 nmol) is evaporated under a stream of argon in a small plastic tube. The dry residue is dissolved in 100 μl of Tris buffer (vortex or sonicate), and the tube is placed in a water bath at 37°. The reaction is started by addition of 10 μl of enzyme solution containing 0.1–5 mg of protein per milliliter. The reaction mixture is incubated for 30 min with gentle shaking. Acetic acid in methanol, 0.11 ml, is added to stop the reaction and precipitate the enzyme. After mixing and sedimenting the precipitate (700 *g*, 5 min) the clear solution is injected into a reversed-phase HPLC column (C_{18} Nucleosil, 250 × 4.6 mm). Methanol–water, 7:3 (v/v)–0.1% acetic acid, adjusted to pH 5.4 with NH_4OH, is used as mobile phase (flow rate: 1 ml/min). A UV light absorption detector, set at 280 nm, is attached to the column effluent and operated at a sensitivity of 0.01 absorbance unit for full-scale deflection of a 10 mV potentiometric recorder. The elution volumes for different leukotrienes C and D are shown in Table I. The conversions of leukotrienes C to D are ca. 90% under these conditions.

In the presence of glutathione, γ-glutamyl transpeptidase catalyzes the

[7] S. Hammarström, R. C. Murphy, B. Samuelsson, D. A. Clark, C. Mioskowski, and E. J. Corey, *Biochem. Biophys. Res. Commun.* **91,** 1266 (1979).

[8] L. Örning, S. Hammarström, and B. Samuelsson, *Proc. Natl. Acad. Sci. U.S.A.* **77,** 2014 (1980).

reaction: leukotriene D + glutathione → leukotriene C + cysteinylglycine.[9] The assay is performed as described above using leukotriene D (0.5 nmol) and glutathione (0.8 μmol) as substrates. The conversion of leukotriene D to C is approximately 70% under these conditions.

Inhibitors. A number of inhibitors of γ-glutamyl transpeptidase have been described. These compounds also inhibit the conversion of leukotrienes C to leukotrienes D. L-Serine–borate complex is a transition-state inhibitor that acts at relatively high concentrations. At equimolar concentrations of L-serine and sodium borate, 50% inhibition of leukotriene D formation (ID_{50}) was observed at 11 m*M* concentration of each constituent.[10] Acivicin [(L-(α*S*,5*S*)-α-amino-3-chloro-4,5-dihydro-5-isoxazolacetic acid] is an irreversible inhibitor of the enzyme that reacts with the glutamyl binding site.[11] Preincubation of the enzyme for 30 min with acivicin (0.25 m*M*) inhibits the conversion of leukotriene C to leukotriene D by more than 90%.

Purification Procedure

Partially purified swine kidney enzyme is commercially available (Sigma Chemical Co.). These preparations are often contaminated with a dipeptidase, which will degrade leukotrienes D further. By using a suitable enzyme to substrate ratio and a suitable incubation time, the latter reaction can be minimized. Although different batches of the commerical enzyme vary, typical conditions are 5 nmol of LTC per 100 μg of enzyme incubated at 37° for 30 min.

A simple purification procedure has been devised by Kozak and Tate.[11] Rat kidney brush border membranes are isolated according to the method of Malathi *et al.*[12] Papain-solubilized membrane proteins are fractionated by $(NH_4)_2SO_4$ precipitation and by chromatography on Sephadex G-150. Renal cortices from male Sprague–Dawley rats are homogenized in 15 volumes of 50 m*M* D-mannitol–2 m*M* Tris-HCl, pH 7.0, using a Potter–Elvehjem-type homogenizer. After addition of 1% (v/v) 1 *M* $CaCl_2$ solution, the mixture is left at 0° for 10 min (stir occasionally) and centrifuged at 3000 *g* for 15 min. The supernatant obtained is centrifuged for 20 min at 43,000 *g*. The second sediment is washed once in the same buffer

[9] S. Hammarström, *J. Biol. Chem.* **256,** 9573 (1981).

[10] L. Örning and S. Hammarström, *J. Biol. Chem.* **255,** 8023 (1980).

[11] E. M. Kozak and S. S. Tate, *FEBS Lett.* **122** 175 (1980).

[12] P. Malathi, H. Preiser, P. Fairclough, P. Mallett, and R. K. Crane, *Biochim. Biophys. Acta* **554,** 259 (1979).

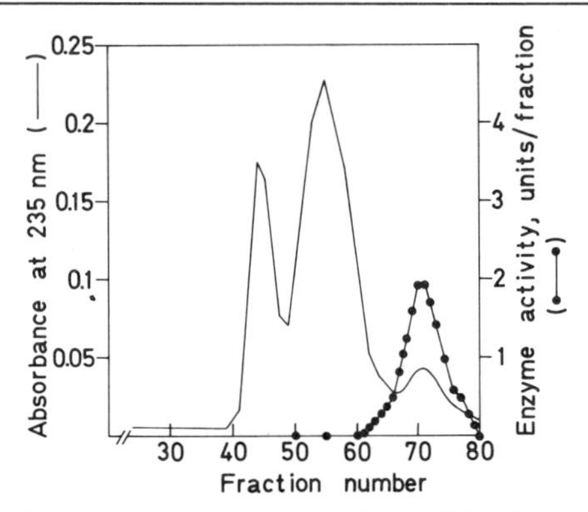

FIG. 1. Sephadex G-150 chromatography of papain-solubilized γ-glutamyl transpeptidase from rat kidney brush border membranes. Column, 1.5 × 90 cm; fractions, 1.3 ml. Modified from Kozak and Tate.[11]

and then resuspended in 10 m*M* phosphate buffer, pH 7.4–0.15 *M* NaCl–10 m*M* 2-mercaptoethanol (1 ml per gram of kidney). Papain (Sigma, 18 units/mg) is dissolved in the same buffer (10 mg/ml) and preincubated at 25° for 30 min. This solution (1 mg of papain per 10 mg of total membrane protein) is added to the resuspended brush border membranes, and the mixture is stirred slowly (25°, 2 hr). After a centrifugation at 43,000 *g* for 30 min, $(NH_4)_2SO_4$ is added to the supernatant (90% saturation). The precipitate, collected by centrifugation (18,000 *g*, 30 min), is dissolved in 50 m*M* Tris-HCl buffer, pH 8.0 (0.1 ml per gram of kidney) and applied to

TABLE II
PURIFICATION OF γ-GLUTAMYL TRANSPEPTIDASE FROM RAT KIDNEY[a]

Fraction	Volume (ml)	Protein[b] (mg)	Transpeptidase activity[c] Total (U)	Specific (U/mg)	Yield (%)
Homogenate (30 g of kidney)	441	3704	9920	2.7	(100)
Brush border membranes	29.5	264	6140	23.3	62
Papain treatment followed by Sephadex G-150 chromatography	2.7	6.5	5150	792	52

[a] From Kozak and Tate.[11]

[b] By the method of O. H. Lowry, N. J. Rosebrough, A. L. Farr, and R. J. Randall, *J. Biol. Chem.* **193,** 265 (1951).

[c] Determined with 1 m*M* L-γ-glutamyl-*p*-nitroanilide and 20 m*M* glycylglycine.

a column of Sephadex G-150 (2.5 × 100 cm), eluted with the same buffer. Fractions containing γ-glutamyl transpeptidase (Fig. 1) are combined and dialyzed against 100 volumes of water for 18 hr. After lyophilization the residue is dissolved in a small volume of water and stored frozen at −20°. The purification is summarized in Table II.

Alternative purification methods and additional properties of γ-glutamyl transpeptidase have been recently described in this series.[13]

[13] A. Meister, S. S. Tate, and O. W. Griffith, this series, Vol. 77, p. 237.

[6] Arachidonic Acid-15-Lipoxygenase from Rabbit Peritoneal Polymorphonuclear Leukocytes

By SHUH NARUMIYA and JOHN A. SALMON

Arachidonic acid + O_2 → 15-hydroperoxy-5,8,11,13-eicosatetraenoic (15-HPETE) acid

—nonenzymic→ 15-hydroxy-5,8,11,13-eicosatetraenoic acid (15-HETE)
15-keto-5,8,11,13-eicosatetraenoic acid (15-ketoETE)
13-hydroxy-14,15-epoxy-5,8,11-eicosatrienoic acid
11,14,15-trihydroxy-5,8,12-eicosatrienoic acid

The enzyme reaction occurs in polymorphonuclear leukocytes as well as other tissues of various species.[1-3] This reaction is distinctly different from that of fatty acid cyclooxygenase (PGH synthase), which also catalyzes peroxide formation at carbon 15 of arachidonic acid, since the lipoxygenase reaction is not inhibited by indomethacin.[4] This enzyme reaction is presumed to be the first step in the biotransformation of arachidonic acid to the 15-series of leukotrienes.[5] It is also suggested that the enzyme reaction is involved in inactivation of slow-reacting substance of anaphylaxis.[6]

[1] P. Borgeat, M. Hamberg, and B. Samuelsson, *J. Biol. Chem.* **251,** 7816 (1976).
[2] M. Hamberg and B. Samuelsson, *Biochem. Biophys. Res. Commun.* **61,** 942 (1974).
[3] P. Borgeat and B. Samuelsson, *Proc. Natl. Acad. Sci. U.S.A.* **76,** 2148 (1979).
[4] O. Radmark, C. Malmsten, and B. Samuelsson, *FEBS Lett.* **110,** 213 (1980).
[5] U. Lundberg, O. Radmark, C. Malmsten, and B. Samuelsson, *FEBS Lett.* **126,** 127 (1981).
[6] P. Sirois, *Prostaglandins* **17,** 395 (1979).

ISBN 0-12-181986-8

Assay Method[7]

Principle. The activity of arachidonic acid-15-lipoxygenase is assayed by measuring the sum of the radioactivities converted from [1-^{14}C]arachidonic acid to 15-HPETE and its decomposition products.

Reagents

Potassium phosphate buffer, 0.5 *M*, pH 7.0

$CaCl_2$, 0.1 *M*

[1-^{14}C]Arachidonic acid, 3 m*M* (15,000 cpm/nmol)

Citric acid, 2 *M*

Procedures. The assay mixture contains 10 μl of potassium phosphate buffer, 5 μl of $CaCl_2$, 10 μl of [1-^{14}C]arachidonic acid, enzyme, and H_2O in a total volume of 0.1 ml. The incubation is carried out at 30° for 2 min and terminated by the addition of 1 ml of methanol. After the insoluble materials have been removed by centrifugation, the supernatant is acidified by the addition of 20 μl of 2 *M* citric acid, and the products are extracted with 5 ml of diethyl ether. The products are chromatographed on silica gel thin-layer (LK5D, Whatman) with diethyl ether–*n*-hexane–acetic acid (60:40:1, v/v/v) and located by autoradiography on Kodak X-ray film. The mobilities of the products relative to that of arachidonic acid are: 15-keto-ETE, 0.87; 15-HPETE, 0.84; 15-HETE, 0.80; two isomers of 13-hydroxy-14,15-epoxyeicosatrienoic acid, 0.57 and 0.50; and 11,14,15-trihydroxy-5,8,12-eicosatrienoic acid, 0.02. Virtually no other HETEs were seen in the enzyme system described below. The silica gel zone corresponding to each product is scraped off, and the radioactivity is determined in a scintillation counter. One unit of enzyme activity is defined as the amount of enzyme that catalyzes the formation of 1 μmol of product in 2 min at 30°.

Purification Procedures

All operations are done at 0–4° unless specified.

Step 1. Crude Extract. Rabbit peritoneal polymorphonuclear leukocytes are collected by the method of Borgeat *et al.*[1] The collected cells are stored at −40° before use. About 28 g of cells are sonicated 15 times for 30 sec in 60 ml of 50 m*M* potassium phosphate buffer, pH 7.0, containing 1 m*M* EDTA. The homogenate is centrifuged successively at 10,000 *g* for 15 min and at 100,000 *g* for 60 min. The final supernatant is termed the crude extract.

[7] S. Narumiya, J. A. Salmon, F. H. Cottee, B. C. Weatherley, and R. J. Flower, *J. Biol. Chem.* **256,** 9583 (1981).

Step 2. Acetone Fractionation. Acetone precooled at −20° is added dropwise to the crude extract to give a final concentration of 40% (v/v). The mixture is stirred at −10° for 10 min and centrifuged at −10° at 5000 *g* for 10 min. The precipitates are dissolved in 10 ml of 10 m*M* potassium phosphate buffer, pH 6.0, and the insoluble materials are spun down by centrifugation at 10,000 *g* for 15 min.

Step 3. CM-Cellulose Column Chromatography. The acetone fraction (0 to 40%) is applied to a column of CM-52 (Whatman; 1.5 × 12 cm) equilibrated with 10 m*M* potassium phosphate buffer, pH 6.0. The column is washed with 30 ml of the same buffer, then a linear gradient elution is performed ranging from 10 to 200 m*M* potassium phosphate, pH 6.0, in a total volume of 80 ml. The enzyme is eluted at approximately 30 m*M* buffer concentration, and the fractions containing the enzyme are pooled and concentrated to 1.0 ml by ultrafiltration (the CM-52 fraction).

Step 4. Gel Filtration on Sephadex G-150. The concentrated CM-52 fraction is applied to a column of Sephadex G-150 (3 × 33 cm) equilibrated with 50 m*M* potassium phosphate buffer, pH 7.0. Elution is carried out with the same buffer at a flow rate of 10 ml/hr, and the active fractions are combined (the G-150 fraction).

Step 5. Hydroxyapatite Column Chromatography. The G-150 fraction is applied on a column of hydroxyapatite (type I, Sigma; 1.5 × 6.0 cm) equilibrated with 50 m*M* potassium phosphate buffer, pH 7.0. The column is washed with the same buffer, then elution is performed by a linear gradient ranging from 50 to 500 m*M* potassium phosphate buffer, pH 7.0, in a total volume of 60 ml. The enzyme is eluted at approximately 350 m*M* buffer concentration. The active fractions are pooled and concentrated by ultrafiltration to about 1 ml (the purified enzyme). A typical purification procedure is summarized in the table.

Purification of Arachidonic Acid-15-Lipoxygenase from Rabbit Leukocytes

Step	Protein (mg)	Total activity[a] (munits)	Specific activity (munits/mg)	Yield (%)	Purification (fold)
1. Crude enzyme	1104	828	0.75	100	1
2. Acetone	504	786	1.56	95	2
3. CM-52	96	254	2.64	31	3.5
4. Sephadex G-150	7	140	20.0	17	27
5. Hydroxyapatite	0.3	56	187	7	250

[a] One unit of enzyme activity is defined as the amount of enzyme that catalyzes the formation of 1 μmol of product in 2 min at 30° under the standard assay conditions.

Properties

Stability and Molecular Characteristics of Enzyme. The purified enzyme loses 30% of its original activity when kept at pH 7.0 for 24 hr, and 60% of the activity when stored at −40° for 1 month. Repeated freezing and thawing causes a considerable loss of activity. The molecular weight of the enzyme is assessed to be 61,000 by gel filtration on a Sephadex G-150 column.

Kinetics and pH Optimum. The enzyme reaction proceeds linearly for 2 min at 30° and for 1 min at 37°. The K_m value for arachidonic acid is determined to be 28 μM, and the enzyme has a pH optimum at pH 6.5.

Inhibitors. The enzyme activity is inhibited by sulfhydryl blocking agents; 30% inhibition is observed at 0.1 m*M* (*p*-chloromercuribenzoate (PCMB), and at 1.0 m*M* iodoacetamide. Among the specific inhibitors of arachidonic acid metabolism, eicosatetraynoic acid and BW755C[8] inhibit the reaction with the respective IC_{50} values of 0.5 μg/ml and 20 μg/ml. Indomethacin has no inhibitory effect at 100 μg/ml.

Reaction Products. 15-HPETE formed in the enzymic reaction decomposes quickly in the incubation mixture in a fashion similar to that reported for other hydroperoxy unsaturated fatty acids.[9,10] The decomposed products isolated on silica gel thin-layer chromatography are derivatized to methyl ester TMS-ethers and identified by gas chromatography–mass spectrometry. Among the decomposed products identified are 15-hydroxy-5,8,11,13-eicosatetraenoic acid, 15-keto-5,8,11,13-eicosatetraenoic acid, two cis/trans isomers of 13-hydroxy-14,15-epoxy-5,8,11-eicosatrienoic acid, and 11,14,15-trihydroxy-5,8,12-eicosatrienoic acid. This decomposition is observed in the reaction using the purified enzyme.

Other Properties. Divalent cations such as calcium, magnesium, and manganese activate the enzyme activity 7- to 10-fold in the reaction with the crude extract. Such dependence on divalent cations is not observed in the enzyme preparation subsequent to step 3 (the CM-52 column chromatography).

[8] G. A. Higgs, R. J. Flower, and J. R. Vane, *Biochem. Pharmacol.* **28,** 1959 (1979).
[9] H. W. Gardner, R. Kleiman, and D. Weisleder, *Lipids* **9,** 696 (1974).
[10] M. Hamberg, *Lipids* **10,** 87 (1975).

[7] Arachidonic Acid-12-Lipoxygenase from Bovine Platelets

By D. H. NUGTEREN

Lipoxygenases catalyze the incorporation of one oxygen molecule into di- or polyunsaturated fatty acids. Reaction occurs at a suitably located 1,4-*cis,cis*-pentadiene system of the fatty acid resulting in a hydroperoxy-trans,cis-conjugated fatty acid:

$$\underset{\text{cis}}{—CH{=}CH}—CH_2—\underset{\text{cis}}{CH{=}CH}— + O_2 \rightarrow —CHOOH—\underset{\text{trans}}{CH{=}CH}—\underset{\text{cis}}{CH{=}CH}—$$

The reaction not only takes place with positional specificity, but is also stereospecific; in other words, the product formed is optically active.

Lipoxygenase was discovered in soybeans in 1932,[1] and similar enzymes were detected later in a large number of plant species. Until relatively recently, the enzyme had been clearly demonstrated only in plants; claims for its presence in animal tissues were confused by the widespread occurrence of hematin catalysis. The enzyme prostaglandin endoperoxide synthase, discovered in 1964 in sheep vesicular glands, has lipoxygenase-like activity: it can convert (in the presence of hydroquinone) certain di-unsaturated fatty acids, which are not suitable for prostaglandin formation, into conjugated monohydroxy acids, e.g., 11-*cis*,14-*cis*-20:2 → 11-L-OH-12-*trans*, 14-*cis*-20:2.[2] However, the first true lipoxygenase of animal origin was discovered in 1974 in blood platelets.[3,4] This enzyme converts arachidonic acid quantitatively into 12-L-hydroperoxy-5-*cis*,8-*cis*,10-*trans*,14-*cis*-eicosatetraenoic acid (12-HPETE), which yields 12-OH-20:4 (12-HETE) after enzymic or chemical reduction. Still more recent is the detection of a 15-HETE-forming lipoxygenase in certain cells derived from leukocytes and the isolation of 5-HETE from basophilic cells as an intermediate (or by product) during the biosynthesis of slow-reacting substance of anaphylaxis (SRS-A; leukotriene).

We describe here the assay, the partial purification, and the properties of arachidonic acid-12-lipoxygenase, which is relatively easily available in large quantities from blood platelets of slaughterhouse animals.

[1] E. André and K. W. Hon, *C. R. Hebd. Seances Acad. Sci.* **194,** 645 (1932).

[2] D. H. Nugteren, D. A. van Dorp, S. Bergström, M. Hamberg, and B. Samuelsson, *Nature (London)* **212,** 38 (1966).

[3] M. Hamberg and B. Samuelsson, *Proc. Natl. Acad. Sci. U.S.A.* **71,** 3400 (1974).

[4] D. H. Nugteren, *Biochim. Biophys. Acta* **380,** 299 (1975).

METHODS IN ENZYMOLOGY, VOL. 86

ISBN 0-12-181986-8

Enzyme Assays

Lipoxygenase activity can be determined in intact platelets or crude, turbid enzyme preparations by incubations with radioactive arachidonic acid. The products are separated by thin-layer chromatography (TLC), and their radioactivity is measured. A simpler and quicker method can be used in more purified, clear enzyme solutions: the increase in UV absorption at 234 nm due to the formation of a conjugated system of double bonds is followed continuously.

Reagents

Buffer: 0.1 *M* Tris-HCl + 0.1 *M* K_2HPO_4, pH 8.2

Citric acid, 0.2 *M* and 2 *M*

TLC Solvent system: chloroform–methanol–acetic acid–water (90:8:1:0.8, v/v/v/v). Plates: Merck Fertigplatten, silica gel 60 F254, 20 cm × 20 cm, 0.25 mm thick (with fluorescence indicator). Spraying: 10% phosphomolybdic acid in 96% alcohol and heating at 120° for about 5 min.

Arachidonic acid (e.g., from Sigma) dissolved in toluene (1–4 mg/ml), [1-^{14}C]Arachidonic acid [50 mCi/mmol from New England Nuclear (NEN)] dissolved in toluene (100 μg/ml). The solutions are stored under nitrogen at −20°. If necessary, the fatty acid is purified by TLC (R_f = 0.80). Just before use, the amounts of cold and/or labeled material needed are pipetted together in one test tube, the solvent is removed by a stream of nitrogen, and the residue is dissolved in buffer.

12-L-Hydroxyeicosatetraenoic acid (12-HETE): This compound can be prepared by large-scale incubation of arachidonic acid with blood platelets; it is useful as a standard and as a reference: 1.5 mg of arachidonic acid are dissolved in 200 ml of buffer, and a suspension of bovine blood platelets (300 mg of protein, equal to about 2×10^{11} platelets) in 20 ml of saline–EDTA (see below) is added. The mixture is incubated aerobically for 60 min at 30° while shaking gently. Subsequently, 10 ml of 2 *M* citric acid are added to give pH 4–5, and the mixture is extracted twice with 250 ml of ether. The extract is evaporated to dryness under vacuum, and the residue is purified by preparative TLC as 16-cm bands on two 0.25-mm silica gel plates. The 12-HETE (R_f = 0.68) can easily be seen as a dark band under a 254 nm UV lamp; this compound is sufficiently separated from arachidonic acid (R_f = 0.80) and the other products 12-L-hydroxyheptadecatrienoic acid (12-OH-17:3) (R_f = 0.60) and thromboxane B_2 (R_f = 0.24). The band is scraped off and eluted twice with 10 ml of ether–methanol (9:1, v/v). The yield is 25–50%.

Incubation with Radioactive Arachidonic Acid. [1-^{14}C]Arachidonic acid (4.5 μg, 15 nmol, about 90,000 dpm) in 2 ml of buffer is incubated aerobically with a suitable amount of enzyme preparation (e.g., 1 mg of protein) at 20° for various periods of time up to 15 min. The reaction is stopped by addition of 1 ml of 0.2 *M* citric acid to give a pH of about 4. The mixture is extracted twice with 4 ml of ether. The extract is evaporated by a stream of nitrogen, and the residue, dissolved in chloroform–methanol (3:1, v/v) is brought on a TLC plate as a 1-cm band. After development, the compounds are detected by phosphomolybdic acid or, alternatively, by autoradiography or a radioactivity scanner. With the TLC system used, the hydroperoxy(HPETE)- and the hydroxy(HETE)- eicosatetraenoic acid have practically the same R_f (0.68), and the sum of these two products is a good measure of the lipoxygenase activity. The radioactive bands are scraped into vials and counted after addition of 1 ml of methanol and 10 ml of 0.4% Omnifluor (NEN). The conversion into HETE in nanomoles is plotted against time of incubation, and from the resulting curve the activity of the enzyme preparation can be calculated.

Spectroscopic Assay. Arachidonic acid solution (9 μg, 30 nmol, in 1.5 ml of buffer) is pipetted into a cuvette (light path 1 cm). The enzyme preparation plus additional buffer is added to an end volume of 2.0 ml. The increase in absorption at 234 nm is followed at 20° for about 15 min. (If the self-absorption of the enzyme solution is too high at 234 nm, the assay can be carried out at 240 nm or 246 nm.) Extinction coefficients of HETE are ϵ_{234} = 27,600; ϵ_{240} = 22,000; ϵ_{246} = 14,200. Thus formation of HETE of 1 nmol per minute results in an increase in absorption (ΔA) per minute, at 234 nm, 0.014; at 240 nm, 0.011; and at 246 nm, 0.007.

Partial Purification of Arachidonate Lipoxygenase

The enzyme is found for the greater part in the 100,000 *g* supernatant after freezing and thawing the platelets three times. Further fractionation is possible with $(NH_4)_2SO_4$, but the enzyme becomes rather labile and additional purification steps were unsuccessful in our hands.

Reagents

- Sodium citrate (3,8%), pH 7.2
- Saline–EDTA: 45 g of NaCl and 18.6 g of disodium EDTA·2 H_2O in 5 liters of distilled water; pH adjusted to 7.2 at 20° with about 20 ml of 2.5 *N* NaOH
- Hypertonic saline: 36 g of NaCl and 14.9 g of disodium EDTA·2 H_2O in 1 liter of distilled water; pH 7.2
- Buffer: 0.1 *M* Tris-HCl; 0.1 *M* K_2HPO_4; pH 8.2

Isolation of Bovine Platelets. Blood from slaughtered cows (8 liters) is collected in a 10-liter plastic flask containing 1 liter of sodium citrate. After thorough mixing, the blood is stored overnight in the cold room at 4°. (This results in a better buffy coat after centrifugation.) All further procedures are carried out at 0–4°, the later steps are followed by microscopy. The blood in 1-liter plastic bottles is centrifuged for 1 hr at 1500 *g* (2500 rpm in an MSE Mistral 4 L, rotor 59563). The clear plasma is cautiously poured off, after which the buffy coat on top of the erythrocytes, containing the leukocytes and the platelets, can be collected. In this manner, 400–500 ml of a mixture of plasma, erythrocytes, leukocytes, and platelets is obtained from 8 liters of whole blood. Erythrocytes are removed from this preparation by an osmotic shock as follows: The volume is brought to exactly 500 ml with saline–EDTA and 1500 ml of distilled water are rapidly added while shaking thoroughly. Precisely 30 sec later, 500 ml of hypertonic saline are added and the solutions are rapidly mixed. Centrifugation for 1 hr at 1500 *g* (see above) results in slightly yellowish pellets consisting of a mixture of platelets and leukocytes (weight ratio about 1:1). The combined pellets are suspended in about 200 ml of saline–EDTA by shaking and gently rubbing with a plastic spoon. The suspension is centrifuged for 10 min at 5000 *g*. The washed cells are separated by several centrifugations in saline–EDTA—using a cooled swing-out low-speed centrifuge—into leukocytes, which precipitate during 5 min at 200 *g*, and platelets, which need 20 min at 4000 *g*. The platelet suspension finally obtained contains 500–900 mg of protein in 40–60 ml of saline–EDTA and is contaminated with less than 10% (by weight) of leukocytes.

Freezing and Thawing. The platelet suspension (300–400 mg of protein) is centrifuged for 20 min at 4000 *g*. The pellet is suspended in 20 ml of buffer in a 100-ml plastic centrifuge tube, and the mixture is slowly frozen by rotating the tube in a Dry Ice–acetone bath. Subsequently, the contents are allowed to thaw in a cold (0–4°) water bath. This freezing and thawing cycle is repeated twice, after which a thick flocculent suspension is obtained.

Centrifugation. The frozen and thawed preparation is centrifuged for 15 min at 1500 *g*. A substantial white pellet is obtained, which is washed with 10 ml of buffer. The combined slightly turbid supernatants are centrifuged for 60 min at 100,000 *g* (30,000 rpm in a Beckman–Spinco L, rotor 50). The clear pink supernatant is saved.

Ammonium Sulfate Precipitation. To the 100,000 *g* supernatant (28 ml) is added 5.6 g of solid $(NH_4)_2SO_4$ (33% saturation). Centrifugation for 10 min at 1000 *g* yields a white pellet that is dissolved in 8 ml of buffer. A clear, almost colorless solution is obtained, containing the bulk of the arachidonate lipoxygenase.

TABLE I
PURIFICATION OF ARACHIDONATE LIPOXYGENASE FROM BOVINE PLATELETS

Fraction	Volume (ml)	Protein (mg)	Specific activity (nmol/min mg^{-1})	Total activity[a] (units)
Intact platelets	17	345	0.6	0.21
After freezing and thawing	22	345	0.4	0.14
100,000 *g* Supernatant	28	105	0.9	0.09
$(NH_4)_2SO_4$, 0–35%	8	40	1.8	0.07

[a] A unit = 1 μmol of substrate converted per minute at 20°.

The results of a typical experiment are given in the table; steps 2–4 were done within 6 hr.

Properties of the Enzyme

The lipoxygenase in intact platelets is rather stable and can be kept for at least 1 week at 0°. However, after breaking the platelets, the enzyme becomes very labile; e.g., the $(NH_4)_2SO_4$ fraction loses half its activity during 24 hr at 0°. After freezing and thawing, the thromboxane synthesizing system is found almost exclusively in the particulate fraction, and part of the lipoxygenase is also recovered in this fraction.[5] As a matter of fact, the soluble arachidonate lipoxygenase is possibly present in aggregated form: gel filtration on agarose 0.5 *M* indicated a molecular weight higher than 300,000[6] and some lipid is still present.

Substrate Specificity. A minimum requirement for every lipoxygenase is the presence of two cis double bonds separated by a methylene group. Regarding the requirement for the position of this double-bond system, arachidonate lipoxygenase has apparently an *n*-9, *n*-12 specificity: 8,11-20:2 is converted but 11,14-20:2 is not.[4,6] This is just the opposite of the specificity of the prostaglandin endoperoxide synthase.[2] Only free fatty acids are converted; the methyl ester of arachidonic acid is not a substrate. More details about the substrate specificity are given by Nugteren.[4,6]

Cofactor Requirements. Apart from molecular oxygen, no further cofactor or other stimulator is required to obtain a maximal reaction rate. It is not yet known whether nonheme iron (as in the case of soybean lipoxy-

[5] P. P. K. Ho, C. P. Walters, and H. R. Sullivan, *Biochem. Biophys. Res. Commun.* **76,** 398 (1977).

[6] D. H. Nugteren *in* "Prostaglandins in Hematology" (M. J. Silver, J. B. Smith, and J. J. Kocsis, eds.), p. 11. Spectrum, New York, 1977.

genase) or perhaps heme is present in the active center of the enzyme. KCN (10^{-3} M) does not inhibit the reaction.

Inhibitors. Arachidonate lipoxygenase is strongly inhibited by eicosatetraynoic acid and other ynoic acids. 5,8,11-Eicosatriynoic acid is to a certain extent a selective inhibitor.[7] Other inhibitory compounds are antioxidants (hydroquinone), epinephrine, and Cu^{2+}.[5,8] Serum albumin prevents conversion because it strongly binds the fatty acid substrate.[4] It has been reported that 15-HETE is a potent and selective inhibitor of platelet lipoxygenase.[9] Anti-inflammatory drugs (aspirin, indomethacin) are not inhibitory.

Occurrence in Various Tissues. 12-HETE acid can be synthesized in many mammalian tissues, such as spleen, lung, skin. Platelets, are present in blood throughout the circulation, and they easily stick to the vessel walls, so that, in principle, we can expect to find them in every organ. Especially lung and spleen are organs known to accumulate platelets, and these are the very organs in which much arachidonate lipoxygenase is found. The nucleated thrombocyte of the chicken contains the enzyme as well,[10] and also the gills of fish can make relatively large quantities of 12-HETE. Arachidonate lipoxygenase is apparently widespread throughout the animal kingdom.

Biological Function. A clear biological function of this 12-lipoxygenase has not yet been found: 12-HETE has chemotactic and chemokinetic properties toward leukocytes,[11] but this is probably not of much specific physiological relevance, because other mono-HETEs, e.g., 5-HETE, are equally active, and leukotriene B (5,12-di-HETE) is 100–1000 times more active as a chemokinetic agent than the mono-HETEs.[12]

[7] S. Hammarström, *Biochim. Biophys. Acta* **487,** 517 (1977).

[8] G. J. Blackwell and R. J. Flower, *Prostaglandins* **16,** 417 (1978).

[9] J. Y. Vanderhoek, R. W. Bryant, and J. M. Bailey, *J. Biol. Chem.* **255,** 5996 (1980).

[10] M. Claeys, E. Wechsung, A. G. Herman, and D. H. Nugteren, *Prostaglandins* **21,** 739 (1981).

[11] S. R. Turner, J. A. Tainer, and W. S. Lynn, *Nature (London)* **257,** 680 (1975).

[12] A. W. Ford-Hutchinson, M. A. Bray, M. V. Doig, M. E. Shipley, and M. J. H. Smith, *Nature (London)* **286,** 264 (1980).

[8] Purification and Assay of PGH Synthase from Bovine Seminal Vesicles

By SHOZO YAMAMOTO

An enzyme preparation of prostaglandin H (PGH) synthase purified to an apparent homogeneity from the microsomes of bovine seminal vesicle, catalyzes two reactions[1] as shown in Fig. 1. One is the fatty acid cyclooxygenase reaction producing prostaglandin G_2 (PGG_2) from arachidonic acid. This is a bisdioxygenase reaction incorporating two molecules of oxygen into arachidonic acid. The other is the prostaglandin hydroperoxidase reaction in which the 15-hydroperoxide of PGG_2 is converted to a hydroxyl group to produce PGH_2. These two enzyme activities have not been resolved in various experimental conditions, the results suggesting that either two separate enzyme activities are attributed to one enzyme protein or two separate enzyme proteins are tightly bound.[1,2] Thus, the preparation with the two enzyme activities is referred to as PG endoperoxide synthetase[1] or PGH synthase.

Assay Methods

Reagents

Tris-HCl, 0.2 *M*, pH 8.0

Hematin, 20 μM: About 20 mg of hematin are dissolved in 25 ml of 0.1 *N* NaOH. Insoluble materials are removed by filtration. A 20-μl portion is mixed with 1.98 ml of 0.1 *N* NaOH. Sodium dithionite (about 1 mg) is added and stirred with a microspatula. Pyridine (0.5 ml) is added. Absorption of pyridine hemochromogen at 557 nm is measured. On the basis of ϵm*M* at 557 nm (34.4), the original hematin solution is diluted to 200 μM with water and stored at $-20°$. For daily use a 20-μM solution is also prepared. Storage at $-20°$ for several months does not cause the loss of cofactor activity with either 20 μM or 200 μM hematin solutions.

Manganese protoporphyrin IX, 40 μM, purchased from Porphyrin Products (Logan, Utah). A solution at 1 m*M* in 0.1 *N* NaOH is prepared on the basis of a molecular weight of 632.58 and diluted to 40 μM with water. Storage at $-20°$ for several months causes no

[1] T. Miyamoto, N. Ogino, S. Yamamoto, and O. Hayaishi, *J. Biol. Chem.* **251,** 2629 (1976).
[2] S. Ohki, N. Ogino, S. Yamamoto, and O. Hayaishi, *J. Biol. Chem.* **254,** 829 (1979).

ISBN 0-12-181986-8

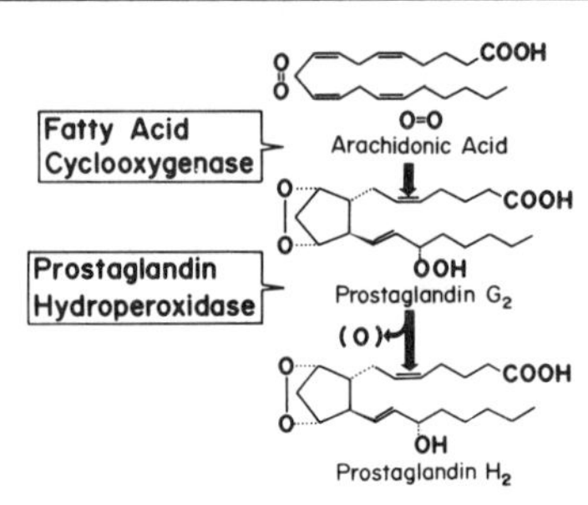

FIG. 1. Reactions catalyzed by fatty acid cyclooxygenase and prostaglandin hydroperoxidase.

loss of cofactor activity with either 40 μM or 1 mM solutions of manganese protoporphyrin IX.

L-Tryptophan, 50 mM. L-Tryptophan is dissolved in water to a concentration of 50 mM by keeping the test tube in hot water. When solutions are stored at $-20°$, the sample must be warmed in hot water to obtain a clear solution.

[1-^{14}C]Arachidonic acid. Radioactive arachidonic acid (55.5 mCi/mmol) is mixed with nonradioactive arachidonic acid to make a solution containing 5 nmol and 100,000 cpm per 5 μl of ethanol.

[1-^{14}C]Prostaglandin G_2, prepared as described by Ohki *et al.*[2] starting with [1-^{14}C] arachidonic acid (50,000 cpm/5 nmol). The final solution is adjusted to a concentration of 5 nmol and 50,000 cpm per 5 μl of diethyleneglycol dimethyl ether or acetone and stored at $-70°$.

Enzyme, prepared as described below.

Fatty Acid Cyclooxygenase Assay. A 100-μl reaction mixture contains 0.2 M Tris-HCl, pH 8.0 (50 μl), 40 μM manganese protoporphyrin IX (10 μl), [1-^{14}C]arachidonic acid (5 μl of ethanol solution containing 5 nmol and 100,000 cpm), and enzyme (usually 5–10 ul) in a 12 × 103 mm test tube. Before the addition of enzyme the reaction mixture is mixed well with a vortex mixer to disperse arachidonic acid. The reaction is carried out with shaking at 24° for 2 min, and terminated by the addition of 0.3 ml of a mixture of ethyl ether–methanol–0.2 M citric acid (30:4:1, v/v/v) precooled to $-20°$. The content of the tube is mixed using a vortex mixer and then kept in an ice bath. About 0.5 g of anhydrous sodium sulfate is added for dehydration of the organic phase. The reaction mixture is gently mixed. The following procedures are performed in a cold room at 4°. A 150-μl portion of the upper organic layer is removed using a graduated capillary tube and placed on a Merck precoated silica gel 60 F254 glass plate (200 × 200 × 0.25 mm) in a width of 1.5 cm at the origin 2 cm above

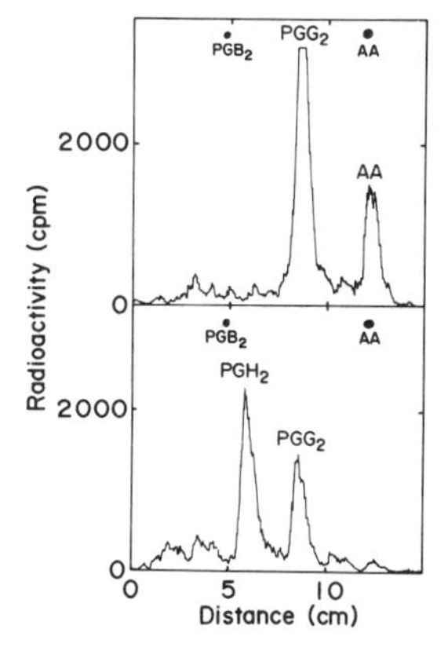

FIG. 2. Thin-layer chromatograms of fatty acid cyclooxygenase (upper) and prostaglandin hydroperoxidase (lower) reactions.

the bottom. The solvent is evaporated under a stream of cold air. The plate is developed in ethyl ether–petroleum ether–acetic acid (85:15:0.1, v/v/v) in a glass tank kept in a deep-freeze at −20°. Development to a height of 17 cm takes about 50 min. Arachidonic acid and PGB_2, if available, are placed on the plate as markers and visualized with iodine vapors. PGG_2 can be located between these two markers. Distribution of radioactivity on the plate can be monitored either by a radiochromatogram scanner or by autoradiography. For the latter method, overnight exposure of the plate to an X-ray film used for routine chest X-ray examination is sufficient to detect the radioactivity utilized for one assay. A typical chromatogram is presented in Fig. 2 (top). Silica gel in the regions corresponding to arachidonic acid and PGG_2 is scraped into scintillation vials, and radioactivity is measured by a scintillation spectrometer in a toluene solution containing 0.03% *p*-bis[2-(4-methyl-5-phenyloxazolyl)]benzene and 0.5% 2,5-diphenyloxazole. The percentage of radioactivity in the section of arachidonic acid or PGG_2 is calculated by dividing the counts per minute value found in each section by the total radioactivity recovered. The amount of PGG_2 (in nanomoles) is then calculated from the added arachidonic acid (5 nmol) multiplied by the percentage of conversion to PGG_2.

The cyclooxygenase activity can also be assayed by following oxygen consumption. The volume of the reaction mixture must be increased depending on the size of the oxygen electrode cell.

Prostaglandin Hydroperoxidase Assay. A 100-μl reaction mixture contains 0.2 *M* Tris-HCl, pH 8.0 (50 μl), 20 μ*M* hematin (10 μl), 50 m*M* L-tryptophan (10 μl), [1-^{14}C]PGG_2 (5 μl of solution containing 5 nmol and 50,000 cpm), and enzyme (usually 5–10 μl). The reaction is started by the addition of PGG_2, and the reaction mixture is mixed with a vortex mixer to disperse PGG_2. After a 1-min incubation at 24°, the reaction is terminated and the products are extracted and subjected to thin-layer chromatography as described above for the cyclooxygenase assay. A typical radiochromatogram is shown in Fig. 2 (bottom).

Enzyme Purification

Bovine seminal vesicles are cleaned by removing fat, muscle, and connective tissues and stored at −70°. Frozen seminal vesicles (400 g) are thawed, cut into small pieces with scissors, and homogenized in a Waring blender for 1 min with one volume of 20 m*M* potassium phosphate buffer at pH 7.3. Then two volumes of the same buffer are added, and 1-min homogenizations are performed three more times. The tissue homogenate is centrifuged at 7,000 *g* (in a Sorvall refrigerated centrifuge Model RC-5 equipped with a rotor Model GSA) for 10 min. The supernatant is further centrifuged at 142,800 *g* (in a Beckman preparative ultracentrifuge equipped with a rotor type 35) for 90 min. The precipitate is suspended in 20 m*M* potassium phosphate, pH 7.3, to a final volume to fill all the centrifuge tubes for one run and centrifuged at 142,800 *g* for 90 min. The washed precipitate is dispersed in one-third the volume of the same buffer (microsomal fraction).

The microsomal fraction is mixed with an equal volume of 20 m*M* phosphate, pH 7.4, containing 2% Tween 20 and 40% glycerol. The mixture is stirred for 30 min in an ice bath and centrifuged at 142,800 *g* for 90 min. The supernatant is applied to a column of Whatman DE-52 (5 × 30 cm) equilibrated with 20 m*M* potassium phosphate, pH 7.4, containing 0.1% Tween 20 and 20% glycerol. The column is washed with the same buffer, and active fractions in the wash are collected and combined (approximately 400 ml), followed by concentration to less than 80 ml with the aid of a Diaflo membrane XM-50 (62 mm in diameter). The concentrated enzyme is subjected to electrofocusing in an LKB column (440 ml). Water at 4° is circulated to cool the column. The enzyme is focused in a 1% LKB Ampholine carrier ampholyte solution (pH 3.5–10) containing 0.1% Tween 20. A glycerol density gradient is established using an LKB Ampholine stirrer motor according to the instruction manual supplied from LKB. The voltage is increased from 400 up to 900 V during the initial 20 hr and then kept at 900 V for about 60 hr. The contents of the column are collected in 5-ml fractions. Absorbance at 280 nm is followed continuously by an ultraviolet detector. The pH value of each fraction is measured using a pH meter. The enzyme appears in fractions in a range of pH 6.6–7.5 (Fig. 3). To obtain a highly purified enzyme preparation, only

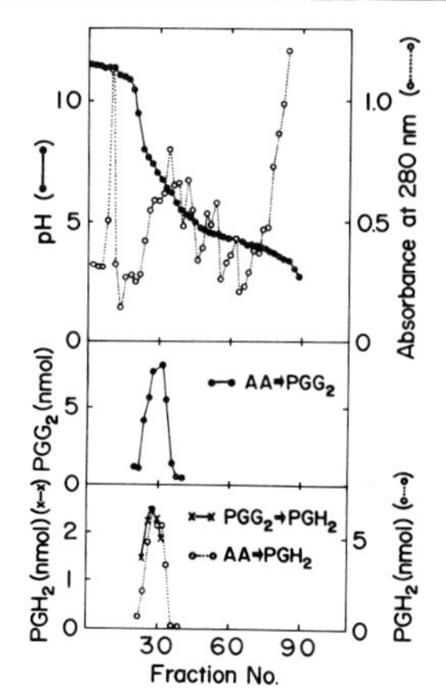

FIG. 3. Isoelectrofocusing pattern of prostaglandin endoperoxide synthase. *Upper:* pH and protein profiles; *middle:* fatty acid cyclooxygenase activity; *lower:* prostaglandin hydroperoxidase activity.

fractions of a major ultraviolet absorbing peak in the pH region are collected. For a better yield, all the active fractions are collected. The enzyme solution is concentrated to less than 30 ml using a Diaflo membrane PM-50 (62 mm in diameter). To remove the contaminating ampholytes, the enzyme is applied to a Sepharose 6B column (3.5 × 50 cm) equilibrated with 20 m*M* potassium phosphate, pH 7.4, containing 0.05% Tween 20 and 20% glycerol. Active fractions (about 80 ml) are collected and concentrated to about 12 ml. Starting with 400 g of bovine seminal vesicles, about 10 mg of the purified enzyme are obtained. The purified enzyme has a specific enzyme activity of about 2.5 μmol/min per milligram of protein determined by the standard assay. The protein concentration is determined by the method of Lowry *et al.*[3] with bovine serum albumin as a standard.

Properties of the Enzyme

Previous papers describe properties of the enzyme in detail.[1,2,4] In this chapter the description is limited to the properties of the enzyme that are closely related to the assay and the purification of the enzyme.

Hematin is required for both the cyclooxygenase and hydroperoxidase activities. The enzyme activities are scarcely detectable without hematin

[3] O. H. Lowry, N. J. Rosebrough, A. L. Farr, and R. J. Randall, *J. Biol. Chem.* **193,** 265 (1951).

[4] N. Ogino, S. Ohki, S. Yamamoto, and O. Hayaishi, *J. Biol. Chem.* **253,** 5061 (1978).

in the assay mixture. The presence of 2 μM hematin in the standard assay mixture gives maximum activities of cyclooxygenase and hydroperoxidase. Manganese protoporphyrin IX replaces hematin in the cyclooxygenase reaction, but not in the hydroperoxidase reaction.[4] The hydroperoxidase reaction requires tryptophan in addition to hematin, and the tryptophan can be replaced by various other compounds, such as guaiacol, epinephrine, hydroquinone, phenol, and serotonin.[2] Some of these compounds inhibit the reaction at higher concentrations, but such an inhibition is not observed with tryptophan, which is used in the routine assays for this reason. Incubation of the enzyme with arachidonic acid produces prostaglandin H_2 rather than G_2 if tryptophan is present in the reaction mixture.

The purified enzyme can be stored at $-70°$ for many months without appreciable loss of activity. Before using the frozen enzyme, preincubation of the enzyme solution at 24° for 15–30 min is recommended to obtain reproducible results. The interaction of the enzyme and hematin (or manganese protoporphyrin) causes rapid inactivation of the enzyme.[4] The enzyme is protected from such inactivation by the presence of tryptophan. Therefore, preincubation of enzyme and hematin for temperature equilibrium should be avoided unless tryptophan is present.

Owing to self-destruction of the enzyme during catalysis, the enzyme reaction slows down and stops 60–90 sec after the start of the reaction.[4] Therefore, the 2-min incubation performed in the routine assay does not allow a precise determination of the initial velocity of reaction.

8,11,14-Eicosatrienoic acid and PGG_1 are also substrates for cyclooxygenase and hydroperoxidase at reaction rates nearly equal to those of arachidonic acid and PGG_2, respectively.

[9] Purification of PGH Synthase from Sheep Vesicular Glands

By F. J. G. VAN DER OUDERAA and M. BUYTENHEK

Prostaglandin H_2(PGH_2) synthase is a membrane-bound glycoprotein catalyzing the conversion of certain polyunsaturated fatty acids into prostaglandin endoperoxides in two distinct reaction steps.[1–4] On using ara-

[1] T. Miyamoto, N. Ogino, S. Yamamoto, and O. Hayaishi, *J. Biol. Chem.* **251,** 2629 (1976).

[2] F. J. van der Ouderaa, M. Buytenhek, D. H. Nugteren, and D. A. van Dorp, *Biochim. Biophys. Acta* **487,** 315 (1977).

[3] M. E. Hemler, W. E. M. Lands, and W. L. Smith, *J. Biol. Chem.* **251,** 5575 (1976).

[4] F. J. van der Ouderaa, F. J. Slikkerveer, M. Buytenhek, and D. A. van Dorp. *Biochim. Biophys. Acta* **572,** 29 (1979).

METHODS IN ENZYMOLOGY, VOL. 86

ISBN 0-12-181986-8

chidonic acid, for example, 15-hydroperoxyprostaglandin H_2 is formed in the first step as a consequence of the cyclooxygenase activity. This compound is converted into PGH_2 catalyzed by the PGG → PGH peroxidase activity of the same enzyme. Sheep vesicular glands are the richest available source of the enzyme.

Assay Methods

Three different procedures for enzyme analysis have been used: (*a*) an assay for the cyclooxygenase activity based on a polarographic oxygen determination[2]; (*b*) an assay for the PGG → PGH peroxidase activity based on a colorimetric procedure[2]; (*c*) analysis of products as resulting from conversion of radioactively labeled arachidonic acid.[1,5]

Generally the cyclooxygenase determination as mentioned under (*a*) is used as a fast specific routine method to measure enzyme activity during the various purification steps of the enzyme. A combination of both the assays for the cyclooxygenase and peroxidase activities has been used to establish that both activities reside in the same molecule[2] and when studying enzyme inhibitors, most of which have been reported to inhibit the cyclooxygenase only.[2-10]

Cyclooxygenase Assay

Reagents

Incubation buffer: 0.1 *M* Tris-HCl, pH 8.0, kept at 25°; air is continuously bubbled through

Arachidonic acid stock solution: 10 mg/ml in toluene kept under nitrogen at −20°; assay solution: 0.5 mg/ml in incubation buffer; freshly prepared each day, kept at room temperature

Hydroquinone, 0.23 m*M* in distilled water, freshly prepared

Hemin (Calbiochem) freshly dissolved in a small volume of 0.01 *M* NaOH, adjusted to pH 8 with concentrated incubation buffer; the final concentration of hemin is 1 μM

The hydroquinone and hemin solutions are kept on ice

Procedure. The cyclooxygenase activity is measured at 25° by monitoring the oxygen consumption as a result of the cyclooxygenation of the

[5] D. H. Nugteren and E. Hazelhof, *Biochim. Biophys. Acta* **326,** 448 (1973).

[6] F. J. van der Ouderaa, M. Buytenhek, and D. A. van Dorp, *Adv. Prostaglandin Thromboxane Res.* **6,** 139 (1980).

[7] M. E. Hemler and W. E. M. Lands, *Lipids* **12,** 591 (1977).

[8] F. J. van der Ouderaa, M. Buytenhek, D. H. Nugteren, and D. A. van Dorp, *Eur. J. Biochem.* **109,** 1 (1980).

[9] W. L. Smith and W. E. M. Lands, *Biochemistry* **11,** 3276 (1972).

[10] P. J. O'Brien and A. Rahimtula, *Biochem. Biophys. Res. Commun.* 70, 832 (1976).

substrate using a Gilson Model K-IC oxygraph equipped with a 2.0-ml thermostatted incubation cell with a YSI 5331 micro Clark oxygen probe covered with standard Teflon membranes (YSI 5539). An aliquot of the enzyme solutions to be analyzed (between 5 and 100 μg of protein) is pipetted into the cell, together with 50 μl of the hydroquinone and 100 μl of the hemin solutions; buffer is added to fill up the cell. The cell is closed, its contents are stirred continuously, and the voltage is applied. When a straight base line is achieved (usually in less than 30 sec) the enzyme reaction is started by injecting 100 μl of arachidonic acid solution through the capillary bore stopper. The cyclooxygenase activity of the sample is calculated from the oxygen consumption curve assuming that a full-scale deflection of the recorder corresponds to 0.26 m*M* oxygen dissolved in the buffer and that 2 mol of oxygen are consumed per mole of arachidonic acid. The linear part of the sigmoid curve of oxygen consumption is used for calculations.

Definition of Unit. The unit of activity is defined as conversion of 1 μmol of arachidonic acid per minute at 25°.

Peroxidase Assay

Reagents (in addition to those stated under Cyclooxygenase Assay)
Tetramethylphenylenediamine (TMPD), 2.4 m*M* in water
Hydrogen peroxide, 9 m*M* in water

Procedure. The PGG → PGH peroxidase activity is determined by measuring the enzyme-catalyzed oxidation of tetramethylphenylenediamine by hydrogen peroxide. The blue reaction product is measured at 610 nm in a double-beam spectrophotometer equipped with stirred cuvettes thermostatted at 25°. To the enzyme solution (2–30 μg of protein) are added: 5 μl of hemin solution, 100 μl of TMPD solution, and buffer up to 3.0 ml. When a stable line is obtained, 100 μl of a hydrogen peroxide solution are added to start the reaction. Under the conditions used, a value of 12,000 liters mol^{-1} cm^{-1} is found for the molar absorption coefficient of the oxidation product of TMPD.

Definition of Unit. One unit of activity is defined as the amount of enzyme required to convert 1 μmol of hydrogen peroxide at 25° in 1 min. Instead of hydrogen peroxide, other substrates, such as linoleic acid 12-hydroperoxide, can be used.

Product Analysis

The procedure for incubations with radioactive arachidonic acid is amply described elsewhere[1,2,5] and in this volume [7]. Compared to the assay procedures for cyclooxygenase and peroxidase described above, product analysis is laborious and quantitative conversion rates of the substrate are difficult to obtain.

Purification Procedure

Prostaglandin H Synthase is an intrinsically membrane-bound enzyme. Solubilization of the enzyme from the membrane is therefore a critical step to arrive at a homogeneous enzyme preparation. The nonionic detergent Tween 20 has been used to solubilize the enzyme. High detergent concentrations, which occur when concentrating enzyme solutions during the purification, appear, however, to cause enzyme inactivation. During the isoelectric focusing step the nondialyzable detergent Tween 20 is therefore exchanged for the dialyzable detergent β-octylglucoside, resulting in an enzyme preparation with a favorable low concentration ($<0.1\%$) of detergent present.

The purification procedure has to be scheduled preferably in such a way that it can be carried out in the shortest possible time in order to minimize losses of enzyme activity during the isolation. In addition, it has been reported that the enzyme is autoinactivated when converting its substrate.[2,9] Insofar as the substrate is made available by phospholipase A_2-catalyzed hydrolysis from the phospholipids, addition of EDTA disodium salt to the homogenization buffer will effectively result in a decreased availability of endogenous substrate. A further way of controlling the rate of inactivation in the first steps of the isolation is addition to the buffer of diethyldithiocarbamate, a reversible inhibitor of the enzyme.[11]

Preparation of Washed Vesicular Gland Microsomes. All procedures are carried out between 0 and 4°. Seminal vesicles from sheep, freshly collected at the slaughterhouse, are trimmed free from fat and connective tissue and stored at −80° before use. Vesicles (250 g) are homogenized six times for 30 sec in a Sorvall Omnimixer in two batches having a total volume of 350 ml of buffer containing 0.05 *M* Tris-HCl (pH 8), 5 m*M* EDTA disodium salt, and 5 m*M* diethyldithiocarbamate (Merck). Sometimes additional homogenization in a Bühler or VirTis homogenizer will be required. The resulting tissue homogenate is centrifuged for 10 min at 13,000 *g* in a Sorvall GSA rotor. The supernatant is filtered through two layers of cheesecloth and subsequently centrifuged for 45 min at 220,000 *g* in a Beckman Ti 45 rotor. The pellet, which is the microsomal fraction, is rehomogenized using a Potter–Elvehjem homogenizer with a Teflon pestle in 270 ml of a buffer containing 0.05 *M* Tris-HCl, pH 8,

[11] L. H. Rome and W. E. M. Lands, *Prostaglandins* **10**, 813 (1975).

0.1 *M* sodium perchlorate, 1 m*M* EDTA disodium salt, and 0.1 m*M* diethyldithiocarbamate and recentrifuged for 45 min at 220,000 *g*.

Solubilization. The perchlorate-washed microsomal pellet consisting of a white fluffy layer and a red tightly packed pellet is subsequently rehomogenized in 210 ml of 0.05 *M* Tris-HCl, pH 8.0, containing 0.1 m*M* EDTA disodium salt and Tween 20 to a final detergent concentration of 1%. This Tween 20-solubilized homogenate is centrifuged for 60 min at 220,000 *g*. The clear supernatant is carefully pipetted off and concentrated to a volume of 30 ml using an Amicon concentration cell (contents 400 ml) equipped with a XM-50 filter or a Bio-Rad BioBeaker 80.

Gel Chromatography I. The concentrate is applied to a column (65 × 5 cm) of Ultrogel ACA-34 equilibrated and eluted at 100 ml/hr with 0.05 *M* Tris-HCl, pH 8, containing 0.1 m*M* EDTA disodium salt, 0.1% Tween 20, and 0.01% sodium azide. The concentration of Tween 20 surrounding the enzyme will drop rapidly during elution, since the elution volume of the Tween 20 micelles differs from that of the enzyme–Tween 20 complex. Eluted fractions of 12 ml are assayed for cyclooxygenase and peroxidase activities. The active fractions are pooled, concentrated, and subsequently washed extensively to lower the ionic strength, using an Amicon cell with a XM-50 filter.

Flat-Bed Isoelectric Focusing. A gel slab ultimately consisting of 4 g of Ultrodex, 10 g of glycerol, 400 mg of β-octylglucoside, 4 ml of Ampholine (40%), pH 5–7, 1.0 ml of Ampholine (40%) pH 3.5–10, and 43 ml of twice-distilled water is cast upon a polyacrylate tray (10.4 × 24 cm) of a Multiphor electrophoresis unit. The gel is cooled at 0° and prefocused for 3 hr at 6 W, using a "constant power" supply. The concentrated eluate of the Ultrogel column (3.5 ml) is carefully pipetted onto the gel. After isoelectric focusing for 16 h at 6 W, the current stabilizes at 2 mA at 1500 V. Thirty fractions are scraped off the plate and suspended in 2 ml of 10% glycerol. The Ultrodex is centrifuged off, and the fractions are analyzed for cyclooxygenase activity. The combined active enzyme is concentrated on an XM-50 filter to 3 ml.

Gel Chromatography II. The concentrate is applied to a column of Ultrogel ACA-34 (60 × 1.5 cm) equilibrated and eluted at 8 ml/hr with 0.05 *M* Tris-HCl, pH 8, containing 0.1 *M* sodium chloride, 0.1% β-octylglucoside, 0.1 m*M* EDTA disodium salt, and 0.01% sodium azide. Fractions (3 ml) of the eluate are monitored for cyclooxygenase activity. The active fractions are pooled and concentrated on a XM-50 filter to a concentration of between 5 and 10 mg/ml.

Comments. The procedure described above has proved to be extremely reproducible in the author's laboratory. A typical example of one of the purification runs is shown in Table I. Compared to procedures pub-

TABLE I
PURIFICATION OF PGH_2 SYNTHASE

Fraction	Total volume (ml)	Total protein (mg)	Cyclooxygenase activity[a] Total (units)	Specific (units/mg of protein)
Supernatant, 13,000 *g*	455	12,000	5400	0.45
Microsomal pellet	420	3,570	5400	1.5
Perchlorate-washed microsomal pellet	210	2,200	5400	2.5
Tween 20-solubilized microsomal fraction	172	774	4600	5.9
Gel chromatography on ACA-34, I	138	130	2000	15
Isoelectric focusing fraction	36	—	1100	—
Gel chromatography on ACA-34, II	20	60	900	15

[a] Cyclooxygenase activity is expressed as micromoles of arachidonate converted per minute at 25°.

lished earlier in which column isoelectric focusing was used,[1-3] the slab gel technique gives considerably improved yields. Specific cyclooxygenase activities of enzyme preparations obtained with the above-mentioned procedure ranged from 12 to 18 units per milligram of protein. The differences in specific activity obtained are presumably caused by the varying quality of the vesicular glands.

Properties

Purity. The pure enzyme obtained showed a single band when analyzed by acrylamide electrophoresis in the presence of 0.1% sodium dodecyl sulfate. Additionally it elutes as a single peak from the Ultrogel ACA-34 column. Throughout the purification both the cyclooxygenase and the PGG → PGH peroxidase activities are essentially copurifying, which indicates that the enzyme catalyzes both reactions.

Storage and Stability. The pure enzyme is preferably stored in small aliquots in 50% glycerol at −80° and should not be thawed more than once. Stored under these conditions, only a marginal loss of activity has been observed over a period of 3 months.

Physical Properties. The most important physical properties of the enzyme are summarized in Table II. The values obtained for the molecular weights of the enzyme–detergent complexes under native conditions excluding the amount of detergent bound, strongly suggest a dimer struc-

TABLE II
PHYSICAL PROPERTIES

Parameter	Procedure or condition chosen	
Polypeptide chain molecular weight	Electrophoresis in the presence of sodium dodecyl sulfate (8, 10, and 12% acrylamide gels)	72,000[a,b]
	Chromatography in the presence of 6 *M* guanidinium hydrochloride	68,000[c]
Molecular weight of the native enzyme (excluding bound detergent)	Chromatography in the presence of 0.1 and 0.3% octylglucoside	132,000[c]
	Chromatography in the presence of 0.1% Tween 20[a]	126,000[a]
	Sedimentation equilibrium in the presence of 0.1% Tween 20	129,000[a]
Detergent bound to the enzyme (w/w)	In 0.1 and 0.3% β-octylglucoside solutions	0.11 g/g protein[c]
	In 0.1% Tween-20 solution	0.69 g/g protein[a]
Stokes' radius of enzyme–Tween complex	In 0.1% Tween 20 solution	53 Å[a]
Sedimentation coefficient ($s_{20,w}$) (S)	In 0.1% Tween 20 solution	7.4×10^{-13}[a]
Absorbance ($A_{280}^{1\%}$)	In 0.1% Tween 20 solution	17.4[a]

[a] van der Ouderaa *et al.*[2]
[b] Hemler *et al.*[3]
[c] van der Ouderaa *et al.*[6]

ture. Additional evidence has been presented to support this.[12] The amount of detergent bound is dependent on the nature of the detergent. It is interesting to note that the enzyme binds approximately 30% more molecules of the bulkier detergent Tween 20 than of β-octylglucoside. This observation obtained from detergent-binding studies[2,6] is in full agreement with the elution characteristics of both types of complexes on Ultrogel ACA-34.[6]

Chemical Properties. The amino acid composition of the pure enzyme is shown in Table III. It was concluded from the orcinol–sulfuric acid method and periodic acid–Schiff staining on acrylamide gels that the enzyme contains carbohydrate.[2] The sugar residues found are included in the table. In order to establish whether the dimer enzyme consists of identical polypeptide chains a quantitative sequenator experiment was carried out.[6,8] Over 70% of the N-terminal amino acid alanine was recovered in

[12] G. R. Roth, C. J. Siok, and J. Ozols, *J. Biol. Chem.* **255,** 1301 (1980).

TABLE III
AMINO ACID COMPOSITION OF PGH$_2$ SYNTHASE[a]

Residue	I[b]	II[c]	III[d]
Asp	40	46	45
Thr	24	30	23
Ser	18	29	19
Glu	63	64	60
Pro	40	52	38
Gly	37	46	39
Ala	25	34	29
Cys[e]	6	14	21
Val	30	30	30
Met	13	13	11
Ile	24	26	24
Leu	55	65	57
Tyr	19	27	20
Phe	35	36	35
Trp	14	15	f
Lys	24	27	25
His	17	18	22
Arg	30	32	29
Man	12	N.D.[f]	16
NAcGlcNH$_2$	5	N.D.[f]	4

[a] Values are expressed as amino acid residues per 70,000.
[b] van der Ouderaa *et al.*[2]
[c] Roth *et al.*[12]
[d] F. van der Ouderaa, unpublished results.
[e] As cysteic acid.
[f] Not determined.

the first step; moreover, a single sequence was found, indicating the presence of identical chains. The following N-terminal amino acid sequence was obtained: Ala-Asp-Pro-Gly-Ala-Pro-Ala-Pro-Val-Asn-Pro.[6,8,12]

Spectral Properties. Participation of a transition metal in the cyclooxygenase and peroxidase reactions is to be expected. Enzyme preparations obtained using the above procedure appeared to contain between 0.1 and 0.3 atom of iron per polypeptide chain[4]; no other metals have been found.[2] However, the pure enzyme requires hemin (ferriprotoporphyrin IX) for its activity.[1-4,7] It has been concluded from spectral and kinetic data that two molecules of hemin per polypeptide chain are bound at high affinity.[4] The restored hemin–enzyme complex has a millimolar absorption coefficient at 408 nm of 61 mM^{-1} cm^{-1} per heme group as measured by difference spectroscopy. The spectra obtained for the restored hemin–enzyme

TABLE IV
KINETIC PROPERTIES OF THE PURE ENZYME

K_m for arachidonic acid (C20:4)	15 μM[a]
K_m for dihomo-γ-linolenic acid (C20:3)	23 μM[a]
K_{aff} for hemin	0.16 μM[b]
K for acetylation by acetyl salicylic acid (37°)	0.3 mM[c]
K_i for indomethacin	4.2 μM[a]

[a] F. van der Ouderaa, unpublished results.
[b] Data of V and [hemin] have been plotted in a Lineweaver–Burk plot (see Fig. 2 of van der Ouderaa.[4]
[c] Calculated from data obtained from a double-reciprocal plot in Fig. 2 of van der Ouderaa *et al.*[8]

complex are typical for a hemoprotein and indicate that the enzyme is a hemoprotein with a noncovalently bound heme group.[4] No nonheme iron has been found. The measured affinity constant for the hemin–protein complex calculated from kinetic measurements is 0.16 μM (Table IV).

Substrate Specificity. Michaelis constants for the cyclooxygenation of both arachidonic acid and 8,11,14-eicosatrienoic acid are shown in Table IV. According to previous experiments on the substrate specificity of the enzyme using sheep vesicular gland microsomes,[13] these fatty acids are the best substrates.

Inhibitors. PGH_2 synthase has been reported to be inhibited by nonsteroidal anti-inflammatory drugs.[14] These inhibitions involve different mechanisms; e.g., acetylsalicyclic acid is capable of irreversibly modifying the enzyme by acetylation of a serine hydroxyl group,[8,12] whereas the inhibitory activity of indomethacin is caused by noncovalent interaction.[15] There is at least a hundredfold difference in activity between both inhibitors (cf. Table IV). Additionally it can be concluded from the literature that the reactivity of the enzyme toward acetylsalicylic acid is variable as well; e.g., the enzyme from sheep vesicular glands has a low reactivity compared to that of other sources, e.g., blood platelets.[16]

[13] C. B. Struyk, R. K. Beerthuis, H. J. J. Pabon, and D. A. van Dorp, *Recl. Trav. Chim. Pays-Bas* **85,** 1233 (1966).
[14] R. J. Flower, *Pharmacol. Rev.* **26,** 33 (1974).
[15] N. Stanford, G. J. Roth, T. Y. Shen, and P. W. Majerus, *Prostaglandins* **13,** 669 (1977).
[16] J. W. Burch, N. Stanford, and P. W. Majerus, *J. Clin. Invest.* **61,** 314 (1978).

[10] Preparation of PGH Synthase Apoenzyme from the Holoenzyme

By BRADLEY G. TITUS and WILLIAM E. M. LANDS

The prostaglandin H (PGH) synthase apoenzyme in the presence of its prosthetic group, heme, catalyzes the initial step in the conversion of arachidonic acid to the prostaglandins. Specifically, PGH synthase catalyzes a bis-dioxygenation with arachidonic acid to form the endoperoxide intermediate, PGG_2, which is subsequently reduced via a hydroperoxidase to PGH_2.

Both the ovine and bovine PGH synthases have been purified to homogeneity.[1-3] Early attempts to prepare ovine PGH synthase provided active enzyme in small yield.[4,5] Our laboratory developed methods for isolating and storing high yields of the active holoenzyme containing the tightly bound heme cofactor.[6] This form of enzyme is conveniently stored and studied without further additions of cofactors. Studies of the heme-free form of the enzyme were hindered because the very high affinity for heme made removal of this cofactor difficult. This chapter describes the preparation of active apo-PGH synthase from the holoenzyme.

Assay Methods

Principle. PGH synthase in the presence of added hematin catalyzes the conversion of arachidonic acid to endoperoxide intermediates. Two moles of oxygen are required for the conversion of 1 mol of fatty acid. The enzymic activity is conveniently determined by monitoring the rate of oxygen uptake during enzyme catalysis. The maximum velocity of the reaction can easily be determined from a plot of the rate of oxygen uptake versus time.

Procedure. Both holo- and apo-PGH synthase activity are determined by placing an oxygen electrode[7] in a 3-ml reaction chamber containing 0.1 *M* Tris-HCl, 40 μM arachidonic acid, 232 μM oxygen, and 0.67 m*M*

[1] M. E. Hemler, W. E. M. Lands, and W. L. Smith, *J. Biol. Chem.* **251,** 5575 (1976).
[2] T. Miyamoto, N. Ogino, S. Yamamoto, and O. Hayaishi, *J. Biol. Chem.* **251,** 2629 (1976).
[3] N. Ogino, S. Ohki, S. Yamamoto, and O. Hayaishi, *J. Biol. Chem.* **253,** 5061 (1978).
[4] M. E. Hemler, C. G. Crawford, and W. E. M. Lands, *Biochemistry* **17,** 1772 (1978).
[5] F. J. van der Ouderaa, M. Buytenhek, D. H. Nugteren, and D. A. van Dorp, *Biochim. Biophys. Acta* **487,** 315 (1977).
[6] M. E. Hemler and W. E. M. Lands, *Arch. Biochem. Biophys.* **201,** 586 (1980).
[7] L. H. Rome and W. E. M. Lands, *Prostaglandins* **10,** 813 (1975).

ISBN 0-12-181986-8

phenol at 30°. The optimum pH for oxygenation is at 8.5 Reactions are initiated upon the addition of PGH synthase (20–70 n*M* in final volume). Oxygen consumption is monitored and recorded versus time. The signal for oxygen consumption can be further processed through a differentiator[8] to give a continuous record of the rate of oxygen uptake. Maximum reaction velocities can be ascertained from the continuous time derivative data. In measurements with the apoenzyme, 1 μM hematin is added to the assay mixture prior to the addition of enzyme. The hematin solution is prepared by dissolving hemin chloride in a few drops of 0.1 *N* NaOH, which is then diluted with 0.1 *M* Tris base and adjusted to pH 8.5 with HCl. Hematin concentrations are determined spectrophotometrically as described below.

The amount of apoenzyme activity present is determined by subtracting the holoenzyme activity (enzyme units observed in the absence of added hematin) from the total activity (units observed in the presence of 1 μM hematin). Upon completion of the cyclooxygenase reaction, excess soybean lipoxygenase (20 μl of a 4 mg/ml solution) should be added to the reaction chamber to oxidize any remaining arachidonic acid and hematin. Hematin appears to stick to the oxygen electrode membranes and is removed by the lipoxygenase-generated peroxides, avoiding contamination of subsequent apocyclooxygenase assays.

Definition of Unit and Specific Activity One unit of cyclooxygenase activity is defined as the amount of enzyme that will catalyze the uptake of 1 nmol of oxygen per minute at 30°. Protein determinations are made by the method of Lowry *et al.*,[9] and enzyme concentrations are based on a molecular weight of 70,000 ds per subunit. Specific activity is expressed as units per milligram of protein.

Spectrophotometric Techniques. Oxidation–reduction states can be monitored at several stages during the conversion of ovine holoenzyme to apoenzyme using a Cary 219 recording spectrophotometer. A modified 1.0-ml Thunberg cuvette with side arm is used (see below) containing the enzyme preparation. The cell has a path length of 10 mm and should be maintained at 2°. The reference solution is 0.1 *M* Tris-HCl, pH 8.5. All spectrophotometric signals should be allowed 5–10 min for stabilization. Fe(III)-PGH synthase has a Soret band at 408 nm that shifts to 427 nm upon reduction with sodium dithionite.

Hematin concentrations are determined by the reduced pyridine hemochrome method[10] using $\Delta\epsilon_{557-540} = 24.5\ \text{m}M^{-1}\ \text{cm}^{-1}$.

[8] H. W. Cook, G. Ford, and W. E. M. Lands, *Anal. Biochem.* **96,** 341 (1979).

[9] O. H. Lowry, N. J. Rosebrough, A. L. Farr, and R. J. Randall, *J. Biol. Chem.* **193,** 265 (1951).

[10] J. E. Falk, "Porphyrins and Metalloporphyrins." Elsevier, Amsterdam, 1964.

Development of Procedure

The procedure to isolate ovine apo-PGH synthase introduced here involves removing the tightly bound heme of purified holo-PGH synthase without concomitant destruction of activity of the apoprotein. We discovered that after anaerobic reduction of holo-PGH synthase with sodium dithionite, the heme moiety is lost upon vigorous addition of oxygen.[11]

During attempts to find the most efficient method of preparing the apo-PGH synthase from holoenzyme, several procedures were employed. In early preparations we attempted to obtain complete conversion of holoenzyme to apoenzyme in one step by adding large amounts of sodium dithionite instead of 1 mg as mentioned below. The amount of holoenzyme converted in one step was essentially the same either with 1 mg or with greater amounts of dithionite.

In subsequent preparations, we employed a procedure whereby the enzyme was anaerobically reduced and oxygenated four times. In these preparations, more vigorous oxygenation for 5 min (instead of 1 min) gave apoenzyme of 90–99% purity. The rate of oxygenation appeared to be the primary factor limiting the conversion of ovine holo-PGH synthase to apoenzyme for any one treatment cycle. The final procedure recommended was based on three cycles in which anaerobic reduction is followed by *vigorous* oxygenation.

Purification Procedure

Holo-PGH synthase is purified to homogeneity from sheep vesicular glands as previously described.[1] All operations are performed at 0–4° unless otherwise stated. The purified holoenzyme (5–10 mg) is placed in the modified Thunberg cuvette, and approximately 1 mg of crystalline sodium dithionite is carefully placed in the side arm to prevent premature mixture with the enzyme preparation. Prior to reducing the Fe(III)-PGH synthase with sodium dithionite, the cuvette is connected to an anaerobic train.[12,13] The cuvette is then alternately evacuated and flushed a total of 10 times with nitrogen that had been passed through reduced BASF catalyst at 150°. (Argon can be substituted for nitrogen.) Under anaerobic conditions, holoenzyme is tipped into the side arm and mixed with the sodium dithionite crystals. To ascertain complete reduction, the reaction can be monitored spectrophotometrically by observing the shift of the Soret

[11] B. G. Titus, R. J. Kulmacz, and W. E. M. Lands, *Arch. Biochem. Biophys.*, in press (1982).

[12] H. Beinert, W. H. Orme-Johnson, and G. Palmer, this series, Vol. 54, p. 111.

[13] C. H Williams, Jr., L. D. Arscott, R. G. Matthews, C. Thorpe, and K. D. Wilkinson, this series, Vol. 62, p. 185.

band at 408 nm [characteristic of Fe(III) enzyme] to 427 nm [characteristic of Fe(II) enzyme]. A stable signal at 427 nm is obtained after repeated shaking to dissolve the dithionite crystals, usually within 5 min.

The resulting Fe(II)-PGH synthase is then flushed vigorously with 100% oxygen for 5 min. The cuvette should be tipped on its side and shaken vigorously to give a larger surface area to volume ratio allowing for better oxygenation. To ascertain complete oxygenation, the enzyme preparation can again be monitored spectrophotometrically by observing the shift of the Soret band optimum wavelength back toward 408 nm, but with a decrease in the A_{408}.

At this point the enzyme preparation contains apoenzyme with some holoenzyme contamination. To measure the fraction of apoenzyme, a 30-μl aliquot can be removed and chromatographed on a microcolumn (0.2 × 1 cm) of Sephadex G-25 previously equilibrated with buffer containing 0.02 *M* phosphate (pH 7.0) and 30% glycerol. Fractions purified on the G-25 microcolumn are collected and assayed for oxygenase activity as described above.

The remaining enzyme preparation not chromatographed is again made anaerobic in a clean cuvette (grease on the cuvette is easily removed in toluene). The holoenzyme in the remaining enzyme preparation is reduced again and then vigorously oxygenated.

The procedure of anaerobic reduction of ovine PGH synthase by sodium dithionite followed by oxygenation is done a total of either three or four times to yield 99% apo-PGH synthase. In a typical preparation, the ratio of apoenzyme units to total enzyme units changed from 0:31 to 25:31 after one cycle of treatment, followed by 27:30 and 21:21 after the second and third cycles.

The final preparation containing apoenzyme is purified on a column of Sephadex G-25 (0.9 × 6 cm) previously equilibrated with buffer containing 0.02 *M* phosphate (pH 7.0) and 30% glycerol. Fractions containing the purified ovine apo-PGH synthase are collected and routinely stored at −70°.

Acknowledgment

This work was supported in part by a grant from the National Science Foundation (PCM 80-15638).

[11] Purification of PGH–PGD Isomerase from Rat Brain

By TAKAO SHIMIZU, SHOZO YAMAMOTO, and OSAMU HAYAISHI

$$\text{Prostaglandin } H_2 \xrightarrow[\text{SH}]{} \text{prostaglandin } D_2$$

An enzyme that isomerized the prostaglandin endoperoxide PGH_2 to the 9-hydroxy-11-keto structure of PGD_2 was first demonstrated in various tissues of rat.[1] Later, bovine serum albumin[2,3] and glutathione *S*-transferase[2] from sheep lung were shown to convert PGH_2 to PGD_2. Two papers[4,5] reported the synthesis of PGD_2 as a predominant product from PGH_2 in the rat brain. This chapter describes the isolation and purification of the specific PGH–PGD isomerase (previously called PGD synthetase) from rat brain. In addition to rat brain, PGH–PGD isomerase activity has been demonstrated in several other tissues, including rat spleen,[6] rat mastocytoma,[7] rat polymorphonuclear cells,[5] and human blood platelets.[8–10]

Assay Method[11]

Principle. The activity of PGH–PGD isomerase is determined by measuring the production of $[1\text{-}^{14}C]PGD_2$ from $[1\text{-}^{14}C]PGH_2$ after separation of substrate and product by thin-layer chromatography (TLC).

Reagents and Materials

Tris-HCl buffer, 0.2 *M*, pH 8.0

$[1\text{-}^{14}C]PGH_2$, 1.25 m*M*; 2.5 nmol (82,500 cpm) dissolved in 2 μl of acetone

[1] D. H. Nugteren and E. Hazelhof, *Biochim. Biophys. Acta* **326,** 448 (1973).

[2] E. Christ-Hazelhof, D. H. Nugteren, and D. A. van Dorp, *Biochim. Biophys. Acta* **450,** 450 (1976).

[3] M. Hamberg and B. B. Fredholm, *Biochim. Biophys. Acta* **431,** 189 (1976).

[4] M. S. Abdel-Halim, M. Hamberg, B. Sjöquist, and E. Änggård, *Prostaglandins* **14,** 633 (1977).

[5] F. F. Sun, J. P. Chapman, and J. C. McGuire, *Prostaglandins* **14,** 1055 (1977).

[6] E. Christ-Hazelhof and D. H. Nugteren, *Biochim. Biophys. Acta* **572,** 43 (1979).

[7] M. M. Steinhoff and B. A. Jakschik, *Biochim. Biophys. Acta* **618,** 28 (1980).

[8] O. Oelz, R. Oelz, H. R. Knapp, B. J. Sweetman, and J. A. Oates, *Prostaglandins* **13,** 225 (1977).

[9] M. Ali, A. L. Cerskus, J. Zamecnik, and J. W. D. McDonald, *Thromb. Res.* **11,** 485 (1977).

[10] H. Anhut, B. A. Peskar, W. Wachter, B. Gräbling, and B. M. Peskar, *Experientia* **34,** 1494 (1978).

[11] T. Shimizu, S. Yamamoto, and O. Hayaishi, *J. Biol. Chem.* **254,** 5222 (1979).

ISBN 0-12-181986-8

Diethyl ether–methanol–0.2 *M* citric acid (30:4:1, v/v/v) precooled to −20°

Thin-layer silica gel glass plate 60 F254 (Merck)

Solvent system: diethyl ether–methanol–acetic acid (90:2:0.1)

Procedures. The incubation mixture contains 25 μl of Tris-HCl, pH 8.0 buffer, 2 μl of [1-^{14}C]PGH_2 (2.5 nmol), and enzyme in a total volume of 50 μl. Reaction is started by the addition of substrate and carried out for 1 min at 24°. The reaction is terminated by the addition of 0.3 ml of a mixture of diethyl ether–methanol–0.2 *M* citric acid (30:4:1) precooled to −20°, and mixed with about 0.5 g of anhydrous sodium sulfate. An aliquot (0.1 ml) of the organic phase is subjected to TLC at −5° for about 50 min. Radioactivities on chromatographic plates are monitored by a Packard radiochromatogram scanner Model 7201. For the quantitative determination of the enzyme activity, the glass plates are kept on a Kodak X-ray film overnight for autoradiography. The silica gel in the region of each visualized radioactive spot is scraped off, and the radioactivity is measured by a Packard liquid scintillation spectrometer Model 3385 in a toluene solution containing 0.03% 1,4-bis[2-(4-methyl-5-phenyloxazolyl)]benzene and 0.5% 2,5-diphenyloxazole. One unit of enzyme activity is defined as the amount of enzyme that produces 1 μmol of PGD_2 per minute at 24°. Specific activity is expressed as the number of units per milligram of protein. Protein concentration is determined according to the method of Lowry *et al.*[12] with bovine serum albumin as a standard.

Purification Procedures[11]

Wistar male rats weighing 250–300 g are killed by decapitation under anesthesia with diethyl ether or intraperitoneal injection of sodium pentobarbital (50 mg/kg). The whole brains, including brainstem, are removed, rinsed thoroughly with ice-cold saline, and stored frozen at −80°. All procedures described below are carried out at 0–4° and centrifugation at 8000 *g* for 10 min, unless stated otherwise.

Step 1. Crude Extracts. Approximately 60 g of rat brains are cut into small pieces and mixed with 300 ml of 20 m*M* potassium phosphate buffer, pH 6.0, containing 0.5 m*M* dithiothreitol. The mixture is homogenized using a Polytron blender homogenizer three times at top speed each for 10 sec. After centrifugation, the supernatant solution is further centrifuged at 105,000 *g* for 60 min. The high speed supernatant solution recovered by decantation is referred to as crude extracts (310 ml).

Step 2. Acetone Fractionation. The same volume of redistilled acetone precooled to −60° is added to the crude extracts and mixed well with a magnetic stirrer at − 10° for 15 min, followed by centrifugation at − 10°. To the supernatant solution is added 1.5 times the volume of acetone, and the mixture is stirred for 15 min at − 10°. After centrifugation at − 10°, the precipitate thus formed is suspended in 20 ml of 10 m*M* potassium phosphate buffer, pH 7.0, containing 0.5 m*M* dithiothreitol. Residual acetone is removed under N_2 stream, and insoluble materials are removed by centrifugation.

Step 3. Chromatography on Phosphocellulose. The acetone fraction is applied to a column of phosphocellulose (Whatman, 2.5 × 5 cm), which has been previously equilibrated with 10 m*M* potassium phosphate buffer, pH 7.0, containing 0.5 m*M* dithiothreitol. The column is washed with the same buffer, and the enzyme passes through the column without being adsorbed.

Step 4. Chromatography on DEAE-Cellulose (DE-52, Whatman). The active fractions from the phosphocellulose column are combined and applied to a column of DE-52 (2.5 × 7 cm), previously equilibrated with 10 m*M* potassium phosphate buffer, pH 7.0, containing 0.5 m*M* dithiothreitol. The column is washed with the same buffer (approximately 180 ml) until the absorbance of the eluate at 280 nm becomes less than 0.05. Then, a linear gradient elution is performed ranging from 10 to 300 m*M* buffer containing 0.5 m*M* dithiothreitol in a total volume of 300 ml. The enzyme activity is eluted at about 90 m*M* buffer concentration.

Step 5. Ammonium Sulfate Fractionation. After the protein concentration of the eluate is adjusted to about 0.8 mg/ml, solid ammonium sulfate is added to 80% saturation (561 g/liter). The precipitate formed after 30 min is collected by centrifugation, and is suspended in a minimum volume of 65%-saturated ammonium sulfate solution prepared in 20 m*M* potassium phosphate, pH 6.0 with the use of a Potter–Elvehjem homogenizer. Insoluble materials are removed by centrifugation.

Step 6. Gel Filtration on Sephadex G-200. The ammonium sulfate fraction (2 ml) is applied to a column of Sephadex G-200 (2 × 30 cm), previously equilibrated with 20 m*M* potassium phosphate butter, pH 6.0, containing 0.1 *M* NaCl and 0.5 m*M* dithiothreitol. The column is washed with the same buffer at a flow rate of about 10 ml/hr. Active fractions (8 ml) are pooled and concentrated to 1 ml with the aid of a Diaflo membrane PM-10. Rechromatography with a smaller column (1 × 20 cm) is carried out when the purification of the enzyme is not satisfactory after the first run.

A typical result of purification is presented in the table. Approximately 1500-fold purification is achieved with a yield of 40%. The final preparation has a specific activity of about 1.8 units per milligram of protein.

TABLE I
PURIFICATION OF PGH-PGD ISOMERASE FROM RAT BRAIN

Step	Volume (ml)	Protein (mg)	Total activity (units)	Specific activity (units/mg of protein)	Yield (%)	Purification (fold)
1. Crude extract	310	1860	2.06	0.0011	100	1
2. Acetone	26	332	1.48	0.0044	72	4
3. Phosphocellulose	66	132	1.36	0.0103	66	9
4. DEAE-cellulose	16	13	1.19	0.0917	58	83
5. Ammonium sulfate	2	4.2	1.22	0.291	59	264
6. Sephadex G-200	2	0.5	0.84	1.68	40	1527

General Properties

About 50% of the total PGH–PGD isomerase activity of the homogenate is found in the cytosol fraction with the highest specific activity[12] (0.0012 unit per milligram of protein). Approximately 20% of the total activity is observed in P_2 fraction where small myelin, synaptosomes, and mitochondria are abundant. The enzyme activity is high in monkey (4.8 milliunits per milligram of protein), mouse (2.8), guinea pig (2.5), and rat (2.4), but rather low in cat (0.3) and rabbit brain (0.8).[13]

The molecular weight of the purified enzyme is estimated to be 80,000–85,000, and the isoelectric point, 5.3. No appreciable absorption of the purified enzyme is found in the visible region. The optimal pH for the reaction is around 8. The V_{max} and K_m values for substrates are respectively, PGG_2, 1.1 μmol/min per milligram of protein, 6 μM; PGH_2, 1.8 μmol/min per milligram of protein, 8 μM. The enzyme does not require glutathione, but is inhibited by sulfhydryl reagents such as 10 μM *p*-chloromercuribenzoate (PCMB) (90% inhibition), 10 μM *N*-ethylmaleimide (80%), and 10 μM iodoacetamide (50%). Optimal pH for the stability is 6. The enzyme is unstable at neutral pH; about 50% of the enzyme activity is lost after 24 hr at 4°. The purified enzyme is essentially free of the glutathione *S*-transferase activity, although crude extracts of rat brain shows significant transferase activity (0.001 unit per milligram of protein as assayed with 1 mM 1-chloro-2,4-dinitrobenzene as a substrate). Glutathione is not required for PGD synthesis, but it is active in preventing in-

[12] O. H. Lowry, N. J. Rosebrough, A. L. Farr, and R. J. Randall, *J. Biol. Chem.* **193,** 265 (1951).

[13] T. Shimizu, N. Mizuno, T. Amano, and O. Hayaishi, *Proc. Natl. Acad. Sci. U.S.A.* **76,** 6231 (1979).

activation of the enzyme. Furthermore, 1-chloro-2,4-dinitrobenzene does not inhibit PGD synthesis. It should also be noted that the specific activity of the highly purified enzyme is higher by three orders of magnitude than those of bovine or rat serum albumin. PGD synthesis by the purified enzyme is not inhibited by 50 μM arachidonic acid. These results indicate that the purified enzyme is clearly distinguishable from glutathione *S*-transferase or serum albumins.

Acknowledgments

This work was supported in part by a Grant-in-Aid for Scientific Research from the Ministry of Education, Science, and Culture of Japan and by grants from the Japanese Foundation on Metabolism and Diseases, Research Foundation for Cancer and Cardiovascular Diseases, Fujiwara Memorial Foundation, and Japan Heart Foundation 1979.

[12] Isolation of PGH–PGD Isomerase from Rat Spleen

By E. CHRIST-HAZELHOF and D. H. NUGTEREN

A key intermediate in prostaglandin biosynthesis is the prostaglandin endoperoxide PGH_2. This labile compound was isolated for the first time in 1973[1]; it is synthesized from arachidonic acid by the enzyme prostaglandin H synthase. When PGH_2 is dissolved in water, it decomposes (isomerizes) almost exclusively into a mixture of PGE_2 and PGD_2 with an E:D ratio of 2.4. This is a spontaneous first-order reaction, and the half-life of the endoperoxide in buffer of pH 7–8 is 5 min at 37° (or about 10 min at 20°). The PGE_2:PGD_2 ratio can be influenced in several ways. For instance, a remarkable increase in the amounts of PGD_2 was observed after addition of cow, sheep, or pig serum albumin.[2] Rabbit and rat albumin are less active. Bovine serum albumin not only alters the E:D ratio, but also increases the velocity of isomerization of the endoperoxide. Another group of binding proteins, the ligandins or glutathione *S*-transferases, can also change the E:D ratio. Sheep transferases isolated from lung or liver produce much PGD_2, together with $PGF_{2\alpha}$, whereas rat transferases give greater amounts of PGE_2 and $PGF_{2\alpha}$.[2]

The enzymic properties of genuine PGH–PGD isomerase isolated from rat spleen are totally different from those of the above-mentioned

[1] D. H. Nugteren and E. Hazelhof, *Biochim. Biophys. Acta* **326,** 448 (1973).

[2] E. Christ-Hazelhof, D. H. Nugteren, and D. A. van Dorp, *Biochim. Biophys. Acta* **450,** 450 (1976).

ISBN 0-12-181986-8

proteins.[3] PGH–PGD isomerase is found not only in rat spleen, but also in rat lung, stomach, intestine,[1] brain,[4] and mast cells.[5] PGD-isomerase activity has also been reported in mouse melanoma cells.[6] In other mammals, it was found among others in gerbil spleen, rabbit lung, guinea pig intestine, monkey intestine, and monkey lung.[3] Mastocytoma patients have highly increased levels of PGD metabolites in their urine.[7]

PGH–PGD isomerase from many tissues turned out to be rather labile, and the activity of the homogenates was sometimes lost within 1 day at 0°, even in the presence of 5 m*M* 2-mercaptoethanol or glutathione. The pure enzyme isolated from rat spleen remained active for about a week at 0° and for several months at −20°. PGD isomerase is the only endoperoxide metabolizing enzyme known to be recovered mainly in the supernatant; all other enzymes are membrane bound. A Japanese group isolated a PGD-forming enzyme from rat brain that has different properties: it has no glutathione requirement, and the molecular weight is 80,000.[8]

Enzyme Assay

Enzyme activity is determined by incubating with radioactive prostaglandin endoperoxide and separating the products by thin-layer chromatography (TLC). Incubations are always done in the presence of 1 m*M* EDTA because otherwise large quantities (up to 40%) of PGF_α may be obtained in the presence of glutathione, even without enzyme.

Preparation of Prostaglandin endoperoxide. For the enzymic or chemical preparation of prostaglandin endopéroxide, several methods have been described[9,10] (see also this volume [13]). We use a lyophilized microsomal preparation of sheep vesicular glands to which an SH-blocker has been added to prevent PGE formation by PGH–PGE isomerase, which is also present. After a short incubation with arachidonic acid, the

[3] E. Christ-Hazelhof and D. H. Nugteren, *Biochim. Biophys. Acta* **572,** 43 (1979).

[4] M. S. Abdel-Halim, M. Hamberg, B. Sjöquist, and E. Änggård, *Prostaglandins* **14,** 633 (1977).

[5] C. W. Parker, B. A. Jakschik, M. G. Huber, and S. F. Falkenheim, *Biochem. Biophys. Res. Commun.* **89,** 1186 (1979).

[6] F. A. Fitzpatrick and D. A. Stringfellow, *Proc. Natl. Acad. Sci. U.S.A.* **76,** 1765 (1979).

[7] L. J. Roberts, B. J. Sweetman, R. A. Lewis, K. F. Austen, and J. A. Oates, *N. Engl. J. Med.* **303,** 1400 (1980).

[8] T. Shimizo, S. Yamamoto, and O. Hayaishi, *J. Biol. Chem.* **254,** 5222 (1979).

[9] N. A. Porter, J. D. Byers, R. C. Mebane, D. W. Gilmore, and J. R. Nixon, *J. Org. Chem.* **43,** 2088 (1978).

[10] R. R. Gorman, F. F. Sun, O. V. Miller, and R. A. Johnson, *Prostaglandins* **13,** 1043 (1977).

prostaglandin endoperoxide is rapidly extracted and isolated by preparative TLC.

A mixture of 1600 μg of hydroquinone, 250 mg of *p*-hydroxymercuribenzoate, and 1 g of lyophilized sheep vesicular gland microsomes (400 mg of protein) in 32 ml of 0.1 *M* Tris, 0.1 *M* potassium phosphate, pH 8.0, is homogenized in a Potter–Elvehjem homogenizer, and the preparation is preincubated at 20° for 10 min. Then 6 mg of [1-^{14}C]arachidonic acid (12×10^6 dpm, New England Nuclear or Amersham) suspended in 32 ml of Tris–phosphate buffer at 30° is added to the preincubated medium; the whole mixture is incubated for 2 min at 25°. The incubate is added to, and immediately extracted with, a precooled (in a −80° bath) mixture of 19 ml of 0.2 *M* citric acid and 160 ml of ethyl acetate. The ethyl acetate layer is removed and dried with Na_2SO_4 (30 min, −80°). After filtration, most of the ethyl acetate is removed under vacuum at low temperature with a rotation evaporator. The residue is diluted with ether and rapidly brought on a silica gel TLC plate (0.5 mm thickness, Kieselgel 60 F254 Fertigplatten, Merck, Darmstadt), as a 12-cm band at 0°. Prostaglandin B_2 (PGB_2) serves as a reference. The TLC plate is immediately developed for 40 min (about 12 cm) at −20° in the solvent system hexane–ether (2:8, v/v). The endoperoxide is rapidly scraped off (PGB_2, which can be seen under a UV lamp, moves somewhat more slowly than the endoperoxide) and added to 10 ml of precooled ether–methanol (9:1, v/v). The extract is transferred to another tube, and solvent up to 20 ml final volume is added. Amounts of 1–2 mg of prostaglandin endoperoxide (based on radioactivity; 15–35% yield) can be obtained following this procedure. The endoperoxide solution is quite stable when kept dry and stored at −70°.

Incubation Procedure and Estimation of Enzyme Activity. Prostaglandin endoperoxide (1.5 μg; 5 nmol, 3000 dpm) is incubated with a suitable amount of enzyme in 1 ml of 0.2 *M* Tris-HCl buffer, pH 8.0, in the presence of 0.5 m*M* glutathione (GSH) and 0.1 m*M* EDTA for 30 min at 30°. This is followed by acidification with 0.2 *M* citric acid until pH about 3, extraction with 2×2 ml of ether, and separation of the products as a 1-cm band on a 0.2-mm silica gel plate (Kieselgel 60 F254 Fertigplatten, Merck, Darmstadt). The solvent system is chloroform–methanol–acetic acid–water (90:6:1:0.7, v/v/v/v). The products, prostaglandins E and D are scraped off and counted in 1.0 ml of methanol + 10 ml of Omnifluor (New England Nuclear Chemicals); the ratio of radioactivity between PGD and PGE is calculated. If for a certain enzyme preparation the D:E ratio is plotted against the amount of protein used, a linear relationship is found until a D:E ratio of about 6.[cf. 3]

We have introduced the following definition for enzyme activity:

Amounts of enzyme that give a D:E ratio between 2.0 and 5.0 under the incubation conditions just described contain 0.20–0.50 milliunits (mU) of PGH–PGD isomerase. According to this definition, it can be estimated that 1 enzyme unit (U) can convert approximately 1 μmol of substrate per minute at 30°.

Purification of PGH–PGD Isomerase

PGH–PGD isomerase is purified from the 150,000 *g* supernatant of rat spleen in the following steps: DEAE-cellulose chromatography, preparative isoelectric focusing, glutathione affinity chromatography, gel filtration. When the whole procedure is done within 4 days, loss of enzyme activity is negligible. All steps are done at 0–4°, always in the presence of 5 m*M* 2-mercaptoethanol. The course of the purification is demonstrated in Fig. 1 and in the table. The determination of PGD isomerase in crude preparations is disturbed by many factors, including proteins, heme, and reducing agents. This may explain the variations found in the total amount of enzyme units in the different steps during the purification procedure.

Homogenization. The solution used is 5 m*M* 2-mercaptoethanol in distilled water. About 40 rat spleens (20 g) are homogenized with 88 ml of 5 m*M* 2-mercaptoethanol in a Bühler homogenizer at maximum speed twice for 15 sec.

Centrifugation. The homogenate is centrifuged for 30 min at 150,000 *g*, and the clear supernatant is decanted. Approximately 80 ml of supernatant are obtained containing about 2.6 g of protein.

DEAE-Cellulose Chromatography. The solutions used are starting buffer (10 m*M* Tris-HCl, 5 m*M* 2-mercaptoethanol, pH 8.0 at 20°) and starting buffer with 0.2 *M* NaCl. The column (Whatman DE-52 cellulose; 14 × 5 cm, 280 ml bed volume) is equilibrated with starting buffer.

The supernatant is applied to the column, and the column is eluted overnight with 600 ml of starting buffer. Subsequently it is eluted with a

TABLE I
PURIFICATION OF PROSTAGLANDIN H–PROSTAGLANDIN D ISOMERASE

Fraction	Total volume (ml)	Total protein (mg)	Total enzyme (units)	Specific activity (units/mg protein)
150,000 *g* supernatant	84	~2600	~0.8	~0.0003
DEAE-cellulose	150	170	0.22	0.0013
Isoelectric focusing	10	40	0.36	0.009
Affinity chromatography	12.5	2.6	1.10	0.42
Gel filtration	13.6	1.3	0.82	0.63

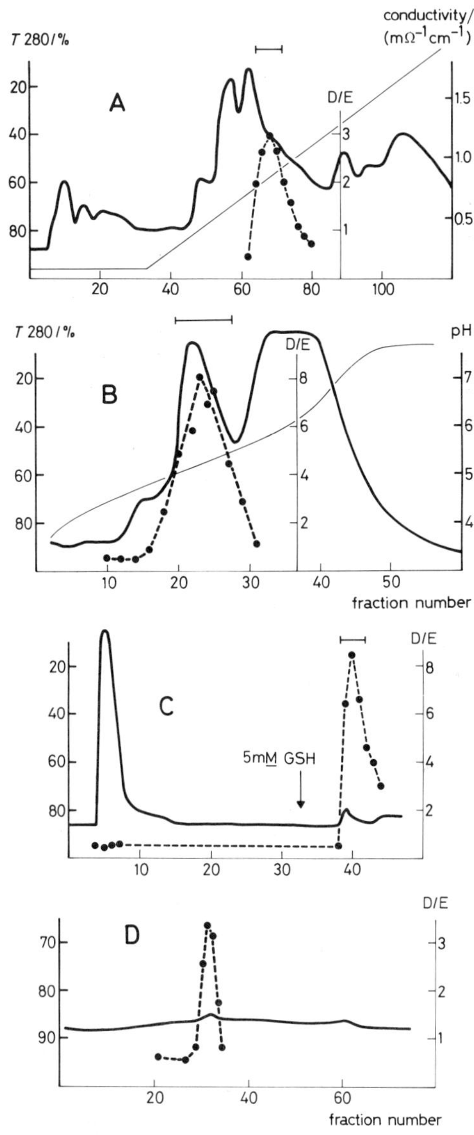

FIG. 1. Purification of prostaglandin endoperoxide PGD isomerase from rat spleen. (A) DEAE-cellulose chromatography; (B) preparative isoelectric focusing; (C) glutathione (GSH) affinity chromatography; (D) gel filtration on Sephadex G-100. The enzymic activity was determined by incubating aliquots of the fractions with 1.5 μg of [1-^{14}C]prostaglandin endoperoxide in 1 ml of 0.2 *M* Tris-HCl (pH 8.0)–1 m*M* GSH–0.5 m*M* EDTA for 30 min at 30°, followed by analysis of the products using thin-layer chromatography. The isomerase activity is expressed as the ratio of counts in PGD_2 to those in PGE_2 (D:E ratio, ●---●). Transmittance (*T*) was measured at 280 nm. —: Fractions used for further purification.

linear gradient made from 1 liter of starting buffer and 1 liter of starting buffer with NaCl. Four fractions per hour (21.5 ml each) are collected; 0.2-ml aliquots of each fraction are used for determination of enzyme activity (see Fig. 1). The fractions containing the bulk of the PGD isomerase activity are combined to give 170 mg of protein in 150 ml.

Concentration and Washing Step. The combined fractions are concentrated and washed with 5 m*M* 2-mercapthoethanol in distilled water over an Amicon filter (Diaflo type PM-10) to an end volume of about 15 ml.

Preparative Isoelectric Focusing. Preparative isoelectric focusing is done as described by O'Brien *et al.*[11] The solutions used are Ampholine 4-6 (LKB) (15 g/liter) with glycerol (100 g/liter), polyacrylamide (150 g/liter), and 5 m*M* 2-mercaptoethanol in distilled water. The column is Sephadex G-15 in distilled water (27 × 2.6 cm).

The column is equilibrated with 160 ml of the Ampholine–glycerol solution, and the protein sample (16 ml in 15 g of Ampholine per liter and 100 g of glycerol per liter) is rinsed into the column. Both ends are then provided with a Millipore filter and with two polyacrylamide plugs polymerized *in situ* as described in detail by O'Brien *et al.*[11] Isoelectric focusing is done for 11 hr at 1250 V. The plugs and the filters are removed, and the column is eluted with 5 m*M* 2-mercaptoethanol (1.3-ml fractions are collected). For each fraction, the pH is measured and 0.02 ml is used to determine the PGD isomerase activity (see Fig. 1). The active fractions contain 40 mg of protein in 10 ml.

Concentration and Washing Step. The buffer is 22 m*M* potassium sodium phosphate, pH 7.0, with 5 m*M* 2-mercaptoethanol. The active fractions are combined and concentrated on an Amicon filter (Diaflo PM-10). They are washed with 22 m*M* potassium sodium phosphate 5 m*M* mercaptoethanol, pH 7.0.

Glutathione Affinity Chromatography. Glutathione affinity chromatography was done as described by Simons and Vander Jagt.[12] The solutions used are 22 m*M* potassium sodium phosphate (pH 7.0) with 5 m*M* 2-mercaptoethanol, and 50 m*M* Tris-HCl–5 m*M* glutathione (pH 9.6). For the column, 100 g of glutathione per liter are coupled at pH 7.0 with epoxy-activated Sepharose 6B (Pharmacia) exactly as described by Simons and Vander Jagt; a 23 × 1.6 cm column is prepared.

The preparation is applied to the column, and the column is washed overnight with 140 ml of 22 m*M* potassium sodium phosphate containing

[11] T. J. O'Brien, H. H. Liebke, H. S. Cheung, and L. K. Johnson, *Anal. Biochem.* **72,** 38 (1976).

[12] P. C. Simons and D. L. Vander Jagt, *Anal. Biochem.* **82,** 334 (1977).

5 m*M* 2-mercaptoethanol (pH 7.0), during which time no enzyme activity is eluted. The elution is continued with 50 m*M* Tris-HCl–5 m*M* glutathione (pH 9.6). From each fraction, 0.02 ml is used for determination of the enzyme activity (see Fig. 1). The bulk of the enzyme activity is present in three fractions (total volume 12.5 ml; 2.6 mg of protein).

Concentration. The combined active fractions are concentrated by ultrafiltration until the volume is reduced to 1.5 ml.

Gel Filtration. The buffer used is 0.1 *M* Tris-HCl with 5 m*M* 2-mercaptoethanol, pH 8.0. The column is Sephadex G-100 (60 cm × 1.6 cm). The enzyme preparation is brought onto the gel filtration column, which is then eluted with buffer (6.7 ml/hr); 1.7-ml fractions are collected, from which 5-μl aliquots are used for determination of enzyme activity. The peak of enzyme activity exactly coincides with a small 280 nm absorbance peak, corresponding to 1.3 mg of protein (see Fig. 1).

Properties of PGH–PGD Isomerase

Determination of Molecular Weight. When the enzyme had been purified as described above, one single band was seen after sodium dodecyl sulfate (SDS)–gel electrophoresis; no impurities could be detected. The mobility of the pure rat spleen PGD isomerase during SDS–polyacrylamide (120 g/liter) slab gel electrophoresis was compared with the migration of three mixtures of reference proteins (molecular weight in parentheses): (A) catalase (60,000), aldolase (40,000), chymotrypsinogen A (25,700) and β-lactoglobulin (18,400); (B) β-galactosidase (130,000), bovine serum albumin (68,000), ovalbumin (43,000), and myoglobin (17,200); (C) pyruvate kinase (57,000) and carbonic anhydrase (29,000). From the results, a molecular weight of 26,000 was calculated for the enzyme. Gel filtration on Sephadex G-100, using the same conditions as described in the purification procedure, indicated a molecular weight of 34,000; reference proteins were: bovine serum albumin (67,000), ovalbumin (45,000), carbonic anhydrase (29,000), soybean trypsin inhibitor (21,500), and myoglobin (17,800). So the enzyme consists of one polypeptide chain with a molecular weight of about 30,000.

Identification of Reaction Product. It was confirmed with gas chromatography–mass spectrometry that PGD_2 was indeed the main product formed on incubation of the purified enzyme with prostaglandin endoperoxide.[3]

Cofactor Requirement Glutathione, cysteine, thioglycolic acid, reduced lipoic acid, dithiothreitol, NADH, and NADPH were tested in 1.0 and 0.2 m*M* concentration. Only in the presence of glutathione was a high yield of PGD_2 obtained; none of the other compounds could stimulate the reaction. The optimum glutathione concentration lies between 10^{-4} and 10^{-3} *M*. Without glutathione, no enzymic PGD_2 formation could be demonstrated.

pH Optimum. Incubations were done between pH 4 and 9.4 in the presence of 1 m*M* glutathione. Optimum conversion was obtained between pH 7 and 8.[3]

Inhibitory Substances. A number of compounds were tested that might inhibit the isomerization either by interference with the substrate binding site or with the glutathione binding site of the enzyme. Only those compounds that can react with free SH groups, such as 15-hydroperoxy-arachidonic acid and *p*-hydroxymercuribenzoate, were found to inhibit the conversion.[3]

[13] Purification of PGH–PGE Isomerase from Sheep Vesicular Glands

By P. MOONEN, M. BUYTENHEK, and D. H. NUGTEREN

For studies on prostaglandin biosynthesis, the microsomal fraction of sheep vesicular glands is more or less the "classical" starting material. The biosynthesis of prostaglandin E (PGE)[1] from certain unsaturated fatty acids is a membrane-bound process that requires two enzymes acting sequentially.[2,3] The first enzyme, which catalyzes the conversion of certain essential fatty acids (e.g., arachidonic acid) into prostaglandin H (PGH) is referred to as PGH synthase. This protein has been fully purified both from bovine and sheep vesicular gland microsomes.[4] The second enzyme needs glutathione (GSH) for its activity and isomerizes the endoperoxy group of PGH to a 9-keto-, 11-hydroxyl arrangement (PGE). This so-called prostaglandin endoperoxide-E isomerase (PGH–PGE isomerase, EC 5.3.99.3) has been partially purified from bovine vesicular gland microsomes.[5] In this contribution our approaches to purification of PGH–PGE isomerase from sheep vesicular gland microsomes are described.

[1] For nomenclature of prostaglandins see N. A. Nelson, *J. Med. Chem.* **17,** 911 (1974).

[2] D. H. Nugteren and E. Hazelhof, *Biochim. Biophys. Acta* **326,** 448 (1973).

[3] T. Miyamoto, S. Yamamoto, and O. Hayaishi, *Proc. Natl. Acad. Sci. U.S.A.* **71,** 3645 (1974).

[4] F. J. van der Ouderaraa and M. Buytenhek, this volume [9].

[5] N. Ogino, T. Miyamoto, S. Yamamoto, and O. Hayaishi, *J. Biol. Chem.* **252,** 890 (1977).

ISBN 0-12-181986-8

Assay Method

Principle. The endoperoxide used as substrate in the assay is not stable in aqueous medium: it has a half-life of about 10 min at pH 7.4 and 20°. The main degradation products are PGE and PGD in a ratio of about 3:1. Therefore, incubations are carried out for only 1 min with different amounts of enzyme and with a fixed amount of ^{14}C-labeled endoperoxide. After quick acidification and extraction with ether, the products formed and the remaining endoperoxide are rapidly separated by silica gel-HPTLC. Quantification of the enzyme activity is possible by scraping off the radioactive bands, counting the radioactivity, and calculating the nanomoles of PGE formed.

Preparation of Substrate (*see also this volume* [12]). Purified prostaglandin endoperoxide synthase[4] (4.5 mg of protein, specific activity 16.9 U/mg) was preincubated with 1 mg of hemin in 16 ml of 0.1 *M* Tris-HCl, pH 8, for 5 min at 0°. [1-^{14}C]Arachidonic acid (Radiochemical Centre, Amersham, 23 × 10^6 dpm) was added to unlabeled arachidonic acid to a total of 9.0 mg. The fatty acid plus 3 mg of hydroquinone were dissolved in a 250-ml Erlenmeyer flask in 40 ml of 0.1 *M* Tris–0.1 *M* phosphate, 1 m*M* EDTA, pH 8.0. The turbid suspension was carefully saturated with air and placed in a water bath at 29°. The enzyme solution was warmed up and subsequently transferred to the fatty acid suspension; the mixture was then incubated for 2.5 min at 29°. The Erlenmeyer flask was shaken manually, and a stream of air was introduced into the flask. The reaction was terminated by pouring the mixture rapidly into 18 ml of 0.2 *M* citric acid plus 100 ml of ethyl acetate cooled in ethanol–solid CO_2, immediately followed by thoroughly shaking. The reaction vessel was cooled so that the water phase just did not freeze. After separation of the two phases, the ethyl acetate phase was immediately decanted and the water phase was rapidly extracted with 25 ml of cooled ethyl acetate. The combined ethyl acetate phases were cooled to −80° for 4 hr to freeze out the water, and then filtered at −80° over a cooled glass filter G3 (15–40 μm).

The precipitate was washed with 25 ml of cooled ethyl acetate. The filtered ethyl acetate was evaporated under reduced pressure to approximately 1 ml and spotted in a cold room as a 10-cm band on a 20 × 20 cm 0.5 mm-thick silica gel 60 F254 plate (Merck); 10 μg of prostaglandin B_1 or B_2 had been spotted beforehand as a reference. The plate was developed without delay at 4° for 1 hr with diethyl ether–methanol–acetic acid (200:1:0.04, v/v/v). After UV detection of PGB_1 (R_f = 0.48), the zone with R_f 0.50–0.70 was scraped off immediately and transferred to 25 ml of a mixture of diethyl ether–methanol (90:10, v/v) at 0°. After mixing and centrifugation, the organic layer was evaporated under a stream of nitrogen to 1 ml. After counting an aliquot of the solution, the total amount of PGH_2 was calculated based on the known specific radioactivity. Then diethyl ether was added to give a stock solution of 200 μg of PGH_2 per milliliter, which was stored at $-80°$. The yield was about 20%.

Enzyme Assay. To 0.4 ml of incubation solution containing 0.1 *M* Tris, 0.1 *M* phosphate, 1 m*M* EDTA, 4 m*M* GSH, pH 7.4, an amount of PGE isomerase activity was added to get a conversion to PGE_2 of between 30 and 70%. The reaction was started by the addition of radioactive PGH_2 (2 μg, 5.7 nmol in 4 μl of ethanol). Incubation was carried out for exactly 1 min at 20°. The reaction was terminated by the rapid addition of 75 μl of 0.2 *M* citric acid, and then 1.2 ml of diethyl ether followed immediately by mixing and cooling at 0°. After centrifugation, the organic phase was separated and evaporated under a stream of nitrogen to approximately 25 μl (not dry). After spotting as a 0.5-cm band on a 10×10 cm (0.1 mm thick) HPTLC plate (Merck) provided with a reference containing $PGF_{2\alpha}$, PGE_2 and PGD_2, the plate was developed immediately with diethyl ether–methanol–acetic acid (90:2:0.5, v/v/v). After drying in air, the plate was sprayed with phosphomolybdic acid in ethanol (100 g/liter) and heated. From the intensity of the blue spots, the conversion into PGE_2 could be estimated. For quantitative determination, the blue spots of PGH, PGD, PGE, and PGF_α were scraped off (R_f values 0.90, 0.55, 0.37, and 0.22, respectively) and the silica gel was transferred to a liquid scintillation vial to which 1.0 ml of methanol and 10 ml of toluene scintillator (Packard) were added. Radioactivity was measured with a liquid scintillation spectrometer. From the result, the amount of PGE_2 formed was calculated.

Calculation of Enzyme Activity. If the amount of PGE_2 formed is plotted as a function of the amount of protein used, curves like the one in Fig.

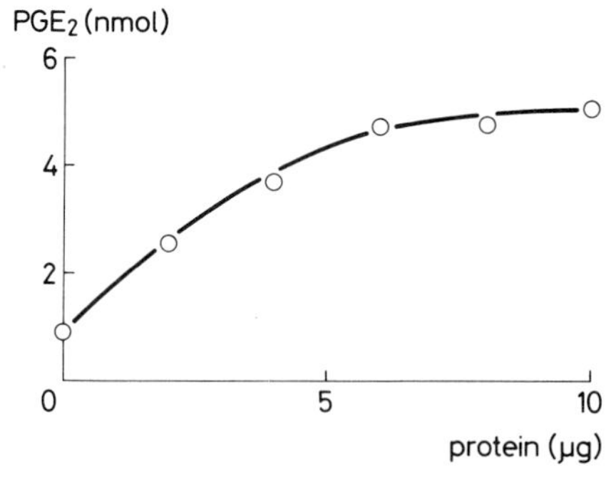

FIG. 1. Production of PGE_2 as a function of the amount of protein in the 1.5% Triton supernatant. Assay method as described in the text.

1 are obtained. A linear relationship is found in a certain range of enzyme concentration and from the slope, the specific activity of the enzyme preparation tested (micromoles PGE_2 formed per minute per milligram of protein at 20°) can be calculated (duration of incubation is always 1 min). If the right amount of enzyme is taken, one or two incubations are sufficient to estimate the specific activity. The enzyme concentration should not be too low because then the accuracy is poor owing to the nonenzymic PGE_2 formation. With excess enzyme a shortage of endoperoxide substrate prevents a reliable determination.

Purification Procedure

The results of the procedures used for the fractionation are summarized in the table. All manipulations were carried out between 0° and 4°.

Step 1. 13,000 g Supernatant. Deep-frozen sheep seminal vesicles (225 g), trimmed free from fat and connective tissue, were homogenized in 75-g portions with 110 ml of 0.05 *M* Tris-HCl, pH 8.0, containing 10 m*M* EDTA disodium salt and 1 m*M* diethyldithiocarbamate using a Sorvall Omnimixer (twice for 30 sec) and a VirTis 45 homogenizer (3 × 10 sec). The homogenate was centrifuged for 20 min at 13,000 *g*.

Step 2. Microsomes. The supernatant was filtered through two layers of cheesecloth and centrifuged for 45 min at 130,000 *g*. The pellets were rehomogenized, using a Potter-Elvehjem homogenizer, in 350 ml of 0.05 *M* Tris-HCl, pH 8.0, containing 0.1 *M* sodium perchlorate, 2 m*M* EDTA, and 0.5 m*M* diethyldithiocarbamate.

Step 3. Perchlorate-Washed Microsomes. The homogenate was centrifuged again for 60 min at 130,000 *g*. The washed microsomes, consist-

PURIFICATION OF PGH–PGE ISOMERASE

Step	Volume (ml)	Protein (mg)	Specific activity (units)
1. 13,000 *g* Supernatant	450	8955	1.0
2. Microsomes	420	2860	1.7
3. Perchlorate-washed microsomes	210	1722	1.7
4. Tween-treated microsomes	140	987	2.2
5. Washed Tween-treated microsomes	140	450	1.3
6. Triton X-100 supernatant	140	231	1.4
7. DEAE-cellulose	30	25	1.7
8. Hydroxyapatite–agarose (unbound protein fractions)	45	9	1.4

ing of a white fluffy layer and a red tightly packed pellet, were then homogenized in 180 ml of 0.05 *M* Tris-HCl, pH 8.0, containing 0.5 m*M* EDTA and 0.1 m*M* diethyl dithiocarbamate.

Step 4. Tween-Treated Microsomes. Twenty-one milliliters of a 10% fresh aqueous Tween 20 (w/v) solution were added to the microsomal suspension to give a final concentration of 1% Tween 20 in a total volume of 210 ml. The Tween 20 homogenate was centrifuged for 60 min at 130,000 *g*. The pellets were rehomogenized in 120 ml of 0.5 m*M* GSH, 0.5 m*M* EDTA adjusted with 1 *M* Na_2CO_3 to pH 8.0.

Step 5. Washed Tween-Treated Microsomes. To the resuspended pellet 14 ml of a 10% Tween 20 solution were added, making a total volume of 140 ml. The homogenate was centrifuged for 60 min at 130,000 *g*. The pellets were homogenized in 0.5 m*M* GSH, 0.5 m*M* EDTA, pH 8, in a total volume of 120 ml.

Step 6. 1.5% Triton Supernatant. By the addition of 21 ml of fresh aqueous 10% Triton X-100 (w/v, Packard) the Triton X-100 concentration was made 1.5%. The mixture was left for 30 min and then centrifuged for 60 min at 130,000 *g*. To the clear supernatant (112 ml), ethylene glycol (28 ml) was added to give a final concentration of 20% in a final volume of 140 ml.

Step 7. Chromatography on DEAE-Cellulose. DEAE-cellulose (Whatman DE-52 microgranular) was treated as described in the Whatman Manual. The DEAE-cellulose column (1.6 × 70 cm) was equilibrated with a solution containing 0.5 m*M* EDTA, 0.5 m*M* GSH, 0.1 m*M* dithiothreitol (DTE, Merck), 0.5% Triton X-100 (w/v), 20% ethylene glycol (v/v) and adjusted to pH 8.0 with 1 *M* Na_2CO_3. After application of the Triton supernatant, elution was continued first with 50 ml of equilibration buffer until the unbound protein had eluted and then with a linear gradient of 250 ml of equilibration buffer and 250 ml of equilibration buffer plus 0.1 *M* potassium phosphate, pH 8.0, made by an LKB Ultrograd, in 24 hr. The flow rate was about 18 ml/hr. The elution profile of this column is presented in Fig. 2. The fractions with a specific activity above 1.3 were combined (30 ml).

Step 8. Chromatography on Hydroxyapatite–Agarose. A column (25 × 1.6 cm) was prepared of hydroxyapatite–agarose (LKB) and equilibrated with 30 m*M* potassium phosphate, pH 7.7, plus 0.1 m*M* EDTA, 0.1 m*M* DTE, 0.5 m*M* GSH, 0.5% Triton X-100, and 20% ethylene glycol. The pooled fractions of the DEAE chromatography were diluted with water made 0.5% in Triton X-100 and 20% in ethylene glycol to give a phosphate concentration of 30 m*M*. After application to the hydroxyapatite–agarose column and washing with 30 ml of starting buffer, the bound protein was eluted stepwise by increasing the phosphate concentration

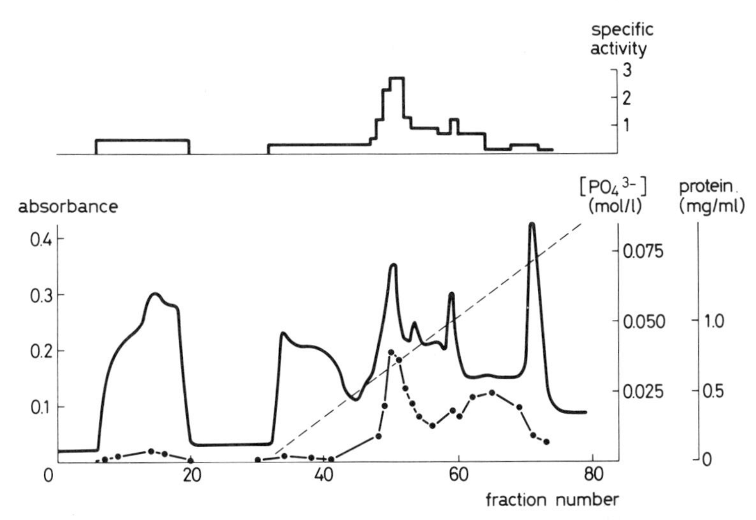

FIG. 2. Elution pattern of the DEAE-cellulose column loaded with the Triton X-100 supernatant. Fractions of 20 min were collected. ——, Absorbance at 280 nm; ----, concentration of PO_4^{3-} of the eluent; ·-·, protein concentration.

from 30 to 60 m*M* (step 1, 40 ml) and from 60 to 100 m*M* phosphate (step 2, 70 ml). The flow rate was about 15 ml/hr. Isomerase activity eluted in two parts: about half of the activity did not bind to the column and the other half eluted in step 1 with 60 m*M* phosphate. The specific activity of the fractions containing the unbound protein was higher than that for the fractions eluted in step 1. However, the specific activity had increased only little with respect to the 13,000 *g* supernatant, probably because of inactivation of the PGE isomerase during the purification procedure.

Properties

Stability. The PGH–PGE isomerase is a highly unstable enzyme; it can be partially protected from inactivation during the purification procedure by addition of thiol compounds in the buffers. In Fig. 3 the time-dependent inactivation at 0° is shown. The half-life of the isomerase activity is about 20 hr at 0°. This rapid inactivation of the isomerase is one of the main reasons why the specific activity does not increase (see the table). Other factors, e.g., lipid requirement and influence of detergent, may also be responsible. A protein fraction of the hydroxyapatite column showed the same lability, indicating that proteolytic digestion is probably not the reason. The inactivation can be suppressed by storage at −80°.

Cofactor Requirement. PGH–PGE isomerase specifically needs gluta-

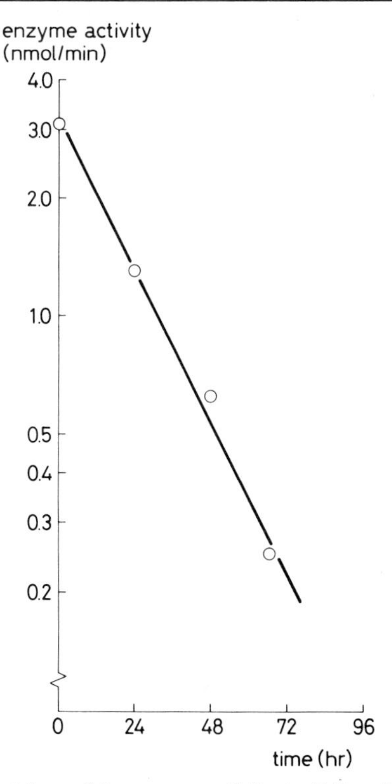

FIG. 3. Time-dependent loss of isomerase activity in Triton X-100 supernatant at 0°. Assay with 8 μg of protein.

thione to exhibit its activity. However, there is no stoichiometric oxidation of glutathione.[5] Blocking reagents for SH groups, such as *p*-hydroxymercuribenzoate and *N*-ethylmaleimide, destroy the isomerase activity even if after incubation with these reagents an excess of glutathione is added.[6]

pH Optimum. The PGH–PGE isomerase present in the Triton X-100 supernatant has a wide pH optimum between pH 5.5 and 7.0. At lower GSH concentrations (0.5 m*M* GSH), the optimum is sharper around pH 7.0.

Molecular Mass and Purity. The unbound hydroxyapatite column fractions show two protein bands with molecular mass between 60,000 and 67,000 daltons after SDS–slab gel electrophoresis on polyacrylamide in the presence of 2-mercaptoethanol (staining with Coomassie Brilliant

[6] M. E. Gerritsen, T. P. Parks, and M. P. Printz, *Biochim. Biophys. Acta* **619,** 196 (1980).

Blue R250). The sensitive silver stain[7] indicated the presence in these fractions of some low molecular mass proteins (10,000 to 30,000), indicating that the purification procedure does not result in a completely homogeneous PGH–PGE isomerase.

[7] B. R. Oakley, D. R. Kirsch, and N. R. Morris, *Anal. Biochem.* **105,** 361 (1980).

[14] Preparation and Assay of Prostacyclin Synthase

By JOHN A. SALMON and RODERICK J. FLOWER

Prostacyclin is the most potent naturally occurring inhibitor of platelet aggregation and is also a powerful vasodilator.[1–3] It is formed by a rearrangement of the prostaglandin endoperoxide PGH_2, the reaction being catalyzed by an enzyme commonly known as prostacyclin synthase (see Fig. 1).

The enzyme is located in mammalian blood vessels, and although prostacyclin is generated by many organs it is often difficult to determine whether this is because all tissues contain vascular elements or whether there are genuine extravascular sources. It is the endothelium that contains the highest amounts of prostacyclin synthase per gram of tissue, but vascular smooth muscle can also synthesize substantial amounts of prostacyclin.

Prostacyclin is unstable, with a half-life of approximately 10 min at physiological temperature and pH; it hydrolyzes quantitatively and nonenzymically to 6-keto-$PGF_{1\alpha}$ (see Fig. 1). Only bioassays permit the assay of prostacyclin itself in biological samples, and therefore the majority of investigators have sought to assess prostacyclin synthase by measuring the formation of 6-keto-$PGF_{1\alpha}$ by radiochemical, radioimmunoassay (RIA), or physicochemical procedures. Many of these assays are fully described elsewhere in this volume, and details will not be repeated here.

Choice of Substrate

Prostacyclin synthase converts prostaglandin endoperoxides to prostacyclin, and therefore the most convenient methods of assessing the en-

[1] S. Moncada, R. J. Gryglewski, S. Bunting, and J. R. Vane, *Nature (London)* **263,** 663 (1976).

[2] R. J. Gryglewski, S. Bunting, S. Moncada, R. J. Flower, and J. R. Vane, *Prostaglandins* **12,** 685 (1976).

ISBN 0-12-181986-8

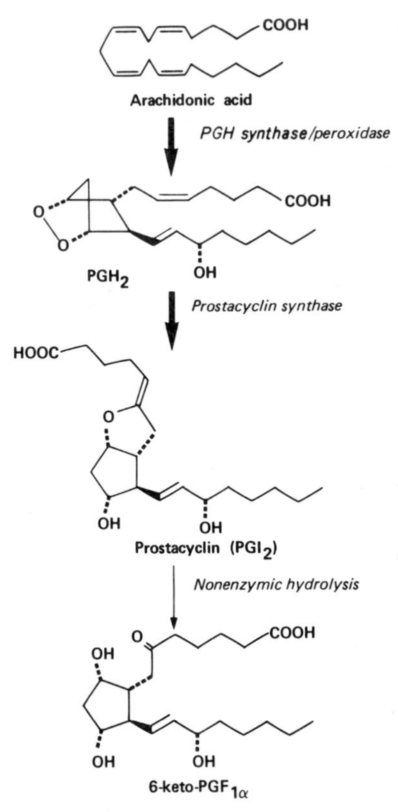

FIG. 1. Biosynthesis and decomposition of prostacyclin.

zyme activity utilize endoperoxides as substrate. The biosynthetic preparation of endoperoxides from arachidonic acid has been described in detail by several authors[4-6] and is also discussed in this volume [49]. Both endoperoxides (i.e., PGG_2 and PGH_2) are substrates for the prostacyclin synthase, but PGG_2 is converted to a mixture of prostacyclin and 15-hydroperoxyprostacyclin; the latter product decomposes to 6,15-diketo-

[3] S. Bunting, R. J. Gryglewski, S. Moncada, and J. R. Vane, *Prostaglandins* **12,** 897 (1976).
[4] F. Ubatuba and S. Moncada, *Prostaglandins* **13,** 1055 (1977).
[5] R. R. Gorman, F. F. Sun, O. V. Miller, and R. A. Johnson, *Prostaglandins* **13,** 1043 (1977).
[6] K. Gréen, M. Hamberg, B. Samuelsson, M. Smigel, and J. C. Frölich, *Adv. Prostaglandin Thromboxane Res.* **5,** 39 (1978).

$PGF_{1\alpha}$ rather than 6-keto-$PGF_{1\alpha}$,[7] and therefore it is more practical to use PGH_2 as substrate.

Other precursors, such as arachidonic acid, could be used, but these would require the presence of a viable PGH synthase (cyclooxygenase) as well as prostacyclin synthase. The conversion of arachidonic acid to endoperoxides is invariably the rate-limiting step, and, if accurate kinetic data are required, it is essential to use an endoperoxide as substrate.

Preparation of Crude "Prostacyclin Synthase"

Reagents

Fresh vascular tissue, approximately 0.5 kg from an abattoir
Liquid nitrogen, 0.5 liter
Tris buffer, 0.05 *M*, pH 7.5

Prostacyclin synthase is abundant in most vascular tissue, and so tissue from many sources could be used as a starting material. We have found that pig arterial vascular tissue is a particularly rich source; it is easily obtained in large amounts from abattoirs and may be stored at −20° until required. The high proportion of elastic and connective tissue in arteries makes this tissue difficult to homogenize, and we have developed a special technique for dealing with this problem.

Batches of about 500 g of frozen tissue are thawed, and the vascular adventitia is removed. The unseparated aortic walls are cut into pieces (2–3 cm in length), then rapidly frozen in liquid nitrogen and immediately pulverized into a fine powder, using a stainless steel pestle and mortar or similar device. The powder is suspended in 0.05 *M* Tris buffer at pH 7.5 (1:4 w/v) and homogenized at high speed using a Polytron homogenizer. The homogenate is initially centrifuged at 1000 *g* for 15 min at 4°, and the resulting supernatant is centrifuged at 10,000 *g* for 15 min at 4°, producing a pellet referred to as the mitochondrial fraction. Finally, this 10,000 *g* supernatant is centrifuged again at 100,000 *g* for 60 min in a refrigerated ultracentrifuge to obtain a pellet termed the microsomal fraction, which contains approximately 50% protein by weight as estimated by the method of Lowry *et al.*[8] The prostacyclin synthase activity resides primarily in this fraction,[9,10] and we usually resuspend it in deionized water

[7] J. A. Salmon, D. R. Smith, and F. Cottee, *Prostaglandins* **17,** 747 (1979).

[8] O. H. Lowry, N. J. Rosebrough, A. L. Farr, and R. J. Randall, *J. Biol. Chem.* **193,** 265 (1951).

[9] S. Moncada, R. J. Grylgewski, S. Bunting, and J. R. Vane. *Prostaglandins* **12,** 715 (1976).

[10] J. A. Salmon, D. R. Smith, R. J. Flower, S. Moncada, and J. R. Vane, *Biochim. Biophys. Acta* **523,** 250 (1978).

and then lyophilize. The average yield is 150 mg of lyophilized microsomal powder per 100 g of aortic tissue. The enzyme activity in the lyophilized powder is maintained for several months when it is stored desiccated at $-20°$.

The above technique is probably the best for preparing microsomes from samples of vasculature, but some other tissues (e.g., rat forestomach) can usually be homogenized successfully without requiring rapid freezing and pulverization.

Further Purification of the Enzyme

For most purposes, the crude enzyme preparation is sufficient, but should further purification be required we can recommend the procedure of Wlodawer and Hammarström.[11] Aortic microsomes are treated with 0.5% Triton X-100, and then the mixture is centrifuged at 100,000 *g* for 60 min in the cold. Most of the prostacyclin synthase activity resides in the supernatant and can be separated on a DEAE-cellulose column. The column is eluted with (*a*) 30 ml of 10 m*M* potassium phosphate buffer (pH 7.4) containing 0.1% Triton X-100; (*b*) 10 ml of 20 m*M* buffer containing 0.1% Triton X-100; and (*c*) 10 ml of 0.2 *M* buffer containing 0.1% Triton X-100. The enriched enzyme is eluted in fraction c and can be concentrated by ultrafiltration.

Enzyme Assays

Radiochemical Assays

For the reasons discussed above, the most useful substrate for assessing prostacyclin synthase activity is the prostaglandin endoperoxide PGH_2. Radioactive PGH_2 permits a straightforward assay of prostacyclin synthase in which the production of labeled prostacyclin or its hydrolysis product, 6-keto-$PGF_{1\alpha}$, is used as a measure of enzyme activity. [1-^{14}C]PGH_2 is ideal for this test and can be synthesized from [1-^{14}C]arachidonic acid (Amersham International, Bucks, England or New England Nuclear, Boston, Massachusetts) (for methods see this volume [49]). The higher specific activity of multilabeled [^{14}C]arachidonic acid (New England Nuclear) is not a significant advantage for most of these assays.

Reagents

Radioactive PGH_2
Lyophilized aortic microsomes prepared as described above
Tris buffer 50 m*M*, pH. 7.4

[11] P. Wlodawer and S. Hammarström, *Febs Lett.* **97**, 32 (1979).

Solvents for extraction and chromatography
Thin-layer chromatography plates

Procedure. For routine assays of prostacyclin synthase using [1-^{14}C]PGH_2 as substrate and aortic microsomes as source of enzyme, the following procedure has been adopted. Lyophilized aortic microsomes (2 mg; equivalent to approximately 1 mg of protein) are gently resuspended in buffer (50 m*M* Tris, pH 7.4) with the aid of a glass–Teflon homogenizer and incubated with [1-^{14}C]PGH_2 (approximately 1.5 nmol with specific activity of 3–6 Ci/mol). After 3 min of incubation at 20°, the protein in the mixture is precipitated with acetone (2 volumes) and the radioactive products are extracted into organic solvents by a procedure described in this volume [59]. The organic extract is concentrated under nitrogen, the residue is reconstituted in 50 μl of chloroform–methanol (2:1, v/v), and this solution is quantitatively applied to a thin-layer chromatography (TLC) plate, which is developed in the organic phase of ethyl acetate–2,2,4-trimethylpentane–acetic acid–water (110:50:20:100, v/v/v/v; upper phase) (see this volume [59]). After development, the radioactive areas on the TLC plate are detected by autoradiography (see Fig. 2) or radiochromatogram scanning; then the relevant zones are scraped off the plate, and the radioactivity in them is determined by conventional liquid scintillation counting. During extraction the prostacyclin formed is quantitatively converted to 6-keto-$PGF_{1\alpha}$, and therefore the amount of [^{14}C]-6-keto-$PGF_{1\alpha}$ produced reflects the conversion of PGH_2 to prostacyclin.

The enzyme activity appears to be relatively insensitive to changes in pH, maximum synthesis of prostacyclin occurring in the range 6.8–7.5. The conversion of endoperoxide to prostacyclin is very rapid at 37° (reaction linear for approximately 30 sec and complete within 1 min), making many experiments impractical; an incubation temperature of 20° (reaction linear for approximately 1 min and complete within 5 min) is our recommended alternative.

Prostacyclin synthase is inhibited by 15-hydroperoxyeicosatetraenoic and other hydroperoxy acids[9,10] (see also Fig. 2), and this can provide useful confirmation of the presence of the enzyme. However, investigators should be aware that these hydroperoxy acids may also inhibit other enzyme systems (e.g., PGH synthase) and also they may not be active inhibitors in tissues or organs, since they may be rapidly degraded by peroxidases.

An alternative radiochemical assay for prostacyclin synthase using 6-[^{3}H]PGH_2 as substrate is described in this volume [17]: In addition there are some other methods of determining synthesis of prostacyclin and/or 6-keto-$PGF_{1\alpha}$, and these are briefly discussed below.

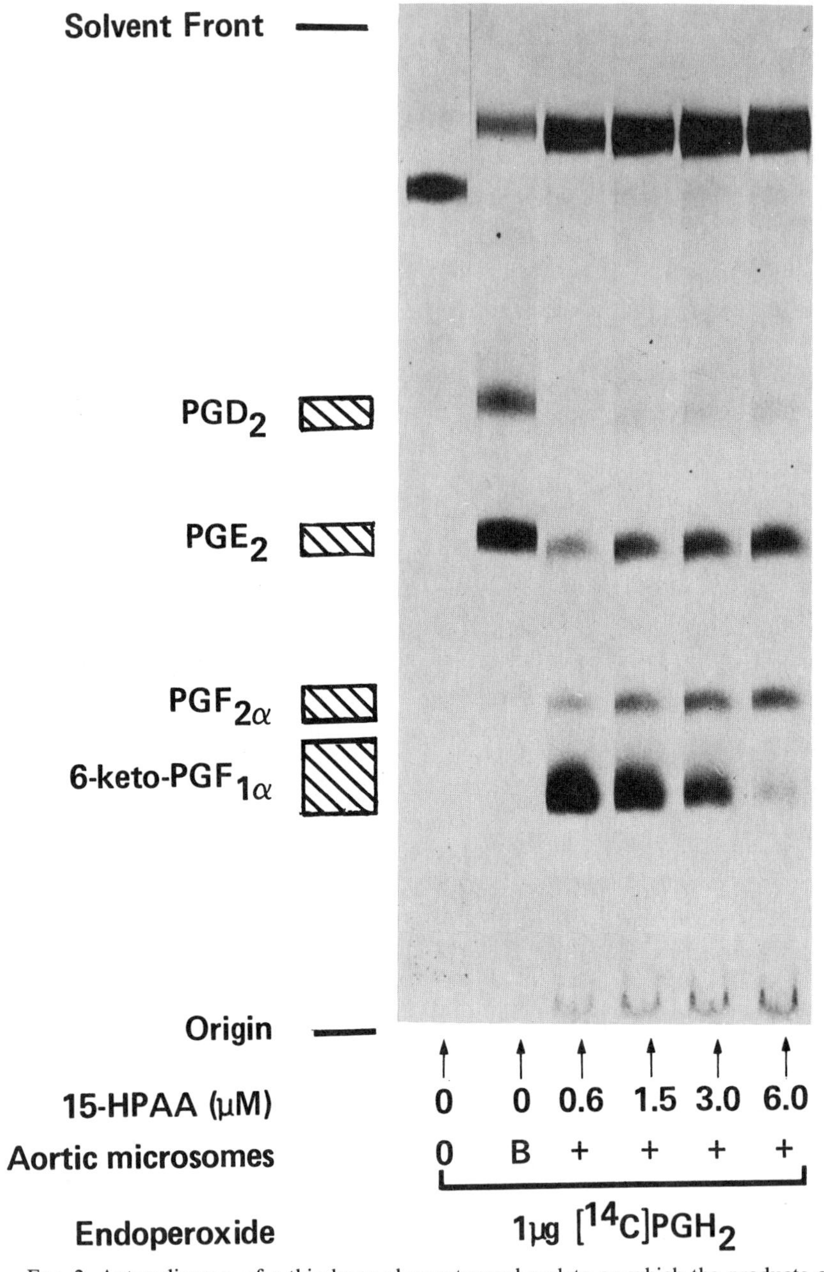

FIG. 2. Autoradiogram of a thin-layer chromatography plate on which the products of incubating 1-[^{14}C]PGH_2 with aortic microsomes have been separated. The lane marked "O" aortic microsomes shows the mobility of unchanged endoperoxide and that labeled "B" shows the products formed during incubation of the endoperoxide with boiled aortic microsomes. All other lanes ("+" aortic microsomes) are obtained after incubation of the endoperoxide with 2 mg of microsomes. Various concentrations (0.6–6 μM) of 15-hydroperoxyeicosatetraenoic acid (15-HPAA) were added to the incubation as indicated. The hatched areas on the left of the autoradiogram indicate the mobility of authentic nonradioactive standards in the same solvent system (system A, Table I in this volume [59]).

Radioimmunoassays

The conversion of exogenous endoperoxides to prostacyclin has been determined by radioimmunoassay of 6-keto-$PGF_{1\alpha}$ (the decomposition product of prostacyclin).[12,13] The development and use of radioimmunoassays for 6-keto-$PGF_{1\alpha}$ are described in this volume [40] and will not be considered further.

Physicochemical Assays

Prostacyclin itself has been determined using high-pressure liquid chromatography (HPLC) with ultraviolet detection.[14,15] However, this technique is probably not applicable for measuring PGI_2 formed from biological samples, since an extraction step would be required and this would convert PGI_2 to 6-keto-$PGF_{1\alpha}$. The separation and measurement of 6-keto-$PGF_{1\alpha}$ by HPLC is described by Eling *et al.* [61].

6-Keto-$PGF_{1\alpha}$ may also be determined by gas–liquid chromatography linked to an electron capture detector or mass spectrometer.[6,16,17]

Bioassay

Smooth Muscle Bioassays. Prostacyclin was originally discovered because it produced a biological profile qualitatively and quantitatively different from that of its precursor, PGH_2 and all other prostaglandins known at that time.[1,3] Initial studies with prostacyclin synthase also employed bioassays; prostacyclin was detected as a relaxant of most spirally cut blood vessels (e.g., rabbit celiac and mesenteric arteries). Prostacyclin also decreases spontaneous activity and reduces the basal tone of rat colon while having a weaker contractile effect than other prostaglandins (e.g., PGE_2) on rat stomach strip. Subsequently, the bovine coronary artery (BCA) was shown to be a valuable addition to the battery of tissues,[18] since it distinguished clearly between prostacyclin (which relaxes the tissue) and PGE_2 (which contracts it). Possibly the single most important advantage of bioassay, especially that of the continuous superfusion system

[12] J. A. Salmon, *Prostaglandins* **15,** 383 (1978).
[13] C. L. Tai and H-H. Tai, *Prostaglandins Med.* **4,** 399 (1980).
[14] G. T. Hill, *J. Chromatogr.* **176,** 407 (1979).
[15] M. A. Wynalda, F. H. Lincoln, and F. A. Fitzpatrick, *J. Chromatogr.* **176,** 413 (1979).
[16] J. A. Salmon and R. J. Flower, *in* "Hormones in Blood" (C. H. Gray and V. H. T. James, eds.), 3rd ed., Vol. 2, p. 237. Academic Press, New York, 1979.
[17] F. A. Fitzpatrick, *Adv. Prostaglandin Thromboxane Res.* **5,** 95 (1978).
[18] G. J. Dusting, S. Moncada, and J. R. Vane, *Prostaglandins* **13,** 3 (1977).

of Vane,[19,20] is its dynamic quality and ability to detect short-lived compounds such as prostacyclin in circulating blood or in the effluent from perfused organs. Specificity is achieved by careful selection of tissues used in the cascade and by addition of antagonists to other biologically active compounds that may be present in the sample and would otherwise compromise the assay. Thus, an assessment of prostacyclin production in a whole anesthetized animal (or perfused organ) can be achieved by determining prostacyclin in circulating blood (or physiological salt solution) by passing the fluid over a combination of rabbit aorta, rabbit celiac (or mesenteric) artery, bovine coronary artery, rat stomach strip, and rat colon.[16,21] However, the slow recovery of the tissues limits the usefulness of the technique for more routine applications, for which the other procedures considered above are preferred.

Antiaggregating Assay. A second bioassay for prostacyclin, based on its ability to inhibit potently platelet aggregation, may have more general application, and the conversion of endogenous or exogenous substrate (either arachidonic acid or PGH_2) can be assessed.

The antiaggregatory activity of prostacyclin is most reliably measured using ADP-induced aggregation in human platelet-rich plasma. Human platelet-rich plasma is recommended because human platelets are more sensitive to prostacyclin than those from most other species. ADP, which is readily available and easy to prepare, is preferred to other inducers of aggregation because blood taken from a volunteer who has recently taken aspirin or other inhibitor of the PGH synthase can still be employed (whereas neither arachidonic acid nor collagen could be used, since these agents induce platelet aggregation by inducing the synthesis of thromboxane A_2, and this would be blocked by aspirin and similar drugs; consequently platelet aggregation would be prevented). In addition, aggregation responses using ADP are more "graded" and permit precise quantitation. The method is as follows.

Citrated human blood (10 ml of 3.15% w/v trisodium citrate solution added per 100 ml of blood) is centrifuged for 15 min at 150 *g* at room temperature to produce platelet rich plasma. Aggregation of the platelets is monitored by observing changes in light transmission in an aggregometer of the type originally described by Born.[22] These instruments can be purchased from several companies, including Payton Associates Ltd., Scarborough, Ontario, Canada and Chrono-log Corporaton, Havertown,

[19] J. R. Vane, *Br. J. Pharmacol.* **23,** 360 (1964).
[20] J. R. Vane, *Br. J. Pharmacol.* **35,** 209 (1969).
[21] S. Moncada, S. H. Ferreira, and J. R. Vane, *Adv. Prostaglandin Thromboxane Res.* **3,** 211 (1978).
[22] G. V. R. Born, *Nature (London)* **194,** 927 (1962).

Pennsylvania; multichannel instruments are available that enable the simultaneous measurement of multiple samples.

The antiaggregatory activity in a sample is assessed using at least two concentrations (a two-point assay) and is compared to authentic prostacyclin. The latter, as the sodium salt, is dissolved in 1 *M* Tris buffer (pH 9.5 at 4°) and freshly diluted in 50 m*M* Tris buffer (pH 8.2 at 4°) when required. The limit of detection is approximately 50 pg of prostacyclin in 50 μl of buffer added to 0.5 ml of platelet-rich plasma, which thus enables samples containing 1 ng of prostacyclin per milliliter to be analyzed.

Criteria for Identification of Prostacyclin

The nature of the antiaggregatory activity should be confirmed as PGI_2 by using at least two of the following criteria: (*a*) unstable in buffer at pH 7.4 after 15 min at 37°; (*b*) unstable at pH 3 after 20 sec incubation, but stable in alkaline buffers, (*c*) activity abolished by 1 min of preincubation of the extract with an antiserum that binds prostacyclin[23] (see also this volume [41]) (*d*) parallel dose-response curves and similar relative potency to authentic prostacyclin in platelet-rich plasma from several species (platelets from different species exhibit differing sensitivities to PGI_2[24]); (*e*) generation of activity inhibited by incubation of the tissue with aspirin-like drugs and 15-hydroperoxyeicosatetraenoic acid.[9]

[23] S. Bunting, S. Moncada, P. Reed, J. A. Salmon, and J. R. Vane, *Prostaglandins* **15,** 565 (1978).

[24] B. J. R. Whittle, S. Moncada, and J. R. Vane, *Prostaglandins* **16,** 373 (1978).

[15] Assay of Prostacyclin Synthase Using $[5,6\text{-}^3H]PGH_2$[1]

By HSIN-HSIUNG TAI and CHARLES J. SIH

Prostacyclin synthase catalyzes isomerization of prostaglandin endoperoxide (PGH_2) to prostacyclin (PGI_2), which is readily hydrolyzed to 6-keto-$PGF_{1\alpha}$ at physiological pH.[2]

[1] This work was supported in part by grants from the National Institutes of Health (GM-25247, AM-09688) and the American Heart Association (78-865) and its Texas Affiliate.

[2] S. Moncada, R. J. Gryglewski, S. Bunting, and J. R. Vane, *Nature, (London)* **263** (1976).

ISBN 0-12-181986-8

PGH_2 → PGI_2 → (H_2O) → 6-Keto-$PGF_{1\alpha}$

PGI_2 inhibits platelet aggregation and relaxes smooth muscle[2] as opposed to thromboxane A_2 (TxA_2), which induces platelet aggregation and contracts smooth muscle.[3] The opposing actions of two different prostaglandin endoperoxide metabolites suggest that the control and balance of the synthesis of these two potent substances may be of great importance to cardiovascular functions and diseases.

A number of methods, including bioassay,[2] radiochromatographic assay,[4] and radioimmunoassay,[5] have been devised to detect the prostacyclin synthase activity. However, these methods lack specificity or require tedious extraction and separation procedures. Since conversion of PGH_2 to PGI_2 entails the loss of hydrogen at C-6, labeling of PGH_2 with tritium at C-6 will provide a simple assay for this enzyme by following the release of C-6 tritium.[6] This presentation describes the preparation of [5,6-^{3}H]arachidonic acid and [5,6-^{3}H]PGH_2 and their use for the assay of prostacyclin synthase from porcine aortic microsomes.

[3] M. Hamberg, J. Svensson, and B. Samuelsson, *Proc. Natl. Acad. Sci. U.S.A.*. **72,** 2994 (1975).

[4] J. A. Salmon, *Prostaglandins* **15,** 383 (1978).

[5] C. L. Tai and H. H. Tai, *Prostaglandins Med.* **4,** 399 (1980).

[6] H. H. Tai, C. T. Hsu, C. L. Tai, and C. J. Sih, *Biochemistry* **19,** 1989 (1980).

Preparation of [5,6-^{3}H]Arachidonic Acid

Principle. [5,6-^{3}H]Arachidonic acid is prepared by 5,6-ditritiation of eicosa-*cis*-8,11,14-trien-5-ynoic acid (5,6-dehydroarachidonic acid), which is synthesized from tetradeca-2,5,8-triyn-1-ol according to the following scheme.[5]

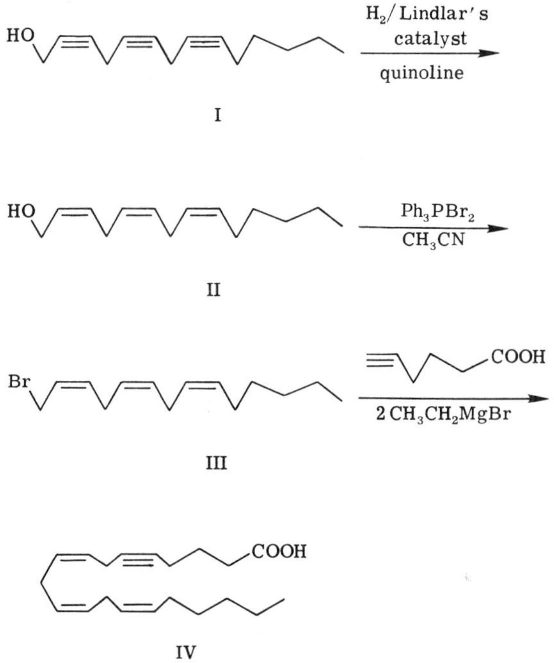

HO–(chain) I —(H_2/Lindlar's catalyst, quinoline)→

HO–(chain) II —(Ph_3PBr_2, CH_3CN)→

Br–(chain) III —(≡–COOH, 2 CH_3CH_2MgBr)→

COOH IV

Reagents

Tetradeca-2,5,8-triyn-1-ol[7]
5-Hexynoic acid (Farchan Chemical Co.)
Lindlar catalyst (Aldrich Chemical Co.)
Silicic acid (Mallinckrodt 2847, 100 mesh)

Procedure

cis-Tetradeca-2,5,8-trien-1-ol (II). Tetradeca-2,5,8-triyn-1-ol (I)[7] (1.067 g, 5.28 mmol in absolute ethanol (30 ml) containing a 5% solution of quinoline in 0.2 ml of absolute ethanol is partially hydrogenated with Lindlar catalyst (0.8 g) after 30 min, 390 ml of H_2 are taken up (theoretical, 387 ml), the catalyst is removed by filtration through a cone of Celite,

[7] J. M. Osbond, P. G. Philpott, and J. C. Wickens, *Chem. Commun.* 2779 (1961).

and the solvent is evaporated on a rotary evaporator. The only product is chromatographed over a silicic acid column (1.5 × 25 cm). Elution of the column with ethyl acetate–hexane (4:6) affords 0.96 g of (II) (colorless oil) with a strong odor: NMR ($CDCl_3$) δ 0.96 (3 H, t, CH_3), 1.30 (6 H, m, CH_2), 2.0 (3 H, m, $C{=}CHCH_2$ and OH), 2.82 (4 H, q, $C{=}CHCH_2CH{=}C$), 4.20 (2 H, d, $C{=}CHCH_2OH$), 5.06–5.74 (6 H, m, vinylic H). The mass spectrum gives *m/e* 208 as the molecular ion.

cis-1-Bromotetradeca-2,5,8-triene (*III*). Into a 200-ml three-necked round-bottom flask is placed a solution of triphenylphosphine (2.42 g, 9.2 mmol) in dry acetonitrile (45 ml) under N_2 in a ice-water bath. Bromine (0.6 ml, 9.3 mmol is added dropwise from a syringe. After addition, a slight excess of Br_2 persists (light yellow) and is removed by adding a small amount of triphenylphosphine. The adduct, triphenylphosphine dibromide, appears as a white precipitate.

After the ice-water bath is removed, a solution of compound (II) (1.9 g, 9.1 mmol) in dry CH_3CN (20 ml) is added dropwise from a dropping funnel. When the addition is complete, the white precipitate disappears. After the reaction mixture is stirred at room temperature for 30 min, the contents are transferred to a 250-ml round-bottom flask and the CH_3CN is evaporated on a rotary evaporator (bath temperature 45°). The residue is dissolved in 50 ml of ether, washed with saturated Na_2CO_3 solution (2 × 50 ml), and dried over $MgSO_4$. After evaporation, the residue (2.5 g) is triturated with hexane to remove the insoluble by-product, triphenylphosphine oxide. The pure product is obtained by chromatography of the hexane fraction over a silicic acid column (1.5 × 25 cm). Elution of the column with 2% ethyl acetate in hexane affords 0.9 g of pure (III): NMR ($CDCl_3$) δ 0.9 (3 H, t, CH_3), 1.32 (6 H, m, CH_2), 2.09 (2 H, m, $C{=}CHCH_2$), 2.75–3.04 (4 H, $C{=}CHCH_2CH{=}C$), 4.02 (2 H,d, $BrCH_2CH{=}C$), 5.06–5.90 (6 H, m, vinylic H). The rest of the column fractions (1.2 g) is further purified by preparative silica gel TLC (Brinkmann) (developed in 3% ethyl acetate in hexane) to yield another 1.02 g (total 1.92 g, 78%).

Eicosa-cis-8,11,14-trien-5-ynoic Acid (*IV*). A solution of 5-hexynoic acid (2 g, 17.5 mmol) in dry tetrahydrofuran (20 ml) is added dropwise to a solution of ethylmagnesium bromide, prepared from magnesium turnings (0.86 g, 35.4 mol) and ethyl bromide (4.4 g, 40 mmol) in dry tetrahydrofuran (30 ml) at 0° under N_2. After the reaction mixture is refluxed for 2 hr, the contents are cooled to 0° and cuprous chloride (125 mg) is added. After 15 min *cis*-1-bromotetradeca-2,5,8-triene (III) (1.9 g, 7 mmol) in dry tetrahydrofuran (15 ml) is added over 5 min. The reaction mixture is refluxed for 4 hr, then another portion of cuprous chloride (100 mg) is added and the contents are refluxed for an additional 14 hr.

The cooled reaction mixture is carefully acidified with 2 *N* H_2SO_4 and extracted with diethyl ether. The ethereal layer is washed with a saturated NH_4Cl solution five times (until free from Cu^{2+} and SO_4^{2+}) and then extracted with three portions of 2 *N* ammonia to which 5% NaCl has been added. The combined aqueous layers are washed once with ether, acidified, and extracted three times with ether. The combined ethereal layers are washed successively with water and saturated NaCl solution and dried over Na_2SO_4. After evaporation of the solvent, the excess 5-hexynoic acid is recovered by Kugelrohr distillation (apparatus from Aldrich Chemical Co.). The product is purified by preparative TLC by using 50% ethyl acetate in hexane as the developing solvent and yielded 558 mg (42.5%) of (IV) as a colorless oil: NMR ($CDCl_3$) δ 0.9 (3 H, t, CH_3), 1.30 (6 H, m, CH_2), 1.80 (2 H, t, $C{\equiv}CCH_2$), 2.01 (2 H, m, $C{=}CHCH_2C$), 2.26 (2 H, m, CH_2CCO_2H), 2.46 (2 H, t, CH_2CO_2H), 2.63–3.06 (6 H, m, $C{=}CHCH_2CH{=}C$ and $C{=}CHCH_2C{\equiv}C$), 5.18–5.60 (6 H, m, vinyl H), 10.82 (1 H, br s, COOH). The mass spectrum for the methyl ester of the product gives *m/e* 316 (M^+) and 301 ($M^+{-}CH_3$).

Catalytic Reduction of Eicosa-8,11,14-trien-5-ynoic Acid with Tritium Gas. Tritiation was performed by New England Nuclear Corp. To 45 mg of (IV) in 3.5 ml of dry petroleum ether (bp 65–67°) is added Lindlar's catalyst (16 mg) and 5% quinoline solution in petroleum ether (0.09 ml). The mixture is stirred under $^3H_2$ (1 atm) at 25° for 20 min. (Owing to solvent evaporation, one cannot measure the amount of $^3H_2$ gas taken up; however, 20 min reaction time should be sufficient). After 20 min, the contents are filtered through Celite, and the catalyst is washed with 1 *N* HCl, water, and saturated NaCl solution and then dried over Na_2SO_4. After filtration, the solvent is evaporated and the desired [5,6-^{3}H]arachidonic acid is purified on a silicic acid–Celite column (1 × 30 cm) by using a 7% diethyl ether–Skelly B elution system.[8] The purified [5,6-^{3}H]arachidonic acid has a specific activity of 120 mCi/mmol.

Assay of Prostacyclin Synthase

Principle. The immediate substrate of prostacyclin synthetase is PGH_2 which is derived from arachidonic acid by the action of prostaglandin endoperoxide synthase.[9] This enzyme is most abundant in sheep seminal vesicular microsomes. [5,6-^{3}H]PGH_2 can be prepared from [5,6-^{3}H]arachidonic acid by incubating with sheep seminal vesicular microsomes for a brief period followed by rapid extraction and silicic acid chromatography

[8] C. J. Sih and C. Takeguchi, *in* "Prostaglandins" (P. W. Ramwell and J. E. Shaw, eds.), Vol. 1, p. 83. Plenum, New York, 1973.

[9] M. Hamberg and B. Samuelsson, *Proc. Natl. Acad. Sci. U.S.A.* **70,** 899 (1973).

of the extract.[9,10] When [5,6-^{3}H]PGH$_2$ is converted into PGI$_2$ by prostacyclin synthase, the tritium at C-6 is lost to the medium as shown in the following scheme.[5] Thus, the progress of this enzymic rearrangement may be monitored by following the rate of tritium release.

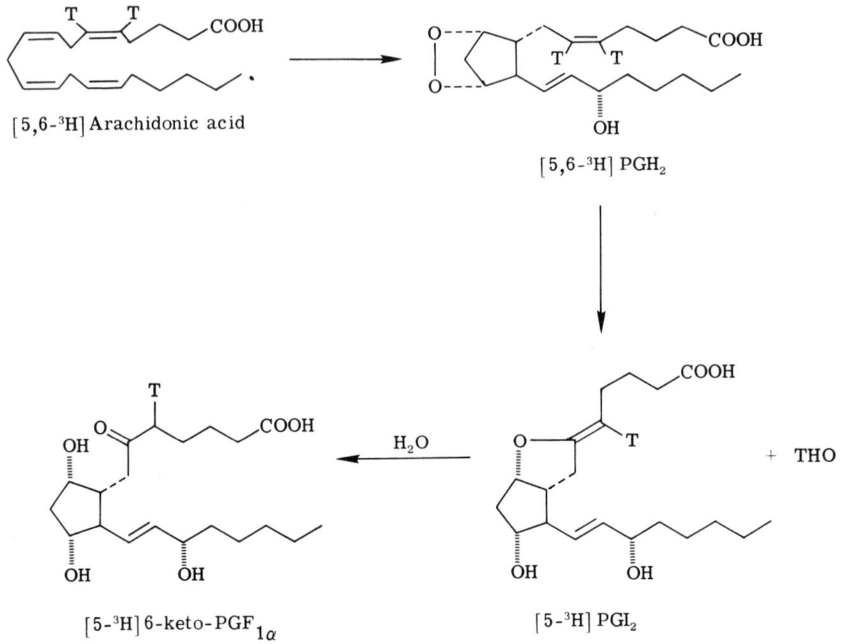

[5,6-^{3}H]Arachidonic acid → [5,6-^{3}H] PGH$_2$ → [5-^{3}H] PGI$_2$ + THO; [5-^{3}H] PGI$_2$ —H_2O→ [5-^{3}H] 6-keto-PGF$_{1\alpha}$

The substrate is tritiated only at C-5 and C-6. There is no known tritium loss at either position during arachidonate transformation except that at C-6, which is unique to prostacyclin synthase and thus confers the desired specificity of the assay. Separation of tritiated H_2O from other labeled substrate and products can be achieved by charcoal adsorption of the latter compounds followed by centrifugation.

Reagents

- Tris-HCl buffer, 0.05 *M*, pH 7.5
- [5,6-^{3}H]Arachidonic acid, diluted with arachidonic acid to a specific activity of 3000 cpm/mmol
- [5,6-^{3}H]PGH$_2$, diluted with PGH$_2$ to a specific activity of 3300 cpm/mmol
- Trichloroacetic acid, 15%
- Charcoal suspension, 10% in H_2O (neutral charcoal, Amend Drug and Chemical Co.)

[10] G. Graff, this volume [49].

DL-Isoproterenol
Human hemoglobin
Porcine aortic microsomes[6]

Procedure

The reaction mixture contains [5,6-^{3}H]PGH_2 (2.8 nmol, 9400 cpm) and porcine aortic microsomes in a final volume of 0.5 ml of 0.05 *M* Tris-HCl buffer, pH 7.5. The reaction is initiated by the addition of labeled PGH_2 and incubated at 22° for 1 min. The reaction is then terminated by the addition of 0.25 ml of cold 15% trichloroacetic acid. The mixture is allowed to stand at 0° for 10 min, then 0.2 ml of 10% charcoal suspension in water is added. After standing for 5 min at 22°, the mixture is centrifuged at 1000 *g* for 5 min. The supernatant is decanted, and the radioactivity is determined in 10 ml of a toluene–Triton X-100 (2:1) mixture containing 0.4% PPO and 0.005% POPOP with a liquid scintillation counter. If [5,6-^{3}H]arachidonic acid is used as the indirect substrate, the reaction mixture contains [5,6-^{3}H]arachidonic acid (20 nmol, 60,000 cpm), DL-Isoproterenol (0.5 μmol), hemoglobin (0.5 nmol), and porcine aortic microsomes in a final volume of 0.5 ml of 0.05 *M* Tris-HCl buffer, pH 7.5. The reaction is initiated by the addition of microsomes, incubated for 5 min at 37°, and terminated by the addition of trichloroacetic acid as before. The radioactivity in the supernatant after centrifugation is determined in the same manner.

Comments

1. The enzyme preparation should be diluted to give no more than 20–20% of the total radioactivity of the labeled substrate released. The linearity of the tritium release versus enzyme protein concentration should be observed.
2. A control sample with acid added at zero time should be run to give the basal unadsorbable organic radioactivity.
3. A kinetic isotope effect is associated with the removal of C-6 tritium in [5,6-^{3}H]PGH_2. Therefore, the percentage conversion of substrate into product is actually higher than the percentage release of substrate radioactivity. The assay can be considered to be an indication of relative enzyme activity.
4. In many tissues, prostaglandin endoperoxide synthase is a rate-limiting enzyme in arachidonate conversion to prostacyclin. When [5,-6-^{3}H]arachidonate is used as an indirect substrate, it is advisable to fortify the system with sheep seminal vesicular microsomes to generate sufficient [5,6-^{3}H]PGH_2 *in situ*.

[16] Partial Purification and Assay of Thromboxane A Synthase from Bovine Platelets

By TANIHIRO YOSHIMOTO and SHOZO YAMAMOTO

Thromboxane A synthase catalyzes the transformation of prostaglandin H_2 (PGH_2) to thromboxane A_2. The 9,11-endoperoxide group of PGH_2 is cleaved, and the two oxygen atoms appear as an epoxide and a constituent of an oxane ring. Thromboxane A_2 is well known as a proaggregatory agent of platelets. The compound is extremely unstable and decomposes with a half-life of approximately 30 sec to thromboxane B_2, which is biologically inactive.[1] The reaction of the enzyme with PGH_2 produces 12-L-hydroxy-5,8,10-heptadecatrienoic acid (HHT) concomitant with thromboxane A_2 formation in a molar ratio of 1-2:1.[2-5] Although the enzyme has not been purified, the production of both thromboxane A_2 and HHT is believed to be catalyzed by the same enzyme.[3-5] Thus, both thromboxane B_2 and HHT are detected when the enzyme acts upon PGH_2 (Fig. 1).

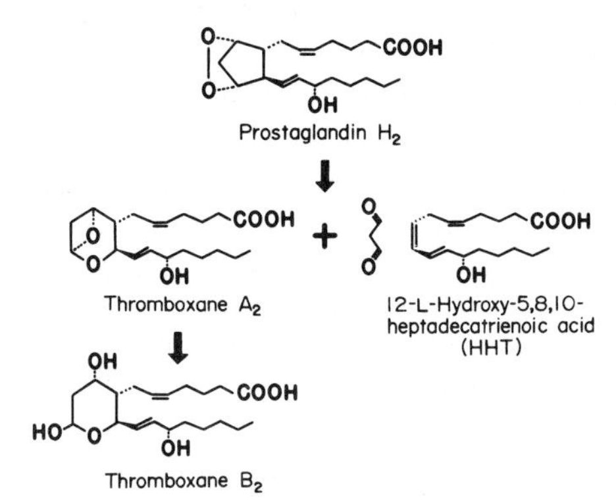

FIG. 1. Reaction catalyzed by thromboxane A synthase.

[1] M. Hamberg, J. Svensson, and B. Samuelson, *Proc. Natl. Acad. Sci. U.S.A.* **72,** 2994 (1975).

[2] T. Yoshimoto, S. Yamamoto, M. Okuma, and O. Hayaishi, *J. Biol. Chem.* **252,** 5871 (1977).

[3] U. Diczfalusy, P. Falardeau, and S. Hammarström, *FEBS Lett.* **84,** 271 (1977).

[4] M. W. Anderson, D. J. Crutchley, B. E. Tainer, and T. E. Eling, *Prostaglandins* **16,** 563 (1978).

[5] A. Raz, D. Aharony, and R. Kenig-Wakshal, *Eur. J. Biochem.* **86,** 447 (1978).

ISBN 0-12-181986-8

Assay Method

With crude enzyme preparations such as whole platelet cells and platelet homogenates, the production of thromboxane B_2 can be detected when arachidonic acid is used as the substrate. However, the use of PGH_2 as substrate is recommended in the assay of thromboxane A synthase itself.

Reagents

Tris-HCl buffer, 0.2 *M*, pH 7.4

[1-^{14}C]Prostaglandin H_2 prepared as described by Yoshimoto *et al.*[2] starting with [1-^{14}C]arachidonic acid (70,000 cpm/5 nmol). The final solution is adjusted to a concentration of 5 nmol and 70,000 cpm per 5 μl of diethylene glycol dimethyl ether or acetone, and stored at −70°.

Procedure. A 100-μl reaction mixture contains 0.1 *M* Tris-HCl, pH 7.4 (50 μl), [1-^{14}C]PGH_2 (5 μl of solution containing 5 nmol and 70,000 cpm), and enzyme. Before the reaction is started by the addition of PGH_2, the other components are placed in a test tube (12 × 103 mm) and equilibrated at 24°. The assay is performed at 24° for 1–5 min and terminated by the addition of 0.3 ml of a mixture of ethyl ether–methanol–0.2 *M* citric acid (30:4:1), which is precooled to −20°. The contents of the test tube are mixed using a vortex mixer and then kept in an ice bath. About 0.5 g of anhydrous sodium sulfate is added for dehydration of the organic phase, and the mixture is mixed gently. The following procedures are performed in a cold room at 4°. A precoated silica gel 60 F254 glass plate for thin-layer chromatography (200 × 200 × 0.25 mm, E. Merck, Darmstadt) is divided into 8 sections of 2.5-cm width. A 150-μl aliquot is removed by a graduated capillary tube from the organic layer of each tube and placed at the origin 2 cm above the bottom of the plate in a width of about 1.5 cm. The solvent is evaporated under a stream of cold air. The developing solvent is a mixture of ethyl ether–petroleum ether–acetic acid (85:15:0.1, v/v/v). The glass tank for thin-layer chromatography together with the solvent is precooled to −20° in a deep freeze, and the plate is developed at this temperature. Development to a height of 17 cm takes about 50 min. Prostaglandin B_2 and thromboxane B_2, if available, are placed as standards and visualized with iodine vapor. Ultraviolet illumination can be used for visualization of PGB_2. Prostaglandin H_2 chromatographs slightly above PGB_2. Distribution of radioactivity on the plate can be monitored either by a radiochromatogram scanner or by autoradiography. For the latter method, overnight exposure of the plate to an X-ray film that is routinely used for chest X-ray examination is sufficient to detect the radioactivity utilized for one assay. A typical chromatogram is shown in Fig. 2. Silica gel in the regions corresponding to thromboxane

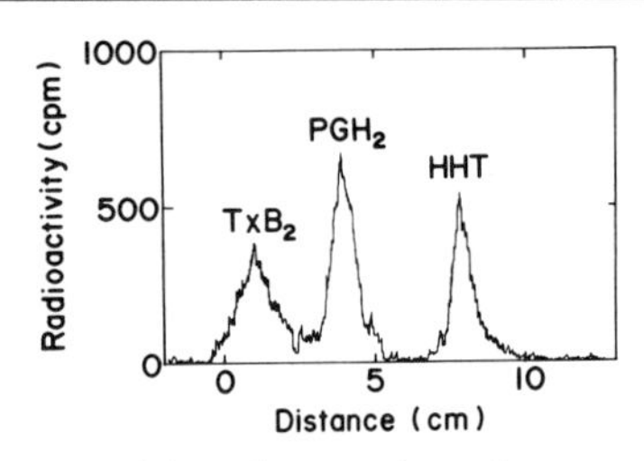

FIG. 2. Radiochromatogram of thromboxane A synthase reaction. Enzyme reaction and thin-layer chromatography were performed as described in the text.

B_2, PGH_2, and HHT is scraped into scintillation vials, and radioactivity is measured using a scintillation spectrometer in a toleune solution containing 0.03% *p*-bis[2-(4-methyl-5-phenyloxazolyl)]benzene and 0.5% 2,5-diphenyloxazole. The percentage of radioactivity in each section is calculated by dividing the counts per minute value found in the section by the total radioactivity recovered. The amount of thromboxane B_2 and HHT in nanomoles is then calculated from the amount of added PGH_2 (5 nmol) multiplied by the percentage conversion to thromboxane B_2 and HHT, respectively.

Enzyme Preparation

At a slaughterhouse, fresh bovine blood (8 liters) is collected in a polyethylene bucket into which an EDTA solution at pH 7.0 is added to give a final concentration of 7.7 m*M*. The blood is transported to the laboratory in an ice-bath. After centrifugation at 200 *g* (a Sorvall refrigerated centrifuge Model RC-5 equipped with a rotor Model GS-3), the upper turbid layer (platelet-rich plasma) is collected by gentle suction into a bottle connected to an aspirator. The platelet-rich plasma is centrifuged at 4000 *g* (a Sorvall refrigerated centrifuge Model RC-5 equipped with a rotor Model GS-3) for 30 min, and the pellet is subjected to hypotonic treatment for 30 sec to induce the lysis of contaminating red cells.

After centrifugation, again at 4000 *g*, the platelet pellet is washed twice with 0.15*M* NaCl containing 10 m*M* EDTA followed each time by centrifugation at 4000 *g* for 20 min. The washed platelets (about 5 g wet weight) are dispersed in 50 ml of 10 m*M* potassium phosphate buffer, pH 7.4. The platelet suspension is sonicated at 20,000 Hz for 1.5 min twice with the aid of a Branson cell disruptor Model 185. The sonicate is centrifuged at 10,000 *g* (a Sorvall refrigerated centrifuge Model RC-5 equipped with a rotor Model SS-34) for 10 min, and the supernatant is further centrifuged at 105,000 *g* (a Beckman preparative ultracentrifuge equipped with a rotor type 40) for 60 min. The precipitate is washed once with

50 ml of 10 m*M* potassium phosphate, pH 7.4, collected by centrifugation at 105,000 *g* for 60 min and then is suspended in the same buffer to a final volume of 15 ml. The protein concentration determined by the method of Lowry *et al.*[6] with bovine serum albumin as a standard is about 7 mg/ml (microsomal fraction).

The microsomal fraction is mixed with one-ninth the volume of 5% Triton X-100 and stirred in an ice-bath for 30 min. The mixture is centrifuged at 105,000 *g* for 60 min. The supernatant solution (43 mg of protein) is applied to a column of Whatman DE-52 (3 × 12.5 cm) equilibrated with 10 m*M* potassium phosphate, pH 7.4, containing 0.1% Triton X-100. The column is washed with the same buffer. Prostaglandin H synthase activity appears in this first wash. When the buffer concentration is increased to 200 m*M*, the thromboxane A synthase activity is detected in the eluate (a total volume of about 60 ml), which is concentrated to 10–12 ml by ultrafiltration using a Diaflo membrane PM-10. The concentrated solution (approximately 2 mg of protein per milliliter) is used as the source of thromboxane A synthase. About 30 μl of enzyme causes approximately 60% conversion of PGH_2 (5 nmol/100 μl) to thromboxane B_2 and HHT under the standard assay conditions, giving a specific enzyme activity of approximately 55 nmol/min per milligram of protein at 24°.

Properties of Enzyme

Primary production of thromboxane A_2 rather than B_2 can be demonstrated by its proaggregatory activity, and the enzyme and PGH_2 can be utilized as a thromboxane A_2 generating system.[2] The enzyme can be stored at −70° without appreciable loss of enzyme activity for several months. The pH optimum of the enzyme reaction is around 7.4. The rate of conversion of PGH_1 is almost equal to that of PGH_2. However, when the enzyme is incubated with PGH_1, thromboxane B_1 is scarcely detectable, and the major product is 12-L-hydroxy-8,10 heptadecadienoic acid.

Addendum

Thromboxane A synthase has also been partially purified from human platelets by essentially identical procedures.[7]

[6] O. H. Lowry, N. J. Rosebrough, A. L. Farr, and R. J. Randall, *J. Biol. Chem.* **193,** 265 (1951).

[7] S. Hammarström and P. Falardeau, *Proc. Natl. Acad. Sci. U.S.A.* **74,** 3691 (1977).

[17] Assay of Thromboxane A Synthase Inhibitors[1]

By HSIN-HSIUNG TAI

Thromboxane synthase catalyzes the conversion of prostaglandin endoperoxide (PGH_2) to an unstable proaggregatory and vasoconstrictive thromboxane A_2 (TxA_2) that is readily hydrolyzed to inactive thromboxane B_2 (TxB_2).[2]

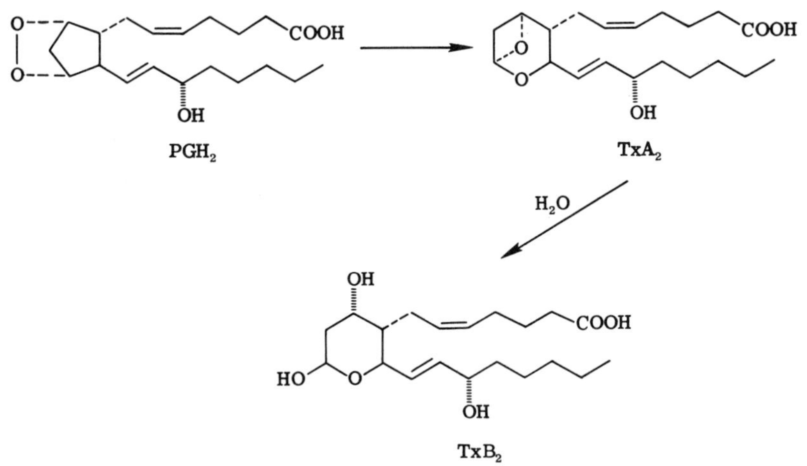

PGH_2 can also be converted by prostacyclin synthase to an antiaggregatory and vasodilatory substance—prostacyclin (PGI_2)—which is also easily hydrolyzed to the relatively inactive 6-keto-$PGF_{1\alpha}$ at physiological pH.[3] Because of the opposing actions of TxA_2 and PGI_2, the control and the balance of the synthesis of these two biologically potent substances are of great importance to cardiovascular functions and diseases. A number of cardiovascular disorders, such as stroke and heart attack, are thought to be related at least in part to an imbalance in the synthesis of TxA_2. Management of these disorders by aspirin has been explored in several clinical trials. The efficacy of aspirin has been found to be variable. One of the arguments against its being effective is that it blocks not only the synthesis of TxA_2 but also that of PGI_2 since aspirin inhibits

[1] This work is supported in part by grants from the National Institutes of Health (GM-25247) and the American Heart Association (78-865).

[2] M. Hamberg, J. Svensson, and B. Samuelsson, *Proc. Natl. Acad. Sci. U.S.A.* **72,** 2994 (1975).

[3] S. Moncada, R. J. Gryglewski, S. Bunting, and J. R. Vane, *Nature (London)* **263,** 663 (1976).

ISBN 0-12-181986-8

PGH_2 synthase (cyclooxygenase), which catalyzes the formation of PGH_2, the precursor, of both TxA_2 and PGI_2.[4] Agents that can selectively inhibit thromboxane synthetase may provide a more useful therapeutic alternative. A number of inhibitors have been discovered and designed and have been shown to be potent and selective in *in vitro* systems.[5-11] The efficacy of these compounds in animal models has also begun to be explored.[12,13] Preliminary results appear to indicate that compounds of similar nature may eventually become potentially valuable antithrombotic and antihypertensive agents. This presentation describes a radioimmunological method for large-scale screening of thromboxane synthase inhibitors and the assessment of the inhibitory potency of these inhibitors.

Assay Methods

Principle. Thromboxane synthase is assayed by monitoring the formation of TxB_2 immunoreactivity from PGH_2 using a specific radioimmunoassay for TxB_2.[14] Compounds that are able to inhibit thromboxane synthase can be assayed by their ability to inhibit the formation of TxB_2 immunoreactivity from PGH_2 catalyzed by human platelet microsomes (which contain very active thromboxane synthase).[15] The relative potency of compounds as thromboxane synthase inhibitors is determined by the relative concentrations (IC_{50}) at which 50% inhibition of the enzyme activity occurs with a given substrate PGH_2 concentration.

Reagents

Tris-HCl buffer, 0.05 *M*, pH 7.5, containing 0.1% gelatin (RIA buffer)
PGH_2[16]
TxB_2

[4] J. R. Vane, *Nature (London)* **231,** 232 (1971).

[5] R. R. Gorman, G. L. Bundy, D. C. Peterson, F. F. Sun, O. V. Miller, and R. A. Fitzpatrick. *Proc. Natl. Acad. Sci. U. S. A.* **74,** 4007 (1977).

[6] R. J. Gryglewiski, A. Zmuda, R. Korbut, E. Krecioch, and K. Rieron, *Nature (London)* **267,** 628 (1977).

[7] S. Moncada, S. Bunting, K. Mullane, P. Throgood, J. R. Vane, A. Raz, and P. Needleman *Prostaglandins* **13,** 611 (1977).

[8] H. H. Tai and B. Yuan, *Biochem. Biophys. Res. Commun.* **80,** 236 (1978).

[9] T. Yoshimoto, S. Yamamoto, and O. Hayaishi, *Prostaglandins* **16,** 529 (1978).

[10] F. F. Sun. *Biochem. Biophys. Res. Commun.* **80,** 236 (1978).

[11] H. H. Tai, C. L. Tai, and N. Lee. *Arch. Biochem. Biophys.* **203,** 758 (1980).

[12] T. Miyamoto, K. Taniguchi, T. Tanouchi, and F. Hirata, *Adv. Prostaglandin Thromboxane Res.* **6,** 443 (1980).

[13] T. Umetsu and K. Sanai, *Thromb. Haemostasis* **39,** 74 (1978).

[14] H. H. Tai and B. Yuan, *Anal. Biochem.* **87,** 343 (1978).

[15] H. H. Tai and B. Yuan, *Biochim. Biophys. Acta* **531,** 286 (1978).

[16] M. Hamberg and B. Samuelsson, *Proc. Natl. Acad. Sci. U. S. A.* **70,** 889 (1973).

TxB_2 antisera[14]
TxB_2–[^{125}I]iodotyrosine methyl ester conjugate[14]
γ-Globulin-coated charcoal suspension (100 ml of RIA buffer containing 1 g of γ-globulin and 3 g of neutral charcoal (Amend Drug and Chemical Co.)
Human platelet microsomes[15]
Thromboxane synthase inhibitors
HCl, 1 *N*
Tris base, 1 *M*

Procedure. Thromboxane synthase assay mixtures contain PGH_2 (5 nmol) varying concentrations of inhibitor including a zero concentration, and human platelet microsomes (50–100 μg of protein) in a final volume of 1 ml of 0.05 *M* Tris-HCl buffer, pH 7.5. The reaction is initiated by the addition of PGH_2 because of the instability of PGH_2 in aqueous solution. Incubation is allowed to proceed for 1 min at 22°. The reactions are terminated by the addition of 0.05 ml of 1 *N* HCl. After neutralization with 1 *M* Tris base, an aliquot (0.1 ml) of the reaction mixture is diluted with 1 ml of RIA buffer for determination of radioimmunoassayable TxB_2.

Radioimmunoassay mixtures contain 0.2 ml of TxB_2 standards (0, 10, 20, 50, 100, 200, 500, 1000 pg) or diluted assay samples, 0.1 ml of diluted TxB_2 antibodies, and 0.1 ml of TxB_2–[^{125}I]iodotyrosine methyl ester conjugate (ca. 10,000 cpm) in RIA buffer. Each standard or sample is run in duplicate. After 1 hr of incubation at room temperature, each tube receives 1 ml of H_2O before addition of 0.2 ml of charcoal suspension and mixing on a vortex mixer. The mixture is centrifuged for 5 min at 1000 *g* after standing for 5 min at room temperature. The supernatant containing the bound hapten is separated from the pellet containing the free hapten adsorbed to charcoal. The radioactivity in the supernatant and the pellet is separately determined by a gamma spectrometer. The bound to total ratio is calculated for each sample, and the concentration of each sample is determined from a standard curve after logit transformation as described by Tai and Chey.[17]

The percentage inhibition of each inhibitor at a series of concentrations is plotted against inhibitor concentrations on a semilog paper. A near-linear relationship should be observed. The concentration of inhibitor that gives 50% inhibition of TxB_2 formation is then determined.

Comments

1. The amount of human platelet microsomes used for assay should be within the linear range of product formation versus protein concentra-

[17] H. H. Tai and W. Y. Chey, *Anal. Biochem.* **74,** 12 (1976).

tions, and the conversion to product should be limited to no more than 20–30% of the total PGH_2 substrate.

2. At least 6 or 7 different concentrations of a inhibitor should be run to obtain a best-fit linear relationship.

3. An alternative method of examining the specificity of the inhibitor is to incubate [1-^{14}C]arachidonic acid with sheep or porcine lung microsomes in the presence of hemoglobin, isoproterenol, and a potential thromboxane synthase inhibitor. Lung microsomes possess high levels of PGH synthase, prostacyclin synthase, and PGH–PGE isomerase in addition to thromboxane synthase.[18] If the inhibitor blocks PGH synthase activity in addition to thromboxane synthase, the formation of other catabolites of PGH_2 will also be inhibited. This will be evident from thin-layer radiochromatographic analysis. If the inhibitor blocks the synthesis of PGI_2 and TxB_2 without affecting the synthesis of their precursor PGH_2, the synthesis of PGE_2 should be stimulated because of less competition in using the substrate. From the relative peak size of each catabolite one should be able to tell if the synthesis of that particular catabolite has been modulated.

4. The ability of thromboxane synthase inhibitors to block platelet aggregation induced by arachidonic acid, ADP, collagen, etc., can also be examined as described.[19]

[18] H. H. Tai, B. Yuan, and A. T. Wu, *Biochem. J.* **170,** 441 (1978).
[19] H. H. Tai, N. Lee, and C. L. Tai, *Adv. Prostaglandin Thromboxane Res.* **6,** 447 (1980).

[18] Purification and Assay of 9-Hydroxyprostaglandin Dehydrogenase from Rat Kidney[1]

By HSIN-HSIUNG TAI and BARBARA YUAN

9-Hydroxyprostaglandin dehydrogenase catalyzes an NAD^+-dependent oxidation of the 9α-hydroxyl group of 15-keto-13,14-dihydro-$PGF_{2\alpha}$ to 15-keto-13,14-dihydro-PGE_2.[2]

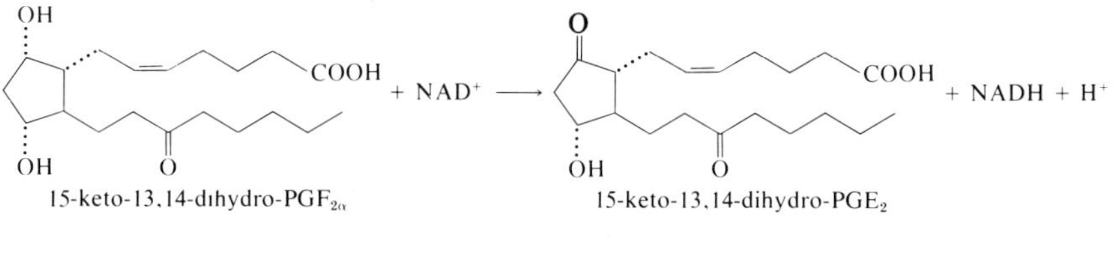

15-keto-13,14-dihydro-$PGF_{2\alpha}$ + NAD^+ → 15-keto-13,14-dihydro-PGE_2 + NADH + H^+

[1] This work was supported in part by a grant from the National Institutes of Health (GM-21588).

ISBN 0-12-181986-8

The enzyme was found to have the highest activity in rat kidney cortex cytosol,[2] although some activity is also detected in rat liver cytosol.[3] The enzyme catalyzes not only the oxidation of 15-keto-13,14-dihydro-$PGF_{2\alpha}$, but also that of 15-keto-$PGF_{2\alpha}$ and $PGF_{2\alpha}$ albeit to a lesser degree.[4] This enzyme thereby provides a mechanism to convert PGF to PGE, particularly at the metabolite level.

Assay Method

Principle. The assay is based on the stereospecific transfer of the tritium label from 15-keto-13,14-dihydro-[9β-^{3}H]$PGF_{2\alpha}$ to lactate by coupling 9-hydroxyprostaglandin dehydrogenase with another A-side specific lactate dehydrogenase as shown in the following scheme.[3]

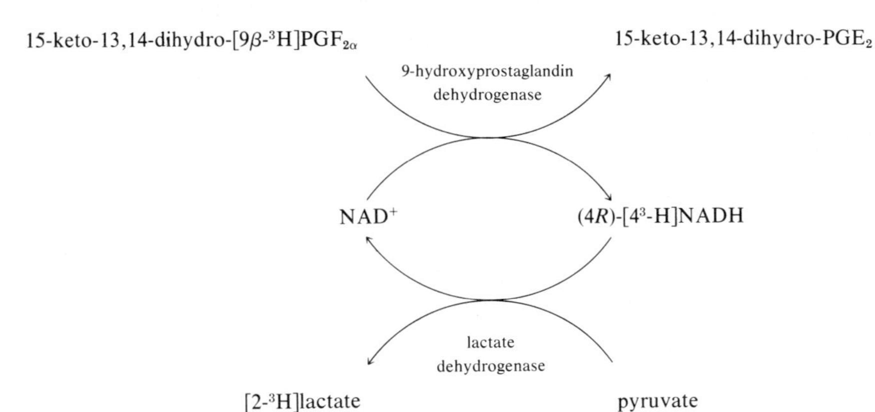

The labeled lactate is separated from labeled 15-keto-13,14-dihydro-$PGF_{2\alpha}$ and NADH by adsorption of the latter two compounds with charcoal followed by centrifugation. The amount of labeled lactate formed that represents the amount of labeled substrate oxidized is quantitated by counting the supernatant in a liquid scintillation counter. The assay is sensitive to oxidation of picomoles of substrate. A similar assay for 15-hydroxyprostaglandin dehydrogenase has been developed.[5]

Reagents

Potassium phosphate buffer, 0.1 *M*, pH 7.5

Lactate dehydrogenase from rabbit muscle (870 U/mg, Sigma Chemical Co.)

---

[2] C. R. Pace-Asciak, *J. Biol. Chem.* **250,** 2789 (1975).

[3] H. H. Tai and B. Yuan, *Anal. Biochem.* **78,** 410 (1977).

[4] B. Yuan, C. L. Tai, and H. H. Tai, *J. Biol. Chem.* **255,** 7439 (1980).

[5] H. H. Tai, *Biochemistry* **15,** 4586 (1976).

9-Hydroxyprostaglandin dehydrogenase[4]
15-Keto-13,14-dihydro-[9β-^{3}H]$PGF_{2\alpha}$[3]
NAD^+
Sodium pyruvate
Charcoal suspension, 10% in H_2O (neutral charcoal, Amend Drug and Chemical Co.)

Procedure. The assay mixture contains 15-keto-13,14-dihydro-[9β-^{3}H]$PGF_{2\alpha}$ (1 nmol, 40,000 cpm), NAD^+ (2 μmol), sodium pyruvate (5 μmol), lactate dehydrogenase (50 μg), and appropriate amounts of crude or purified 9-hydroxyprostaglandin dehydrogenase in a final volume of 1 ml of 0.1 *M* potassium phosphate, pH 7.5. The reaction is allowed to proceed at 37° for 10 min and is terminated by the addition of 0.3 ml of a 10% charcoal suspension with vigorous mixing. After standing for 5 min at room temperature, the reaction mixture is centrifuged at 1000 *g* for 5 min. The supernatant is decanted, and the radioactivity is determined by liquid scintillation counting in 10 ml of a toluene–Triton X-100 (2:1) mixture containing 0.4% PPO and 0.005% POPOP.

Comments

1. A linearity of product formation versus protein concentration and time should first be established to determine the amount of enzyme and incubation time to be used for the assay.
2. The sensitivity of the assay can be decreased by increasing the amount of unlabeled substrate added to the assay mixture.
3. A control sample with charcoal added at zero time should be run to give the basal unadsorbable radioactivity.
4. The assay can be used to quantitate 15-keto-13,14-dihydro-$PGF_{2\alpha}$ by constructing a standard curve based on the fact that increasing amount of unlabeled 15-keto-13,14-dihydro-$PGF_{2\alpha}$ added to the labeled substrate results in decrease of radioactivity released. Purification of 15-keto-13,14-dihydro-$PGF_{2\alpha}$ may be necessary to avoid interference with competing substances. The assay can be sensitive in the picomole range.

Purification of 9-Hyrdoxyprostaglandin Dehydrogenase[4]

Preparation of Crude Enzyme Extract. One hundred grams of frozen rat kidney (Pel Freez Biologicals) are homogenized in 300 ml of 10 m*M* potassium phosphate buffer, pH 7.5, containing 1 m*M* EDTA and 0.2 m*M* dithiothreitol (buffer A) in a Waring blender for 2 min. The homogenate is centrifuged at 10,000 *g* for 10 min, and the supernatant is further centrifuged at 105,000 *g* for 1 hr. The final supernatant is designated S_{105}.

Ammonium Sulfate Fractionation. Ammonium sulfate is slowly added to the S_{105} to give 45% saturation (258 g/liter). After stirring for 30 min at 4° the precipitate is removed by centrifugation at 10,000 *g* for 10 min. The clear supernatant is brought to 70% saturation with ammonium sulfate (additional 156 g/liter), stirred for 30 min, and centrifuged at 10,000 *g* for 10 min. The precipitate is collected and dissolved in 38 ml of buffer A. Any insoluble material is removed by centrifugation at 10,000 *g* for 10 min.

Sephadex G-100 Chromatography. The above clear supernatant is applied to a Sephadex G-100 column (5 × 100 cm) equilibrated with buffer A. The column is developed with the same buffer, and fractions of 13 ml are collected. Active fractions are combined, and ammonium sulfate is added to 80% saturation (516 g/liter). After stirring for 30 min, the precipitate is collected by centrifugation at 10,000 *g* for 10 min and dissolved in 10 ml of buffer A. This solution is desalted by passing through a Sephadex G-25 column (2.5 × 27 cm) equilibrated with buffer A.

TEAE-Cellulose Chromatography. The above desalted fraction is applied to a TEAE-cellulose column (2.2 × 19 cm) equilibrated with buffer A. The column is developed with the same buffer, and fractions of 8.6 ml are collected. The active fractions appearing behind the major protein peak are combined, lyophilized, and redissolved in 8.4 ml of buffer A. This solution is desalted using a Sephadex G-25 column (2.5 × 27 cm) equilibrated with buffer A.

Affi-Gel Blue Affinity Chromatography. The above desalted solution is applied to an Affi-Gel Blue column (1.5 × 10 cm) equilibrated with buffer A. The column is washed with the same buffer until no protein appears in the washing. The enzyme is then eluted with buffer A containing 3 m*M* NAD^+. Fractions of 4.2 ml are collected. The active fractions are pooled and stored in aliquots at −80°. A summary of the results of the purifica-

PURIFICATION OF 9-HYDROXYPROSTAGLANDIN DEHYDROGENASE FROM RAT KIDNEY

Fraction	Activity[a] (units)	Protein (mg)	Specific activity (units/mg)	Yield (%)	Purification (fold)
S_{105}	42.26	4058.8	0.01	100	1
Ammonium sulfate (45–70%)	40.51	1380.6	0.03	95.9	3
Sephadex G-100	32.28	157.4	0.21	76.4	21
TEAE-cellulose	17.86	10.6	1.68	42.3	168
Affi-Gel Blue	12.77	1.4	9.12	30.2	912

[a] One unit of enzyme activity is defined as the amount of enzyme that will catalyze the oxidation of 1 nmol of substrate per minute at 37°. Calculation of the amount of substrate oxidized is based on the assumption that no kinetic isotope effect is involved in the removal of 9β-tritium during oxidation.

tion of 9-hydroxyprostaglandin dehydrogenase is shown in the table. The overall procedure represents a purification of 912-fold with 30% recovery.

Properties[4]

Homogeneity. The final preparation shows a single band upon electrophoresis on a 7.5% polyacrylamide gel at pH 8.4 and on a polyacrylamide gel in 0.1% SDS.

Molecular Weight. The native enzyme consists of a single polypeptide chain with molecular weight of 33,000.

Reversibility. The enzyme appears to catalyze irreversible oxidation at pH 7.5.

Substrate Specificity. The enzyme catalyzes the oxidation of 15-keto-13,14-dihydro-$PGF_{2\alpha}$, 15-keto-$PGF_{2\alpha}$, and $PGF_{2\alpha}$ with relative rates of 9.1, 5.9, and 1.4 nmol/min per milligram, respectively, with K_m values of 0.33, 3.0, and 5.0 μM, respectively. The enzyme does not catalyze the oxidation of 6-keto-$PGF_{1\alpha}$, nor does it catalyze the oxidation of 15(*S*) hydroxyl group of any prostaglandins.

Coenzyme Specificity. The oxidation rate is 10 times faster with NAD^+ (1 m*M*) than with $NADP^+$ (1 m*M*). The activity with the NAD^+ analog, thionicotinamide-NAD^+ (1 m*M*), is 1.5 times greater than with NAD^+ (1 m*M*). The apparent K_m for NAD^+ is 0.11 m*M*.

pH Optimum. The pH optimum is in the range of 8.5–9.0.

Inhibitors. The enzyme is inhibited 70% and 50% by 1 m*M* sodium mersalyl and *p*-hydroxymercuribenzoate, respectively. The enzyme is also inhibited by PGA_2, PGB_2, PGD_2, PGE_2, arachidonic acid, linoleic acid, triiodothyroacetic acid, and indomethacin in the low micromolar range[4]; PGE_2 is a competitive inhibitor, and other substances are noncompetitive inhibitors with respect to prostaglandin substrate. The purified enzyme can be stored at −80° for a year without appreciable loss of activity.

[19] Purification of PGD_2 11-Ketoreductase from Rabbit Liver

By Patrick Y-K. Wong

Nugteren and Hazelhof[1] first described the formation of PGD_2 from endoperoxides in 1973. Subsequently, PGD_2 was found to be a significant product of arachidonic acid cascade in many tissue and cell types.[2] Sev-

[1] D. H. Nugteren and E. Hazelhof, *Biochim. Biophys. Acta* **326,** 448 (1973).

ISBN 0-12-181986-8

eral groups of investigators have demonstrated the transformation of PGH_2 in brain homogenates[3,4] and neuroblastoma cells.[5] The enzyme, PGH–PGD isomerase, which catalyzed the conversion of PGH_2 to PGD_2, has been purified to homogeneity and clearly distinguished from that of glutathione *S*-transferase.[6] It has been demonstrated that PGD_2 is released by platelets during aggregation.[7] PGD_2 has also been found to be a potent inhibitor of platelet aggregation, with a potency only less than that of prostacyclin (PGI_2) and its biologically active metabolite, 6-keto-PGE_1.[8,9] PGD_2, like PGI_2 and 6-keto-PGE_1, is not metabolized by the lung 15-hydroxyprostaglandin dehydrogenase *in vitro*.[10–12] Thus, PGD_2, like PGI_2 and 6-keto-PGE_1, is a potential circulating antithrombotic agent. Hensby[13] first reported the conversion of PGD_2 and PGE_2 to $PGF_{2\alpha}$ in sheep blood. Ellis *et al.*[14] demonstrated the formation of $PGF_{2\alpha}$ (cyclopentane-1,3-diol structure) type metabolites after infusion of PGD_2 *in vivo*, suggesting the presence of a unique enzymic system in the metabolism of PGD_2 *in situ*.

The presence of PGD_2 11-ketoreductase in rabbit liver was confirmed by Reingold and co-workers.[15] They demonstrated that this enzyme was present in the highest concentration in rabbit liver. Thus, reports on the estimation of the level of $PGF_{2\alpha}$ in various tissues, as measured by the F-type metabolites by radioimmunoassay or by gas chromatography and mass spectrometry, may have to be reexamined in order to account for the possible contribution from PGD_2. Further, the loss of biological activity of PGD_2 on passage through the liver by the action of 11-ketoreductase

[2] B. Samuelsson, *Annu. Rev. Biochem.* **47,** 997 (1978).

[3] M. S. Abdel-Halim, M. Hamberg, B. Sjöquist, and E. Anggård, *Prostaglandins* **14,** 633 (1977).

[4] F. F. Sun, J. P. Chapman, and J. C. McGuire, *Prostaglandins* **14,** 1055 (1977).

[5] T. Shimizu, N. Mizuro, T. Amano, and O. Hayaishi, *Proc. Natl. Acad. Sci. U.S.A.* **76,** 6231 (1979).

[6] T. Shimizu, S. Yamamoto, and O. Hayaishi, *J. Biol. Chem.* **254,** 5222 (1979).

[7] O. Oelz, R. Oelz, H. R. Knapp, B. J. Sweetman, and J. A. Oates, *Prostaglandins* **13,** 225 (1977).

[8] G. DiMinno, M. J. Silver, and G. DeGaetano, *Br. J. Haematol.* **43,** 637 (1979).

[9] P. Y.-K. Wong, J. C. McGiff, F. F. Sun, and W. H. Lee, *Eur. J. Pharmacol.* **60,** 245 (1979).

[10] M. F. Ruckrich, W. Schlegel, and A. Jung, *FEBS Lett.* **68,** 59 (1976).

[11] F. F. Sun, S. B. Armour, V. R. Bockstanz, and J. C. McGuire, *Adv. Prostaglandin Thromboxane Res.* **1,** 163 (1976).

[12] P. Y.-K. Wong, W. H. Lee, C. P. Quilley, and J. C. McGiff, *Fed. Proc., Fed. Am. Soc. Exp. Biol.* **40,** 2001 (1981).

[13] C. N. Hensby, *Prostaglandins* **8,** 369 (1974).

[14] C. K. Ellis, M. D. Smigel, J. A. Oates, O. Oelz, and B. J. Sweetman, *J. Biol. Chem.* **254,** 4252 (1979).

[15] D. F. Reingold, A. Kawasaki, and P. Needleman, *Biochim. Biophys. Acta* **659,** 179 (1981).

may convert a potent inhibitor of platelet aggregation (PGD_2) to a potent vasoconstrictor ($PGF_{2\alpha}$) that has no antithrombotic activity.

These reports strongly suggest the presence of an enzyme that can reduce the 11-keto group of PGD_2 to an 11-hydroxy group. In this chapter, we describe the purification of PGD_2 11-ketoreductase activity from rabbit liver to apparent homogeneity. Thus, PGD_2-11-ketoreductase may be a major enzymic pathway for the metabolism of endogenous PGD_2.

Purification Method

Reagents and Materials

[5,6,8,9,12,14,15(n)-^{3}H]-PGD_2 (specific activity 100 ci/mmol) was a generous gift from Dr. David Ahern of New England Nuclear. The purity of the [^{3}H]PGD_2 was established by thin-layer chromatography (TLC plate, 0.25 mm thick, 20 × 20 cm silica gel H precoated plastic sheet, Brinkmann, New Jersey) in two solvent systems: (A) ethyl acetate–acetic acid (99:1, v/v); (B) organic phase of the mixture of isooctane–ethyl acetate–acetic acid–water (25:55:10:50, v/v). PGD_2 and $PGF_{2\alpha}$ were kindly supplied by Dr. John Pike and Dr. Udo Axén of the Upjohn Company. $NADP^+$, NADPH, and glucose-6-phosphate dehydrogenase were purchased from Sigma Chemical Co. (St. Louis, Missouri).

Procedure

Step 1. Preparation of Crude Extract. New Zealand white male rabbits (2–3 kg) were anesthetized with sodium pentobarbital (25 mg/kg). After midline laparotomy, the liver was exposed. The portal vein was cannulated, and the liver was flushed with 200–300 ml of 50 m*M* Tris-HCl buffer, pH 7.4, containing 0.1 m*M* dithiothreitol (buffer I). After the liver was free of blood, it was removed and cut into thin slices. Approximately 50 g of thin slices were homogenized in 5 volumes of ice-cold buffer I with a Polytron homogenizer operated at top speed for 1 min. The homogenization process was repeated twice. The final homogenate was centrifuged at 8000 *g* for 20 min. The supernatant was centrifuged further at 105,000 *g* for 60 min in a Beckman Model L-75 centrifuge using a type W-28 rotor. The supernatant is referred to as fraction I.

Step 2. Ammonium Sulfate Fractionation. All purification steps were carried out at 4° in the cold room. Ammonium sulfate (12 g) was added slowly to fraction I to produce 30% saturation. After stirring for 30 min at 4°, the solution was centrifuged at 9000 *g* for 30 min. The clear supernatant was brought to 60% saturation with additional ammonium sulfate (25 g). After stirring for 60 min at 4°, the solution was centrifuged at

9000 *g* for 30 min. The precipitate was collected and resuspended in 20 ml of buffer I and dialyzed three times with 100 volumes of buffer I. This fraction (30–60% of ammonium sulfate) was referred to as fraction II.

Step 3. DEAE-Sephadex Chromatography. Fraction II was applied onto a DEAE-Sephadex column (A-50-120, Sigma Co., Missouri) (4 × 30 cm),previously equilibrated with buffer I. The column was washed with 500 ml of buffer I, then the enzyme was eluted with a stepwise gradient of various concentrations of NaCl (0.2, 0.3, 0.4, and 0.5 *M*) in 500 ml of buffer I. Fractions containing 11-ketoreductase activity were pooled and concentrated by an Amicon ultrafiltration cell using a 10,000 dalton cutoff (PM-10) membrane. The concentrated fraction was referred to as fraction III.

Step 4. DEAE-Cellulose Chromatography. A DEAE-cellulose column (4 × 30 cm) (DE-52, preswollen, Whatman, LTD, Springfield Mill, Kent, England) was equilibrated with buffer I. Fraction III was loaded onto the column and was first washed with 500 ml of buffer I. The enzyme was then eluted by a step gradient of NaCl in 500 ml of buffer I (0.1 *M* to 0.4 *M*). The enzyme activity was eluted as one single peak by buffer I with 0.1 *M* NaCl. Active fractions were pooled and immediately concentrated to 10 ml by using an ultrafiltration cell as described above.

Step 5. Isoelectric Focusing and Gel Electrophoresis. Active enzyme fractions obtained from the DEAE-cellulose column were applied to the top of a sucrose gradient analytical focusing column (LKB) with 1% Ampholine (pH 3.5 to 10). Focusing was started at 250 V, and the voltage was gradually increased to 400 V within 2 hr. Total focusing time was 16 hr. Fractions were collected, and PGD_2 11-ketoreductase activity was monitored by the radiometric assay method as described.

Electrophoresis was performed in an LKB electrophoresis. A sodium dodecyl sulfate (SDS)–polyacrylamide slab gel that contained a 5 to 15% linear acrylamide gradient was prepared as described by Laemmli.[16] Electrophoresis was performed at 2 mA per gel until the marker dye (bromophenol blue) approached the end of the gel. Protein bands were stained with Coomassie Brilliant Blue. The molecular weight of PGD_2 11-ketoreductase was estimated by comparison to enzymes and proteins of known molecular weight on the same gel, according to the method of Wallace *et al.*[17]

Step 6. Radiometric Assay of PGD_2 11-ketoreductase. The PGD_2 11-ketoreductase activity was assayed in a mixture containing $[^3H]PGD_2$ (500,000 dpm), 5 μM unlabeled PGD_2; 0.5–0.9 ml of the enzyme fractions

[16] U. K. Laemmli, *Nature (London)* **227,** 680 (1970).

[17] R. W. Wallace, T. J. Lynch, A. A. Tallant, and W. Y. Cheung, *J. Biol. Chem.* **254,** 377 (1978).

eluted from the columns or 10–25 μg of protein of the purified enzyme fraction obtained from isoelectric focusing column and buffer I to a final volume of 1 ml. Reactions were carried out by incubating the mixture at 37° for 60 min with constant shaking. The reaction was terminated by the addition of 1 *N* HCl to bring the pH to 3.0. The reaction mixture was extracted with ethyl acetate (3 ml) twice, and the lipid extracts were dried under a stream of N_2. The residues were redissolved in 100 μl of dry acetone, applied onto TLC plates, and developed twice in solvent system B using authentic PGD_2 and $PGF_{2\alpha}$ as standards. Radioactive zones on the TLC plates were located by a radiochromatogram scanner (Packard radiochromatogram scanner, Model 7320). Zones corresponding to $PGF_{2\alpha}$ and PGD_2 were cut out and suspended in 10 ml of 0.4% Omnifluor and 20% Triton X-100 toluene liquid scintillation fluid and counted in a Beckman L-75 liquid scintillation counter. The observed counts per minute were converted to disintegrations per minute and then converted to picomoles of $PGF_{2\alpha}$. The specific activity of the enzyme was expressed as picomoles of $PGF_{2\alpha}$ formed per hour per milligram of protein. Protein concentration was determined by the method of Lowry *et al.*[18] using bovine serum albumin as standard.

Step 7. Platelet Aggregation Studies. Blood was drawn from volunteers who had not taken any aspirin-like drugs for the previous 10 days. Nine parts of whole blood were mixed with one part of 3.8% sodium citrate to a total volume of 5 ml. The blood was centrifuged at 150 *g* for 10 min, then the platelet-rich plasma was removed with a siliconized pipette. Platelet-poor plasma was prepared by centrifuging the remaining blood at 12,000 *g* for 10 min. The final platelet count in platelet-rich plasma was adjusted to 2×10^8 per milliliter with platelet-poor plasma. Platelet aggregation studies were performed with 0.5 ml of platelet-rich plasma stirred at 1200 rpm at 37° in a dual-channel Payton aggregation module. The aggregation response (increased light transmission) was transcribed on a linear recorder (Payton Associates, Buffalo, New York).

Purity and Properties of PGD_2 11-Ketoreductase

The percentage yield and the increases in specific activity during purification of PGD_2 11-ketoreductase from rabbit liver are shown in Table I. As can be seen by the yield and specific activity, little activity was lost from the cytoplasmic fractions during ammonium sulfate fractionation. After chromatography on two DEAE columns and isoelectric focusing, approximately 0.2% of the total cytoplasmic protein but 24% of the PGD_2

[18] O. H. Lowry, N. J. Rosebrough, A. L. Farr, and R. J. Randall, *J. Biol. Chem.* **193,** 265 (1951).

TABLE I
PURIFICATION OF PROSTAGLANDIN D_2 11-KETOREDUCTASE FROM RABBIT LIVER

Fraction	Total protein (mg)	Total activity (pmol/60 min)	Specific activity (pmol/60 min mg^{-1} protein)	Purification (fold)	Percentage recovery
I. Crude extract	2502	648	0.25	1	100
II. Ammonium sulfate precipitate (30–60%)	1987	630	0.32	1.26	97.2
III. DEAE-Sephadex	192	554	2.89	11.5	85.5
IV. DEAE-cellulose	53	315	5.90	23.5	48.6
V. Isoelectric focusing	6	153	25.52	101.7	23.6

11-ketoreductase activity remained. In our purification procedure, the enzyme was first separated by the DEAE-Sephadex column from other proteins by a step gradient of NaCl (0.2 to 0.5 *M*) in buffer I (Fig. 1). When the enzyme fraction was further purified by DEAE-cellulose column chromatography, the enzyme activity was eluted with buffer I as one single peak. The active enzyme fractions from DEAE-cellulose column were fo-

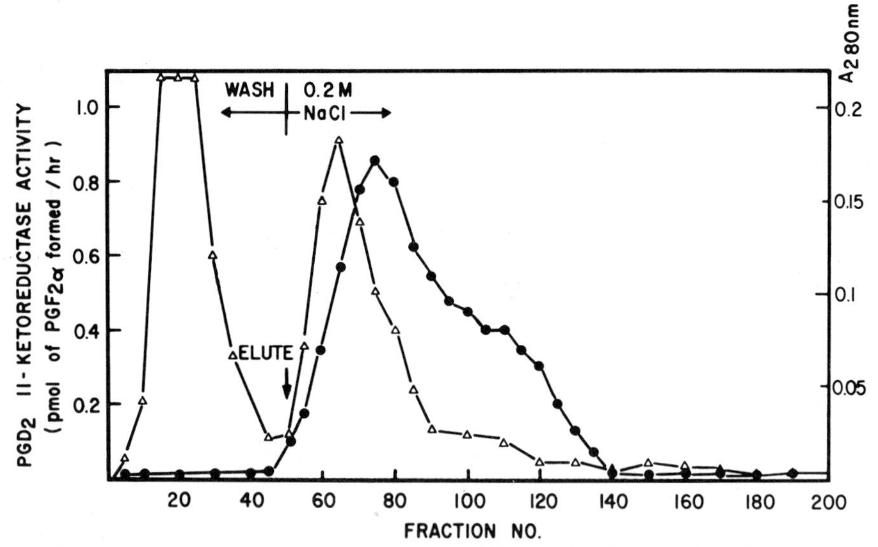

FIG. 1. DEAE-Sephadex column chromatograph. The desalted fraction of ammonium sulfate (30 to 60%) precipitation was applied to a DEAE-Sephadex column (4 × 30 cm) previously equilibrated with buffer I. The enzyme was eluted with 0.1 *M* NaCl in buffer I; 10-ml fractions were collected, and the PGD_2 11-ketoreductase activity was assayed by the radiometric assay (●—●). Protein concentration was monitored by absorbance at 280 nm (△—△).

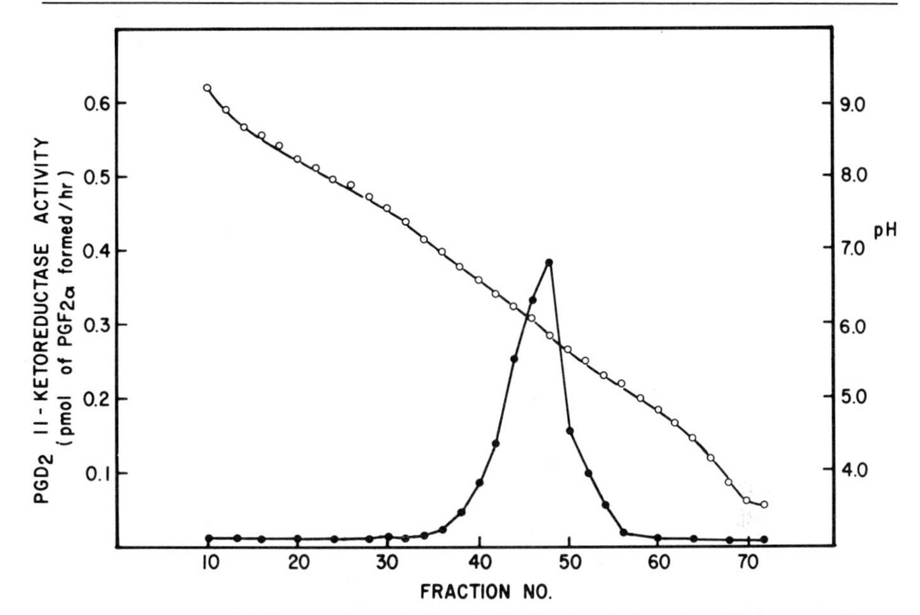

FIG. 2. Purification of PGD_2 11-ketoreductase by isoelectric focusing. The enzyme fractions obtained from DEAE-cellulose column chromatography were concentrated by ultrafiltration. Approximately 20 mg of protein were electrofocused. PGD_2 11-ketoreductase activity (●—●), was determined on every second fraction by radiometric assay.

cused using an analytical isoelectric focusing column (LKB) containing a pH gradient of 3.5 to 10.0. The enzyme activity was found to focus with a fraction having a pH value of 5.8, indicating the isoelectric point (p*I*) of the enzyme (Fig. 2).

Enzyme activity was found to be dependent on time and protein concentration and to have a pH optimun of 7.5. The effect of reduced pyridine nucleotides was measured at a constant concentration of NADPH, or using an NADPH-generating system that contained the following mixture: $NADP^+$ (2 m*M*), glucose 6-phosphate (3.5 m*M*), and 2 units of glucose-6-phosphate dehydrogenase. The activity of PGD_2 11-ketoreductase activity was found to be highest with NADPH and NADPH generating system (Table II). NADPH had lower activity; with NAD^+ and $NADP^+$ as cofactors, no activity was detectable. The apparent K_m of the enzyme was estimated to be 200 μM. This was in agreement with that reported by Reingold *et al.*[15] SDS–gel electrophoresis of the active enzyme fraction obtained from isoelectric focusing revealed one major band with a molecular weight estimated to be 66,000, as compared to proteins with known molecular weight on the same gel (Fig. 3).

The conversion of $[^3H]PGD_2$ to $[^3H]PGF_{2\alpha}$ was measured by radiomet-

TABLE II
COENZYME SPECIFICITY OF PGD_2 11-KETOREDUCTASE IN RABBIT LIVER

Enzyme	Coenzyme (4 m*M*)	Specific activity (pmol/mg protein)
Fraction IV	NAD+	ND[a]
	NADP+	ND
	NADH	1.63
	NADPH	5.13
	NADPH (GS)[b]	4.83
Boiled fraction IV	NADPH	ND

[a] ND, not detectable.
[b] NADPH (GS), NADPH-generating system.

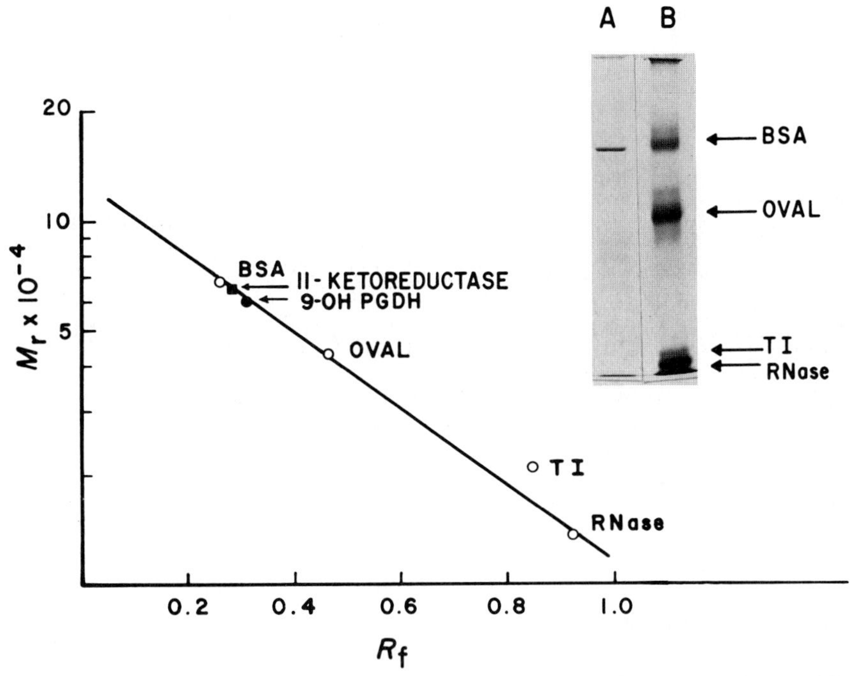

FIG. 3. Polyacrylamide gel electrophoresis of the purified PGD_2 11-ketoreductase. The purified enzyme obtained from isoelectric focusing (20 μg) was applied to sodium dodecyl sulfate polyacrylamide gel (A). A slab gel with a 5 to 15% linear polyacrylamide gradient was prepared according to the method of Laemmli.[16] Protein was stained with Coomassie Brilliant Blue. The molecular weight was calibrated with bovine serum albumin (BSA, M_r = 67,000), ovalbumin (OVAL, M_r = 42,000), soybean trypsin inhibitor (TI, M_r = 21,000), and ribonuclease A (RNase, M_r = 13,700) (B). A indicated that the molecular weight of the PGD_2 11-ketoreductase in rabbit liver was estimated to be 66,000.

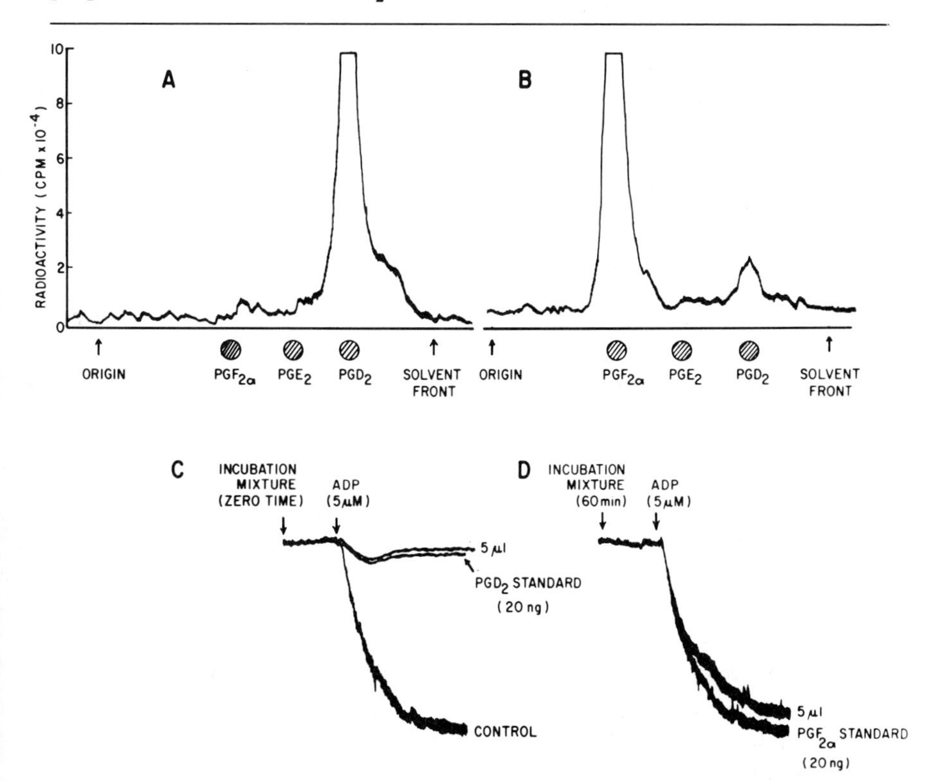

FIG. 4. (A, B) Radiochromatograph scan of the radiometric assay of PGD$_2$ 11-ketoreductase. (C, D) Bioassay of the PGD$_2$ 11-ketoreductase on platelet aggregation. (A, B) Purified enzyme was incubated with [^{3}H]PGD$_2$ (500,000 dpm, 5 nmol), NADPH (4 m*M*) and buffer I to a volume of 1 ml. After incubation at 37° for 0 min (A), and 60 min (B), the reactions were terminated by acidification and extraction. The radioactive products extracted were separated by TLC. (C, D) The enzymes were assayed as described above, except that unlabeled PGD$_2$ (14 nmol) was added to the incubation mixture before the addition of the enzymes. After incubation at 37° for 0 min (C), and 60 min (D), the reactions were terminated and extracted as described. The extracted products were resuspended in Tris buffer (pH 7.5, 50 m*M*); 5-μl aliquots of each suspension (equivalent to 20 ng of PGD$_2$) were tested for their ability to inhibit ADP-induced platelet aggregation as compared to authentic PGD$_2$ and PGF$_{2\alpha}$ standards.

ric method; concomitantly, the assay extracts were monitored for their biological activity on platelet aggregation. Since PGD$_2$ is a potent inhibitor of platelet aggregation, the transformation of PGD$_2$ to PGF$_{2\alpha}$ resulted in the loss of the anti-platelet aggregatory activity of PGD$_2$ after incubation with the purified enzyme (Fig. 4C and D). The loss of anti-platelet aggregatory activity coincides with the conversion of [^{3}H]PGD$_2$ to [^{3}H]PGF$_{2\alpha}$, the latter being inactive on platelet aggregation[19] (Fig. 4, A–D).

[19] A. J. Marcus, *J. Lipid Res.* **19,** 793 (1978).

[20] Isolation and Properties of an NAD^+-Dependent 15-Hydroxyprostaglandin Dehydrogenase from Human Placenta

By JOSEPH JARABAK

15-Hydroxyprostaglandin + $NAD^+ \rightleftarrows$ 15-ketoprostaglandin + NADH + H^+

The first step in the biological inactivation of prostaglandins involves oxidation of the 15-hydroxyl group to a ketone.[1] In many tissues this reaction is catalyzed by either an NAD^+-dependent or an $NADP^+$-dependent enzyme. The activity of the former enzyme is particularly high in human placenta, making this tissue a good source of the NAD^+-dependent enzyme.[2,3]

Assay Method

The enzyme activity is measured at 25 ± 0.5° spectrophotometrically in a cuvette with a l-cm light path. The cuvette contains 3 ml of aqueous solution consisting of 290 μmol of Tris-chloride at pH 9.0, 1.36 μmol of NAD^+, and 114 nmol of PGE_1 added in 0.02 ml of ethanol. The blank cuvette contains no PGE_1 but is identical otherwise. The reaction is started by the addition of enzyme, and NADH production is followed by measuring the change in absorbance at 340 nm. One unit of enzyme activity is defined as the amount of enzyme that catalyzes the production of 1 μmol of NADH per minute under the conditions of the assay.

Purification Procedure

The procedure to be described involves minor modifications of an earlier one.[4] Three normal human term placentas are used for the purification. Steps 1 and 2 are performed on individual placentas. Unless specified otherwise, all operations are performed at below 4°.

Step 1. Homogenization. A placenta is chilled immediately after delivery. The villous tissue is dissected from the membranes, rinsed in cold tap water, and homogenized. Portions of from 75 to 100 g are homogenized at top speed for 2 min in a Waring blender with two volumes of buffer A (20% glycerol, v/v, 5 m*M* potassium phosphate, and 1 m*M* EDTA, at a

[1] E. Änggård and B. Samuelsson, *Ark. Kemi* **25,** 293 (1966).
[2] J. Jarabak, *Proc. Natl. Acad. Sci. U. S. A.* **69,** 533 (1972).
[3] J. Jarabak, *Am. J. Obstet. Gynecol.* **138,** 534 (1980).
[4] S. S. Braithwaite and J. Jarabak, *J. Biol. Chem.* **250,** 2315 (1975).

ISBN 0-12-181986-8

final pH of 7.0). The homogenate is centrifuged at 10,000 *g* for 45 min, and the precipitate is discarded.

Step 2. Ammonium Sulfate Precipitation. Solid ammonium sulfate is added to the supernatant solution from the preceding step (0.41 g/ml) slowly and with continuous stirring. The pH is maintained at about 7 by the addition of 3 *N* NH_4OH. After standing for 2 hr[5] the solution is centrifuged at 10,000 *g* for 30 min. The supernatant solution is discarded. The precipitate is suspended in 50–75 ml of buffer B (50% glycerol, v/v, 5 m*M* potassium phosphate, and 1 m*M* EDTA, at a final pH of 7.0), dialyzed for 24 hr against two changes of at least 20 volumes each of buffer B, and stored at −20°.

Step 3. Dialysis against Buffer A. The resuspended ammonium sulfate precipitates are combined and dialyzed for 24 hr against two changes of at least 20 volumes each of buffer A and then centrifuged for 30 min at 10,000 *g*. The precipitate is discarded.

Step 4. DEAE-Cellulose Chromatography. The supernatant solution from the preceding step is applied to a 4 × 45 cm column of DEAE-cellulose (high capacity, Bio-Rad) that has been equilibrated with buffer C (20% glycerol, v/v, 40 m*M* potassium phosphate, and 1 m*M* EDTA, at a final pH of 7.0). The column is washed with buffer C until the absorbance of the column eluate at 280 nm is below 0.2 and then with a linear gradient: chamber I contains 1 liter of buffer C; chamber II contains 1 liter of 20% glycerol, v/v, 400 m*M* potassium phosphate, and 1 m*M* EDTA at a final pH of 7.0. The elution of the enzyme starts at a conductivity of 2.2 $m\Omega^{-1}$ and is completed at 6.0 $m\Omega^{-1}$. Fractions having the highest specific activity are combined, solid ammonium sulfate is added (0.41 g/ml), and the suspension is allowed to stand for 1–2 hr before centrifugation at 10,000 *g* for 30 min. The supernatant solution is discarded. The precipitate is suspended in 20 ml of buffer B, dialyzed against buffer B, and stored at −20°.

Step 5. Sephadex G-100 Gel Filtration. The enzyme from the preceding step is applied to a 5 × 94 cm column of Sephadex G-100 that has been equilibrated with a buffer of 20% glycerol, v/v, 10 m*M* potassium phosphate, 1 m*M* EDTA, at a final pH of 7.0. The column is developed with the equilibrating buffer. The fractions having the highest specific activity are combined, dialyzed against buffer B, and stored at −20°.

Step 6. NAD–Hexane–Agarose Affinity Chromatography. A 2.5 × 6 cm column of NAD-hexane-agarose (NAD·AG) is equilibrated with a buffer containing 20% glycerol, v/v, 25 m*M* potassium phosphate, and 1 m*M* EDTA, at a final pH of 7.0. The enzyme from the preceding step is

[5] If the precipitate is allowed to stand for more than 10 hr, 30–60% of the enzyme activity is lost.

diluted to achieve this glycerol and salt concentration and is applied to the column immediately after dilution. The column is then washed with the equilibration buffer until the absorbance of the eluate at 280 nm is less than 0.002. Then the column is washed with buffer D (20% glycerol, v/v, 200 m*M* potassium phosphate, and 1 m*M* EDTA, at a final pH of 7.0). This wash is important to elute some protein impurities, but, since it also elutes a small amount of the enzyme, it is continued only until the absorbance of the eluate at 280 nm ceases to decline. Then the enzyme is eluted from the column with 30 ml of buffer D containing 390 μmol of NADH followed by buffer D containing no additions. The fractions containing the highest enzyme activity are combined, dialyzed against buffer B, concentrated by ultrafiltration in a Diaflo cell to 3–5 ml, and stored at −20°.

Step 7. DEAE-BioGel A Chromatography. Even after exhaustive dialysis, the enzyme eluted from NAD·AG contains appreciable amounts of NADH. The NADH is not adsorbed to DEAE-BioGel A and may be removed from the enzyme in this manner. A 1.2 × 3.5 cm DEAE-BioGel A column is equilibrated with buffer B, and the enzyme from the preceding step is applied to it. The column is then washed with buffer B until the absorbance of the eluate at 340 nm has dropped to 0.000 (generally 400–600 ml of buffer B are required). The enzyme is eluted with a buffer containing 20% glycerol, v/v, 25 m*M* potassium phosphate, and 1 m*M* EDTA

PURIFICATION OF NAD$^+$-DEPENDENT 15-HYDROXYPROSTAGLANDIN DEHYDROGENASE FROM HUMAN PLACENTA

Step	Volume (ml)	Total activity (units)	Specific activity (units/mg protein[a]) × 10^3	Yield (%)
1. Centrifuged homogenate	2560[b]	191	1.85	100
2. Ammonium sulfate precipitation	373	87.4	2.23	46
3. Supernatant after dialysis	510	70.0	2.92	37
4. DEAE-cellulose	16	45.4	49.8	24
5. Sephadex G-100	55	29.2	950	15
6. NAD·AG[c]	60	19.7	—	10
7. DEAE-BioGel A	9.4	13.8	1750	7

[a] During most of the purification it is assumed that a solution containing 1 mg of protein per milliliter has an absorbance at 280 nm of 1.0 in a cuvette of 1-cm path length. In the final step of the purification, the protein concentration is calculated by the formula $1.5 \times A_{280} - 0.75\ A_{260}$ = protein concentration (mg/ml). E. Layne, this series, Vol. 3, p. 447.

[b] Volume of the supernatant solution after centrifugation of the homogenate from 910 g of placental tissue.

[c] NAD–hexane–agarose.

at a final pH of 7.0. Peak fractions are combined, concentrated to about 5 ml by ultrafiltration, dialyzed against buffer B, and stored at $-20°$.

The enzyme resulting from this purification is apparently homogeneous by several criteria.[4] Typical results of this purification are summarized in the table.

Properties

Stability. The placental NAD^+-dependent 15-hydroxyprostaglandin dehydrogenase is quite labile. Since placental homogenates and partially purified fractions both lose more than 90% of their NAD^+-dependent 15-hydroxyprostaglandin dehydrogenase activity when stored at 4° for 24 hr in 0.1 M potassium phosphate at pH 7.0,[2] it is unrealistic to attempt to purify this enzyme in the absence of a stabilizing agent. Partial stabilization may be achieved with the addition of 2-mercaptoethanol, dithiothreitol, or NAD^+, but none of these is as effective as glycerol.[2] At relatively high protein concentrations (7.7 mg/ml), storage at 4° in 20% glycerol completely protects the enzyme for 72 hr. For that reason, steps 1–6 are performed at 4° in buffers containing 20% glycerol. At lower protein concentrations (0.05 mg/ml) there are significant losses of activity within 24 hr under these same conditions. Since buffers containing 50% glycerol fully protect the enzyme activity for at least 24 hr at low protein concentrations and a temperature of 4°, such a buffer is used for the last step in the purification. Storage in buffers containing 50% glycerol at $-20°$ provides the best long-term protection; samples stored in this manner retain more than 90% of their original activity after 7 months.

Pyridine Nucleotide Specificity. NAD^+ and thionicotinamide adenine dinucleotide are cofactors for the enzyme, but $NADP^+$ and acetylpyridine dinucleotide are not. The activity of the enzyme with thionicotinamide adenine dinucleotide is only 5 to 33% of its activity with NAD^+, depending on the substrate used: PGA_1, 5%; PGE_2, 20%; PGE_1, 21%; $PGF_2\alpha$, 33%.[4] The Michaelis constant for NAD^+ is 30.8 μM at pH 7.0 and 29.1 μM at pH 9.0 with PGE_1 as substrate.[6] The Michaelis constant for NADH is 15.6 μM at pH 7.0 with 15-keto-PGE_1 as substrate.[6]

Substrate Specificity. The maximum velocity of oxidation of various prostaglandins (expressed as moles of pyridine nucleotide reduced per minute per mole of enzyme at 25°) is as follows: PGE_2, 110; $PGF_{2\alpha}$, 107; PGE_1, 98; PGI_2, 80; PGA_1, 72; 6-keto-$PGF_{1\alpha}$, 53; $PGF_{1\alpha}$, 37; and PGD_1, 3.9.[7] Thromboxane B_2, 15-keto-$PGF_{1\alpha}$ and PGB_1 are not substrates for the enzyme.[7] The Michaelis constant for PGE_1 is 1.33 μM at pH 7.0 and

[6] J. Jarabak and S. S. Braithwaite, *Arch. Biochem. Biophys.* **177**, 245 (1976).

[7] J. Jarabak and J. Fried, *Prostaglandins* **18**, 241 (1979).

4.8 μM at pH 9.0 with NAD^+ as cofactor.[6] The Michaelis constant for 15-keto-PGE_1 is 30.1 μM at pH 7.0 with NADH as cofactor.[6]

Kinetic Mechanism.[6] The enzymic reaction proceeds by a single displacement mechanism; the addition of reactants is ordered with the pyridine nucleotide binding first. At pH 7.0 a kinetically significant ternary complex is formed, whereas at pH 9.0 the ternary complex is not significant (Theorell–Chance mechanism).

Inhibitors.[6] At high substrate concentrations there is formation of unreactive complexes between the 15-hydroxyprostaglandin and both the free enzyme and the enzyme·NADH complex. The inhibition caused by various prostaglandins (e.g., PGB_1) and prostaglandin analogs (e.g., prostanoic acid) may be explained by the formation of similar unreactive complexes. In addition, certain prostaglandin analogs, arachidonic acid and ethacrynic acid, also affect the activity of the enzyme by causing its irreversible inactivation.

Equilibrium Constant.[4]

$$K_{eq} = \frac{[H^+][\text{15-keto-PGE}_1][\text{NADH}]}{[\text{PGE}_1][\text{NAD}^+]}$$

The average value for the K_{eq} is $(6.4 \pm 0.2) \times 10^{-8}$ M when NAD^+ and PGE_1 are initial components of the reaction mixture and $(6.6 \pm 0.8) \times 10^{-8}$ M when NADH and 15-keto-PGE_1 are present initially.

Ultraviolet Absorption Spectrum.[4] The enzyme has a typical protein ultraviolet absorption spectrum; the molar absorbance at 280 nm in buffer B is 41.6×10^3.

Activation Energy.[4] The activation energy for the enzyme-catalyzed reaction is 9900 cal per mole.

Molecular Weight.[4] When sodium dodecyl sulfate disc electrophoresis is used to estimate the molecular weight, a value of 42,000 is obtained. A value of 51,500 is obtained when gel filtration on Sephadex G-100 is performed.

Acknowledgments

This work was supported by grants from the National Institutes of Health (HD-07045) and the Louis Block Foundation.

[21] Assay of NAD^+-Dependent 15-Hydroxyprostaglandin Dehydrogenase Using (15*S*)-[15-^3H]PGE_2[1]

By HSIN-HSIUNG TAI

NAD^+-dependent 15-hydroxyprostaglandin dehydrogenase catalyzes the oxidation of 15(*S*)-hydroxyl group of prostaglandins to 15-ketoprostaglandins.[2]

$$PGE_2 + NAD^+ \longrightarrow \text{15-keto-}PGE_2 + NADH + H^+$$

This reaction is considered to be both the initial and major route for transformation of biologically active prostaglandins to inactive metabolites.[3] The enzyme has been found in most animal tissues examined.[4,5] Its activity can vary during tissue development,[6] hormonal treatment,[7] and pathological conditions.[8-11]

Although virtually all primary prostaglandins except PGB can be oxidized by this enzyme, PGE_2 is known to be the best substrate.[12,13] Numerous methods for assaying the enzyme activity using PGE_2 as a substrate have been developed. However, few of these methods can be used reliably for determination of enzyme activity with sufficient sensitivity in

[1] This work was supported in part by a grant from the National Institutes of Health (GM-21588)

[2] E. Änggård and B. Samuelsson, *J. Biol. Chem.* **239,** 4097 (1964).

[3] E. Änggård, *Acta Physiol. Scand.* **66,** 509 (1966).

[4] E. Änggård, C. Larsson, and B. Samuelsson, *Acta Physiol. Scand.* **81,** 396 (1971).

[5] H. H. Tai, *Biochemistry* **15,** 4586 (1976).

[6] C. R. Pace-Asciak, *J. Biol. Chem.* **250,** 2795 (1975).

[7] N. A. Alam, P. Russell, T. W. Tabor, and Moulton, B. C., *Endocrinology* **98,** 859 (1976).

[8] J. M. Armstrong, G. J. Blackwell, R. J. Flower, J. C. McGift, K. M. Mullane, and J. R. Vane, *Nature (London)* **260,** 582 (1976).

[9] C. R. Pace-Asciak, *Nature (London)* **263,** 510 (1976).

[10] C. J. Limas and C. Limas, *Am. J. Physiol.* **233,** 487 (1977).

[11] H. H. Tai, B. Yuan, and M. Sun, *Life Sci.* **24,** 1275 (1979).

[12] J. Nakano, E. Änggård, and B. Samuelsson, *Eur. J. Biochem.* **11,** 386 (1969).

[13] D. T. Kung-Chao and H. H. Tai, *Biochim. Biophys. Acta* **614,** 1 (1980).

ISBN 0-12-181986-8

crude enzyme extracts. We have developed a rapid and sensitive assay using (15*S*)-[15-^{3}H]PGE$_2$ as a substrate and following the release of tritium as a result of oxidation of secondary 15(*S*)-hydroxyl group.[4] The assay is applicable to either crude or purified enzyme preparations and can be sensitive in the picomole range of substrate oxidation.

Preparation of (15*S*)-[15-^{3}H]PGE$_2$

Principle. (15*S*)-[15-^{3}H]PGE$_2$ is synthesized enzymically by coupling two B-side specific dehydrogenases, β-D-galactose dehydrogenase and 15-hydroxyprostaglandin dehydrogenase, in the presence of β-D-[1-^{3}H]galactose, 15-keto-PGE$_2$, and NAD$^+$.[4] The tritium of β-D-[1-^{3}H]galactose can be readily transferred to 15-keto-PGE$_2$ via (4*S*)-[4-^{3}H]NADH, resulting in the formation of (15*S*)-[15-^{3}H]PGE$_2$ as shown in the following scheme.

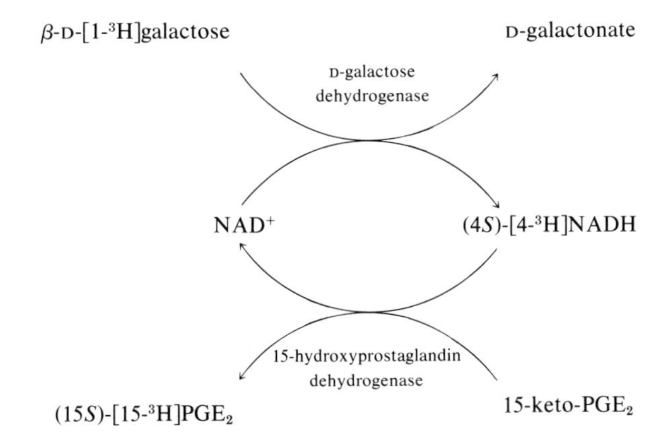

The final product is extracted from the reaction mixture and purified by thin-layer chromatography.

Reagents

Potassium phosphate buffer, 0.1 *M*, pH 6.0
β-D-Galactose dehydrogenase (7.6 U/mg, Sigma Chemical Co.)
15-Hydroxyprostaglandin dehydrogenase (0.65 U/mg partially purified through Sephadex G-100)[13]; if Affi-Gel Blue is used NADH should be removed by dialysis or the Affi-Gel Blue column should be eluted using 1 *M* NaCl instead of 1 m*M* NADH
15-Keto-PGE$_2$
NAD$^+$
D-[1-^{3}H]Galactose (New England Nuclear)
Silica gel G precoated plate (Brinkmann)

Ethyl acetate
Isooctane
Acetic acid

Procedure. The reaction mixture contains 15-keto-PGE_2 (50 nmol), NAD^+ (10 nmol), D-[1-^{3}H]galactose (3.52 nmol, 50 μCi), β-D-galactose dehydrogenase (100 μg), and 15-hydroxyprostaglandin dehydrogenase (180 μg) in a final volume of 1 ml of 0.1 *M* potassium phosphate buffer adjusted to pH 6.0 with 1 *N* acetic acid. The reaction is allowed to proceed for 3 hr at 37° and terminated by acidification to pH 3.0 with 1 *N* HCl. The reaction mixture is extracted twice with 3 ml of ethyl acetate. The combined extract is evaporated under a stream of N_2 at 40°. The residue is taken up in 100 μl of acetone and spotted on a silica gel G plate (2 × 20 cm). The plate is developed in the organic layer of the solvent system ethyl acetate–acetic acid–isooctane–water (11:2:5:10) to a height of 12.5 cm. The (15*S*)-[15-^{3}H]PGE_2 can be localized by scanning with a Berthold thin-layer radioscanner or by cochromatography of an authentic standard of PGE_2 on a separate plate followed by exposure of this plate in iodine vapor to localize PGE_2. Labeled PGE_2 is scraped off the plate and extracted from the gel twice with 1-ml portions of ethanol. Each extraction is followed by centrifugation at 1000 *g* to precipitate fine gel particles. The combined extract, which contains (15*S*)-[15-^{3}H]PGE_2, is kept at −20°. The specific activity should be comparable to that of D-[1-^{3}H]galactose.

Comments

1. The same procedure can be used for preparation of (15*S*)-[15-^{3}H]$PGF_{2\alpha}$ and other prostaglandins provided that 15-keto prostaglandins are available.
2. (15*S*)-[15-^{3}H]PGA_2 can be prepared from (15*S*)-[15-^{3}H]PGE_2 by acetic acid–phosphoric acid dehydration.[5]
3. Preparation of (15*S*)-[15-^{3}H]$PGF_{2\alpha}$ can alternatively be achieved by tritiated sodium borohydride reduction of 15-keto-$PGF_{2\alpha}$. The desired epimer can be purified by thin-layer chromatography. Preparation of (15*S*)-[15-^{3}H]PGE_2 by chemical reduction is not practical, since purification of the desired epimer from several other epimers is required.

Assay of 15-Hydroxyprostaglandin Dehydrogenase

Principle. The assay is based on the transfer of tritium from (15*S*)-[15-^{3}H]PGE_2 to glutamate via (4*S*)-[4-^{3}H]NADH by coupling 15-hydroxyprostaglandin dehydrogenase with another B-side-specific glutamate dehydrogenase as shown in the following scheme.[5]

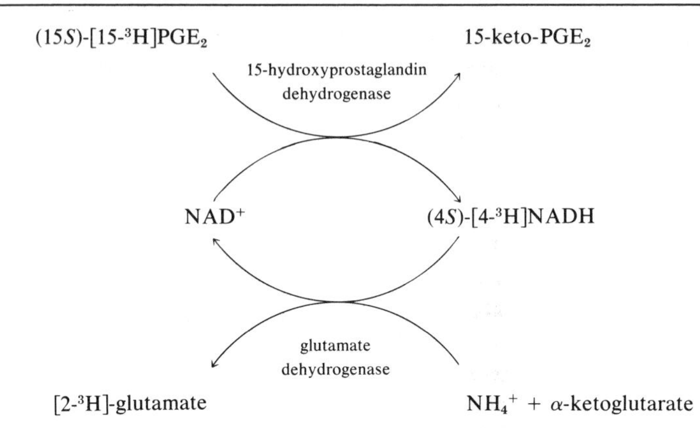

The labeled glutamate is separated from labeled PGE_2 and NADH by adsorption of the latter two labeled compounds with charcoal followed by centrifugation. The amount of labeled glutamate formed, which represents the amount of labeled substrate oxidized, is quantitated by counting the supernatant in a liquid scintillation counter. A similar assay for 9-hydroxyprostaglandin dehydrogenase has been developed.[14]

Reagents

Potassium phosphate buffer, 0.1 *M*, pH 7.5
(15*S*)-[15-^{3}H]PGE_2, diluted with PGE_2 to give a specific activity of 20,000 cpm/nmol
NAD^+
NH_4Cl
Sodium α-ketoglutarate
Glutamate dehydrogenase (51 U/mg, Sigma Chemical Co.)
15-Hydroxyprostaglandin dehydrogenase
Charcoal suspension, 10%, in H_2O (neutral charcoal, Amend Drug and Chemical Co.)

Procedure. The assay mixture contains (15*S*)-[15-^{3}H]PGE_2 (1 nmol, 20,000 cpm), NAD^+ (1 μmol), NH_4Cl (5 μmol), sodium α-ketoglutarate (1 μmol), glutamate dehydrogenase (100 μg), and an appropriate amount of crude or purified 15-hydroxyprostaglandin dehydrogenase in a final volume of 1 ml of 0.1 *M* potassium phosphate, pH 7.5. The reaction is initiated by adding an aliquot of the 15-hydroxyprostaglandin dehydrogenase preparation (see below) and allowed to proceed at 37° for 10 min. The reaction is terminated by the addition of 0.3 ml of a 10% charcoal suspension followed by mixing on a vortex mixer. The mixture is centrifuged at 1000 *g* for 5 min after standing for 5 min at room temperature. The super-

[14] H. H. Tai and B. Yuan, *Anal. Biochem.* **78,** 410 (1977).

natant is decanted into the scintillation vial, and the radioactivity present in labeled glutamate is determined in 10 ml of a toluene–Triton X-100 (2:1) mixture containing 0.4% PPO and 0.005% POPOP with a liquid scintillation counter.

Comments

1. The enzyme preparation should be diluted to the extent that (*a*) the amount of tritium release is not more than 20–30% of the total radioactivity of labeled substrate; and (*b*) linearity exists between product formation and added 9-hydroxyprostaglandin dehydrogenase.
2. The sensitivity of the assay can be increased or decreased by changing the amount of unlabeled PGE_2 added to the labeled PGE_2.
3. A control sample with charcoal added at zero time should be run to give the basal unadsorbable radioactivity.
4. The assay can be used to quantitate PGE_2 by constructing a standard curve based on the fact that increasing amount of unlabeled PGE_2 added to the labeled substrate results in decrease of radioactivity released. This assay can modified to detect picomoles of PGE_2.

[22] Microassay Procedure for NAD^+-Dependent 15-Hydroxyprostaglandin Dehydrogenase

By CLINTON N. CORDER

Principle. The assay procedure is based upon the determination of the amount of substrate [prostaglandin E_1 (PGE_1)] lost during a reaction with NAD^+-dependent 15-hydroxyprostaglandin dehydrogenase (NAD^+-PGDH) in the tissue sample.[1-3] The reaction is carried out in a small reaction volume at 37° for the specified time, then terminated. The amount of unreacted PGE_1 is measured in a coupled enzyme system. The initial reaction may be carried out on nanogram quantities of freeze-dried tissue in a reaction volume of 0.1 μl suspended in oil. Conversely, the microassay may be applied to larger sample sizes in a 1-ml reaction volume. The quantitative histochemical assay consists of seven steps. The procedure is described for NAD^+-PGDH in kidney.

[1] J. T. Wright, C. N. Corder, and R. Taylor, *Biochem. Pharmacol.* **25**, 1669 (1976).
[2] J. T. Wright and C. N. Corder, *Prostaglandins* **17**, 431 (1979).
[3] J. T. Wright and C. N. Corder, *J. Histochem. Cytochem.* **27**, 657 (1979).

ISBN 0-12-181986-8

Reagents and Materials

All chemicals and enzymes used in the assays should be of the highest purity available.

Alcohol dehydrogenase, yeast (EC 1.1.1.1) (300 U/mg) treated with activated charcoal to remove pyridine nucleotides[4]

Malic dehydrogenase, pig heart (EC 1.1.1.37) (1200 U/mg) treated as for alcohol dehydrogenase

Glutamic-oxaloacetate transaminase, pig heart (EC 2.6.1.1) (200 U/mg).

NAD^+-15-hydroxyprostaglandin dehydrogenase, beef lung (EC 1.1.1.141) (600 mmol kg^{-1} hr^{-1}).[5] Alternatively, it may be prepared from porcine lung.[6] Care must be taken to ensure that this preparation is free of nonspecific NADH-oxidase activity. The enzyme is standardized in the fluorometer with 110 μM PGE_1 in step 1 reagent and sufficient PGDH to reduce 0.05–0.2 μM NAD^+ per minute.

NAD^+, 10 m*M* stock solution in double-distilled water is acid treated to remove NADH and standardized at weekly intervals.[7,8]

PGE_1, 10 m*M* stock solution in ethanol, standardized in the fluorometer with beef lung PGDH (7–10 μg/ml) by allowing the reaction of 2–6 μM PGE_1 to go to completion in the standard PGDH reaction medium

Bovine serum albumin (10% solution), dialyzed 4 hr at 4° against double-distilled water

Light mineral oil, washed with acid and alkali[7]

Reagents for Microassay

Step 1 reagents

Reagent A: potassium phosphate, 100 m*M*, pH 7.5; mercaptoethanol, 2 m*M*; ethylenediaminetetraacetic acid, 2 m*M*; bovine serum albumin, 0.2 mg/ml; NAD^+, 0.7 m*M*; PGE_1, 0.01 m*M*. Components 1–4 may be stored at 4° in solution. Add NAD^+ and PGE_1 within 1 hr of intended use.

Reagent B: Same as reagent A except that PGE_1 is omitted

Step 2 reagent: phosphoric acid, 0.2 *N* (concentration must be adjusted to lower pH of step 2 to pH 3–3.5)

Step 3 reagent: Sodium hydroxide, 0.3 *N* (concentration must be adjusted to raise pH of step 3 to pH 8.0)

[4] T. Kato, S. J. Berger, J. Carter, and O. H. Lowry, *Anal. Biochem.* **53,** 86 (1973).

[5] F. M. Matschinsky, D. Shanahan, and J. Ellerman, *Anal. Biochem.* **60,** 188 (1974).

[6] E. Änggård and B. Samuelsson, this series, Vol. 14, p. 215.

[7] O. H. Lowry, J. V. Passonneau, and M. K. Rock, *J. Biol. Chem.* **236,** 2756 (1961).

[8] M. Ciotti and N. Kaplan, this series, Vol. 3, p. 890.

Step 4 reagent: Tris-HCl, 100 mM, pH 8.0, ethylenediaminetetraacetic acid, 2 mM; mercaptoethanol, 2 mM; bovine serum albumin, 0.6 mg/ml; NAD^+, 2.8 mM (acid-treated to removed NADH); NAD^+-15-hydroxyprostaglandin dehydrogenase, 7–10 μg/ml. Components 1–4 may be stored at 4° in solution for several days. Add NAD^+ and enzyme within 1 hr of intended use.

Step 5 reagent: sodium hydroxide, 0.3 N (concentration must be adjusted to a concentration to raise pH of step 5 to pH 11.0–12.0)

Step 6 reagents (cycling system)

Reagent A: Tris-HCl, 100 mM, pH 8.0; ethanol, 300 mM; oxaloacetate, 2 mM; mercaptoethanol, 2 mM; bovine serum albumin, 0.2 mg/ml; alcohol dehydrogenase, 140 μg/ml; malic dehydrogenase, 43 μg/ml

Reagent B: sodium hydroxide, 1 N

Step 7 reagent

Reagent A: 2-amino-2-(hydroxymethyl)-1,3-propanediol, 40 mM, pH 9.9; NAD^+, 0.2 mM; glutamate, 10 mM

Other reagents for microassays: 0.3 M Na_3PO_4–0.3 M K_2HPO_4

Reagents for Macroassay

Step 2 reagent: 0.4 N H_3PO_4

Step 4 reagent: 1 M Tris base; 2 mM ethylenediaminetetraacetic acid; 2 mM merceptoethanol; 0.2 mg bovine serum albumin per milliliter. Adjusted to pH 8.0 with HCl

Homogenizing Medium

100 mM potassium phosphate, pH 7.5, 200 mM NaCl, 2 mM mercaptoethanol, and 1% glycerol

Tissue Preparations

Histochemical: Tissues to be assayed histochemically for PGDH are prepared by standard techniques for quantitative histochemistry. These techniques are beyond the scope of this monograph, but are adequately described in the literature.[3,9–12] Briefly, tissue (brain, muscle, liver, kidney, etc.) is quick frozen with Dry Ice-cooled Freon, sectioned into 20-μm-thick sections in a cryostat at −18°, and freeze dried. Considerable PGDH activity will be lost if the tissue is allowed to thaw before it is freeze dried. There is no loss of activity otherwise. Histochemical samples are taken under a dissection microscope, then weighed on a quartz-fiber microbalance.

[9] O. H. Lowry and J. V. Passonneau, "A Flexible System of Enzymatic Analysis." Academic Press, New York, 1972.

[10] T. Brannan, C. N. Corder, and M. Rizk, *Proc. Soc. Exp. Biol. Med.* **148,** 714 (1975).

[11] C. N. Corder, M. L. Berger, and O. H. Lowry, *J. Histochem. Cytochem.* **22,** 1034 (1974).

[12] C. N. Corder, J. G. Collins, T. S. Brannan, and J. Sharma, *J. Histochem. Cytochem.* **25,** 1 (1977).

The sample is transferred to the reaction medium (step 1) suspended in "light" mineral oil in an "oil-well rack" (Teflon block with appropriate wells to contain the reactions). Then steps 1–6 are carried out in the "oil-well," whereas step 7 is in 10 × 75 mm glass fluorometer tubes.

Nonhistochemical: Tissues may be taken fresh and homogenized, or frozen tissue collected for histochemical assay may be powdered and homogenized.

Fresh tissue: Homogenize the tissue in 3 volumes of homogenizing medium cooled on ice. PGDH will be stable for 3–6 hr at 4° but cannot be frozen without significant loss of activity.

Frozen tissue: Powder the frozen tissue at −40° with a mortar and pestle. Weigh a desired portion of powder without it thawing. Homogenize it in 3 volumes of ice cold homogenizing medium. There should be no loss of PGDH with this procedure if the powder thaws directly in the homogenizing medium. The enzyme will lose some activity upon freezing the homogenate. The enzyme will be stable in the homogenate at 4° for 6–12 hr. If one wants to do more assays on the enzyme, one must make new homogenates from the frozen powder. *Note:* Neither of these homogenates can be used in the radioactive thin-layer chromatography assay owing to interference by mercaptoethanol.[2]

Fluorometer

Any standard photofluorometer will suffice, although one should use an instrument with proven reliability and stability in a useful range of 0.01 to 8 μM NADH fluorescence.[13,14] Regular 10 × 75 mm glass culture tubes can be used, if they are screened for optical uniformity.[9,13] The tubes should be specially cleaned.

Microassay Procedures for 1–100 ng of Tissue

Step 1. Load the wells with 0.1 μl of reagent A and others with reagent B (no PGE_1). It is recommended that one assay four samples from a histological specimen in reaction A and four in B. In some cases, one may need to use 8–10 samples if there is unacceptable variability. Place the suspended medium at 37°. Start the reactions by introducing the samples at 20-sec intervals. Allow the reactions to proceed for 90 min. There should be 3–8 μM PGE_1 remaining, but if it is determined that less than 3 μM PGE_1 remains in reagent A at the end of the reaction, new samples from the histological specimens should be assayed for 30 min, if necessary

[13] H. B. Burch, this series, Vol. 3, p. 946.

[14] D. Laurence, this series, Vol. 4, p. 174.

using less sample. The tissue concentration should remain within a range equivalent to 0.03 to 1.0 mg of protein per milliliter to ensure reaction linearity with tissue concentration. Quadruplicate standards will be reagent A (precisely standardized for PGE_1) and blanks. This step will have tissue PGDH assayed in reagent A, in reagent B (tissue blanks), and in standards (reagent A) and blanks (reagent B).

Step 2. Sequentially at 20-sec intervals, stop the reactions with 0.1 μl of step 2 reagent. Heat the rack for 10 min at 65° in a sand bath to destroy all enzyme activity and newly formed NADH.

Step 3. Add 0.1 μl of step 3 reagent at 37°.

Step 4. Add 0.1 μl of step 4 reagent at sequential 20-sec intervals and allow to react at 37°.

Step 5. After 45 min, add 0.1 μl of step 5 reagent at sequential intervals; at the end, heat the rack at 65° for 15 min to destroy NAD^+, leaving NADH.

Step 6. Add 8.5 μl of step 6 cycling reagent A at sequential 20-sec intervals. After reaction at 37° for 45 min, sequentially add 0.7 μl of 1 *N* NaOH (reagent B) at 20-sec intervals. Heat at 65° for 15 min. This reaction should be adjusted to 5000–6000 cycles.[3] The various parameters of cycling have been adequately described in references cited in this volume and in this series.[15]

Step 7. Add 8.5 μl of the step 6 final reaction to 1.0 ml of step 7 malate assay reagent. Record the fluorescence, then start the reaction with 5 μg of malic dehydrogenase and 2 μg of glutamic-oxaloacetate transaminase per milliliter and follow the reaction to completion (approximately 20 min).

Calculation of PGDH Activity

A. Calculate the net fluoresence developed in step 7 for
 Net fA1: Reagent A, step 1 (PGE_1 standard)
 Net fA2: Reagent B, step 1 (reagent blank)
 Net fA3: Tissue in reagent A, step 1; average of four samples per histological substructure.
 Net fA4: Tissue in reagent B, step 1; average of four samples per histological substructure.

B. Standardization of histochemical procedure

$$\frac{\mu M \text{ PGE}_1 \text{ in step 1}}{\text{fA1} - \text{fA2}} = \mu\text{mol/liter/f(unit)}$$

Example: 0.025 μmol of PGE_1 per liter per fluorescent unit in the original 0.1 μl.

[15] O. H. Lowry and J. V. Passonneau, this series, Vol. 6, p. 792.

C. Tissue blank (millimoles equivalent per kilogram dry weight)

$$\frac{(fA4 - fA2)(B)(10^{-3})}{\text{kg tissue/liter}} = \text{mmol kg}^{-1}$$

Example: 1.10 mmol kg^{-1}

D. NAD^+-PGDH activity

{fA1-fA2} − {fA3-fA2} = f for 90 min

(f_{90}) (B) (10^{-3}) ÷ kg tissue/liter = mmol kg^{-1} in 90 min But a certain amount of PGE_1 lost was apparently added back by the tissue blank (C), therefore the PGE_1 lost is greater than the f_{90} would indicate. Therefore, the increase due to C is added.

Example: 14.5 mmol·kg^{-1} before correction. Add C (e.g., 1.1 mmol kg^{-1}) giving 15.6 mmol·kg in 90 min. This is expressed as mmol·kg·hr; therefore multiply by 60/90: 10.4 mmol·kg·hr.

E. For other histochemical samples in the assay, use the same values for A and B. However, C and D will be done in quadruplicate on each substructure.

Microassay Procedure for (0.1–0.6 μg Tissue)

The reactions are similar for steps 1–4, except that the reaction volume is approximately 0.7 μl: one may use larger reaction volumes with subsequent adjustment in other aliquots, but the tissue protein concentration should be in the range of 0.03–1.0 mg/ml, unless studies on the tissue being assayed indicate other permissible concentrations.

Step 5. Add 4 μl of 0.3 *M* Na_3PO_4–0.3 *M* K_2HPO_4 at sequential intervals and then heat at 65° for 15 min.

Step 6. Place 100 μl of step 6 cycling reagent to specially cleaned 10 × 75 mm fluorometer tubes; except that this should contain 85 μg of malic dehydrogenase per milliliter so as to give approximately 2000 cycles in 30 min. Add 4.2 μl from step 5 at 15-sec intervals and allow to react at 37° for 30 min. Stop the reactions sequentially at 90° for 15 min.

Step 7. Add 1 ml of step 7 reagent to the reactions in the fluorometer tubes in step 6. Record the fluorescence, then add enzymes as in step 7 of the microassay.

Macroassay Procedure (for 0.1–1 mg of Tissue)

This procedure is similar to that for the microassays, except that steps 5–7 are omitted. The entire reaction is carried out in 10 × 75 mm fluorometer tubes. Reagent blanks and standards are as before. Tissue blanks and reactions are done in duplicate. One can also do the step 1 reaction in a volume of several milliliters and take aliquots at timed intervals for study of kinetics of the reaction.

Step 1. Load the tubes with 0.3 ml of step 1 reagents A and B, respectively. Add 5 μl of tissue homogenate and incubate the reaction at 37° for 45 min, then cool on ice. The tissue protein concentration should be in a range of 0.08–1.0 mg/ml. Depending upon the tissue and time linearity of the samples being analyzed, variations may be necessary.

Step 2. Add 0.3 ml of 0.4 *N* H_3PO_4. Final pH must be 3.0–3.5. Heat for 20 min in a 60° water bath and cool on ice.

Step 3. Add 0.3 ml of 0.3 *N* NaOH to a final pH of 7–9.

Step 4. Add 0.3 ml of step 4 macroassay reagent. Record the fluorescence at 37°. Add 10 μg of beef lung NAD-PGDH and follow the reaction to completion: sufficient PGDH should be added to complete the reaction with a half-time of 3–5 min for 10 μM PGE_1 in step 1. Activity is calculated as for the microassays and expressed as millimoles of PGE_1 lost per kilogram of tissue protein per hour (mmol kg^{-1} h^{-1}). Protein was assayed using the Folin phenol reagent.[16]

Comments

The substrate loss assay is readily applicable to micro- and macroassays. It has been verified to be specific for NAD^+-PGDH in the kidney. It is applicable to other tissues having significant levels of NADH-oxidase. Other enzymes that may interfere are at a relatively high dilution. Because tissues from various organs may have characteristics different from the kidney, it is suggested that extensive applications of this method be preceded with a verification of the method on a macroassay using the 15-keto-PGE_1 chromophore method or the radiometric assay or other appropriate methods.[2,3] At tissue concentrations greater than 1.0 mg/ml, activity decreases owing to rapid depletion of PGE_1, and also to interference from unidentified factors. Tissue levels less than 0.03 mg/ml generally do not have enough NAD^+-PGDH to exceed the limit of sensitivity of the assay, about 0.4 μM PGE_1. The maximum level of PGE_1 in the step 1 reaction is 16 μM, but the level of 10 μM is more practical. Although the pH optimum of kidney NAD^+-PGDH is about 9.2, this assay will not give reproducible results above pH 8.0 owing to instability of PGE_1 in alkaline pH, therefore the enzyme is assayed at pH 7. PGE_1 loss due to chemical instability is 10% per hour at pH 9.6 and 30% per hour at pH 10.5. The usual precautions and techniques of microassay in "oil wells" are necessary for reproducible results.

[16] O. H. Lowry, N. J. Rosenbrough, A. L. Farr, and R. J. Randell. *J. Biol. Chem.* **193,** 265 (1951).

[23] Purification of $NADP^+$/NADPH-Dependent 15-Hydroxyprostaglandin Dehydrogenase and Prostaglandin 9-Ketoreductase from Porcine Kidney[1]

By HSIN-HSIUNG TAI and DAVID GUEY-BIN CHANG

$NADP^+$-dependent 15-hydroxyprostaglandin dehydrogenase (also known as type II 15-hydroxyprostaglandin dehydrogenase) was first known to catalyze the oxidation of 15(*S*)-hydroxyl group of prostaglandins to 15-ketoprostaglandins.[2]

$$PGF_{2\alpha} + NADP^+ \longrightarrow \text{15-keto-}PGF_{2\alpha} + NADPH + H^+$$

Subsequent studies showed that the same enzyme not only catalyzes reversible oxidoreduction of prostaglandins at C-15, but also affects reversible oxidoreduction of prostaglandins at C-9.[3,4] These findings lead to the conclusion that the previously recognized $NADP^+$-dependent 15-hydroxyprostaglandin dehydrogenase and prostaglandin 9-ketoreductase activities are in fact alternate activities of a single enzyme protein. The enzyme was shown to exist in multiple forms[3-6] and was thought to control the PGE:PGF ratio and the duration of biologically active prostaglandins.[7] The diversified functions of the enzyme reflect the versatility of the enzyme and the expression of each enzyme activity may depend on the availability of each particular prostaglandin substrate and the ratio of NADPH:$NADP^+$. We and others have isolated two homogeneous forms of this enzyme from porcine kidney and human placenta, respectively.[3,4] This presentation describes our method of purification and some properties of the enzyme from porcine kidney.

[1] This work was supported in part by grants from the National Institutes of Health (GM-25247) and the American Heart Association (78-865).

[2] S. C. Lee and L. Levine, *J. Biol. Chem.* **250,** 548 (1975).

[3] Y. M. Lin, and J. Jarabak, *Biochem. Biophys. Res. Commun.* **81,** 1227 (1978).

[4] D. G.-B. Chang and H. H. Tai, *Biochem. Biophys. Res. Commun.* **99,** 745 (1981).

[5] S. C. Lee and L. Levine, *J. Biol. Chem.* **250,** 4549 (1975).

[6] A. Hassid and L. Levine, *Prostaglandins* **13,** 503 (1977).

[7] S. C. Lee, S. S. Pong, D. Katzen, K. Y. Wu, and L. Levine, *Biochemistry* **14,** 142 (1975).

ISBN 0-12-181986-8

Assay Method

Principle. The assay of 15-hydroxyprostaglandin dehydrogenase activity can be done by a spectrophotometric method based on the conversion of PGE to 15-keto-PGE, which forms a chromophore (E_m = 27,000 M^{-1} cm^{-1} at λ_{max} = 500 nm) after alkalinization of the reaction mixture.[8] The activity can also be assayed spectrofluorometrically by following the increase in fluorescence of NADPH at 460 nm with excitation at 340 nm. Alternatively, the assay of 15-hydroxyprostaglandin dehydrogenase and prostaglandin 9-ketoreductase activities can be carried out by radioimmunological methods. The production of 15-keto-$PGF_{2\alpha}$ from $PGF_{2\alpha}$ (15-hydroxyprostaglandin dehydrogenase) or the formation of $PGF_{2\alpha}$ from PGE_2 (prostaglandin 9-ketoreductase) can be determined quantitatively by specific radioimmunoassays for 15-keto-$PGF_{2\alpha}$[9] or $PGF_{2\alpha}$,[9] respectively.

Reagents

Potassium phosphate buffer, 0.1 *M*, pH 7.5
Tris-HCl, 0.05 *M*, pH 7.5, containing 0.1% gelatin (RIA buffer)
NaOH, 2 *N*
$NADP^+$
NADPH
PGE_2
$PGF_{2\alpha}$
15-Keto-$PGF_{2\alpha}$
$PGF_{2\alpha}$ antisera[9]
15-Keto-$PGF_{2\alpha}$ antisera[9]
$PGF_{2\alpha}$–[^{125}I]iodotyrosine methyl ester conjugate[10]
15-Keto $PGF_{2\alpha}$–[^{125}I]iodotyrosine methyl ester conjugate[10]
Charcoal suspension: 100 ml of RIA buffer containing 1 g of bovine γ-globulin and 3 g of Norit A) and enzyme

Procedures

Spectrophotometric or Spectrofluorometric Method. The assay mixture contains PGE_2 (56.8 nmol), $NADP^+$ (200 nmol), and enzyme in a final volume of 1 ml of 0.1 *M* potassium phosphate, pH 7.5. The reaction is allowed to proceed at 37° for 40 min and then is terminated by the addition of 0.1 ml of 2 *N* NaOH. The transient chromophore generated is quantitated by measuring the absorption at 500 nm. If the enzyme activity is assayed spectrofluorometrically, the increase in fluorescence of NADPH at

[8] E. Änggård, C. Larsson, and B. Samuelsson, *Acta Physiol. Scand.* **81,** 396 (1971).
[9] J. C. Cornette, K. L. Harrison, and K. T. Kirton, *Prostaglandins* **5,** 155 (1974).
[10] H. H. Tai and B. Yuan, *Anal. Biochem.* **87,** 343 (1978).

460 nm with excitation at 340 nm is measured. The concentration of NADPH is determined by assuming that $E_m = 6200\ M^{-1}\ cm^{-1}$ at $\lambda_{max} =$ 340 nm.[11]

Radioimmunological Method. The assay mixture contains PGE_2 (14 nmol, for prostaglandin 9-ketoreductase) or $PGF_{2\alpha}$ (14 nmol, for 15-hydroxyprostaglandin dehydrogenase), NADPH or $NADP^+$ (500 nmol), and enzyme in a final volume of 0.2 ml of 0.1 *M* potassium phosphate buffer, pH 7.5. The reaction is allowed to proceed at 37° for 10 min and then is terminated by boiling for 2 min. After removal of denatured proteins by centrifugation at 1000 *g* for 10 min, the supernatant is diluted and assayed for immunoreactive $PGF_{2\alpha}$ or 15-keto-$PGF_{2\alpha}$. The radioimmunoassay mixture contains 0.2 ml of standards or diluted sample, 0.1 ml of diluted $PGF_{2\alpha}$ or 15-keto-$PGF_{2\alpha}$ antisera, and 0.1 ml of the respective labeled hapten in 0.05 *M* Tris-HCl, pH 7.5, containing 0.1% gelatin. Each sample is run in duplicate. After 1 hr of incubation at room temperature, each tube receives 1 ml of H_2O before addition of 0.2 ml of charcoal suspension and mixing in a vortex mixer. After standing for 5 min at room temperature, the mixture is centrifuged for 5 min at 1000 *g*. The supernatant containing the bound hapten is separated from the pellet containing the free hapten. The radioactivity in the supernatant and the pellet is separately determined by a gamma spectrometer. The ratio of cpm bound to the total cpm is calculated for each sample, and the concentration of each sample is determined from a standard curve after logit transformation as described by Tai and Chey.[12]

Purifiction of $NADP^+$/NADPH-Dependent 15-Hydroxyprostaglandin Dehydrogenase and Prostaglandin 9-Ketoreductase from Porcine Kidney[4]

Preparation of Crude Enzyme Extract. All steps are performed at 4°. Porcine kidney (500 g) is homogenized in 1 liter of 10 m*M* Tris-HCl buffer, pH 7.5, containing 1 m*M* EDTA (buffer A) in a Waring blender for 2 min. The homogenate is centrifuged at 40,000 *g* for 2 min. The supernatant is designated S_{40}.

Acetone Fractionation. To S_{40} is added slowly prechilled acetone (−68°) to 33% (v/v) with stirring for 10 min. The precipitate is removed by centrifugation at 40,000 *g* for 15 min. Acetone is further added to the supernatant to 43% with stirring for 10 min, and the precipitate is again removed by centrifugation at 40,000 *g* for 15 min. The supernatant is then brought to an acetone concentration of 67% and stirred for 10 min before

[11] B. L. Horecker and A. Kornberg, *J. Biol. Chem.* **175,** 385 (1949).
[12] H. H. Tai and W. Y. Chey, *Anal. Biochem.* **74,** 12 (1976).

centrifugation at 40,000 *g* for 15 min. The precipitate is dissolved in 50 ml of buffer A and designated as the acetone fraction.

Sephadex G-100 Chromatography. The acetone fraction is immediately applied to a Sephadex G-100 column (6 × 100 cm) equilibrated with buffer A. The column is eluted with the same buffer, and fractions of 13 ml are collected. The active fractions as determined by spectrophotometric method are pooled and concentrated to 20 ml using an Amicon ultrafiltration PM-10 membrane.

TEAE-Cellulose Chromatography. The concentrated material is applied to a TEAE-cellulose column (1.5 × 25 cm) equilibrated with buffer A. The column is eluted with 10 m*M* potassium phosphate, pH 7.5, containing 1 m*M* EDTA (buffer B), and fractions of 13 ml are collected. The active fractions are pooled and concentrated to 20 ml by Amicon ultrafiltration.

Isoelectric Focusing. The concentrated fraction from TEAE-cellulose chromatography is isoelectrofocused in a sucrose density gradient using 2.5% Ampholine (pH 4–6) in a 110-ml LKB column as described by Vesterberg *et al.*[13] Water at 3° is circulated to cool the column. Electrofocusing is performed at 1600 V for 16 hr. The contents of the column are collected in 1-ml fractions. Two peaks of activity appear and are designated form I and form II with p*I* values of 5.8 and 4.8, respectively. Active fractions from both peaks are concentrated separately and passed through a Sephadex G-50 (1 × 15 cm) column equilibrated and eluted with 50 m*M* potassium phosphate, pH 7.5, containing 1 m*M* EDTA. The active fractions are concentrated by lyophilization and dialyzed against buffer B for 24 hr. The enzyme is stored in aliquots at −80°. A summary of the results of the purification of 15-hydroxyprostaglandin dehydrogenase (15-PGDH) with its concurrent activities of prostaglandin 9-ketoreductase (9-PGKR), 9-hydroxyprostaglandin dehydrogenase (9-PGDH), and prostaglandin 15-ketoreductase (15-PGKR) is shown in the table.

Properties[14]

Homogeneity. The final preparations of both enzyme forms show single bands upon sodium dodecyl phosphate–polyacrylamide gel electrophoretic analysis.

Molecular Weight. Both forms of the native enzyme contain a single polypeptide chain with a molecular weight of 29,000, although minor differences in amino acid composition are noted.

[13] O. Vesterberg, T. Wadstrom, K. Vesterberg, H. Svensson, and B. Malugren, *Biochim. Biophys. Acta* **133,** 435 (1967).

[14] D. G.-B. Chang, and H. H. Tai, *Arch. Biochem. Biophys.* **214,** 464 (1982).

PURIFICATION OF 9-PGKR (I), 9-PGDH (II), 15-PGKR (III), AND 15-PGDH (IV) FROM PORCINE KIDNEY[a]

Fraction	Protein (mg)	Enzyme activity (units)				Specific activity (units/mg)				Purification (fold)				Recovery (%)			
		I	II	III	IV	I	II	III	IV	I	II	III	IV	I	II	III	IV
I. Crude extract	9360																
II. Acetone fraction (43–67%)	862.5	70	104	24	49	0.08	0.12	0.03	0.06	1	1	1	1	100	100	100	100
III. Sephadex G-100	176.8	83	120	25	55	0.47	0.68	0.14	0.31	5.8	5.7	5	5.5	118	115	104	112
IV. TEAE-cellulose	71.4	58	82	22	41	0.81	1.15	0.31	0.57	10	9.6	11	10.1	83	79	91	84
isoelectric focusing																	
V. Form I (p*I* 5.8)	2.9	22	33.1	7.7	19.8	7.5	11.3	2.6	6.8	93	94	93	119	31	32	32	40
VI. Form II (p*I* 4.8)	2.0	18.6	27.4	6.3	15	9.3	13.6	3.1	7.5	115	115	111	132	27	26	26	30

[a] Activities of I and IV were assayed as described in Methods. Activities of II and III were assayed using $PGF_{2\alpha}$ and 15-keto-$PGF_{2\alpha}$ as substrate and following the production of PGE_2 and $PGF_{2\alpha}$, respectively, as described for assay of activities of I and IV. One unit of enzyme activity was defined as that amount of enzyme which catalyzed the formation of 1 nmol of product per minute under the standard assay conditions.

Reversibility. The enzyme is able to catalyze reversible $NADP^+$/NADPH-dependent oxidoreduction of prostaglandins at C-9 and C-15 at pH 7.5.

Substrate Specificity. For 15-hydroxyprostaglandin dehydrogenase activity, either form of the enzyme utilizes all kinds of prostaglandins as a substrate, PGB_1 being the best substrate. For prostaglandin 9-ketoreductase activity, prostaglandins having 9-keto group can serve as substrate, PGA_1-GSH being the best substrate. K_m values for various prostaglandins have been determined.[15] The enzyme does not catalyze oxidoreduction of prostaglandins at C-11.

Coenzyme Specificity. $NADP^+$/NADPH are preferred coenzyme for oxidoreduction of prostaglandins at C-9 and C-15. K_m values for $NADP^+$ and NADPH are 1.4 and 100 μM, respectively.

Stability and Storage. The enzyme can be stored at $-80°$ for 1 year without appreciable loss of activity.

[15] D. G.-B. Chang and H. H. Tai, *Biochem. Biophys. Res. Commun.* **101,** 898 (1981).

[24] Isolation of $NADP^+$-Dependent PGD_2-Specific 15-Hydroxyprostaglandin Dehydrogenase from Swine Brain

By TAKAO SHIMIZU, KIKUKO WATANABE, HIDEKADO TOKUMOTO, and OSAMU HAYAISHI

$$\text{Prostaglandin } D_2 + NADP^+ \rightarrow \text{15-ketoprostaglandin } D_2 + NADPH + H^+$$

The first step in the metabolic inactivation of prostaglandins is the oxidation of the 15-hydroxy group.[1-4] Two types of 15-hydroxyprostaglandin dehydrogenase have been described (types I and II).[5-9] However, PGD_2 was a poor substrate for these enzymes in terms of V_{max} and K_m values.[8,10-12] Ellis *et al.*[13] demonstrated that the major urinary metabolites of

[1] E. Änggård and B. Samuelsson, *J. Biol. Chem.* **239,** 4097 (1964).
[2] E. Änggård and B. Samuelsson, *Ark. Kemi* **25,** 293 (1966).
[3] S. S. Braithwaite and J. Jarabak, *J. Biol. Chem.* **250,** 2315 (1975).
[4] J. Nakano, E. Änggård, and B. Samuelsson, *Eur. J. Biochem.* **11,** 386 (1969).
[5] S.-C. Lee and L. Levine, *J. Biol. Chem.* **250,** 548 (1975).
[6] L. Kaplan and L. Levine, *Arch. Biochem. Biophys.* **167,** 284 (1975).
[7] S.-C. Lee, S.-S. Pong, D. Katzen, K.-Y. Wu, and L. Levine, *Biochemistry* **14,** 142 (1975).
[8] J. Jarabak and J. Fried, *Prostaglandins* **18,** 241 (1979).
[9] H. S. Hansen, *Prostaglandins* **12,** 647 (1976).
[10] F. F. Sun, S. B. Armour, V. R. Bockstanz, and J. C. McGuire, *Adv. Prostaglandin Thromboxane Res.* **1,** 163 (1976).

METHODS IN ENZYMOLOGY, VOL. 86

ISBN 0-12-181986-8

PGD_2 injected to monkey intravenously include the 15-keto derivatives of PGD_2. We discovered an NADP-linked PGD dehydrogenase in swine brain and purified it partially.[14] The enzyme may be responsible for the metabolic inactivation of PGD_2, which is biosynthesized[15] and may play some important functional roles in the brain.[16,17]

Assay Method[14]

Principle. The activity of PGD dehydrogenase is determined spectrometrically by measuring the increase in absorbance at 415 nm, which is characteristic of the enolate anion form of 15-keto-PGD_2 formed under alkaline conditions.[14,18]

Reagents

Tris-HCl buffer, 0.5 *M*, pH 9.0
PGD_2, 20 m*M* (ethanol solution)
NADP (Sigma), 20 m*M*
Ethanol

Procedures. The sample cuvette contains 100 μl of Tris-HCl buffer, pH 9.0, 50 μl of NADP, 5 μl of PGD_2 (ethanol solution), and the enzyme in a total volume of 0.5 ml. In a reference cuvette are included all reagents except PGD_2, which is replaced by 5 μl of ethanol. The initial velocity of increase in absorbance at 415 nm (or 340 nm, the absorbance of NADPH) is determined. The enzyme activity is calculated using molecular extinction coefficients of 35,000 M^{-1} cm^{-1} or 6,200 M^{-1} cm^{-1}, at 415 or 340 nm, respectively. One unit of the enzyme activity is defined as the amount that produces 1 μmol of 15-keto-PGD_2 per minute at 24°. The specific activity is expressed as the number of units per milligram of protein. Protein is determined according to the method of Lowry *et al.*[19] using bovine serum albumin as a standard.

[11] M. F. Rückrich, W. Schlegel, and A. Jung, *FEBS Lett.* **68,** 59 (1976).
[12] H. Ohno, Y. Morikawa, and F. Hirata, *J. Biochem.* **84,** 1485 (1978).
[13] C. K. Ellis, M. D. Smigel, J. A. Dates, O. Oelz, and B. J. Sweetman, *J. Biol. Chem.* **254,** 4152 (1979).
[14] K. Watanabe, T. Shimizu, S. Iguchi, H. Wakatsuka, M. Hayashi, and O. Hayaishi, *J. Biol. Chem.* **255,** 1779 (1980).
[15] T. Shimizu, S. Yamamoto, and O. Hayaishi, *J. Biol. Chem.* **254,** 5222 (1979).
[16] T. Shimizu, N. Mizuno, T. Amano, and O. Hayaishi, *Proc. Natl. Acad. Sci U. S. A.* **76,** 6231 (1979).
[17] K. Kondo, T. Shimizu, and O. Hayaishi, *Biochem. Biophys. Res. Commun.* **98,** 648 (1981).
[18] R. L. Jones and N. H. Wilson, *J. Chem. Soc., Perkin Trans. I* 209 (1978).
[19] O. H. Lowry, N. J. Rosebrough, A. L. Farr, and R. J. Randall, *J. Biol. Chem.* **193,** 265 (1951).

Purification Procedures

All procedures are carried out at 0–4° and centrifugation is at 8000 *g* for 10 min, unless stated otherwise.

Step 1. High Speed Supernatant. Approximately 250 g of swine brains are cut into small pieces and mixed with 3 volumes of 10 m*M* potassium phosphate buffer, pH 7.0, containing 0.5 m*M* dithiothreitol (buffer A). The mixture is homogenized using a Polytron blender homogenizer five times at top speed for 10 sec each. After centrifugation, the supernatant fraction is centrifuged at 105,000 *g* for 60 min. The high speed supernatant solution is recovered by decantation.

Step 2. Ammonium Sulfate Fractionation. Solid ammonium sulfate is slowly added to the solution to attain 40% saturation (243 mg/ml); the preparation is mixed well for 40 min at 0° with a magnetic stirrer. After centrifugation, solid ammonium sulfate is further added to obtain 60% saturation (132 mg/ml). The precipitate thus formed is dissolved in a minimum volume of buffer A and dialyzed overnight against three changes of 100 volumes of buffer A.

Step 3. Blue Sepharose Chromatography. After removal of insoluble materials by centrifugation, the sample is applied to a column of Blue Sepharose 6B (2 × 10 cm), previously equilibrated with buffer A. The column is washed with buffer A and then buffer A containing 0.1 *M* KCl[20]

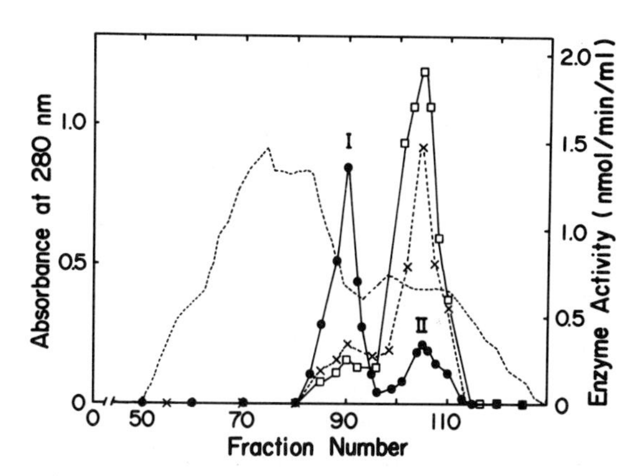

FIG. 1. Separation of 15-hydroxy-PGD dehydrogenase (peak I) and type II (peak II) dehydrogenase by Sephadex G-100 gel filtration. About 15 mg of protein are applied to a column of Sephadex G-100 (2.5 × 48 cm), and 1-ml fractions are collected. Enzyme activities are determined by the increase in the absorbance at 415 nm (●—●, PGD_2) or 340 nm (×---×, PGE_2; □—□, $PGF_{2\alpha}$). Absorbance at 280 nm is determined in every second fraction (----).

[20] Type II enzyme (peak II in Fig. 1) is eluted in part in this fraction.

PURIFICATION OF PGD DEHYDROGENASE FROM SWINE BRAIN

Step	Volume (ml)	Total protein (mg)	Total activity (milliunits)	Specific activity (milliunits of protein)	Yield (%)	Purification (fold)
High speed supernatant	520	3110	28.0	0.009	100	1
Ammonium sulfate	20	808	25.9	0.032	93	3.7
Blue Sepharose	2.5	20.3	16.6	0.820	59	96.8
Sephadex G-100	3.8	4.8	8.6	1.795	31	207.0

until the absorbance of the eluate at 280 nm decreases to less than 0.05. The enzyme is then eluted with buffer A containing 1 *M* KCl.

Step 4. Sephadex G-100 Chromatography. The sample obtained in the preceding step (about 50 ml) is concentrated to about 2.5 ml with the aid of Diaflo membrane PM-10 and applied to a column of Sephadex G-100 (2.5 × 50 cm), previously equilibrated with buffer A. Elution is carried out with the same buffer at a flow rate of 5 ml/hr. The dehydrogenase activities elute in two peaks (Fig. 1). The first peak at around fraction 90 exhibits much greater activity toward PGD_2 than prostaglandins E_2, F_2, $F_{2\alpha}$, and B_2, and will be referred to as 15-hydroxy-PGD dehydrogenase. The second peak of activity, which eluted at around fraction 105, oxidizes PGE_2 $PGF_{2\alpha}$ much more effectively than PGD_2. PGB_2 can serve as the best substrate for peak II enzyme. Furthermore, the second peak contains PGE 9-ketoreductase activity, which reportedly copurifies with the so-called type II (i.e., $NADP^+$-dependent) dehydrogenase. A typical result of purification is summarized in the table. Overall purification is about 200-fold with a yield of 30%.

Properties of the Enzyme

The molecular weight of the enzyme is estimated to be 52,000–62,000 by gel filtration on Sephadex G-100,[21] and the isoelectric point, pH 7.8. The enzyme utilizes NADP exclusively as a cofactor. The K_m values for NADP and PGD_2 are 0.5 μM and 70 μM, respectively. The relative rates of reaction as assayed in the presence of 200 μM NADP 200 μM PGs at pH 9 are PGD_2, 100; PGD_1, 25; PGD_3, 10; PGE_2, 6; and $PGF_{2\alpha}$, 10. The A and B series of PGs cannot serve as substrates. The partially purified enzyme contains 15-keto-Δ^{13}-prostaglandin reductase activity, but is essen-

[21] The molecular weight of about 28,000 was obtained by sodium dodecyl sulfate–polyacrylamide gel electrophoresis (H. Tokumoto *et al.*, unpublished data).

tially free of NADPH-linked PGE 9-ketoreductase and 11-keto-PGD reductase activities.[22] The enzyme is quite unstable at the low protein concentrations (less than 10 mg/ml). Sulfhydryl reagents such as dithiothreitol, 2-mercaptoethanol, and reduced glutathione protect the enzyme from inactivation.

Identification of Reaction Sequence and Product

[1-^{14}C]PGD_2 is biosynthesized by incubation of [1-^{14}C]PGH_2 with PGH–PGD isomerase[16] (0.8 unit per milligram of protein). When the partially purified enzyme is incubated with [1-^{14}C]PGD_2 in the presence of 200 μM NADP and 1 mM dithiothreitol and the ethereal extracts are subjected to thin-layer chromatography [silica gel 60 F254 (Merck)], the major radioactive peak comigrates with authentic 13,14-dihydro-15-keto-PGD_2 (R_f value = 0.47 in a solvent system of diethyl ether–methanol–acetic acid, 90:2:0.1, v/v/v). A small but significant amount of radioactivity is also detected, diffusely distributed between R_f 0.35 and 0.41. In this region, an ultraviolet quenching is observed. When extracted from a silica gel plate, the material in the region shows an absorbance maximum at 415 nm at alkaline pH values, which is characteristic of the enolate anion of 15-keto-PGD_2. The product with an R_f value of 0.47 was extracted and treated with diazomethane, methoxyamine, and silylating reagents and was subjected to gas chromatography–mass spectrometry. A single peak appeared with a retention time of 2.8 min and gave a mass spectrum essentially identical with that obtained with the corresponding

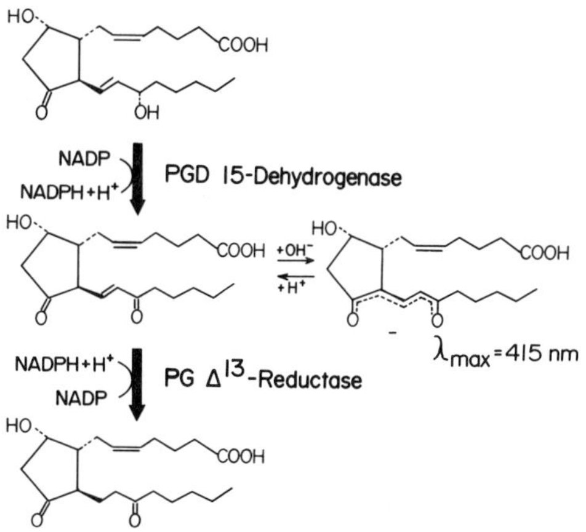

FIG. 2. A probable reaction sequence of PGD_2 metabolism in swine brain.

[22] K. Watanabe, T. Shimizu, and O. Hayaishi, *Biochem. Int.* **2,** 603 (1981).

derivative of the authentic 13,14-dihydro-15-keto-PGD_2. The definitive identification of 15-keto-PGD_2, a primary product, has not so far been successful, because it is enzymically and chemically unstable.[14,18] On the basis of spectrophotometric observation,[14] we propose that PGD_2 is first converted to 15-keto-PGD_2, an unstable intermediate, followed by the reduction of C13-C14 double bond by an NADPH-linked Δ^{13}-reductase which is copurified with $NADP^+$-linked 15-hydroxy-PGD dehydrogenase (Fig. 2).

Acknowledgments

This work was supported in part by a Grant-in-Aid for Scientific Research from the Ministry of Education, Science, and Culture of Japan and by grants from the Japanese Foundation on Metabolism and Diseases, Research Foundation for Cancer and Cardiovascular Diseases, Fujiwara Memorial Foundation, and Japan Heart Foundation 1979.

[25] Isolation and Properties of an $NADP^+$-Dependent PGI_2-Specific 15-Hydroxyprostaglandin Dehydrogenase from Rabbit Kidney

By JEFFREY M. KORFF and JOSEPH JARABAK

$$PGI_2 + NADP^+ \rightleftarrows \text{15-keto } PGI_2 + NADPH + H^+$$

$NADP^+$-dependent 15-hydroxyprostaglandin dehydrogenases have been identified in a number of tissues.[1] Those that have been studied in sufficient detail catalyze the oxidoreduction at either the 9- or the 15-position of prostaglandins, the reduction of the 9-keto group being more rapid than oxidation of the 15-hydroxyl.[2,3] Several forms of this enzyme have been isolated from rabbit kidney.[4] The following text describes the purification and characteristics of one of these rabbit kidney enzymes, which is relatively specific for PGI_2.

Assay Method

The enzyme activity is measured spectrophotometrically in a cuvette with a 1-cm light path at 25 ± 0.5°. The cuvette contains 3 ml of aqueous solution consisting of 290 μmol of sodium pyrophosphate at pH 10.3, 1.25

[1] S.-C. Lee and L. Levine, *J. Biol. Chem.* **250,** 548 (1975).
[2] A. Hassid and L. Levine, *Prostaglandins* **13,** 503 (1977).
[3] Y.-M. Lin and J. Jarabak, *Biochem. Biophys. Res. Commun.* **81,** 1227 (1978).
[4] J. Korff and J. Jarabak, *Prostaglandins* **20,** 111 (1980).

ISBN 0-12-181986-8

μmol of $NADP^+$, and 267 nmol of PGI_2 added in 0.01 ml of sodium pyrophosphate. The blank contains no PGI_2 but is identical otherwise. The reaction is started by the addition of enzyme, and NADPH production is followed by measuring the change in absorbance at 340 nm. One unit of enzyme activity is defined as the amount of enzyme that catalyzes the production of 1 μmol of NADPH per minute under the conditions of the assay.

Histochemical Staining of Polyacrylamide Gels. To each 20 ml of 0.1 *M* sodium pyrophosphate, pH 10.3, are added 1 mg of nitroblue tetrazolium, 10 mg of $NADP^+$, and, in dim light, 0.1 mg of phenazine methosulfate. Each gel is incubated for 1 hr at room temperature in tubes protected from light that contain 4.6 ml of this solution or 4.6 ml of this solution and 150 μg of PGI_2.

Purification Procedure

All of the purification steps are performed at 4–9°.

Step 1. Homogenization. Frozen rabbit kidneys (812 g) are thawed and minced. Portions of 50–60 g are homogenized with four volumes of buffer A (5 m*M* potassium phosphate, pH 7.0, and 1 m*M* EDTA) in a Waring blender at top speed for 2 min. A 10-ml portion of the homogenate is centrifuged at 30,000 *g* for 60 min, and the supernatant solution is assayed for enzyme activity. The remainder of the homogenate is centrifuged at 9000 *g* for 75 min, and the supernatant solution is collected.

Step 2. Ammonium Sulfate Fractionation. Solid ammonium sulfate (0.231 g/ml) is slowly added with continuous stirring to the supernatant solution from step 1, while the pH is maintained at about 7 by dropwise addition of 3 *N* NH_4OH. The resulting suspension is allowed to stand for 1 hr and then is centrifuged at 9000 *g* for 60 min. The precipitate is discarded, and solid ammonium sulfate (0.123 g/ml) is added slowly and with continuous stirring to the supernatant solution while maintaining the pH at about 7 with the dropwise addition of 3 *N* NH_4OH. After standing for 1 hr, the resulting suspension is centrifuged at 9000 *g* for 60 min. The precipitate is resuspended in a minimal volume of buffer A and dialyzed against 100 volumes of the same buffer for 24 hr.

Step 3. DEAE-Cellulose Chromatography. The dialyzed material from the preceding step is applied to a 40 × 400 mm column of DEAE-cellulose that has been equilibrated with buffer A. The column is washed with buffer A until the absorbance at 280 nm of the eluate falls below 0.4, and then a linear gradient is started: chamber I contains 1000 ml of buffer A and chamber II contains 1000 ml of 5 m*M* potassium phosphate, pH 7.0, 1 m*M* EDTA, 0.4 *M* KCl. The enzyme is eluted at a buffer conductivity of

1.5–5.0 mΩ^{-1}, and the fractions having the highest specific activity are pooled.

Step 4. Mātrex Gel Blue A Chromatography. The pooled fractions from the preceding step are chromatographed in two portions on a 25 × 95 mm column of Mātrex Gel Blue A that has been equilibrated with buffer A. After the enzyme is applied, the column is washed with buffer A and then developed stepwise with buffer A containing increasing concentrations of KCl. All of the enzyme is adsorbed to the column and is eluted with buffer A containing 0.75 *M* KCl. The fractions with highest activity are pooled and concentrated to less than 20 ml by ultrafiltration on a Diaflo UM-10 membrane.

Step 5. Sephadex G-100 Gel Filtration. The enzyme solution from the preceding step is applied to a 50 × 940 mm column of Sephadex G-100 that has been equilibrated with 10 m*M* potassium phosphate, pH 7.0. The column is developed with the same buffer.

Step 6. Hydroxyapatite Chromatography. The fractions with highest specific activity from step 5 are combined and applied to a 20 × 30 mm column of hydroxyapatite[5] which has been equilibrated with 5 m*M* potassium phosphate buffer, pH 7.0. The column is washed with 50 m*M* potassium phosphate buffer, pH 7.0, until the absorbance of the eluate at 280 nm falls to 0.015, then the enzyme is eluted with 100 m*M* potassium phosphate buffer, pH 7.0.

PURIFICATION OF $NADP^+$-DEPENDENT PGI_2-SPECIFIC 15-HYDROXYPROSTAGLANDIN DEHYDROGENASE FROM RABBIT KIDNEY

Step	Volume (ml)	Total activity (units)	Specific activity (units/mg protein[a]) $\times 10^2$	Yield (%)
1. Centrifuged homogenate	3305[b]	11.2[c]	0.015	100
2. Ammonium sulfate precipitate	239	5.65	0.032	50
3. DEAE-cellulose	291	4.87	0.184	43
4. Mātrex Gel Blue A	490	3.23	1.47	29
5. Sephadex G-100	59	1.71	12.1	15
6. Hydroxyapatite	44	1.46	19.9	13

[a] Protein determinations were based on the assumption that a solution containing 1 mg of protein per milliliter has an absorbance at 280 nm of 1.0 in a cuvette of 1-cm light path.

[b] Volume of the supernatant solution after centrifugation of the homogenate from 812 g of rabbit kidney.

[c] This activity was determined in an ultracentrifuged sample, as described in the text, because the supernatant solution after centrifugation at 9000 *g* was too turbid to assay spectrophotometrically.

[5] R. L. Anaker and V. Stoy, *Biochem. Z.* **330,** 141 (1958).

The enzyme resulting from this purification migrates as a single diffuse band on polyacrylamide disc gel electrophoresis performed at pH 7.0 or 8.3 in 5, 7.5, or 15% gels. This band corresponds to the migration of enzymic activity as determined by histochemical staining. Typical results of this purification are summarized in the table.

Properties

Stability. The enzyme is stable for months when frozen, but it loses activity slowly when repeatedly thawed and frozen.

Cofactor and Substrate Specificity. Unlike other $NADP^+$-dependent 15-hydroxyprostaglandin dehydrogenases that have been described, the rabbit kidney PGI_2 dehydrogenase does not use NAD^+ as a cofactor. It catalyzes the oxidation of the 15-hydroxyl group in a number of prostaglandins (PGI_2, PGB_1, PGE_2, $PGF_{2\alpha}$, and 6-keto $PGF_{1\alpha}$). PGI_2 is a much better substrate for the enzyme than any of the prostaglandins tested to date, as indicated by the following relative rates of oxidation: PGI_2 (89 μM), 100%; PGB_1 (40 μM), 8%; PGE_2 (94 μM), 4%; $PGF_{2\alpha}$, (70 μM), 2.2%; 6-keto $PGF_{1\alpha}$ (90 μM), 1.8%. The rabbit kidney enzyme also catalyzes the reduction of the 9-keto group in PGE_2, but this occurs at a rate less than 1% that of PGI_2 oxidation.[6]

Kinetic Constants.[6] The apparent K_m of PGI_2 is 278 μM, and that of $NADP^+$ is 13.4 μM.

Molecular Weight.[6] The molecular weight of the enzyme is 62,000 when estimated by gel filtration on Sephadex G-100.

Reversibility of the Reaction. It is presumed that the reactions is reversible, but this has not been established experimentally because 15-keto-PGI_2 has not been available.

Equilibrium Constant. Both PGI_2 and 15-keto-PGI_2 are rapidly hydrolyzed in aqueous solutions[7] to 6-keto-$PGF_{1\alpha}$ and 6,15-diketo-$PGF_{1\alpha}$, respectively. This makes determination of the equilibrium constant unreliable.

Acknowledgments

This work has been supported by grants from the National Institutes of Health (HD-07045 and AM-07011).

[6] J. Korff and J. Jarabak, *Prostaglandins,* **21,** 719 (1981).
[7] M. J. Cho and M. Allen, *Prostaglandins* **15,** 943 (1978).

[26] Purification and Assay of 15-Ketoprostaglandin Δ^{13}-Reductase from Bovine Lung

By HARALD S. HANSEN

The initial steps in the metabolism of E- and F_α-type prostaglandins[1,2] are catalyzed by two enzymes: 15-hydroxyprostaglandin dehydrogenase and 15-ketoprostaglandin Δ^{13}-reductase.[3,4] The same metabolic pathway may also exist for PGD_2,[5,6] PGI_2,[7,8] and TxA_2,[9] whereas the hydrolytic breakdown products (6-keto-$PGF_{1\alpha}$ and TxB_2) of the latter two compounds seem to be poor substrates for the 15-hydroxyprostaglandin dehydrogenase.[10,11] The dehydrogenase and reductase are responsible for the formation of the initial major metabolite, 13,14-dihydro-15-ketoprostaglandin, found in human,[12] bovine,[13] and rat[14] plasma. Of the two enzymes, 15-hydroxyprostaglandin dehydrogenase (EC 1.1.1.141) and 15-ketoprostaglandin Δ^{13}-reductase, the reductase has the highest activity in bovine lung homogenates,[15] whereas the opposite is found in human placenta homogenates.[16]

15-Ketoprostaglandin Δ^{13}-reductase catalyzes the following reaction.

[1] E. Änggård and C. Larsson, *Eur. J. Pharmacol.* **14,** 66 (1971).
[2] M. Hamberg and B. Samuelsson, *J. Biol. Chem.* **246,** 1073 (1971).
[3] E. Änggård and B. Samuelsson, *J. Biol. Chem.* **239,** 4097 (1964).
[4] H. S. Hansen, *Prostaglandins* **12,** 647 (1976).
[5] C. K. Ellis, M. D. Smigel, J. A. Oates, O. Oelz, and B. J. Sweetman, *J. Biol. Chem.* **254,** 4152 (1979).
[6] K. Watanabe, T. Shimizu, S. Iguchi, H. Wakatsuka, M. Hayashi, and O. Hayaishi, *J. Biol. Chem.* **225,** 1779 (1980).
[7] F. F. Sun and B. M. Taylor, *Biochemistry* **17,** 4096 (1978).
[8] J. Korff and J. Jarabak, *Prostaglandins* **20,** 111 (1980).
[9] W. Dawson, J. R. Boot, A. F. Cockerill, D. N. B. Mallen, and D. J. Osborn, *Nature (London)* **262,** 699 (1976).
[10] C. R. Pace-Asciak, M. C. Carrara, and Z. Domazet, *Biochem. Biophys. Res. Commun.* **78,** 115 (1977).
[11] L. J. Roberts, B. J. Sweetman, and J. A. Oates, *J. Biol. Chem.* **253,** 5305 (1978).
[12] M. Hamberg and B. Samuelsson, *J. Biol. Chem.* **246,** 6713 (1971).
[13] H. Kindahl, L. E. Edqvist, A. Bane, and E. Granström, *Acta Endocrinol.* **82,** 134 (1976).
[14] K. Bukhave and H. S. Hansen, *Biochim. Biophys. Acta* **489,** 403 (1977).
[15] H. S. Hansen, *Biochim. Biophys. Acta* **574,** 136 (1979).
[16] J. Jarabak, *Am. J. Obstet. Gynecol.* **138,** 534 (1980).

METHODS IN ENZYMOLOGY, VOL. 86

ISBN 0-12-181986-8

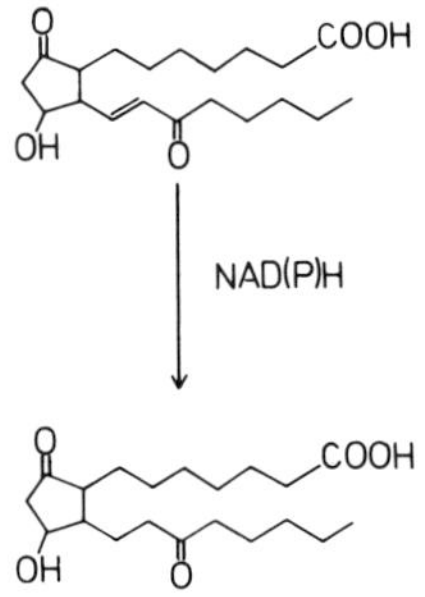

15-Ketoprostaglandin Δ^{13}-reductase has been isolated in varying degrees of purity from chicken heart,[17] human placenta,[18] and bovine lung.[15] The procedure reported here describes the purification to homogeneity of the enzyme from bovine lung by the use of affinity chromatography on a 2′,5′-ADP-Sepharose column. The purified enzyme has been used for obtaining monospecific anti-15-ketoprostaglandin Δ^{13}-reductase immunoglobulins[19] for immunoelectrophoretic procedures.

Assay Methods

Reduction of 15-Keto-PGE_1. 15-Keto-PGE_1 forms a labile red chromophore in alkaline solution.[20] This property is the basis for routine detection of the enzyme activity during the purification. The reaction mixture contains 0.1 *M* sodium phosphate, pH 7.4, 1 m*M* EDTA, 1 m*M* mercaptoethanol, 14.2 μM 15-keto-PGE_1 (10 μg), and 1 m*M* NADH. The reaction is initiated by the addition of enzyme to give a final total volume of 2.0 ml. After incubation at 44°, usually 45 min, the concentration of the remaining substrate is quantified after addition of 0.5 ml of 2 *N* NaOH solution by measuring the maximum absorption at 500 nm, reached ca. 1 min after alkali addition. Under these conditions the maximal molar extinction coefficient is 37.4 cm^{-1} mM^{-1}. The value of the maximal molar extinction coefficient varies according to the incubation conditions.[4]

Disappearence of 15-keto-PGE_1 is not an absolute specific assay for 15-ketoprostaglandin Δ^{13}-reductase activity, since this can also be due to

[17] S. C. Lee and L. Levine, *Biochem. Biophys. Res. Commun.* **61,** 14 (1974).
[18] C. Westbrook and J. Jarabak, *Biochem. Biophys. Res. Commun.* **66,** 541 (1975).
[19] H. S. Hansen, O. Norén, and H. Sjöström, unpublished results (1980).
[20] E. Änggård and B. Samuelsson, this series, Vol. 14, 215 (1969).

reversability of the 15-hydroxyprostaglandin dehydrogenase-catalyzed reaction.[21,22]

Formation of 13,14-Dihydro-15-keto-PGE$_1$. The initial rate of formation of 13,14-dihydro-15-keto-PGE$_1$ from 15-keto-PGE$_1$ is determined by the use of 5,6-$^3H_2$-labeled 15-keto-PGE$_1$.[15] Preparation of the radioactive prostaglandin metabolites is described by Hansen and Toft.[23] The reaction mixture contains 6000 cpm of ^{3}H-labeled 15-keto-PGE$_1$ appropriate amounts of 15-keto-PGE$_1$ and coenzyme (NADH or NADPH), respectively, and 0.1 *M* sodium phosphate (pH 7.4), 1 m*M* EDTA, 1 m*M* mercaptoethanol. The reaction is initiated by addition of enzyme (total incubation volume, 0.5 ml), and the incubation is stopped by adding 1.5 ml of cold acetone (0–4°). Then 400 cpm of ^{14}C-labeled 13,14-dihydro-15-keto-PGE$_1$ and 10 μg of 13,14-dihydro-15-keto-PGE$_1$ are added as internal standard and carrier, respectively. Several incubations with varied incubation time are performed in order to obtain a progress curve. The samples can at this stage be left overnight when stored at −20°. The precipitated protein is removed by centrifugation, and the acetone is evaporated at 35° under a stream of N_2.

The aqueous phase is brought to pH 5 with 50 μl acetic acid, and the prostaglandins are extracted with two 1-ml portions of ethyl acetate. After evaporation with N_2 the prostaglandins are separated by TLC (DC-Alufolie Kieselgel GF-254, Merck) in two developments using $CHCl_3/CH_3OH/CH_3COOH$ (95:1:5, v/v/v) as solvent. With this system the following R_f values are obtained: PGE$_1$ 0.08; 13,14-dihydro-PGE$_1$, 0.15; 15-keto-PGE$_1$, 0.33; and 13,14-dihydro-15-keto-PGE$_1$, 0.41. The spots are visualized by gentle spraying with 10% (w/v) phosphomolybdic acid in ethanol, followed by heating to 120° for 10 min. The spots containing 13,14-dihydro-15-keto-PGE$_1$ are cut from the plate, and radioactivity is estimated by liquid scintillation spectrometry. In some incubations of tissue homogenates[1,2] a 15-ketoreductase activity is present (probably due to 15-hydroxyprostaglandin dehydrogenase), and in such cases the activity of the 15-ketoprostaglandin Δ^{13}-reductase may be calculated from the sum of 13,-14-dihydro-PGE$_1$ and 13,14-dihydro-15-keto-PGE$_1$. Under the present conditions and using bovine lung homogenates 13,14-dihydro-PGE$_1$ is not formed in significant amounts. It has been shown that 13,14-dihydro-15-keto-PGE$_1$ is unstable under some circumstances.[24,25] However, under

[21] M. Yamazaki and M. Sasaki, *Biochem. Biophys. Res. Commun.* **66,** 255 (1975).

[22] S. S. Braithwaite and J. Jarabak, *J. Biol. Chem.* **250,** 2315 (1975).

[23] H. S. Hansen and B. S. Toft, *Biochim. Biophys. Acta* **529,** 230 (1978).

[24] E. Granström, M. Hamberg, G. Hansson, and H. Kindahl, *Prostaglandins* **19,** 933 (1980).

[25] F. A. Fitzpatrick, R. Aguirre, J. E. Pike, and F. H. Lincoln, *Prostaglandins* **19,** 917 (1980).

the conditions of the present assay there is no significant breakdown of 13,14-dihydro-15-keto-PGE_1.

Formation of 13,14-dihydro-15-keto-$PGE_{2\alpha}$ from 15-keto-$PGE_{2\alpha}$ can also be estimated by the use of radioimmunoassay.[17]

Oxidation of NAD(P)H. The disappearance of the coenzyme NADH or NADPH measured by spectrophotometry has been used to estimate enzyme activity[18,27] (see Table II). Disappearance of coenzyme is not a specific assay for the same reasons described above for the disappearance of substrate 15-ketoprostaglandin. Furthermore, the use of high concentrations of coenzyme restricts the sensitivity of the assay, which measures small changes in absorbance against a high background.

The assay mixture as used in Table II (see later) contains 0.1 *M* sodium phosphate buffer (pH 7.4), 1 m*M* EDTA, 1 m*M* mercaptoethanol, 100 μM NADH, and 3 μg of purified enzyme in a 3-ml cuvette with a 1-cm light path. Total volume is 2.00 ml, and incubations are performed at 37°. The reaction is initiated with the addition of appropriate amounts of prostaglandin in 10 μl of ethanol. Only ethanol is added to the blank cuvette. The oxidation of the coenzyme is followed by the decrease in absorbance at 340 nm using a molar extinction coefficient of 6.2 cm^{-1} mM^{-1}.

Purification Procedure

Step 1. All operations in the purification procedure are performed at 4°. Bovine lung (typically 70–100 g) is sliced and homogenized in 2 volumes (v/w) of buffer A (0.1 *M* sodium phosphate, pH 7.4, 1 m*M* EDTA, 1 m*M* mercaptoethanol) using a Waring blender. The homogenate is centrifuged 30 min at 20,000 *g*. The resulting supernatant is centrifuged 60 min at 107,000 *g*. The precipitate is discarded.

Step 2. Solid ammonium sulfate (43.2 g) is added to 140 ml of supernatant solution to achieve 50% saturation. After 3 hr this solution is centrifuged at 20,000 *g* for 30 min, and the precipitate is discarded. Solid ammonium sulfate (13.8 g) is added to the supernatant solution (137 ml) to achieve 65% saturation. The solution is stirred overnight, then centrifuged at 20,000 *g* for 30 min. The precipitate (50–65% fraction) is redissolved in 150 ml of buffer B (0.05 *M* sodium phosphate, pH 6.0) and stirred for 40 min. Solid ammonium sulfate (30 g) is added to achieve 35% saturation. After 4 hr this solution is centrifuged at 20,000 *g* for 30 min; the precipitate is discarded. Solid ammonium sulfate (23.5 g) is added to

[26] B. Weeke, *in* "A Manual of Quantitative Immunoelectrophoresis. Methods and Applications" (N. H. Axelsen, J. Krøll, and B. Weeke, eds.), pp. 37–46. Universitets-forlaget, Oslo, 1973.

the supernatant solution (162 ml) to achieve 58% saturation. This solution is stirred overnight, then centrifuged at 20,000 *g* for 30 min. The precipitate (35–58% fraction) is redissolved in 230 ml of buffer C (0.08 *M* sodium phosphate, pH 5.8). Insoluble material is precipitated by centrifugation at 20,000 *g* for 30 min and discarded. The supernatant is concentrated to 110 ml using an Amicon cell 202 and a PM-10 filter.

Step 3. The concentrated supernatant is applied to a CM-Sephadex C-50 column (5 × 49 cm) equilibrated in buffer D (buffer C containing 1 m*M* EDTA and 1 m*M* mercaptoethanol). The column is washed with buffer D, and the enzyme is eluted by an NaCl gradient, 0 to 250 m*M*. In the actual case shown in Table I, a sequence of two linear gradients of NaCl, 0 to 50 m*M* (460 ml total volume) and 50 to 250 m*M* (300 ml total volume), is applied, using an LKB Ultrograd (flow rate, 114 ml/hr; 9.4-ml fractions). Before beginning the 50 to 250-ml gradient, the column is washed with 100 ml of buffer containing 50 m*M* NaCl. Fractions containing enzyme activity are pooled, and the solvent buffer is changed to buffer A either by dialysis or by using a concentration cell. The use of a two-step gradient relative to a one-step linear gradient results in a higher degree of purification.

Step 4. The enzyme solution from step 3 is applied on a 2′,5′-ADP–Sepharose 4B column (1.6 × 8.4 cm) equilibrated in buffer A. The column is washed in buffer A, and the enzyme is eluted with a linear gradient of 0 to 600 m*M* NaCl (total volume of 90 ml; flow rate, 21 ml/hr; 2.6-ml fractions). The enzyme is concentrated on a Amicon cell. Generally the enzyme is stored in buffer A containing the NaCl from the gradient (approximately 240 m*M* NaCl). Occasionally the purified enzyme is gel filtered and stored in buffer A with no NaCl; this has no apparent effect on the stability of the enzyme.

The affinity column step gives 5- to 6-fold purification. The enzyme can be eluted by $NADP^+$, which, however, has to be removed before assay for enzyme activity because of to its strong inhibitory properties.[15] This affinity chromatography may also be useful in separation of 15-keto-prostaglandin Δ^{13}-reductase and NAD^+-dependent 15-hydroxyprostaglandin dehydrogenase, since the latter is not bound to the column.

This purification procedure (summarized in Table I) comprises two column steps and takes about 3 days. The yield is 10–20%, and the enzyme is purified to homogeneity.

Properties

Purity and Stability. Polyacrylamide gel electrophoresis with and without SDS revealed only one protein, which also could be stained for

TABLE I
PURIFICATION OF 15-KETOPROSTAGLANDIN Δ^{13}-REDUCTASE FROM BOVINE LUNG[a]

Step	Total activity (mU)	Total protein (mg)	Specific activity (mU/mg)	Yield (%)
107,000 *g* supernatant[b]	7808	3670	2.13	100
$(NH_4)_2SO_4$ fractionation	2048	164	12.48	26
CM-Sephadex	1664	8.78	190	21
2′,5′-ADP-Sepharose	1224	1.15	1067	16

[a] One unit is 1 μmol of 13,14-dihydro-15-keto-PGE_1 formed per minute at 37° with 2 m*M* NADH and 85.2 μM 15-keto-PGE_1.
[b] Material from 81 g of lung tissue.

enzyme activity.[15] Immunization of a rabbit with the purified enzyme preparation resulted in formation of monospecific antibodies, as seen by rocket immunoelectrophoresis[26] of a bovine lung homogenate (Fig. 1). The immunization schedule was as follows: A rabbit was immunized with purified enzyme, previously mixed with an equal volume of Freund's incomplete adjuvant. The animal was injected intracutaneously every second week with 100 μl of the mixture (approximately 60 μg of enzyme protein). One week after the third injection, the rabbit was bled (40 ml). Boosters were given every sixth week, followed by new bleedings.

The purified 15-ketoprostaglandin Δ^{13}-reductase is stable for more than a month at 4° in buffer A. Stored at −80°, it is stable for more than a year.

Molecular Weight and Isoelectric Point. Two molecular weights, 56,000 and 39,500, were found by gel filtration and SDS–gel electrophoresis.[15] It is at present not known whether the difference in the molecular weights obtained by the two methods is due to the existence of dimer–monomer or to an asymmetric enzyme molecule. The latter property results in an overestimation of the molecular weight by the gel filtration procedure. Only one N-terminal amino acid has been found, valine.[15] By the use of the same two methods, molecular weights of 68,500 and 35,500, respectively, have been found for the enzyme from human placenta.[27] The p*I* value for the bovine lung enzyme is 7.8.[15]

Substrates. The enzyme can use both NADH and NADPH as coenzymes, however with different K_m values: 91 μM for NADH and 8 μM for NADPH.[15] The V_{max} value for NADPH is less than half that for NADH.[15] The enzyme from human placenta uses only NADH as coenzyme.[18,27] With near saturated concentrations (600 μM) of NADH, K_m

[27] C. Westbrook and J. Jarabak, *Arch. Biochem. Biophys.* **185,** 429 (1978).

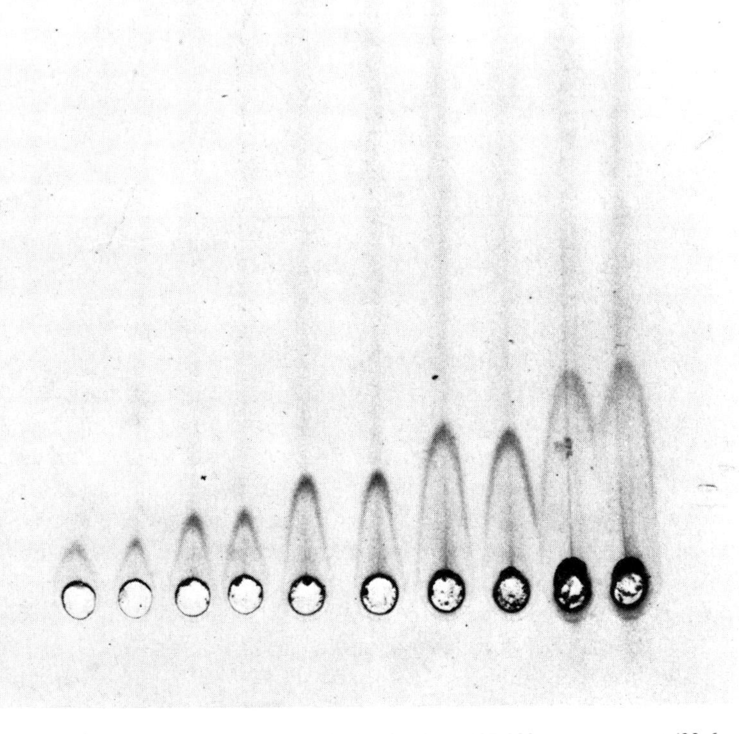

FIG. 1. Rocket immunoelectrophoresis[26] of a bovine lung 107,000 g supernatant (33.6 mg of protein per milliliter, and 67.2 μg of 15-ketoprostaglandin Δ^{13}-reductase protein per milliliter) in different dilutions (v/v) with electrophoresis buffer. Wells of 10 μl, 1–10 from left: $\frac{1}{16}$ dilution, $\frac{1}{8}$ dilution, $\frac{1}{4}$ dilution, $\frac{1}{2}$ dilution, and undiluted in double estimation. The immunoglobulin G concentration (purified on a protein A–Sepharose column) in the gel was 30 $\mu g/cm^2$. Electrophoresis was carried out at pH 8.6 using a 0.02 M barbital buffer with 3 V/cm overnight. Stained with Coomassie Brilliant Blue R.

(15-keto-PGE_1) is 10 μM.[15] Other prostaglandins have been tested as substrates using a NADH concentration close to the K_m value (Table II).

The purified enzyme does not to any significant degree catalyze the reverse reaction, i.e., formation of 15-keto-PGE_1 from 13,14-dihydro-15-keto-PGE_1 using either NAD^+ or $NADP^+$ as coenzyme.

Inhibitors. p-Chloromercuribenzoic acid and $NADP^+$ are inhibitory. The latter observation was the reason for using the 2′,5′-ADP ligand in the affinity chromatography.

TABLE II
SUBSTRATES FOR BOVINE LUNG 15-KETOPROSTAGLANDIN Δ^{13}-REDUCTASE[a]

	Apparent K_m (μM)	Apparent relative V_{max}
15-Keto-PGE_1	30	100
15-Keto-PGE_2	62	181
15-Keto-$PGF_{1\alpha}$	184	251
6,15-Diketo-$PGF_{1\alpha}$	258	54
PGE_2	Not substrate	—
PGD_2	Not substrate	—
6-Keto-$PGF_{1\alpha}$	Not substrate	—
PGA_1	Not substrate	—
PGB_1	Not substrate	—
TxB_2	Not substrate	—

[a] The reaction mixture is described under assay methods.

Acknowledgments

The study was supported by a research grant from the Danish Fat Research Foundation. Birgit Nielsen is thanked for skillful technical assistance.

[27] Isolation and Properties of a 15-Ketoprostaglandin Δ^{13}-Reductase from Human Placenta

By JOSEPH JARABAK

$$\text{15-Ketoprostaglandin} + \text{NADH} + \text{H}^+ \rightarrow \text{13,14-dihydro-15-ketoprostaglandin} + \text{NAD}^+$$

The second step in the biological inactivation of prostaglandins involves reduction of the 13-14 double bond of 15-ketoprostaglandins.[1] Although the enzyme catalyzing this reaction has not been studied as extensively as the 15-hydroxyprostaglandin dehydrogenase, several of the studies that have been performed indicate that differences exist between the 15-ketoprostaglandin Δ^{13}-reductases isolated from different species and organs.[2,3] Human placenta is a good source for this enzyme even though it is present in smaller quantities than the NAD^+-dependent 15-hydroxyprostaglandin in placental tissue.[4]

[1] E. Änggård and C. Larsson, *Eur. J. Pharmacol.* **14,** 66 (1971).
[2] C. Westbrook and J. Jarabak, *Biochem. Biophys. Res. Commun.* **66,** 541 (1975).
[3] S.-C. Lee and L. Levine, *Biochem. Biophys. Res. Commun.* **61,** 14 (1974).
[4] J. Jarabak, *Am. J. Obstet. Gynecol.* **138,** 534 (1980).

ISBN 0-12-181986-8

Assay Method

Since the reaction catalyzed by the NAD^+-dependent 15-hydroxyprostaglandin dehydrogenase is reversible, both this enzyme and the 15-ketoprostaglandin Δ^{13}-reductase reduce 15-ketoprostaglandins in the presence of NADH. Because the dehydrogenase activity is greater than the reductase activity in placental tissue, it is not possible to assay the reductase accurately until the dehydrogenase has been removed, i.e., before step 3 in the purification. The enzyme activity is measured spectrophotometrically in a cuvette with a 1-cm light path at 25 ± 0.5°. The cuvette contains 3 ml of aqueous solution consisting of 30 μmol of potassium phosphate at pH 7.0, 680 nmol of NADH and 114 nmol of 15-keto-PGE_2 added in 0.02 ml of 95% ethanol. The reaction is initiated by the addition of enzyme. The blank cuvette contained no 15-keto-PGE_2 or ethanol. Enzyme is included in the blank only when NADH oxidase activity is present in the enzyme preparation, i.e., before step 4 in the purification. Oxidation of NADH is followed by measuring the decrease in absorbance at 340 nm. One unit of enzyme activity is defined as the amount of enzyme that catalyzes the oxidation of 1 μmol of NADH per minute under the conditions of the assay.

Purification Procedure[5]

Six normal-term human placentas are used for this purification. Steps 1 and 2 are performed on individual placentas. All operations are performed at below 4° unless indicated otherwise.

Step 1. Homogenization. A placenta is chilled immediately after delivery. The villous tissue is dissected from the membranes, rinsed in cold tap water and homogenized in 75–100-g portions at top speed in a Waring blender for 2 min with 2 volumes of buffer A (20% glycerol, v/v, 5 m*M* potassium phosphate, and 1 m*M* EDTA, at a final pH of 7.0). The homogenate is centrifuged at 10,000 *g* for 45 min, and the precipitate is discarded.

Step 2. Ammonium Sulfate Precipitation. Solid ammonium sulfate is added slowly, with continuous stirring, to the supernatant solution (0.298 g/ml). The pH is maintained at about 7 by the addition of 3 *N* NH_4OH. After 1 hr this solution is centrifuged at 10,000 *g* for 1 hr and the precipitate is discarded. Solid ammonium sulfate is added slowly, with continuous stirring, to the supernatant solution (0.197 g/ml), while maintaining the pH at 7 with the addition of 3 *N* NH_4OH. After 1 hr this solution is

[5] C. Westbrook and J. Jarabak, *Arch. Biochem. Biophys.* **185,** 429 (1978).

centrifuged for 1 hr at 10,000 *g*, and the precipitate is suspended in 40–50 ml of buffer B (50% glycerol, v/v, 5 m*M* potassium phosphate, and 1 m*M* EDTA, at a final pH of 7.0). This material is stored for no longer than 24 hr before performance of the next step.

Step 3. DEAE-Cellulose Chromatography. Ammonium sulfate precipitates from three placentas are combined and dialyzed for 40 hr against two 12-liter changes of buffer A, then centrifuged at 39,000 *g* for 15 min. The precipitate is discarded. If the conductivity of the supernatant solution is greater than 0.8 mΩ^{-1}, cold 20% glycerol is added with stirring until the conductivity is below that value. This solution is then applied to a 4 × 45 cm DEAE-cellulose column (Bio-Rad, high capacity) that has been equilibrated with buffer A, and the column is washed with 1 liter of buffer A. All the material that does not bind to the column and that has an A_{280} greater than 0.6 (a value that is less than 1% of the absorbance of the solution applied) is collected and used for the next purification step within 24 hr.

Step 4. Blue Sepharose Chromatography. The enzyme from two DEAE columns (six placentas) is combined for this step and applied to a 2.5 × 10.5 cm Blue Sepharose column (prepared as described by Westbrook and Jarabak[5]) that has been equilibrated with a 10 m*M* potassium phosphate buffer, pH 7.0, containing 1 m*M* EDTA. After the enzyme has been applied, the column is successively washed with 200 ml of equilibration buffer; 500 ml of buffer containing 10 m*M* potassium phosphate, pH 7.0, 1 m*M* EDTA, 0.05 *M* KCl; and 200 ml of 10 m*M* potassium phosphate, pH 7.0, 1 m*M* EDTA, 0.1 *M* KCl. The enzyme is eluted from the column with a linear gradient formed from 400 ml of 10 m*M* potassium phosphate, pH 7.0, 1 m*M* EDTA, 0.1 *M* KCl; and 400 ml of 10 m*M* potassium phosphate, pH 7.0, 1 m*M* EDTA, 1.0 *M* KCl. Fractions of the highest specific activity are combined, and the salt concentration is reduced on a Diaflo UM-10 membrane by repeated concentration and dilution with 10 m*M* potassium phosphate buffer, pH 7.0, until the conductivity of the solution is 0.3 mΩ^{-1}.

Step 5. Hydroxyapatite Chromatography. The enzyme from the preceding step is applied to a 2.5 × 10 cm hydroxyapatite column that has been equilibrated with 10 m*M* potassium phosphate buffer, pH 7.0. The hydroxyapatite was prepared according to Anaker and Stoy.[6] The column is then washed with 200 ml of the same buffer, and the enzyme is eluted with a linear gradient formed from 500 ml of 10 m*M* potassium phosphate, pH 7.0, and 500 ml of 160 m*M* potassium phosphate, pH 7.0.

Step 6. CM-Cellulose Chromatography. The enzyme solution from the

[6] R. L. Anaker and V. Stoy, *Biochem. Z.* **330,** 141 (1958).

PURIFICATION OF 15-KETOPROSTAGLANDIN Δ^{13}-REDUCTASE FROM HUMAN PLACENTA

Step	Volume (ml)	Total activity (units)	Specific activity (units/mg protein[a]) × 10^3	Yield (%)
1. Centrifuged homogenate	4430[b]	—	—	—
2. Ammonium sulfate precipitation	690	—	—	—
3. DEAE-cellulose	1070	1.64	0.029	100
4. Blue Sepharose	20	1.49	10.6	91
5. Hydroxyapatite	30	0.520	94.2	32
6. CM-cellulose	30	0.265	425	16

[a] During most of the purification it is assumed that a solution containing 1 mg of protein per milliliter has an absorbance at 280 nm of 1.0 in a cuvette of 1-cm path length. In the final step of the purification, the protein concentration is calculated by the formula $1.5 \times A_{280} - 0.75\, A_{260}$ = protein concentration (mg/ml) (E. Layne, this series, Vol. 3, p. 447).

[b] Volume of the supernatant solution after centrifugation of the homogenate from 1998 g of placental tissue.

preceding step is adjusted to pH 6.0 by dropwise addition of 1 *N* HCl, and its salt concentration is reduced on a Diaflo UM-10 membrane by repeated concentration and dilution with cold distilled water until the conductivity is 0.5 $m\Omega^{-1}$. This solution is applied to a 2 × 7 cm CM-cellulose column (Bio-Rad, Cellex-CM) that has been equilibrated with 5 m*M* potassium phosphate, pH 6.0, 1 m*M* EDTA. The column is washed with 80 ml of this buffer, and then the enzyme is eluted with a linear gradient formed from 200 ml of 5 m*M* potassium phosphate, pH 6.0, 1 m*M* EDTA, and 200 ml of 200 m*M* potassium phosphate, pH 6.0, 1 m*M* EDTA. The fractions of highest enzymic activity were combined, concentrated by vacuum dialysis, dialyzed against 10 volumes of buffer B, and stored at −20°.

Results of a typical purification are summarized in the table. Disc gel electrophoresis of the purified enzyme reveals only minor protein impurities, containing less than 5% of the protein in the sample.

Properties

Stability.[5] The purified enzyme is stable for at least 6 months when stored at −20° in buffer B. It is also stable for several weeks at 4° in the absence of glycerol.

Pyridine Nucleotide Specificity.[5] Both NADH and reduced nicotinamide hypoxanthine dinucleotide are cofactors for the enzyme, while NADPH and reduced 3-acetylpyridine adenine dinucleotide are not. At a

15-keto PGE_2 concentration of 179 μM, the apparent K_m for NADH is 31.9 μM, and that for the hypoxanthine dinucleotide is 63.1 μM.

Substrate Specificity.[5] All the 15-ketoprostaglandins that were tested are substrates for the enzyme. At an NADH concentration of 166 μM, the following apparent Michaelis constants are obtained: 15-keto PGA_1, 7.5 μM; 15-keto-PGB_1, 11.2 μM; 15-keto-PGE_1, 3.9 μM; 15-keto-PGE_2, 12.7 μM; 15-keto-$PGF_{2\alpha}$, 10.8 μM. None of the 15-hydroxyprostaglandins tested is a substrate. Furthermore, substances for various double-bond reductases, progesterone, cortisone, orotic acid, fumaric acid, and uracil are not reduced by the enzyme.

Reversibility of the Enzymic Reaction.[5] No evidence has been obtained to indicate that the enzymic reaction is reversible. The reaction continues until one of the substrates is completely consumed, even if the products are added in excess. After prolonged incubation of NAD^+ and 13,14-dihydro-15-keto-PGE_2 with the enzyme at pH 6, 7, 8, or 9, no NADH or 15-keto-PGE_2 can be demonstrated.

Activation Energy.[5] The activation energy of the enzyme is 12,100 cal per mole.

Effect of Sulfhydryl Reagents.[5] Although neither *N*-ethylmaleimide nor iodoacetate inhibited the enzyme at concentrations as high as 1 mM, *p*-chloromercuribenzoate was a very effective inhibitor, causing 85% inhibition at a concentration of 10^{-7} M.

Kinetic Mechanism.[5] The kinetic mechanism is a rapid-equilibrium random mechanism. The substrate and product inhibition is compatible with the formation of "dead-end" complexes.

Inhibition by NADPH and Cibacron Blue.[5] Although NADPH is not a cofactor for the enzyme, it is a competitive inhibitor with respect to NADH and it has a K_i of 10.7 nM. Cibacron Blue 3G-A, the dye that gives Blue Sepharose its color, is also a competitive inhibitor with respect to NADH. Its K_i is 12.5 nM. Thus both of these compounds are bound to the enzyme with a greater affinity than its cofactor, NADH, whose dissociation constant is 200 μM.

Acknowledgments

This work has been supported by grants from the National Institutes of Health (HD-07045 and PHS 5-T05-GM 01939) and from the Louis Block Foundation.

[28] Measurement of Prostaglandin ω-Hydroxylase Activity

By WILLIAM S. POWELL[1]

Omega (ω) oxidation is an important pathway in prostaglandin (PG)[2] metabolism. A large proportion of the urinary metabolites of PGs have been oxidized in the ω or ω-1 positions.[3] Semen from primates contains high concentrations of 19-hydroxy-PGs. Since some of these metabolites have biological activities comparable to those of the corresponding unmetabolized compounds,[4] ω-hydroxylation may not result in the biological inactivation of PGs. In most cases, however, ω-hydroxylation of PGs appears to follow other pathways of metabolism, such as oxidation of the 15-hydroxyl group, which does result in biological inactivation.

The major sites of ω-oxidation of PGs in most species appear to be the liver and the kidney.[5] The liver has both PG 19- and PG 20-hydroxylase activities. In most species, the kidney, especially the cortex, has PG 20-hydroxylase activity and, in some cases, also has PG 19-hydroxylase activity.[5,6] These enzymes are found in microsomal fractions and require NADPH as a cofactor. They appear to be related to cytochrome P-450, since they are inhibited by cytochrome P-450 inhibitors such as metyrapone[7] and SKF-525A.[5,7] 20-Hydroxy metabolites are further oxidized to the corresponding aldehydes[8] and ω-carboxylic acids[9,10] in the presence of either cytosolic fractions and NAD^{+}[10] or microsomal fractions.[8,9]

The rabbit is distinct from other species in that it has a wider distribution of PG 20-hydroxylase, which is also found in the lung and the uterus in this species. PG 20-hydroxylase is induced in lung, liver, kidney, and uterus as a result of pregnancy in the rabbit and is present in appreciable

[1] The author is a scholar of the conseil de la recherche en santé du Québec.

[2] The abbreviations used are: PG, prostaglandin; Tx, thromboxane, ODS, octadecylsilyl; HPLC, high-pressure liquid chromatography; TLC, thin-layer chromatography; GC–MS, gas chromatography–mass spectrometry.

[3] B. Samuelsson, E. Granström, K. Gréen, and M. Hamberg, *Ann. N. Y. Acad. Sci.* **180,** 138 (1971).

[4] C. H. Spilman, K. K. Bergstrom, and A. D. Forbes, *Prostaglandins* **13,** 795 (1977).

[5] W. S. Powell, *Prostaglandins* **19,** 701 (1980).

[6] J. Navarro, D. E. Piccolo, and D. Kupfer, *Arch. Biochem. Biophys.* **191,** 125 (1978).

[7] D. Kupfer, *Pharmacol. Ther.* **11,** 469 (1980).

[8] W. S. Powell, *Biochim. Biophys. Acta* **575,** 335 (1979).

[9] W. S. Powell and S. Solomon, *J. Biol. Chem.* **253,** 4609 (1978).

[10] D. Kupfer, J. Navarro, G. K. Miranda, D. E. Piccolo, and A. Theoharides, *Arch. Biochem. Biophys.* **199,** 228 (1980).

METHODS IN ENZYMOLOGY, VOL. 86

ISBN 0-12-181986-8

amounts in the placenta.[5] The mechanism of induction, at least in the lung, appears to be mediated by progesterone.[5,11] In the hamster and the rat, on the other hand, both hepatic PG 19- and 20-hydroxylase activities are depressed during pregnancy.[11]

Prostaglandin ω-hydroxylase activity can be measured by determining the amounts of 19- and 20-hydroxy-PGs formed after incubation of microsomal fractions with PGs in the presence of NADPH. The hydroxylated products can be quantitated as discussed below by gas chromatography–mass spectrometry (GC–MS) using selected ion monitoring, by thin-layer chromatography (TLC) or high-pressure liquid chromatography (HPLC) using radioactivity or UV detectors, or by GC.

Materials

[9β-^{3}H]PGF$_{2\alpha}$ (ca. 16 Ci/mmol) was from Amersham/Searle, Oakville, Ontario. [9β-^{3}H]PGF$_{2\alpha}$ of a low specific activity (ca.1 Ci/mmol) was synthesized by reduction of PGE$_2$ with sodium boro[^{3}H]hydride as described in the literature.[12] [3,3,4,4-^{2}H]PGE$_2$, [3,3,4,4-^{2}H]PGF$_{2\alpha}$, and nonlabeled PGs and TXB$_2$ were kindly supplied by Dr. J. E. Pike of the Upjohn Company, Kalamazoo, Michigan. Cartridges containing octadecylsilyl (ODS) silica (SEP-PAK C$_{18}$ cartridges) were obtained from Waters Associates, Milford, Massachusetts. New Zealand white rabbits were purchased from Canadian Hybrid Farms, Stanstead, Quebec.

Synthesis of Labeled 19- and 20-Hydroxy-PG Standards

Deuterated hydroxy-PGs are required as internal standards for analysis by GC–MS using selected ion monitoring. Tritium-labeled hydroxy-PGs can be used as recovery markers in double-label experiments and as standards for chromatography using radioactivity detectors.

20-Hydroxy-[3,3,4,4-^{2}H]PGF$_{2\alpha}$. This substance is prepared by incubating a microsomal fraction from pregnant rabbit lungs (25 days of gestation) with [3,3,4,4-^{2}H]PGF$_{2\alpha}$.[11] Lungs (10 g) are minced in 3 volumes of 0.05 *M* Tris-HCl, pH 7.4, containing 0.25 *M* sucrose. The mixture is homogenized in an ice-water bath with a VirTis homogenizer (6 periods of 10 sec with 1 min in between to allow for cooling). The homogenate is centrifuged at 8000 *g* for 10 min at 4°, and the supernatant is removed and recentrifuged at 100,000 *g* for 60 min at 2°. The pellet is resuspended in 0.05 *M* Tris-HCl, pH 7.4 (3 volumes) and recentrifuged as described

[11] W. S. Powell, *J. Biol. Chem.* **253,** 6711 (1978).

[12] E. Granström and B. Samuelsson, *Eur. J. Biochem.* **10,** 411 (1969).

above. The final pellet is resuspended in 0.05 *M* Tris-HCl, pH 7.4, (1 volume per gram of tissue).

The microsomal fraction prepared in this way is incubated with a mixture of [3,3,4,4-^{2}H]PGF$_{2\alpha}$ (500 μg) and [9β-^{3}H]PGF$_{2\alpha}$ (10^8 dpm, 1 μg) in the presence of NADPH (1 m*M*) for 15 min at 37°. The incubation is terminated by the addition of ethanol (2 volumes).

The purification procedure reported originally[11] involved extraction with ethyl acetate, chromatography on an open column of silicic acid, and finally TLC on silicic acid. This can be simplified by extraction with ODS silica. The incubation mixture containing ethanol is centrifuged at 400 *g* for 10 min. The supernatant is concentrated under vacuum to the original volume of the incubation (10 ml), followed by the addition of water (40 ml). The mixture is acidified and passed through an ODS silica cartridge, which is eluted as described below. The 20-hydroxy[3,3,4,4-^{2}H]PGF$_{2\alpha}$ is purified by HPLC on a silicic acid column with benzene–ethyl acetate–acetonitrile–methanol–acetic acid (30 : 40 : 30 : 2 : 0.5) as the mobile phase (see Table II).

20-Hydroxy-[3,3,4,4-^{2}H]PGE$_2$. This substance is synthesized from [3,3,4,4-^{2}H]PGE$_2$ by a procedure[5] exactly analogous to that described above for the synthesis of 20-hydroxy[3,3,4,4-^{2}H]PGF$_{2\alpha}$.

19-Hydroxy-[9β-^{3}H]PGF$_{2\alpha}$. 19-Hydroxy-PGE$_2$ (500 μg, Upjohn Co.) in methanol (0.2 ml) is mixed with sodium boro[^{3}H]hydride (50 mCi) in methanol (0.3 ml). The mixture is kept in an ice bath for 15 min and then at room temperature for 60 min. Water (9.5 ml) is added, and the mixture is acidified to a pH of ca 3 and passed through an ODS silica cartridge, which is eluted as described below. The products (19-hydroxy-[9β-^{3}H]PGF$_{2\alpha}$ and 19-hydroxy-[9α-^{3}H]PGF$_{2\beta}$) are purified by HPLC on a silicic acid column with benzene–ethyl acetate–acetonitrile–methanol–acetic acid (30 : 40 : 30 : 2 : 0.5) as the mobile phase (see Table II, below).

20-Hydroxy-[9β-^{3}H]PGF$_{2\alpha}$. [9β-^{3}H]PGF$_{2\alpha}$ (100 μCi, ca 35 μg), synthesized as described in the literature,[12] is incubated for 15 min at 37° in the presence of NADPH (1 m*M*) with a microsomal fraction prepared from the lungs of a pregnant rabbit as described above. The incubation is terminated with ethanol (2 ml), and the mixture is extracted using ODS silica. A similar procedure can be used to prepare 20-hydroxy-[1-^{14}C]PGF$_{2\alpha}$.

Extraction of Hydroxy-PGs

19-Hydroxy-PGs and 20-hydroxy-PGs can be extracted from biological media using cartridges of ODS silica[13] as described in this volume [58].

[13] W. S. Powell, *Prostaglandins* **20,** 947 (1980).

Incubation mixtures of small volumes (up to 2 ml) are terminated by the addition of ethanol (2 volumes). Water is added to give a final concentration of ethanol of 10%. The mixtures are centrifuged at 400 *g* for 10 min, and the supernatants are acidified to a pH of ca 3 and passed through a cartridge of ODS silica (SEP-PAK C_{18} cartridge), which has been prepared by treatment with ethanol and water.[14] The ODS silica is eluted, using a syringe, with 10% ethanol in water (20 ml), petroleum ether (20 ml), and methyl formate (10 ml). The methyl formate fraction, which contains the ω-hydroxy-PGs, is evaporated under a stream of nitrogen, and the residue is dissolved in a solvent suitable for subsequent analysis by TLC or HPLC.

Incubations containing larger volumes are terminated with 2 volumes of ethanol and centrifuged at 400 *g* for 10 min. The supernatant is concentrated under vacuum to the original volume of the incubation mixture and diluted with 4 volumes of water (it is assumed that after concentration under vacuum the concentration of ethanol will not exceed 50%). The mixture is then acidified and passed through a cartridge containing ODS silica, which is eluted as described above.

Chromatography of Prostaglandin ω-Oxidation Products

Thin-Layer Chromatography. Table I lists the R_f values on silicic acid of the free acids and the methyl esters of some of the ω-oxidation products of PGB_2, PGE_1, PGE_2, $PGF_{2\alpha}$, 13,14-dihydro-15-oxo-$PGF_{2\alpha}$ and TxB_2. The 19-hydroxy metabolites of each of these substances has the same R_f value as the corresponding 20-hydroxy derivative, both for the free acids and the methyl esters. In general, the free acids of the ω-carboxy derivatives have R_f values slightly higher than those of the corresponding ω-hydroxy compounds. There is a greater difference between the R_f values of the methyl esters of ω-hydroxy and ω-carboxy metabolites (Table I).

High-Pressure Liquid Chromatography. Table II shows the retention times of $PGF_{2\alpha}$, 19-hydroxy-$PGF_{2\alpha}$, 19-hydroxy-$PGF_{2\beta}$ and 20-hydroxy-$PGF_{2\alpha}$ on normal-phase, reversed-phase, and argentation HPLC. The free acids of 19- and 20-hydroxy-$PGF_{2\alpha}$ are not separated by HPLC on a silicic acid column using the solvent systems described in Table II. These conditions are therefore useful for applications requiring the copurification of 19- and 20-hydroxy-PGs in a single fraction (e.g., analysis by GC–MS).

[14] ODS silica must be wetted with an organic solvent before use. This can be accomplished by passing ethanol (20 ml) through the ODS silica cartridge. The ethanol is then removed by eluting with water (20 ml). ODS silica cartridges can be re-used a number of times provided the samples being extracted are not too large. In this case we also wash the cartridge with 80% ethanol prior to 100% ethanol.

TABLE I
CHROMATOGRAPHIC PROPERTIES OF SOME ω-OXIDATION PRODUCTS OF PGs AND TxB_2[a]

	R_f values		C values[d] of methyl esters	
Compound	Solvent 1[b]	Solvent 2[c]	Me_3Si derivatives	O-Methyloxime Me_3Si derivatives
19-Hydroxy-PGB_2	—	0.18 (2%)	26.3	—
20-Hydroxy-PGB_2	—	0.18 (2%)	27.1	—
PGE_1	0.34	0.42 (4%)	—	24.1, 24.6
19-Hydroxy-PGE_1	0.08	0.17 (6%)	—	26.1, 26.6
20-Hydroxy-PGE_1	0.08	0.17 (6%)	—	26.9, 27.4
ω-Carboxy-PGE_1	ca. 0.08	0.42 (6%)	—	26.9, 27.4
PGE_2	0.34	0.46 (4%)	—	24.0, 24.5
19-Hydroxy-PGE_2	0.08	0.15 (6%); 0.42 (10%)	—	26.1, 26.6
20-Hydroxy-PGE_2	0.08	0.15 (6%); 0.42 (10%)	—	26.8, 27.3
ω-Carboxy-PGE_2	0.16	0.40 (6%)	—	26.6, 27.2
$PGF_{2\alpha}$	0.21	0.39 (6%)	24.1	—
19-Hydroxy-$PGF_{2\alpha}$	0.05	0.10 (6%) 0.27 (10%)	26.2	—
20-Hydroxy-$PGF_{2\alpha}$	0.05	0.10 (6%); 0.27 (10%)	27.0	—
ω-Carboxy-$PGF_{2\alpha}$	0.10	0.23 (6%)	26.9	—
TxB_2	0.43	0.37 (2%)	24.7	—
19-Hydroxy-TxB_2	0.12	0.25 (6%)	26.8	—
20-Hydroxy-TxB_2	0.12	0.25 (6%)	27.5	—
ω-Carboxy-TxB_2	0.17	0.34 (6%)	27.4	—
13,14-Dihydro-15-oxo-$PGF_{2\alpha}$	0.52	0.64 (4%)	24.4	—
13,14-Dihydro-20-hydroxy-15-oxo-$PGF_{2\alpha}$	0.11	0.17 (4%)	27.4	27.2
ω-Carboxy-13,14-dihydro-15-oxo-$PGF_{2\alpha}$	0.18	0.40 (4%)	27.2	27.1

[a] Taken from Powell and Solomon[9] and Powell.[8]

[b] Thin-layer chromatography of underivatized PGs and Txs was carried out on glass plates precoated with silica gel (E. Merck, Darmstadt, FRG) using diethyl ether–methanol–acetic acid (100:6:1) as solvent.

[c] Thin-layer chromatography of the methyl esters of PGs and Txs was carried out on glass plates coated with silica gel G (E. Merck) in the laboratory. Various concentrations of methanol (indicated in parentheses) in diethyl ether were used as solvents. The R_f values of substances using these plates were slightly higher than those when precoated plates were used.

[d] Gas chromatography was carried out using a 6-ft column of 1.5% OV-101 and a temperature of 260°. C values were calculated as described in the text.

TABLE II
HIGH-PRESSURE LIQUID CHROMATOGRAPHY (HPLC) RETENTION TIMES OF SOME PROSTAGLANDIN F (PGF) DERIVATIVES[a]

	Silicic acid[b]			ODS silica[c]		
Compound	B	A:B (3:7)	B_M	C:D (1:1)	C:D (9:11)	Ag^+ [d]
$PGF_{2\alpha}$	5.8	10.1	6.2	—	42.1	13.2
19-Hydroxy-$PGF_{2\alpha}$	11.9	30.8	15.9	22.1	11.6	18.8
19-Hydroxy-$PGF_{2\beta}$	18.7	—	26.2	15.5	8.7	11.9
20-Hydroxy-$PGF_{2\alpha}$	11.9	30.8	14.1	12.1	7.0	16.8

[a] HPLC was carried out using a Milton Roy minipump with a Berthold radioactivity monitor. The retention times of [9β-^{3}H]$PGF_{2\alpha}$, 19-hydroxy-[9β-^{3}H]$PGF_{2\alpha}$, 19-hydroxy-[9α-^{3}H]$PGF_{2\beta}$ and 20-hydroxy-[1-^{14}C]$PGF_{2\alpha}$ are given in minutes.

[b] Normal-phase HPLC was carried out on a column (35 × 0.46 cm) of Partisil (particle size, 5 μm), which was purchased from Alltech Associates, Deerfield, Illinois. Solvent A: hexane–benzene–acetic acid, 50:50:0.5; solvent B; benzene–ethyl acetate–acetonitrile–methanol–acetic acid, 30:40:30:2:0.5; solvent B_M: benzene–ethyl acetate–acetonitrile–methanol, 30:40:30:2. The flow rate was 2 ml/min.

[c] Reversed-phase HPLC was carried out on a column (30 × 0.46 cm) of μBondapak C_{18}, purchased from Alltech Associates. Solvent C: water–acetic acid, 99.9:0.1; solvent D: methanol–acetic acid, 99.9:0.1. The flow rate was 1.5 ml/min.

[d] Argentation HPLC was carried out using a silver ion-loaded cation exchange column, prepared as described in this volume [63]. The mobile phase was methanol–acetic acid, 99.8:0.2. The flow rate was 1.5 ml/min.

The methyl esters of 19- and 20-hydroxy-$PGF_{2\alpha}$ are separated by HPLC on silicic acid using benzene–ethyl acetate–acetonitrile–methanol (30:40:30:2, v/v/v/v) as the mobile phase.

Mixtures of $PGF_{2\alpha}$ and its 19- and 20-hydroxy metabolites can be completely resolved by argentation HPLC using methanol–acetic acid (99.8:0.2) as mobile phase. This solvent suppresses polar–polar interactions between the solute and the stationary phase, and consequently solutes are retained mainly on the basis of olefin–Ag^+ interactions (see this volume [63]). Thus it is not surprising that 19-hydroxy-$PGF_{2\beta}$ has a retention time shorter than that of $PGF_{2\alpha}$ under these conditions. We have not tested other mobile phases (see this volume [63]) for the separation of the above standards by argentation HPLC.

Of the three types of HPLC listed in Table II, reversed-phase HPLC gives the best separation of 19- and 20-hydroxy-$PGF_{2\alpha}$. These compounds are completely separated using a mobile phase consisting of 45% solvent C (water–acetic acid (99.9:0.1) and 55% solvent D (methanol–acetic acid

(99.9:0.1) (Table II). Kupfer *et al.*[15] have used an ODS silica column with a mobile phase consisting of water–acetonitrile–acetic acid (80:20:1) to separate 19- and 20-hydroxy-PGB_2 by HPLC using a UV detector.

Gas Chromatography. The *C* values of the methyl ester-trimethylsilyl ether or methyl ester-trimethylsilyl ether-*O*-methyloxime derivatives of PGE_1, PGE_2, $PGF_{2\alpha}$, 13,14-dihydro-15-oxo-$PGF_{2\alpha}$ and TxB_2 and some of their ω-oxidation products on a 1.5% OV-101 column are shown in Table I. *C* values were calculated from standard curves obtained from the retention times (t_R) of a series of saturated fatty acid methyl esters chromatographed under conditions identical to those used for the PG or Tx derivatives.[16] The 19- and 20-hydroxy derivatives of PGs and TxB_2 are all well separated from one another by GC. The *C* values of the 19-hydroxy derivatives are all 2.0–2.1 units higher than for the corresponding unmetabolized compounds, whereas the *C* values of the 20-hydroxy metabolites are 2.8–3.0 units higher.[8,9] The *C* values of the ω-carboxy derivatives of PGs and TxB_2 are the same or slightly lower than those of the corresponding 20-hydroxy derivatives.[8,9]

Analysis of 19- and 20-Hydroxy-PGs by GC–MS Using Selected Ion Monitoring

The activities of PG 19- and 20-hydroxylases with either PGE_2[5] or $PGF_{2\alpha}$[11] as substrate can be determined by GC–MS. The amounts of hydroxylated products formed are measured using either 20-hydroxy[3,3,4,4-^{2}H]PGE_2 or 20-hydroxy[3,3,4,4-^{2}H]$PGF_{2\alpha}$ as internal standards. Microsomal fractions are prepared as described above, except that tissues are homogenized either with a ground-glass homogenizer (tissues that are difficult to homogenize, such as lung or uterus) or a Potter–Elvehjem homogenizer (most other tissues). The final pellet is, in most cases, resuspended in 0.5–1.0 volume per gram of tissue of 0.05 *M* Tris-HCl, pH 7.4. If the tissue in question is especially rich in PG ω-hydroxylase activity, the microsomal fraction should be resuspended in a larger amount of buffer (e.g., 20 volumes per gram of tissue for lung microsomes from pregnant rabbits near term), so that the percentage conversion to hydroxylated products is not greater than ca 10%.

Microsomal fractions (0.5 ml) are incubated with an excess (0.113 m*M*) of PGE_2 or $PGF_{2\alpha}$ and NADPH (2 m*M*) for 15 min at 37°. Incubations are terminated by the addition of ethanol (1 ml). Mixtures of

[15] D. Kupfer, J. Navarro, and D. E. Piccolo, *J. Biol. Chem.* **253,** 2804 (1978).

[16] K. Gréen, *Chem. Phys. Lipids* **3,** 254 (1969).

either 20-hydroxy-[3,3,4,4-^{2}H]PGF$_{2\alpha}$ (0.5–1.0 μg) and 20-hydroxy-[9β-^{3}H]PGF$_{2\alpha}$ (100,000–200,000 dpm) or 20-hydroxy-[3,3,4,4-^{2}H]PGE$_2$ (0.5–1.0 μg) and 20-hydroxy-[5,6,8,11,12,14,15-^{3}H]PGE$_2$ (100,000–200,000 dpm) are added, and the mixtures are extracted using ODS silica as described above. After evaporation of the methyl formate, the residue is dissolved in 0.1 ml of methanol and PGs are methylated by the addition of ethereal diazomethane. The 19- and 20-hydroxy derivatives of either PGE$_2$ or PGF$_{2\alpha}$ are purified as a single fraction by TLC on silicic acid using diethyl ether–methanol (9:1) as the mobile phase. Alternatively, the products could be purified prior to methylation by HPLC on a silicic acid column using 100% B as the mobile phase (Table II).

After purification, the methyl esters of 19- and 20-hydroxy-PGF$_{2\alpha}$ are converted to their Me$_3$Si derivatives by treatment with trimethylsilylimidazole in pyridine (10 μl of Tri-Sil Z, Pierce Chemical Co., Rockford, Illinois) for 5 min at 60°. The methyl esters of 19- and 20-hydroxy-PGE$_2$ are converted to the corresponding trimethylsilyl ether derivatives of PGB$_2$ by treatment with Tri-Sil Z (8 μl) for 5 min at 60°, followed by piperidine (8 μl) for an additional 5 min at 60°. 19-Hydroxy-PGB$_2$ (t_R, 2.7 min), 20-hydroxy-PGB$_2$ (t_R, 3.5 min), 19-hydroxy-PGF$_{2\alpha}$ (t_R, 2.6 min) and 20-hydroxy-PGF$_{2\alpha}$ (t_R, 3.4 min) are analyzed by GC–MS using selected ion monitoring with a 16-inch column of OV-101 (1.5%) at a temperature of 215°. The ions monitored are 423 and 427 (M-159-90)[17] for 19- and 20-hydroxy-PGF$_{2\alpha}$ and 321 and 325 (M-159-28)[17] for 19- and 20-hydroxy-PGB$_2$. To confirm further the identities of the substances being measured, ions of higher mass [508 (M) for hydroxy-PGB$_2$ and 582 (M-90) for hydroxy-PGF$_{2\alpha}$] can also be monitored. Figure 1A shows the standard curves obtained from samples containing various amounts of unlabeled 19- and 20-hydroxy-PGF$_{2\alpha}$ along with 20-hydroxy-[3,3,4,4-^{2}H]PGF$_{2\alpha}$ (0.5 μg). Figure 1B shows a standard curve obtained in a similar fashion for 19- and 20-hydroxy-PGE$_2$.

Figure 2 shows the effect of diluting microsomal fractions from pregnant rabbit lung and liver with buffer to give different protein concentrations. It is apparent that the amount of 20-hydroxy-PGF$_{2\alpha}$ formed is proportional to the enzyme concentration in the incubation mixture. The formation of 20-hydroxy-PGF$_{2\alpha}$ is linear with time up to at least 15 min, the incubation time used for the assays.[11]

[17] The ions at m/e 159 correspond the terminal pentyl side chain (i.e., carbons 16–20) of the 19- and 20-hydroxy derivatives of PGB$_2$ and PGF$_{2\alpha}$. The ions at m/e 90 and m/e 28 correspond to trimethylsilanol and carbon monoxide, respectively.

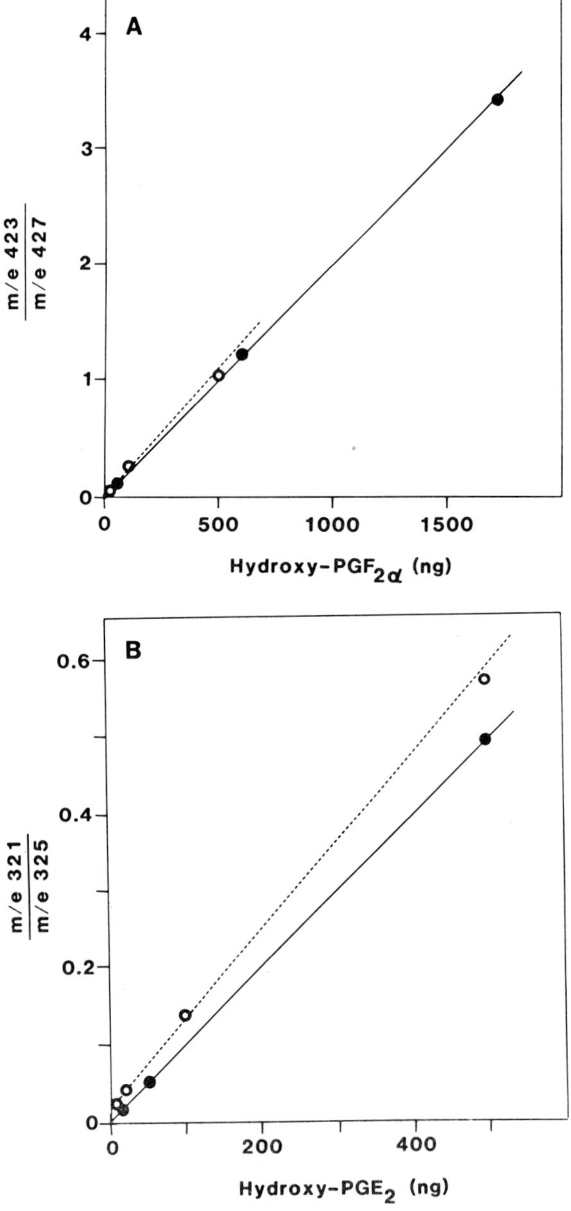

FIG. 1. (A) Standard curves for the quantitation of 19-hydroxy-$PGF_{2\alpha}$ (○····○) and 20-hydroxy-$PGF_{2\alpha}$ (●—●) by gas chromatography–mass spectrometry (GC–MS) using selected ion monitoring. Mixtures of 20-hydroxy-[3,3,4,4-^{2}H]$PGF_{2\alpha}$ (0.5 μg) and varying amounts of 19(*R*)-19-hydroxy-$PGF_{2\alpha}$ or 20-hydroxy-$PGF_{2\alpha}$ were analyzed by GC–MS as described in the text. Since relatively small amounts of 19-hydroxy-$PGF_{2\alpha}$ were formed by the tissues that we investigated, the range of the standard curve for this compound is smaller than that for 20-hydroxy-$PGF_{2\alpha}$. (B) Standard curves for the quantitation of 19-hydroxy-PGE_2 (●—●) and 20-hydroxy-PGE_2 (○····○) by GC–MS using selected ion monitoring. Mixtures of 20-hydroxy-[3,3,4,4-^{2}H]PGE_2 (0.9 μg) and varying amounts of 19(*R*)-19-hydroxy-PGE_2 and 20-hydroxy PGE_2 were analyzed by GC–MS as described in the text. Figure 1B was taken from Powell.[5]

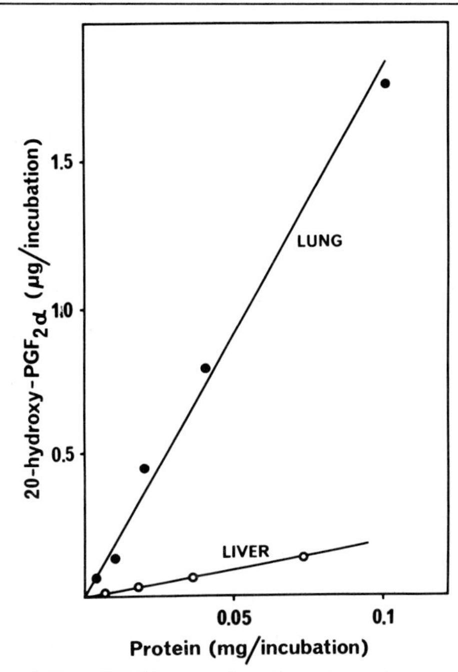

FIG. 2. 20-Hydroxylation of $PGF_{2\alpha}$ as a function of protein concentration. Microsomal fractions from the lungs (●) and liver (○) of a pregnant (25 days of gestation) rabbit were diluted to give various concentrations of microsomal protein. Aliquots (0.5 ml) of the microsomal fractions were incubated for 15 min at 37° with $PGF_{2\alpha}$ (0.113 m*M*) in the presence of NADPH (2 m*M*). 20-Hydroxy-$PGF_{2\alpha}$ was measured by gas chromatography–mass spectrometry as described in the text. Taken from Powell.[11]

Analysis of PG ω-Hydroxylation by TLC or HPLC Using Radioactive Substrates

Prostaglandin ω-hydroxylase activity can be determined by measuring the amount of radioactive substrate converted into radioactive hydroxy-PGs.[8,15] We have used this method to investigate the specificities of PG 20-hydroxylase from pregnant rabbit lung and liver, using tritiated derivatives of PGE_2, $PGF_{2\alpha}$, 15-methyl-$PGF_{2\alpha}$, 16,16-dimethyl-$PGF_{2\alpha}$, and 17-phenyl-18,19,20-trinor-$PGF_{2\alpha}$, as well as [1-^{14}C]TxB_2 as substrates.[8] Microsomal fractions are incubated with 0.06–0.3 m*M* ^{3}H-labeled substrate and 1 m*M* NADPH for 15 min at 37°. Incubations are terminated with ethanol, and the products are extracted and analyzed by TLC using diethyl ether–methanol–acetic acid (100:6:1) as solvent. The TLC plates are scanned using a radiochromatogram scanner, and the radioactive bands are scraped off. The silicic acid is eluted with methanol–diethyl ether

(1 : 1; 5 ml) which is collected in counting vials. The solvent is evaporated and the radioactivity in each fraction is determined by liquid scintillation counting. When we used this procedure we expressed our results in terms of the percentage of recovered radioactive products that were hydroxylated. For more accurate results, it would be possible to incubate microsomal fractions with [1-^{14}C]PGF$_{2\alpha}$ and use 19-hydroxy-[9β-^{3}H]PGF$_{2\alpha}$ and/or 20-hydroxy-[9β-^{3}H]PGF$_{2\alpha}$ as internal standards to correct for recovery. If it were desired to measure both 19- and 20-hydroxylase activities by this method, these products would have to be separated by reversed-phase HPLC (Table II). Alternatively, [9β-^{3}H]PGF$_{2\alpha}$ could be used as substrate and 20-hydroxy-[1-^{14}C]PGF$_{2\alpha}$, synthesized as described above for the corresponding ^{3}H-labeled compound, could be used as an internal standard.

Other Methods for Measurement of ω-Hydroxy-PGs

We have used GC with a flame ionization detector to measure the ω-hydroxylation of the nonradioactive PG analogs, 13,14-didehydro-PGF$_{2\alpha}$ and *ent*-13,14-didehydro-15-epi-PGF$_{2\alpha}$, by lung and liver microsomes from pregnant rabbits.[8] This method cannot be recommended for measuring low levels of these products, however, owing to interference from unrelated contaminants. Reversed-phase HPLC with a UV detector has been used to measure 19-hydroxy-PGE$_1$ and 19-hydroxy-PGE$_2$ in human semen after conversion to their PGB derivatives, which have λ_{max} values of 278 nm.[18]

Choice of Substrate for Measuring PG ω-Hydroxylase Activity

PG 19- and 20-hydroxylases appear to be different enzymes.[7,11] PG 20-hydroxylases from pregnant rabbit lung and liver also have different properties with regard to inhibition by SKF-525A[5] and substrate specificity.[5,8] PGE$_2$ and PGF$_{2\alpha}$ are metabolized at equal rates by the lung enzyme, whereas PGE$_2$ is metabolized at about twice the rate of PGF$_{2\alpha}$ by the liver enzyme.[5]

An advantage of using PGE$_1$ or PGE$_2$ as substrates is that these compounds, along with their ω-hydroxy metabolites, can be converted to their PGB derivatives, which can be detected using a UV detector.[15,18] On the other hand, ^{3}H-labeled PGF$_{2\alpha}$ and its 19- or 20-hydroxylated metabolites can be synthesized inexpensively by reduction of the corresponding PGE

[18] D. E. Piccolo and D. Kupfer *in* "Biological/Biomedical Applications of Liquid Chromatography II" (G. L. Hawk, ed.), p. 425. Dekker, New York, 1979.

compounds with sodium boro[^{3}H]hydride or by incubation of [^{3}H]$PGF_{2\alpha}$ with pregnant rabbit lung microsomes as described above. The tritiated metabolites can then be used as recovery markers and to indicate the positions of HPLC peaks using a radioactivity monitor. The separation of 19- and 20-hydroxy-$PGF_{2\alpha}$ by reversed-phase HPLC (Table II) appears to be considerably better than that of 19- and 20-hydroxy-PGB_2.[15]

One problem with the use of PGEs as substrates for ω-hydroxylation reactions is that a number of side reactions can take place.[8] The major products formed when PGE_1 is incubated with liver microsomes from male rabbits in the presence of NADPH, for example, are $PGF_{1\alpha}$ and 8-iso-PGE_1. Smaller amounts of 19- and 20-hydroxy-PGE_1 are formed. Similar results were obtained with PGE_2. On the other hand, the only products detected with $PGF_{2\alpha}$ as substrate were 19- and 20-hydroxy-$PGF_{2\alpha}$.[8]

Acknowledgments

These studies were supported by grants from the Medical Research Council of Canada, the Quebec Heart Foundation, and the Quebec Thoracic Society.

[29] Receptors for PGI_2 and PGD_2 on Human Platelets

By ADELAIDE M. SIEGL

Platelets are small, anucleate disks that, in humans, are released into the circulation at a rate (per day) of 35,000 per microliter of blood. Platelets are intimately involved in homeostasis, playing such important roles as effecting the initial arrest of bleeding by both the production of vasoconstrictive substances and platelet plug formation; stabilization of the hemostatic plug by providing a catalytic membrane surface on which thrombin formation can occur; mediation of clot retraction; maintenance of endothelial integrity; and promotion of wound healing through the secretion of growth factors. On the pathological side, platelets play important roles in atherogenesis, inflammation, and transplant rejection. The modification of platelet reactivity, therefore, has important physiological and clinical implications.

Platelets can be stimulated toward aggregation and clot formation by a wide variety of substances. Endogenous stimulatory substances include thrombin, ADP, epinephrine, arachidonic acid, collagen, and prostaglandin endoperoxides. Platelet aggregation can be inhibited by agents that act as local anesthetics and by agents that increase levels of cyclic AMP (cAMP). PGI_2, PGD_2, and PGE_1 have been demonstrated to be the most

 ISBN 0-12-181986-8

potent endogenous inhibitors of platelet aggregation and appear to act via stimulation of platelet adenylate cyclase. PGI_2 has been of particular interest, since it is produced by endothelial cells and its production can be stimulated by many of the factors, such as thrombin and prostaglandin endoperoxides, that cause platelet aggregation. PGD_2 has been shown to be produced by platelets during platelet aggregation. These two prostaglandins, therefore, may act as feedback regulators of platelet reactivity. For that reason there has been a great deal of interest in the measurement and characterization of receptors for these prostaglandins on human platelets. Indeed, by the end of 1979, specific binding sites for $[^3H]PGI_2$,[1] $[^3H]PGD_2$,[2,3] and $[^3H]PGE_1$[4] had been characterized both with intact platelets and with platelet membranes. The present chapter summarizes the methods used in these studies and their results.

Methodology for Binding Assays

Binding of a radiolabeled prostaglandin to platelet suspensions or membranes should be measured as a function of time and as a function of concentration. In order to be physiologically relevant, the binding of a radiolabeled prostaglandin should, ideally, be measured over the 10^{-10}–10^{-5} M range. The conditions of incubation appear to be relatively unimportant except that, as will be discussed later, direct contact with glass should be avoided.

The procedure chosen for separation of ligand bound to receptor from free ligand is perhaps the most critical aspect of binding studies where the suspected affinity for the receptor is in the 10^{-9} to 10^{-6} M range. Over this range the rate of dissociation of the hormone from the receptor varies widely. Bennett[5] has determined that, for an accurate determination of the dissociation constant, the time required for separation of the bound from the free ligand must be less than one-seventh the $t_{1/2}$ for dissociation of the bound ligand. That means, therefore, that the allowable separation time ranges from 102 sec for a hormone–receptor interaction with a K_D of 10^{-9} M to 0.01 sec for a K_D of 10^{-6} M. Microcentrifugation is the separation procedure of choice because its extremely short separation time per-

[1] A. M. Siegl, J. B. Smith, M. J. Silver, K. C. Nicolaou, and D. Ahern, *J. Clin. Invest.* **63,** 215 (1979).

[2] A. M. Siegl, J. B. Smith, and M. J. Silver, *Biochem. Biophys. Res. Commun.* **90,** 291 (1979).

[3] B. Cooper and D. Ahern, *J. Clin. Invest.* **64,** 586 (1979).

[4] A. I. Schafer, B. Cooper, D. O'Hara, and R. I. Handin, *J. Biol. Chem.* **254,** 2914 (1979).

[5] J. P. Bennett, Jr., *in* "Neurotransmitter Receptor Binding" (H. I. Yamura, S. J. Enna, and M. J. Kuhar, eds.), p. 57. Raven, New York, 1978.

mits measurement of a wide range of dissociation constants. A drawback of microfuge separation is the trapping of a significant amount of plasma containing free ligand in the pellet. Corrections were made for this in studies on prostaglandin binding by including ^{14}C-labeled sucrose as an extracellular marker to assess plasma space in the pellet. Double-channel counting of the tritium and carbon-14 with correction for crossover was not sufficiently precise because of the low level of bound ligand. Therefore the Packard 306 oxidizer, which enables complete separation of ^{3}H and ^{14}C labels, was utilized.

The filtration method, although widely used, may yield inaccurate determinations with ligands having dissociation constants greater than 10^{-8} *M*, since a small but significant length of time is required for the separation of bound from free ligand. A second potential problem is the possibility of a large degree of nonspecific binding of ligand to the filter. In some cases, ligand bound to filter can represent more than 60% of the total bound radioactivity, making difficult any measurement of small differences in specific binding. Another potential problem with the filtration method when it is applied to intact platelets is that the glass fibers in the filters may cause platelet activation with consequent alteration of platelet membranes.

A third method that has been used for separation of ligand bound to platelets from free ligand is column chromatography.[cf 6] The time required for separation, however, is approximately 2 min, which is too slow for the accurate determination of dissociation constants greater than 10^{-9} *M*. The incubations were performed by Schillinger and Prior[6] in buffer at pH 8.0 as opposed to plasma or buffer at pH 7.4. Although high pH stabilizes $[^{3}H]PGI_2$, binding characteristics may be altered at such a nonphysiological pH.

In conclusion, microfuge centrifugation is the best separation procedure for studies with $[^{3}H]PGI_2$, $[^{3}H]PGD_2$, and $[^{3}H]PGE_1$, where binding constants are in the range of 10^{-9}–10^{-6} *M*.

Analysis of Binding Data

In order to determine the dissociation constant (K_D) and the number of binding sites (B_{max}) for $[^{3}H]PGI_2$, $[^{3}H]PGE_1$ and $[^{3}H]PGD_2$, the data obtained from concentration-dependent binding experiments must be subjected to Scatchard analysis.[7] The derivation of this analysis may be found

[6] E. Schillinger and G. Prior, *Biochem. Pharmacol.* **29,** 2297 (1980).

[7] G. Scatchard, *Ann. N.Y. Acad. Sci.* **5,** 660 (1950).

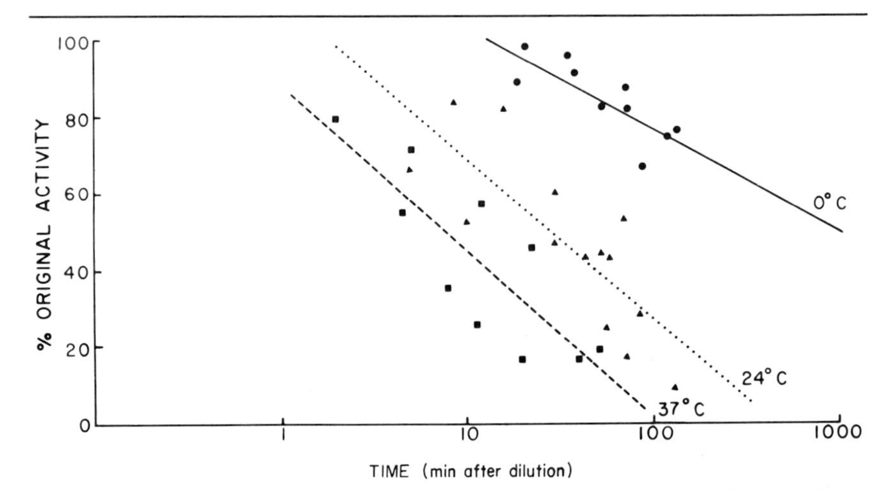

FIG. 1. Stability of PGI_2 in plasma at different temperatures. PGI_2 was diluted in plasma (final concentration 10 μM) and tested for its ability to inhibit ADP-induced aggregation. Lines were drawn using all data points obtained. Incubation at 37°, $t_{1/2}$ equaled 7.7 min ($p < 0.001$), $N = 2$ experiments (■----■). Incubation at 24°, $t_{1/2}$ equaled 28.7 min ($p < 0.001$), $N = 4$ experiments (▲----▲). Incubation at 0°, $t_{1/2}$ equaled 1031 min ($p < 0.001$), $N = 4$ experiments (●—●). Taken from Siegl.[8]

in Bennett.[5] The final equation is

$$B/F = (B_{max} - B)/K_D \qquad (1)$$

Knowledge of free ligand (F) at the end of the incubation period is necessary for the accurate utilization of this method. In these experiments the bound ligand (B) was less than 1% of the total radioactivity; therefore it was not necessary to correct the concentration of ligand (F) for the bound ligand. $[^3H]PGD_2$ and $[^3H]PGE_1$ are both stable under the conditions used for these assays. $[^3H]PGI_2$, however, is fairly labile under these conditions. It was therefore necessary to construct a curve for the decay of PGI_2 in plasma (Fig. 1). For this purpose, a known initial concentration of PGI_2 was incubated in plasma at different temperatures for various periods and its activity with respect to inhibition of platelet aggregation as induced by ADP (10 μM) was then measured. From a known dose-response curve for PGI_2, the actual concentration of PGI_2 after various periods of preincubation could be determined and a line describing the decay of PGI_2 as a function of time could be constructed. This plot was then used to calculate the remaining free concentration of $[^3H]PGI_2$ under such incubation conditions.

For a single class of binding sites, plotting B/F vs B will yield a straight line with a slope equal to $1/K_D$ and an x intercept equal to B_{max}

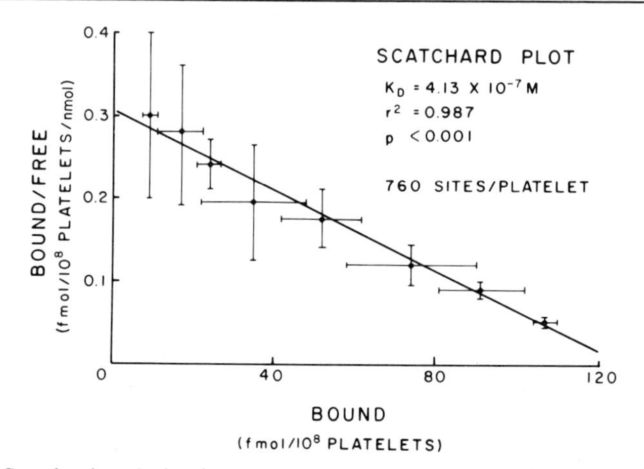

FIG. 2. Scatchard analysis of concentration-dependent [^{3}H]PGD$_2$ binding. Platelet-rich plasma was incubated with [^{3}H]PGD$_2$ in concentrations from 30 to 2000 n*M*. Binding was measured with ^{14}C-labeled sucrose as the internal marker. Each point is plotted ± SEM and represents the mean of duplicate determinations on 3–5 subjects. Platelet counts for these subjects were 5.6 ± 0.83 (range 3.72–5.82) × 10^8/ml. Reprinted from Siegl *et al.*[2] with permission of the publisher.

(cf. [^{3}H]PGD$_2$ binding, Fig. 2). In cases where the ligand binds to two independent sites or where negative cooperativity is involved, plotting B/F vs B will yield a hyperbola (cf Fig. 3, [^{3}H]PGI$_2$ binding). The determination of K_D and B_{max} from such a curve is complex but may be done by a number of methods.

The first step in analysis of a hyperbola is to construct the hyperbola that best fits the data points. Since a general equation for a hyperbola is

$$C = x^n y \tag{2}$$

taking the logarithim of both sides gives

$$\log C = n \log x + \log y \tag{3}$$

or

$$C^1 = n \log x + \log y \tag{4}$$

$$\log y = -n \log x + C^1 \tag{5}$$

Plotting log y (or B/F) vs log x (or B) will therefore yield a straight line (Fig. 4). For [^{3}H]PGI$_2$ and [^{3}H]PGE$_1$ binding [8] the linear regression pro-

[8] A. M. Siegl, doctoral dissertation. Identification of prostaglandin receptors on human platelets: Measurements of ligand binding, stimulation of adenyl cyclase and inhibition of platelet aggregation. Department of Pharmacology, Thomas Jefferson University, Philadelphia, Pennsylvania, 1980.

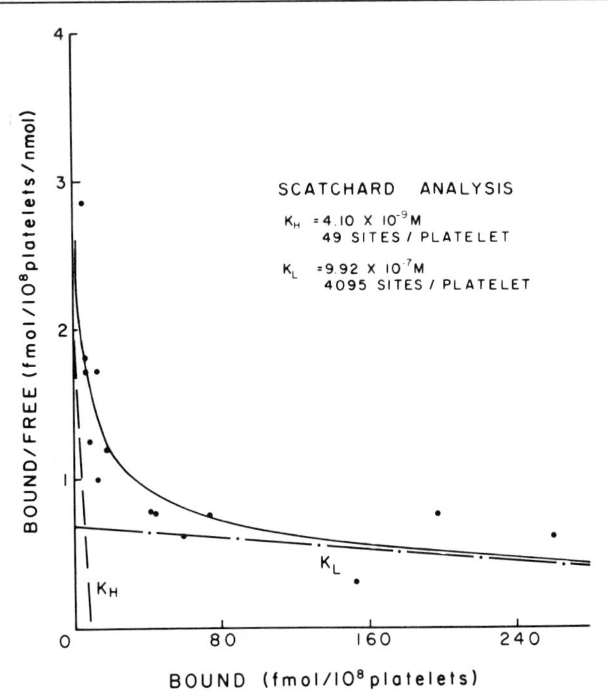

FIG. 3. Scatchard analysis of concentration-dependent [^{3}H]PGI_2 binding. Platelet-rich plasma was incubated with [^{3}H]PGI_2 in initial concentrations from 4 to 518 nM, and binding was measured with ^{14}C-labeled sucrose as the internal marker. The smooth curve represents the hyperbolic plot derived from the log–log transformation (Fig. 4), and the straight lines represent the partial plots used to obtain K_H and K_L. Each point represents the mean of duplicate determinations of between 4 and 12 subjects. Average percentage standard deviation for these points was 40% (range 23–54%). Figure is similar to that found in Siegl *et al.*[1]

gram of a TI-55 calculator (Texas Instruments, Houston, Texas) was used to derive the straight line with the best fit. This was then used to draw the best-fit hyperbola through these data points.

Ohnishi *et al.*[9] have demonstrated that it is possible to construct a hyperbolic Scatchard plot from two partial plots that are straight lines. Using a nonlinear equation describing a hyperbolic Scatchard plot, they resolved the above equation into its component partial plots by repetitive numerical analysis of the two legs of the hyperbola using a computer program. The same process may be mimicked, however, by repetitive, iterative (i.e., first low binding area, than high binding area) geometric con-

[9] T. Ohnishi, E. J. Masero, H. A. Borkand, and B. Palyai, *Biophys. J.* **12,** 1251 (1972).

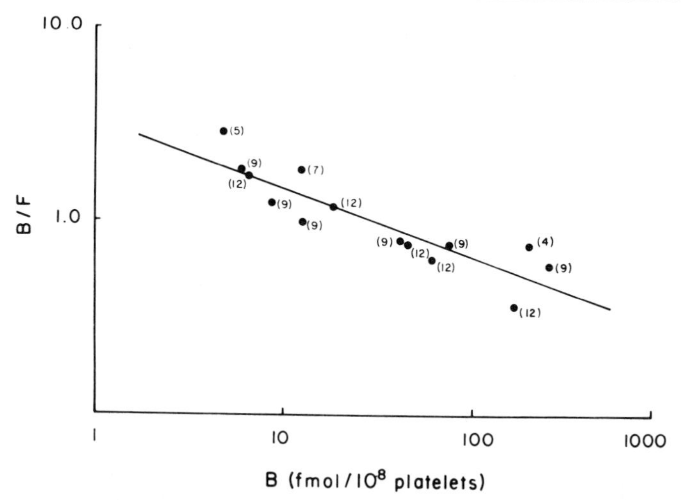

FIG. 4. Log–log transformation of Scatchard analysis binding data for [^{3}H]PGI$_2$. Each of the data points from Fig. 3 was plotted on double-log paper. The straight line represents the best-fit line derived through a linear regression program ($r^2 = 0.8801, p < 0.001$). This line was then used to form the hyperbola in Fig. 3 by extrapolation of the linear regression program. The numbers in parentheses represent the number of subjects used for each point. Taken from Siegl[8]

struction of the partial plots (see Fig. 3 for final iterative plot of that hyperbola).

The possibility that [^{3}H]PGI$_2$ binding (Fig. 3) reflected negative cooperativity was tested by measuring the rate of dissociation of [^{3}H]PGI$_2$ bound to a concentrated suspension of platelets during dilution by buffer in the presence or the absence of a high concentration of unlabeled PGI$_2$. If negative cooperativity was present, then one would expect dissociation in the presence of a high concentration of unlabeled PGI$_2$ to be faster than dissociation by plain buffer.[10] This was not the case. Therefore, it was concluded that [^{3}H]PGI$_2$ binding represented interaction with two classes of binding sites.

Under circumstances where binding is found to reflect a high-affinity, specific site and a lower-affinity, nonspecific site, the Scatchard plot will also appear to be a hyperbola. Log–log transformation of the data, however, will not yield a straight line and, hence, no true hyperbola can be drawn. Once the nonspecific binding is subtracted, however, the Scatchard plot should reflect the parameters of specific binding.

[10] P. DeMeyts, *J. Supramol. Struct.* **4,** 241 (1976).

Criteria for Relevance of Binding Data

Hormone binding is generally believed to reflect hormone–receptor interaction when the binding can be shown to be saturable, reversible, specific, and parallel in time course and concentration to the effects produced by the hormone.[11] For [^{3}H]PGI_2,[1] [^{3}H]PGD_2,[2,3] and [^{3}H]PGE_1,[4] the binding has been shown to be complete within 2–3 min and to be rapidly reversible with the addition of a large excess of the appropriate unlabeled prostaglandin. This time course agrees well with both the onset of inhibition of aggregation and the increases in platelet cAMP that may be observed with these prostaglandins. The binding of [^{3}H]PGD_2 has been shown to be saturable while binding of [^{3}H]PGI_2, and [^{3}H]PGE_1 does not appear to saturate, perhaps owing to the large number of low-affinity sites and the limitations of the methods. The specificity of binding of [^{3}H]PG has been tested in two complementary ways. In the studies performed by Siegl *et al.*[1,2] the platelets were equilibrated with the labeled prostaglandin before the addition of unlabeled prostaglandins. This technique measures the ability of prostaglandins to displace the bound [^{3}H]PG. For [^{3}H]PGI_2 binding to the high-affinity site $PGI_2 >> PGE_1 >> PGF_{1\alpha}$, 6-keto-$PGF_{1\alpha} > PGD_2$, PGE_2, $PGF_{2\alpha}$.[1] For [^{3}H]-PGD_2 binding, $PGD_2 >> PGE_1$, PGI_2, PGE_2, TxB_2, $PGF_{2\alpha}$.[2] These results indicate that both the high-affinity PGI_2 binding site and the PGD_2 binding site are specific. In the studies performed by Schafer *et al.*[4] and Cooper and Ahern[3] the ability of prostaglandins to compete with [^{3}H]PGE_1 or [^{3}H]PGD_2 was measured and, except for the fact that PGI_2 was a more potent competitor for [^{3}H]PGE_1 binding than PGE_1, the results were essentially identical with those of Siegl *et al.*[1,2]

A summary of the dissociation constants and number of binding sites obtained by Scatchard analysis by a number of investigators may be seen in Table I. It should be noted that there is a discrepancy not only between the results obtained in different laboratories using the same system, but also from the same laboratory using different systems. As shown in Table II, the data obtained using intact platelets appear to be the most relevant to the receptors involved in activation of adenylate cyclase.

The final and most important criterion that should be met before concluding that a hormone binding site is a physiologically relevant receptor is that there be some correspondence between the K_D for binding and the EC_{50} for some early biochemical event. As seen in Table II, the binding sites for [^{3}H]PGI_2, [^{3}H]PGE_1, and [^{3}H]PGD_2 meet this criterion. There is a relatively good agreement between the K_D values for the [^{3}H]PGD_2 binding site and lower-affinity [^{3}H]PGI_2 and [^{3}H]PGE_1 binding sites and the

[11] O. Hechter, *Adv. Exp. Med. Biol.* **96,** 1 (1975).

TABLE I
SUMMARY OF [^{3}H]PG BINDING SITES ON HUMAN PLATELETS

[^{3}H]PG	System	K_D	Sites
[^{3}H]PGD_2	Intact platelets	4.1×10^{-7} *M*	126 fmol/10^8 platelets[a]
		5.4×10^{-8} *M*	35 fmol/10^8 platelets[b]
[^{3}H]PGI_2	Intact platelets	4.1×10^{-9} *M*	8.13 fmol/10^8 platelets[c]
		9.9×10^{-7} *M*	680 fmol/10^8 platelets
	Platelet lysate	2.1×10^{-8} *M*	160 fmol/mg protein[c]
		7.1×10^{-6} *M*	39,269 fmol/mg protein
		6.3×10^{-8} *M*	360 fmol/mg protein[d]
		3.2×10^{-7} *M*	2820 fmol/mg protein
[^{3}H]PGE_1	Intact platelets	9.5×10^{-9} *M*	2 fmol/10^8 platelets[c]
		1.6×10^{-6} *M*	282 fmol/10^8 platelets
	Platelet membranes	6.6×10^{-9} *M*	24 fmol/mg protein[e]
		1.9×10^{-6} *M*	450,000 fmol/mg protein

[a] Data taken from Siegl *et al.*[2]
[b] Data taken from Cooper and Ahern.[3]
[c] Data taken from Siegl.[8]
[d] Data taken from Schillinger and Prior.[6]
[e] Data taken from Schafer *et al.*[4]

EC_{50} values for stimulation of platelet adenyl cyclase reported by Tateson *et al.*[12] The nature of the higher-affinity binding sites for [^{3}H]PGI_2 and [^{3}H]PGE_1 is not known at this time.

In summary, the results indicate the presence of specific binding sites on human platelets for PGI_2, PGD_2, and PGE_1, which correspond to receptors linked to adenylate cyclase. Furthermore, the data strongly suggest that PGI_2 and PGE_1 interact with the same set of binding sites.

Assay for Binding of [^{3}H]PGI_2 and [^{3}H]PGD_2

Materials

[9-^{3}H (N)]PGI_2 (>97% pure, specific activity 12.6 Ci/mmol); [5,6,7,9,-12,14,15-^{3}H (N)]PGD_2 (>99% pure, specific activity 100 Ci/mmol); [5,6-^{3}H (N)]PGE_1 (>99% pure, specific activity 89.5 Ci/mmol); [1-^{14}C]$PGF_{1\alpha}$ (98% pure, specific activity 45 mCi/mmol); and [U-^{14}C]sucrose (>99% pure, specific activity between 1 and 5 mCi/mmol) were obtained from New England Nuclear Corp., Boston, Massachusetts.

Nonradioactive prostaglandins were kindly donated by Dr. John Pike of the Upjohn Company, Kalamazoo, Michigan. Most prostaglandins may now be obtained commercially through Upjohn Diagnostics (Kala-

[12] J. E. Tateson, S. Moncada, and J. R. Vane, *Prostaglandins* **13,** 389 (1977).

TABLE II
CORRELATION OF [^{3}H]PG BINDING AND INCREASES IN CYCLIC AMP

[^{3}H]PG	K_D intact platelets	EC_{50} increase of platelet cAMP[c]
[^{3}H]PGI$_2$	4.1×10^{-9} M[a]	—
	9.9×10^{-7} M[a]	1.8×10^{-7} M
[^{3}H]PGE$_1$	9.5×10^{-9} M[a]	—
	1.6×10^{-6} M[a]	1×10^{-6} M
[^{3}H]PGD$_2$	4.1×10^{-7} M[b]	5×10^{-7} M

[a] Data taken from Siegl.[8]
[b] Data taken from Siegl *et al.*[2]
[c] Data taken from Tateson *et al.*[12]

mazoo, Michigan) or Sigma Chemical Co. (St. Louis, Missouri). Nonradioactive PGI$_2$ was a gift from Dr. K. C. Nicolaou, Department of Chemistry, University of Pennsylvania, Philadelphia, Pennsylvania, but may be obtained through Sigma Chemical Co. (St. Louis, Missouri). Dibutyl phthalate was obtained from Eastman Kodak Company (Rochester, New York). Whatman GF/C filters (2.4 cm in diameter) were purchased from Fisher Scientific (Pittsburgh, Pennsylvania).

Solutions

Stock solutions of 0.1 M EDTA (pH 7.4), 3.8% trisodium citrate, and 0.1 M Na_2CO_3 (pH >9.5) were kept at 4°. Stock solutions of prostaglandins dissolved in absolute ethanol (with exception of PGI$_2$) were stored at −20°. PGI$_2$ was dissolved in 0.1 M Na_2CO_3 and kept at −20°. Under these conditions all prostaglandins were stable for 3 months if kept on ice after removal from freezer.

Tris-saline-glucose–EDTA (TSG–EDTA) buffer was composed of 134 mM sodium chloride, 5 mM D-glucose, and 1 mM EDTA buffered to pH 7.4 with 15 mM Tris-HCl. Lysing buffer for the preparation of platelet membranes was composed of 1.54 mM Tris-HCl, 6 mM EDTA, and distilled H_2O at a pH of 7.4. The incubation buffer for binding studies with platelet membranes consisted of 40 mM Tris-HCl, 10 mM $MgCl_2$, and 5 mM EDTA at pH 7.4.

Methods

Preparation of Human Platelets. Between 40 and 150 ml of blood were collected by venipuncture (19-gauge, thin-walled, butterfly needle) of healthy, drug-free donors into a siliconized or plastic syringe containing one-tenth volume of 3.8% trisodium citrate. The blood was then centrifuged at 200 *g* for 15 min at room temperature. This procedure separated the blood into three layers: a top layer of platelet-rich plasma, which was removed to a separate container via a siliconized pipette; a small middle layer rich in white cells; and a large bottom layer of red cells. Platelet density was determined using a Coulter counter, Coulter Electronic Corporation (Hialeah, Florida). The platelet-rich plasma so obtained had an average platelet density of 4×10^8 platelets/ml, less than 5% red blood cells, and less than 0.0005% white cells.

It is especially important when working with platelets to keep them in either plastic or siliconized glass containers, since contact with glass surfaces will activate even anticoagulated platelets and platelet aggregation will be the result.

For studies where larger volumes of platelets are required, fresh platelet concentrates anticoagulated with acid–citrate–dextrose (ACD) obtained from platelet pheresis donors may be used. Platelets obtained in this manner were always washed by the following procedure: platelet-rich plasma was chilled on ice before the addition of one-tenth volume 0.1 *M* EDTA (final concentration 10 m*M* EDTA). This was then centrifuged at 6000 *g* for 15 min at 4°.

The plasma was discarded, and the platelet pellet was resuspended in an equal volume of TSG–EDTA buffer. This platelet suspension was then centrifuged at 6000 *g* for 10 min at 4°. The preceding step was repeated twice. After the final washing the platelet pellet was either resuspended in TSG–EDTA to the desired platelet density for use in studies involving washed platelets or treated to produce platelet membranes.

Platelets free of plasma may also be obtained using albumin density centrifugation as described by Walsh *et al.*[13] or albumin density centrifugation followed by gel filtration using Sepharose 2B as described by Mustard *et al.*[14] Prostaglandin binding to platelets obtained through either of these procedures was essentially the same as that to platelets maintained in PRP; therefore PRP was used in the majority of these studies.

Binding Studies—Intact Platelets. In each of these experiments 0.2 μCi of ^{14}C-labeled sucrose or 0.05 μCi of [^{14}C]PGF$_{1\alpha}$ was added per milliliter as an internal marker for the extracellular space. [^{14}C]PGF$_{1\alpha}$ does not bind to platelets. Appropriate concentrations of [^{3}H]PGI$_2$, [^{3}H]PGE$_1$, or [^{3}H]PGD$_2$ containing unlabeled prostaglandin to dilute the preparation to a known specific activity were added to 1 ml of either

[13] P. N. Walsh, D. C. B. Mills, and J. G. White, *Br. J. Haematol.* **36,** 281 (1977).

[14] J. F. Mustard, D. W. Perrig, N. G. Ardhie, and M. A. Packham, *Br. J. Haematol.* **22,** 193 (1972).

platelet-rich plasma or washed platelets; unless otherwise noted the sample was incubated at room temperature for 5 min.

For measurement of concentration-dependent binding approximately 40,000 cpm of [^{3}H]PG diluted to the desired concentration was added per milliliter of plasma. The concentration ranges examined were 1.5–1120 n*M* for PGI_2, 30–2000 n*M* for PGD_2, and 2.2–504 n*M* for PGE_1. Time-dependent binding was measured from 0 to 15 min at room temperature. Displacement of bound radioactivity with a hundredfold excess of unlabeled prostaglandin was measured from 0 to 15 min also. Experiments in which the ability of various unlabeled prostaglandins to displace [^{3}H]PG was measured were performed as follows: The [^{3}H]PG was equilibrated with PRP for 4 min, then either buffer or unlabeled PG was added and the PRP was equilibrated an additional 4 min. Results were expressed as a percentage of the counts per minute bound to the control samples at the end of the 8-min incubation period. The range of concentrations of unlabeled prostaglandins used was 3×10^{-8} *M* to 3×10^{-5} *M*. Prostaglandins were added in either ethanol or 0.1 *M* Na_2CO_3. The maximum final concentration of ethanol used was 0.5% v/v. Incubations were terminated by centrifuging 1-ml samples in an Eppendorf microfuge (Brinkmann Instrument Co., Westbury, New York) at 15,000 *g* for 2 min. The supernatant was rapidly removed with a Pasteur pipette. The tube was inverted and allowed to drain for 5 min; the pellet was removed with a cotton swab. The pellets and samples of the supernatant were oxidized in a Packard 306 oxidizer (Packard Instrument Company, Downers Grove, Illinois). The ^3H_2O and $^{14}CO_2$ produced by this process were determined by liquid scintillation counting. Efficiency of combustion was 99%, and counting efficiencies were determined by combustion and liquid scintillation counting of known amounts of radioactive standards. Bound radioactivity per 10^8 platelets was determined after correction for background, extracellular space, and platelet count utilizing a tape-programmable SR-52 calculator (Texas Instruments Co., Houston, Texas) and the following equation:

$$[(A - B \times C)/D]/\text{platelet count} = \text{cpm } [^3\text{H}]\text{PG bound per } 10^8 \text{ platelets}$$

where A = (cpm [^{3}H]PG/pellet) − background, B = (cpm [^{14}C]sucrose or [^{14}C]$PGF_{1\alpha}$ − background, C = (cpm [^{3}H]PG/μl supernatant) − background, and D = (cpm [^{14}C]sucrose or [^{14}C]$PGF_{1\alpha}$/μl supernatant) − background.

No nonspecific binding was observed using this method. In some experiments bound [^{3}H]PG was expressed as femtomoles per 10^8 platelets using the known specific activity after correction for counting efficiency. Each of these experiments was performed in duplicate.

Similar experiments were performed in which the platelets were spun through a small (0.2 ml) layer of dibutyl phthalate to separate the platelets from plasma. This process was discontinued, however, when it was found that there was essentially no difference in either the amount of binding or the size of the extracellular space from samples prepared as described above. In addition it was found that the residue of oil left on the pellet interfered with the functioning of the oxidizer.

Preparation of Platelet Lysates. A platelet lysate preparation adapted from Tsai and Lefkowitz[15] proved to be satisfactory. Washed platelets were prepared as described above except that after the final washing the platelet pellet was resuspended in one-fifth the initial volume of lysing buffer and frozen in liquid nitrogen. This frozen suspension was allowed to thaw at room temperature and was then homogenized with 20 strokes of a Teflon serrated-tip tissue grinder (A. H. Thomas Co., Philadelphia, Pennsylvania). The homogenized suspension was then centrifuged at 39,000 *g* for 12 min at 4°. The resulting pellet was washed twice with lysing buffer. The final pellet was resuspended to a protein concentration between 1 and 2 mg/ml as measured by a Lowry assay[16] using bovine serum albumin standard. This was an uncharacterized crude membrane preparation.

Binding Studies—Platelet Membranes. Aliquots (200 μl) of platelet membrane suspensions prepared as described above were incubated with the appropriate concentrations of [^{3}H]PG for 2.5 min (for concentration ranges, see Binding Studies—Intact Platelets). At the end of that time the membrane suspensions were placed on Whatman GF/C filters in a Millipore 3025 Sampling Manifold (Millipore Corporation, Bedford, Massachusetts) and rapidly filtered under vacuum (at least 20 in. of Hg). The filters were rapidly rinsed with 5 ml of ice-cold incubation buffer. This did not produce any change in the amount of specific binding. Control filters were run under identical conditions without membranes to determine the amount of nonspecific binding. This was linear with the amount of radioactivity added. Specific binding was equal to the total binding minus the nonspecific binding and equaled approximately 50–60% of the total radioactivity on the filters.

After rinsing with 5 ml of ice-cold incubation buffer, the filters were air dried and placed in glass liquid scintillation vials to which 15 ml of ACS II were added. These vials were counted in a Packard Model 3003 Tri-Carb scintillation spectrometer (Packard Instrument Co., Downers

[15] B. S. Tsai and R. J. Lefkowitz, *Mol. Pharmacol.* **16,** 61 (1979).

[16] O. H. Lowry, N. J. Rosebrough, A. L. Farr, and R. J. Randall, *J. Biol. Chem.* **193,** 265 (1951).

Grove, Illinois). Counting efficiency for tritium under these conditions was 31%.

Results are expressed as femtomoles of [^{3}H]PG bound per milligram of protein utilizing the known specific activity of the [^{3}H]PG solutions.

Acknowledgments

A. M. S. was supported during this work by an Advanced Predoctoral Fellowship from the Pharmaceutical Manufacturers Association Foundation.

[30] Distribution of PGE and $PGF_{2\alpha}$ Receptor Proteins in the Intracellular Organelles of Bovine Corpora Lutea

By Ch. V. Rao and S. B. Mitra

Previous data [1-4] and beliefs have been that only the outer cell membranes of target tissues contain receptors for prostaglandin (PG) E and $F_{2\alpha}$. More recent data from our laboratory demonstrated, however, that a variety of intracellular organelles, as well as outer cell membranes of bovine corpora lutea, contain these receptors. This contribution outlines the methodology and approach that we used in obtaining these data.

Selection of Tissue

A large amount of luteal tissue is required for the isolation of adequate quantities of various highly purified subcellular organelles. This is due to severe and sometimes unaccountable organelle losses during these fractionation and purification procedures. Bovine corpora lutea meet this requirement because they are easily obtainable in abundance from many slaughterhouses (each corpus luteum weighs 4–7 g) at little or no cost. An added advantage with bovine corpora lutea is that those obtained throughout the entire gestation period (they are functional[5-7]) can be used. This reduces the time required to obtain the needed amount of tissue. The disadvantage of using bovine corpora lutea, however, is that

[1] M. Smigel and S. Fleischer, *Biochim. Biophys. Acta* **332,** 358 (1974).
[2] W. S. Powell, S. Hammarström, and B. Samuelsson, *Eur. J. Biochem.* **61,** 605 (1976).
[3] I. Grunnet and E. Bojesen, *Biochim. Biophys. Acta* **419,** 365 (1976).
[4] M. E. Carsten and J. D. Miller, *Arch. Biochem. Biophys.* **204,** 404 (1980).
[5] F. Stormshak and R. E. Erb, *J. Dairy Sci.* **44,** 310 (1961).
[6] R. C. Mills and M. C. Morrissette, *J. Reprod. Fertil.* **22,** 435 (1970).
[7] C. V. Rao, *Fertil. Steril.* **26,** 1185 (1975).

ISBN 0-12-181986-8

many of the physiological studies that one wishes to conduct simply cannot be undertaken because they are prohibitively expensive.

We have repeatedly experienced that subcellular organelles isolated from bovine corpora lutea collected during the early spring of every year bind very little or none of the added [^{3}H]PGs. The reasons for this finding or whether this problem occurs in other areas are unknown.

Buffers

Homogenizing buffer consisted of 10 m*M* Tris-HCl, pH 7.3, containing 250 m*M* sucrose and 1 m*M* Ca^{2+}. The buffer used for washing the isolated organelles contained the above ingredients without the sucrose. Heavy sucrose solutions were made in 10 m*M* Tris-HCl, pH 7.3, containing 1 m*M* Ca^{2+}; critical tables were used in making those based on density. The composition of homogenizing buffer was altered in an effort to establish optimal composition, i.e., increasing Tris-HCl concentration to 50 m*M* and addition of dithiothreitol and/or gelatin to final concentrations of 1 m*M* and 0.1%, respectively. None of these alterations worked very well because they either interfered with the clear-cut separation of various subcellular organelles or greatly reduced [^{3}H]PGs binding to the organelles, or both.

Isolation of Various Subcellular Organelles

Bovine corpora lutea used in the present studies were collected in a slaughterhouse located within a 10-min drive (5 miles) from our laboratory. The ovaries were removed approximately 10 min after the animals were killed. Corpora lutea were dissected free from remaining ovarian tissue, placed in homogenizing buffer at 4°, and brought to the laboratory once a sufficient amount of tissue was collected. Typically this amounted to about 150 g of luteal tissue.

Corpora lutea were minced and homogenized in approximately 30 ml of homogenizing buffer per corpus luteum, at 4° with a Polytron homogenizer (PCU-110) using three 10-sec bursts at a setting of 6 interspersed with a resting period of about 1 min to avoid generating any local heat. The homogenizations were usually completed within 4–5 hr after the ovaries were removed from the animals. However, on several occasions, corpora lutea had to be frozen overnight at −20° and homogenization accomplished the next morning because of time constraints. When homogenization by Polytron was compared with that done by a motor-driven glass–Teflon homogenizer, plasma membranes yield and their gonadotropin binding capacity were similar. The use of a Polytron, however, saved a lot of time.

The homogenates were filtered through four layers of surgical gauze to remove debris. The filtrates were then subjected to differential and various sucrose density (sedimentation or flotation) gradient centrifugations to obtain nuclear membranes,[8] plasma membranes,[9] mitochondria,[10,11] lysosomes,[10,11] rough endoplasmic reticulum,[12] and heavy, medium, and light Golgi.[12] A flow chart (Fig. 1) detailing the isolation scheme is provided to supplement the above information and for the reader's convenience. All the centrifugations and procedures were conducted at 4°.

Procedures recommended in the literature [i.e., administration of detergent to animals prior to removal of tissues (which was expected to aid complete separation of lysosomes from mitochondria[11]), acute intoxication of animals with ethanol prior to sacrifice (which was expected to aid the separation of Golgi[12]), and addition of potassium citrate (10%) in the isolation of nuclear membranes[8]] were deliberately omitted. The reasons for these omissions were that detergent and ethanol administration is impractical with cows and would not have influenced lysosomes and Golgi isolation from bovine corpora lutea, and potassium citrate addition resulted in a drastic reduction of $[^3H]PGF_{2\alpha}$ specific binding to nuclear membranes.

After fractionation was finished, all the organelles were washed once with washing buffer, resuspended in homogenizing buffer (resuspension volume varied with the organelle), and stored in 1-ml aliquots at −20°. The protein content in an aliquot of the organelles was determined by the Folin-phenol reagent method of Lowry *et al.*[13] using bovine serum albumin as the standard. If the optical density reading of any sample is higher than the standard curve range (10–100 μg), dilution of the sample after color development to give a lower optical density should not be done since it underestimates protein by about 30%.

Table I shows that recoveries of various organelles with respect to homogenate protein are very meager. It should be mentioned in this context that the recoveries of crude pellets (for example 400, 9000, and 100,000 *g* pellets) or relatively crude organelles isolated therefrom, are usually good and add up to approximately 100%. It is only when attempts are made to achieve a high degree of purity of organelles that the recoveries become poor and unaccountable. This is common in all studies of this type and is due to losses that occur during numerous transfers, handlings, etc., in the procedures.

[8] D. M. Kashing and C. B. Kasper, *J. Biol. Chem.* **244,** 3786 (1969).

[9] D. Gospodarowicz, *J. Biol. Chem.* **248,** 5050 (1973).

[10] A. Trouet, *Arch. Int. Physiol. Biochim.* **72,** 698 (1964).

[11] F. Leighton, B. Poole, H. Beufay, P. Baudhin, J. W. Caffey, S. Fowler, and C. de Duve, *J. Cell. Biol.* **37,** 482 (1968).

[12] J. H. Ehrenreich, J. J. M. Bergeron, P. Siekevitz, and G. E. Palade, *J. Cell. Biol.* **59,** 45 (1973).

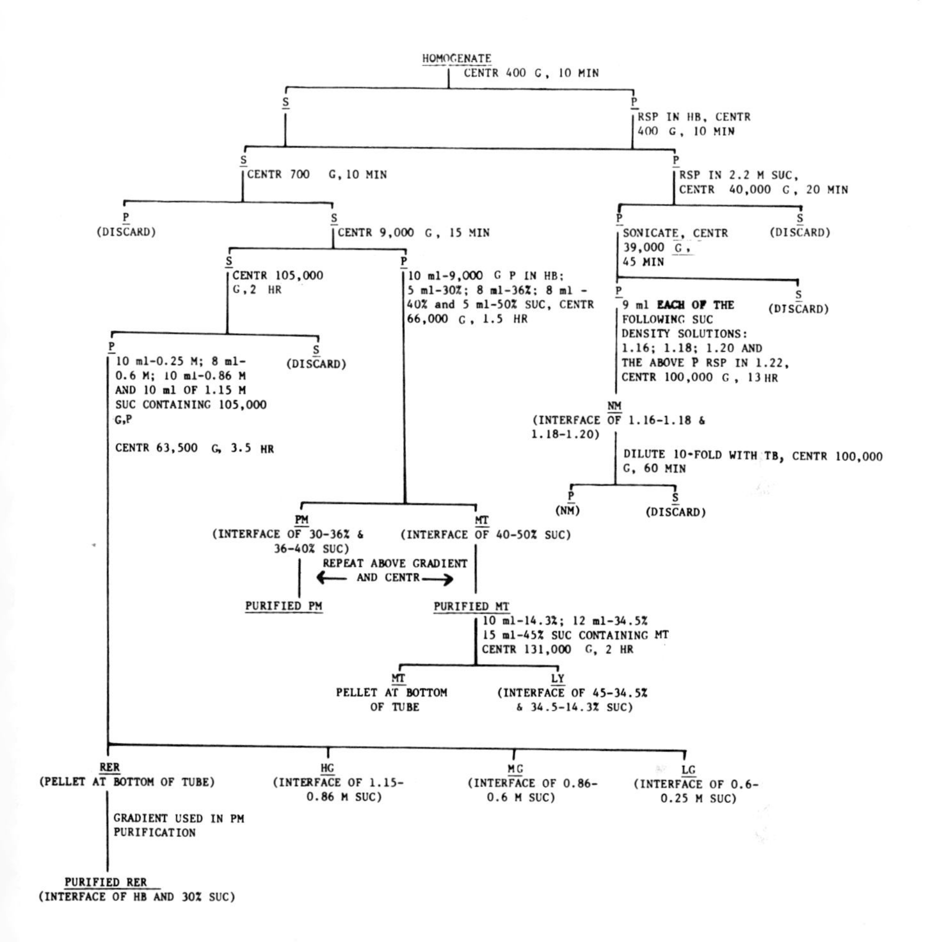

FIG. 1. Flow chart for the preparation of various subcellular organelles from bovine corpora lutea. The following abbreviations are used: CENTR, centrifuge; P, pellet; S, supernatant; SUC, sucrose; RSP, resuspend; HB, homogenizing buffer; TB, 10 m*M* Tris-HCl buffer, pH 7.3, containing 1 m*M* Ca^{2+}; NM, nuclear membranes; MT, mitochondria; LY, lysosomes; PM, plsma membranes; RER, rough endoplasmic reticulum; HG, heavy Golgi; MG, medium Golgi; LG, light Golgi. Sonication was conducted at 4° after adjusting the protein concentration to approximately 1.2 mg/ml, using a Sonifer cell disrupter (Model W40, Heat Systems Ultrasonic, Inc.) for 10 sec at a setting of 6.5. Sonication had no effect on plasma membranes binding of $[^3H]PGE_1$ and $[^3H]PGF_{2\alpha}$.

[13] O. H. Lowry, N. J. Rosebrough, A. L. Farr, and R. J. Randall, *J. Biol. Chem.* **193,** 265 (1951).

TABLE I

RECOVERIES OF VARIOUS SUBCELLULAR ORGANELLES, 5′-NUCLEOTIDASE ACTIVITY, AND [^{3}H]PGs SPECIFIC BINDING[a]

Organelle	Percent recovery with respect to homogenate protein	5′-Nucleotidase (nmol P_i released/min/mg of protein)	[^{3}H]PGE_1 (fmol bound/mg of organelle protein)	[^{3}H]$PGF_{2\alpha}$ (fmol bound/mg of organelle protein)
Nuclear membranes	5.0 ± 1.4	47.5 ± 4.8	34.3 ± 3.8	16.8 ± 1.8
Plasma membranes	1.5 ± 0.3	624.8 ± 16.8	264.0 ± 16.6	184.5 ± 13.1
Lysosomes	1.4 ± 0.0	26.4 ± 3.2	67.7 ± 6.2	98.0 ± 5.2
Rough endoplasmic reticulum	0.3 ± 0.2	ND[b]	38.8 ± 4.9	26.9 ± 5.2
Heavy Golgi	0.393 ± 0.055	41.0 ± 4.0	257.4 ± 20.5	104.8 ± 3.1
Medium Golgi	0.024 ± 0.000	ND	169.8 ± 18.7	46.6 ± 3.1
Light Golgi	0.006 ± 0.000	ND	114.6 ± 9.3	ND

[a] Mitochondria free from lysosomes do not exhibit intrinsic PG binding. Although lower than nuclear membranes, chromatin does appear to contain binding sites for PGs. Smooth endoplasmic reticulum also seems to contain these binding sites. There is no good evidence to date whether or not cytosol of bovine corpora lutea contain PG binding sites.

[b] ND, not detectable.

Subcellular fractionation of bovine corpora lutea slices incubated at 4°, 22°, or 38°, for 1 min to 4 hr resulted in different separation patterns and apparent yields of various subcellular organelles.[14] This implies, therefore, that one needs to reevaluate purity and identity of organelles by marker enzymes, when fractionation is attempted on incubated tissues.

Four to five full days were required for complete fractionation and isolation of all the subcellular organelles. Two Sorvall RC-2B refrigerated centrifuges with SS-34 rotors and one Beckman L5-65 preparative ultracentrifuge with SW-27 rotor and new or used (no more than 4 times and thoroughly washed after each use) nitrocellulose tubes were used during this period. Various steps were planned so that the centrifuges were in constant use during the day and sometimes at night. The time required for fractionation can be somewhat shortened by the simultaneous use of two ultracentrifuges.

The procedures used in these studies for isolation and purification of various subcellular organelles were essentially the same (with some modifications) as those described by others.[8-12] Thus, we have reproduced many of these procedures in our laboratory over the last 4 years with the exception of lysosomes. After our initial success in isolating intact lysosomes,[15] we have never been able to repeat it, even after multiple changes in the procedure, and always ended up with lysed lysosomes.

In our experience, achieving a high degree of purity of various organelles is not only a function of good centrifugation strategy, but also critically depends on the care and handling during the procedures. To illustrate this point, one should remove organelles from the top sucrose interfaces first, not attempt to remove them to the extent that would overlap with other interfaces, and use fresh syringes and needles (which we used because it gave us better control) to remove different organelles. Purity will be severely compromised despite an excellent centrifugation strategy if sufficient attention is not paid to the above and other seemingly minor handling details.

Marker Enzymes Assayed and Their Use in Evaluation of Purity of Subcellular Organelles

The use of marker enzymes to assess the purity of subcellular organelles should be more sensitive than morphological assessment when they remain active during the isolation procedures.

A prerequisite for using marker enzymes for assessment of purity is that the selected enzyme(s) should be found in only one organelle. In real-

[14] S. Mitra, C. V. Rao, M. T. Lin, and F. R. Carman, Jr., *Biochim. Biophys. Acta* (submitted for publication).

[15] S. Mitra and C. V. Rao, *Arch. Biochem. Biophys.* **185,** 126 (1978).

TABLE II
LIST OF MARKER ENZYMES ASSAYED

Marker enzyme	Organelle for which enzyme is used as a marker
5′-Nucleotidase[a]	Plasma membranes
NAD pyrophosphorylase[b]	Nuclei
Cytochrome *c* oxidase[c]	Mitochondria
Acid phosphatase[d]	Lysosomes
NADH cytochrome *c* reductase[e]	Rough endoplasmic reticulum, nuclear membranes
Galactosyltransferase[f]	Light, medium, and heavy Golgi
Glucose-6-phosphate dehydrogenase[g]	Cytosol

[a] P. Emmelot and C. J. Bos, *Biochim. Biophys. Acta* **120,** 369 (1966).
[b] A. Kornberg, *J. Biol. Chem.* **182,** 779 (1950).
[c] S. J. Cooperstein and A. Lazarow, *J. Biol. Chem.* **189,** 665 (1951).
[d] P. R. N. Kind and E. J. King, *J. Clin. Pathol.* **7,** 322 (1954).
[e] H. R. Mahler, this series, Vol. 2, p. 688.
[f] M. Treloar, J. M. Sturgess, and M. A. Moscarello, *J. Biol. Chem.* **249,** 6628 (1974). Although this enzyme activity was higher in Golgi, it was found in all the other organelles of bovine corpora lutea; it does not appear to be due to Golgi contamination.
[g] T. J. Kelly, E. D. Nielson, R. B. Johnson, and G. S. Vestling, *J. Biol. Chem.* **212,** 545 (1955).

ity, this prerequisite may not be strictly met for all the enzymes, primarily because it is difficult to resolve unequivocally whether small amounts of enzymes found in inappropriate organelles are a reflection of low-level contamination or intrinsic.

Except for galactosyltransferase, all the enzymes used reasonably met the prerequisite for their use as markers (see Table II). We do not know what enzyme could be used as a marker for Golgi of bovine corpora lutea.

The nuclear membrane has no unique enzymic composition, as it is a morphological continuum of rough endoplasmic reticulum. Therefore the relative enrichment of rough endoplasmic reticulum marker in nuclear membranes with respect to nuclei, dramatic decrease of a marker (enzyme that is not present in nuclear membranes, for example, NAD pyrophosphorylase) in nuclear membranes as compared to nuclei and other marker enzymes data, were the only means we had to evaluate the purity of nuclear membranes.

Other enzymes measured were: succinic dehydrogenase (specific for mitochondria); lactate dehydrogenase (for cytosol); β-glucuronidase, alkaline phosphatase, and *N*-acetyl-β-D-glucosaminidase (the latter was more reliable for lysosomes). Some enzymes found to be unsatisfactory were Mg^{2+} ATPase (which was not specific for the plasma membrane) and glucose-6-phosphatase (not detected in the endoplasmic reticulum). The following changes and/or modifications were made in the enzyme assays:

25 μM α,β-methylene ADP, a specific inhibitor of 5′-nucleotidase,[16,17] was used to determine P_i release from 5′-AMP by nonspecific membrane phosphatases. This release was found to be minimal. β-Glycerophosphate, instead of phenyl phosphate, was used as the substrate in acid phosphatase assay since this enzyme of lysosomal origin is specific for this substrate. ^{3}H-labeled instead of ^{14}C-labeled UDP galactose (38 pmol) as the donor and ovomucoid instead of *N*-acetylglucosamine as the acceptor (ovomucoid worked better than even microsomal protein; phosphotungstic acid precipitation of ovomucoid after the reaction) were used in the galactosyltransferase assay. Triton X-100 was present in the acid phosphatase (0.1%) and galactosyltransferase (0.4%) assays. Although various enzyme assay components were added in buffers recommended in the original reference sources, organelles were added in our homogenizing buffer. In some cases, time and volume of incubation, concentration of assay components, and amount of organelle protein used were also at variance with the reference sources. All these changes, however, were validated with respect to linearity of enzyme reactions. Appropriate blanks were run in all the enzyme assays, and corrections were made for these values. Various enzyme specific activities were calculated from linear rate data obtained from two different organelle protein concentrations (data for only 5′-nucleotidase are presented here, and our original papers contain the data for other enzymes).

Determination of [^{3}H]PGE$_1$ and [^{3}H]PGF$_{2\alpha}$ Specific Binding

[^{3}H]PGs purchased from commercial sources were routinely checked for radiochemical purity on thin-layer sheets of silica gel without gypsum (5 × 20 cm; Brinkmann Instrument Co.) and purified, if there were more than 5% contaminants, by thin-layer chromatography on silica gel G plates (Analtech) using a solvent system of benzene–*p*-dioxane–acetic acid (20 : 10 : 1, v/v) without liners in the tanks. We consider this to be a very important step regardless of assurances of purity by the supplier.

[^{3}H]PGs were diluted in redistilled ethanol, and 100-μl aliquots (final concentration during incubation was 10^{-9} M, which is equivalent to the K_d of binding to plasma membranes) were pipetted into 12 × 75 mm disposable glass tubes. The tubes for nonspecific binding received 1 μg of corresponding unlabeled PGs in 10 μl of ethanol. The ethanol was dried under a stream of nitrogen, and then 0.1 ml of organelles containing 20–100 μg of protein) in homogenizing buffer were added to the tubes, gently

[16] R. M. Burger and J. M. Lowenstein, *J. Biol. Chem.* **245,** 6274 (1970).
[17] M. K. Gentry and R. A. Olson, *Anal. Biochem.* **64,** 624 (1975).

vortexed, and incubated for 1 hr at 38° ([^{3}H]PGE$_1$) or for 2 hr at 22° ([^{3}H]PGF$_{2\alpha}$). These incubations were done on all the organelles of the same batch at the same time.

After incubation, 1.0 ml of 10 m*M* Tris-HCl, pH 7.0, at 4° was added to each tube, and the contents were poured onto Metricel filters (0.20–0.45 μm pore size) positioned on multiplace Millipore manifolds. The tubes were rinsed once with 1.0 ml of the Tris buffer, and finally the filters were washed with 10 ml of the same buffer at 4°. The time taken for the entire separation procedure varied depending on the number of tubes to be filtered (usually 10 min for 30 tubes). The filters were removed, cut into halves, and placed in a vial containing 10 ml of scintillation fluid (toluene–25 X Permaflour–Triton X-100 (18.3:1:6, v/v/v). The vials were capped and agitated several times over a 30-min period until the filters became transparent. The vials were then vigorously vortexed and counted in a liquid scintillation counter with an efficiency of 25% for ^{3}H.

The counts per minute in nonspecific binding tubes were subtracted from those in the total binding tubes to obtain specific binding. The differences in magnitude of specific binding among different subcellular organelles was primarily due to variations in the total binding. The femtomoles of [^{3}H]PGs bound per milligram of organelle protein were calculated using the molecular weight of 354 for PGs.

Assessment of Whether Binding of [^{3}H]PGs was Intrinsic and/or Due to Contamination

The binding data in Table I show that all the intracellular organelles specifically bound both [^{3}H]PGE$_1$ and [^{3}H]PG$_{2\alpha}$ (except light Golgi, which did not bind [^{3}H]PGF$_{2\alpha}$). Since plasma membranes were thought to be the primary or exclusive site of these receptors, the question arose whether intracellular organelle [^{3}H]PGs binding could be attributable to plasma membrane contamination. To answer this question, we measured 5′nucleotidase activity in the same subcellular organelles that were used for specific binding of [^{3}H]PGs. Table I shows that rough endoplasmic reticulum and medium and light Golgi had no detectable 5′-nucleotidase activity, yet they all bound [^{3}H]PGs. Nuclear membranes, lysosomes, and heavy Golgi exhibited disproportionately high binding of [^{3}H]PGs as compared to their 5′-nucleotidase activity (ratios of [^{3}H]PGs bound:5′-nucleotidase activity were higher than for plasma membranes). This indicates that these intracellular organelles contain more binding sites for PGs than can possibly be explained by plasma membrane contamination. In this context, the possibility of plasma membrane contamination that is not faithfully reflected by 5′nucleotidase activity should be considered. Indeed we feel, on the basis of several logical grounds and observations, that it is very unlikely.[15,18,19]

The above approach was also used (using appropriate marker enzymes for appropriate organelles) to evaluate the possibility of cross contamination (except for Golgi because of the lack of a specific marker) among different subcellular organelles, explaining the binding observed in the intracellular organelles. But such a possibility included for Golgi also seems very unlikely.

Using repeated freezing and thawing (5–20 times using a Dry Ice–acetone bath) and Alamethacin (10 μg/100 μg of organelle protein), an antibiotic ionophore that permeabilizes biological membranes,[20] we have been able to demonstrate latency of binding of [^{3}H]PGs in the lysosomes and Golgi, respectively, but not in plasma membranes treated similarly. The use of 0.1% Triton X-100 to reveal latency did not help, perhaps owing to solubilization of PG receptor sites. The losses of [^{3}H]PG binding from nuclear and plasma membranes following washing with 10 m*M* Tris-HCl, pH 7.0, containing 0.1% Triton X-100 were different. The properties of $PGF_{2\alpha}$ binding to some of the intracellular organelles differed from those of plasma membranes as well as from other intracellular organelles.[23] All of the above events should not have been observed if simple contamination with plasma membranes (or cross contamination) were responsible for intracellular organelles binding of [^{3}H]PGs.

Applicability of Our Approach to Other Systems

The methodology and approach used in the present studies should be applicable to other tissues and receptor systems as long as the amount of tissue available is adequate. Fractionation procedures have previously been described for small amounts of tissue.[21,22] However, these procedures do not generally yield all the subcellular organelles, separation of various organelles is not complete, and organelles purity is somewhat compromised.

Caution should be exercised in the use of crude pellets (for example 400, 9000, 100,000 *g*) for assessing the subcellular distribution of receptors. This is because purified organelles isolated from such crude pellets can give qualitatively as well as quantitatively different results from those obtained by using just the crude pellets.

[18] S. Mitra and C. V. Rao, *Arch. Biochem. Biophys.* **191,** 331 (1978).

[19] C. V. Rao and S. Mitra, *Biochim. Biophys. Acta* **584,** 454 (1979).

[20] G. Roy, *J. Membr. Biol.* **24,** 71 (1975).

[21] T. J. Peters and C. A. Seymour, *Biochem. J.* **174,** 435 (1978).

[22] C. V. Rao, S. Mitra, J. Sanfilippo, and F. R. Carman, Jr., *Am. J. Obstet. Gynecol.* **139,** 655 (1981).

[23] C. V. Rao, S. Mitra, and F. R. Carman, Jr., *Biol. Reprod.* **20,** Suppl. IIIA (1979).

There are few previous studies in which the subcellular distribution of PG receptors has been investigated.[1-4] In these studies, however, neither the isolation of all the subcellular organelles nor the critical assessment of the intrinsic nature of intracellular organelles binding was accomplished.

Acknowledgments

The excellent help of Mr. F. R. Carman, Jr. in binding studies is gratefully acknowledged. This work was supported by Grants HD09577 and HD15177 from the National Institutes of Health.

[31] A Receptor for Prostaglandin $F_{2\alpha}$ from Corpora Lutea

By SVEN HAMMARSTRÖM

Prostaglandin $F_{2\alpha}$ is a luteolytic hormone in a number of subprimate mammals.[1] It is produced by the uterus at the end of the estrous cycle and is transported via a venoarterial shunt mechanism to the corpus luteum of the ovary.[2] Interaction of $PGF_{2\alpha}$ with a receptor in the corpus luteum[3,3a] leads to a decrease in luteotropic hormone-stimulated cyclic AMP levels in the ovary. This reduces progesterone secretion, causes corpus luteum regression, and initiates a new estrous cycle.[4]

Determination of $PGF_{2\alpha}$ Receptors

Preparation of Tritium-Labeled $PGF_{2\alpha}$.[5] Prostaglandin E_2 (6 mg, dissolved in 0.5 ml of methanol and kept at 0°) is mixed with NaB^3H_4 (1 mg, ca 8 Ci/mmol, dissolved in 0.5 ml of methanol at 0°) and left at 0° for 15 min and at 22° for 60 min. The mixture is then chilled to 0°, and 5 mg of unlabeled $NaBH_4$ are added and allowed to react under the same conditions. Two volumes of water are added plus HCl to give a pH of 2–3. The products are extracted with diethyl ether and separated by reversed-phase partition chromatography on 4.5 g of Hyflo Supercel coated with 30 ml of chloroform–isooctanol, 1:1 (v/v) equilibrated with 300 ml of mobile phase (methanol–water, 57:93, v/v). The elution volumes for

[1] E. W. Horton and N. L. Poyser, *Physiol. Rev.* **56,** 595–651 (1976).
[2] J. R. Goding, J. A. McCracken and D. T. Baird, *J. Endocrinol.* **39,** 37 (1967).
[3] W. S. Powell, S. Hammarström, and B. Samuelsson, *Eur. J. Biochem.* **41,** 103 (1974).
[3a] W. S. Powell, S. Hammarström, and B. Samuelsson, *Eur. J. Biochem.* **56,** 73 (1975).
[4] J. P. Thomas, L. J. Dorflinger, and H. R. Behrman, *Proc. Natl. Acad. Sci. U.S.A.* **75,** 1344 (1978).
[5] E. Granström and B. Samuelsson, *Eur. J. Biochem.* **10,** 411 (1969).

ISBN 0-12-181986-8

[9α-^{3}H]PGF$_{2\beta}$ and [9β-^{3}H]PGF$_{2\alpha}$ are 78–100 and 100–150 ml, respectively.[5] Alternatively, reversed-phase high-performance liquid chromatography can be used to separate the diastereoisomeric products.[6] The purity of the products can be determined by thin-layer chromatography (silica gel G; diethyl ether–methanol, 97:3, v/v; R_f for PGF$_{2\alpha}$ methyl ester = 0.25) or by gas–liquid radiochromatography (*C* value on SE-30 for PGF$_{2\alpha}$ methyl ester, tris-*O*-trimethylsilyl derivative is 24.5). To determine the specific radioactivity, a known amount (e.g., 0.5 μCi) of the labeled product is mixed with a known amount of deuterium-labeled PGF$_{2\alpha}$ (e.g., 3 nmol of [$^2H_4$]PGF$_{2\alpha}$). The degree of isotope dilution caused by the addition of radioactive product is determined by multiple ion analysis of the methyl ester, tri-*O*-acetyl derivative.[7]

Membrane-Bound Receptor Preparations. Corpora lutea from sheep or bovine corpora lutea, estimated to be 6–16 days in the estrous cycle, are dissected free from ovarian stromal tissue while still frozen. The luteal tissue is thawed in 4 volumes of 0.01 *M* Tris-HCl buffer, pH 7.5 (4°) and homogenized (20 strokes) using a Potter–Elvehjem-type homogenizer. After centrifugation (1000 *g*, 20 min, 4°) the supernatant is passed through four layers of gauze bandage and recentrifuged (35,000 *g*, 45 min, 4°). The sediment is resuspended in Tris-HCl buffer (1.25 ml per gram of corpus luteum) and used for binding experiments.

Binding Assays. Tritium-labeled PGF$_{2\alpha}$ is dissolved in Tris-HCl buffer, pH 7.5 at concentrations between 0.2 and 2 μg/ml. Unlabeled PGF$_{2\alpha}$ is dissolved in ethanol at a concentration of 1.5 mg/ml. These solutions can be stored at −20° for at least a month. Generally, four or five different concentrations of [^{3}H]PGF$_{2\alpha}$, in the range indicated above, are prepared. Separation of macromolecule-bound and free radioactivity is performed on 2-ml columns of Sephadex G-50, fine. The gel is swollen in 0.01 *M* Tris-HCl buffer, pH 7.5, and packed in 2-ml disposable plastic syringes fitted with filter paper at the bottom. Filter papers are also placed on top of the columns, which are washed with a few milliliters of buffer prior to use and kept at 4°. The columns can be prepared 8–12 hr in advance, provided they are washed shortly before use.

Resuspended receptor preparation (0.2 ml), [^{3}H]PGF$_{2\alpha}$ (10 μl) and ethanol (3.5 μl) or unlabeled PGF$_{2\alpha}$ (3.5 μl) are incubated with gentle shaking at 23° for 120 min. The incubation mixtures are put on ice, diluted with 0.2 ml of cold buffer, and immediately applied to Sephadex columns. As soon as the samples have entered the gel, 0.01 *M* Tris-HCl buffer, pH 7.5 (0.85 ml) is added. The combined eluate (0.95 ml; "macromolecule

[6] Column: C_{18} Nucleosil (250 × 4.6 mm); solvent: methanol–water, 6:4 (v/v) plus 0.01% acetic acid (1 ml/min); retention times: PGF$_{2\beta}$, 25 min; PGF$_{2\alpha}$, 30 min.

[7] K. Gréen, E. Granström, B. Samuelsson, and U. Axén, *Anal. Biochem.* **54,** 434 (1973).

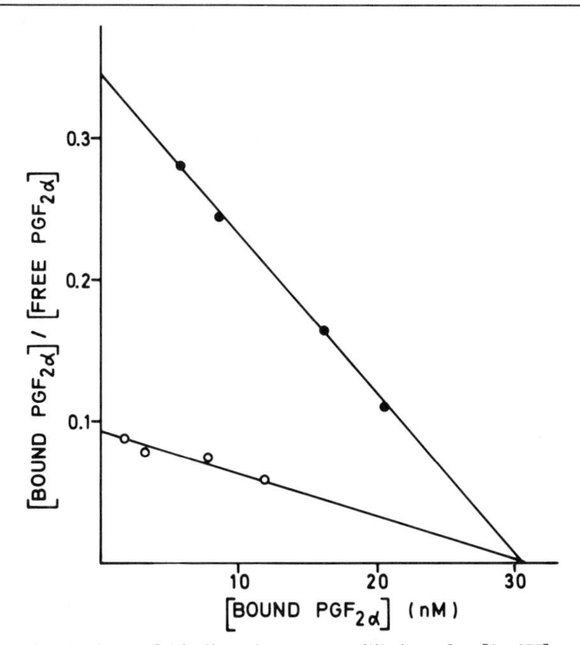

FIG. 1. Scatchard plots of binding data at equilibrium for [9β-^{3}H]prostaglandin $F_{2\alpha}$ (●—●) (2, 4, 10, and 20 ng) and [9β-^{3}H]prostaglandin $F_{2\alpha}$ (2, 4, 10, and 20 ng) plus prostaglandin E_2 (500 ng) (○—○). Volume, 0.2 ml. From Powell *et al.*[3]

bound $PGF_{2\alpha}$") is collected in a scintillation vial. The column is then eluted with an additional 3.5 ml of buffer, and the second eluate is collected in the same way as the first. The radioactivity in the two fractions (bound and free $PGF_{2\alpha}$, respectively) is determined by liquid scintillation counting after the addition of 10 ml of Instagel to each scintillation vial. Binding assays are done in duplicates to quadruplicates for total binding (no unlabeled $PGF_{2\alpha}$ added) and in duplicates for nonspecific binding (presence of unlabeled $PGF_{2\alpha}$). Data on specific binding (total minus nonspecific binding) are plotted according to Scatchard[8] for determination of receptor concentration and dissociation constant (Fig. 1).

Properties of $PGF_{2\alpha}$ Receptors

Affinity and Specificity. Binding of $PGF_{2\alpha}$ to membrane preparations from corpora lutea is saturable and reversible, and no metabolic alteration of the ligand takes place under the assay conditions described above. Fatty acids, steroids, nucleotides, luteotropic and follicle-stimulating hor-

[8] G. Scatchard, *Ann. N.Y. Acad. Sci.* **51,** 660 (1949).

mone do not interfere with the binding reaction.[3,3a] Dissociation constants determined by Scatchard plot analyses of binding data are of the order of 50 nM for bovine and 100 nM for ovine preparations. A somewhat lower value (28 nM) is obtained from the rate constants for association and dissociation.[3a]

The affinities of other prostaglandins and prostaglandin analogs can be determined using unlabeled compounds. Binding assays for Scatchard plot analyses are performed as described above. In one set of experiments, various concentrations of tritium-labeled $PGF_{2\alpha}$ are incubated in the presence or the absence of unlabeled $PGF_{2\alpha}$ in the usual way. In other sets of experiments a fixed concentration of the unlabeled prostaglandin or prostaglandin analog is added in addition to [^{3}H]$PGF_{2\alpha}$ (plus or minus

TABLE I
EFFECTS OF STRUCTURAL MODIFICATIONS OF PROSTAGLANDIN $F_{2\alpha}$ ON THE DISSOCIATION CONSTANT FOR BINDING TO BOVINE CORPUS LUTEUM RECEPTOR

Structural modification	Increase in K_d (K_d analog/K_d $PGF_{2\alpha}$)
CO_2H → CH_2OH	108
CH_2CO_2H → CO_2H	52
5,6-CH=CH → 5,6-CH_2CH_2	40
9α-OH → 9β-OH[a]	146
→ 9-oxo	54
11α-OH → 11-oxo	9.2
13,14-CH=CH → 13,14-CH_2CH_2	4
→ 13,14-C≡C	4.2
15-OH → 15-oxo	180
→ 15-epi-OH[b]	12
→ 15-OAc	9.2
→ 15-OMe	1.2
15-H → 15-Me	1.3
15-Pentyl → 15-(1,1-dimethylpentyl)	1.7
16-Butyl → 16-(*m*-chlorophenoxy)	0.40
→ 16-(1-butenyl)	3.4
17-Propyl → 17-phenyl	0.54
20-H → 20-ethyl	2.4
Inversion of all asymmetric centers (carbons 8, 9, 11, 12, and 15)[c]	6200
Inversion of all asymmetric centers except carbon-15[c]	30

[a] Ovine corpora lutea.
[b] The reference compound is 15-methyl-PGE_2 methyl ester.
[c] The reference compound is 13,14-didehydro-$PGF_{2\alpha}$.

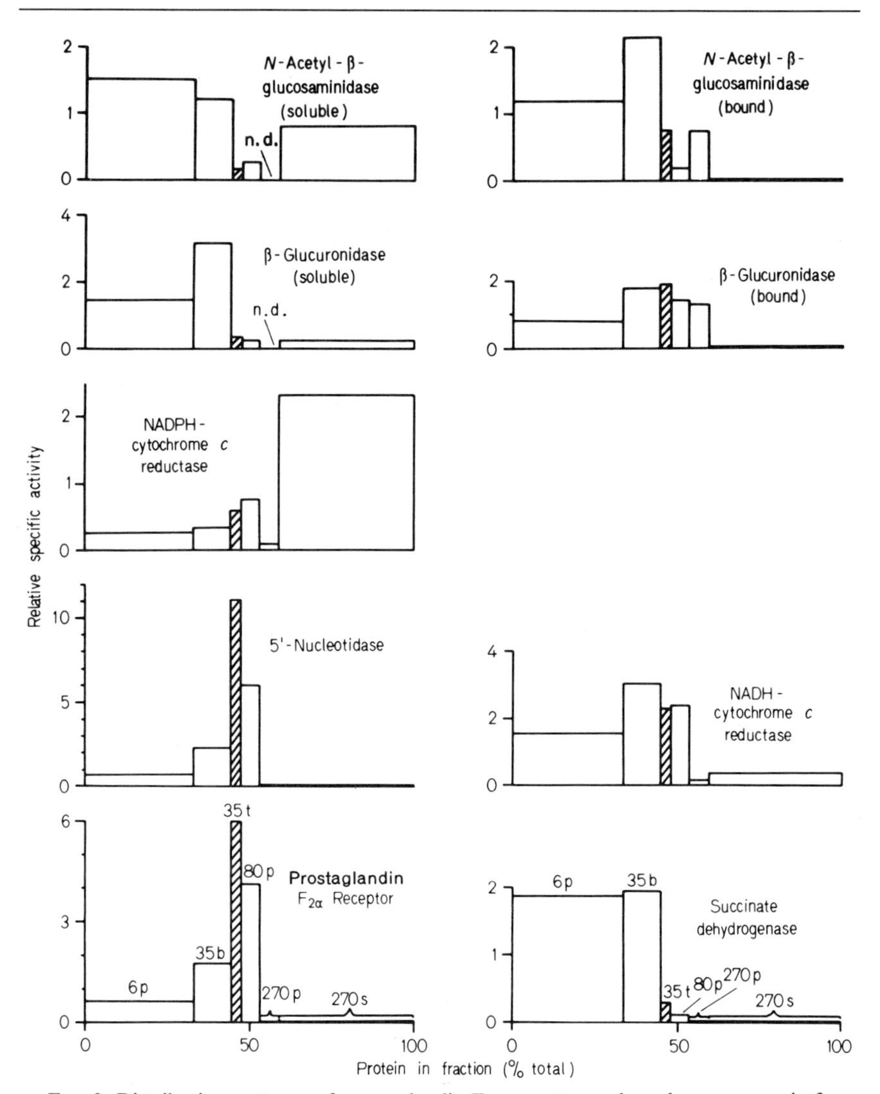

FIG. 2. Distribution patterns of prostaglandin $F_{2\alpha}$ receptor and marker enzymes in fractions obtained from homogenates of bovine corpora lutea by differential centrifugation. Designations: 6p, 80p, 270p indicate 6,000, 80,000, and 270,000 *g* pellets; 35b, 35t indicate bottom and top layers of 35,000 *g* pellet; 270s, 270,000 *g* supernatant. From Powell *et al.*[9]

unlabeled $PGF_{2\alpha}$). The unlabeled compound under investigation causes a decrease in the apparent dissociation constant, determined by Scatchard plots (Fig. 1). The dissociation constant for the unlabeled compound is calculated from the apparent dissociation constant, the dissociation constant for $PGF_{2\alpha}$, and the concentration of the unlabeled compound. Table

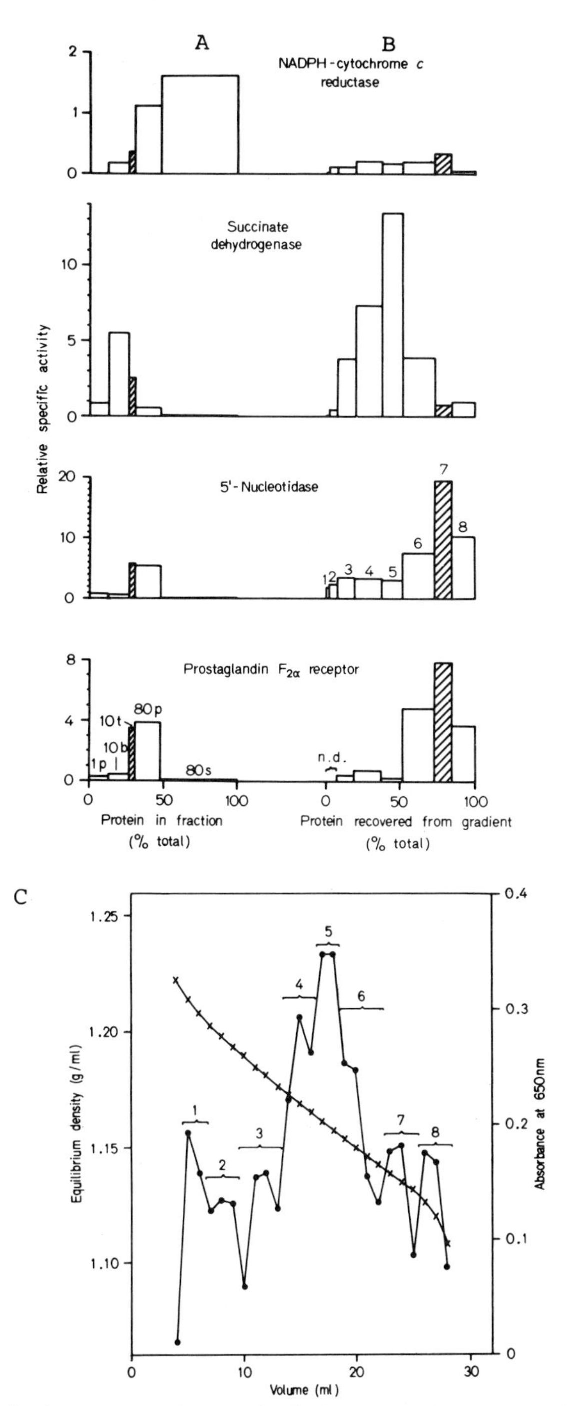

FIG. 3. Distribution patterns of prostaglandin $F_{2\alpha}$ and marker enzymes in fractions obtained from homogenates of bovine corpora lutea by differential (A) and density gradient (B) centrifugations. (C) Density gradient fractionation of the top layer of the 10,000 *g* pellet (10t) obtained by differential centrifugation; (B) shows the analytical results using fractions 1–8 from the gradient. Designations as in Fig. 2. From Powell *et al.*[9]

TABLE II
PHYSIOCHEMICAL PROPERTIES OF TRITON X-100-SOLUBILIZED PROSTAGLANDIN $F_{2\alpha}$ RECEPTOR[a]

R_s (Å)	$s_{20,w} \times 10^{13}$ (cm/sec · dyne)	ν (cm³/g)	% Triton (w/w)	$MW_{complex}$ (g/mol)	$MW_{receptor}$ (g/mol)	Triton bound (mol/mol complex)	f/f_0
63	4.6	0.78	26	144000	107000	58	1.6

[a] From Kyldén and Hammarström.[10]

I summarizes the effects of various structural modifications of $PGF_{2\alpha}$ on the dissociation constant.

Subcellular Distribution.[9] Subcellular fractions prepared from homogenates of corpora lutea have been used to compare the distributions of marker enzymes with that of $PGF_{2\alpha}$ receptor (Fig. 2). The results show that the distribution of the receptor is the same as that of the plasma membrane marker, 5′-nucleotidase. On the other hand, marker enzymes for mitochondria (succinate dehydrogenase), endoplasmic reticulum (NADPH–cytochrome *c* reductase) and lysosomes (soluble *N*-acetyl-β-glucosaminidase and β-glucuronidase) have distinctly different distribution patterns. The receptor is purified 6-fold (over the homogenate) in the top layer of the 35,000 *g* sediment (fraction 35t in Fig. 2). A 32-fold purification can be achieved by a combination of differential and sucrose density gradient centrifugations (Fig. 3). In this case, the top layer of the 10,000 *g* sediment (10t; 4-fold purification of receptor) is fractionated on a sucrose density gradient. Fraction 5 contains 32-fold purified receptor and 120-fold purified 5′-nucleotidase, suggesting that the enzyme is more stable than the receptor to purification.

Physical Properties.[10] The binding of $PGF_{2\alpha}$ to the receptor is inhibited by low concentrations of detergents. However, preformed hormone–receptor complex is stable to much higher detergent concentrations and can be solubilized by, e.g., deoxycholate or Triton X-100. By using tritium-labeled $PGF_{2\alpha}$, a convenient marker for the hormone–receptor complex is obtained. The molecular size of the complex is determined by a combination of Sepharose 6B chromatography and density gradient centrifugations on sucrose–H_2O and sucrose–^2H_2O gradients. The former analysis gives the Stokes' radius, and the latter gives the sedimentation coefficient

[9] W. S. Powell, S. Hammarström, and B. Samuelsson, *Eur. J. Biochem.* **61,** 605 (1976).
[10] U. Kyldén and S. Hammarström, *Eur. J. Biochem.* **109,** 489 (1980).

and the partial specific volume. From these parameters, the molecular weight of the receptor and the amount of detergent bound to the receptor can be calculated (Table II).[10]

Acknowledgment

The work was supported by a grant from the World Health Organization.

Section II

Immunochemical Assays of Enzymes and Metabolites

[32] Characteristics of Rabbit Anti-PGH Synthase Antibodies and Use in Immunocytochemistry

By WILLIAM L. SMITH and THOMAS E. ROLLINS

Prostaglandin derivatives are produced by virtually all mammalian tissues and organs.[1] In addition, the active forms (PGD, PGE, PGF_α, PGI, and TxA) are converted to biologically inactive products quite rapidly. For example, PGE and PGF_α are catabolized by the lung to a mixture of 15-ketoprostaglandin, and 15-keto-13,14-dihydroprostaglandins during a single passage through the circulation,[2] and the unstable TxA_2 ($t_{1/2}$ = 30 sec at 37°[3]) is probably hydrolyzed nonenzymically to the inactive TxB_2 in blood. This combination of ubiquitous synthesis and rapid catabolism has led to the concept that prostaglandins are local hormones as distinct from classical, circulating hormones and act at or near their sites of synthesis. Using this local hormone concept, it is possible to rationalize quite easily events such as the interplay in hemostasis between TxA_2 formed by platelets and PGI_2 synthesized by the vasculature[4] and the role in natriuresis of PGE_2 synthesized by renal collecting tubules.[5,6]

Because of the local character of prostaglandin action, defining the cellular sites of prostaglandin biosynthesis can provide important insights into the function of prostaglandins. This is particularly true in organs, such as kidney, brain, intestine, that contain numerous, functionally heterogeneous cell types.[7] A convenient and specific method for determining the location of prostaglandin synthesis in tissues and organs is to localize the PGH synthase by immunocytochemistry. The rationale for this approach is that cells that synthesize prostaglandins generally do so beginning with endogenous arachidonate and, thus, must be able to convert arachidonic acid to PGH_2 via the PGH synthase. In this chapter we describe (*a*) procedures for preparing conventional rabbit antibodies against PGH synthase[7]; (*b*) the properties of these antisera, particularly in terms of species cross-reactivities[7]; (*c*) the preparation of IgG and Fab fragments from the sera[8]; (*d*) the use of antiserum for immunoprecipitation of

[1] E. J. Christ and D. A. van Dorp, *Biochim. Biophys. Acta* **270,** 537 (1972).
[2] S. H. Ferreira and J. R. Vane, *Nature* (*London*) **216,** 868 (1967).
[3] M. Hamberg, J. Svensson, and B. Samuelsson, *Proc. Natl. Acad. Sci. U.S.A.* **72,** 2994 (1975).
[4] S. Moncada and J. R. Vane, *N. Engl. J. Med.* **300,** 1142 (1979).
[5] J. B. Stokes and J. P. Kokko, *J. Clin. Invest.* **59,** 1099 (1977).
[6] J. B. Stokes, *J. Clin. Invest.* **64,** 495 (1979).
[7] W. L. Smith and G. P. Wilkin, *Prostaglandins* **13,** 873 (1977).
[8] W. L. Smith and T. G. Bell, *Am. J. Physiol.* **235,** F451 (1978).

ISBN 0-12-181986-8

PGH synthase[7]; and (*e*) the use of anti-PGH synthase serum and IgG for immunocytofluorescence[7-9] and immunoelectron microscopy.[10]

Preparation of Rabbit Anti-PGH Synthase Serum

Female New Zealand white rabbits (2–3 kg) are inoculated intramuscularly with an emulsion prepared by sonication of a mixture of 0.5 ml of 0.1 *M* Tris-chloride, pH 7.4, containing 100 μg of purified sheep vesicular gland PGH synthase[11] and 0.5 ml of complete Freund's adjuvant (GIBCO). Intramuscular innoculations are made at four sites on the back: (*a*) behind the front shoulders on both sides of and about 2 cm from the backbone; and (*b*) in front of the hips on both sides of the backbone. Subsequent inoculations are performed 7 and 14 days later with emulsions made from 0.5 ml of 0.1 *M* Tris-chloride, pH 7.4, containing 50 μg of pure PGH synthase[11] and 0.5 ml of incomplete Freund's adjuvant (GIBCO). One month later, rabbits are inoculated again with 50 μg of purified enzyme in incomplete Freund's adjuvant. One week later rabbits are bled (30–40 ml) from the marginal ear vein into silanized 50-ml glass centrifuge tubes. A wooden applicator stick is inserted into the tube, the blood is allowed to stand overnight at 24°, and most of the clot is removed by withdrawing the stick. The sample is centrifuged at 1000 *g* for 2 min at 24°, and the supernatant (serum) is withdrawn with a Pasteur pipette. The serum is made to 0.02% in NaN_3, partitioned into 5-ml aliquots, and stored in screw-top tubes at −20°. Rabbits can be bled at biweekly or monthly intervals for about 6 months with no appreciable loss of antibody titer. Approximately 15 ml of serum are obtained from each bleeding. It is important to bleed the rabbits several times prior to immunization so that approximately 50 ml of preimmune serum are available.

Rabbit anti-PGH synthase sera when tested by standard Ouchterlony double-diffusion analyses show single lines of precipitation with samples of both solubilized sheep vesicular gland microsomes and pure PGH synthase (each containing 0.25–0.50 unit of cyclooxygenase activity); a zone of identity is present between the lines formed with microsomal and purified PGH synthases.[7] When tested by immunoelectrophoresis, the purified cyclooxygenase shows an extended, but continuous, arc of precipitation that is due to the presence of multiple isoelectric forms of the enzyme.[7]

We have been uniformly successful in raising rabbit immunoglobulins

[9] R. L. Huslig, R. L. Fogwell, and W. L. Smith, *Biol. Reprod.* **21,** 589 (1979).
[10] T. E. Rollins and W. L. Smith, *J. Biol. Chem.* **255,** 4872 (1980).
[11] M. E. Hemler, W. E. M. Lands, and W. L. Smith, *J. Biol. Chem.* **251,** 5575 (1976).

against PGH synthase. However, we have been unable in several attempts to prepare antibodies against sheep vesicular gland PGH synthase in either goats or sheep using standard protocols.

Isolation of Rabbit IgG

For many purposes (e.g., immunofluorescence, immunoprecipitation) it is unnecessary to purify the IgG from the serum. However, when the IgG concentration needs to be carefully controlled or when Fab fragments are to be prepared, rabbit IgG can be purified by chromatography on protein A-Sepharose.[9] All steps are performed at 4°. Protein A-Sepharose CL-4B (1.5 g; Pharmacia Fine Chemicals) is swelled overnight in 0.1 *M* sodium phosphate, pH 7.0, and poured into a 5-ml disposable plastic syringe with a Luer-Lok fitting on the tip and a small wad of glass wool at the bottom to prevent loss of the Sepharose beads. The column is equilibrated by washing with three volumes of 0.1 *M* sodium phosphate, pH 7.0 Serum (10 ml) is applied to the column at a slow flow rate (ca 10 ml/hr), and the column is then washed with 20 volumes of equilibration buffer (ca 1 hr) or until the absorbance of the eluent at 280 nm falls below 0.1. The IgG fraction is eluted with 4–5 ml of 1 *M* acetic acid, and the eluent is quickly adjusted to pH 7–8 with concentrated NH_4OH and then dialyzed at 4° against 100 volumes of 0.1 *M* sodium phosphate, pH 7.0. Approximately 70 mg of IgG ($\epsilon_{280}^{1\%}$ = 1.4[12]) are isolated from 10 ml of serum. The IgG is pure as judged by comparative Ouchterlony double-diffusion analysis against goat anti-rabbit whole serum (Miles Laboratories, Inc.) and goat anti-rabbit IgG (Miles) performed in 1.5% Bacto-agar containing 0.02% NaN_3.[9,10] It is best to isolate IgG from serum immediately before use, since precipitates are often observed in solutions of IgG after storage at −20°, particularly with solutions containing IgG at concentrations greater than 2 mg/ml.

Preparation of Fab Fragments

Fab fragments are useful for immunocytochemistry on cells such as lymphocytes that have Fc receptors or when it is necessary to minimize problems with penetration of large IgG molecules into cells and tissues immobilized with chemical fixatives. IgG (5 mg/ml) is incubated at 24° for 4 hr with gentle stirring in 0.1 *M* sodium phosphate, pH 7.0, containing 0.002 *M* EDTA, 0.01 *M* cysteine, and papain (Worthington Biochemical Corporation; 0.5 mg/100 mg IgG). Papain-digested IgG is dialyzed at 4°

[12] M. H. Freedman, A. L. Grossberg, and D. Pressman, *Biochemistry* **1,** 1941 (1968).

three times for 12 hr each against 500 volumes of 0.02 *M* sodium acetate, pH 6.0. The dialyzate is centrifuged at 10,000 *g* for 10 min to remove any precipitated Fc fragments. The supernatant is applied to a column (1 × 10 cm) of CM cellex (Bio-Rad Laboratories, Inc.) equilibrated with 0.02 *M* sodium acetate, pH 6.0, and fractions containing Fab fragments are eluted with starting buffer.[13] After dialysis against 0.1 *M* sodium phosphate, pH 7.4, the immunochemical purity of isolated Fab fragments is tested by Ouchterlony double-diffusion analyses against goat anti-rabbit IgG (Miles Laboratories, Inc.). With purified Fab a single line showing partial identity with IgG is observed.

Immunoprecipitation of PGH Synthases

Sheep vesicular glands or other tissues (1 g) are homogenized with a Polytron in 10 ml of 0.1 *M* Tris-chloride, pH 8.0, containing 20 m*M* diethyldithiocarbamate. The homogenate is centrifuged at 10,000 *g* for 10 min, and microsomes are prepared from the resulting supernatant by centrifugation at 100,000 g for 60 min. The microsomal pellet is resuspended by homogenization in 1 ml of 0.1 *M* Tris-chloride, pH 8.0, containing 20 m*M* diethyldithiocarbamate and 1% Tween 20 and again centrifuged at 100,000 *g* for 60 min. The resulting solubilized enzyme preparation is assayed for cyclooxygenase activity at 37° using a Yellow Springs Instruments Model 53 oxygen monitor.[9,11] The assay mixture contains 3 ml of assay buffer containing 0.1 *M* Tris-chloride, pH 8.0, 100 μM arachidonic acid, 1 m*M* phenol, 2 n*M* bovine hemoglobin, and 5 m*M* EDTA; reactions are initiated by the addition of enzyme.[11] The specific activity of the solubilized enzyme from sheep vesicular gland is routinely 5–10 μmol of O_2 consumed per minute per milligram of protein when assayed at 37°.

To perform immunoprecipitation reactions, aliquots of solubilized PGH synthase (0–0.15 units) are added to tubes containing 0.025 ml of either anti-PGH synthase or control preimmune serum, and the samples are made to a final volume of 0.5 ml with 0.1 *M* Tris-chloride, pH 8.0, containing 20 m*M* diethyldithiocarbamate and 0.1% Tween 20. After incubation overnight at 4°, the mixtures are centrifuged at 2000 *g* for 15 min, and the supernatants are withdrawn and assayed for cyclooxygenase activity. The results of one experiment with solubilized sheep vesicular gland microsomes is shown in Fig. 1. On the average 1 ml of rabbit anti-PGH synthase serum will precipitate 2.0 units of cyclooxygenase activity at the immunochemical equivalence point. This corresponds to approxi-

[13] D. H. Campbell, *in* "Methods in Immunology: A Laboratory Text for Instruction and Research" (D. H. Campbell, ed.), pp. 224–233. Benjamin, New York, 1970.

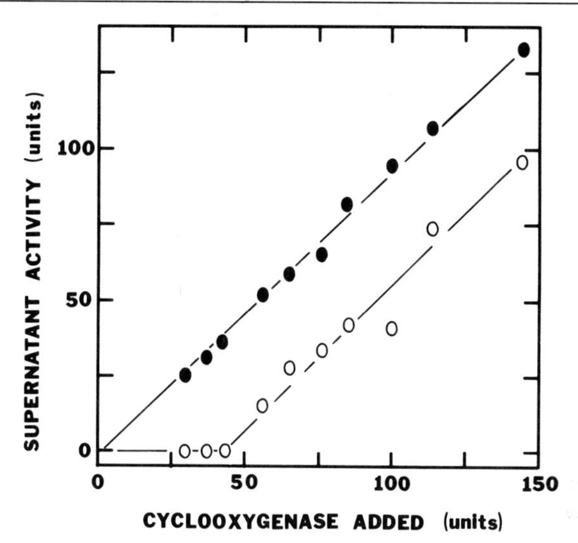

FIG. 1. Precipitation of sheep vesicular gland PGH synthase (cyclooxygenase) with rabbit anti-PGH synthase serum. Rabbit anti-PGH synthase serum (○—○) or control preimmune serum (●—●) was incubated for 24 hr at 4° as described in the text with the indicated amounts of cyclooxygenase activity. Immunoprecipitates were removed by centrifugation for 15 min at 2000 *g*, and the supernatant was assayed for cyclooxygenase activity as described in the text. From Smith and Wilkin.[7]

mately 70 μg of purified enzyme. If one estimates that there are three antigenic determinants on the enzyme, this means that there are approximately 200 μg of IgG directed against the enzyme per milliliter of serum. Since we routinely obtained 7 mg of IgG per milliliter of serum, about 3% of the serum IgG is anti-PGH synthase IgG.

Rabbit anti-PGH synthase serum (prepared using the purified sheep vesicular gland enzyme as the immunogen) was capable of precipitating quantitatively cyclooxygenase activity from solubilized microsomes prepared from sheep vesicular gland, bovine seminal vesicles, rabbit kidney medulla, Swiss mouse 3T3 cells, human platelets, guinea pig lung, dog spleen, and sheep uterus.[7,9,10] Thus, at least two antigenic determinants are shared between the sheep PGH synthase and synthases from other species. It should also be noted that (*a*) PGH synthase that has been inactivated by aspirin treatment is as reactive with the rabbit anti-PGH synthase as the native enzyme[7]; and (*b*) addition of anti-PGH synthase serum to cyclooxygenase assay mixtures has no appreciable effect on enzyme activity. Thus, none of the antibody molecules appear to be directed against the active site.

Immunocytofluorescence Using Anti-PGH Synthase Serum

One convenient method for determining which cells within a tissue can form prostaglandin endoperoxides (and, thus, in general, which cells form prostaglandins) is to test tissues sections for PGH synthase antigenic reactivity using indirect immunofluorescence staining. The method to be described has been useful in tissues from sheep, cow, guinea pig, pig, rabbit, rat, and dog including kidney,[7,8] uterus,[9] heart,[14] intestine,[15] and brain.[16] Fresh tissue to be sectioned is cut in 0.5–1.0 cm^3 blocks with a razor blade and then mounted in 5% gum tragacanth on cork cylinders (2 mm thick) cut from cork stoppers. The tissue, gum, and cork are immersed for 1 min in 2-methylbutane precooled to −70° in a Dry-Ice–acetone bath, and the frozen sample is then transferred to a cryotome (−20°). Freezing in liquid N_2 tends to cause splits and cracks in the tissues that interfere with subsequent sectioning. Sections (10 μm) are cut on a Lab-Tek cryotome, mounted on glass cover slips, placed in cover-slip carriers and then desiccated in the presence of anhydrous $CaSO_4$ for 1 hr under water aspiration. The cover slips are placed tissue side up on two layers of cheesecloth backing, and overlayed with 0.3 ml of diluted serum [1:10–1:100 with 0.1 *M* sodium phosphate, pH 7.0 (washing buffer)]. The samples are covered with the top of a plastic petri dish to prevent dehydration and incubated for 30 min. The cover slip is then washed by applying 0.5 ml of washing buffer 3–4 times, decanting after each wash into a beaker covered with a nylon mesh. The mesh prevents cover slips from dropping into the beaker during washing. Fluorescein isothiocyanate (FITC)-labeled goat anti-rabbit IgG (Miles) diluted with 20 volumes of washing buffer (ca 0.4 ml) is then applied to each cover slip; the sections are incubated another 30 min and then washed as previously. The cover slips are mounted tissue side down in glycerol and viewed under a fluorescence microscope outfitted with filters appropriate for visualizing fluorescein fluorescence. Rabbit kidney is a convenient tissue for testing the anti-PGH synthase serum. At dilutions of 1:50, one observes specific staining of collecting tubules, vascular endothelial cells, and, perhaps most characteristically, the parietal layer of Bowman's capsule (Fig. 2). These latter cells do not stain appreciably in the renal cortex of other animals. The pattern of intracellular staining obtained with anti-PGH synthase is characteristic. Fluorescence occurs throughout the cytoplasm, and an intense circle of stain-

[14] W. L. Smith and P. Needleman, *in* "Prostaglandins and the Microcirculation" (G. Kaley, ed.), Raven, New York, in press.

[15] D. M. Bebiak, E. R. Miller, R. L. Huslig, and W. L. Smith, *Fed. Proc., Fed. Am. Soc. Exp. Biol.* **38,** 884 (1979).

[16] W. L. Smith, D. I. Gutekunst, and R. H. Lyons, Jr., *Prostaglandins* **19,** 61 (1980).

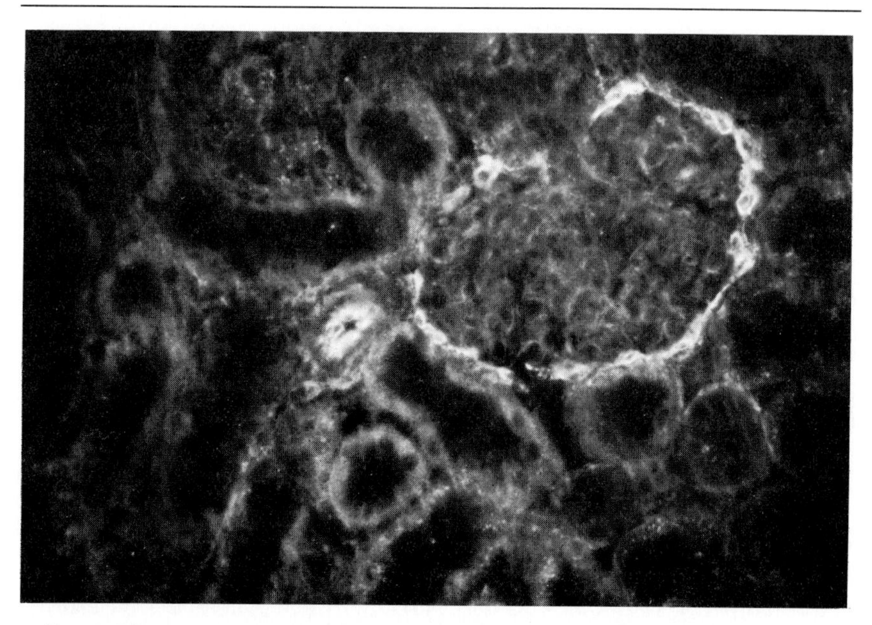

FIG. 2. Fluorescence photomicrograph of a glomerulus and adjoining arteriole in a section of rabbit renal cortex stained using anti-PGH synthase serum (1:20 dilution), then fluorescein isothiocyanate-labeled goat anti-rabbit IgG (1:20 dilution). G, Glomerulus; P, parietal layer; AE, arterial endothelial cells. From Smith and Bell.[8]

ing is often observed around the nucleus[8-10]; this latter staining can be visualized by moving the microscope objective slightly in and out of focus.

Two sets of controls are performed.[17,18] One set of sections is stained substituting preimmune for immune serum; the other set is stained using immune serum that has been adsorbed with microsomal PGH synthase. Adsorption is performed by mixing 5 units of cyclooxygenase activity per milliliter of undiluted serum, incubating at 24° for 5 min, and then diluting the serum sample for staining. In control samples, no staining should be observed other than that seen with FITC-labeled goat anti-rabbit IgG.

Use of frozen sections from fresh tissue is well suited for localization of PGH synthase. Antigenic reactivity is destroyed by fixation with acetone (2 min, 0°), glutaraldehyde, picric acid, and formalin solutions. Para-

[17] P. Petrusz, M. Sar, P. Ordonneau, and P. DiMeo, *J. Histochem. Cytochem.* **24,** 1110 (1976).

[18] D. F. Swaab, C. W. Pool, and F. W. van Leeuwen, *J. Histochem. Cytochem.* **25,** 388 (1977).

formaldehyde fixation (4% for 2 hr at 24°) can be used but provides no substantial benefits for light microscopy.

Immunoelectron Microscopy Using Anti-PGH Synthase IgG

Maintenance of cellular ultrastructure in immunoelectron microscopy requires prior fixation of cell or tissue samples. Fixation causes two problems with respect to intracellular antigens: (*a*) destruction of antigenic determinants and thus loss of immunoreactivity; and (*b*) sealing of membranes, thereby preventing penetration of labeling reagents. The following protocol was developed to circumvent these difficulties in staining for PGH synthase in Swiss mouse 3T3 cells.[10] Cells grown in monolayer culture are used to minimize problems of reagent penetration normally associated with the use of whole tissue or even suspended cells. Cells ($10^6/150\ mm^2$) are seeded in polystyrene culture dishes (35 × 10 mm). Cells in each dish are washed twice with 2–3 ml of 0.1 *M* sodium phosphate, pH 7.4, containing 0.9% NaCl (phosphate-buffered saline, PBS) to remove excess media. Cells are then fixed for 4 hr at 24° in a freshly prepared solution of 10 m*M* $NaIO_4$, 75 m*M* lysine, 2% paraformaldehyde, and 37m*M* sodium phosphate, pH 7.4,[19] containing 0 to 0.05% Tween 20. Fixed cells are subjected to three 15-min washes with 3 ml of phosphate-buffered saline and then overlayed with 1.5 ml of a solution of IgG isolated from either rabbit anti-PGH synthase or preimmune sera at final concentration of IgG of 5–50 μg/ml in phosphate-buffered saline. The dilution of IgG to be used in this procedure needs to be determined empirically.[20] All washes and dilutions of antisera are performed with 0.1 *M* sodium phosphate, pH 7.2, containing 0.9% NaCl. After incubation for 2.5 hr at 24°, IgG-treated cells are washed four times, 15 min each time. The cells are then incubated for 2 hr at 24° with goat anti-rabbit IgG (1:10 dilution, Miles) and subsequently washed for 15 min four times. Peroxidase rabbit anti-peroxidase-soluble (PAP) complex (1:50 dilution; Miles) is added to the cells, which are then incubated for 2 hr at 24° and washed as above. Washed cells are incubated for 10 min with 1 ml of a freshly prepared solution containing 0.3 m*M* 3,3′-diaminobenzidine and 0.3 m*M* H_2O_2 in 0.05 *M* Tris-chloride, pH 8.0.[21]

Cells stained in this manner with IgG isolated from anti-PGH synthase serum and then viewed under the light microscope show diffuse brown

[19] I. W. McLean and P. K. Nakane, *J. Histochem. Cytochem.* **22,** 107 (1974).

[20] G. C. Moriarty, C. M. Moriarty, and L. A. Sternberger, *J. Histochem. Cytochem.* **21,** 825 (1973).

[21] R. C. Graham and M. K. Karnovsky, *J. Histochem. Cytochem.* **14,** 291 (1966).

staining throughout the cytoplasm growing more intense nearer the nucleus; a discrete, sharply defined brown ring of staining occurs around the nucleus.[10] Only diffuse light brown staining is seen in control samples. Unless clear differences are seen between experimental and control samples at the light microscope level, it is fruitless to proceed to the electron microscope. In this regard, the time of staining with 3,3′-diaminobenzidine and H_2O_2 is important. It is often useful to prepare four experimental and four control dishes of cells and then stain for 2, 5, 10, or 20 min with the peroxidase substrates and process only the samples from the time points showing the clear distinctions between immune and control staining for electron microscopy.

The following protocol is used to prepare stained cell samples for electron microscopy.[10] The stained cells are subjected to four 15-min washes with 2 ml of 0.1 *M* sodium cacodylate, pH 7.4, and then postfixed in 2 ml of 1% OsO_4 prepared in 0.1 *M* sodium cacodylate, pH 7.4, for 2 hr at 24°. The cells are washed with 2 ml of 0.1 *M* sodium cacodylate, pH 7.4, and then dehydrated by washing with 2 ml of 50, 70, 80, 90, and 100% solutions of ethanol (three times each, 5 min each time). During the third treatment with 100% ethanol (1–2 ml), the cells are removed from culture dishes by rubbing gently with a rubber policeman, transferred in ethanol using a disposable pipette to 4-dram screw-top vials, and then collected by centrifugation at 500 *g* for 2 min. Cell pellets are resuspended by agitation with propylene oxide (0.5 ml) and then mixed with an equivalent volume of Epon–Araldite resin (Epon 812–Araldite 502–dodecenyl succinic anhydride, 25:20:60, v/v/v; Electron Microscopy Sciences) and agitated for 4 hr at 24° on a Labquake (Bolab Incorporated). Extra resin (0.5 ml) is then added to provide a ratio of resin to propylene oxide of 2, and the mixture is agitated overnight. The cells are then collected by centrifugation and resuspended in Epon–Araldite resin to which 0.024 volume of 2,4,6-tri(dimethylaminomethyl)phenol (Electron Microscopy Sciences) had been added and mixed thoroughly (30 min on Labquake). The samples are transferred to the tip of long-nosed plastic Beem capsules and incubated for 72 hr at 60°. The hardened resin is sectioned, and the sections are examined and photographed using a Phillips Model 201 transmission electron microscope. The sections are not counterstained with heavy metals. Kodak EM4463 film is used for photography. With mouse 3T3 cells, electron-dense staining was associated only with the endoplasmic reticulum and nuclear membranes in cells stained using anti-PGH synthase-IgG; no organelle-specific, electron-dense deposits were observed in cells stained using IgG isolated from rabbit preimmune serum or immune IgG that had been preadsorbed with purified PGH synthase.[10] It is useful to establish criteria for positive and negative staining (e.g., uniform nuclear membrane staining) and then to examine at random 400–500 cells in sections stained with immune and control IgG. When this was done in the case of 3T3 cells, approximately 90% of the cells in immune sections showed clear evidence of positive staining,[10] whereas in control sections, partial positive staining was noted in 10% of the cells.

Acknowledgments

This work was supported in part by U.S. Public Health Service NIH Grants HD10013 and AM22042 and by an Established Investigatorship (W. L. S.) from the American Heart Association.

[33] Radioimmunoassay for PGH Synthase

By GERALD J. ROTH

Prostaglandin H (PGH) synthase is a ubiquitous enzyme, found in a variety of species[1] and tissues.[2] As shown in studies of human endothelial cells in culture, the enzyme is continually produced by new protein synthesis,[3] and enzyme content may vary under different physiological or pathological conditions. Also, as shown in patients with "aspirin-like" defects of platelet function, PGH synthase may be congenitally absent or abnormal in certain individuals.[4] Therefore, a relatively rapid and convenient assay for enzyme protein, namely, a radioimmunoassay (RIA), may prove to be useful in physiological and clinical studies.[5,6]

Principle

The described radioimmunoassay for PGH synthase is a conventional precipitation-inhibition assay. The enzyme to be measured competes with [^{125}I]PGH synthase for a limiting amount of anti-PGH synthase antibody. After incubation, antibody-bound [^{125}I]PGH synthase is precipitated, and the amount of ^{125}I in the precipitate is determined.

[1] B. Samuelsson, M. Goldyne, E. Granström, M. Hamberg, S. Hammarström, and C. Malmsten, *Annu. Rev. Biochem.* **47,** 997 (1978).
[2] W. L. Smith and G. P. Wilkin, *Prostaglandins* **13,** 873 (1977).
[3] J. C. Hoak, R. L. Czervionke, G. L. Fry, and J. B. Smith, *Fed. Proc., Fed. Am. Soc. Exp. Biol.* **39,** 2606 (1980).
[4] H. J. Weiss, *Semin. Hematol.* **17,** 228 (1980).
[5] H. H. Tai, C. L. Tai, and W. L. Smith, *Fed. Proc., Fed. Am. Soc. Exp. Biol.* **39,** 292a (1980).
[6] G. J. Roth, *Circulation* **62,** 403a. (1980).

ISBN 0-12-181986-8

Materials

Antigen, [^{125}I]PGH Synthase

PGH synthase is purified to near homogeneity in the acetylated, acetyl-^{3}H-labeled form as described in this volume [51]). The material is used to standardize the assay, using 0–100 ng amounts of the enzyme to give a standard curve of precipitation-inhibition. The RIA works equally well with enzyme in the acetylated or nonacetylated form. The acetylated ^{3}H-labeled form is used as standard, since the amount of enzyme present can be measured directly by its ^{3}H content.

To prepare the ^{125}I-labeled form of the enzyme, pure PGH synthase from sheep vesicular gland is iodinated with ^{125}I according to the lactoperoxidase method of Thorell and Johansson.[7] Carrier-free Na^{125}I (0.5 mCi, 230 pmol, Amersham) is mixed with 200 pmol of pure PGH synthase (0.05 ml in 0.01 *M* sodium phosphate, pH 7.0, containing 0.05% sodium dodecyl sulfate (SDS), azide free). Lactoperoxidase (10 μg in 0.005 ml, 0.1 *M* sodium phosphate, pH 7.0; approximately 60 units/mg, Sigma) and hydrogen peroxide (220 nmol in 0.005 ml of water) are added, and the mixture is agitated for 1 min at 24°. The reaction is stopped by dilution with 0.5 ml of sodium phosphate, pH 7.4, containing 1% SDS, followed by 0.1 ml of normal rabbit serum containing 5% SDS. The preparation is dialyzed against borate saline buffer (0.036 *M* borate, 0.154 *M* NaCl, pH 7.8) containing 0.1% SDS; it is then subjected to gel filtration on agarose (BioGel A1.5M, Bio-Rad, Richmond, California) in a buffer of 0.01 *M* sodium phosphate, pH 7.0, 0.05% SDS, 0.02% sodium azide. Protein-bound ^{125}I elutes from the column in an identical fashion to [acetyl-^{3}H]PGH synthase. Fractions containing ^{125}I are pooled and used as the ^{125}I-labeled antigen in the RIA. SDS–polyacrylamide gel electrophoretic analysis of the final material shows that all of the ^{125}I comigrates with [acetyl-^{3}H]PGH synthase, indicating that the ^{125}I label is bound to the enzyme.

The iodination reaction generally gives 70% incorporation of ^{125}I into protein, and at least 95% of the protein-bound ^{125}I is acid-precipitable. The ^{125}I-labeled antigen is stored at −70°. When used in the RIA, the antigen is diluted to contain approximately 10,000 cpm per 0.05 ml, and 0.05 ml is used in each assay.

Anti-PGH Synthase Antiserum

Rabbits were immunized at 2-week intervals with 0.1 mg of pure PGH synthase from sheep vesicular glands. Antigen was solubilized in 0.05%

[7] J. I. Thorell and B. G. Johansson, *Biochim. Biophys. Acta* **251,** 363 (1971).

SDS and emulsified in Freund's adjuvant; complete adjuvant was used for the first injection, and incomplete adjuvant for subsequent injections. Rabbits were bled at intervals, and serum was tested for anti-PGH synthase activity. A known amount of PGH synthase was mixed with various amounts of antiserum or nonimmune rabbit serum as control, incubated for 24 hr at 0°, and then centrifuged to remove enzyme-antibody precipitate. The supernatant was assayed for remaining oxygenase activity, and serum samples with the highest enzyme-precipitating activity were used in the RIA. The most effective antiserum was obtained from a rabbit that had been immunized for almost 4 months. This antiserum would both precipitate and inactivate the enzyme.[6]

Antiserum was also analyzed by double-diffusion analysis in agar as shown in Fig. 1. The antiserum was apparently monospecific for PGH synthase, since the double-diffusion analysis shows a line of identity between the pure enzyme and a crude preparation of the enzyme in microsomes (Fig. 1). The antiserum cross-reacts with the enzyme from human platelet microsomes, producing a line of partial identity with the enzyme from sheep vesicular gland. Apparently, the human enzyme lacks an antigenic determinant(s) present in the sheep enzyme.

Serum was obtained from clotted blood, heated at 56° for 30 min, and stored at −20°. Initial experiments were performed to determine how much antiserum was needed to precipitate a given amount of ^{125}I-labeled antigen (about 10,000 cpm). Relatively large amounts of antiserum (1:100 to 1:1000 dilutions) gave 80–90% precipitation, while relatively small amounts of serum (1:10^6 dilution) or nonimmune serum gave 10–15% precipitation. An antiserum dilution was chosen that gave about 50% precipitation of the ^{125}I-labeled antigen. The actual dilution varied slightly with different antigen preparations, but ranged between a 1:3000 and a 1:8000 dilution. For use in the RIA, 0.05 ml of the antiserum dilution was used in a final assay volume of 0.5 ml, giving a final antiserum dilution of 1:30,000 to 1:80,000.

The "Nonspecific" Precipitating System

A "nonspecific" or "second" precipitating system is needed to precipitate the small amount of PGH synthase–anti-PGH synthase immune complex formed in the assay. This system consists of nonimmune rabbit IgG and an antiserum directed against rabbit IgG. The appropriate amounts of these materials are determined by testing various proportions for their ability to provide maximum ^{125}I precipitation in the presence of anti-PGH synthase antiserum and minimum ^{125}I precipitation in the presence of nonimmune serum. I found that 16 μg of rabbit IgG prepared by

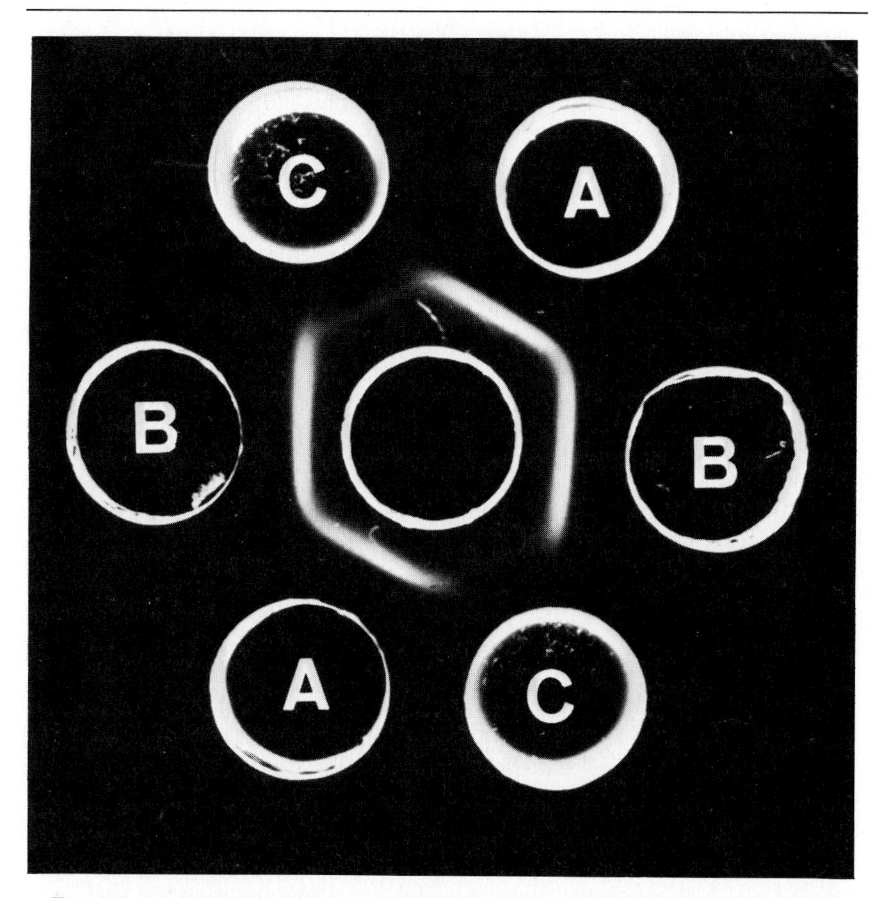

FIG. 1. Immunodiffusion analysis of anti-PGH synthase antiserum. The center well contained 0.01 ml of antiserum diluted 1:3. Wells A contained 100 μg of sheep vesicular gland microsomes. Wells B contained 5 μg of pure sheep PGH synthase. Wells C contained about 1.5 mg of human platelet microsomes. The diffusions were performed in 1% agar containing 0.1% sodium dodecyl sulfate at pH 8.4.

caprylic acid precipitation[8] and a 1:8 dilution of a sheep anti-rabbit IgG antiserum gave satisfactory results.

Sample Preparation

To date, the enzyme has been assayed from sheep vesicular gland, human platelets, and human foreskin fibroblasts. The preparation method

[8] M. Steinbuch and R. Audran, *Arch. Biochem. Biophys.* **134,** 279 (1969).

is essentially the same for each tissue. The material is first concentrated into a pellet by centrifugation and then solubilized in an SDS-containing buffer by sonication. The preparation of human platelets for assay is described as an example. Citrated venous blood (10 ml, anticoagulated with 0.2 ml of 0.5 *M* citric acid–sodium citrate, pH 5.0) is centrifuged (250 g, 10 min, 24°) to give about 3 ml of platelet-rich plasma. The platelet-rich plasma is centrifuged (1500 *g*, 15 min, 24°) to give a platelet pellet containing approximately 2×10^9 platelets. The plasma overlying the pellet is aspirated off and discarded. The platelet pellet is solubilized by sonication in 0.25 ml of 0.025 *M* sodium phosphate, pH 7.4, containing 2% SDS. Platelet protein is determined by Lowry protein assay using bovine serum albumin as standard, and platelet number is estimated by assuming 1.5 mg of protein per 10^9 platelets. The solubilized platelet material is diluted 1:40 with assay buffer that does not contain SDS. The detergent is omitted in order to lower the SDS concentration from 2% to 0.05%. Subsequent dilutions of the platelet material are made in the assay buffer containing 0.05% SDS, and aliquots (0.05 ml) of the dilutions are added to the RIA as test material.

Procedure

The assay buffer is composed of 0.025 *M* sodium phosphate, pH 7.4, containing 10 mg of bovine serum albumin per milliliter, 0.05% SDS, and 0.02% sodium azide. All dilutions of the various components are made in this buffer except for the initial dilution of test material.

For the initial incubation, use 0.25 ml of assay buffer. Add 0.05 ml of anti-PGH synthase antiserum, approximately 1:5000 dilution; then add 0.05 ml of an appropriate dilution of test or standard material. Mix and incubate for 2 hr at 37°.

For the ^{125}I-labeled antigen incubation, add 0.05 ml of [^{125}I]PGH synthase (10,000 cpm). Mix and incubate for 2 hr at 37°.

For incubation with "nonspecific" precipitating system, add 0.05 ml of nonimmune rabbit IgG (16 μg); then add 0.05 ml of sheep anti-rabbit IgG antiserum (1:8 dilution). Mix and incubate for 18 hr at 4°. Centrifuge (1500 *g*, 15 min, 24°) and discard the supernatant. Assay the pellet for ^{125}I.

The assay is performed using an SDS-containing buffer with a relatively high protein content, 10 mg of bovine serum albumin per milliliter. Detergent is required to solubilize the antigen, PGH synthase, which is a membrane-bound protein. The assay is more sensitive when SDS is present, compared to the nonionic detergent Tween 20. This finding may be due to the fact that the rabbit antiserum was raised to the SDS-solubilized enzyme as antigen. The albumin that is added to the buffer serves to "buffer" the detergent and to minimize nonspecific adsorption of protein to surfaces.

The assay involves an initial incubation of a limiting amount of antiserum with a noniodinated test or standard antigen. The "preincubation" step prior to addition of the ^{125}I-labeled antigen gives increased sensitivity to the assay. The ^{125}I-labeled antigen is then added; after a second incubation, a relatively large amount of the "nonspecific" precipitating system is added to precipitate the relatively small amount of antigen and anti-PGH synthase antibody complex formed in the assay.

For each assay, a standard curve is determined by duplicate assays containing 10, 5, 2.5, 1.0, and 0.5 ng of standard antigen (Fig. 2). A positive control is included without antigen, giving about 50% precipitation of the ^{125}I-labeled antigen, and a negative control is performed with nonimmune rabbit serum replacing the anti-PGH synthase antiserum, giving about 15% precipitation. For assay of test material, triplicate assays of several dilutions of the test material (for example, 1:80, 1:160, and 1:320 dilutions of the platelet preparation described above) are performed. The dilutions of the test material are chosen as those that give precipitation-inhibition comparable to 0.5–5.0 ng amounts of standard. The actual measurement of PGH synthase in the test material is made by comparing the extent of precipitation-inhibition by test material to that seen with the standards (Fig. 2). The results of the assays are averaged and expressed

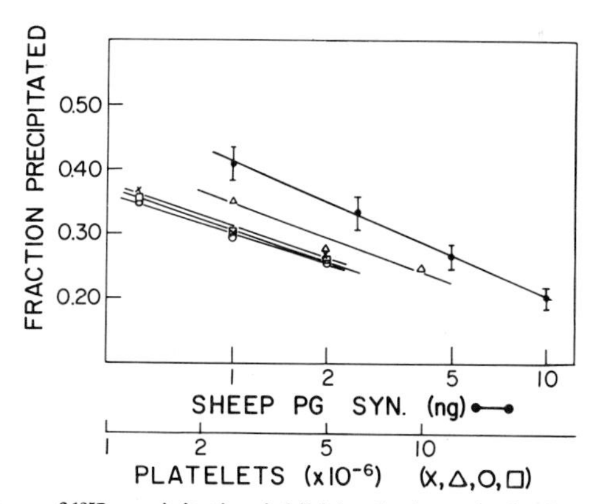

FIG. 2. Comparison of ^{125}I precipitation-inhibition by human platelets and by sheep vesicular gland PGH synthase. The standard precipitation line plots the fraction of total ^{125}I precipitated versus the amount of sheep PGH synthase added (1, 2.5, 5, or 10 ng). The fraction of total ^{125}I precipitated in the presence of doubling dilutions of platelets (corresponding to 1.25, 2.5, and 5.0 × 10^6 platelets) is plotted for comparison according to the separate horizontal abscissa for platelets. Platelet dilutions from four separate donors (×, △, ○, □) were tested.

as nanograms of PGH synthase per milligram of protein. Human platelets contain about 1200 ng of PGH synthase/10^9 platelets, corresponding to 800 ng per milligram of protein. Human foreskin fibroblasts grown in cell culture contain about 300 ng of enzyme per milligram of protein.

Discussion

A number of assays are available to measure PGH synthase. The most direct assays are activity measurements such as those described using thin-layer chromatography (TLC) for identification of prostaglandin products or O_2 electrode measurement of O_2 uptake. Also, the enzyme can be measured by its ability to interact selectively with radioactive aspirin and undergo a site-specific acetylation reaction, giving rise to acetyl-^{3}H-labeled enzyme (this volume [51]). However, PGH synthase is a labile enzyme which rapidly loses activity in broken cell preparations, making both activity and aspirin-acetylation measurements inaccurate. The advantages of the RIA of the enzyme lie in the ability to assay the enzyme by its immunological reactivity regardless of its catalytic activity. In addition, the RIA requires much lower amounts of enzyme for accurate measurement. For example, the enzyme can be quantitated by RIA using 2×10^7 platelets, whereas at least 10-fold more platelets are needed for aspirin-acetylation or TLC assays. Finally, the RIA is an easier assay in terms of the laboratory effort required to obtain an accurate result.

The assay shows good reproducibility. Replicate assay on successive days using a single sample agree within 10%, and the assay results by RIA show good agreement with an independent enzyme assay using [acetyl-^{3}H]aspirin.[6]

When applying the RIA to various tissues, one can validate the accuracy of the measurement by doing preliminary enzyme measurements using TLC analysis of PG products or using [acetyl-^{3}H]aspirin as described earlier. Assuming that these assays give consistent results, one can utilize the RIA as an accurate and convenient measurement of the enzyme protein.

Acknowledgments

This work was supported by NIH Grant HL20974, a Grant-in-Aid from the American Heart Association with funds contributed in part by the Connecticut Heart Association, and the University of Connecticut Research Foundation. G. J. R. is an Established Investigator of the American Heart Association.

[34] Monoclonal Antibodies against PGH Synthase: An Immunoradiometric Assay for Quantitating the Enzyme

By DAVID L. DEWITT, JEFFREY S. DAY, JOHN A. GAUGER, and WILLIAM L. SMITH

PGH synthase catalyzes the formation of the prostaglandin endoperoxide PGH_2 from arachidonic acid and oxygen.[1–4] Fluctuations in tissue concentrations of PGH synthase have been shown to occur in rat Graafian follicles in response to luteinizing hormone,[5] in the ovine[6] and guinea pig[7] uterus during the estrous cycle, and in the hydronephrotic rabbit kidney in response to perfusion *ex vivo*.[8] The diminished production of PGI_2 by arteries of rabbits subjected to high-fat diets[9] and the increased PGE_2 formation by kidneys of rabbits fed low-salt diets[10] could also be due to alterations in the levels of PGH synthase. These observations from several laboratories indicate that changes in PGH synthase levels can play a role in regulating the rates of prostaglandin formation in both normal and pathological situations. There are a number of methods for assaying PGH synthase enzyme activity including polarographic assays,[6] use of radioactive fatty acid substrates and measurement of labeled products,[7] and the use of unlabeled fatty acids and measurement of products by radioimmunoassays.[5] There does, however, exist a need for methods to quantitate changes in enzyme protein levels. In this chapter, we describe the preparation of four different monoclonal antibodies against the PGH synthase and the use of iodinated monoclonal antibodies for quantitating PGH synthase protein concentrations in tissue extracts by immunoradiometric assay. This immunoradiometric assay is $10^3–10^4$ times as sensitive as the most accurate and sensitive polarographic assay for enzyme activity.

[1] M. E. Hemler, W. E. M. Lands, and W. L. Smith, *J. Biol. Chem.* **251,** 5575 (1976).

[2] F. J. G. van der Ouderaa, M. Buytenhek, D. H. Nugteren, and D. A. van Dorp, *Biochim. Biophys. Acta* **487,** 315 (1977).

[3] G. J. Roth, N. Stanford, J. W. Jacobs, and P. W. Majerus, *Biochemistry* **16,** 4244 (1977).

[4] T. Miyamoto, N. Ogino, S. Yamamoto, and O. Hayaishi, *J. Biol. Chem.* **251,** 2629 (1976).

[5] M. R. Clark, J. M. Marsh, and W. J. LeMaire, *J. Biol. Chem.* **253,** 7757 (1978).

[6] R. L. Huslig, R. L. Fogwell, and W. L. Smith, *Biol. Reprod.* **21,** 589 (1979).

[7] N. L. Poyser, *J. Reprod. Fertil.* **56,** 559 (1979).

[8] A. R. Morrison, H. Montz, and P. Needleman, *J. Biol. Chem.* **253,** 8210 (1978).

[9] A. Dembinska-Kiec, T. Gryglewski, A. Zmuda, and R. J. Gryglewski, *Prostaglandins* **14,** 1025 (1977).

[10] R. A. K. Stahl, A. A. Attallah, D. L. Bloch, and J. B. Lee, *Am. J. Physiol.* **237,** F334 (1979).

ISBN 0-12-181986-8

Preparation of Monoclonal Antibodies against PGH Synthase

Tissue Culture Media

Incomplete HT (hypoxanthine–thymidine) medium: 68.88 g of Dulbecco's modified Eagle medium [DMEM; 4.5 g of glucose per liter (GIBCO)]; 18.5 g of $NaHCO_3$; 0.5 g of penicillin (Sigma); 0.5 g of streptomycin (Sigma); 0.068 g of hypoxanthine (Sigma), 0.01 g of glycine; 0.0195 g of thymidine (Sigma). All reagents are dissolved in 4.5 liters of glass-distilled, deionized water, filtered through a sterile Millipore filter, and stored in ten 500-ml bottles at 4°.

Complete HT medium: 450 ml of incomplete HT medium; 50 ml of fetal bovine serum (KC Biologicals, Inc.); 50 ml of horse serum (Flow Laboratories); 50 ml NCTC 109 medium (Microbiological Associates); 6 ml of 200 m*M* glutamine (glutamine stored at −20°).

Complete HAT (hypoxanthine–aminopterin–thymidine) medium is complete HT medium containing 1 μM aminopterin (Sigma).

Procedure. Monoclonal antibodies can be prepared in any laboratory equipped for mammalian cell culture by personnel with some training in tissue culture and basic immunological techniques. However, since myeloma–spleen cell hybridization procedures are still as much artistic as scientific, it is most efficient and least expensive to visit a laboratory performing this type of work prior to embarking on the preparation of monoclonal antibodies. In addition to the description presented here, we also recommend that the reader refer to volumes that give further technical details.[11,12]

Myeloma–spleen cell fusions are performed by modification of a method of Galfré *et al.*[13] Four- to six-week-old female C57BL mice or Swiss albino mice are immunized intraperitoneally at 2-week intervals with 20 μg of PGH synthase purified from sheep vesicular gland[1] suspended in an emulsion prepared by sonicating and vortexing a mixture containing 0.1 ml of 0.1 *M* Tris-chloride, pH 7.4, and 0.1 ml of complete Freund's adjuvant (GIBCO). Three days after the third inoculation, the mice are killed by cervical dislocation and their spleens are removed under sterile conditions (see Mishell and Shugi[11] for details). The spleens are placed in a 60-mm sterile petri dish containing 5 ml of incomplete HT medium containing 20 m*M* HEPES, pH 7.6. Spleens are cut into pieces (1

[11] B. B. Mishell and S. M. Shugi, eds., "Selected Methods in Cellular Immunology." Freeman, New York, 1980.

[12] D. M. Weir, ed., "Handbook of Experimental Immunology," Vol. 1. Blackwell, Oxford, 1979.

[13] G. Galfré, S. C. Howe, C. Milstein, G. W. Butcher, and J. C. Howard, *Nature (London)* **266,** 550 (1977).

cm^3) with small scissors and then teased apart with tweezers to release the lymphocytes (all equipment used in handling the spleens is sterilized prior to use). The liquid is transferred to a 50-ml culture tube and vortexed. The large tissue fragments are allowed to settle briefly. The supernatant containing the lymphocytes is transferred to another culture tube and then centrifuged at 1500 *g* for 5 min. Red blood cells in the pellet are removed by hypotonic lysis: 5.0 ml of 0.2% saline are added, and the sample is allowed to stand for 30 sec; then 5.0 ml of 1.6% saline is added, and the sample is incubated for an additional 30 sec. Next, 10 ml of incomplete HT medium containing 20 m*M* HEPES, pH 7.6 is added, and the remaining spleen cells are collected by centrifugation and resuspended in 5 ml of incomplete HT medium containing 20 m*M* HEPES, pH 7.6.

The mouse myeloma strain[14] SP2/0-Ag14 obtained from the Cell Distribution Center of the Salk Institute, La Jolla, California, is grown in Dulbecco's modified Eagle medium (DMEM, GIBCO) containing 10% fetal bovine serum (KC Biologicals, Inc.) and 100 mg of both penicillin and streptomycin per liter at 37° under a water-saturated 10% CO_2 atmosphere. SP2 myeloma cells (1 to 5 $\times$ 10^6), which had been washed and resuspended in incomplete HT medium containing 20 m*M* HEPES pH 7.6, are mixed with 1 to 5 $\times$ 10^7 of the isolated splenic lymphocytes. Cell numbers are determined using a hemacytometer. The cell mixture is collected by centrifugation at 1000 *g* for 5 min in a sterile silanized glass centrifuge tube [glass tubes were silanized by immersion for 5 min in 2% dimethyldichlorosilizane (Sigma) in hexane, then washing the glassware first with hexane and then methanol prior to autoclaving]. After careful removal of the supernatant, 1 ml of the fusion solution is added and the cells are shaken from the bottom of the test tube gently so as not to break up the clumps. The fusion solution is prepared immediately prior to use by autoclaving 7 ml of polyethylene glycol 1000 (Baker) in a 50-ml plastic culture tube and then diluting the liquid (maintained at 37°) with 1 ml of dimethyl sulfoxide and enough incomplete HT medium containing 20 m*M* HEPES, pH 7.6, to make the final volume 20 ml. During the ensuing 3 min, the fusing cell mixture is gradually diluted with 3 ml of incomplete HT medium containing 20 m*M* HEPES, pH 7.6; then, over a period of 6 min, 12 ml of complete HT medium is added. Finally, the cells are collected by centrifugation, resuspended in 48 ml of complete HT medium, and dispensed in 0.5-ml aliquots into 2–24-well Costar 3524 tissue culture plates. After 24 hr, 1 ml of complete HAT medium is added to each well. Half of the medium is removed and replaced with fresh complete HAT medium 2 and 4 days thereafter; 14–21 days after the cell fusion, when the media

[14] M. Shulman, C. D. Wilde, and G. Kohler, *Nature (London)* **276**, 269 (1978).

from those wells with growing hybridomas begin to acidify (turn yellow), aliquots of media are removed to test for the presence of anti-PGH synthase antibody.

Selection of Hybridomas Producing Antibody to PGH Synthase

The protein-A-bearing Cowen I strain of *Staphylococcus aureus* kindly provided by Dr. Ronald Patterson of Michigan State University is grown in Trypticase soy broth soybean–casein digest medium (BBL Microbiology Systems) and isolated and attenuated as described by Kessler.[15] Formaldehyde-fixed, heat-killed *S. aureus* cells are stored at −80° as a 10% cell suspension in 10 m*M* HEPES, pH 7.5, containing 150 m*M* NaCl.

Protein A-bearing *S. aureus* cells bind many subclasses of mammalian IgG; however, mouse IgG_1, which constitutes about 30% of the total IgG in mouse serum, is not bound by *S. aureus* cells at pH 7.5. To circumvent this limitation, *S. aureus* cells are first absorbed with rabbit anti-mouse IgG. This rabbit anti-mouse IgG–*S. aureus* complex precipitates all mouse IgG molecules and is prepared as follows. Aliquots (1 ml) of the 10% *S. aureus* cell suspensions are defrosted and washed by sequential centrifugation and resuspension: (*a*) twice with 1 ml of 0.1 *M* Tris-chloride, pH 8.0, containing 5% bovine serum and 1% Tween 20 (w/v); (*b*) once with 1 ml of 0.1 *M* Tris-chloride, pH 8.0, containing 1% Tween 20 (w/v) and 80 μg of rabbit anti-mouse IgG (Miles Laboratories); and (*c*) twice with 1 ml of 0.1 *M* Tris-chloride, pH 8.0, containing 1% Tween 20 (w/v). Finally the *S. aureus* cells are resuspended in 1 ml of 0.1 *M* Tris-chloride, pH 8.0, containing 20 m*M* diethyldithiocarbamate, 1 m*M* phenol, and 0.1% Tween 20 (w/v).

For assay, 0.1 ml of the rabbit anti-mouse IgG–*S. aureus* suspension is mixed with 1 ml of medium removed from each well with growing hybridomas. The sample is vortexed and then centrifuged for 2 min at 500 *g* on a desk-top centrifuge. A volume of solubilized sheep vesicular gland microsomes (ca 5 μl) containing approximately 100 units of cyclooxygenase activity[16,17] is added to the *S. aureus* pellet along with enough 0.1 *M* Tris-chloride, pH 8.0, containing 20 m*M* diethyldithiocarbamate, 1 m*M*

[15] S. W. Kessler, *J. Immunol.* **117,** 1482 (1976).

[16] One unit of cyclooxygenase activity is defined as the amount of enzyme that will catalyze the consumption of 1 μmol of O_2 per minute at 37° when assayed in 3 ml of reaction buffer containing 0.1 *M* Tris-chloride, pH 8.0, 100 μ*M* arachidonic acid, 1 m*M* phenol, 2 n*M* bovine hemoglobin, and 5 m*M* EDTA.

[17] A description of the preparation of microsomes and indirect immunofluorescence staining is presented in this volume [32].

phenol, and 0.1% Tween 20 to give a final volume of approximately 0.1 ml. The mixture is vortexed and then centrifuged to pellet the cells. The supernatant is removed, and the resulting pellet is resuspended in 0.1 ml of the starting buffer. Both the supernatant and pellet are assayed for cyclooxygenase activity. Precipitation of cyclooxygenase activity is taken as evidence that at least some of the hybridoma cells present in the test well are producing anti-PGH synthase antibodies.

Cells from wells yielding a positive response in the immunoprecipitation test are cloned in soft agar. Sterile stock agar solution is made by autoclaving 1.25 g of Seaplaque agarose (FMC Company) in 25 ml of phosphate-buffered saline, pH 7.4. While still hot, 10 ml of stock agarose are added to 90 ml of complete HT medium previously warmed to 37°. Five milliliters of this complete HT-agarose is overlaid on a feeder layer of 5×10^4 Swiss mouse 3T3 fibroblast (ATCC CCL 92) seeded 15–30 min beforehand onto a 100-mm tissue culture plate (Corning). The agarose is allowed to gel at 4° for 5 min and then returned to the incubator to warm to 37°. Next, 1 ml of complete HT–agarose (37°), containing 2×10^3 hybridoma cells, is dripped evenly over the already poured, solidified agarose, and the new agarose ia allowed to harden at 4° before returning the tissue culture plate to the incubator. After 7–10 days, the clones, visible by the naked eye as small spots in the agar, are ready to transfer to liquid media. Clones are removed using Eppendorf or Pipetman pipettors with sterile plastic pipette tips (0–200 μl size). Working in a positive-pressure tissue culture hood, the cells are easily viewed under a dissecting microscope. Each clone is sucked up with a pipettor (50 μl setting) and transferred to 1 ml of complete HT medium present in each well of Costar 24 well plates. Usually 24–48 clones (1–2 plates) are selected. Practice on extra plates is highly advisable until one can successfully remove single clones without disturbing adjacent clones. When the clones have grown almost to confluency (10–12 days), the medium is tested for production of antibody. Normally, two positive clones derived from each well are propagated. The resulting cells are frozen for permanent storage in liquid N_2 and for subsequent production of antibody. To freeze the cells, 5 to 10 $\times 10^6$ hybridoma cells are suspended in 1 ml of fetal calf serum containing 50 μl of DMSO, transferred to a freezing vial (Nunc), placed at −80° overnight, and then transferred to liquid N_2. To thaw the cells, the liquid is rapidly brought to 37° and pipetted into 20 ml of complete HT medium. The cells are then collected by centrifugation at 24°, resuspended in complete HT medium, and transferred to culture dishes.

Four cloned hybridoma lines (*cyo*-1, *cyo*-3, *cyo*-5, and *cyo*-7) secreting anti-PGH synthase antibody were prepared from two separate fusions. An additional hybridoma (*2c3*), which secretes an IgG_2 that does not inter-

TABLE I
ANALYSIS OF MONOCLONAL ANTIBODIES AGAINST PGH_2 SYNTHASE[a]

Antibodies produced by hybridoma line	Subclass[b]	Species cross-reactivities[c]	Antigenic determinant on sheep[d]
cyo-1	IgG_{2b}	Positive: sheep, bovine, human, rat Negative: guinea pig, rabbit, mouse, dog	Site 1
cyo-3	IgG_1	Positive: sheep, bovine, human, guinea pig, rabbit Negative: rat, mouse, dog	Site 3
cyo-5	IgG_{2b}	Positive: sheep, bovine, human, guinea pig, rabbit Negative: rat, mouse, dog	Site 5
cyo-7	IgG_2	Positive: sheep, bovine, human, guinea pig, rabbit Negative: rat, mouse, dog	Site 3
2c3	IgG_2	Negative with all species	No reaction

[a] From DeWitt *et al.*[18]

[b] Determined on the basis of Ouchterlony double-diffusion analyses with rabbit anti-mouse IgG_1, IgG_{2a}, and IgG_{2b} antisera and elution profiles from protein A-Sepharose[19] as described in the text.

[c] Determined by immunoprecipitation of solubilized cyclooxygenase activity from sheep vesicular gland, rat small intestine, rabbit renal medulla, guinea pig renal medulla, bovine seminal vesicle, and human platelet microsomes, and/or by immunofluorescent staining using rat, dog, mouse, guinea pig, and rabbit kidneys.[17,18]

[d] Deduced on the basis of species cross-reactivities and reactivities in immunoradiometric assay.

act with the PGH synthase, was used as a negative control in the immunoradiometric assay described below. The properties of the immunoglobulins secreted by each of the hybridoma lines are summarized in Table I.[18]

Preparation of IgG-Free Fetal Calf Serum and Cell Culture Medium

Fetal calf serum (100 ml) is adjusted to pH 8.2 and applied to a protein A–Sepharose CL-4B column (Pharmacia Fine Chemicals; ca 1×5 cm) equilibrated with 0.1 *M* sodium phosphate, pH 8.0. The eluent is collected, and the bovine protein absorbed to the column is removed by washing the column with 2–3 volumes of 0.1 *M* sodium citrate, pH 3.5. The column is then reequilibrated with 0.1 *M* sodium phosphate, pH 8.0, and the entire procedure is repeated. After three or four passages of fetal calf serum through the column, no protein (i.e., A_{280}) is found to elute with

[18] D. L. DeWitt, T. E. Rollins, J. S. Day, J. A. Gauger, and W. L. Smith, *J. Biol. Chem.*, **256,** 10375 (1981).

0.1 *M* sodium citrate, pH 3.5. The pH 8.0 eluent is designated as IgG-free fetal calf serum. The medium used for growth of hybridoma lines for isolation of mouse IgG is complete HT medium, but contains 20% IgG-free fetal calf serum and no horse serum.

Purification of Mouse IgG_1 from Hybridoma Culture Media

Chromatography is performed at 4°. Media (IgG-free as above) obtained by growing *cyo*-3 (or other hybridomas) to confluency is adjusted to pH 8.2, cooled at 4°, and applied to a protein A–Sepharose CL-4B column (1 × 5 cm). Absorbed material is eluted stepwise using 0.1 *M* buffers of pH 8.0 (sodium phosphate), pH 6.0, pH 4.5, and pH 3.5 (sodium citrate).[19] Anti-PGH synthase activity in each fraction is monitored by measuring the ability of the fraction when mixed with rabbit anti-mouse IgG–*S. aureus* complexes (see above) to precipitate cyclooxygenase activity. IgG_1 secreted by hybridoma line *cyo*-3 is eluted at pH 6.0; IgG_{2b} secreted by *cyo*-1, *cyo*-5, *cyo*-7, and *2c3* is eluted at pH 3.5 and adjusted to pH 7–8 with concentrated NH_4OH. Fractions containing IgG_1 (*cyo*-3) are pooled, dialyzed overnight against 0.125 *M* sodium borate, pH 8.4, and stored at −80°. In one typical experiment in which *cyo*-3 was grown to confluency, 9 mg of IgG_1 (as determined by the absorbance at 280 nm (A_{280} = 1.4 for 1 mg of IgG/ml[20]) was isolated from 120 ml of media; 1 μg of isolated IgG_1 when bound to 0.1 ml of the rabbit anti-mouse IgG–*S. aureus* cell suspension is able to bind 0.05 unit (ca 0.6 μg) of cyclooxygenase activity.

Radioiodination of IgG_1 (*cyo-3*)

IgG_1 is radioiodinated essentially as described by Bolton and Hunter.[21] For routine iodinations, an aliquot containing 0.4 mCi of Bolton and Hunter's reagent (Amersham) is evaporated under a gentle stream of dry N_2 in a 6 × 50 mm test tube. IgG_1 (40 μg, *cyo*-3) in 0.01–0.05 ml of 0.125 *M* sodium borate, pH 8.4, is added, and the sample incubated at 4° for 15 min with frequent agitation. Unreacted iodinating reagent is destroyed by the addition of 0.5 ml of 0.2 *M* glycine in the reaction buffer followed by a 10-min incubation at 4°. Products are separated by chromatography (4°) on a column of BioGel P-30 (Bio-Rad Laboratories; 1 × 7 cm) equilibrated with 0.05 *M* sodium phosphate, pH 7.4, containing 2.5 mg of gelatin per milliliter and 0.02% NaN_3. Aliquots of each fraction (ca 0.5 ml) are

[19] P. L. Ey, S. J. Prowse, and C. R. Jenkin, *Immunochemistry* **15,** 429 (1978).

[20] M. H. Freedman, A. L. Grossberg, and D. Pressman, *Biochemistry* **7,** 1941 (1968).

[21] A. E. Bolton and W. M. Hunter, *Biochem. J.* **133,** 529 (1973).

counted in a gamma counter, and those fractions eluting at the void volume are pooled and stored at $-80°$ in small aliquots (0.2 ml) containing 12–20 μCi. The percentage of starting ^{125}I incorporated into IgG_1 by this procedure ranges from 60 to 80%; most (ca 80%) of the ^{125}I present in IgG_1 coelectrophoreses with the heavy chain on SDS-gel electrophoresis. Prior to use of $[^{125}I]IgG_1$ for immunoradiometric assays, samples are thawed and diluted with 4.8 ml of 0.05 *M* sodium phosphate, pH 7.4, containing 2.5 mg of gelatin per milliliter. After thawing a tube, unused material is discarded.

Immunoradiometric Assay of PGH Synthase

A 10% (w/v) suspension of attenuated *S. aureus* cells are washed by centrifugation once in 0.1 *M* Tris-chloride, pH 7.4, containing 5% (w/v) bovine serum albumin and 1% Tween 20, and then twice in 0.1 *M* Tris-chloride, pH 7.4, containing 20 m*M* diethyldithiocarbamate and 1% Tween 20 (assay buffer; prepared daily) followed by resuspension to 10% (w/v) *S. aureus* in the assay buffer. Equal volumes of washed *S. aureus* suspensions and media from IgG_2-producing hybridoma lines (*cyo*-1, *cyo*-5, *cyo*-7, or *2c3*) are mixed, allowed to stand 15 min at 24°, and then centrifuged. Pellets are resuspended and washed twice in the assay buffer and finally resuspended to give a 10% *S. aureus* suspension.

The interactions involved in the immunoradiometric assay for PGH synthase are illustrated in Fig. 1. Solubilized microsomes[16] containing 5×10^{-4} units of PGH synthase (i.e., cyclooxygenase) activity are diluted into 0.1 ml of assay buffer. Various amounts of this diluted material (0–1.5 mU) are added to 0.1 ml of *S. aureus*–IgG_2 complex suspensions (sufficient to bind 10^{-3} unit of cyclooxygenase activity) in 6×50 mm glass test tubes. Finally, 0.01 ml of $[^{125}I]IgG_1$ (*cyo*-3), containing 50,000 ^{125}I cpm originally (this amount is not altered to compensate for decay of the ^{125}I) is added. Assays are performed in duplicate, then incubated at 4° overnight. Pellets are collected by centrifugation (1500 *g* for 10 min at 24°), washed once in 0.2 ml of assay buffer, and recentrifuged. The supernatants are removed by aspiration, and the tubes containing the cell pellets inserted into vials and counted using a Beckman Biogamma γ-counter.

A positive, linear relationship between precipitated ^{125}I and purified PGH synthase[1] exists over the range of 1.35×10^{-6} to 1.35×10^{-5} cyclooxygenase units (0.15–1.5 ng; slope $\approx$250 cpm/unit) when using either the *cyo*-1- or *cyo*-5-IgG_{2b}–*S. aureus* cells as precipitating complexes (Fig. 2); however, minimal ^{125}I above control levels is bound using the *cyo*-7-IgG_2–*S. aureus* complex. The observation that binding of $[^{125}I]IgG_1$ (*cyo*-3) cannot occur to enzyme that is bound to *S. aureus* cells via the

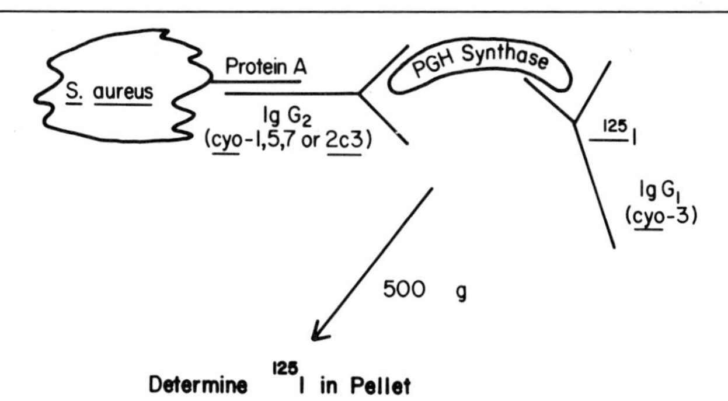

FIG. 1. Illustration of the interactions involved in the immunoradiometric assay for PGH synthase.

IgG_2 secreted by *cyo*-7 suggests that *cyo*-3 and *cyo*-7 secrete immunoglobulins directed against the same site. This conclusion is further supported by the observation that adsorption of intact sheep vesicular gland microsomes with an excess of IgG_2 secreted by *cyo*-7 prevents subsequent binding of $[^{125}I]IgG_1$ (*cyo*-3) to the microsomes; in contrast, adsorption of mi-

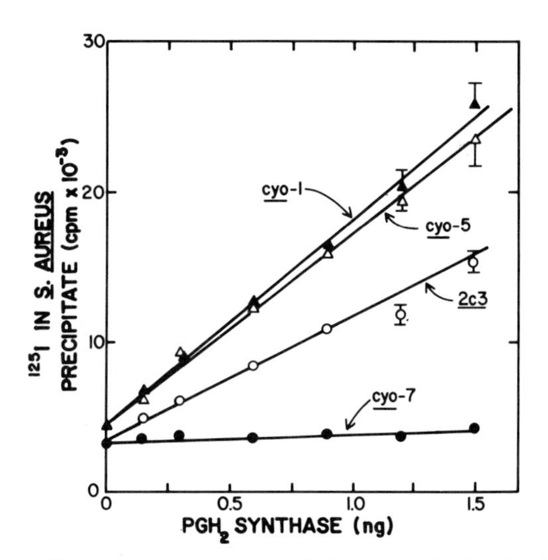

FIG. 2. Immunoradiometric assay using purified sheep vesicular gland PGH synthase.[1] IgG_{2b}–*Staphylococcus aureus* complexes were prepared as described in the text with $[^{125}I]IgG_1$ (*cyo*-3) and incubated overnight at 4° with various amounts of the purified PGH synthase. The cell pellets were collected and washed; cell-bound radioactivity was quantitated. The *S. aureus* precipitating complexes were with IgG_2 secreted by *cyo*-1 (▲—▲), *cyo*-5 (△—△), *cyo*-7 (●—●), or *2c3* (○—○).

crosomes with IgG_{2b} molecules secreted by *cyo*-1 and *cyo*-5 does not interfere with [^{125}I]IgG_1 (*cyo*-3) binding.

Curiously, substantial precipitation of [^{125}I]IgG_1 occurs with the *2c3*-IgG_2–*S. aureus* complex (Fig. 2) or with *S. aureus* cells alone (not shown) indicating that small amounts of PGH synthase are bound to *S. aureus* cells in the absence of an intervening antibody. However, the fact that no binding of [^{125}I]IgG_1 occurs with enzyme bound to the *cyo*-7-IgG_2–*S. aureus* complex indicates that the enzyme preferentially binds via antibody directed against it when that antibody is present on the *S. aureus* cells. The fact that PGH synthase can bind *S. aureus* cell directly is of no practical disadvantage in the immunoradiometric assay since the *cyo*-7-IgG_2–*S. aureus* complex provides a negative control.

As shown in Fig. 3, the purified PGH synthase and the enzyme from detergent solubilized sheep vesicular gland microsomes behave similarly when using *cyo*-5-IgG_2–*S. aureus* precipitating complexes. The PGH synthases from solubilized microsomes prepared from human platelets and from guinea pig and rabbit kidneys are considerably less reactive on a per

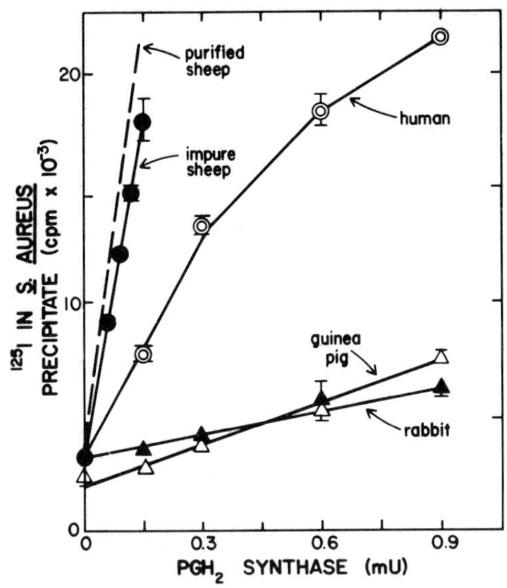

FIG. 3. Comparison of the reactivities of PGH synthases from different species in the immunoradiometric assay. IgG_2 (*cyo*-5)–*Staphyloccus aureus* precipitating complexes were used in all the experiments. Control using IgG_2 (*cyo*-7)–*S. aureus* cells gave a slope of zero. Experiments were performed as described in the text and the legend to Fig. 2 with solubilized microsomes[17] prepared from sheep vesicular glands (●—●), human platelets (○—○), guinea pig renal medulla (△—△), and rabbit renal medulla (▲—▲). A dashed line indicating the slope obtained with the purified sheep enzyme is also shown.

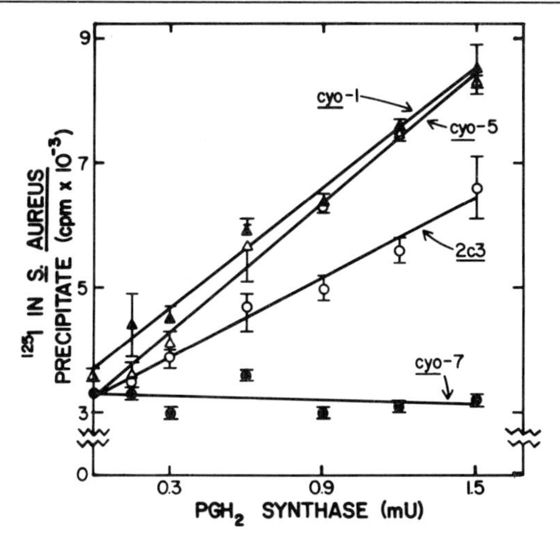

FIG. 4. Immunoradiometric assay using PGH synthase solubilized from rabbit renal medulla microsomes. Incubations were performed as described in the text and the legend to Fig. 2 except that rabbit kidney medulla microsomes solubilized with 1% Tween 20 were substituted for the sheep vesicular gland enzyme.

unit basis than the sheep enzyme; whether these differences represent species differences in antigen–antibody interactions or reflect species differences in PGH synthase turnover numbers is not known. However, as illustrated in detail in Fig. 4 for the rabbit kidney enzyme (qualitatively similar results are noted for the human platelet and guinea pig kidney enzyme), the immunoradiometric assay is useful for quantitating small amounts of cyclooxygenase from a variety of tissues. For purposes of day-to-day standardization, we have found it useful to generate a standard curve using pure PGH synthase (e.g., see Fig. 3). Moreover, it is important to include pure PGH synthase as an internal standard in some tissue samples to determine recovery of antigenic activity.

TABLE II

FACTORS INFLUENCING IMMUNORADIOMETRIC ASSAY FOR SYNTHASE FROM DETERGENT SOLUBILIZED SHEEP VESICULAR GLAND MICROSOMES

No effect on PGH synthase reactivity	Prevents reactivity of PGH synthase
1. Inactivation of enzyme with aspirin	1. Heating of enzyme at 60° for 10 min
2. Self-catalyzed destruction of enzyme	2. > 0.01% SDS in assay buffer
3. 0.1 *M* or 1 *M* NaCl in assay buffer	3. ≥2 *M* urea in assay buffer
4. 0, 0.1, 1% Tween 20 in assay buffer	
5. Allowing enzyme to stand 48 hr at 4°	

In an attempt to optimize the radiometric assay, a number of factors were investigated to determine their influence on the reactivity of both the purified and solubilized microsomal sheep PGH synthase in the assay. These results are summarized in Table II.

Acknowledgments

This work was supported in part by U.S. Public Health Service NIH Grants Nos. HD10013 and AM22042, by a Grant-in-Aid from the Michigan Heart Association, and by an Established Investigatorship (W. L. S.) from the American Heart Association.

[35] Monoclonal Antibodies against PGI_2 Synthase: An Immunoradiometric Assay for Quantitating the Enzyme

By DAVID L. DEWITT and WILLIAM L. SMITH

In this chapter we describe the preparation of two hybridoma lines producing monoclonal antibodies against two different antigenic sites on the PGI_2 synthase enzyme molecule. The hybridoma lines were derived from mice immunized with a partially purified preparation of PGI_2 synthase from bovine aorta. Using these antibodies we have developed an immunoradiometric assay with which to quantitate PGI_2 synthase protein concentrations. This assay is 50–100 times more sensitive than conventional radiochromatographic enzyme activity assays.

PGI_2 Synthase Purification

Bovine aorta obtained fresh at slaughter is frozen immediately on Dry Ice, then stored at $-80°$. The abdominal region of the aorta beginning 15–20 cm from the heart is easiest to homogenize. Frozen aorta is chopped into small pieces (ca. 2 cm^3) with a hammer and homogenized in 2–3 volumes of ice-cold 0.1 *M* Tris-chloride, pH 8.0, containing 1×10^{-4} *M* Fluribiprofen (or any PGH synthase inhibitor at an inhibitory concentration) with a Polytron (Brinkmann) homogenizer. Care should be taken to maintain the buffer temperature below 5° during homogenization. The homogenate is centrifuged at 10,000 *g* for 10 min, and the resulting supernatant is centrifuged for 35 min at 200,000 *g* to collect the microsomal pellet. When stored at $-80°$ the pellets retain their PGI_2 synthase activity for 2–3 months.

Further purification of PGI_2 synthase is performed using a modifica-

ISBN 0-12-181986-8

PARTIAL PURIFICATION OF PGI_2 SYNTHASE

Step	Protein concentration (mg/ml)	Specific activity (nmol of 6-keto-$PGF_{1\alpha}$ min^{-1} mg^{-1} protein)	Recovery of activity (%)	Purification (fold)
1. 10,000 *g* supernatant of homogenized aorta	4.5	12	100	—
2. Microsomal suspension (0.1 *M* Tris-chloride, pH 8.0)	3.9	23	45	2
3. 20,000 *g* supernatant of solubilized microsomes	1.05	94	41	7.8
4. Eluent from DE-52 column	0.73	110	24	9.2

tion of the method of Wlodawer and Hammarström.[1] Microsomal pellets (0.5 g) from 25–30 g of tissue are resuspended with a glass homogenizer in 10 ml of 0.1 *M* Tris-chloride, pH 8.0, containing 1×10^{-4} *M* Fluribiprofen. This homogenate is centrifuged at 200,000 *g* for 35 min. The washed pellet is resuspended in 10 ml of 10 m*M* sodium phosphate, pH 7.4, containing 0.5% Triton X-100 (Calbiochem) and again centrifuged at 200,000 *g* for 35 min. The supernatant is removed, and the solubilized PGI_2 synthase is applied to a DE-52 cellulose (Whatman) column (2 × 8 cm) equilibrated with 10 m*M* sodium phosphate, pH 7.4, containing 0.1% Triton X-100; the column is then washed with 60 ml of the equilibration buffer. PGI_2 synthase is eluted with 0.2 *M* sodium phosphate, pH 7.4, containing 0.1% Triton X-100. The specific activity of the PGI_2 synthase at this step ranges from 100 to 225 with an average of 150 units per milligram of protein per minute. [One unit of activity is defined as that amount of enzyme that will catalyze the formation of 1 nmol of PGI_2 per minute under standard assay conditions (described below).] This represents a purification of approximately 10-fold. A summary of a typical purification is presented in the table.

Immunization Protocol

Partially purified PGI_2 synthase (through De-52 chromatography) is used for immunization of outbred 4 to 6-week-old female ICR swiss white mice (Harlan Laboratories). Approximately 250 μg of protein (average

[1] P. Wlodawer and S. Hammarström, *FEBS Lett.* **97**, 32 (1979).

specific activity of 150 units per milligram of protein) in 0.2 ml of 0.2 *M* phosphate, pH 7.4, containing 0.1% Triton X-100 is emulsified by sonication with 200 μl of complete Freund's adjuvant (GIBCO) and injected intraperitoneally. After 2–4 weeks the mice are again inoculated using PGI_2 synthase emulsified in incomplete Freund's adjuvant. Three days after the second booster, mice are killed by cervical dislocation. Spleens are removed asceptically, and blood is collected to test for anti-PGI_2 synthase activity. Spleen cells from all mice are fused, but only those hybridomas from mice found to have anti-PGI_2 synthase activity in their serum are screened for anti-PGI_2 synthase activity.

Fusion of Mouse Spleen Cells

Spleen cells (1 to 5 × 10^7) from mice inoculated with PGI_2 synthase are fused with 1 to 5 × 10^6 HGPRT-negative SP2/0-Ag14 mouse myeloma cells as described in this volume[2] with the following modifications. After fusion, cells are suspended in 90 ml of complete HT medium and distributed into six 96-well tissue culture plates (Costar). After 24 hr, 150 μl of complete HAT medium (complete HT medium containing 1 μ*M* aminopterin) are added to each well. Two and four days thereafter, 150 μl of medium is removed from each well and replaced with 150 μl of fresh complete HAT medium. When the medium in a well with growing hybridomas begins to turn yellow (12–15 days after fusion), 200 μl of the spent medium are removed to test for anti-PGI_2 synthase antibody. The medium is replaced with 200 μl of complete HT medium.

Assay for Monoclonal Antibodies against PGI_2 Synthase

Staphylococcus aureus cells conjugated with rabbit anti-mouse IgG (Miles) can bind and precipitate all subclasses of mouse IgGs. When mixed with medium containing mouse anti-PGI_2 synthase antibody, the newly formed *S. aureus*–rabbit anti-mouse IgG–mouse IgG complex will precipitate solubilized PGI_2 synthase. This precipitate can be assayed for PGI_2 synthase activity. Immunoglobulin classes other than IgG are not detected by this method.

Staphylococcus aureus (Cowan strain I) is grown and attenuated by the method of Kessler as described by DeWitt *et al.*[2] The rabbit anti-mouse IgG–*S. aureus* complexes are prepared by washing, collecting (by centrifugation), and resuspending 5 ml of a 10% *S. aureus* cell suspension (w/v) as follows: (*a*) twice with 5 ml of 0.1 *M* Tris-chloride, pH 8.0, con-

[2] D. L. DeWitt, J. S. Day, J. A. Gauger, and W. L. Smith, this volume [34].

taining 5% (w/v) bovine serum albumin and 0.5% (v/v) Triton X-100; (*b*) once with 5 ml of 0.1 *M* Tris-chloride, pH 8.0, containing 0.5% (v/v) Triton X-100 and 250 μl of rabbit anti-mouse IgG (Miles Laboratories; ca. 2.5 mg of IgG per milliliter); and (*c*) once with 5 ml of 0.1 *M* Tris-chloride, pH 8.0, containing 0.5% (v/v) Triton X-100. Finally, the rabbit anti-mouse IgG–*S. aureus* is resuspended in 5 ml of the last buffer.

To assay for the presence of PGI_2 synthase antibody, either 50 μl of serum from mice immunized with PGI_2 synthase preparations or 200 μl of medium from a well containing a growing hybridoma (after the medium turns yellow) is mixed with 0.1 ml of the rabbit anti-mouse IgG–*S. aureus* suspension. The mixture is vortexed and centrifuged at 1500 *g* on a desk-top centrifuge, and the supernatant is removed by aspiration. After resuspending the cell pellet in 0.5 ml of 0.1 *M* Tris-chloride, pH 8.0, containing 0.5% Triton X-100, solubilized PGI_2 synthase (ca 15 units) is added. The PGI_2 synthase used in this screening assay is obtained from solubilizing microsomes as described above, but prior to chromatography on DE-52. The mixture is vortexed briefly and again the *S. aureus* cells are pelleted by centrifugation at 1500 *g* for 5 min. The supernatant is removed by aspiration, and the pellet is resuspended a second time in 1 ml of 0.1 *M* Tris-chloride containing 0.5% Triton X-100. PGI_2 synthase activity is then assayed as described below.

Assay of PGI_2 Synthase

For PGI_2 synthase assay, $[^3H]PGH_2$ is synthesized[3] using [5,6,8,9,11,12,14,15-^{3}H-(N)] arachidonic acid (New England Nuclear 62.2 Ci/mmol) diluted to a specific activity of 3.7 μCi/mg (ca 1000 cpm/nmol) with unlabeled arachidonic acid (NuChek Prep). To assay, 30 nmol of $[^3H]PGH_2$ in 10 μl of dry-acetone is evaporated under a stream of N_2 in a test tube and enzyme (or resuspended *S. aureus* complex) in a total of 1 ml of 0.1 *M* Tris-chloride, pH 8.0, containing 0.5% Triton X-100 is added to initiate the reaction. After incubation for 1 min at 24°, the reaction is stopped by adding 7 ml of chloroform–methanol (1:1, v/v); then 1.8 ml of 0.02 *M* HCl and 3 ml of chloroform are added, and the mixture is mixed vigorously. The aqueous layer is removed by aspiration and the organic layer is evaporated with N_2. Finally, the residue is redissolved in 100 μl of chloroform and applied quantitatively to a 250 μm silica gel G thin-layer chromatography plate (Analtech) along with authentic 6-keto-$PGF_{1\alpha}$ and $PGF_{2\alpha}$ standards. The plate is developed twice to a height of 20 cm using the upper layer of ethyl acetate–isooctane–acetic acid–H_2O

[3] G. Graff, this volume [49].

(110:50:20:100, v/v/v/v). After drying the plate in air, the standards are visualized with I_2 vapor. The region corresponding to 6-keto-$PGF_{1\alpha}$ is marked, and the rest of each vertical lane is divided into three or four regions. All areas are then scraped into individual scintillation vials, and samples are counted in 7 ml of Bray's solution.[4] The percentage of counts chromatographing with authentic 6-keto-$PGF_{1\alpha}$ is determined, and the nanomoles of PGI_2 formed are calculated (nanomoles of $[^3H]PGH_2$ in the reaction mixture times the percentage of radioactivity chromatographing with 6-keto-$PGF_{1\alpha}$). Under maximal conditions as much as 85% of PGH_2 can be converted to PGI_2.

It must be noted that this assay does not yield a well-defined rate. One reason is that the substrate concentration is not saturating at all times, since in a typical assay 40–60% of substrate is converted to product. Also the reaction rate is not linear over the entire assay time, possibly because PGI_2 synthase is inactivated during the reaction.[5,6] However, the assay does provide a convenient estimation of relative enzyme activities and is useful for purification and for screening sera and hybridoma media for antibodies to PGI_2 synthase.

Hybridomas Producing Monoclonal Antibodies to PGI_2 Synthase

The medium from two of 75 hybridoma-containing wells from a single fusion was found to precipitate PGI_2 synthase activity, and cells from these wells were cloned to form two anti-PGI_2 synthase antibody-producing cell lines, *isn*-1 and *isn*-3. Both hybridomas secrete a mouse IgG_1 as determined by Ouchterlony double diffusion using subclass specific antisera. Each monoclonal antibody will cause the immunoprecipitation of an iodinated protein of molecular weight equal to 50,000 from iodinated, solubilized bovine aortic microsomes.

The two hybridoma lines were grown in media free of bovine IgG, and pure mouse IgG_1 was isolated by chromatography on protein A–Sepharose (Pharmacia). IgG_1 (*isn*-3) was iodinated using Bolton–Hunter reagent. The procedures for isolation and iodination of mouse IgG are described in detail in this volume.[2]

$[^{125}I]IgG_1$ (*isn*-3) is precipitated after incubation with nonsolubilized bovine aortic microsomes and centrifugation at 200,000 *g* for 30 min. Unlabeled IgG_1 (*isn*-3) competes with $[^{125}I]IgG_1$ (*isn*-3) for binding to the mi-

[4] G. A. Bray, *Anal. Biochem.* **2,** 279 (1960).

[5] J. A. Salmon, D. R. Smith, R. J. Flower, S. Moncada, and J. R. Vane, *Biochim. Biophys. Acta* **523,** 250 (1978).

[6] K. Watanabe, S. Yamamoto, and O. Hayaishi, *Biochem. Biophys. Res. Commun.* **87,** 192 (1979).

crosomes. In contrast, IgG_1 (*isn*-1) has no effect on the binding of [^{125}I]IgG_1 (*isn*-3) indicating that the IgG_1 molecules secreted by *isn*-3 and *isn*-1 bind different antigenic sites. Since the two antibodies bind different sites, they can be used in a double-antibody radioimmunometric assay similar to the one developed for the PGH synthase.[2]

A complex of IgG_1 (*isn*-1) bound to *S. aureus* was prepared by adding 100 μg of purified IgG_1 (*isn*-1) in 0.1 *M* sodium citrate, pH 7.0, to 1 ml of a 10% suspension of rabbit anti-mouse IgG–*S. aureus* cell complex. After 5 min the *S. aureus* cells are pelleted by centrifugation and resuspended in 1 ml of 0.1 *M* Tris-chloride, pH 8.0, containing 0.5% Triton X-100. This IgG_1 (*isn*-1)–*S. aureus* complex is stable for at least 2 weeks when stored at 4°.

A standard curve for quantitating PGI_2 synthase protein is generated as follows. Aliquots of solubilized bovine aortic microsomes (containing 0–0.05 unit of PGI_2 synthase; microsomes solubilized in 0.1 *M* Tris-chloride, pH 8.0, containing 0.5% Triton X-100) are added to 6 × 50 mm glass test tubes each containing 100,000 cpm of [^{125}I]IgG_1 (*isn*-3) and allowed to stand for 30 min at 24°; next, 10 μl of the IgG_1 (*isn*-1)–*S. aureus* complex are added, and the tubes are vortexed and centrifuged immediately at 1500 *g* for 10 min at 24°. Any long delay (>5 min) before centrifugation will increase the degree of nonspecific precipitation of ^{125}I to unacceptable levels. After centrifugation, the supernatant is removed by aspiration and the pellets are washed once in 0.5 ml of the solubilization buffer. The washed cell pellets present in the 6 × 50 mm test tubes are placed in vials and counted on a Beckman Biogamma counter.

A positive linear relationship between precipitated ^{125}I and added PGI_2 synthase activity exists over the range of 0.005–0.05 unit of activity (Fig. 1). The slope is equal to 300,000 cpm precipitated per unit of PGI_2 synthase. This assay is approximately 50–100 times more sensitive than the enzyme activity assays. When IgG_1 (*isn*-3)–*S. aureus* complexes are used as controls for IgG_1 (*isn*-1)–*S. aureus* cells, no ^{125}I above background is precipitated.

This immunoradiometric assay provides a simple, sensitive, and highly specific method for quantitating PGI_2 synthase. The method should be useful for measuring changes in PGI_2 synthase protein concentrations in tissues during physiological stresses such as aging[7] and the development of atherosclerosis.[8] The antibodies should also be of value in the im-

[7] R. S. Kent, B. B. Kitchell, D. G. Shand, and A. R. Whorton, *Prostaglandins* **21,** 483 (1981).

[8] A. Dembinska-Kiec, T. Gryglewski, A. Zmuda, and R. J. Gryglewski, *Prostaglandins* **14,** 1025 (1977).

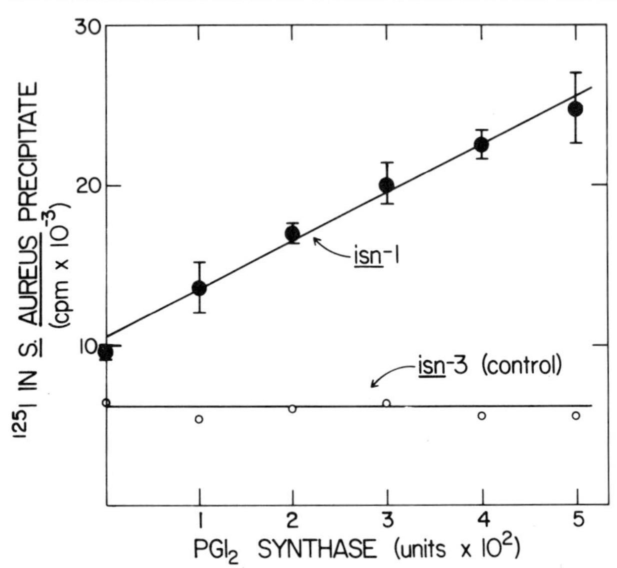

FIG. 1. Immunoradiometric assay of PGI_2 synthase using solubilized bovine aortic microsomes. PGI_2 synthase was incubated with [^{125}I]IgG_1 (*isn*-3) for 30 min, then *Staphylococcus aureus* cells conjugated to IgG_1 secreted by either *isn*-1 (●—●) or *isn*-3 (○—○) was added. The *S. aureus* were pelleted by centrifugation and washed, and precipitated ^{125}I was quantitated. Results are averages of triplicates (●—●); error bars ± SD.

munocytochemical localization of PGI_2 synthase at both the cellular and subcellular levels.

Acknowledgments

This work was supported in part by U.S. Public Health Service NIH Grant No. HD10013, by a Grant-In-Aid from the Michigan Heart Association, and by an Established Investigatorship (W. L. S.) from the American Heart Association.

[36] Radioimmunoassay and Immunochromatography of 12-L-Hydroxyeicosatetraenoic Acid

By RICHARD A. MORGAN and LAWRENCE LEVINE

Metabolism of arachidonic acid occurs via two enzymic systems: cyclooxygenase pathways, which produce the prostaglandins, prostacyclin, and thromboxane, and lipoxygenase pathways, which mediate the formation of hydroxyeicosatetraenoic acids and leukotrienes. The cyclooxy-

ISBN 0-12-181986-8

genase products, covalently linked to carrier molecules, are immunogenic,[1] and radioimmunoassays for their measurement have been developed.[2] Leukotrienes are also immunogenic; a leukotriene that contains a cysteinylglycine function (leukotriene D_4), when conjugated to bovine albumin, elicited antibodies to leukotriene D_4, and a radioimmunoassay for leukotrienes has been developed.[3] Although lipids are poor immunogens,[4,5] hydroxyarachidonic acids, which are unsaturated and contain one or more hydroxyl functions, also elicit specific antibodies; immunization with a conjugate of 12-hydroxyeicosatetraenoic acid (12-L-HETE) and human albumin has produced in rabbits antibodies directed toward 12-L-HETE,[6] and immunization with a conjugate of 15-HETE and bovine albumin has elicited in rabbits antibodies directed toward 15-HETE.[7]

Here we describe the production and the radioimmunoassay for 12-HETE. We also present two high-performance liquid chromatography procedures to separate mono- and dihydroxy acids from other arachidonic acid metabolites.

Preparation of Conjugate

12-L-HETE, purified by and obtained from T. J. Carty, was conjugated to human albumin using the method of Bauminger.[8] 12-L-HETE (8 mg) was dissolved in *N,N*-dimethylformaide (100 μl). To this, 3 mg of dicyclohexylcarbodiimide and 3.5 mg of *N*-hydroxysuccinimide were added, and the mixture was stirred for 30 min at 25°. The *N,N'*-dicyclohexylurea precipitate was removed by centrifugation, and the supernatant fluid was added to 500 μl of 0.1 *N* $NaHCO_3$ containing 12.5 mg of human albumin. The reaction solution was stirred for 2 hr at 4° followed by dialysis in the cold against five changes of 2 liters of phosphate buffer (0.15 *M* NaCl, 0.005 *M* sodium phosphate, pH 7.5). After dialysis the final conjugate volume of 2.5 ml was stored at −20°.

[1] L. Levine and H. Van Vunakis, *Biochem. Biophys. Res. Commun.* **47,** 888 (1972).

[2] E. Granström, *Prostaglandins* **15,** 3 (1978).

[3] L. Levine, R. A. Morgan, R. A. Lewis, K. F. Austen, A. Marfat, and E. J. Corey, *Proc. Natl. Acad. Sci., U.S.A.* **78,** 7692 (1981).

[4] M. M. Rapport and L. Graf, *Prog. Allergy* **13,** 273 (1969).

[5] C. R. Alving, *in* "The Antigens" (M. Sela, ed.), Vol. IV, p. 1. Academic Press, New York, 1977.

[6] L. Levine, I. Alam, H. Gjika, T. J. Carty, and E. J. Goetzl, *Prostaglandins* **20,** 923 (1980).

[7] D. H. Hwang and R. W. Bryant, International Symposium on Leukotrienes and Other Lipoxygenase Products, Florence, Italy (June 10–12, 1981).

[8] S. Bauminger, U. Zor, and H. R. Lindner, *Prostaglandins* **4,** 313 (1973).

Immunization

Two milligrams of the conjugate in complete Freund's adjuvant were used to immunize a rabbit. Injections were made intramuscularly in the hind limbs and in the toe pads. After 2 weeks, and in the 3 following weeks, 40 ml of blood were collected in 0.1 *M* EDTA (final EDTA concentration was 0.01 *M*) and centrifuged to obtain the plasmas. The plasmas were stored at −20°. Three months after the first injection, the rabbit was boosted via the same procedure with 3.0 mg of the conjugate again in complete Freund's adjuvant. One week later, and for several weeks following, plasmas were collected and stored at −20°. A second, and final, boost was given 6 months after the primary injection, and again plasmas were obtained over successive weeks. Antibodies directed toward 12-HETE were detected after the primary immunization and increased after each booster injection.

Radioimmunoassay (RIA)

A double-antibody technique was used to separate free antigen from antibody-bound antigens in solution.[9] Assays were carried out in 3.5-ml polypropylene test tubes (No. 535, Walter Sarstedt, Inc., Princeton, New Jersey). All dilutions were made in Tris buffer (0.01 *M* Tris, 0.14 *M* NaCl, pH 7.5, containing 0.1% gelatin). Approximately 12,000 cpm of [^{3}H]12-HETE (New England Nuclear, Boston, Massachusetts), 100 μl of a 1:900 dilution of the immune plasma, and known concentrations of ligand (for standard curve construction) or the unknown sample were added to the test tubes in a total volume of 300 μl, with mixing (100 μl of a 1:900 dilution of normal rabbit plasma was used to obtain nonspecific binding). The reaction mixtures were incubated for 1 hr at 37°, after which carrier rabbit IgG (in the form of 100 μl of a 1:25 dilution of normal rabbit plasma) was added with mixing. One hundred microliters of goat-plasma anti-rabbit IgG (previously titered to be at equivalence or in antibody excess with respect to the rabbit IgG antigen) were added with mixture and incubated at 4° for 16–18 hr. The immune precipitates were collected by centrifugation at 1500 *g* for 60 min. The supernatant fluids were decanted, and the tubes were kept inverted in test tube racks lined with paper towels. The sides of the test tube above the precipitates were wiped dry with folded strips of filter paper (Whatman No. 1) with care not to disturb the precipitates during the wiping procedure. The immune precipitates were then

[9] H. Van Vunakis and L. Levine, *in* "Immunoassays for Drugs Subject to Abuse" (S. J. Mulé, I. Sunshine, M. Brande, and R. E. Willette, eds.), p. 23. CRC Press, Cleveland, Ohio, 1974.

dissolved in 200 μl of 0.1 *N* NaOH. Scintillation fluid (2.5 ml) was added, and the test tubes were capped with push-in polyethylene stoppers (Walter Sarstedt, Inc., Princeton, New Jersey), mixed, and counted for radioactivity.

Serological Specificity

The [^{3}H]12-HETE anti-12-HETE binding was inhibited most effectively by 12-HETE (50% inhibition with 1.4 ng of 12-HETE). The other hydroxy fatty acids tested also reacted with the anti-12-HETE, but only 5% or less as effectively as 12-HETE (see the table). The 12-OH function between two double bonds appears to be an immunodominant function. Because of this cross-reactivity with the anti-12-HETE, a direct quantitation of 12-HETE levels cannot be made unless it is known that the sample in question contains only 12-HETE. This ambiguity was circumvented by separating the hydroxy fatty acids by high-performance liquid chromatography (HPLC).

High-Performance Liquid Chromatography—Normal Phase

For the resolution of cyclooxygenase products and hydroxyeicosatetraenoic acids two Waters Model 6000A pumps (Waters Associates, Milford, Massachusetts), Model 660 solvent programmer, Model U6K injector, and a Waters μ-Porasil column (30 cm × 3.9 mm) were used. The solvent system consisted of hexane–acetic acid (125:1, v/v solvent A) eluted isocratically for 11 min followed by linear gradient from solvent A

SEROLOGIC SPECIFICITY OF THE [^{3}H]12-HETE ANTI-12-HETE REACTION

Ligand	Percent serologic activity
12-L-HETE	100[a]
12-L-Hydroxyheptadecatrienoic acid (HHT)	5
Leukotriene B_4	5
5-HETE	5
8+9-HETE	3.3
11-HETE	3.3
15-HETE	1.3
Leukotriene D_4	0.8
Leukotriene C_4	0.2
Leukotriene E_4	0.1
PGE_1, E_2, $F_{1\alpha}$, 6-K-$F_{1\alpha}$, TxB_2	<1

[a] 1.4 ng of 12-L-HETE inhibited binding 50%.

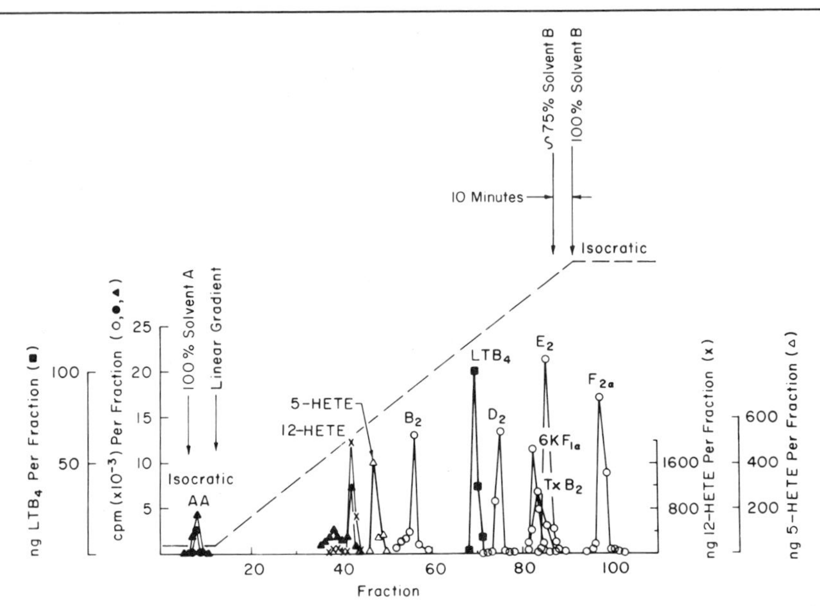

FIG. 1. Separation by high-performance liquid chromatography (normal phase) of arachidonic acid (AA), 12-hydroxyeicosatetraenoic acid (12-HETE), 5-HETE, LTB_4, PGB_2, PGD_2, 6-keto-$PGF_{1\alpha}$ (6-K-$F_{1\alpha}$), thromboxane B_2 (TxB_2), PGE_2, and $PGF_{2\alpha}$. The standard arachidonic acid metabolites were radiolabeled with ^3H (●, ○) and ^{14}C (▲). [^{14}C]12-L-HETE, 5-HETE, and LTB_4 were also measured serologically by inhibition of the binding of antibodies to 12-L-HETE. From A. Rigas and L. Levine, *Prostaglandins and Medicine* **7**, 217 (1981).

to chloroform–methanol–acetic acid (125:5:1, v/v/v; solvent B) for the total time of 76 min. Final elution with 100% solvent B (attained over 10 min) was run isocratically for total time of 100 min. The flow rate for the entire program was set at 1 ml/min per tube. All solvents were obtained from Fisher Chemicals, Fairlawn, New Jersey (C_6H_{14}, $CHCl_3$, CH_3OH HPLC grade, acetic acid reagent grade).

Resolution of lipoxygenase products 12-HETE, 5-HETE, and LTB_4, as well as several cyclooxygenase products by this normal-phase HPLC is shown in Fig. 1. The retention times of 12-HETE and its quantitation were obtained by RIA using purified 12-HETE and a calibration curve generated with the purified 12-HETE, whereas the retention times of 5-HETE and LTB_4 and their quantitation were obtained with standard 5-HETE and LTB_4 preparations and from calibration curves generated with the heterologous ligands. Methods for identification and quantification of the cyclooxygenase products have been reported in this series.[10] The

[10] I. Alam and L. Levine, this series, Vol. 73, p. 275.

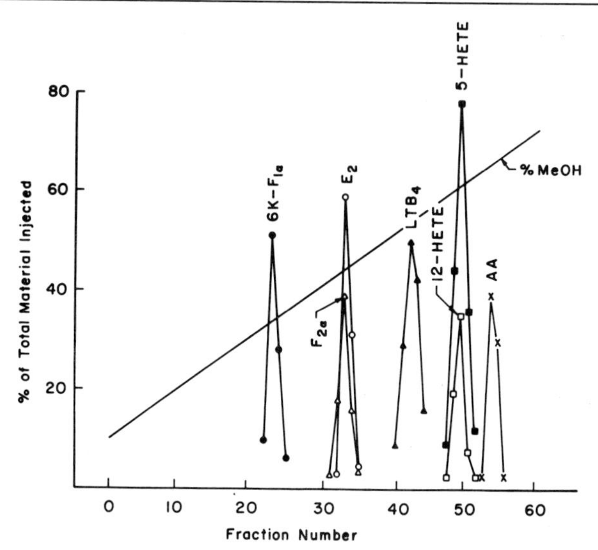

FIG. 2. Separation by high-performance liquid chromatography (reversed phase) of 6-keto-$PGF_{1\alpha}$, $PGF_{2\alpha}$, PGE_2, LTB_4, 12-HETE, 5-HETE, and arachidonic acid. The arachidonic acid was radiolabeled. The other arachidonic acid metabolites were assayed serologically. The LTB_4, the 12-HETE, and the 5-HETE were measured with the antiserum to 12-HETE.

mono- and dihydroxyeicosatetraenoic acids could be resolved by this normal-phase HPLC procedure, but the leukotrienes containing amino acid residues were adsorbed, but could not be eluted. To resolve the leukotrienes and the cyclooxygenase products, a procedure that utilized the HPLC in the reverse phase mode was used.

High-Performance Liquid Chromatography—Reversed Phase

Reversed-phase HPLC was performed as described by Parker *et al.*[11] The Waters Model 6000A pumps (Waters Associates, Milford, Massachusetts), Model 660 solvent programmer, Model U6K injector, and a C_{18} fatty acid analysis column 3.9 × 300 mm were used. A sample run consisted of a linear gradient from 100% solvent A to 100% solvent B for 100 min at a flow rate of 1 ml/min collecting 1-ml fractions. Solvent A consisted of 93.4% 0.01 *M* phosphate buffer, pH 7.4, 6% MeOH (HPLC grade Fisher Chemicals, Fairlawn, New Jersey), 0.6% *t*-amyl alcohol (Aldrich, Milwaukee, Wisconsin). Solvent B was 99.4% MeOH with 0.6% *t*-amyl alcohol. Fractions were collected in the same tubes used in the ra-

[11] C. W. Parker, S. F. Falkenhein, and M. M. Huber, *Prostaglandins* **20,** 863 (1980).

dioimmunoassay. For analysis of individual fractions by RIA the alcohols were evaporated by blowing N_2 over each fraction. The chromatographic properties of some standard arachidonic acid metabolites on the reversed-phase HPLC system are shown in Fig. 2. The reversed-phase system does not separate the mono-HETEs; their resolution is achieved by normal-phase HPLC (Fig. 1). However, this reversed-phase system resolves leukotrienes C_4 D_4 E_4 (see this volume [37]).

Acknowledgments

This work was supported by Grants GM-27256 and CA-17309 from the National Institutes of Health. This is publication 1370 from the Department of Biochemistry, Brandeis University, Waltham, Massachusetts 02254. L. L. is a Research Professor of Biochemistry of the American Cancer Society (Award PRP-21).

[37] Radioimmunoassay of the 6-Sulfido-Peptide-Leukotrienes and Serologic Specificity of the Anti-Leukotriene D_4 Plasma

By LAWRENCE LEVINE, ROBERT A. LEWIS, K. FRANK AUSTEN, and E. J. COREY

The constituents of slow-reacting substance of anaphylaxis (SRS-A) generated with calcium ionophore or immunologically have been defined by comparison with totally synthetic structures as 5(*S*)-hydroxy,-6(*R*)-*S*-glutathionyl-7,9-*trans*,11,14-*cis*-eicosatetraenoic acid (leukotriene C_4, LTC_4)[1,2] and its 6-sulfidocysteinylglycine (leukotriene D_4, LTD_4)[3-5] and 6-sulfidocysteine (leukotriene E_4, LTE_4)[6] analogs. The limited sensitivity and reproducibility of the bioassay of SRS-A leukotrienes[3,7] and the fur-

[1] R. C. Murphy, S. Hammarström, and B. Samuelsson, *Proc. Natl. Acad. Sci. U.S.A.* **76,** 4275 (1979).

[2] E. J. Corey, D. A. Clark, G. Goto, A. Marfat, C. Mioskowski, B. Samuelsson, and S. Hammarström, *J. Am. Chem. Soc.* **102,** 1436 (1980).

[3] R. A. Lewis, K. F. Austen, J. M. Drazen, D. A. Clark, A. Marfat, and E. J. Corey, *Proc. Natl. Acad. Sci. U.S.A.* **77,** 3710 (1980).

[4] H. R. Morris, G. W. Taylor, P. J. Piper, and J. R. Tippins, *Nature* (*London*) **285,** 104 (1980).

[5] L. Örning, S. Hammarström, and B. Samuelsson, *Proc. Natl. Acad. Sci. U.S.A.* **77,** 2014 (1980).

[6] R. A. Lewis, J. M. Drazen, K. F. Austen, D. A. Clark, and E. J. Corey, *Biochem. Biophys. Res. Commun.* **96,** 271 (1980).

[7] R. P. Orange and K. F. Austen, *in* "Methods in Immunology and Immunochemistry" (C. A. Williams and M. W. Chase, eds.), Vol. 5, p. 145. Academic Press, New York, 1976.

METHODS IN ENZYMOLOGY, VOL. 86

ISBN 0-12-181986-8

ther limitation in sensitivity as well as in numbers of samples conveniently assessed by detection of absorbance at 280 nm after high-performance liquid chromatography (HPLC)[3] in the reversed phase prompted the development of a 6-sulfido-peptide-leukotriene class-specific antiserum for use in a radioimmunoassay of the class and of its components after their resolution by reversed-phase HPLC.[8]

Preparation of Leukotrienes and Leukotriene Analogs

The preparative routes to LTC_4, LTD_4, LTE_4, and their respective 11-*trans* analogs[2,6,9] Δ^{14}-dihydro-LTC_4[2]; 6-*epi*-LTD_4[10]; the 12-*S*-glutathionyl position isomer of LTC_4[11]; the 11-*S*-glutathionyl,12-hydroxy analog of LTC_4[12,13]; the deamino, homocysteinyl, and D-penicillamyl analogs of LTD_4[14]; LTB_4[15]; and 5-HETE[16,17] have been described. The 7-*cis*-$\Delta^{9,11,14}$-hexahydro analogs of LTC_4 and LTD_4 and the 7-*trans*-$\Delta^{9,11,14}$-hexahydro analog of LTC_4 were synthesized from the corresponding LTA_4 analogs.[2,13]

Production of Immunogen and Antibodies

The dimethyl ester of *N*-trifluoroacetyl LTD_4[2,6] was treated with 2.5 equivalents of lithium hydroxide in dimethoxyethane–water (4:1) at 23° for 1.5 hr to yield the corresponding C-1 monoacid (eicosanoid carboxyl-free, glycine carboxyl remaining as methyl ester), which was purified by silica gel thin-layer chromatography (TLC) with methylene chloride–methanol (9:1) and was obtained in a 56% yield. Reaction of this monoacid with 4 equivalents of triethylamine and 2 equivalents of isobutyl chloroformate in a minimum of dry dimethoxyethane at −25° to −30° for 15 min produced the mixed anhydride of the (Gly)-monomethyl ester

[8] L. Levine, R. A. Morgan, R. A. Lewis, K. F. Austen, D. A. Clark, A. Marfat, and E. J. Corey, *Proc. Natl. Acad. Sci. U.S.A.* **78,** 7692 (1981).

[9] E. J. Corey, D. A. Clark, A. Marfat, and G. Goto, *Tetrahedron Lett.* **21,** 3143 (1980).

[10] E. J. Corey, and G. Goto, *Tetrahedron Lett.* **21,** 3463 (1980).

[11] E. J. Corey, and D. A. Clark, *Tetrahedron Lett.* **21,** 3547 (1980).

[12] E. J. Corey, A. Marfat, and G. Goto, *J. Am. Chem. Soc.* **120,** 6607 (1980).

[13] J. M. Drazen, R. A. Lewis, K. F. Austen, M. Toda, F. Brion, A. Marfat, and E. J. Corey, *Proc. Natl. Acad. Sci. U.S.A.* **78,** 3195 (1981).

[14] R. A. Lewis, J. M. Drazen, K. F. Austen, M. Toda, F. Brion, A. Marfat, and E. J. Corey, *Proc. Natl. Acad. Sci. U.S.A.* **78,** 4579 (1981).

[15] E. J. Corey, A. Marfat, J. Munroe, K. S. Kim, P. B. Hopkins, and F. Brion, *Tetrahedron Lett.* **22,** 1077 (1981).

[16] E. J. Corey, J. O. Albright, A. E. Barton, and S. Hashimoto, *J. Am. Chem. Soc.* **102,** 1435 (1980).

[17] E. J. Corey and S. Hashimoto, *Tetrahedron Lett.* **22,** 299 (1981).

of *N*-trifluoroacetyl LTD_4 and the isobutyl ester of carbonic acid, as analyzed by TLC (as above), with a 90% estimated yield. To this solution at −25° was added bovine albumin (Sigma Chemical Co., St. Louis, Missouri; crystallized and lyophilized, globulin-free) in deionized distilled water (with a 50:1 molar ratio of leukotriene to protein) as a 40 mg/ml solution.

The reaction mixture was gradually warmed to 0° over 1 hr and stirred at 0° for an additional 1.5 hr. Methanol (3 volumes) and 0.15 *M* aqueous potassium carbonate (~100 equivalents based on leukotriene) were added, and the solution was stirred at 23° for 1 hr to effect hydrolytic cleavage of *N*-trifluoroacetyl and ester functions. After neutralization with glacial acetic acid, the mixture was chromatographed on Sephadex G-25 with 10% methanol in water used as eluent. The fractions containing the conjugate (by ultraviolet absorbance detection) were combined and dialyzed against distilled water for 15 hr at 0°. The resulting solution of leukotriene–bovine albumin conjugate was analyzed by ultraviolet absorbance and after correction of protein absorbance at 280 nm (0.5 mg/ml, $A = 0.26$) it was determined that the molar ratio of LTD_4:protein was 7.

A 5-month-old New Zealand white rabbit responded to an intramuscular injection of 500 μg of LTD_4-bovine albumin conjugate in complete Freund's adjuvant followed by a subcutaneous injection 3 weeks later with 250 μg of LTD_4–bovine albumin in incomplete Freund's adjuvant and was boosted intramuscularly and subcutaneously at 3 months with 250 μg of LTD_4-bovine albumin in incomplete Freund's adjuvant. Blood was collected in acid–citrate–dextrose and centrifuged at 400 *g*, and the plasma was separated.

Preparation of 11-trans-[^3H]LTC_4

14,15-Ditritiated leukotriene A_4 (LTA_4) was prepared by the catalytic tritiation (Lindlar Pd-Pb catalyst) of 14,15-dehydro-LTA_4 (synthesized in a manner analogous to LTA_4 itself).[2] The reaction of 14,15-ditritiated LTA_4 with glutathione yielded both LTC_4 and 11-*trans*-LTC_4, which were resolved by C_{18} reversed-phase HPLC.[9]

Radioimmunoassay (RIA)

The radioimmunoassays (RIA) were performed in 3.5-ml polypropylene test tubes (No. 535; Walter Sarstedt, Inc., Princeton, New Jersey). The diluent for all reagents was Tris buffer (0.01 *M* Tris, pH 7.4, containing 0.14 *M* NaCl and 0.1% gelatin). Appropriately diluted immune rabbit plasma to LTD_4-bovine albumin, 11-*trans*-[^{3}H]LTC_4 (40 Ci/mmol, New England Nuclear), and either standard compounds or unknown samples were added to the test tubes in a total volume of 300 μl with mixing after each addition; the reactions were incubated at 37° for 60 min. A 100-μl portion of goat anti-rabbit IgG plasma (previously titrated to equivalence or antibody excess with respect to the rabbit IgG antigen in the immune plasma) was added, the mixture was shaken, and the rabbit IgG-goat anti-rabbit IgG complex was precipitated overnight at 4°. The immune precipitates were centrifuged at 1500 *g* for 60 min at 4°, and the supernatant fluids were decanted. The precipitates were dissolved in 200 μl of 0.1 *N* NaOH, and 2.5 ml of scintillation fluid were added to each tube. The tubes were stoppered with polyethylene push-in stoppers (Walter Sarstedt, Inc.), thoroughly mixed, and analyzed in a liquid scintillation counter (Beckman, LS7500, Fullerton, California) with an efficiency of 63%.

Coprecipitation of 11-*trans*-[^{3}H]LTC_4 with the anti-leukotriene rabbit IgG–goat anti-rabbit IgG complex increased with increasing amounts of rabbit anti-leukotriene plasma. Of the 5880 cpm of 11-*trans*-[^{3}H]LTC_4 added, 1340 cpm were bound by 9.5 μl of immune rabbit plasma and goat anti-rabbit IgG. After incubation of the ligand and rabbit nonimmune plasma for 60 min at 37° followed by addition of the second antibody, 200 cpm were bound; thus, specific binding to antibodies was 20%. The combining sites of the antibodies in 9.5 μl of immune plasma were saturated with 2222 cpm of 11-*trans*[^{3}H]LTC_4. At a counting efficiency of 63%, 9.5 μl of the immune plasma contains 0.02 pmol of 11-*trans*-[^{3}H]LTC_4 binding IgG molecules, and each milliliter of undiluted rabbit plasma contains 2.1 pmol (0.32 μg) of specific antibody. An average association constant, K_0, was calculated to be 2.8×10^9 mol^{-1} at 37°.

Fifty percent inhibition of 11-*trans*-[^{3}H]LTC_4-anti-LTD_4 binding by the synthetic SRS-A leukotrienes LTC_4, LTD_4 and LTE_4 occurred in the 500 pg range (Table I) indicating an assay sensitivity one to two logs greater than that afforded by biological or physical assay methods.

The immunodominant role of the 6(*R*)-sulfido-peptide domain of LTC_4, LTD_4 and LTE_4 is demonstrated by the lack of reactivity of the antiserum with arachidonic acid, PGE_1, PGA_1, 5-HETE and the relative lack of reactivity with LTB_4, which possesses a 5(*S*)-hydroxyl and a triene structure, but lacks the sulfido-peptide subunit. As the antibody recognized LTC_4, LTD_4, and LTE_4 equally well, the C-terminal glycine of LTC_4 or LTD_4 and the N-terminal glutamic acid of LTC_4 do not block access of the antibody to the immunodominant cysteinyl domain. Further, the exact structure of the 6-sulfidocysteine was not discriminated by the anti-LTD_4, since the homocysteinyl and deamino analogs of LTD_4 were inhibitors equally effective to LTD_4. The importance of the position

TABLE I
INHIBITION OF 11-*trans*-[^{3}H]LTC_4 ANTI-LTD_4 BINDING[a]

Ligand	Mass required for 50% inhibition (ng)
LTC_4	0.33
11-*trans*-LTC_4	0.39
Δ^{14}-Dihydro-LTC_4	0.51
LTD_4	0.66
11-*trans*-LTD_4	0.23
LTE_4	0.72
11-*trans*-LTE_4	0.41
Deamino-LTD_4	0.24
Homocysteinyl-LTD_4	0.60
D-Penicillamyl-LTD_4	3.7
6-*epi*-LTD_4	11.0
5-Hydroxy-12-*S*-glutathionylleukotriene analog	25.0

[a] Data are taken from Levine *et al.*[8] The following compounds inhibited 0% with 10 ng: 11-*S*-glutathionyl-12-hydroxyleukotriene analog; 7-*cis*-hexahydro-LTC_4; 7-*cis*-hexahydro-LTD_4; 7-*trans*-hexahydro-LTC_4. The following compounds inhibited less than 25% with 100 ng: LTB_4; 5-hydroxyeicosatetraenoic acid (5-HETE); glutathione; cystinylbisglycine; arachidonic acid; PGE_1; PGA_1.

and spatial orientation of the 6-sulfidocysteinyl function to the serologic activity of the rabbit anti-LTD_4 was demonstrated by the decreased activity of the 6-*epi*-LTD_4 (5% of LTD_4) and the 5-hydroxy, 12-*S*-glutathionylleukotriene analog (2% of LTC_4), and the apparent lack of activity (<5%) of the 11-*S*-glutathionyl–12-hydroxyleukotriene analog. That the integrity of the lipid is required for serologic activity was demonstrated by the lack of activity of glutathione, cystinylbisglycine, and 7-*cis*- and 7-*trans*-hexahydro-LTC_4 (<5%).

Immunochromatography

Since the antibodies do not differentiate among the various 6-sulfidopeptide-containing leukotrienes, these compounds must be resolved from one another before identification and individual quantitation by RIA. A single chromatographic separation on C_{18} reversed-phase HPLC allows the resolution of each of the SRS-A leukotrienes and some additional 5-lipoxygenase products, namely 5-HETE and LTB_4, from one another and from the PGH synthase products.[18]

The Waters Model 6000A pumps, Model 660 solvent programmer,

[18] R. A. Morgan and L. Levine, this volume [36].

TABLE II
COMBINED HIGH-PERFORMANCE LIQUID CHROMATOGRAPHY AND RADIOIMMUNOASSAY (RIA) OF SOME LIPOXYGENASE PRODUCTS

Arachidonic acid metabolite	Retention time (min)[a]	Percent recovery
LTC_4	35	48[b]
11-*trans*-LTC_4	36	44,[b] 52[c]
LTD_4	38	70[b]
11-*trans*-LTD_4	41	44[b]
LTE_4	43	60[b]
11-*trans*-LTE_4	47	90[b]
LTB_4	43	50[d]
5-HETE	50	78[d]
12-HETE	50	38[b]

[a] Mean (SD = 2 min).
[b] By RIA.
[c] By absorbance at 280 nm.
[d] By RIA using heterologous reaction with anti-12-HETE.

Model U6K injector, and a 3.9 × 300 mm C_{18} fatty acid analysis column (Waters Associates, Milford, Massachusetts) were used for reversed-phase HPLC. Each sample was run on a linear gradient program from 100% solvent A to 100% solvent B for 100 min at a flow rate of 1 ml/min, and 1-ml fractions were collected. Solvent A consisted of 93.4% 0.01 *M* phosphate buffer, pH 7.4, 6% methanol (HPLC grade, Fisher Chemical Co., Fairlawn, New Jersey), and 0.6% *t*-amyl alcohol (Aldrich, Milwaukee, Wisconsin); solvent B was 99.4% methanol with 0.6% *t*-amyl alcohol.

The reversed-phase HPLC retention times and recoveries of 9 oxidative products of lipoxygenases, after chromatography of radiolabeled and/or unlabeled reference compounds, are presented in Table II as identified by RIA and/or ultraviolet absorbance. Recoveries of compounds were calculated as a percentage of the mass, assessed spectrophotometrically at their respective ultraviolet absorbance maxima, applied to HPLC. Separation of the lipoxygenase products from one another was suitable for quantitation. Recoveries of the leukotrienes and other lipoxygenase products by RIA (44–90%).[19,20] The combined RIA–HPLC method allows detection of about 0.2 ng of LTC_4, LTD_4, and LTE_4, in compari-

[19] I. Alam and L. Levine, this series, Vol. 73, p. 275.
[20] I. Alam, K. Ohuchi, and L. Levine, *Anal. Biochem.* **93,** 339 (1979).

son to 5–10 ng, 1–2 ng, and 5–10 ng, respectively, by combined HPLC and bioassay,[13] and 25–30 ng of each leukotriene by reversed-phase HPLC with detection of absorbance at 280 nm.

Acknowledgments

This work was supported in Boston by Grants AI-07722, AI-10356, HL-17832, and RR-05669 from the National Institutes of Health and a grant from the Lillia Babbitt Hyde Foundation; in Cambridge by the National Science Foundation; and in Waltham by GM-27256 from the National Institutes of Health. L. L. is an American Cancer Society Research Professor of Biochemistry (Award PRP-21), and R. A. L. is the recipient of an Allergic Diseases Academic Award (AI-00399).

[38] Problems of PGE Antisera Specificity

By FERNAND DRAY, SUZANNE MAMAS, and BISTRA BIZZINI

Prostaglandins E (PGEs) are particularly unstable because of the existence of a β-hydroxyketone moiety. In both mild acid or alkaline pH, PGE is converted to PGA loss of a proton at C-10 results in a negative charge, expulsion of the hydroxyl group at C-11 and formation of a C-10–C-11 double bond. At pH > 10, PGA undergoes internal rearrangement to form the relatively stable PGB[1,2] (C-8–C-12 double bond). The same compounds PGA and PGB can be produced under enzymic conditions.[3-6] A dehydrase that converts PGE to PGA was detected in human serum, and two isomerases were measured in sera of various species: one converting PGA to PGC (C-11–C-12 double bond) and the other converting PGC to PGB.[6] The structure and ultraviolet (UV) absorption (λ_{max}) for these compounds are shown in Fig. 1.

Between 1971 and 1976 efforts were made to raise specific antibodies in animals immunized against PGE_1 or PGE_2 coupled through a peptidic bond to an antigenic carrier. The results were generally negative because the antisera recognized essentially PGA or PGB.[7-9] In a few cases, the results were apparently positive; i.e., the serum of such animals con-

[1] S. Bergström, R. Ryhage, B. Samuelsson, and J. Sjövall, *J. Biol. Chem.* **238,** 3555 (1963).
[2] N. H. Andersen, *J. Lipid Res.* **10,** 320 (1969).
[3] R. L. Jones, *Biochem. J.* **119,** 64p (1970).
[4] H. Polet and L. Levine, *Biochem. Biophys. Res. Commun.* **45,** 1169 (1971).
[5] E. Horton, R. Jones, C. Thompson, N. Poyser, *Ann. N.Y. Acad. Sci.* **180,** 351 (1971).
[6] H. Polet and L. Levine, *J. Biol. Chem.* **250,** 351 (1975).
[7] R. M. Zusman, B. V. Caldwell, and L. Speroff, *Prostaglandins* **2,** 41 (1972).
[8] L. Levine, R. M. Gutierrez Cernosek, and H. Van Vunakis, *J. Biol. Chem.* **246,** 6782 (1971).

ISBN 0-12-181986-8

Prostaglandin	Structure		λ_{max} (nm)	ϵ_m
PGE_2	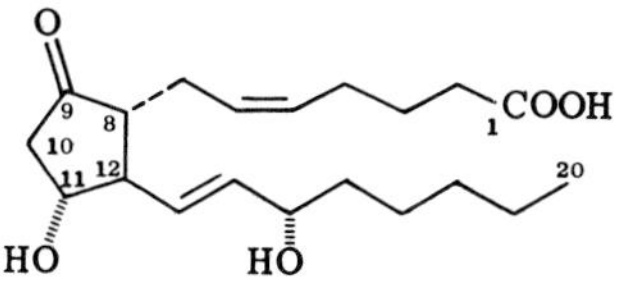 [structure: 9-oxo, 11,15-dihydroxy prosta-5,13-dienoic acid; COOH, 1, 20, 8, 9, 10, 11, 12, HO, HO]		No absorption	—
		Acid/base or dehydrase ↓		
PGA_2	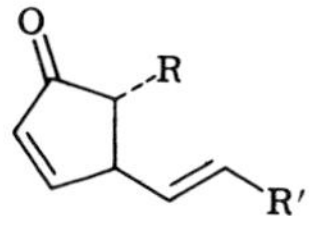 [structure: O, R, R′]		224	9030
PGC_2	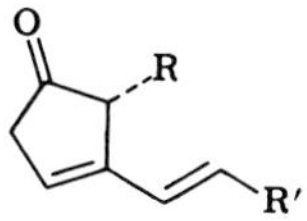 [structure: O, R, R′]	Base pH > 10 or isomerases	[228–234–243][a]	$16,000 < \epsilon_m < 21,000$
PGB_2	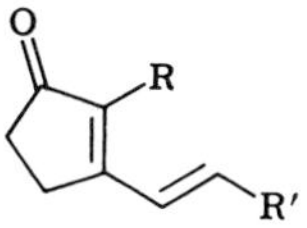 [structure: O, R, R′]	↓	284	23,400

[a] Values identical to those given for PGC_1.[24]

FIG. 1. Ultraviolet spectra of transformation products of PGE_2. λ_{max} (nm) in 0.02 *M* CO_3Na_2, 10% ethanol solution; ϵ_m: molar extinction coefficient.

tained relatively specific PGE antibodies that were convenient for developing a corresponding radioimmunoassay (RIA), but these assays, when controlled, showed that PGA and/or PGB antibodies were also present and that these were in the majority and had higher binding affinities.[10–18]

[9] A. Raz, M. Schwartzman, R. Kenig-Wakshal, and E. Perl, *Eur. J. Biochem.* **53,** 145 (1975).
[10] M. Jaffe, J. W. Smith, W. T. Newton, and C. W. Parker, *Sciences* **171,** 494 (1971).
[11] S.-C. Yu, L. Chang, and G. Burke, *J. Clin. Invest.* **51,** 1038 (1972).
[12] S.-C. Yu and G. Burke, *Prostaglandins* **2,** 11 (1972).
[13] W. Jubiz, J. Frailey, C. Child, and K. Bartholomew, *Prostaglandins* **2,** 471 (1972).
[14] W. Stylos, L. Howard, E. Ritzi, and R. Skarnes, *Prostaglandins* **6,** 1 (1974).
[15] A. Jobke, B. A. Peskar, and B. M. Peskar, *FEBS Lett.* **37,** 192 (1973).
[16] S. Bauminger, U. Zor, and H. R. Lindner, *Prostaglandins* **4,** 313 (1973).
[17] E. M. Ritzi and W. A. Stylos, *Prostaglandins* **8,** 55 (1974).
[18] P. Christensen and P. P. Leyssac, *Prostaglandins* **11,** 399 (1976).

TABLE I
PGE_1 ANTISERA[a]

Hapten	Carrier	Coupling agent	Coupling procedure	^3H Tracer(s) SA (Ci/mM) (*PG)	Binding data[b]	IC_{50} PGE[c] (pg)	References
PGE_1-poly-L-lysine	Succinilated KLH	EDC	—	E_1 87.5	$*B_1 \gg *A_1 \gg *E_1$	Null	Levine *et al.*[8]
				A_1 81.0			
				B_1 81.0			
PGE_1	KLH	Triethylamine ethylchloroformate	0.1 *M* $NaHCO_3$, 10°, 1 hr	E_1 87.5	$(*E_1)\ \underset{105}{A_1} \geqslant \underset{100}{E_1} > \underset{80}{E_2}$	>150	Jaffe *et al.*[10]
PGE_1	HSA	Ethylchloroformate		E_1 88.0	R No. 21-3 $(*E_1)\ E_1 > E_2 > A_1 \gg A_2 > B_1 > B_2$	>150	Yu *et al.*[11]; Yu and Burke[12]
				A_1 88.0	R No. 22-4 $(*B_1)\ B_1 \geqslant A_1 > E_2 > E_1 > B_2$		
				B_1 88.0	For each rabbit, data given only for one tracer.		
PGE_1	Porcine IgG	EDC		E_1 87.0	$(*E_1)\ \underset{100}{E_1} > \underset{25}{E_2} > \underset{5}{B_1} \ggg A_1 = \underset{<0.1}{A_2}$	≃150	Jubiz *et al.*[13]
				A_1 81.3	No data with $*A_1$ or $*B_1$		

PGE_1	Bovine Tg	EDC	Solubilization with trypsin	E_1	110.0	$(^*E_1)\ \underset{100}{E_1} > \underset{15}{E_2} \ggg A_1 = A_2 = B_1 = B_2$ (<0.1) No data with $^*A_1$ or $^*B_1$	≃100	Stylos *et al.*[14]
PGE_1	BSA	EDC	Hapten dissolved in DMF; pH 6.6	E_1	55.0	$(^*E_1)\ B_1 > A_1 > E_1 > E_2$	Null	Raz *et al.*[9]
PGE_1	Bovine Tg	EDC	Solubilization with trypsin			$(^*E_1)\ \underset{100}{E_1} > \underset{12}{E_2} \gg A_1 = B_1$ (<1) No data with $^*A_1$ or $^*B_1$	≃500	Raz *et al.*[9]
PGE_1	BSA	EDC	0.02 *M* Na_2CO_3, 10% ethanol, pH 5.5, 22° overnight	E_1 B_1 A_1	110.0 53.0 46.0	S No. 144: $^*E_1 \gg {^*B_1} \gg {^*A_1}$ $(^*E_1)\ E_1 \ggg A_1 > B_1$ (K_a = 1.9 $10^{10}M^{-1}$) $(^*B_1)\ B_1 \ggg A_1 > E_1$ (K_a = 4.10^9M^{-1}) R No. C1: $^*E_1 \geqslant {^*A_1} \gtrsim {^*B_1}$ $(^*E_1)\ E_1 = B_1 > E_2 > A_1 > B_2 > A_2$ R No. C4: $^*B_1 \gg {^*A_1} > {^*E_1}$	39 100	Maclouf *et al.*[20] Maclouf *et al.*[20]

[a] Key to abbreviations: BSA, bovine serum albumin; HSA, human serum albumin; KLH, keyhole limpet hemocyanin; Tg, thyroglobulin; EDC, 1-ethyl-3-(3-dimethylaminopropyl)carbodiimide-HCl; DMF, *N,N*-dimethylformamide, SA, specific activity; R, rabbit; S, sheep.

[b] Binding data are successively given as follows: the binding with each tracer E_1, A_1, and B_1, the order of the cross-reaction for each one, the percentage of cross-reaction, and the association constant (K_a).

[c] IC_{50}, corresponds to the amount of PGE required to displace 50% of $[^3H]$PGE.

The difficulties incurred to obtain antisera containing antibodies raised almost exclusively against the PGE structure may explain why since 1976 no new data have been published.

We met the same difficulties in spite of a large program of immunizations against PGE_1 and mainly PGE_2, which included 64 rabbits, 3 sheep, 3 goats, and 4 guinea pigs. In only two animals (rabbit No. 79585 immunized against PGE_2-BSA and sheep No. 144 immunized against PGE_1-BSA), the PGE antibodies were in the majority and had high binding affinities allowing very sensitive RIAs[19–21] to be developed. In all other animals, the immune response was anti-PGB and/or anti-PGA.

The main binding properties of PGE antisera published in the literature or found by our group are given in Table I for PGE_1 and in Table II for PGE_2.

In this presentation, some factors that may alter the PGE structure during immunogen preparation or immunization will be discussed and some suggestions will be made as to methods that may be used to reduce these alterations or even to prevent them if a stable PGE derivative is used for coupling[22,23] (Table III).

Preparation and Storage of the Immunogen

As dehydration of the 11-hydroxyl position with formation of A and B series prostaglandins can occur under various chemical conditions, it was important to evaluate the degree of such alterations due to the coupling procedures and to the conditions of purification and storage of the immunogen.

Coupling Procedure. To prepare an immunogen with a prostaglandin structure as hapten, the carboxyl group of the prostaglandin was engaged in a peptidic bond with amino groups of various antigenic carriers. Either carbodiimide or mixed anhydride methods were used and will be described with PGE_2 as an example.

Carbodiimide Method. PGE_2(10 mg) and $[^3H]PGE_2$ (4×10^6 dpm) were dissolved in 10 ml of an aqueous solution of 0.02 *M* Na_2CO_3 and 10% ethanol. The pH was adjusted at 5.5. Then 1-ethyl-3-(3-dimethylaminopropyl)carbodiimide-HCl (EDC) (10 mg) and BSA (20 mg) were added successively. The pH was maintained between 5.4 and 5.6 for 1 hr

[19] F. Dray, B. Charbonnel, and J. Maclouf, *Eur. J. Clin. Invest.* **5,** 311 (1975).

[20] J. Maclouf, J. M. Andrieu, and F. Dray, *FEBS Lett.* **56,** 273 (1975).

[21] F. Dray, *in* "Prostaglandins, Prostacyclin, and Thromboxanes Measurement. Development in Pharmacology" (J. M. Boeynaems and A. G. Herman, eds.), Vol. 1, p. 17. Nijhoff, The Hague, 1980.

[22] F. A. Fitzpatrick and G. L. Bundy, *Proc. Natl. Acad. Sci. U.S.A.* **75,** 2689 (1978).

[23] F. Dray, unpublished observations.

TABLE II
PGE_2 ANTISERA[a]

Hapten	Carrier	Coupling agent	Coupling procedure	^3H Tracer(s)	SA (Ci/m*M*) (*PG)	Binding data[b]	IC_{50} PGE[c] (pg)	References
PGE_2	BSA	EDC	Hapten dissolved in DMF	A_1		$(*A_1)\ A_2 > A_1 > E_2 > E_1 > B_2$, No data with $*E_2$, $*A_2$ and $*B_2$	Null	Zusman *et al.*[7]
PGE_2	BSA	EDC	22° overnight	E_1	68.5	$(*E_1)\ \underset{100}{E_1} > \underset{15}{E_2} > A_2 \gg A_1 = B_1 = \underset{<0.1}{B_2}$	≃200	Jobke *et al.*[15]
				A_1	68.5	$(*A_1)\ A_2 > A_1 > B_2 \gg E_2$		
				B_1	68.5	$(*B_1)\ B_2 > A_2 > E_2$		
PGE_2	BSA	DCC *N*-Hydroxy-succinimide	Hapten dissolved in DMF	E_2	160.0	$(*E_2)\ E_2 > E_1 > A_1 > A_2 > B_1 = B_2$	>100	Bauminger *et al.*[16]
				B_2	160.0	$(*B_2)\ B_2 \gg E_2$ No data with $*A_2$		
PGE_2	KLH	EDC		E_2	130.0	$(*E_2)\ E_2 > E_1 \gg A_1 = A_2 = B_1 = B_2$ No data with $*A_2$ and $*B_2$	1000	Ritzi and Stylos[17]
PGE_2	BSA	EDC		E_2	130.0	$(*E_2)\ B_2 > A_2 > E_2 > E_1$		
PGE_2	Bovine Tg	EDC		E_2	139.0	$(*E_2)\ \underset{100}{E_2} > \underset{19}{E_1} \gg A_2 = \underset{<1}{B_2}$ No data with $*B_2$ and $*A_2$	≃500	Raz *et al.*[9]
PGE_2	Hen IgG	4-Methylmorpholine isobutyl chlorocarbonate	Solvent: dioxane, 4°	E_2	160.0	$(*E_2)\ \underset{100}{E_2} > \underset{14}{E_1} \gg \underset{3}{A_2} > \underset{<1}{A_1} > \underset{<1}{B_2}$ No data with $*A_2$ and $*B_2$	≃200	Christensen and Leyssac[18]
PGE_2	BSA	EDC	0.02 *M* Na_2CO_3, 10% ethanol, pH 5.5, 22° overnight	E_2	160.0	R No. 79585: $*E_2 = *B_2 \gg *A$	10	Dray[21]
				B_2	114.0	$(*E_2)\ \underset{100}{E_2} \ggg \underset{0.3}{A_2} > \underset{0.01}{B_2}\ (K_a = 8.2 \times 10^{10} M^{-1})$		
				A_2	160.0	$(*B_2)\ \underset{100}{B_2} \ggg \underset{0.08}{A_2} > \underset{0.03}{E_2}\ (K_a = 1.8 \times 10^{9} M^{-1})$		
						R No. 109: $*B_2 \gg *A_2 > *E_2$		

[a] Key to abbreviations: BSA, bovine serum albumin; Tg, thyroglobulin; KLH, keyhole limpet hemocyanin; EDC, 1-ethyl-3-(3-dimethylaminopropyl)carbodiimide-HCl; DCC, *N,N*-dicyclohexylcarbodiimide; DMF, *N,N*-dimethylformamide; SA, specific activity; R, rabbit.

[b] Binding data are successively given as follows: the binding with each tracer E, A, and B, the order of the cross-reaction for each one, the percentage of cross-reaction, and the association constant (K_a).

[c] IC_{50} corresponds to the amount of PGE required to displace 50% of [^{3}H]PGE.

TABLE III

ANTI-PGE_2 ANTIBODIES OBTAINED WITH A MODIFIED HAPTEN.[a]

Hapten	Carrier	Coupling agent	Coupling procedure	^3H Tracer(s) SA (Ci/mmol) (*PG)		Binding data[b]	IC_{50} PGE^c (pg)	References
9-Deoxy-9-methylene-$PGF_{2\alpha}$	BSA	EDC	Hapten dissolved in DMF; pH 5.5 25° overnight	E_2	117.0	$(^*E_2)\ \underset{100}{E_2} = E_1 \gg \underset{4.9}{A_1} > \underset{2.8}{A_2} > B_1 = \underset{0.1}{B_2}$ $K_a = 1.5 \times 10^9 M^{-1}$	>100	Fitzpatrick and Bundy[22]
	KLH		Hapten dissolved in DMF; pH 7.0 25° overnight	E_2	117.0	$(^*E_2)\ \underset{100}{E_2} > \underset{90}{E_1} \gg A_1 = A_2 \underset{<1}{=} B_1 = B_2$ No data with $^*A_2$ and $^*B_2$	20	Fitzpatrick and Bundy[22]
PGE_2-9-methoxime	BSA	EDC	Hapten dissolved in DMF; pH 5.5 22° overnight	E_2	160.0	$^*E_2 \ggg {}^*B_2 = {}^*A_2$	50	Dray[23]
				A_2	160.0	$(^*E_2)\ \underset{100}{E_2} > \underset{21}{E_1} \ggg A_2 \underset{<0.2}{=} B_2$		
				B_2	114.0	$K_a = 5.8 \times 10^9\ M^{-1}$		

[a] Key to abbreviations: BSA, bovine serum albumin; KLH, keyhole limpet hemocyanin; EDC, 1-ethyl-3-(3-dimethylaminopropyl)carbodiimide-HCl; DMF, *N,N*-dimethylformamide; SA, specific activity.

[b] Binding data are successively given as follows: the binding with each tracer E_2, A_2, and B_2, the order of the cross-reaction for each one, the percentage of cross-reaction, and the association constant (K_a).

[c] IC_{50}, corresponds to the amount of PGE required to displace 50% of [^3H]PGE.

TABLE IV
EVOLUTION OF ULTRAVIOLET SPECTRA DURING COUPLING[a]

Compounds	Concentration of total PGs (nmol/ml)	Concentration of prostaglandins at different wavelengths (nmol/ml)		
		283 nm (PGB_2)	243 nm (PGC_2)	224 nm (PGA_2)
PGE_2	2800	3.6	0	0
PGE_2 + EDC + BSA				
Immediately	2800	3.6	100	0
At the end of coupling	2800	10	100	0
PGE_2-BSA after dialysis	594	24	100	0
PGE_2-BSA after 1 month's storage at −20°	594	78	364	0
PGB_2	3000	3000	0	0
PGB_2-BSA after dialysis	180	180	0	0

[a] All optical densities have been measured against reference solutions containing all the reagents except prostaglandins, in buffer solution 0.02 *M* CO_3Na_2, pH 5.5; 10% ethanol.

at 25°. The mixture was allowed to stand overnight at 4° and then dialyzed against distilled water at 4° for 24 hr.

Mixed Anhydride Method. PGE_2 (28 mg) and [^{3}H]PGE_2 (4×6^{10} dpm) were dissolved in dioxane (1 ml), and 36 μl of tri-*n*-butylamine were added. The solution was cooled at + 10°, and 21.5 μl of isobutyl chloroformate were added, and the mixture was cooled for 20 min at 4° (the solution solidified). The mixture was then thawed and added to a cooled and stirred solution of 70 mg of BSA in 3.4 ml of water–dioxane mixture (v/v, 1:1). The solution was adjusted at pH 7 with 0.1 *N* NaOH, stirred for 1 hr at 4°, then dialyzed against distilled water for 24 hr.

Stability of PGE_2 (Table IV). This was estimated immediately after dialysis in both compartments by immunoreactivity and UV absorption (e.g., see Fig. 1) and 1 month after storage of the immunogen at − 20° by UV absorption only. In any case the addition of tritiated PGE_2 before coupling allowed us to estimate the amount of "PG" material found after coupling.

PGE_2 immunoreactivity of serial dilutions of the first dialysis water was compared to that of equivalent amounts of standard PGE_2. The exact superposition of both dose-response curves argued against any significant alteration of the structure of PGE_2 molecule. Furthermore the identical UV spectrum for the "PG" material and the standard PGE_2 strengthened this assertion.

The same procedure was applied to the immunogen PGE_2-BSA. Dose-response curves were established for PGE_2 standard and corresponding dilutions of the immunogen, respectively, using [^{3}H]PGE_2 as tracer and PGE_2 antiserum from which BSA antibodies had been removed: the relative inhibition (IC_{50}) was only 11%. Furthermore, the UV spectrum of the immunogen, established against BSA, showed the appearence of two small but significant peaks at 283 and 243 nm. The first peak corresponded to PGB_2, as shown in an identical experiment using PGB_2 as hapten, and represented 2.3 residues per molecule of BSA. The second one at 243 nm may correspond to the very unstable PGC (intermediate structure between PGA and PGB[6,24]) and represented 3 residues per molecule of BSA. As the epitope density calculated from radioactivity was 18 residues per molecule of BSA, it means that at least one-third of PG material coupled to the carrier was found altered after coupling and before addition of adjuvant.

This percentage of alteration was found after coupling by the carbodiimide method. Less alteration (20%) was found immediately after coupling by the mixed anhydride method, but several coupling reactions would have to be performed with either method before giving any significance to this difference.

Nature of the Carrier

Immunizations against PGE were made using various proteins as antigenic carriers. Bovine serum albumin was claimed to be a factor that altered the PGE structure[9] and other proteins, such as keyhole limpet hemocyanin (KLH), bovine thyroglobulin (bTg) and porcine or hen immunoglobulins (pIg or hIg) have been considered less drastic (Tables I–III). Used as carrier, BSA corresponded generally to fraction V of Cohn, crystallized, electrophoretically pure, but not cleared of associated molecules such as ions and fatty acids, and tending to give unforeseen properties to this macromolecule; it is interesting to mention that PGA and PGC isomerase activities were found in BSA solutions.[6]

A systematic study was undertaken to evaluate the degree of *in vitro* alteration of the PGE structure in the presence of various proteins for 24 hr at 4° (coupling temperature) and at 37° (body temperature). The alteration of the structure was followed by modifications of the UV spectrum. The alteration was relatively important at 37° in the presence of most proteins, particularly of albumins. It is probable that this process may also take place in the body of the recipient animal before it is recognized by the lymphocyte. During that time, endogenous materials, such as

[24] R. L. Jones, *J. Lipid Res.* **13,** 511 (1972).

proteins, particularly specific enzymes such as PGE dehydrase and PGA and PGC isomerases, can add their own effects to those previously described.

In conclusion, the exclusive noxious role of albumins is not evident, and the results of numerous immunizations against PGE_2 coupled to various proteins such as BSA, KLH, Tg, or IgG show that the question remains open. Indeed, except in one animal (rabbit No. 79585) which produced both specific PGE_2 and PGB_2 antibody populations, all animals (whatever the carrier) produced specific PGB_2 or to a lesser degree PGA_2 antibody populations. This good responder to PGE_2 was immunized against PGE_2 coupled to BSA. BSA was again used as carrier in the case of PGE_1, and both specific PGE_1 and PGB_1 antibody populations were raised in sheep No. 144.

Choice of Animals

PGE dehydrase and PGA and PGC isomerase activities were found in the blood of animals and seem to differ from one species to another.[4] It is postulated that these enzyme activities subsist in spite of the fact that their substrate binds to macromolecules. If that is the case, then one could reduce these endogenous factors using species with very low enzyme activities: sheep ($n = 2$), goats ($n = 2$), rabbits ($n = 5$) and guinea pigs ($n = 4$) were chosen because of the large difference in isomerase activities, guinea pigs have been the less active enzyme. All animals were immunized with the same preparation of PGE_2-BSA immunogen and under the same conditions in order to reduce and to render uniform the *in vitro* influences. All animals produced essentially PGB_2 or to a lesser extent PGA_2 and very few PGE_2 antibodies. These negative results show that the "species" factor was not determining. On the contrary, it is clear that animals from the same species, not genetically selected, respond differently to the same immunogen. Six rabbits, in another experiment, were immunized against PGE_2-BSA under identical conditions: three of them produced essentially PGB_2 antibody populations, and three last specific PGE_2 and PGB_2 antibodies, but only one (rabbit No. 79585) with a higher affinity of PGE_2 antibodies. As the degree of *in vitro* alteration of PGE_2 was identical for each animal, one can suppose that this unique good responder had low enzymic activities (dehydrase or isomerase) and/or a high immunological recognition for PGE_2 epitopes.

Comments and Proposals

It has been shown that the immunological machinery of animals from various species working against haptened immunogens prepared with

PGE molecule produced, with few exceptions, antibody populations that recognized essentially PGB or PGA, but not PGE. Even if an anti-PGE response appeared, it was never exclusive but more often a minority in comparison to anti-PGB or anti-PGA populations.

It has been shown also that a lot of factors could occur *in vitro* and *in vivo* to alter the structure of PGE differently according to whether the factor was or was not coupled to the carrier. All these factors will determine the nature and the density of various and unknown epitopes presented to competent immunocytes.

Here are some suggestions if, in spite of this long story of failure, one persists in immunizing animals against the PGE hapten.

1. Select a protein as carrier that alters as little as possible the PGE structure in solution at 37° and after 1 or 2 days of contact.
2. Choose the less drastic coupling procedure for the hapten (pH, temperature, solvent, time elapsed until lyophilization, etc.).
3. Distribute the immunogen solution in small fractions appropriate for all immunizations, and lyophilize them for storage.
4. Calculate by immunoreactivity and UV spectrum the percentage of initial PGE as epitope before the first immunization. Do not proceed to immunization if this percentage is too low.
5. Immunize a lot of animals previously selected for their low PGE dehydrase and PGA and PGC isomerase activities.[6]
6. Establish the binding parameters of antisera for $[^3H]$PGE, $[^3H]$PGA and $[^3H]$PGB, respectively.
7. In the case of a successful immunization, i.e., an important but not exclusive production of PGE antibody populations, it is possible to withdraw the PGA or PGB antibody populations after purification of the antiserum by affinity chromatography using PGA or PGB–AH-Sepharose 4B.
8. Satisfying all these obligations is only a part of success when polyclonal antibodies are raised. An alternative and more recent way exists—the production of monoclonal antibodies. The screening of hydridomas would allow the recognition of antibodies produced against the unaltered PGE molecule.

Another approach, which may be more successful, consists in stabilizing PGE by a minor modification of the structure introduced before coupling, inasmuch as the antibody populations recognize specifically and essentially PGE and its stabilized derivative. For this purpose, two procedures were developed. In the first one,[22] the stable derivative 9-deoxy-9-methylene-$PGF_{2\alpha}$ was coupled to BSA or KLH. Specific PGE_2 antibody populations were raised in rabbits with an affinity as shown in Table III. However, the total success of this procedure cannot be defended since no binding data were obtained with $[^3H]PGA_2$ and $[^3H]PGB_2$, respectively. In the second procedure, another stable derivative—PGE_2-9-methoxime—was coupled to BSA using the carbodiimide method. Ten rabbits were immunized, and the binding parameters were evaluated for eventual PGE, PGA, or PGB antibody populations using corresponding tritiated PG. All animals produced essentially PGE_2 antibody populations, and Table III gives binding data concerning the best bleed from the best responder (rabbit No. 12079). The sensitivity was inferior to that obtained previously (rabbit No. 79585), but it is possible to increase the sensitivity after methoxime transformation of both radioactive and nonradioactive competitors.

[39] Enzyme Immunoassay of $PGF_{2\alpha}$

By YOKO HAYASHI and SHOZO YAMAMOTO

Enzyme immunoassay of prostaglandin (PG) $F_{2\alpha}$ has been developed as a replacement of radioimmunoassay. In this method the antigen $PGF_{2\alpha}$ is labeled with an enzyme (β-galactosidase or alkaline phosphatase) rather than radioactivity, and the immune complex is measured by assaying enzyme activity. Since the application of this method to biological materials is now under investigation, this chapter describes the assay procedures with authentic $PGF_{2\alpha}$.

β-Galactosidase-Linked Immunoassay

Preparation of Anti-$PGF_{2\alpha}$

$PGF_{2\alpha}$ is conjugated with bovine serum albumin by the mixed anhydride reaction using isobutyl chloroformate.[1] The conjugate (1 mg with respect to protein) is dissolved in 1 ml of water and emulsified with an equal volume of Freund's complete adjuvant. A male New Zealand white rabbit weighing about 2 kg receives 1 ml of the immunogen by subcutaneous injections of five 0.2-ml portions along the dorsal surface. The immunization is performed every second week for about 3 months, in all 7 times. Blood is collected through the ear marginal vein for the titration of antiserum by radioimmunoassay. Serum from a nonimmunized rabbit is used as a control. With this immunization schedule, rabbits develop antisera that can be used at a final dilution of about 1 to 3 × 10^6. About 100–150 ml

[1] B. F. Erlanger, F. Borek, S. M. Beiser, and S. Liebermann, *J. Biol. Chem.* **234,** 1090 (1959).

ISBN 0-12-181986-8

of blood are collected from one rabbit through a polyethylene tubing inserted into the carotid artery.

Enzyme Labeling of $PGF_{2\alpha}$

β-Galactosidase from *Escherichia coli* (a specific activity of 30 μmol/min per milligram of protein at 25° with lactose as substrate) is a product of Boehringer (Mannheim). A carboxyl group of $PGF_{2\alpha}$ is linked to an amino group of enzyme by the mixed anhydride method.[1] $PGF_{2\alpha}$ (6 mg) is dissolved in 100 μl of dioxane. [9-^{3}H]$PGF_{2\alpha}$ (0.45 μg, 13.8 μCi) is added to follow the extent of coupling of $PGF_{2\alpha}$ to enzyme. Tri-*n*-butylamine (2.5 μl) and isobutyl chloroformate (1.3 μl) are added to the $PGF_{2\alpha}$ solution. The mixture is stirred for 30 min at 10–12° (below this temperature the solution is frozen), and then added dropwise each time in a 10-μl aliquot to β-galactosidase (0.35 mg) dissolved in a mixture of dioxane (0.35 ml) and 0.5% sodium bicarbonate (0.35 ml). The mixture is stirred for 2 hr at 4°. The solution is transferred to a Visking dialysis tube (8/32), and dialyzed against 500 ml of 50% dioxane at 4° for 2 hr to remove unconjugated $PGF_{2\alpha}$. Dialysis is continued against 2 liters of 0.1 *M* sodium phosphate, pH 7.0, for about 40 hr at 4° with two changes of buffer. After dialysis the solution is centrifuged for 15 min at 1200 *g*. A portion (100 μl) of the supernatant is taken for counting of radioactive $PGF_{2\alpha}$ and estimation of enzyme protein by its absorbance at 280 nm. Based on the molecular weight of the enzyme (518,000)[2] and the specific radioactivity of [9-^{3}H]$PGF_{2\alpha}$ (4.9×10^5 cpm/μmol), the number of $PGF_{2\alpha}$ molecule linked to one molecule of enzyme can be estimated. After extensive dialysis to remove free $PGF_{2\alpha}$, approximately 100 molecules of $PGF_{2\alpha}$ are coupled to 1 molecule of enzyme. Since β-galactosidase of *E. coli* contains only 95–111 lysine residues,[3] part of the $PGF_{2\alpha}$ molecules may be bound to the enzyme by nonspecific adsorption.

A solution of conjugate is diluted to an enzyme concentration of 10 μg of protein per milliliter with buffer A described below and stored at 4°. The recoveries of enzyme protein and the enzyme activity are 65% and 28%, respectively. The K_m value (approximately 0.4 m*M*) of the enzyme for 4-methylumbelliferyl-β-D-galactoside as substrate is unchanged by the conjugation reaction. V_{max} values of native and conjugated enzyme are 879 and 243 μmol/min per milligram of protein, respectively, under the conditions described below. With this enzyme preparation, the reaction proceeds linearly for 1.5 hr and the enzyme activity increases in linear proportion to the enzyme protein in a range of 0.03–0.13 ng. The antigeni-

[2] H. Sund and K. Weber, *Biochem. Z.* **337,** 24 (1963).

[3] K. Wallenfels and R. Weil, *in* "The Enzymes" (P. D. Boyer, ed.), 3rd ed., Vol. 7, p. 617. Academic Press, New York, 1972.

city of enzyme-labeled $PGF_{2\alpha}$ is about 40% that of free antigen when comparing the amount of free and enzyme-labeled $PGF_{2\alpha}$ required for 50% competition in radioimmunoassay. The $PGF_{2\alpha}$–enzyme conjugate is stable for over a year with regard to both enzyme activity and antigenicity.

Reagents

Buffer A: 10 mM sodium phosphate, pH 7.0, containing 0.1 M NaCl, 1 mM $MgCl_2$, 0.1% NaN_3, and 0.1% ovalbumin

Buffer B: ovalbumin omitted from buffer A

β-Galactosidase-labeled $PGF_{2\alpha}$ dissolved in buffer A; equivalent to 0.5 pmol of $PGF_{2\alpha}$/ml and 2.6 ng of enzyme/ml

Standard $PGF_{2\alpha}$ dissolved in buffer B (0.06, 0.12, 0.23, 0.47, 0.94, 1.9, 3.8, 7.5, 15, and 30 pmol/50 μl)

Rabbit antiserum to $PGF_{2\alpha}$ (first antibody), diluted in buffer A by 10^6-fold

Goat antiserum to rabbit IgG (second antibody), diluted in buffer B by 5-fold

Normal rabbit serum (carrier), diluted in buffer A by 200-fold

Reagents for enzyme assay

- 4-Methylumbelliferyl-β-D-galactoside, 0.1 mM (substrate solution): 4-methylumbelliferyl-β-D-galactoside (M_r 338.3) at 15 mM in N,N'-dimethylformamide (0.3 ml) is added to buffer A (44.7 ml).
- 4-Methylumbelliferone, 0.25 and 0.5 μM (standard for fluorometry): 4-Methylumbelliferone (M_r 176.2) at 25 mM in N,N'-dimethylformamide is diluted with buffer A to 0.25 and 0.5 μM, respectively.
- Glycine–NaOH buffer, 0.1 M, pH 10.3 (stopping solution)

Assay Procedures

The sequential saturation technique[4] is used for competition between enzyme-labeled and free $PGF_{2\alpha}$. First antibody (100 μl) and standard $PGF_{2\alpha}$ (50 μl) are preincubated at 4° for 10 min, and the enzyme-labeled $PGF_{2\alpha}$ (50μl) is added. After incubation for at least 2 hr at 26°, diluted normal rabbit serum (50 μl) and second antiserum (50 μl) are added in that order. The mixture is allowed to stand for more than 16 hr at 4° and then centrifuged at 1200 g at 4° for 15 min. Immunoprecipitate is washed with 1 ml of buffer B by centrifugation and suspended in 0.3 ml of the substrate solution. After 1 hr at 30° the enzyme reaction is terminated by the addition of 2.5 ml of 0.1 M glycine-NaOH buffer, pH 10.3. Fluorescence in-

[4] A. Zettner and P. E. Duly, *Clin. Chem.* **20,** 5 (1974).

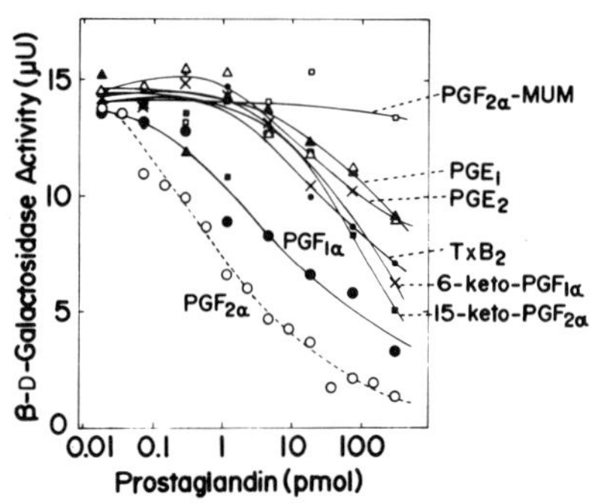

FIG. 1. Competition of β-galactosidase-linked $PGF_{2\alpha}$ with free $PGF_{2\alpha}$ and other PGs. $PGF_{2\alpha}$ (○---○), $PGF_{1\alpha}$ (●), PGE_1 (▲), PGE_2 (△), thromboxane B_2 (•), 6-keto-$PGF_{1\alpha}$ (×), 15-keto-$PGF_{2\alpha}$ (■), and $5\alpha,7\alpha$-dihydroxy-11-ketotetranorprosta-1,16-dioic acid (□). MUM, major urinary metabolite.

tensity of the reaction product is determined against a freshly prepared standard of 0.25 or 0.50 M 4-methylumbelliferone using a Hitachi spectrofluorometer Model MPF 2A. The wavelengths used for excitation and emission are 360 and 450 nm, respectively. One unit of β-galactosidase is defined as the amount of enzyme that hydrolyzes 1 μmol of 4-methylumbelliferyl-β-D-galactoside per minute under the standard assay conditions. A blank value is obtained by the addition of buffer A in place of first antibody.

Calibration Curve

A calibration curve obtained by the procedure described above is shown in Fig. 1. The detection limit, which is defined as the amount of unlabeled $PGF_{2\alpha}$ causing 10% displacement of the precipitation of enzyme-labeled $PGF_{2\alpha}$, is approximately 0.03 pmol. The cross-reactivity (%) of the antibody with another PG is calculated from the amount of the compound necessary to give 50% competition; $PGF_{1\alpha}$ (11%), PGE_1 (0.1%), PGE_2 (0.1%), 6-keto-$PGF_{1\alpha}$ (0.4%), thromboxane B_2 (0.6%), $5\alpha,7\alpha$-dihydroxy-11-ketotetranorprosta-1,16-dioic acid (main urinary metabolite of $PGF_{2\alpha}$, <0.1%), 15-keto-$PGF_{2\alpha}$ (0.9%).

Alkaline Phosphatase-Linked Immunoassay

A similar method using alkaline phosphatase has also been developed.[5] The enzyme activity in immunoprecipitate is measured spectrophotometrically. As shown in Fig. 2, the detection range is about 1–300

[5] Y. Hayashi, T. Yano, and S. Yamamoto, *Biochim. Biophys. Acta* **663,** 661 (1981).

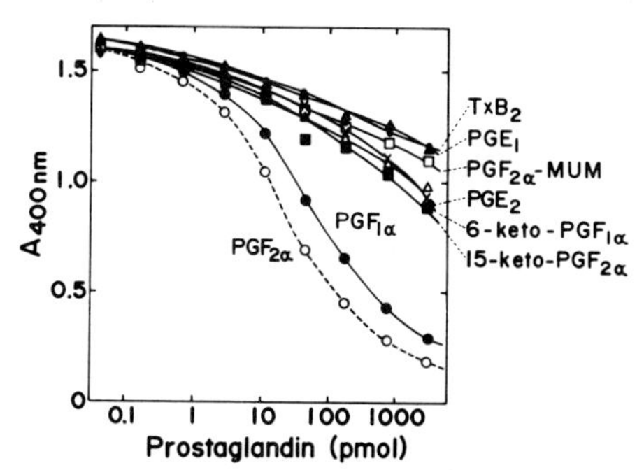

FIG. 2. Competition of alkaline phosphatase-linked PGF$_{2\alpha}$ with free PGF$_{2\alpha}$ and other PGs. Symbols are as in Fig. 1.

pmol. This method is less sensitive than the β-galactosidase-linked immunoassay. A detailed description of the alkaline phosphatase method is given by Hayashi *et al.*[5]

[40] A Radioimmunoassay for 6-Keto-PGF$_{1\alpha}$

By JACQUES MACLOUF

The high sensitivity of radioimmunoassay is often needed for the quantitation of prostaglandins occurring at very low concentrations. Unfortunately, the method itself has been discredited unfairly because of its frequent use without certain indispensable precautions and controls. Nevertheless, with the proper controls, radioimmunoassay has proved to be adapted to prostaglandin analysis. In this chapter the development of a radioimmunoassay for 6-ketoprostaglandin F$_{1\alpha}$ (6-keto-PGF$_{1\alpha}$, the hydrolysis product of prostaglandin I$_2$ (PGI$_2$; prostacyclin) is described. The excellent review by Granström and Kindahl[1] is recommended as a comprehensive assessment of prostaglandin radioimmunoassay.

Preparation of Conjugate and Immunization

The antiserum was raised against 6-keto-PGF$_{1\alpha}$ coupled to bovine serum albumin. Low molecular weight substances need this covalent attachment to an immunogenic molecule to elicit the production of anti-

[1] E. Granström and H. Kindahl, *Adv. Prostaglandin Thromboxane Res.* **5**, 119 (1978).

ISBN 0-12-181986-8

bodies.[1] The conjugate[2] was prepared by dissolving 6-keto-$PGF_{1\alpha}$ (24.3 μmol = 9 mg + 2 × 10^6 dpm [^{3}H]6-keto-$PGF_{1\alpha}$) in 0.5 ml of acetonitrile–water (95:5, v/v); it was then diluted in 8.5 ml of distilled water. After adjusting the pH to 5.5, 93.7 μmol (18 mg) of 1-ethyl-3-(3-diethylamino-propyl)carbodiimide-HCl (Sigma, St. Louis, Missouri) were added and the mixture was allowed to react for 1 hr at room temperature. The pH variation was checked intermittently during the first 30 min. Bovine serum albumin, 300 nmol (18 mg) was then added to the prostaglandin–carbodiimide solution, which was gently stirred until total dissolution of the albumin.

The mixture was kept at room temperature for 2 hr, and then overnight at 4° without stirring. It was then transferred into a dialysis bag and extensively dialyzed for 24 hr against several changes of large volumes of distilled water (pH 7.4) at 4°. The content of the dialysis bag was separated into different aliquots, 2 mg/ml each, that were kept at −80° until use. The number of hapten molecules fixed per molecule of albumin was determined by the measurement of protein-associated radioactivity after dialysis. Approximately 9–10 molecules of prostaglandin were coupled per mole of albumin.

The immunization was carried out with rabbits. It is preferable to immunize 4–6 rabbits to increase the chances of getting a high-titer antibody with good specificity and to increase the probability of survival for responders. About 0.50 mg of conjugate in 1 ml of water was emulsified with 1 ml of Freund's complete adjuvant (Difco Laboratories, Detroit, Michigan). The emulsion was injected intradermally into a rabbit at multiple sites in the flank area (5–10 points).[3] After 2 months, the first booster was administered by the same route, and the next boosters were done every 2–3 months according to the same protocol. Blood was collected weekly commencing 1 week after the first booster. The blood, in normal glass tubes, was clotted at 25° overnight, and the serum was retained. After addition of sodium azide (0.02% w/v) to each bleeding, the serum was stored at 4°.

Preparation of the Iodinated Ligand

Coupling Procedure

Reagents

6-Keto-$PGF_{1\alpha}$ (a kind gift of Dr. John Pike, Upjohn Co., Kalamazoo, Michigan)

[2] F. Dray, B. Charbonnel, and J. Maclouf, *Eur. J. Clin. Invest.* **5,** 311 (1975).

[3] J. Vaitukaitis, J. B. Robbins, E. Nieschlag, and G. T. Ross, *J. Clin. Endocrinol.* **33,** 988 (1971).

1-Ethyl-3-(5-dimethylaminopropyl)carbodiimide-HCl (Sigma) Histamine free base, crystalline (Sigma)

[2-^{14}C]-Histamine, HCl 32 mCi/mmol (CEA, Saclay, France)

Thin-layer chromatography (TLC) plate: silica gel 60 without fluorescent indicator, layer thickness 0.25 mm (Merck Darmstadt, Germany)

Pauly's reagent: Mix 0.5% sulfanilic acid in 1 *N* HCl (w/v), 10 ml; $NaNO_2$ 5% in water (w/v), 5 ml. Wait 1 min, then add 10 ml of a 10% solution of Na_2CO_3 in water (w/v); CO_2 bubbles should appear. This reagent must be prepared fresh for use.

n-Butanol, acetic acid

Coupling of Prostaglandin to Histamine. The 6-keto-PGF$_{1\alpha}$ histamide was prepared using the carbodiimide method.[4] In a small conical plastic tube, 7.3 μmol (2.7 mg) of 6-keto-PGF$_{1\alpha}$ dissolved in 50 μl of acetonitrile–water (95:5, v/v) were mixed with 31.25 μmol (6 mg) of 1-ethyl-3-(3-dimethylaminopropyl)carbodiimide in 150 μl water at pH 5.5. After a 1-hr incubation at room temperature, 30 μmol (3.3 mg) of histamine in 100 μl of water at pH 5.5 containing 7.3×10^6 dpm of [^{14}C]histamine were added to the mixture. After incubation overnight at 4°, the compounds were separated by TLC in *n*-butanol–acetic acid–water (75:10:25, v/v/v). The products are spotted on a continuous 16-cm line on a 20 × 20 cm plate. The profile of the radioactivity is shown in Fig. 1. The histamide conjugate can be localized by several procedures.

The simultaneous use of [^{14}C]histamine and [^{3}H]6-keto-PGF$_{1\alpha}$ identifies the zone containing the histamide derivative by the distinctive comigration of both ^{3}H and ^{14}C radioactivity. Alternatively, the reaction mixture may be applied as one continuous 14-cm line for recovery of histamide, flanked by two additional 0.5-cm zones applied exclusively for localizing histamide, histamine, and prostaglandin. After solvent migration, the area of the plate designated for recovery is masked by another glass plate, and the unmasked corridor containing one 0.5-cm zone is sprayed with Pauly's reagent to detect substances with an imidazole ring. After the sprayed area is heated with a hot-air gun, three imidazole-positive zones should appear corresponding to free histamine, a reaction product, and authentic 6-keto-PGF$_{1\alpha}$–histamide. The other unmasked corridor, containing the second 0.5-cm zone, is then sprayed with phosphomolybdic acid (3.5% w/v, spray reagent, E. Merck), and heated. Two molybdate-positive, brown-purple spots should appear corresponding to prostaglandin free acid and its histamide derivative. Identical migration and localization of a radioactive zone in both test corridors identifies the histamide of 6-keto-PGF$_{1\alpha}$ ($R_f = 0.3 \pm 0.05$) (see Fig. 1). The 14-cm re-

[4] J. Maclouf, M. Pradel, P. Pradelles, and F. Dray, *Biochim. Biophys. Acta* **431,** 139 (1976).

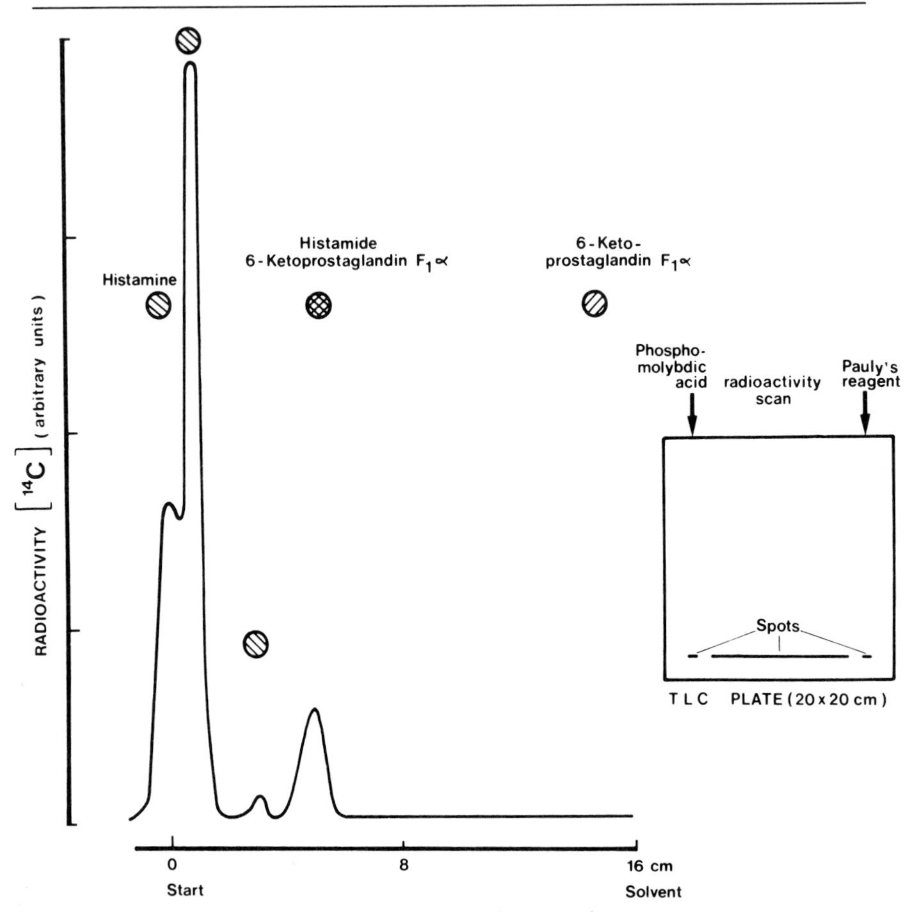

FIG. 1. Thin-layer radiochromatogram of the [^{14}C]histamine products after coupling to 6-keto-$PGF_{1\alpha}$.

covery zone containing this prostaglandin histamide is scraped and extracted twice with 1.5 ml of ethanol–water, 70:30, v/v. After centrifugation or filtration to remove silica, the supernatant can be stored at $-30°$ for at least 2 years without evident difficulty.

Iodination

All the operations are done using micropipettes equipped with disposable plastical tips.

Reagents

Histamide derivative: Into one small plastic conical cube, transfer an amount of the ethanolic solution containing 7 nmol of the histamide derivative. Evaporate to dryness under nitrogen, and dissolve in 10 μl of phosphate buffer.

Potassium phosphate buffer: 100 ml of K_2HPO_4, 0.5 *M*, + 18 ml of KH_2PO_4, 0.5 *M*, pH 7.4

Chloramine-T, 2.5 mg per milliliter of phosphate buffer

Sodium metabisulfite, 16 mg per milliliter of phosphate buffer

$Na^{125}I$, 2000 Ci/mmol (The Radiochemical Centre, Amersham, England)

Thin-layer chromatography (TLC) plate, 10 × 20 cm (same type as for coupling)

Solvent: chloroform–methanol, 90:10 (v/v) or 80:20 (v/v)

Procedure. Iodination is carried out using the chloramine-T procedure.[5] Na ^{125}I, 300 uCi (3 μl) is transferred into a small conical plastic tube containing 7 nmol of the 6-keto-$PGF_{1\alpha}$–histamide dissolved in 10 μl of phosphate buffer. Chloramine-T, 2 μl (5 μg), is added and mixed with the plastic tip of the pipette. After 20 sec, the reaction is stopped by the addition of 2 μl of sodium metabisulfite (32 μg). The reaction mixture is purified by TLC (the zone width is about 1 cm). Radioactive zones are detected by autoradiography (approximately 5–10 min of film contact with the plate). The major immunoreactive zone (see profile on Fig. 2), $R_f = 0.10$, is scraped and extracted with ethanol–water (70:30, v/v). The second solvent system gives an R_f at 0.47. Attention should be paid to scrape exactly at the zone of the tracer (about 3 × 15 mm) in order to exclude possible contamination by unlabeled 6 keto-$PGF_{1\alpha}$-histamide. Some less abundant immunoreactive products probably correspond to 6-keto in equilibrium with its acetal forms.[6] These zones were not isolated. The specific radioactivity of the tracer equals that of ^{125}I, i.e., 2000 Ci/mmol.

The iodination sequence takes approximately 3–4 hr, including TLC migration, autoradiography, etc., every 2 months. We found that the tracer gave reliable results for that time when stored in at −30°.

Radioimmunoassay Procedures

Incubation Conditions

Using ^3H Tracer

6-Keto-$PGF_{1\alpha}$ labeled 5, 8, 9, 11, 12, 14, 15-[^3H]-(N); 100 Ci/mmol purchased from New England Nuclear (Boston, Massachusetts)

Gelatin-Tris-HCl buffer, pH 7.4: 0.1% gelatin, 0.9% NaCl, 0.01 *M* Tris

[5] W. Hunter and F. Greenwood, *Nature (London)* **194,** 495 (1962).

[6] C. R. Pace-Asciak and M. Nashat, *Biochim. Biophys. Acta* **487,** 495 (1977).

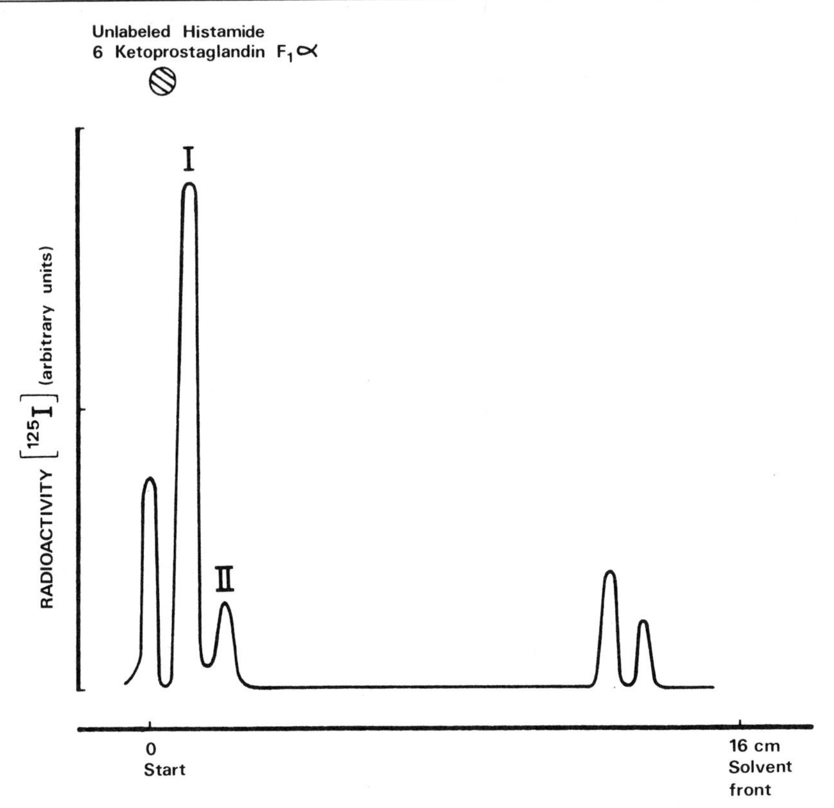

FIG. 2. Thin-layer radiochromatogram of the ^{125}I-labeled products after iodination of 6-keto-$PGF_{1\alpha}$–histamide (chloroform–methanol, 90:10 (v/v).

Incubation volume, 0.6 ml: (a) 200 μl gelatin-Tris-buffer; (b) 200-μl standard samples or unknown (standards, respectively, 6.25, 12.5, 25, 50, 100, 200, 400, 800 pg/tube); (c) 100 μl [^{3}H]-6-keto-$PGF_{1\alpha}$ (9000 dpm); (d) 100 μl of antibody

The incubation proceeds at room temperature for 2 hr; then overnight at 4°.

Separation of bound from free ligand was done by adding to each sample 0.6 ml of cold dextran-coated charcoal at 0° (neutral norit 1%, Serva dextran-70,000, 1% in water (w/w/v).[7] After careful mixing, the tubes were centrifuged for 15 min at 3000 *g* at 4° in a large-capacity centrifuge. The supernatant was transferred into a scintillation vial with 10 ml of scintilla-

[7] B. V. Caldwell, S. Burstein, W. A. Brock, and L. Speroff, *J. Clin. Endocrinol. Metab.* **33**, 171 (1971).

TABLE I
EXAMPLE OF A SETUP FOR THE RADIOIMMUNOASSAY OF 6-KETO-$PGF_{1\alpha}$ USING ^{125}I TRACER[a]

Tube number	Name	Comment
1	Total activity	Tracer solution only (0.1 ml)
2		
3	Blank	All reagents except antibody or standard (0.7 ml)
4		
5	Initial maximal binding	All reagents except standard
6		
7	1.25	
8		
9	2.5	
10		
11	5	
12		
13	7.5	
14		
15	10 standard tubes (pg/tube)	All reagents
16		
17	15	
18		
19	20	
20		
21	40	
22		
23	Sample tubes	10–300 μl of pure or diluted biological sample adjusted with buffer to the final 0.7-ml total volumes, 2–4 dilutions for the same sample).

[a] For all other details, see text.

tion fluid (Atomilit, New England Nuclear), and antibody-bound radioactivity was then counted in a Packard Tri-Carb 7785 counter.

Using [^{125}I]Histamide-Prostaglandin Tracer. The incubation procedure derived from Granström *et al.*[8] was performed in 5-ml plastic tubes.

Tris-HCl buffer, 0.05 *M*, pH 7.4, containing 10^{-3} *M* EDTA

Incubation volume, 0.7 ml (Table I): (a) 200 μl of 0.5% bovine γ-globulin (Cohn fraction II, Sigma Co.); when stored at 4°, this solution tends to precipitate and it should be mixed again prior to dispensing; (b) 300 μl of standard sample or unknown (standards 1.25–40 pg/tube: see Table I); (c) 100 μl of [^{125}I]histamide–6-keto-$PGF_{1\alpha}$ (12,000 dpm); (d) 100 μl of antibody

[8] E. Granström, H. Kindahl, and B. Samuelsson, *Anal. Lett.* **9,** 611 (1976).

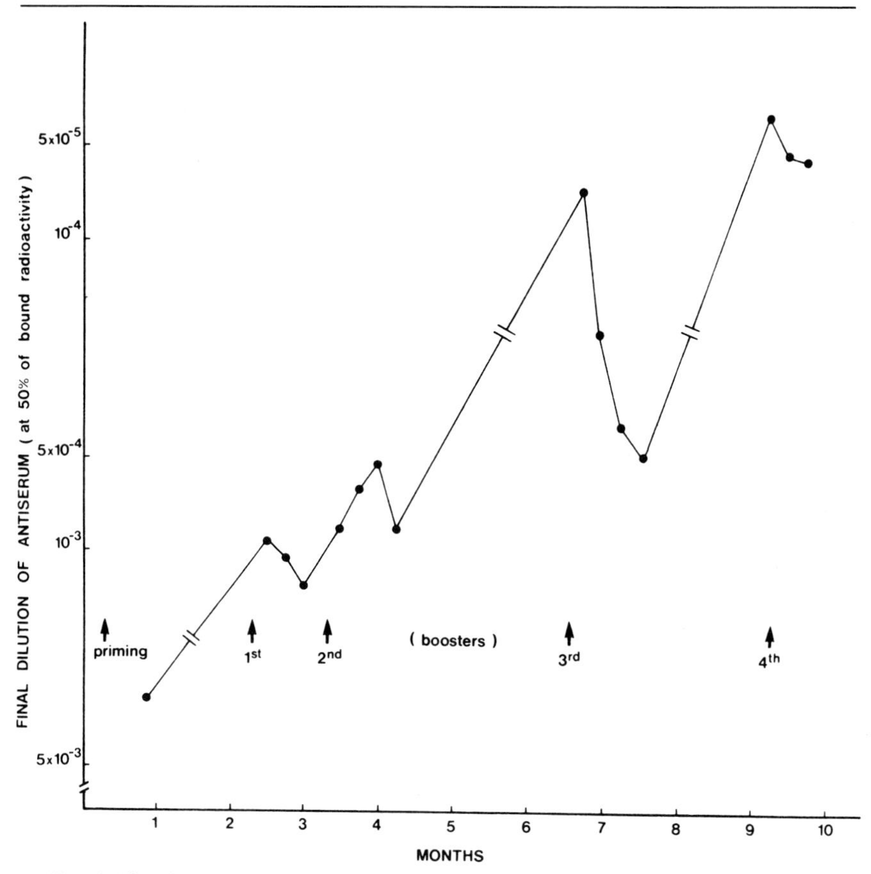

FIG. 3. Titration of a 6-keto-$PGF_{1\alpha}$ antiserum in the time course of immunization. Tracer, [^{125}I]histamide derivative; titer expressed as final dilution that can bind 50% of the added radioactivity.

After mixing, the incubation proceeds at room temperature for 1 hr, then overnight at 4°.

Separation of bound from free ligand was done by adding to each sample, except for total radioactivity, 0.7 ml of an ice-cold 25% w/v solution of polyethylene glycol 4000 in water at 0°C. After careful mixing, tubes were centrifuged at 4° for 30 min at 3000 *g* in a high sample-capacity centrifuge. A white pellet corresponding to the precipitated γ-globulins should be visible. The supernatant was removed by suction under vacuum, and the antibody precipitate labeled with [^{125}I]histamide-prostaglandin was counted in its tube in a gamma spectrometer (Beckman type 7000).

Determination of Antibody Titer: Selection of Dilution

Doubling dilutions of antiserum (1:100, 1:200 . . . 1:25,600) are prepared with 0.05 *M* Tris buffer. A series of duplicate test tubes are prepared containing buffer (300 μl), bovine γ-globulin (200 μl), diluted antiserum (100 μl), and a constant amount of radiolabeled antigen (100 μl, 12,000 cpm; [^{125}I]histamide of 6-keto-$PGF_{1\alpha}$) or [5, 8, 9, 11, 12, 14, 15-^{3}H]-6-keto-$PGF_{1\alpha}$). Several tubes labeled "non-bound" contain buffer (400 μl), bovine γ-globulins (200 μl), and radiolabeled antigen (100 μl, 12,000 cpm) but no antiserum. After incubation, separation of antibody-bound radioactivity, and gamma or beta counting, a dilution curve is established by plotting the percentage of the total radioactivity bound to the antibody (ordinate) versus the antibody dilution (abscissa). This is done for each bleeding. A profile of the antiserum titer of one rabbit is presented in Fig. 3. This curve shows the dilution of antiserum required to bind 50% of the total radioactivity (^{125}I tracer) as a function of the time course of immunization. As expected, the titer increased after each booster. One should select a dilution of the antiserum that binds 50 $\pm$ 10% of the labeled ligand in the absence of any unlabeled ligand for use in the quantitative radioimmunoassay. Dilution curves from the same bleeding are shown using ^{3}H or ^{125}I tracer (Fig. 4). The curves depict (bound radioactivity/total added radioactivity) $\times$ 100 ($B/T \times 100$) versus the logarithm of the antiserum dilution. Since the incubations are performed in different volumes, we have reported final dilutions. As expected from its higher specific radioactivity, the use of ^{125}I tracer allows a greater dilution of antiserum at 50% of binding. We use 1:12,000 and 1:70,000 (final dilutions), respectively, for ^{3}H and ^{125}I tracer.

Standard Curves and Cross-Reactivities

Figure 5 represents the standard curves obtained with the different tracers in terms of (bound radioactivity/bound radioactivity at zero dose) $\times$ 100 ($B/B_0 \times 100$) versus the logarithm of the dose. The amounts of 6-keto-$PGF_{1\alpha}$ giving 50% inhibition of binding for each tracer were, respectively, 125 fmol (46 pg) for ^{3}H tracer and 9.1 fmol (3.4 pg) for ^{125}I tracer.

Cross-reactions of the antiserum with different prostaglandins and metabolites have been studied using the [^{125}I]histamide tracer. The results are presented in Table II. Cross-reactivities have been calculated from the amount of the heterologous compounds required to displace 50% of the bound [^{125}I]histamide from the binding sites compared to the amount of 6-keto-$PGF_{1\alpha}$ necessary to achieve the same displacement. The closest cross-reacting heterologous prostaglandins were $F_{1\alpha}$ and $F_{2\alpha}$. These pos-

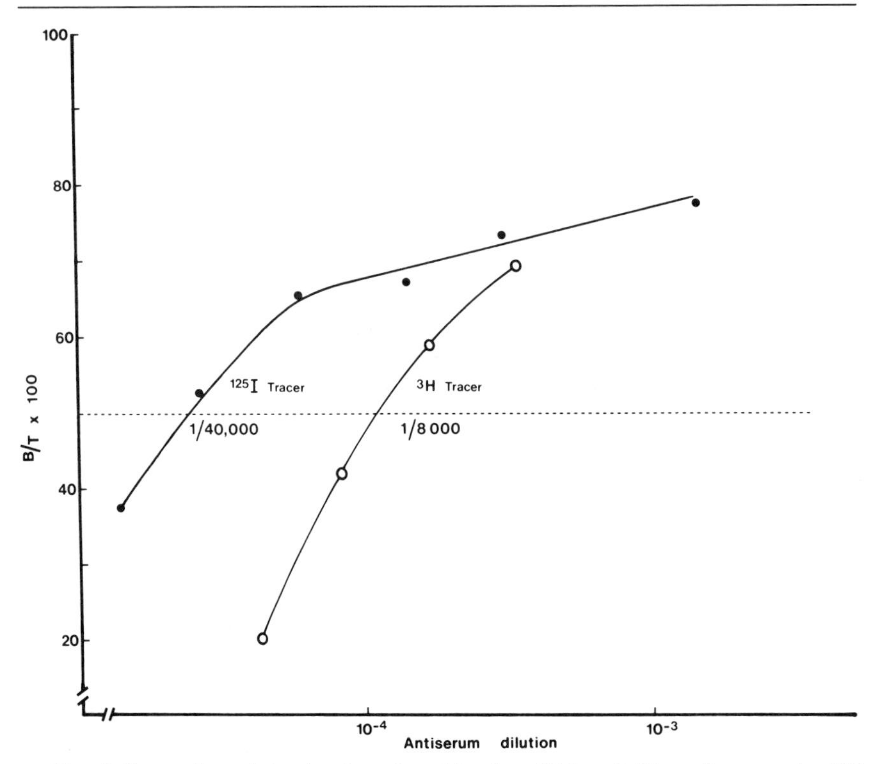

FIG. 4. Comparison of the titration of one bleeding of 6-keto-PGF$_{1\alpha}$ antiserum using ^{125}I tracer (●) and ^{3}H tracer (○). Dilutions are expressed as final volume. ^{3}H results are available by courtesy of Dr. G. Folco.

sess a cyclopentane ring identical to that of 6-keto-PGF$_{1\alpha}$. All other compounds that we have tested presented only minor cross-reactions ($\leq 1\%$).

The determination of cross-reactivities is not an absolute criterion of the specificity:

1. The cross-reaction determined only at 50% displacement does not indicate how parallel are the curves for heterologous ligands. The validity of possible corrections for interferences requires that cross-reactivity be known at every point of the standard curve.
2. In biological samples, unknown or untested substances might be present at sufficiently high concentrations to cause nonspecific interferences. Such is the case for fatty acids.

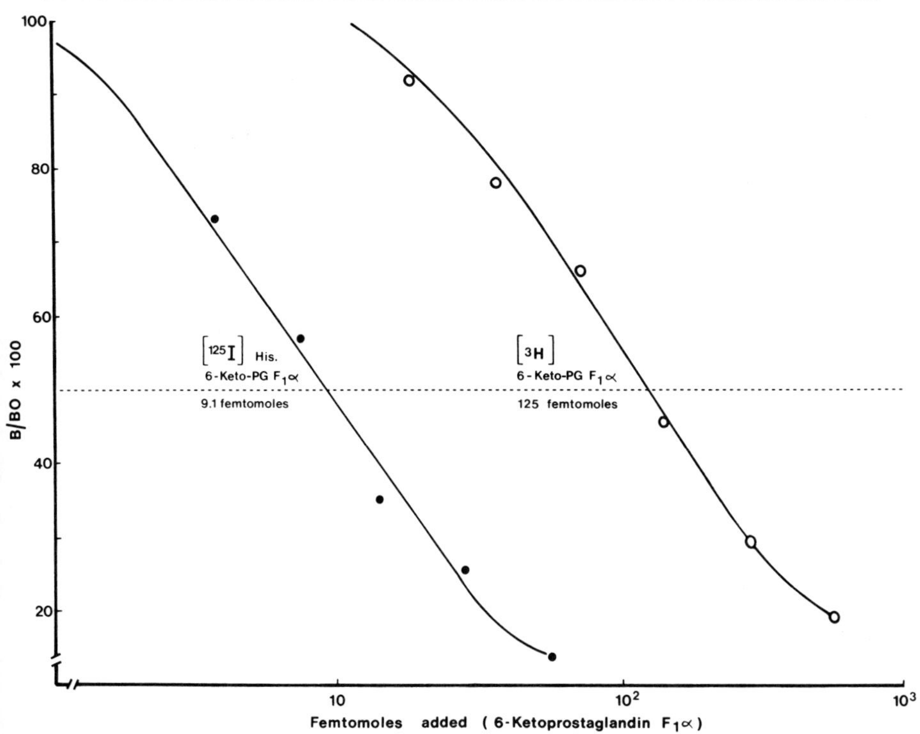

FIG. 5. Standard curves of 6-keto-$PGF_{1\alpha}$ antiserum obtained with ^{125}I tracer (●) and ^{3}H tracer (○). With both tracers, the final dilution of antiserum yielded 40% initial binding. ^{3}H results are available by courtesy of Dr. G. Folco.

TABLE II
SPECIFICITY OF THE RABBIT 6-KETO-$PGF_{1\alpha}$ ANTISERUM

Inhibitor	Relative cross-reaction (%)
6-Keto-$PGF_{1\alpha}$	100
$PGF_{1\alpha}$	3
PGF_2	1.2
PGE_1	1
PGE_2	0.5
PGD_1	0.4
6,15-Diketo-$PGF_{1\alpha}$	0.3
Thromboxane B_2	0.1
PGD_2	0.07

Precision of Measurements of 6-keto-$PGF_{1\alpha}$

Reproducibility of the assay was tested by 15 replicate determinations of the response ($B/B_0 \times 100$) for different standard doses (1.25–40 pg). The mean coefficient of variation was 6.3%. This variation reflects pipetting, mixing, and other errors only within the same assay. To analyze interassay reproducibility, we highly recommend the preparation of biological material containing 6-keto-$PGF_{1\alpha}$ and its distribution into 50–100 separate aliquots for storage at −80°. In each assay that will be run throughout 1 year, one aliquot from the pool will be thawed and analyzed. This constitutes an additional control complementary to the blank, the initial binding, or the reproducibility of the standard dose-response curve. Such a program implies that 6-keto-$PGF_{1\alpha}$ will be stable in the presence of the biological matrix and that samples are kept at −80°.

Serial dilutions (1:2) of unknown samples in each experiment should be prepared and analyzed to check nonspecific binding within the sample matrix. The introduction of multiple dilutions, in duplicate, strongly confirms the specificity of an assay whenever the results of the assay of these dilutions parallel the standard curve. An example is shown in Fig. 6.

Concluding Remarks

The establishment of a reliable and specific radioimmunoassay for each oxygenated metabolite of arachidonic acid could be a valuable source of biochemical, physiological, or pharmacological information. Before applying such a method to the measurement of substances active in the picomolar range, several aspects dealing mainly with the specificity should be verified. A specific assay system should (*a*) be able to estimate dilutions of a "pure" sample of the substance intended for measurement; (*b*) be uninfluenced by related materials (precursors, metabolites, drugs, etc.) at concentrations that exist in typical biological samples; (*c*) be able to reflect changes in the concentration of the substance induced by suitable physiological or pharmacological manipulations. For instance, thrombin should release prostacyclin, and therefore 6-keto-$PGF_{1\alpha}$, from endothelial cells; this production should be inhibited if the stimulation occurs in the presence of nonsteroidal anti-inflammatory drugs. (*d*) The system should also give quantitative results that agree with quantitations by other methods, when these can be applied (biological activity of precursor prostacyclin or measurement of 6-keto-$PGF_{1\alpha}$ by gas chromatography–mass spectrometry). Technically, point *d* may be the most difficult to check. Furthermore, when using a commercial source of antiserum,

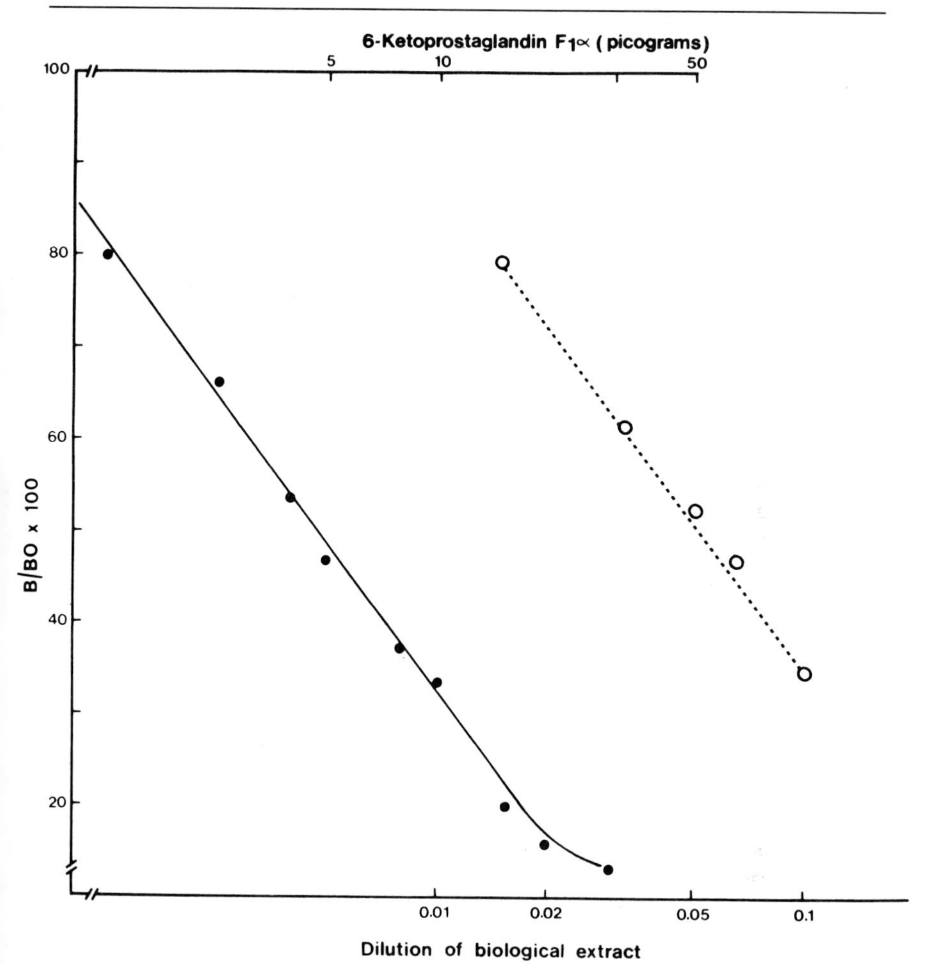

FIG. 6. Comparison of the inhibition of binding of [^{125}I]histamide–6-keto-PGF$_{1\alpha}$ antiserum by standard 6-keto-PGF$_{1\alpha}$ (●) and by serial dilutions of endothelial cell culture supernatant (○).

these controls may seem expensive although they should be prerequisites for any investigation.

We favor the use of ^{125}I tracer because it has a high specific radioactivity; it minimizes quenching problems; it allows fast handling of the assay because incubation and counting are performed in the same tube; and compared to ^{3}H tracers its higher sensitivity allows the assay of minute amounts of diluted biological material, therefore decreasing possible interference with nonspecific substances. The preparation of a new tracer every 2 months should not be a problem, since the 6-keto-$PGF_{1\alpha}$–histamide conjugate is a stable stock product that will be used for subsequent iodinations. The iodination itself is a very fast process when one has already performed it three or four times.

Acknowledgments

The author is indebted to Dr. G. Folco (Instituto di Farmacologia, Milan, Italy) for providing the data with ^3H tracer.

[41] A Radioimmunoassay for Thromboxane B_2

By F. A. FITZPATRICK

Radioimmunoassay of thromboxane B_2 can reflect the concentration of its unstable, biologically active precursor, thromboxane A_2[1].Granström *et al.*[2] developed the first radioimmunoassay for thromboxane B_2. Other groups subsequently developed similar assays.[3–8]

Choice of Materials

Preferences for certain reagents are evident from Table I. Bovine serum albumin was the favored protein immunogen. Synthetic thromboxane B_2 (Upjohn Diagnostics, Kalamazoo, Michigan), and 5,6,8,9,11,12,-14-[^3H]thromboxane B_2 (Catalog No. NET-603, New England Nuclear, Boston, Massachusetts) have supplanted the biosynthetic preparations cited in Table I. The commercial availability of pure TxB_2 and high specific activity [^3H]TxB_2 has improved and normalized the sensitivity of these assays. The typical mass required to displace 50% of the bound ligand is now approximately 30 ± 10 pgs. Sensitivities were not improved

[1] M. Hamberg, J. Svensson, and B. Samuelsson, *Proc. Natl. Acad. Sci. U.S.A.* **72,** 2994 (1975).

[2] E. Granström, H. Kindahl, and B. Samuelsson, *Anal. Lett.* **9,** 611 (1976).

[3] F. Fitzpatrick, R. Gorman, J. McGuire, R. Kelly, M. Wynalda, and F. F. Sun, *Anal. Biochem.* **82,** 1 (1977).

[4] H. Anhut, W. Bernauer, and B. Peskar, *Eur. J. Pharmacol.* **44,** 85 (1977).

[5] V. Ferraris, J. B. Smith, and M. J. Silver, *Thromb. Haemostasis* **38,** 20 (1977).

[6] H. Sors, P. Pradelles, F. Dray, M. Rigaud, J. Maclouf, and P. Bernard, *Prostaglandins* **16,** 277 (1978).

[7] H. Tai and B. Yuan, *Anal. Biochem.* **87,** 343 (1978).

[8] L. Viinika and O. Ylikorkala, *Prostaglandins* **20,** 759 (1980).

ISBN 0-12-181986-8

TABLE I
COMPARATIVE FEATURES OF THROMBOXANE B_2 RADIOIMMUNOASSAYS

Protein immunogen[a]	Thromboxane B_2	Conjugation reaction[b]	Radiolabeled ligand	Cross-reaction with PGD_2 (%)	Sample volume	Separation procedure	Reference[c]
BSA	Biosynthetic	CDI	$[^3H]TxB_2$[d]	11.8	≤0.5 ml direct	Globulin precipitation	Ganström *et al.*[2]
KLH	Synthetic (Upjohn)	EDAC	$[^3H]TxB_2$	1.0	≤0.1 ml direct	Charcoal sequestration	Fitzpatrick *et al.*[3]
BSA	Synthetic (Upjohn)	CDI	$[^3H]TxB_2$	13.3	≤0.1 ml direct	Charcoal sequestration	Anhut *et al.*[4]
BSA	Synthetic (Upjohn)	CDI	$[^3H]TxB_2$	0.05	≤1 ml direct	Globulin precipitation	Ferraris *et al.*[5]
BSA	Synthetic (Upjohn)	EDAC	$[^{125}I]$Histamide of TxB_2	0.15	≤0.5 ml direct	Globulin precipitation	Sors *et al.*[6]
BSA	Synthetic (Upjohn)	EDAC	$[^{125}I]$Tyrosine methyl ester of TxB_2	0.03	≤0.2 ml direct	Charcoal sequestration	Tai and Yuan[7]
BSA	Synthetic (Upjohn)	EDAC	$[^3H]TxB_2$	NT	1–2 ml prior to extraction	Charcoal sequestration	Viinikka and Ylikorkala[8]

[a] BSA, bovine serum albumin; KLH, keyhole limpet hemocyanin.
[b] CDI, *N,N'*-carbonyldiimidazole; EDAC, 1-ethyl-3(3-dimethylaminopropyl)carbodiimide·HCl.
[c] Superscript numbers refer to text footnotes.
[d] Sensitivities currently are 30 ± 10 pg for 50% B/B_0 using commercially available $[^3H]TxB_2$.

dramatically when iodinated forms of TxB_2 were substituted for 5,6,8,9,-11,12,14-[^{3}H]TxB_2. There was an equal preference for water-soluble carbodiimide as for *N,N'*-carbonyldiimidazole as a coupling reagent. Both reagents are reliable, reproducible, and simple to use. Rabbits were the preferred animal. Prostaglandin D_2 was the major interfering compound in all the methods. There were significant differences between them in this regard, but they appear to be independent of procedural details. There was an equal preference for charcoal sequestration as for γ-globulin precipitation to separate free ligand from antibody-bound ligand. For trace levels (<100 pg of TxB_2 per milliliter) γ-globulin precipitation is more compatible with direct analyses of larger volumes.

Reagents

Thromboxane B_2 (Upjohn Diagnostics, Kalamazoo, Michigan, catalog No. 468911)

5,6,8,9,11,12,14-[^{3}H]thromboxane B_2 (New England Nuclear, Boston, Massachusetts; catalog No. NET-603), specific activity $\geq$100 Ci/mmol

Stock buffer (1.0 *M* PBS): 1.0 *M* potassium phosphate buffer, pH 7.4, containing 0.9% w/v NaCl and 0.1% w/v NaN_3. Store at 4° for 60 days.

Working buffer (0.1 *M* PBSG): 0.1 *M* potassium phosphate buffer, pH 7.4, containing 0.9% w/v NaCl, 0.01% NaN_3, 0.1% gelatin. Dilute stock buffer 50.0 to 500.0 with sterile water containing 0.9% w/v NaCl. Add 500 mg of bacteriological grade gelatin. Warm under hot water to dissolve gelatin. *Prepare fresh daily. Do not store and re-use.*

Charcoal suspension: Add 2.50 ± 0.10 g of Norit A (Sigma, St. Louis, Missouri) and 0.25 ± 0.01 g of Dextran T70 (Pharmacia, Sweden) to a 600-ml beaker. Add 500 ml of 0.1 *M* PBSG warmed to about 40–45° under a hot water tap. Stir continuously with a Teflon-coated magnetic bar to suspend the charcoal evenly. *Prepare fresh daily. Do not store and re-use.* It is convenient to fill 100 scintillation vials with the ingredients (2.5 g of Norit A and 0.25 g of dextran T70) for a daily unit (500 ml) of charcoal suspension. Cap securely and store at 25° indefinitely.

[^{3}H]TxB_2 solution: Add 1.0 μCi of 5,6,8,9,11,12,14-[^{3}H]thromboxane B_2 (150 Ci/mmol) to 50.0 ml of 0.1 *M* PBSG. Mix thoroughly. Store at 4° for 7–10 days. The solution should contain about 6000 dpm/100 μl.

TxB_2 antiserum solution: Dilute thromboxane B_2 antiserum 1:10,000 in 0.1 *M* PBSG. See the accompanying procedure for determining the proper antiserum dilution. This antiserum is commercially available (Upjohn Diagnostics, Kalamazoo, Michigan, catalog No. 401760).

Preparation of Immunogenic Conjugate

We use keyhole limpet hemocyanin, rather than albumin, as an immunogenic protein, because rabbits have responded more quickly when inoculated with KLH conjugates, in our experience.

1. Dissolve 10 mg of thromboxane B_2 in 1.0 ml of anhydrous dimethyl formamide (Pierce Chemical Co., Rockford, Illinois; catalog No. 20672).
2. Dissolve 20 mg of keyhold limpet hemocyanin (Calbiochem, La Jolla, California; catalog No. 374805) in 6.0 ml of distilled water. Adjust the pH to 7.0 ± 0.2.
3. Add 10 mg of water-soluble carbodiimide, 1-ethyl-3(3-dimethylaminopropyl)carbodiimide·HCl (Bio-Rad Laboratories, Richmond, California) to the protein solution. Immediately, add the dimethylformamide solution containing thromboxane B_2. It is advisable to add this in ten successive 100-μl portions at 30-second intervals. Stir the resulting mixture for 1 hr at 25°. After 1 hr, add an additional 10 mg of water-soluble carbodiimide; stir and allow the mixture to react for 12 hr at 25°.
4. Wash a dialysis membrane, 8 mm in diameter, 100 cm in length (Union Carbide, Chicago, Illinois), in running water for 2 hr. Transfer the contents of the reaction mixture to a 25-ml volumetric flask. Dilute to the mark with distilled water. Sonicate to disperse the protein evenly. Seal one end of the dialysis membrane; transfer the contents of the volumetric flask to the dialysis membrane; and then seal the other end securely so that only a small 5–10 mm air pocket exists. Dialyze for 24 hr against 4 liters of distilled water at 4°.
5. After dialysis, transfer the contents of the membrane to a clean 50-ml round-bottom flask and lyophilize. Weigh the protein to monitor recovery. Ordinarily this will exceed 80% of the starting mass of the protein (20 mg). Values below 75% recovery usually indicate mechanical losses during dialysis. Values greater than 100% recovery usually indicate incomplete dialysis.
6. Store the conjugate desiccated under Ar or N_2 at −20°.

Inoculation of Rabbits

1. Using a high speed micro blender, mix 2.0 mg of conjugate with 1.0 ml of sterile water containing 0.9% w/v NaCl. Add 1.0 ml of Freund's complete adjuvant (Calbiochem, La Jolla, California; catalog No. 344289) and emulsify thoroughly.

2. Shave the backs of at least two male, albino New Zealand rabbits to produce a rectangular bare patch. Swab with ethanol–water, 70:30, v/v. Use a sterile, disposable 1-ml plastic syringe with a 26 gauge $\frac{1}{2}$-in. needle to inject each rabbit, intradermally, at about 20 sites with a total of 1 ml of emulsion. Each rabbit will receive about 500 μg of conjugate because of incomplete expulsion of the emulsion from the syringe and "leakage" from the injection site.

3. Forty-five days later, emulsify 0.5 mg of conjugate, as above, and inject each rabbit with 1.0 ml of emulsion. Starting 2 weeks after the first booster injection, bleed the rabbits via their central ear artery to collect antiserum.

Determination of Antiserum Titer/Assay Dilution

1. Prepare serial dilutions of antiserum in 0.1 *M* PBSG: 1:100, 1:500, 1:1000, 1:2000, 1:5000, 1:10,000, 1:20,000, 1:50,000.

2. Prepare a solution of [^{3}H]TxB_2 in 0.1 *M* PBSG to contain about 6000 dpm/0.1 ml

3. Add 0.10 ml of each antiserum dilution (1:100 . . . 1:50,000) to 12 × 75 millimeter tubes (triplicate). Add 0.10 ml of [^{3}H]TxB_2 solution to each tube. Add 0.10 ml of 0.1 *M* PBSG to each tube. The final volume is 0.30 ml/tube. To three 12 × 75 millimeter tubes labeled NSB (nonspecific binding) add 0.10 ml of [^{3}H]TxB_2 and 0.20 ml of 0.1 *M* PBSG. *Do not add antibody to the nonspecific binding tubes.*

4. Incubate these tubes for 1 hr at 25°; then for 20–24 hrs at 4°. Finally, incubate the tubes on an ice bath (0°) for 1 hr prior to the charcoal sequestration step to separate antibody-bound and free [^{3}H]TxB_2.

5. Prepare a charcoal suspension in 0.1 *M* PBSG containing 5 mg/ml Norit A and 0.50 mg/ml dextran T70 (see Reagents). Chill the charcoal suspension to 0° on an ice bath. Transfer 1.0 ml of chilled charcoal suspension to each tube. Incubate the charcoal and the tube contents at 0° for *exactly 11.0 ± 1.0 mins*. Centrifuge the tubes in a refrigerated centrifuge (4°) at 1200 *g* for 15 mins. Remove 1.0 ml of supernatant and, by scintillation counting, determine the amount of [^{3}H]TxB_2 bound to the antibody as a function of antibody dilution. Sample calculations are shown in Table II.

6. (a) To monitor the antibody response, plot the dilution that binds 50% of [^{3}H]TxB_2 as a function of time after inoculation (day 0). Improvements in the antibody response are reflected by equivalent binding at higher dilutions; deterioration of the antibody response is reflected by equivalent binding at a reduced dilution (or reduced binding at a fixed dilution). Boost animals only when the antibody response has declined significantly. For example, if the animals bound 50% of [^{3}H]TxB_2 at a

TABLE II
TYPICAL RESULTS FROM A TITER DETERMINATION

Anti-TxB_2				
Initial dilution	Final dilution	Tube[a] (dpm/ml)	Bound[b] (dpm/ml)	Bound[c] (%)
1:100	1:300	5100	5000	83.3
1:500	1:1500	4900	4800	80.0
1:1000	1:3000	4600	4500	75.0
1:5000	1:15,000	4100	4000	66.7
1:10000	1:30,000	3100	3000	50.0
1:20000	1:60,000	2100	2000	33.3
1:50000	1:150,000	600	500	8.3
NSB	NSB	100	0	0

[a] Total (dpm/ml) per tube = 6000.
[b] Bound (dpm/ml = tube (dpm/ml) − NSB (dpm/ml). NSB, nonspecific bound.
[c] Percent bound = Bound (dpm/ml)/Total (dpm/ml).

1:10,000 dilution, do not boost until their titer declines to 1:5000 for an equivalent binding.

(b) In the radioimmunoassay procedure, the antiserum *must* be used at a dilution that binds 40–50% of $[^3H]TxB_2$, in the absence of any other competitive binding substance.

(c) The specific activity of $[^3H]TxB_2$ affects the binding. The incubation conditions, the concentration of charcoal suspension, and the sequestration conditions can affect the binding of $[^3H]TxB_2$. Empirical adjustments may be required.

Preparation of Standards

1. Weigh 10.00 ± 0.10 mg of TxB_2 into a 10.0-ml volumetric flask. Fill to the mark with 0.1 *M* PBSG. Sonicate to dissolve the TxB_2. The concentration of this stock solution is 1.0 mg/ml. Mix thoroughly.

2. Transfer 5.0 ml of the stock solution to a 50.0 ml volumetric flask. Fill to the mark with 0.1 *M* PBSG. Sonicate and mix thoroughly. The concentration of this dilution A is 0.10 mg/ml.

3. Repeat Step 2 to produce two further dilutions: B = 0.01 mg TxB_2/ml and C = 0.001 mg TxB_2/ml or 1000 ng/ml.

4. Label screw-cap polyethylene tubes (50 ml) in sequence: 15,14, 13 . . . 3, 2, 1. Pipette 20.0 ml of 0.1 *M* PBSG to each tube. Add 20.0 ml of dilution C to tube No. 15. Cap securely and mix. Tube No. 15 now contains 500 ng of TxB_2 per milliliter. Repeat this serial geometric dilution process by transferring 20.0 ml of tube No. 15 to tube No. 14. Mix and

repeat. The final concentrations (ng/ml) will be: 500 (No. 15), 250 (No. 14), 125 (No. 13), 62.5 (No. 12), 31.25 (No. 11), 15.6 (No. 10), 7.8 (No. 9), 3.9 (No. 8), 1.95 (No. 7), 0.98 (No. 6), 0.49 (No. 5), 0.24 (No. 4), 0.12 (No. 3), 0.06 (No. 2), 0.03 (No. 1). Prepare several tubes labeled 0. These contain only 0.1 *M* PBSG; however, it is useful to store these under the identical conditions encountered by the standards.

Radioimmunoassay Procedure

1. To appropriately labeled 12 × 75 mm tubes, add sample (0.10 ml) or standard (0.10 ml); [^{3}H]TxB$_2$ solution (0.10 ml ~ 6000 dpm); and TxB$_2$ antiserum (0.10 ml) diluted 1:10,000. To several tubes labeled NSB (nonspecific bound), add sample (0.10 ml) or standard (0.10 ml); [^{3}H]TxB$_2$ solution (0.10 ml ~ 6000 dpm); and 0.1 *M* PBSG (0.10 ml). *Do not add antiserum to the nonspecific bound tubes.* Incubate the tubes for 1 hr at 25°; then for 20–24 hrs at 4°. Finally incubate the tubes on an ice bath (0°) for 1 hr prior to the charcoal sequestration step to separate antibody bound and free [^{3}H]TxB$_2$. Table III depicts a representative plan.

2. Separate the antibody-bound from the free [^{3}H]TxB$_2$ by charcoal sequestration. Prepare a charcoal suspension (see Reagents) and chill it to 0° on an ice bath. Add 1.0 ml of chilled charcoal suspension to each sample or standard tube. Incubate the charcoal and the tube contents for 11.0 ± 1.0 min. Immediately centrifuge the tubes in a refrigerated centrifuge (4°) at 1200 *g* for 15 min. *Note:* An Eppendorf pipette (1.0 ml) or Gilson adjustable pipette (0–2.0 ml) with a disposable tip is preferred. The bore of the tip should be widened to ~3–4 mm to prevent clogging. A novice can add charcoal suspension to not more than 24 tubes in 1 min; an experienced worker can add charcoal suspension to 36–48 tubes in 1 min with acceptable reliability and reproducibility. Attempts to exceed these limits will lower the precision of the analysis.

3. After centrifugation to sediment the charcoal, use a pipette with a disposable tip to remove 1.0 ml of supernatant (total volume ~1.3 ml) from each tube and transfer it to a scintillation vial. *Note: We strongly advise against decanting* supernatant into a scintillation vial. The variability of fluid adhesion to the glass tubes during decantation is $\geq \pm 10\%$. In contrast, the variability of quantitative transfer of 1.0 ml of supernatant by pipette is $\leq \pm 3\%$.

Add 10.0 ml of scintillation fluid [ACS, Amersham, Arlington Heights, Illinois]. Determine, by scintillation counting, the dpm of [^{3}H]TxB$_2$ bound to the antibody as a function of the standard concentration (pg/0.10 ml). Table IV depicts some representative results.

TABLE III
REPRESENTATIVE EXPERIMENTAL PLAN FOR THROMBOXANE B_2 RADIOIMMUNOASSAY

Tube no.	Volume of standard per assay tube (ml)	Mass of TxB_2 standard per assay tube (pg)	Volume of $[^3H]TxB_2$ (6000 dpm) per assay tube (ml)	Volume of antiserum per assay tube, 1:10,000 dilution (ml)	No. of assay tubes
Standards					
0	0.10	0	0.10	0.10	6
1	0.10	3	0.10	0.10	3
2	0.10	6	0.10	0.10	3
3	0.10	12	0.10	0.10	3
4	0.10	24	0.10	0.10	3
5	0.10	48	0.10	0.10	3
6	0.10	98	0.10	0.10	3
7	0.10	195	0.10	0.10	3
8	0.10	390	0.10	0.10	3
9	0.10	780	0.10	0.10	3
10	0.10	1560	0.10	0.10	3
NSB-std	0.10	0	0.10	Replace with 0.10 ml 0.1 *M* PBSG	6
NSB-sample[a]	0.10	Irrelevant	0.10		6
Samples					
A	0.10	Unknown	0.10	0.10	2 or more
B	0.10	Unknown	0.10	0.10	2 or more
C	0.10	Unknown	0.10	0.10	2 or more
↓	0.10	Unknown	0.10	0.10	2 or more
X	0.10	Unknown	0.10	0.10	2 or more
Y	0.10	Unknown	0.10	0.10	2 or more
Z	0.10	Unknown	0.10	0.10	2 or more

[a] It is advisable to determine the nonspecific binding (NSB) in the presence of the sample matrix. Ideally it will equal the nonspecific binding of the standard curve.

4. Plot $(\text{bound}/\text{bound}_0) \times 100$ [ordinate] vs picograms [abscissa] using multicycle log-linear. All calculations are performed using only the linear region of this curve (~85% B/B_0 to ~20% B/B_0). For unknowns, calculate:

$$\frac{\text{dpm/``ml unknown''} - \text{dpm/ml NSB}}{\text{dpm/ml ``0''} - \text{dpm/ml NSB}}$$

TABLE IV
REPRESENTATIVE RESULTS OF A TxB_2 STANDARD CURVE[a]

Number of tubes (n)	Tube label	Dpm/ml per tube[b]	Dpm/ml bound	$(Bound/bound_0)^c \times 100$	Concentration (pg TxB_2/tube)
3	0	3214 ± 4	3096	100 ± 0.14	0
3	1	3039 ± 57	2921	94.4 ± 1.8	5.5
3	2	2730 ± 48	2612	84.4 ± 1.6	11.0
3	3	2172 ± 34	2054	66.4 ± 1.1	22.2
3	4	1542 ± 31	1424	46.0 ± 1.0	44.4
3	5	1000 ± 31	882	28.5 ± 1.0	88.7
3	6	614 ± 23	496	16.0 ± 0.8	177.5
3	7	382 ± 8	264	8.5 ± 0.2	355.0
3	8	258 ± 8	140	4.5 ± 0.3	710
6	NSB	118 ± 7	0	—	—
5	Total added: 6030 ± 120				

[a] All values = mean ± standard deviation. Notebook 15126-FAF-pp. 71–77: Thromboxane B_2 from monocytes.

[b] This corresponds to the $[^3H]TxB_2$ that remained bound to the antibody after the charcoal sequestration removed free (unbound) $[^3H]TxB_2$. Note that the assay was carried out under equilibrium conditions with an initial $[^3H]TxB_2$ content of 6030 dpm per tube; consequently, the maximal binding in the "0" tube was approximately 50% (3096/6030) of the added $[^3H]TxB_2$. The charcoal sequestration step is never absolutely efficient. The inclusion of a nonspecific bound tube provides a measure of the actual efficiency of sequestration, and it allows a correction for this factor. For example, to calculate dpm/ml bound, subtract 118 dpm/ml from each tube. The standard deviation of triplicate tubes should not exceed ±5%. For skilled workers, ±1–3% is common.

[c] It is convenient to normalize all values to the maximal binding, the "0" tube. $Bound_0$ is defined as 3096 dpm/ml, and all other degrees of binding are expressed as a percentage of this value.

If Unknown A contained 1000 dpm/ml

$$\frac{1000 \text{ dpm/ml} - 118 \text{ dpm/ml}}{3096 \text{ dpm/ml} - 118 \text{ dpm/ml}} = \frac{882}{2978} = 29.6\% \ B/B_0$$

From the calibration curve, this corresponds to 98 pg of TxB_2. Since the unknown volume was fixed at 0.1 ml, the unknown concentration was 980 pg/ml.

Ordinarily, all calculations are performed on a computer. There are scores of programs to manipulate data. It is beyond the scope of this chapter to consider all of them. We prefer a Hewlett-Packard desk-top computer, or one of equivalent capacity, with a commercially available RIA program "Radioimmunoassay" in BASIC language for four different standard curve models: raw data, and logit, probit, or arcsin transformations. All elements of this program are based on the mathematical treatments developed elsewhere.

5. The TxB_2 content has been determined in samples of various composition. The main applications have been for TxB_2 released into the fluid media during platelet aggregations; during growth or stimulation of cell cultures, *in vitro;* during tissue homogenizations. In some cases TxB_2 has been measured in plasma following an *in vivo* provocation to cellular thromboxane synthetase. It is important to recall that TxB_2 derives from the hydrolysis of TxA_2, but TxA_2 may not always hydrolyze exclusively.

Synopsis

1. The radioimmunoassay for TxB_2 can be reliable, accurate, precise, and convenient. Table V shows the reproducibility of standard curves over a 9-month interval, using the same standards, with daily preparation of fresh [^{3}H]TxB_2 and antibody solutions. These results indicate that standards of TxB_2 (0–100 ng/ml) are stable for 9 months in 0.1 *M* PBSG stored at 4°. Provided identical conditions are maintained from day to day, one should be able to attain equivalent reproducibility. Table V indicates both intra- and interassay variability.

2. Various controls are advised when using a TxB_2 radioimmunoassay. First, whenever possible, specific thromboxane synthase inhibitors, such as imidazole, alkylimidazoles, 9,11-iminoepoxyprosta-5,13-dienoic acid, should be included in the experimental design. These agents should suppress TxB_2 concentrations in a dose-dependent fashion without suppressing the concentration of prostaglandins or other metabolites of arachidonic acid. Second, nonsteroidal anti-inflammatory agents, such as indomethacin, aspirin, or ibuprofen, should suppress TxB_2 concentrations and the concentrations of other products of PGH synthase.

3. It is important to recognize that TxB_2 forms by the hydrolysis of its unstable precursor, TxA_2. In samples containing albumin, TxA_2 does not necessarily hydrolyze with a 100% yield of TxB_2. Some TxA_2 will bind, covalently, to proteins or even biological membranes.[9–12] To maximize the hydrolytic transformation of TxA_2 into TxB_2, we advise a mild acid

[9] F. Fitzpatrick and R. Gorman, *Prostaglandins* **14,** 881 (1977).

[10] J. Maclouf, H. Kindahl, E. Granström, and B. Samuelsson, *Eur. J. Biochem.* **109,** 561 (1980).

[11] A. Wilson, H. Kung, M. Anderson, and T. Eling, *Prostaglandins* **18,** 409 (1979).

[12] T. Eling, A. Wilson, A. Chaudari, and M. W. Anderson, *Life Sci.* **21,** 245 (1977).

TABLE V
COMPARATIVE CALIBRATION CURVES FOR TYPICAL THROMBOXANE B_2 ASSAYS OVER A 9-MONTH SPAN FROM JULY, 1980 TO MARCH, 1981[a]

TxB_2 (pg/tube)	(Bound/bound$_0$ ± standard deviation) × 100					
0	100 ± 1.5	100 ± 1.3	100 ± 1.8	100 ± 0.7	100 ± 1.2	100 ± 0.4
5	94.6 ± 2.2	94.6 ± 0.8	96.4 ± 1.2	94.3 ± 1.6	88.8 ± 2.5	92.4 ± 3.6
11	82.3 ± 2.8	86.8 ± 1.1	88.4 ± 1.1	88.0 ± 2.2	72.9 ± 1.7	84.4 ± 1.8
22	60.8 ± 0.9	67.2 ± 0.8	69.1 ± 1.4	69.2 ± 1.4	57.5 ± 0.3	64.2 ± 2.7
44	38.8 ± 1.6	46.2 ± 0.4	45.6 ± 0.8	46.5 ± 0.4	38.2 ± 2.8	42.4 ± 0.6
89	21.7 ± 1.5	26.0 ± 1.4	26.6 ± 0.3	26.9 ± 0.2	22.5 ± 0.7	24.7 ± 1.0
178	21.7 ± 1.5	15.8 ± 0.6	14.9 ± 1.2	13.8 ± 0.6	12.8 ± 1.1	13.3 ± 0.7
355	6.1 ± 0.0	7.8 ± 0.3	7.4 ± 0.2	7.7 ± 0.5	6.2 ± 0.8	7.6 ± 0.9
Logit slope	−2.845	−2.6075	−2.6509	−2.5602	−2.5602	−2.5087
Correlation coefficient	0.998	0.997	0.995	0.996	0.999	0.998
Date of assay	07-03-80	09-22-80	08-23-80	09-26-80	09-28-80	10-01-80

[a] Values represent the mean ± standard deviation of three replicate standard tubes.

quench (final pH 3 ± 1) for samples where such covalent bonding is likely.

4. Everyone contemplating the analysis of thromboxane B_2 or other arachidonic acid metabolites should first familiarize himself with the chemical characteristics of these compounds, their enzymic transformations, and their susceptibility to artifact formation. The general problems of prostaglandin radioimmunoassay have been completely elaborated in a review by Granström.[13]

Acknowledgments

The author thanks Linda Missias and William Bothwell for their expert assistance and contributions to the preparation of this manuscript.

[13] E. Granström, *Adv. Thromboxane Prostaglandin Res.* **5,** 217 (1978).

TABLE V—*Continued*

(Bound/bound$_0$ ± standard deviation) × 100				Mean ± SD	Coefficient of variation
100 ± 2.0	100 ± 0.7	100 ± 1.3	100 ± 2.3	100	—
94.9 ± 1.8	95.9 ± 0.9	94.4 ± 1.9	93.1 ± 0.5	93.9 ± 2.1	±2.2
86.0 ± 2.2	86.1 ± 1.0	84.4 ± 1.6	85.2 ± 4.5	84.5 ± 4.4	±5.2
67.2 ± 1.2	68.6 ± 1.4	66.4 ± 1.1	65.7 ± 1.9	65.6 ± 3.8	±5.8
44.4 ± 1.5	46.6 ± 0.5	46.0 ± 1.0	43.7 ± 1.6	43.8 ± 3.1	±7.1
25.4 ± 1.0	27.9 ± 0.1	28.5 ± 1.0	26.6 ± 0.6	25.7 ± 2.2	±8.6
13.7 ± 0.9	14.6 ± 0.6	16.0 ± 0.8	14.7 ± 1.3	14.1 ± 1.4	±9.9
7.5 ± 0.3	7.4 ± 0.1	8.5 ± 0.2	7.7 ± 1.0	7.4 ± 0.7	±9.5
−2.5948	−2.6882	−2.5158	−2.8638		
0.996	0.998	0.999	0.997		
10-06-80	11-17-80	01-26-81	03-10-81		

[42] Iodinated Derivatives as Tracers for Eicosanoid Radioimmunoassays

By FERNAND DRAY

The immune-competitive assays for prostaglandins (PGs) were developed largely during the 1970s and have been applied progressively to most stable eicosanoids, including prostanoid[1–3] as well as nonprostanoid[4] derivatives. The eicosanoids are present in most biological systems at low concentrations (10^{-9} to 10^{-11} *M*), thus requiring extremely sensitive detection techniques. The commercially available tritiated prostanoids have up to 8 atoms of tritium incorporated into their molecular structures; this high specific radioactivity has already allowed various radioimmunoassays (RIA) to be developed. However, the quantity of sample needed is often too large, and more sensitive assays would be useful.

First attempts[5] using PG covalently bound to bacteriophage T_4 as a

[1] L. Levine and H. Van Vunakis, *Biochem. Biophys. Res. Commun.* **41,** 1171 (1970).
[2] F. Dray and B. Charbonnel, *Eur. J. Clin. Invest.* **5,** 311 (1975).
[3] E. Granström and H. Kindahl, *Adv. Prostaglandin Thromboxane Res.* **1,** 81 (1976).
[4] L. Levine, this volume [36].
[5] F. Dray, E. Maron, S. A. Tillson, and M. Sela, *Anal. Biochem.* **50,** 339 (1972).

ISBN 0-12-181986-8

tracer yielded a highly sensitive assay for $PGF_{2\alpha}$ ($\simeq$ 1 pg), but the application to plasma was difficult owing to nonspecific binding.[6]

On the other hand, we succeeded, using radioiodinated derivatives, in developing very sensitive and reliable RIAs, which could be applied to any eicosanoid.[7]

Preparation of Radioiodinated Derivatives

All the eicosanoids have a free carboxyl group. This acidic function has been used for the formation of a peptide bound either with the free amino groups of the antigenic carrier for the preparation of the immunogen or with the free amino group of a substance that can be iodinated.

Coupling to the Substance to be Iodinated

The strategy was to select a substance that on the one hand had a low molecular weight in order to preserve the immunological recognition of the hapten by the antibody site, and on the other hand can be easily iodinated (e.g., substances with a phenol or imidazole ring). Three compounds were tested: tyrosyl methyl ester (TME), tyramine (Tyr), and histamine (His).

Soluble Carbodiimide Method. This method has been applied to most eicosanoids, and the coupling reaction between 6-keto-PGE_1, a metabolite of prostacyclin and histamine is described as a model. 6-Keto-PGE_1 (2.5 mg; 7.1 μmol) in ethanol (50 μl); and 1-ethyl-3-(3-dimethylaminopropyl)carbodiimide-HCl (EDCI; (4 mg; 20.86 μmol) in 50 μl of distilled H_2O were mixed together. After inactivation at room temperature, the following reaction was carried out in the dark: 2.5 mg (22.5 μmol) of histamine in 100 μl of distilled H_2O were added and the pH was adjusted to pH 6; 5 μCi [^{3}H]histamine in 5 μl was then added. After incubation overnight at room temperature, the reaction mixture was purified by thin-layer chromatography (TLC) on silica gel in *n*-butanol–acetic acid–water (75:10:25, v/v/v). The imidazole ring of histamine was localized on the plate using Pauly's reagent (0.5% sulfanilic acid in 1 *N* HCl–5% aqueous $NaNO_2$–10% aqueous Na_2CO_3, 2:1:2, v/v/v; heat plate for a few minutes; phenolic rings yield a pink color), and [^{3}H]histamine was localized by radiochromatography (Kodak film). Quantitative estimation of the coupling reaction was performed by isotopic dilution. The coupled prostanoid (R_f 0.5) was scraped from the plate and extracted twice with eth-

[6] J. M. Andrieu, S. Mamas, and F. Dray, *Prostaglandins* **6,** 15 (1974).

[7] J. Maclouf, M. Pradel, P. Pradelles, and F. Dray, *Biochim. Biophys. Acta* **431,** 139 (1976).

anol–water (70%, v/v). The solution was distributed in small aliquots (1–5 nmol) and lyophilized.

DCCI Method. The coupling efficiency was low for some eicosanoids, such as PGD_2, so a more gentle reaction in organic solvent was used: 1.35 mg (3.82 μmol) of PGD_2 in 0.3 ml of dichloromethane at 4°, 1.24 mg of *N,N*-dicyclohexylcarbodiimide (DCCI) in 3 μl of dichloromethane, 3.53 mg (19.17 μmol) of histamine-HCl in 0.15 ml of dimethylformamide, and 5.3 μl of triethylamine were mixed together in a test tube. The reaction mixture was stirred, and 5 μCi of $[^3H]$histamine in 5 μl were added. After overnight incubation at 4° with stirring, the reaction mixture was purified as previously described.

Iodination Procedure

Chloramine-T Method. The eicosanoid derivative was labeled with ^{125}I using the chloramine-T method.[8] To a conical vial containing the lyophilized eicosanoid derivative were added successively 10 μl of 0.5 *M* phosphate buffer, pH 7.4, 2 μl of $Na^{125}I$ (100 mCi/ml), and 4 μl of chloramine-T (3 mg per milliliter of 0.05 *M* phosphate buffer). After 20 sec, the reaction was stopped with sodium metabisulfite (32 μg). The labeled derivative was purified by TLC on silica gel in chloroform–methanol–water, 80:20:2, v/v/v).

Radioactive spots were located by autoradiography and eluted from the plate with ethanol. Binding properties were evaluated after serial dilutions; the radioactive tracer was then distributed (1 μCi) into vials and lyophilized. This iodination procedure usually gives a monoiodinated derivative, and the specific radioactivity reaches, after purification, the theoretical maximum (i.e., 2000 Ci/mmol).

Iodogen Method. Iodogen was introduced as an ideal reagent for protein iodination,[9] and we first adapted this method for the iodination of some eicosanoids that can be altered by the chloramine-T method (e.g., PGD_2) or are available in small amounts. In a plastic conical tube, 2 μg of iodogen [1,3,4,6-tetrachloro-3α,6α-diphenylglycoluril (Pierce)] was added in 20 μl of methylene chloride. The solvent was evaporated under N_2. Then 0.35 nmol of PGD_2 in 32.5 μl of 0.5 *M* phosphate buffer, pH 7.4, was added; after 10 min at room temperature, 0.35 nmol (5 μl) of $Na^{125}I$ was added. After 10 min at room temperature the mixture was purified by TLC on silica gel in chloroform–methanol–water (80:20:2, v/v/v). Elution was carried out in acetonitrile–H_2O (1:1, v/v). Films of iodogen react rapidly in the solid phase with an aqueous mixture of ^{125}I and the His

[8] W. H. Hunter and F. C. Greenwood, *Nature (London)* **194,** 495 (1962).

[9] P. Y. Fraker and J. C. Speck, *Biochem. Biophys. Res. Commun.* **80,** 849 (1978).

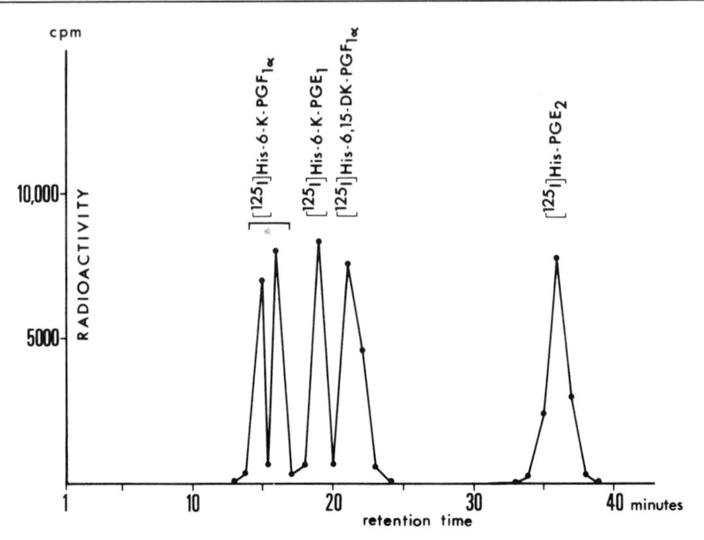

FIG. 1. High-pressure liquid chromatography of some prostanoid [^{125}I]histamine derivatives. These compounds were previously purified by thin-layer chromatography. The interconvertible 6-keto-PGF$_{1\alpha}$ [^{125}I]His isomers that have the same immunoreactivity can be mixed for the assay. Experimental conditions: instrument from Waters Associates; packing, μBondapak C_{18}; column, 3.9 mm × 300 mm; flow rate, 1 ml/min, isocratic elution; solvent system, acetonitrile–water–acetic acid, 30:70:0.1, v/v/v).

derivative, to yield the ^{125}I labeled His derivative. These reactions eliminate the reduction step that is required when chloramine-T is used. Furthermore, the iodogen requires less PGD$_2$. The iodogen technique is now used for iodinating all His derivatives of eicosanoids.

Storage of Iodinated Derivatives. The immunoreactive properties of the tracers after TLC purification were unchanged for about 1 month; they then lost their immunoreactivity at an unpredictable rate. To extend the period of immunostability, we have added another purification step using high-pressure liquid chromatography (HPLC). The radioactive eluates were lyophilized and then tested regularly for binding properties; 4 months after labeling, the immunoreactive properties of five iodinated derivatives selected for this test were unchanged. Figure 1 shows a reconstituted HPLC profile of these tracers. The 6-keto-PGF$_{1\alpha}$ derivatives gave two interconvertible isomers. They were bracketed together because of their identical binding properties.

Radioimmunoassays

Table I shows the procedures we have developed for tritiated or iodinated tracers. Although the dextran-coated charcoal separation of bound from free ^{125}I fractions gave identical results in terms of binding, polyeth-

TABLE I
RADIOIMMUNOASSAYS OF EICOSANOIDS: COMPETITIVE TECHNIQUE USING TRITIATED OR IODINATED TRACER[a]

	^{3}H Tracer	^{125}I Tracer
Immunological reaction, incubation overnight at 4°	1. Tracer, 100 μl, 7,500 dpm [^{3}H]eicosanoid, 15,000 dpm [^{125}I]His-eicosanoid 2. Sample, 100 μl, unknown or standard 3. Antiserum, 100 μl, appropriate dilutions for binding of the tracer between 35 and 50%	
Separation of free and bound fractions	1. Incubation 12 min at 0° in dextran-coated charcoal 1 ml 2. Centrifugation 15 min at 4° at 2000 *g*	1. Precipitation at 0° with polyethylene glycol (PEG) 500 μl (mixing) 2. Centrifugation 15 min at 4° at 2000 *g*
Radioactivity counting	Of supernatant (bound fraction) in a scintillation counter	Of pellet (bound fraction) in a gamma counter
All reagents diluted in:	0.1 *M* saline phosphate buffer (0.1 *M* sodium phosphate containing 0.9% NaCl) pH 7.4; gelatin, 0.1%	0.05 *M* saline phosphate buffer, pH 7.4; bovine γ-globulin 0.3%

[a] Dextran-coated charcoal: NORIT A. (250 mg)–Dextran T 20 (25 mg) for 100 ml of buffer; PEG 6000: 25 g/100 ml of distilled water.

ylene glycol (PEG) was selected for this tracer because it is more convenient for counting bound fractions and is more reproducible.

Binding Parameters

Binding of Tracers. The maximum binding of iodinated tracers was tested in the presence of a large excess of selected antisera and showed in all cases a binding (bound/total radioactivity) equal to or above 80% after TLC purification and more than 90% after HPLC purification for the five prostanoids tested. The dilution of each antiserum adjusted to obtain 40% binding was always greater with iodinated than tritiated tracers: the degree of difference depends on various factors, which will be analyzed further.

Sensitivity. In the first assays, we compared the performances obtained with His, Tyr, and TME derivatives; we selected the His derivative for its stability. When a comparison of dose-response curves using iodinated or tritiated tracer was made, the curves were generally parallel and more sensitive with the iodinated tracer. In a few cases, the sensitivity

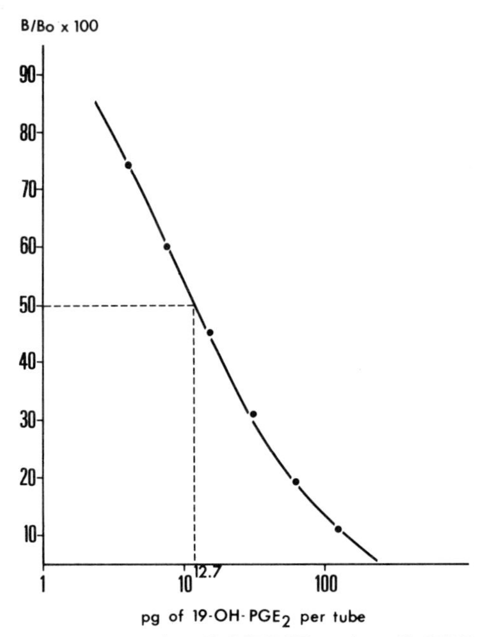

FIG. 2. Dose response curve for 19-OH-PGE_2 using 19-OH-PGE_2-BSA antiserum (R.14.A.) and 19-OH-PGE_2-[^{125}I]His as tracer.

was not improved, as in the case of $PGF_{2\alpha}$; or the curves were not parallel, as in the case of PGE_1. These observations were obviously valid only with the selected antisera. In some cases, the tritiated tracer was not available commercially or had been obtained enzymically with a low specific radioactivity. Figure 2 illustrates with the case of 19-OH-PGE_2, a major prostanoid of human seminal fluid, the general application of iodinated tracers for the assay of natural as well as synthetic eicosanoid derivatives.

Table II shows some binding parameters (final dilution of antisera, sensitivity) for eight prostanoids using tritiated or iodinated tracers and for five prostanoids using only iodinated tracers.

Specificity. Comparative studies have been carried out to compare the specificity of the antisera according to whether the radioactive tracer was an iodinated or a tritiated derivative. In general, a similar order of inhibition is found with the same antiserum and various related compounds, with a few exceptions, namely PGE_1 and 13,14,-diketo-PGE_2. This is likely to be due to alterations induced by blocking the carboxyl group of the tracer.[8]

TABLE II

RELATIVE AFFINITIES OF RELATED PROSTANOIDS FOR ANTISERA RAISED IN RABBITS (EXCEPT FOR DHK-$PGF_{2\alpha}$, RAISED IN SHEEP) AGAINST 11 PROSTANOIDS OF SERIES 2 COUPLED TO BOVINE SERUM ALBUMIN[a]

	Antisera										
Inhibitor	PGE_2	PGD_2	$PGF_{2\alpha}$	DHK-PGE_2	DHK-$PGF_{2\alpha}$	TxB_2	6-K-$PGF_{1\alpha}$	6,15-DK-$PGF_{1\alpha}$	6-K-PGE_1	19-OH-PGE_2	19-OH-$PGF_{2\alpha}$
PGE_1	15.0	0.05	0.1	<0.1	<0.1	<0.1	5.6	<0.8	<1.0	—	—
PGE_2	100.0	0.05	1.0	<0.1	<0.1	<0.1	2.2	<0.3	<1.0	—	—
$PGF_{1\alpha}$	<0.1	0.05	7.0	<0.1	<0.1	<0.1	18.0	<0.3	<0.1	<0.1	2.0
$PGF_{2\alpha}$	0.8	0.05	100.0	<0.1	4.0	<0.1	11.0	<0.3	<0.1	<0.1	4.0
PGD_1	<1.0	16.00	<0.3	<0.1	<0.1	<0.1	1.5	<0.3	<0.1	—	—
PGD_2	<1.0	100.00	<0.3	<0.1	<0.1	<0.1	0.5	<0.2	<0.1	—	<0.1
DHK-PGE_2	<0.5	<0.05	<0.5	100.0	7.0	<0.1	<0.1	1.4	<0.1	—	—
DHK-$PGF_{2\alpha}$	<0.1	<0.05	<0.1	<0.1	100.0	<0.1	<0.1	3.5	<0.1	—	—
TxB_2	<0.1	0.90	<0.1	<0.1	<0.1	100.0	<0.1	<0.1	<0.1	<0.1	<0.1
6-K-$PGF_{1\alpha}$	2.0	<0.05	11.0	<0.1	0.3	<0.1	100.0	1.0	1.7	<0.1	<0.1
6,15-DK-$PGF_{1\alpha}$	<0.3	<0.10	<0.3	1.4	3.5	<0.3	<0.1	100.0	4.3	—	—
6-K-PGE_1	—	—	—	—	—	—	—	—	100.0	—	—
19-OH-PGE_2	23.0	—	<0.2	—	—	—	—	—	—	100.0	<0.1
19-OH-$PGF_{2\alpha}$	—	—	2.0	—	—	—	—	—	—	<0.1	100.0

[a] See Granström and Kindahl.[3] Values were calculated on the basis of quantity (pg) necessary for 50% tracer displacement. Tritiated tracers were used for the first 6 antisera and iodinated tracers for the remaining antisera. 6-Keto-$PGF_{1\alpha}$, 6,15-diketo-$PGF_{1\alpha}$ and 6-keto-PGE_1 are PGI_2 metabolites. K = keto; DK = diketo; DHK = dihydroketo.

Interest of Iodinated Tracers

Thermodynamic Considerations

Competitive radioimmunoassays are based on competition between unlabeled (S) and labeled (*S) molecules of the ligand for the same antibody sites (R). At equilibrium, the association constants are

$$K_a = \frac{(S)(R)}{(SR)} \quad \text{for the unlabeled system}$$

$$^*K_a = \frac{(^*S)(R)}{(^*SR)} \quad \text{for the labeled system}$$

In the case of tritiated tracer, there is no difference between the structures of both competitors and $K_a \neq {}^*K_a$. The sensitivity of the assay will increase, the greater K_a and the specific activity of the tracer.

In the case of iodinated tracers, the structures for both competitors are different and are recognized differently by the same antibody sites; so, $K_a \neq {}^*K_a$.

On the one hand, the blockage of the carboxyl group in a peptidic bond increases the structural analogy between the tracer and immunogen:

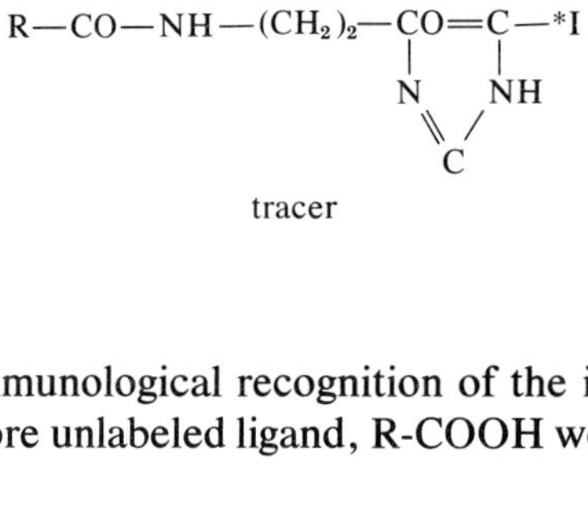

```
R—CO—NH—(CH2)2—CO=C—*I                                   |
                |   |                                    CO
                N   NH                                   |
                 \\ /              R—CO—NH—(CH2)2—CH
                  C                                      |
        tracer                                           NH
                                                         |
                                          immunogen
```

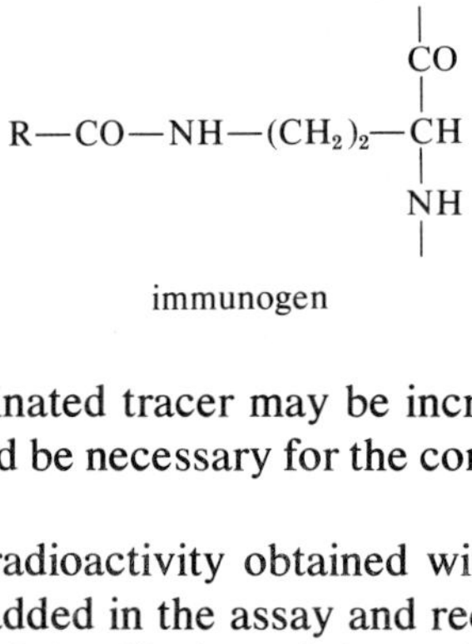

The immunological recognition of the iodinated tracer may be increased and more unlabeled ligand, R-COOH would be necessary for the competition.

On the other hand, the high specific radioactivity obtained with ^{125}I makes negligible the weight of the tracer added in the assay and requires less antiserum for binding. The increase of the dilution of the antiserum may involve antibody populations with low capacity but high affinity; the effect will be an increase in sensitivity.

Finally, we are in the presence of contradictory effects, and the resulting effect cannot be predicted but can only be observed by analysis of dose-response curves.

Another way of increasing the sensitivity of competitive assays with iodinated tracer is to block the carboxyl group of the ligand, in order to

TABLE III
COMPARISON OF THE SENSITIVITY OF PGE_1 AND PGE_2 RADIOIMMUNOASSAYS WITH DIFFERENT TRACERS AND INHIBITORS

		PGE_1		PGE_2	
Tracer[a]	Inhibitor[a]	Final dilution[b]	Sensitivity[c] (pmol/ml)	Final dilution[b]	Sensitivity[c] (pmol/ml)
[^{3}H]PG	PG	1:45,000	0.30	1:75,000	0.051
[^{3}H]PG-His	PG	1:75,000	4.51	1:75,000	0.043
[^{3}H]PG-ME	PG	1:75,000	4.61	1:75,000	0.047
[^{3}H]PG-ME	PG-ME	1:75,000	1.40	1:75,000	0.050
^{125}I-Labeled PG-His	PG	1:300,000	0.17	1:150,000	0.025
^{125}I-Labeled PG-His	PG-His	1:300,000	0.11	1:150,000	0.024
^{125}I-Labeled PG-His	PG-Lys	1:300,000	0.06	1:150,000	0.019
^{125}I-Labeled PG-His	PG-ME	1:300,000	0.09	1:150,000	0.028

[a] His, histamine; ME, methyl ester; Lys, lysine.
[b] Final dilutions of the antisera are expressed for 40% initial binding (B_0).
[c] The sensitivity is given for 50% displacement of B_0.

TABLE IV
PROSTANOID RADIOCOMPETITIVE ASSAYS[a]

		Hapten coupled to BSA				
		Final dilution ($\times 10^{-3}$)			Sensitivity	
Antisera		^{3}H		^{125}I	^{3}H	^{125}I
PGE_1	(S.144.B)[b]	1:45	*90*[c]	1:300	32	18
PGE_2	(R.6.79585)[b]	1:15	*160*	1:25	6	2
$PGF_{1\alpha}$	(R.5758-58A)	1:5	*90*	1:60	50	30
$PGF_{2\alpha}$	(R.5.G)	1:25	*178*	1:40	16	14
DHK-PGE_2	(R.8788)	1:3	*70*	1:9	298	95
DHK-$PGF_{2\alpha}$	(S.229.B1)	1:30	*85*	1:240	24	10
PGD_2	(R.4079.A)	1:3	*100*	1:12	80	20
TxB_2	(R.3643.A)	1:24	*125*	1:270	14	6
6-K-$PGF_{1\alpha}$	(R.40056.A)	1:45	*100*	1:178	39	13
6,15-Diketo-$PGF_{1\alpha}$	(R.44.A)	—	—	1:15	—	25
6-K-PGE_1	(R.100.A)	—	—	1:3.6	—	100
19-OH-PGE_2	(R.14.A)	—	—	1:120	—	13
19-OH-$PGF_{2\alpha}$	(R.16.A)	—	—	1:150	—	164

[a] Final dilutions of antisera and sensitivities for each assay using ^{3}H and/or ^{125}I-tracers. The sensitivity (pg/tube) is given for 50% displacement of B_0 (cpm bound of the tracer in the absence of the unlabeled prostanoid).
[b] R, rabbit; S, sheep.
[c] Specific activity (Ci/mmol) is indicated in italics for each tritiated tracer and estimated to be 2000 Ci/mmol for the iodinated tracers.

mimic the structure of the tracer.[11] Table III shows the effect on the sensitivity of the PGE_1 and the PGE_2 assays when ligand transformations were made. Three derivatives for PGE_1 and PGE_2 were synthesized: methyl esters and lysyl and histamine derivatives. The sensitivity was improved in both cases when a peptidic bond was formed; only in the PGE_1 assay was the sensitivity improved by the methyl ester derivative. Blocking the carboxyl group in a peptidic bond would be the best solution. However, it is difficult completely to bind the ligand using the carbodiimide method, in spite of the complete esterification with methanol in diazomethane.

Practical Considerations

The foregoing data suggest that ^{125}I-labeled radioligands have an important role to play in radioimmunocompetitive assays for prostaglandins and for eicosanoids in general. The [^{125}I]histamine derivatives are readily prepared by the iodogen procedure, and the introduction of purification of tracers on HPLC improves their immunostability and reduces the frequency of labeling. The increase of sensitivity obtained with the iodinated tracers reduces the volume of biological sample used and opens new possibilities for investigation.

[11] J. Maclouf, H. Sors, P. Pradelles, and F. Dray, *Anal. Biochem.* **87,** 169 (1978).

[43] Radioimmunologic Determination of 15-Keto-13,14-dihydro-PGE_2: A Method for Its Stable Degradation Product, 11-Deoxy-15-keto-13,14-dihydro-11β,16ξ-cyclo-PGE_2

By ELISABETH GRANSTRÖM, F. A. FITZPATRICK, and HANS KINDAHL

Many assays have been developed for the initial metabolite of $PGF_{2\alpha}$, 15-keto-13,14-dihydro-$PGF_{2\alpha}$, and have been successfully applied to a number of biological studies (This volume [44]). However, very few quantitative methods exist for the corresponding PGE_2 metabolite, 15-keto-13,14-dihydro-PGE_2, and consequently much less is known about the roles of PGE_2 in the body. Two gas chromatographic–mass spectrometric methods exist[1,2] (see also this volume [67]), and a few attempts

[1] B. Samuelsson and K. Gréen, *Biochem. Med.* **11,** 298 (1974).
[2] W. C. Hubbard and J. T. Watson, *Prostaglandins* **12,** 21 (1976).

ISBN 0-12-181986-8

have also been made to develop radioimmunoassays for the compound.[3-5]

The chemical instability of 15-ketodihydro-PGE_2,[6-8] however, renders it very unsuitable as a target for measurements, particularly using radioimmunoassay, which normally involves long incubations in aqueous solution. Among the products that are formed during decomposition of the labile PGE_2 metabolite,[6-8] a bicyclic product, 11-deoxy-15-keto-13,14-dihydro-11β, 16ξ-cyclo-PGE_2, was found to be stable and particularly suitable as a parameter for radioimmunoassay purposes.

The present chapter describes the chemical instability of 15-keto-13,14-dihydro-PGE_2 under various conditions, the development of a radioimmunoassay for the stable bicyclic degradation product, and conditions for the induced quantitative conversion of 15-ketodihydro-PGE_2 in biological samples into this product prior to measurements.

Chemical Instability of 15-Keto-13,14-dihydro-PGE_2

15-Keto-13,14-dihydro-PGE_2 is rapidly degraded in aqueous solution into several other compounds (Fig. 1). First, the well known enolization of the C-9 keto group leads to the formation of the corresponding 8-iso metabolite. Second, and most important, a rapid dehydration occurs from the sensitive β-ketol structure in the ring, leading initially to the corresponding PGA_2 metabolite, 15-keto-13,14-dihydro-PGA_2. This is the major product around or below neutral pH. The slightly electron-deficient C-11 of this compound may subsequently undergo a nucleophilic attack by, for example, sulfhydryl groups, which may lead to the appearance of water-soluble compounds particularly in biological samples (see below).

However, 15-ketodihydro-PGA_2 may also undergo a different fate: cyclization into a bicyclic product, 11-deoxy-15-keto-13,14-dihydro-11β,16ξ-cyclo-PGE_2.[6-8] The reaction is very rapid at high pH, but is noticeable with time also at physiological pH.

This cyclization product (Fig. 1) contains at least two epimeric sites, at C-8 and C-16; furthermore, the cyclization of the PGA_2 derivative might also involve a transient formation of the corresponding PGC_2 compound,

[3] K. Gréen and E. Granström, *in* "Prostaglandins in Fertility Control" (S. Bergström, ed.), Vol. 3, p. 55. WHO Research and Training Centre on Human Reproduction, Karolinska Institutet, Stockholm, 1973.

[4] B. A. Peskar, A. Holland, and B. M. Peskar, *FEBS Lett.* **43,** 45 (1974).

[5] L. Levine, *Prostaglandins* **14,** 1125 (1977).

[6] E. Granström and H. Kindahl, *Adv. Prostaglandin Thromboxane Res.* **6,** 181 (1980).

[7] F. A. Fitzpatrick, R. Aguirre, J. E. Pike, and F. H. Lincoln, *Prostaglandins* **19,** 917 (1980).

[8] E. Granström, M. Hamberg, G. Hansson, and H. Kindahl, *Prostaglandins* **19,** 933 (1980).

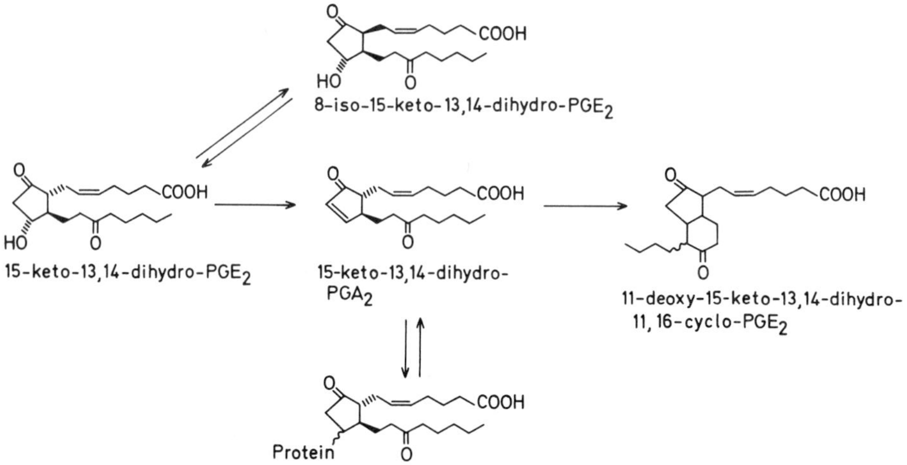

8-iso-15-keto-13,14-dihydro-PGE_2

15-keto-13,14-dihydro-PGE_2

15-keto-13,14-dihydro-PGA_2

11-deoxy-15-keto-13,14-dihydro-11,16-cyclo-PGE_2

Protein

FIG. 1. Structures of products and tentative pathways in the degradation of 15-keto-13,14-dihydro-PGE_2. Reproduced from E. Granström *et al.*[8]

resulting in additional epimeric sites at C-11 and C-12. High-performance liquid chromatography (HPLC) analysis of a mixture of degradation products revealed at least four epimers of this compound (Fig. 2). Under certain conditions the proportions between formed epimers is constant and reproducible. The radioimmunoassay described in this chapter was developed using a racemic mixture of these epimers for the preparation of the immunogen.

Development of the Radioimmunoassay

Preparation of the Bicyclic Derivative for Coupling

15-Keto-13,14-dihydro-PGE_2, 3 mg, was dissolved in 1.2 ml of methanol, and 0.2 ml of 1 M NaOH was added. After 2 hr at room temperature the solution was diluted with water, and the products were extracted with ether after prior acidification. HPLC analysis (straight-phase HPLC using Nucleosil 5; solvent system, hexane–isopropanol–acetic acid, 95:5:0.01, v/v/v) revealed a total conversion of the PGE_2 metabolite into the four epimers of the bicyclic compound. No 15-ketodihydro-PGA_2 was detected after this treatment.

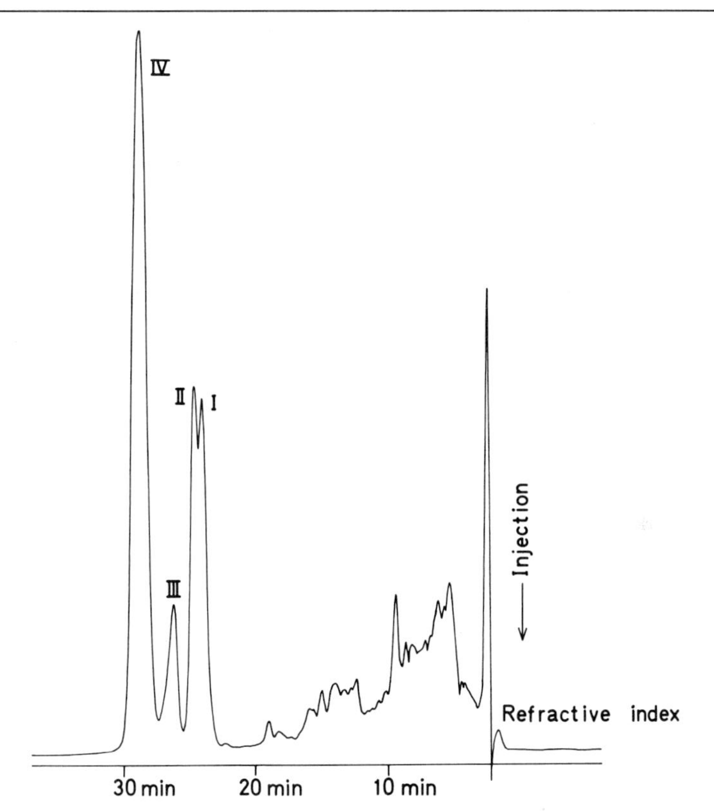

FIG. 2. Reversed-phase HPLC (solvent system: methanol–water–acetic acid, 60:40:0.01, v/v/v) of the bicyclic products formed by alkali treatment of 15-ketodihydro-PGE_2.

Coupling Procedure and Immunization

These steps were performed essentially as described in this volume [44], describing a radioimmunoassay for 15-ketodihydro-$PGF_{2\alpha}$. Briefly, 2.5 mg of the bicyclic compound were dissolved in 0.5 ml of dimethylformamide, converted into a reactive imidazolide using 5 mg of *N,N'*-carbonyldiimidazole,[9] and coupled to amino groups on bovine serum albumin, 10 mg in 0.75 ml of H_2O. After dialysis and lyophilization, portions of the conjugate were emulsified using Freund's complete adjuvant and injected into a rabbit at weekly intervals. After four injections the animal was bled and the antiplasma was prepared.

[9] U. Axén, *Prostaglandins* **5**, 45 (1974).

Preparation of the Labeled Ligand

[5,6,8,11,12,14-$^3H_6$(n)]15-Keto-13,14-dihydro-PGE_2 was purchased from The Radiochemical Centre, Amersham, England, with a specific activity of 80 Ci/mmol. This compound was treated with 1 *M* NaOH as described above for the unlabeled compound to induce dehydration and cyclization. The resulting reaction mixture was extracted after acidification and subjected to straight-phase HPLC as described above. To get a better separation between the epimers, the bicyclic compounds from this HPLC chromatography were isolated and subjected to a reversed-phase HPLC (solvent system: methanol–water–acetic acid, 60:40:0.01, v/v/v). Four peaks of radioactivity appeared (Fig. 2). The material in each was collected separately and tested for affinity for the antibody preparation. All of them were bound by the antibody, but to different degrees (see below).

The chemical conversion and extraction proceed with quantitative formation and recovery of [^{3}H]11-deoxy-13,14-dihydro-15-keto-11β,16ξ-cyclo-PGE_2; however, only 75% ± 5% of the total radioactivity is recovered. The remainder converts into tritiated water, formed by the predictable loss of an acidic, exchangeable tritium from the C-8 and C-14 positions of the [5,6,8,11,12,14-^{3}H]15-keto-13,14-dihydro-PGE_2, under alkaline conditions. The loss of tritium reduces the specific activity of the recovered [5,6,11,12-^{3}H]11-deoxy-13,14-dihydro-15-keto-11β,16ξ-cycloprostaglandin E_2; however, it does not reduce the purity or utility of the radiolabeled ligand prepared for the assay.

Properties of the Antiplasma

The titer of the antiplasma was very high, and somewhat variable depending on which labeled ligand was used. The best ligand was the material in peak IV from the HPLC chromatography: using this material as the tracer, the antiplasma could be used in a final dilution of 1:200,000. Probably the antibody was mainly directed against this epimer. The titers obtained when the other labeled isomers (I, II, or III) were employed as tracers ranged between 1:50,000 and 1:120,000. In all experiments described below, either isomer IV alone or a racemic mixture of [^{3}H]11-deoxy-13,14-dihydro-15-keto-11β,16ξ-cyclo-PGE_2 was employed as the labeled ligand. The dilution of antibody in the latter case was 1:75,000. No differences were seen in the results obtained using either method.

Specificity was also very high for this antibody. Cross-reactions with other structurally related compounds were below 0.1% for most compounds, including the primary prostaglandins, TXB_2, and 15-ketodihydro-$PGF_{2\alpha}$. After a rapid incubation with 15-ketodihydro-PGE_2 as the unlabeled ligand, a cross-reaction of 4% was recorded: it is not likely, however, that this figure represents the true cross-reaction with the PGE_2 metabolite, as dehydration of this compound occurs rapidly during radioimmunoassay conditions.

Radioimmunoassay Procedure

The crucial point in this assay is the prior quantitative conversion of all 15-ketodihydro-PGE_2 present in the sample into the bicyclic derivative before analysis. The practical details of the remainder of the procedure, i.e., reagents, equipment, and incubation procedure, are essentially the same as those described for the assay of 15-ketodihydro-$PGF_{2\alpha}$ (this volume [44]).

However, this assay differs from the earlier described ones in at least one fundamental respect. As briefly mentioned above, the intermediary 15-ketodihydro-PGA_2, which appears transiently during the conversion of 15-ketodihydro-PGE_2 into the bicyclic product, is a highly reactive compound and may be trapped by proteins and other substances containing nucleophilic groups.[7,8] Thus, direct assay of, for example, plasma samples is not possible to perform in this case: about 50% of the desired product will then escape detection because of this trapping (see below, Analysis of Plasma and Other Complex Biological Samples; cf. also Fig. 3).

Thus, the practical details of this radioimmunoassay of the bicyclic derivative differ somewhat for complex biological samples (containing proteins, mainly albumin, or other sulfhydryl-containing compounds) and for samples without these reactive substances.

Analysis of Aqueous Samples without Proteins

Analysis of protein-free samples and samples devoid of sulfhydryl or other nucleophilic reagents is comparatively simple: only alkali treatment is required before radioimmunoassay. Perfusates from organs, eluates from chromatographic columns, chemical reaction mixtures, etc., may fall into this category. It is advisable, however, first to run a test experiment using [^{3}H]15-ketodihydro-PGE_2, followed by chromatographic analysis of the products, to establish that conversion into the bicyclic derivative is really complete under the conditions employed. The predictable loss of ^{3}H from C-8 and C-14 during the conversion step must, however, be taken into consideration when judging the result of the experiment.

Analysis of Plasma and Other Complex Biological Samples

In samples containing proteins, particularly albumin, 15-ketodihydro-PGE_2 is degraded even more rapidly than in buffer of the same pH. Figure

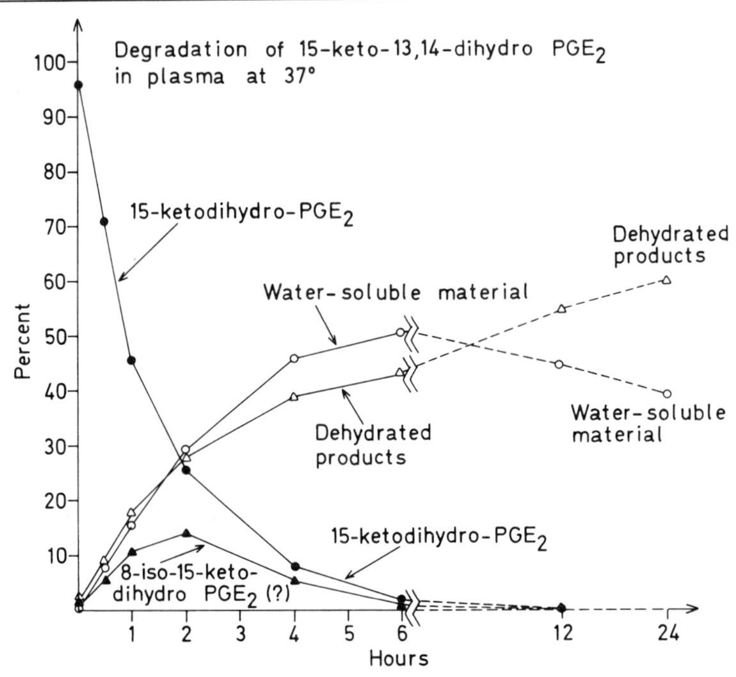

FIG. 3. Degradation of [^{3}H]15-keto-13,14-dihydro-PGE$_2$ in plasma at 37°. Reproduced from Granström *et al.*[8]

3 shows the fates of the PGE$_2$ metabolite in human plasma at +37°. As can be seen, its half-life was only about 45 min in this environment. Over 40% of the material originally present was eventually converted into nonpolar dehydration products. HPLC analysis of this material revealed that it was a mixture of 15-ketodihydro-PGA$_2$ and the cyclization product, the latter accumulating with time.

The remainder of the material was highly polar, and although some of the tritium in this fraction was due to the above-mentioned exchange from C-8 and C-14, most of the material was identified as albumin adducts.[7,8] As this fraction represented about 50% of the 15-ketodihydro metabolite originally present in the sample, it cannot be neglected in quantitative studies. Many attempts were made to induce the dehydration and cyclization of 15-ketodihydro-PGE$_2$ without the accompanying formation of the polar adducts; however, regardless of reaction conditions, the formed amounts of this water-soluble fraction always approached 50%.

Three different approaches can be used in coping with this problem: (1) removal of the protein fraction prior to the alkali treatment; (2) hydrolysis of the formed protein adducts after alkali treatment; (3) blocking the reactive group(s) on the protein prior to alkali treatment, which might at least theoretically prevent formation of water-soluble adducts. All these approaches have been employed in this radioimmunoassay; however, attempts to block the reactive groups on the protein were not successful, and this approach was abandoned.

1. Removal of Proteins prior to Alkali Treatment

Plasma samples, 0.5 ml, were subjected to addition of two volumes of acetone. This precipitated almost all proteins. The protein precipitate was removed by centrifugation, and the acetone was allowed to evaporate spontaneously at 37°. The remaining water was removed by lyophilization; 100 μl of 0.05 *M* NaOH were added to the dry residues, and the solutions were left overnight. After this treatment, 0.2 ml of the radioimmunoassay buffer was added, which contained 112.5 mg of serum albumin per milliliter. This gave a final concentration of albumin close to the physiological one. At this stage the albumin may safely be added to the samples: the bicyclic product which now has been formed is not reactive. It is in fact necessary to reconstitute the samples with respect to albumin concentration prior to radioimmunoassay, for reasons that were discussed in this volume [44]. For the standards, isomer IV from a preparation of the unlabeled bicyclic compound (Fig. 2) was diluted in the radioimmunoassay buffer in a geometric series from 8, 16, . . . to 256 pg/0.1 ml buffer. The standard tubes, containing 0.1 ml of each concentration in duplicates, were subjected to the same treatment as the plasma samples (see above).

The samples and standards were then carried through the radioimmunoassay in the normal way: addition of 0.2 ml of a 0.5% bovine γ-globulin solution, 0.1 ml of an antibody preparation (working dilution, 1:30,000), and 0.1 ml of a tracer solution (5000 dpm/0.1 ml buffer of isomer IV of the tritium-labeled bicyclic compound). The pH of the incubation mixture at this stage was 8.5, which is slightly higher than normal owing to the prior addition of NaOH, but still acceptable for the radioimmunoassay. Incubation was done for 12–18 hr overnight at room temperature, after which the antibody-bound fraction was precipitated using 25% polyethylene glycol. For details, see this volume [44].

2. Hydrolysis of Formed Protein Adducts after Alkali Treatment

Analytical Procedure: Chemical Formation of 11-Deoxy-13,14-dihydro-15-keto-11β,16ξ-cyclo-PGE$_2$. Analysis requires a 24-hr, 37° incubation of samples and standards buffered at pH 10.5 to transform 15-keto-13,14-dihydro-PGE$_2$ into 11-deoxy-13,14-dihydro-15-keto-11β,16ξ-cyclo-

PGE_2.[7,8] Standards of 15-keto-13,14-dihydro-PGE_2 ranging from 0.05 to 10 ng/ml were prepared by geometric dilutions of a 100 ng/ml stock solution. All standards, prepared in methanol–water (70:30, v/v), were stored at −20°. Aliquots (0.1 ml) of each standard were added to plastic-capped 12 × 75 mm tubes, and evaporated to dryness under nitrogen; then 20 μl of 1.0 *M* Na_2CO_3 and 0.5 ml of 2% w/v human albumin in isotonic saline were added, sequentially. Selected samples and standards were checked to verify that their pH was 10.5 ± 0.5; then all tubes were capped securely and incubated at 37° for at least 24 hr. After the quantitative transformation of 15-keto-13,14-dihydro-PGE_2 into 11-deoxy-13,14-dihydro-15-keto-11β,16ξ-cyclo-PGE_2, the alkaline solutions were neutralized (pH 7.4 ± 0.6) by the addition of exactly 30 μl of 1.0 *M* KH_2PO_4. Neutralized samples and standards were stored frozen before analysis. Reconstitution of the standards with an albumin solution is recommended when peripheral plasma samples are being analyzed; other reconstitution media may be appropriate for some samples. It is advisable to expose the standards to a matrix that resembles the projected sample matrix.

Incubation of 0.5-ml samples with 20 μl of 1.0 *M* Na_2CO_3 for 24 hr at 37° facilitated the reproducible, controlled, and rapid transformation of 15-keto-13,14-dihydro-PGE_2 and all its degradation products into a form suitable for analysis regardless of their distribution within the sample or its albumin content. For all samples and standards, the pH varied only slightly around 10.5. Neutralization with a fixed, 30-μl volume of 1.0 *M* KH_2PO_4 was also reproducible; the pH varied only slightly around 7.4. At pH 10 and 37°, the formation of the bicyclic derivative was complete after 24 hr (see the table).

INFLUENCE OF INCUBATION CONDITIONS ON RADIOIMMUNOASSAY

	Bound/bound$_0$ × 100[a,b]		
15-Ketodihydro-PGE_2 (pg/tube)	a	b	c
24	89.6 ± 1.3	83.8 ± 2.1	82.5 ± 3.8
48	74.7 ± 1.4	68.1 ± 1.1	65.1 ± 2.1
96	48.6 ± 1.6	41.4 ± 0.3	42.0 ± 0.8
190	26.8 ± 0.6	23.4 ± 0.3	23.9 ± 1.1
380	14.6 ± 0.4	12.2 ± 0.6	12.0 ± 0.1

[a] Conditions: (a) 20 hr of incubation at 37°, 0.10 *M* Na_2CO_3, human albumin 25 mg/ml in 0.9% w/v NaCl; (b) 36 hr of incubation at 37°, 0.10 *M* Na_2CO_3, human albumin 25 mg/ml in 0.9% w/v NaCl; (c) 96 hr of incubation at 37°, 0.10 *M* Na_2CO_3, in 0.9% w/v NaCl, albumin *absent*. After alkaline hydrolysis of samples (0.5 ml) according to the conditions cited above (a, b, or c), neutralization and assay was carried out as described.

[b] Values represent the mean ± SD for $n = 3$ assay tubes.

It was inappropriate to mix intact radiolabeled [5,6,8,11,12,14-^{3}H]15-keto-13,14-dihydro-PGE_2 with standards or samples prior to the alkaline transformation step. Attempted mutual transformation of tritiated and standard ligands into 11-deoxy-13,14-dihydro-15-keto-11β,16ξ-cyclo-PGE_2, followed by radioimmunoassay of the mixture, always gave a high, nonspecific binding, ranging from 15% to 25% of the total radioactivity added. Initially, we attributed this to irreversible covalent binding between albumin and [^{3}H]15-keto-13,14-dihydro-PGA_2; however, similar results were obtained in protein-free solutions. During radioactivity balance studies, we isolated tritiated water as the source of the nonspecific effect. The loss of tritium from the C-8 position of [5,6,8,11,12,14-^{3}H]15-keto-13,14-dihydro-PGE_2 under alkaline conditions, and tritium exchange with labile hydrogens of water or other sample components is predictable.[10] Since neither globulin precipitation nor charcoal sequestration removed tritiated water, it registered as nonspecific binding. Preparation and isolation of authentic [^{3}H]11-deoxy-13,14-dihydro-15-keto-11β,16ξ-cyclo-PGE_2, with no exchangeable protons, and its use in the radioimmunoassay eliminated this problem. Under these conditions, nonspecific effects were reduced to levels typical of prostaglandin radioimmunoassays.

Radioimmunoassay with Polyethylene Glycol 6000 Precipitation. After its quantitative formation, 11-deoxy-13,14-dihydro-15-keto-11β,16ξ-cyclo-PGE_2 was measured by radioimmunoassay. A solution of the racemic mixture of isomers of [^{3}H]11-deoxy-13,14-dihydro-15-keto-11β,16ξ-cyclo-PGE_2 containing 3000 dpm/0.1 ml was prepared in 0.10 *M* PBS. Antiplasma against 11-deoxy-13,14-dihydro-15-keto-11β,16ξ-cyclo-PGE_2 was diluted 1:10,000 in 0.1 *M* PBS containing 2% w/v human γ-globulin. Under typical assay conditions, this dilution of antiplasma bound 50% of the labeled ligand in the absence of competitive displacement by nonradiolabeled ligand. For the radioimmunoassay, neutralized (pH 7.4 $\pm$ 0.6) samples or standards (0.55 ml), [^{3}H]11-deoxy-13,14-dihydro-15-keto-11β,16ξ-cyclo-PGE_2 solution (0.1 ml, 3000 dpm), and antiplasma (0.1 ml, 1:10,000) were mixed in 12 $\times$ 75 mm plastic-capped tubes. The total volume was 0.75 ml at this stage. Tubes were capped securely, vortexed, and incubated at 37° for 1 hr and then at 4° for 24 hr. At equilibrium, the antibody-bound and free ligand were separated by precipitating γ-globulin with 0.75 ml of a chilled (4°) aqueous solution of 25% w/v polyethylene glycol and then centrifugation at 300 *g* at 4° for 1 hr to compact the γ-globulin pellet.[11] The supernatant was aspirated and dis-

[10] E. Daniels, W. Krueger, F. Kupiecki, J. Pike, and W. Schneider, *J. Am. Chem. Soc.* **90,** 5894 (1968).

[11] D. Van Orden and D. Farley, *Prostaglandins* **4,** 215 (1973).

carded; and the γ-globulin pellet was reconstituted in 2.0 ml of 0.10 *M* PBS. The reconstituted sample (2.0 ml) was mixed with 10 ml of ACS aqueous counting scintillant, and antibody-bound [^{3}H]11-deoxy-13,14-dihydro-15-keto-11β,16ξ-cyclo-PGE_2 was determined by scintillation counting. Data were calculated with a Hewlett-Packard Model 9830A computer, using a logit transformation program to linearize the relationship between the logarithmic concentration of 15-keto-13,14-dihydro-PGE_2 and the percentage of bound antigen. Standard curve presentations and statistics were calculated according to Rodbard *et al.*[12]

The interassay coefficients of variation ranged from ±8.1% to ±14.6% for 0.5-ml samples containing 12–386 pg of 15-keto-13,14-dihydro-PGE_2.The intraassay coefficients of variation of the method for six experiments with $n = 3$ tubes per assay ranged from ±1.7% to ±8.9% standard deviation, with a mean of ±4.2% for 0.5-ml samples containing 12–386 pg. Analysis of normal human plasma fortified with known amounts of 15-keto-13,14-dihydro-PGE_2 showed a linear proportionality ($r = 0.998$) between the mass found and the mass added from 12–386 pg, with a recovery of 97.1 ± 8.7% (mean ± standard deviation, $n = 3$), ranging from 85.4% to 106.3%. Although the recovery was quantitative, we characterized the assay further to assure that covalent binding between the dehydration product, 15-keto-13,14-dihydro-PGA_2, and albumin did not influence the results.[8] When standards were reconstituted in solutions containing 0–40 mg of albumin per milliliter and incubated at pH 10, 37°, for 24 hr, before assay, the calibration curves were equivalent and independent of the albumin concentration. Covalent binding between albumin and the dehydration products of 15-keto-13,14-dihydro-PGE_2 is maximal after 6 hr at 37°, pH 7.4, according to earlier studies.[8] Therefore, plasma samples or standards reconstituted in 40 mg/ml albumin were incubated at 37°, pH 7.4, for 6 hr to maximize covalent binding prior to the addition of 20 μl of 1.0 *M* sodium carbonate. When these calibration curves and samples were compared to an identical set with no preincubation, the results were equivalent, indicating that covalent binding was reversible.

Radioimmunoassay with Dextran-Coated Charcoal. Sequestration with dextran-coated charcoal is a common alternative to globulin precipitation as a technique for separating antibody bound from free ligand.[13] For this technique, 0.5-ml samples and standards were incubated at 37° for 24 hr with 20 μl of 1.0 *M* Na_2CO_3 as before. Neutralized samples or standards (0.1 ml), and [^{3}H]11-deoxy-13,14-dihydro-15-keto-11β,16ξ-cyclo-PGE_2 (3000 dpm/0.1 ml of PBS) were mixed with antiplasma (0.1 ml), at a

[12] D. Rodbard, W. Bridsom, and P. Rayford, *J. Lab. Clin. Med.* **74,** 770 (1969).
[13] M. Binoux and W. Odell, *J. Clin. Endocrinol. Metab.* **36,** 303, 1973.

dilution (1:25,000 in PBS) adjusted to equal a final dilution of 1:75,000, for the 0.3-ml assay volume. Samples were vortexed and incubated as before. At equilibrium, the antibody-bound and free ligand were separated by incubation with 1.0 ml of dextran-coated charcoal suspension (5.0 mg of Norit A and 0.50 mg of dextran T70 in 1.0 ml of PBSG) for 12 min at 0°. After centrifugation (300 *g*, 15 min) the antibody-bound [^{3}H]11-deoxy-13,14-dihydro-15-keto-11β,16ξ-cyclo-PGE_2 in the supernatant was measured by scintillation counting. Calculations were performed as described above.

3. Blocking the Reactive Groups on the Proteins prior to Alkali Treatment

Sulfhydryl groups are the most likely reactants in a chemical reaction of this type.[7,8] Therefore, plasma was preincubated with various sulfhydryl blocking agents that are known to act by different mechanisms, prior to addition of [^{3}H]15-keto-13,14-dihydro-PGE_2 and alkali treatment. The following blockers were tried in varying concentrations: dithiobisnitrobenzoic acid, *N*-ethylmaleimide, iodoacetic acid, and *p*-hydroxymercuribenzoate. Success was very limited, however, and although we may not have found the optimal reaction conditions, this approach has been abandoned.

Application of the Radioimmunoassay

Analysis of Protein-Free Samples

Figure 4 shows the results from one biological study resulting in essentially protein-free samples. A normal guinea pig lung was perfused with Krebs solution at a rate of 3 ml/min. PGE_2 (0.2, 1.0, and 10.0 μg) was injected into the pulmonary artery at hourly intervals, and the perfusate was collected in 3-ml fractions. Aliquots, 0.1 ml, of the collected fractions were alkali-treated overnight and then diluted 50- to 200-fold in the radioimmunoassay buffer to obtain the proper pH for assay. Other aliquots were instead assayed for content of PGE_2.[14] With these assay methods, distinct peaks of PGE_2 as well as its 15-ketodihydro metabolite were found to emerge in the perfusate after PGE_2 injection (Fig. 4).

That the measured metabolite peaks do not simply reflect cross-reaction with the large amounts of unconverted PGE_2 appearing in the perfusate can be deduced from two different observations. First, there is an

[14] J. Å. Lindgren, H. Kindahl, and S. Hammarström, *FEBS Lett.* **48**, 22 (1974).

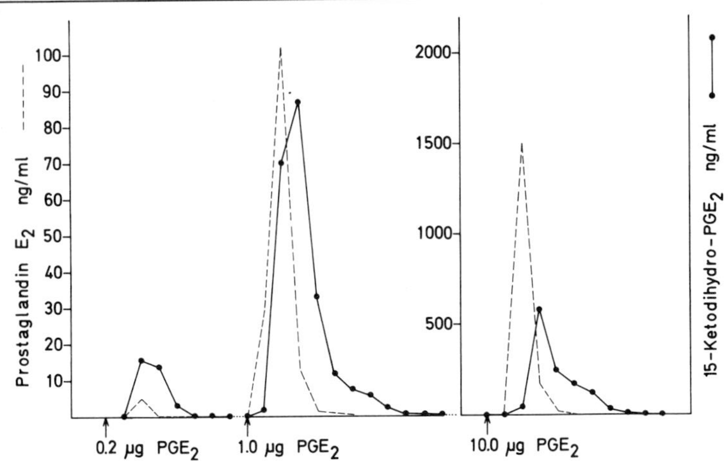

FIG. 4. Appearance of 15-keto-13,14-dihydro-PGE_2 (measured by radioimmunoassay after conversion into bicyclic derivatives) (●—●) and unmetabolized PGE_2 (measured by radioimmunoassay according to Lindgren *et al.* [14]) (----) after injection of 0.2, 1.0, and 10.0 μg of PGE_2 into the pulmonary artery of a perfused, normal guinea pig lung.

inverse relationship between peak heights of PGE_2 and the metabolite after the first (low dose) and last (high dose) injections, which would not have been possible if the registered metabolite peaks had been caused simply by cross-reaction with PGE_2. Almost total conversion into the metabolite was obviously the case when the low dose was given; however, when 10 μg of PGE_2 were injected, the metabolic capacity of the lung was quite inadequate to cope with this large amount of prostaglandin, and approximately 75% of the injected dose escaped degradation in this case. Second, a slight but distinct lag of the metabolite peaks can be seen compared with the PGE_2 peaks. This lag is barely suggested after the lowest dose of the prostaglandin, but quite pronounced after the highest dose.

Analysis of Samples Containing Proteins

Dogs were given intravenous bolus injections of PGE_2, and blood samples were collected. A dose-dependent increase in plasma 15-keto-13,14-dihydro-PGE_2 concentrations was seen. Ten minutes after administration of 30, 10, or 3 μg of PGE_2 per kilogram, its plasma metabolite concentrations were 34.1 ± 8.4, 6.8 ± 0.6, and 2.0 ± 0.2 ng/ml (mean ± SEM, $n = 4$). The metabolite disappeared from the circulation rapidly with half-lives of 9.9 ± 0.2, 9.2 ± 0.3, and 7.8 ± 0.8 min (mean ± SEM, $n = 4$) for the respective doses noted above. Similar studies with PGE_1

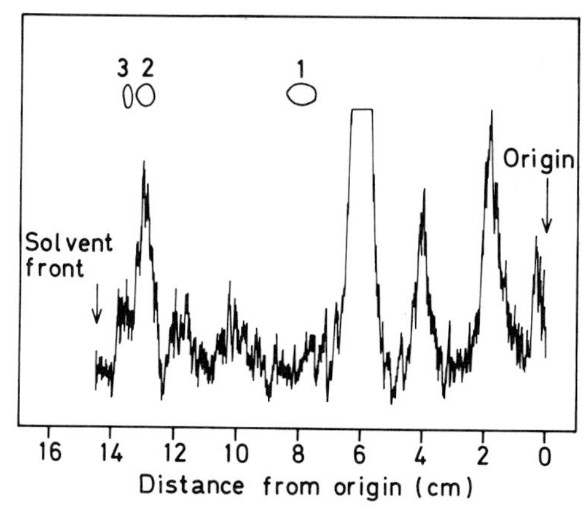

FIG. 5. Thin-layer chromatography of rabbit plasma collected 15–20 min after intravenous injection of [^{3}H]PGE$_2$. Spots on the plate are 1, PGE$_2$; 2, 15-keto-13,14-dihydro-PGE$_2$; and 3, the dehydration products of 15-keto-13,14-dihydro-PGE$_2$ (i.e., 15-ketodihydro-PGA$_2$ and the bicyclic product; no separation is seen between these substances in this TLC system).

were possible because of the predictable cross-reaction between the bicyclic derivatives of 15-keto-13,14-dihydro-PGE$_2$ and 15-keto-13,14-dihydro-PGE$_1$. Since the latter compound is not normally present in mammals, measurements of endogenous 15-keto-13,14-dihydro-PGE$_2$ are still accurate. Basal concentrations of 48 $\pm$ 31 pg/ml (mean $\pm$ SD, n = 15) in human plasma accord with concentrations measured by gas chromatography–mass spectrometry.[1]

15-Keto-13,14-Dihydro-PGE$_2$ as a PGE$_2$ Parameter in Biological Samples

Assay of 15-ketodihydro-PGE$_2$ may be carried out either monitoring the compound as such, which may be done using gas chromatographic–mass spectrometric methods (see this volume [67]), or after its conversion into a final, stable degradation product as described in this chapter. This latter approach is preferable when using radioimmunoassay, which is always performed in environments deleterious for the unstable metabolite, but may naturally also be employed in GC–MS methodology.

Regardless of type of method and of approach to the problem, one might question the suitability of the compound as a PGE$_2$ parameter. The very complex metabolic fates of prostaglandins are not known in detail. Several studies indicate, for example, that PGE compounds may be extensively converted into PGF compounds.[15-17] It is even possible that this pathway may dominate. Figure 5 shows a thin layer chromatogram of a plasma sample obtained 20 min after intravenous injection of [5,6,8,11,12,14,15-$^3H_7$]PGE_2 to a rabbit. No radioactivity was found at the R_f value of PGE_2, which indicated complete conversion. However, very little radioactivity was found corresponding to the expected "major" metabolite, 15-ketodihydro-PGE_2, or its degradation products. The material in the major chromatographic peak was tentatively identified as a PGF product instead (E. Granström, unpublished observation).

Acknowledgments

This study was supported by grants from the Swedish Medical Research Council, project No. 03P-5804 and 03X-05915, the Swedish Council for Forestry and Agriculture Research, the World Health Organization, and Harald och Greta Jeanssons Stiftelse.

[15] M. Hamberg and U. Israelsson, *J. Biol. Chem.* **246,** 5107 (1970).
[16] M. Hamberg and M. Wilson, *Adv. Biosci.* **9,** 39 (1973).
[17] C. A. Leslie and L. Levine, *Biochem. Biophys. Res. Commun.* **52,** 717 (1973).

[44] Radioimmunoassay of the Major Plasma Metabolite of $PGF_{2\alpha}$, 15-Keto-13,14-dihydro-$PGF_{2\alpha}$

By ELISABETH GRANSTRÖM and HANS KINDAHL

In studies on the biological roles of $PGF_{2\alpha}$ *in vivo,* its initial metabolite, 15-ketodihydro-$PGF_{2\alpha}$ (Fig. 1), is often the best parameter: it has a longer half-life in the circulation than the parent compound, it occurs in higher concentrations, and no artifactual formation of the compound takes place during collection and handling of the sample.[1] Many assay methods have been developed for this metabolite and successfully applied to a large number of biological studies.[2,3] Radioimmunoassay is at present the most widely used method for this compound.

[1] E. Granström and B. Samuelsson, *Adv. Prostaglandin Thromboxane Res.* **5,** 1 (1978).
[2] B. Samuelsson, M. Goldyne, E. Granström, M. Hamberg, S. Hammarström, and C. Malmsten, *Annu. Rev. Biochem.* **47,** 997 (1978).
[3] E. Granström and H. Kindahl, *Adv. Prostaglandin Thromboxane Res.* **5,** 119 (1978).

ISBN 0-12-181986-8

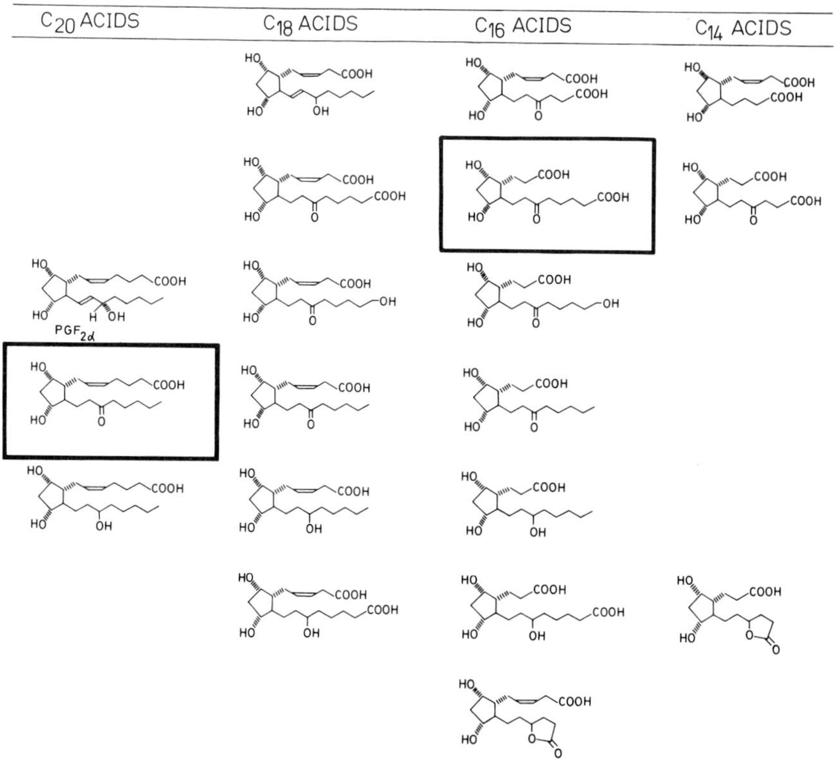

FIG. 1. Structures of metabolites of $PGF_{2\alpha}$. Framed compounds are major metabolites in the human as well as some other species.

Development of the Radioimmunoassay

In 1972 when we developed the first radioimmunoassay for this metabolite,[4] all reagents had to be prepared and standardized in our own laboratory. Since then, however, all necessary reagents have become commercially available, including several preparations of antibody and of labeled ligand of high specific activity. Even complete radioimmunoassay kits for this metabolite are now available. However, the quality of the commercial preparations, particularly the antibody, varies considerably, and scientists may therefore prefer to prepare their own reagents.

[4] E. Granström and B. Samuelsson, *FEBS Lett.* **26,** 211 (1972).

Preparation of Antigen

Since prostaglandins are not immunogenic in themselves because of their low molecular weight, they have to be coupled to larger molecules prior to injection into animals. This can be achieved in a number of ways,[5-7] normally by the formation of a peptide bond between the carboxyl group of the prostaglandin and an amino group on a protein. A convenient procedure employs *N,N'*-carbonyldiimidazole as the coupling reagent[7] and bovine serum albumin as the carrier protein:

15-Keto-13,14-dihydro-$PGF_{2\alpha}$, 10 mg, is dissolved in 2 ml of dry dimethylformamide and stirred for 10 min at room temperature under N_2. *N,N'*-Carbonyldiimidazole, 5 mg, is added, and the mixture is stirred for 20 min. This solution of the reactive prostaglandin imidazolide is then added dropwise to a solution of 40 mg of bovine serum albumin in 3 ml of H_2O and kept with stirring for 6 hr. The conjugate is then dialyzed, first against dimethylformamide–water (2 : 3, v/v), then against several changes of water. The dialyzed preparation is divided into eight equal portions (sufficient for the immunization of two rabbits) and lyophilized.

Immunization

For each injection, one portion of the lyophilized conjugate is dissolved in 2 ml of distilled water and emulsified with 2 ml of complete Freund's adjuvant. A simple method to prepare a satisfactory emulsion is repeated aspiration of the mixture into a syringe and ejection through a medium thin needle into a medical flask. This is repeated until the emulsion is white, stable, and viscous. For dissolution of the conjugate, distilled water should be employed instead of saline, because the presence of salts counteracts the formation of emulsion with the mineral oil of the adjuvant. For the best results, it is essential that the emulsion remain stable for long periods in the animal, not only during the injection. The slight hypotonicity of the small volume of water needed is of no importance for the animal.

Injection is done at multiple sites in the flank area of rabbits, subcutaneously, intradermally, and/or intramuscularly. We avoid foot pad injection, because of the intense pain it causes the animal. Injections are given weekly 3 times, after which antibodies are normally detectable. After a booster injection, perhaps after a rest period of 1–2 months, the antibody

[5] K. Kirton, J. Cornette, and K. Barr, *Biochem. Biophys. Res. Commun.* **47,** 903 (1972).
[6] B. Caldwell, S. Burstein, W. Brock, and L. Speroff, *J. Clin. Endocrinol. Metab.* **33,** 171 (1971).
[7] U. Axen, *Prostaglandins* **5,** 45 (1974).

titer rises sharply and may remain elevated for several months. We generally collect a large volume of blood (50 ml blood from an adult rabbit, about 4 kg body weight) when the titer reaches its maximum, which generally occurs 7–10 days after the boost. The marginal ear veins are best suited for bleeding in this animal: the easiest way is by free flow of blood through a needle after dilatation of the vessel with xylene. No negative pressure or any particular equipment is necessary even for this relatively large volume of blood.

Plasma or serum may be isolated. In this particular case, it does not seem to matter which preparation is preferred: 15-ketodihydro-$PGF_{2\alpha}$ is not formed during blood coagulation, and thus, apart from the endogenous amounts already present in the blood, no additional amounts can be formed to occupy the antibody binding sites. This aspect, on the other hand, has to be carefully taken into consideration when dealing with substances that are produced during clotting, such as TxB_2, the primary prostaglandins, 12-HETE, and other compounds.[2]

The antiserum or antiplasma is then stored at $-20°$ and is stable for many years.

Properties of the Antibody

When employing radioimmunoassay it must be kept in mind that several prostaglandins may occur simultaneously in the sample, and the antibody may not completely distinguish between these. Before setting up a radioimmunoassay routine it is thus necessary to run a series of standard curves with such related compounds instead of the proper antigen. At least those substances that can be expected to occur in the sample should be tested.

Most prostaglandin antibodies recognize structures far from the coupling site well, i.e., structures in the ring and in the ω side chain. For example, this particular antibody against 15-ketodihydro-$PGF_{2\alpha}$ displays low cross-reactions with the corresponding 15-hydroxy compound, 13,14-dihydro-$PGF_{2\alpha}$ (4%) and with the corrresponding PGE_2 metabolite (1.7%) (Fig. 2). Even a minor alteration in those parts of the molecule, such as the introduction of the Δ^{13} trans double bond, is recognized by the antibody: cross-reaction with the unsaturated metabolite, 15-keto-$PGF_{2\alpha}$, is only 16%.

On the other hand, structures close to the coupling site, i.e., the carboxyl, are generally not well recognized by the antibody, and cross-reactions with compounds modified only at the carboxyl group may be close to 100%, such as methyl esters. It is also often seen that antibodies directed against prostaglandins of the 2-series also accept the 1-series as

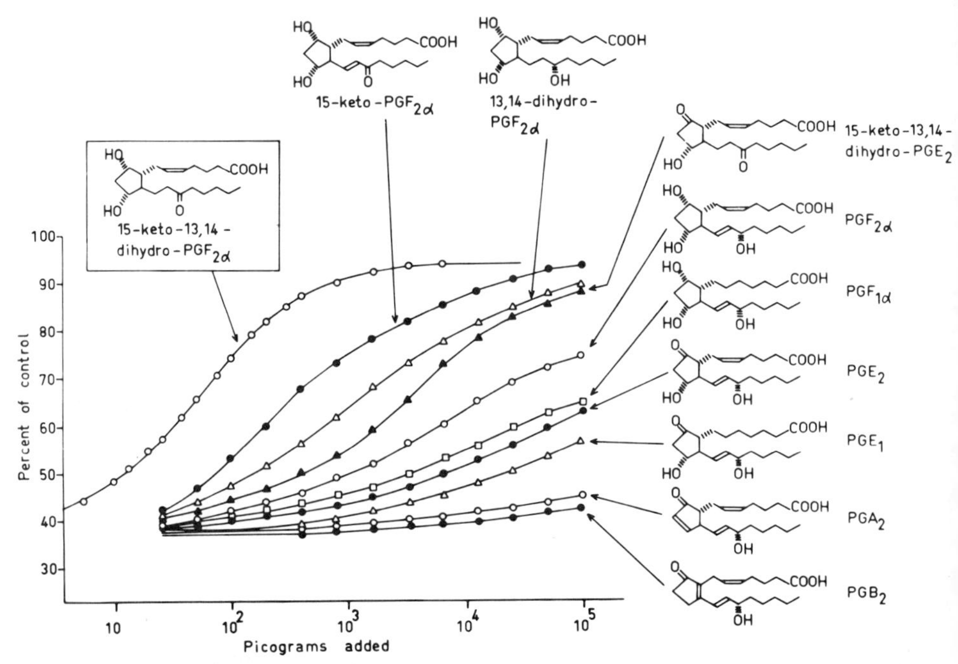

FIG. 2. Properties of an antiserum raised against a bovine serum albumin conjugate of 15-keto-13,14-dihydro-$PGF_{2\alpha}$.

ligands. In this particular case cross-reaction with 15-ketodihydro-$PGF_{1\alpha}$ was 42% (not shown in Fig. 2). Even values close to 100% have often been reported in the literature.

Labeled Ligand

The labeled ligand should have the highest possible specific activity. For 15-ketodihydro-$PGF_{2\alpha}$ the preparation with the largest number of tritium atoms is [5,6,8,9,11,12,14(n)-$^3H_7$]-15-ketodihydro-$PGF_{2\alpha}$, which can be obtained from either The Radiochemical Centre, Amersham, England, or New England Nuclear, Dreieich, West Germany. The specific activity ranges from 60 to 100 Ci/mmol. The conditions normally employed in radioimmunoassay (i.e., aqueous solutions at pH 7.4 or slightly more alkaline) probably lead to loss of the tritium at C-14 due to enolization, so the actual specific activity of the compound may be somewhat lower; however, it seems to be sufficient for most radioimmunoassay purposes.

Purity should always be checked: this can be done by thin-layer chromatography or radio-gas liquid chromatography.[8] Although this compound is very stable, some decomposition inevitably takes place with time, leading to decreased binding (see below).

Even higher specific activities can be obtained with iodinated tracers[9,10] (see also this volume [42]).

Radioimmunoassay Procedure

Reagents

Radioimmunoassay buffer: 0.05 *M* Tris-HCl buffer, pH 7.8, containing 1 m*M* EDTA

Bovine γ-globulin, 0.5% solution in the above buffer.

Antibody, suitable dilution (see below) in the above buffer.

Labeled ligand, "tracer," suitable concentration: 50,000 dpm/ml (5000 dpm/0.1 ml, which is the amount added to each assay tube)

Standard: 15-ketodihydro-$PGF_{2\alpha}$, unlabeled; solutions in the radioimmunoassay buffer ranging from about 20 pg/ml to 2560 pg/ml (= 2, 4, 8, 16, etc., in geometric series, to 256 pg per 0.1 ml, which is the amount added to the standard curve series of tubes).

Polyethylene glycol with an average molecular weight of at least 4000, 25% solution in distilled H_2O.

Equipment

Disposable glass tubes, 10 × 80 mm

Rack suitable for large number of tubes

Hand pipettes with disposable tips, e.g., MLA (Medical Laboratory Automation, Inc. Mount Vernon, New York), Eppendorff (Gerätebau Netheler and Henz, Hamburg, Germany), Gilson, Biopette (Schwarz-Mann, Orangeburg, New York); and/or

Repeating dispenser systems, e.g., Hamilton (Bonaduz, Switzerland) gastight syringes with repeating dispenser unit

Automatic diluter and dispenser, e.g., LKB (LKB-Produkter, Stockholm) Ultrolab System Diluter 2075.

Shaking or vortexing equipment, preferably one that can shake the whole test tube rack

Dispenser for polyethylene glycol solution (Cornwall syringe, Becton-Dickinson, Rutherford, New Jersey), suitable for volumes less than 1 ml

[8] E. Granström, *Eur. J. Biochem.* **27,** 462 (1972).

[9] S. Ohki, T. Hanyu, K. Imaki, N. Nakasawa, and F. Hirata, *Prostaglandins* **6,** 137 (1974).

[10] J. Maclouf, M. Pradel, P. Pradelles, and F. Dray, *Biochim. Biophys. Acta* **431,** 139 (1976).

Refrigerated centrifuge with large sample capacity, suitable for the above tube size

Scintillation counter with equipment for automatic efficiency counting

Computer for calculations

Incubation Procedure

A common setup of the radioimmunoassay tubes is as follows:

Tubes 1 through 5: "Total radioactivity" tubes (T), or "zero binding" tubes (no antibody, but tracer present).

Tubes 6 through 8: "Maximal binding" tubes (B_0) (antibody plus tracer present).

Tubes 9 through 26: Standard tubes (giving the B-values): increasing amounts of the unlabeled compound in a geometrical series, e.g., 0, 2, 4, 8, 16, . . . 256 pg per tube and in duplicate for each amount. Antibody and tracer are present. The reason for the inclusion of two extra "maximal binding" tubes (Nos. 9 and 10) is discussed below in the section dealing with common problems encountered.

Tubes 27 through 31: Biological standards, i.e., some samples from previously analyzed plasma pools as a routine control of the current interassay variation.

Tubes 32 through *n*: Sample tubes, at least in duplicates and in two or more different dilutions, giving the U values (unknowns).

In some situations additional tubes have to be included: for example, if extraction of the plasma is performed, tubes giving information about the procedural blank and extraction recovery must also be assayed. Some antibodies are sensitive to variations in the protein concentration, and it may be necessary to extend the standard curve tubes to include also series containing various amounts of protein in addition to the unlabeled compound (see this volume [45] for an example.)

A convenient protocol for a radioimmunoassay of unextracted plasma is shown in the table. The reagents are added in the order shown, starting with the buffer, and with careful mixing after each of the last three additions prior to the incubation.

In this particular assay, plasma volumes of as much as 0.5 ml may be assayed without disturbing interferences from other plasma constituents (i.e., analysis of plasma in different dilutions gives a perfectly linear response). This is not always the case and must be carefully checked (see this volume [45] for an example.)

After the addition of the labeled ligand (giving a final volume of 0.7 ml in all tubes), the tubes are left for incubation. The optimal time and temperature for this step must be found for each assay system. In this particu-

PROTOCOL EMPLOYED FOR RADIOIMMUNOASSAY OF 15-KETO-13-14-DIHYDRO-$PGF_{2\alpha}$ IN PLASMA[a]

Tube no.	Buffer (ml)	Standards or samples (ml)	Bovine γ-globulin 0.5% (ml)	Antibody, dil. 1:4,000 (ml)	Labeled ligand, 50,000 dpm/ml (ml)	Polyethylene glycol, 25% (ml)
1	0.4	—	0.2	—	0.1	0.7
2	0.4	—	0.2	—	0.1	0.7
3	0.4	—	0.2	—	0.1	0.7
4	0.4	—	0.2	—	0.1	0.7
5	0.4	—	0.2	—	0.1	0.7
6	0.3	—	0.2	0.1	0.1	0.7
7	0.3	—	0.2	0.1	0.1	0.7
8	0.3	—	0.2	0.1	0.1	0.7
9	0.2	0.1 (= 0 pg)	0.2	0.1	0.1	0.7
10	0.2	0.1 (= 0 pg)	0.2	0.1	0.1	0.7
11	0.2	0.1 (= 2 pg)	0.2	0.1	0.1	0.7
12	0.2	0.1 (= 2 pg)	0.2	0.1	0.1	0.7
13	0:2	0.1 (= 4 pg)	0.2	0.1	0.1	0.7
14	0.2	0.1 (= 4 pg)	0.2	0.1	0.1	0.7
·	·	·	·	·	·	·
·	·	·	·	·	·	·
·	·	·	·	·	·	·
23	0.2	0.1 (= 128 pg)	0.2	0.1	0.1	0.7
24	0.2	0.1 (= 128 pg)	0.2	0.1	0.1	0.7
25	0.2	0.1 (= 256 pg)	0.2	0.1	0.1	0.7
26	0.2	0.1 (= 256 pg)	0.2	0.1	0.1	0.7
27	—	0.5 (aliquots	—	0.1	0.1	0.7
28	—	0.5 from	—	0.1	0.1	0.7
·	·	· plasma	·	·	·	·
·	·	· pool)	·	·	·	·
31	—	0.5	—	0.1	0.1	0.7
32	0.2	0.1 (Plasma	0.2	0.1	0.1	0.7
33	0.2	0.1 sample	0.2	0.1	0.1	0.7
34	0.1	0.2 No. 1)	0.2	0.1	0.1	0.7
35	0.1	0.2	0.2	0.1	0.1	0.7
36	—	0.5	—	0.1	0.1	0.7
37	—	0.5	—	0.1	0.1	0.7
38	0.2	0.1 (Plasma	0.2	0.1	0.1	0.7
39	0.2	0.1 sample	0.2	0.1	0.1	0.7
40	0.1	0.2 No. 2)	0.2	0.1	0.1	0.7
41	0.1	0.2	0.2	0.1	0.1	0.7
42	—	0.5	—	0.1	0.1	0.7
43	—	0.5	—	0.1	0.1	0.7
·	·	·	·	·	·	·
·	·	·	·	·	·	·
·	·	·	·	·	·	·

[a] After addition of the labeled ligand, the tubes are incubated for 12–18 hr at +4°.

lar case, incubation at room temperature or at +4° give essentially the same results; nor does the incubation time seem to be critical (the results obtained after 3 hr or 48 hr of incubation do not differ significantly for this antibody). For practical purposes it may be convenient to leave the tubes overnight in the cold.

The next step is the separation of the free and the antibody-bound fractions. Of all methods tried,[3] the polyethylene glycol method gave the most reliable results. It is comparatively insensitive to influences from interfering factors, and the results have always compared well with those from gas chromatography–mass spectrometry methods. The tubes are cooled on ice if previously incubated at room temperature; otherwise +4° is sufficient. Cold polyethylene glycol, 0.7 ml of a 25% solution in water, is added to each tube using a rapid dispensing system, for example, a Cornwall syringe with a refill unit. The conventional hand pipettes with disposable tips are unsuitable for this step because of the high viscosity of the polyethylene glycol solution.

The tubes are vigorously vortexed, for at least 15 sec and preferably twice because of the viscosity of the precipitating agent, and centrifuged in the cold at 1400 *g* for 1 hr. The formed grayish-white pellet contains the precipitated γ-globulins including the antibody-bound fraction of the prostaglandin metabolite, whereas the free fraction is found in the supernatant. The 0.5% γ-globulin solution previously added (see above and see the table) is necessary as a carrier for the minute amounts of γ-globulin from the highly diluted antiserum. If this carrier is omitted, precipitation of the antibody by polyethylene glycol will not be complete.[11] For analysis of plasma volumes exceeding about 0.3 ml no extra addition of γ-globulins is necessary.

We find it most convenient to assay the supernatant for radioactivity. One milliliter is pipetted off using an LKB diluter into scintillation vials, followed by 1 ml of water, which is also added by the diluter. As scintillation cocktail, 10 ml of Instagel (Packard, Downers Grove, Illinois) can be used. This type of cocktail has a high capacity for water and proteins. An emulsion is initially formed that becomes clear after standing in the cold for 24 hr.

Calculations

Scintillation Counting. The yellow color of unextracted plasma gives about 10% reduction in counting efficiency in the scintillation counter as compared to the standard vials without plasma. This figure varies among

[11] B. Desbuquois and G. Aurbach, *J. Clin. Endocrinol.* **33,** 732 (1971).

different individuals and species and can be even higher, for example, in hemolytic samples. It is thus very important that the counted values for counts per minute are corrected for quenching. This can easily be done if the scintillation counter is equipped with automatic quench correction facilities, such as automatic external standard channels ratio. The relation between the standard channels ratio and the efficiency can be linearized, and the values for disintegrations per minute calculated from the following mathematical formula:

$$\text{dpm} = (\text{cpm} \cdot k)/(\text{ratio} - a) \qquad (1)$$

where k = slope of the calibration curve for the instrument and a = intercept with ordinate.

Calculation of Standard Curves and Unknown Samples. Only analyses of the free fractions (supernatants after polyethylene glycol precipitation) are discussed below.

The maximal binding of the labeled ligand is expressed as B_0/T, and the decreasing binding in the series of standard tubes as B/T. The B/T values increase with increasing concentration of the unlabeled ligand, and finally, after total displacement of the bound radioactivity, approach 1. In our assay this normally occurs around or somewhat above 200 pg of the unlabeled ligand (final antiserum dilution in the assay tube, 1:28,000; maximal binding about 0.50).

Several methods exist for plotting the standard curve. For manual calculations, B_0/T and B/T are plotted versus the log dose of the amount of the unlabeled ligand (log X). This results in a sigmoid curve (cf. Fig. 2), from which the amounts of the metabolite can be found from their corresponding U/T values.

For computerized handling of data a convenient method to linearize the standard curve is to make a logit transformation[12] of $(B - B_0)/(T - B_0)$ $(= Y)$ versus $\log_e$ dose (or log dose). This gives

$$\text{logit}\,(Y) = \log_e [Y/(1 - Y)] \qquad (2)$$

There is a linear correlation between logit (Y) and $\log_e$ dose:

$$\text{logit}\,(Y) = a_1 + k_1 \cdot \log_e X \qquad (3)$$

where X = the amount of unlabeled prostaglandin in a standard tube, k_1 = slope of the line, and a_1 = intercept with the ordinate. [Note: k_1 and a_1 should not be confused with the corresponding variables from Eq. (1).]

[12] D. Rodbard, W. Bridson, and P. Rayford, *J. Lab. Clin. Med.* **74,** 770 (1969).

Equation (3) can be rearranged to

$$\log_e X = [\text{logit}\ (Y) - a_1]/k_1 \tag{4}$$

This equation can then easily be used for the calculations of the unknowns.

Assessment of Antibody Titer

Before setting up routine radioimmunoassay runs, the optimal titer of the antibody must be established. This should always be done also with the commercial preparations, as experience shows that the recommended procedure from the manufacturer is often far from optimal.

The titer test can be done by analysis of binding capacity of serial dilutions of the antibody in the assay buffer. One-tenth milliliter of dilutions ranging from 1:20 to about 1:20,000 in geometric series is added to a series of assay tubes containing 0.3 ml of the assay buffer, 0.2 ml of bovine γ-globulin, and 0.1 ml of tracer. The five T tubes should also be included in the series to obtain information about the total radioactivity per tube. Incubation, precipitation, and radioactivity analysis is performed as described above. A suitable antibody titer is one that gives a B_0/T value of about 0.5, at least within the limits 0.4–0.6. A maximum binding higher or lower than this may affect the sensitivity and/or precision of the assay.

Application of the Radioimmunoassay

The assay is very suitable for detection of increases in the production of $PGF_{2\alpha}$, such as occur during luteolysis in many species.[13–15] This is probably the most common application of radioimmunoassays for this metabolite. Figure 3 shows an example of this: the luteolytic period of a heifer was studied with hourly samples of peripheral blood, which were assayed for progesterone and 15-ketodihydro-$PGF_{2\alpha}$. The prostaglandin release was found to occur during 2–3 days as rapid pulses with a duration of 1–5 hr prior to and during luteolysis, which was indicated by decreasing levels of progesterone.

The sensitivity of the assay is high enough, however, also for studies of decreased prostaglandin production, such as can be seen after aspirin ingestion. The study shown in Fig. 4 illustrates this: the basal levels of the PGF metabolite ranged between 40 and 50 pg/ml prior to the study, de-

[13] A. J. Peterson, R. J. Fairclough, E. Payne, and J. F. Smith, *Prostaglandins* **10,** 675 (1975).

[14] H. Kindahl, L. E. Edqvist, A. Bane, and E. Granström, *Acta Endocrinol.* (*Copenhagen*) **82,** 134 (1976).

[15] H. Kindahl, L. E. Edqvist, E. Granström, and A. Bane, *Prostaglandins* **11,** 871 (1976).

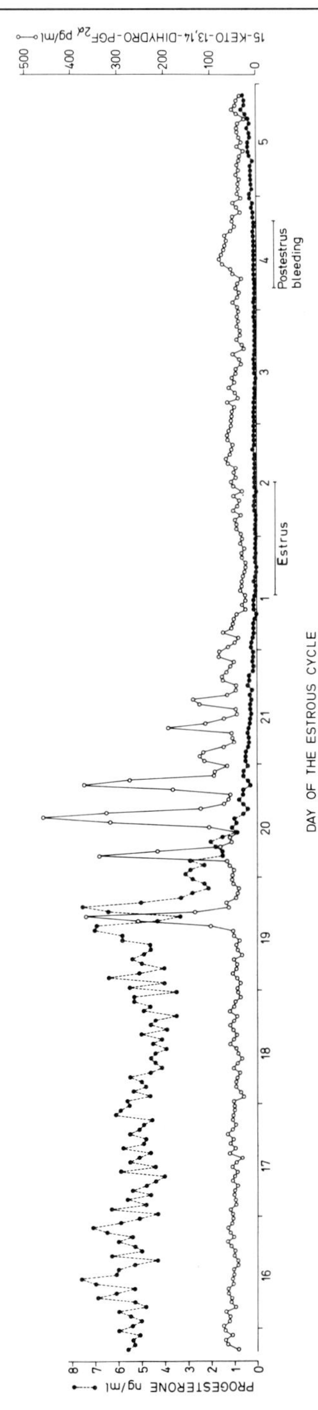

FIG. 3. Peripheral plasma levels of 15-keto-13,14-dihydro-$PGF_{2\alpha}$ (○——○) and progesterone (●---●) during luteolysis, estrus, and the early postestrous period in a heifer. Reproduced from H. Kindahl *et al.*[15]

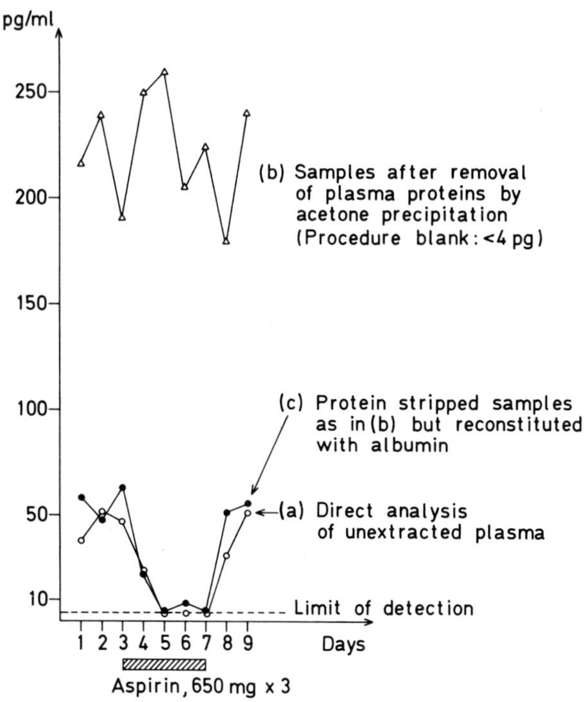

FIG. 4. Radioimmunoassay of "15-keto-13,14-dihydro-PG$F_{2\alpha}$ levels" in peripheral plasma of a human volunteer before, during, and after treatment with aspirin, 650 mg × 3 for 4 days. The plasma samples were analyzed in three different ways: direct analysis of unextracted plasma (curve a, ○——○); analysis of samples corresponding to 0.5 ml of plasma after precipitation of albumin and other proteins with acetone (curve b, △——△); and analysis of such protein-stripped samples (again corresponding to 0.5 ml of plasma after reconstitution to the physiological concentration of albumin (curve c, ●——●).

creased to <4 pg/ml (limit of detection) during aspirin medication, and returned to normal after discontinuation of the drug.

Figure 4 also illustrates another important matter. The problem of extraction of samples prior to assay has been subject to some controversy. Some investigators feel that unprocessed plasma contains too large amounts of other compounds to be assayed directly. The binding of low molecular weight substances such as prostaglandins to albumin has also been thought to interfere with the antigen–antibody binding (see Granström and Kindahl[3] and references therein).

However, our opinion is that an extraction step prior to assay introduces more problems than it solves[3,16]: impurities from solvents, columns, evaporation procedures, etc., interfere strongly with the antigen–antibody binding in the subsequent radioimmunoassay. Furthermore, and most important, the presence of albumin is generally no drawback: it rather exerts a protective influence because it extensively binds other low molecular weight substances that might otherwise interfere. If albumin is removed by a precipitation step, all these compounds are liberated, for example, free fatty acids, which occur in plasma in very high concentrations compared to prostaglandins. Their interference in the radioimmunoassay is thus greatly enhanced.

Figure 4 demonstrates this protective effect of albumin. The plasma samples were first analyzed directly, i.e., 0.5 ml unextracted plasma as usual (series a). The results show the expected sequence of events (curve a). A second and third series of 0.5 ml aliquots from the same plasma samples (series b and c, respectively) were then subjected to addition of two volumes of acetone, which almost completely precipitated the proteins. The samples were centrifuged, the supernatants were transferred to new assay tubes, and the acetone was allowed to evaporate spontaneously. The remaining water was removed by lyophilization. Two samples of 0.5 ml of distilled water were processed in the same way to give information about the procedure blank. After evaporation the samples were dissolved in either 0.3 ml of the assay buffer (series b) or 0.3 ml of a human serum albumin solution (75 mg/ml of assay buffer, series c). To both of these series the normal amount of necessary γ-globulin was added. Series c had thus been reconstituted to a final concentration of albumin close to the physiological one. Both series were then carried through the normal radioimmunoassay procedure.

The protein-free samples (series b) now showed a totally different picture compared with the series of unextracted plasma (curve b vs a). The values did not display any tendency to decrease following aspirin ingestion, and the measured "levels" were unrealistically high and fluctuated irregularly throughout the experiment. Obviously, these readings have nothing to do with prostaglandin levels.

The third series, on the other hand, which had first been stripped of proteins but had then been reconstituted with respect to the albumin concentration, showed essentially the same picture as the unextracted series: small variations in the low, basal levels; a pronounced decrease following

[16] E. Granström, *in* "Radioimmunoassay of Drugs and Hormones in Cardiovascular Medicine" (A. Albertini, M. Da Prada, and B. A. Peskar, eds.), p. 229. Elsevier/North-Holland, Amsterdam, 1979.

aspirin ingestion, and then return to normal after the aspirin treatment was stopped.

Thus, albumin seems to exert a protective action by its capacity to bind a large number of potentially interfering substances, such as plasma fatty acids. At least this particular radioimmunoassay definitely requires the presence of physiological amounts of albumin in the samples. We have also seen the same phenomenon in other radioimmunoassays, although it may not necessarily hold true for all antibodies. It must also be mentioned in this context that albumin may, in fact, even exert a harmful influence on certain prostaglandins, such as 15-keto-13,14-dihydro-PGE_2 (see this volume [43]).

Validation of the Radioimmunoassay

Four parameters must always be evaluated for each assay, viz. specificity, sensitivity, precision, and accuracy. Some common approaches for such studies in prostaglandin radioimmunoassay have been described earlier.[3]

Of great importance is establishment of linearity; i.e., the results obtained from analysis of different volumes of a sample must give the same result when expressed in amounts per unit volume. Lack of linearity in such an experiment can generally be seen also as lack of parallelism of two standard curves, one of which is run in the presence of a certain amount of the biological sample. It is mandatory to look for such phenomena, as their presence will strongly influence the final results.

One of the most important aspects when evaluating the properties of a prostaglandin radioimmunoassay is the comparison of obtained data with those from an independent method. The superior method in this field is quantitative gas chromatography–mass spectrometry.[17] However, very few laboratories have such facilities, and most workers in the field have to rely on other methods.

Comparison with published values is now often possible, since abundant data, obtained from reliable quantitative methods, have accumulated in the literature over the years. For example, most mass spectrometric studies agree that basal levels of 15-keto-13,14-dihydro-$PGF_{2\alpha}$ in peripheral plasma are in the order of magnitude of a few tens of picograms per milliliter (about 20–60 pg/ml). If a radioimmunoassay gives values considerably outside this range the reason must be sought. Obviously, how-

[17] K. Gréen, M. Hamberg, B. Samuelsson, M. Smigel, and J. C. Frölich, *Adv. Prostaglandin Thromboxane Res.* **5**, 39 (1978).

ever, basal values within this range are no guarantee that the assay is valid and reliable.

Obtained data must also compare well with the known biochemistry, physiology, and pharmacology of prostaglandins. For example, if measured levels are not influenced by known inhibitors or stimulators of prostaglandin production, it is not likely that they reflect the true situation (cf. Fig. 4, curve b).

Design of the Study

It must be stressed that all that radioimmunoassay does is to measure the inhibition of the binding of labeled antigen by the antibody. Ideally, this inhibition is caused by unlabeled molecules of the substance under study. However, a vast number of other substances may also inhibit this binding: structurally related compounds by cross-reaction with the antibody binding sites, and unrelated compounds by nonimmunologic inhibition. These latter factors are particularly important, since their inhibitory effects on the antigen–antibody binding may be strong (cf. Fig. 4) and their presence in the sample easily overlooked.[3,16] A decreased antigen–antibody binding obviously will result in an increased radioactivity in the free fraction, and unless the scientist is aware of the possibility of a nonimmunologic cause for the inhibition, the phenomenon will inevitably be interpreted as "high prostaglandin levels." There are many examples of such misinterpretation of data in the prostaglandin literature.[3]

Since the presence of unidentified, nonspecifically interfering substances can never be completely ruled out in a biological sample (extracted or not), attempts to measure *absolute* levels of a certain compound will always be hazardous. We normally try to increase the reliability of our radioimmunoassays by avoiding measurements of single samples: the preferred design of a study in our laboratory is to collect a series of samples during a time when a change in the prostaglandin level can be expected, and the measurements are aimed only at following these *changes*. Figures 3, 4, and 5 (see later) show some examples of such studies, and several more are presented in this volume [45].

Common Practical Problems Encountered in Radioimmunoassay Practice

Once optimal conditions have been established, the radioimmunoassay can generally be used for some time without further checking. However, with time certain alterations may appear that seriously affect the outcome of the assay. This is often the case also with the commercial radioimmunoassay kits, where the procedure recommended in the manual may be far from optimal. The following section deals with the detection and remedies of some of the most common problems in radioimmunoassay.

Decreased Maximal Binding

This can be caused by decomposition of either tracer or antibody, most commonly the tracer. Even if the compound is chemically stable, radiation damage inevitably occurs with time in labeled preparations. The purity of the tracer should thus be checked regularly. If necessary (> 10% decomposition) it must be purified, preferably using high-performance liquid chromatography (HPLC). A suitable chromatographic system for 15-ketodihydro-$PGF_{2\alpha}$ is reversed-phase HPLC, using Nucleosil 5 C_{18} and solvent system methanol–water–acetic acid (65:35:0.01, v/v/v). If some other chromatographic procedure must be employed, care must be taken to avoid introduction of, e.g., solvent impurities with this step. This is another reason for decreased maximal binding, and is best avoided by purification of large amounts of the tracer simultaneously: the influence of possible impurities can then be almost completely removed by the high dilution necessary prior to assay.

Decomposition of the antibody is rare: it can easily be detected by a renewed titer test.

Decomposition of Standard

Decomposition of the unlabeled compound is deleterious for the interpretation of the results. This problem is not often encountered with 15-ketodihydro-$PGF_{2\alpha}$, since the compound is chemically very stable. Should decomposition occur, however, it can be detected as a shift to the right of the standard curve; i.e., what is believed to be higher amounts than expected are necessary to achieve displacement of the tracer. To avoid this phenomenon, which is difficult to detect in its early stages, the purity of the standard stock solution should be checked regularly, using, e.g., thin-layer chromatography and gas–liquid chromatography.[8]

Impurities Introduced into the Standard Dilutions

The table shows our normal protocol for radioimmunoassay of the prostaglandin metabolite. Two tubes in the standard curve, Nos. 9 and 10, do not contain any unlabeled ligand and thus actually give additional information about the maximal binding. These tubes are included for the following reason. If the standard dilutions are not made fresh every day—which is not necessary when working with a chemically stable compound such as 15-ketodihydro-$PGF_{2\alpha}$, and would rather increase the interassay variation—and the different standard solutions are thus used repeatedly, it is recommended to include also a "standard" consisting of buffer alone that is exposed to the same number of pipettings as the rest of the standards. Should contamination occur from the repeated pipettings, for example by dissolution of some component from the plastic tips,[3,16] this is easily detected as differences in the values from these tubes and the true B_0 values (from tubes 6 through 8, in which fresh buffer is used). When this happens, the whole series of standards must be discarded and new dilutions prepared.

Insufficient Sensitivity of the Assay

When analyzing samples expected to contain very low amounts of the compound, the sensitivity of the assay may not be sufficient. The limit of detection of a radioimmunoassay can be considerably improved in several ways. First, it may be possible to use a more diluted solution of the antibody. The maximal binding then decreases somewhat, but the slope of the standard curve becomes more favorable and allows the determination of smaller amounts (cf. Fig. 10 of Granström and Kindahl[3]). Second, if possible, a tracer of higher specific activity should be used. If such a preparation cannot be obtained, an alternative is to reduce the amount of tracer added and compensate for the decreased radioactivity by prolonged counting times. Otherwise the precision of the assay will suffer.

Using either of these alternatives we have always been able to find a suitable radioimmunoassay system for biological samples of interest. In our opinion the above solutions are better alternatives than extraction of larger volumes of the biological material, with the subsequent evaporation of solvents and analysis of the concentrated sample. This procedure will inevitably introduce impurities[16] and may thus not increase the sensitivity.

15-Keto-13,14-dihydro-$PGF_{2\alpha}$ as a $PGF_{2\alpha}$ Parameter in Plasma

15-Ketodihydro-$PGF_{2\alpha}$ in peripheral plasma has now been used for many years as an indicator of $PGF_{2\alpha}$ production in the body. This metabolite replaced the parent compound as the target of choice when the drawbacks of $PGF_{2\alpha}$ measurements became apparent.[1] The short biological half-life of $PGF_{2\alpha}$ was one reason why this compound was considered unsuitable for measurements. However, even the half-life of its initially formed metabolite, 15-ketodihydro-$PGF_{2\alpha}$, is only about 10 min, and unless very frequent blood samples are collected, short spikes of $PGF_{2\alpha}$ release may be missed completely.[1] Extended studies on the metabolism of

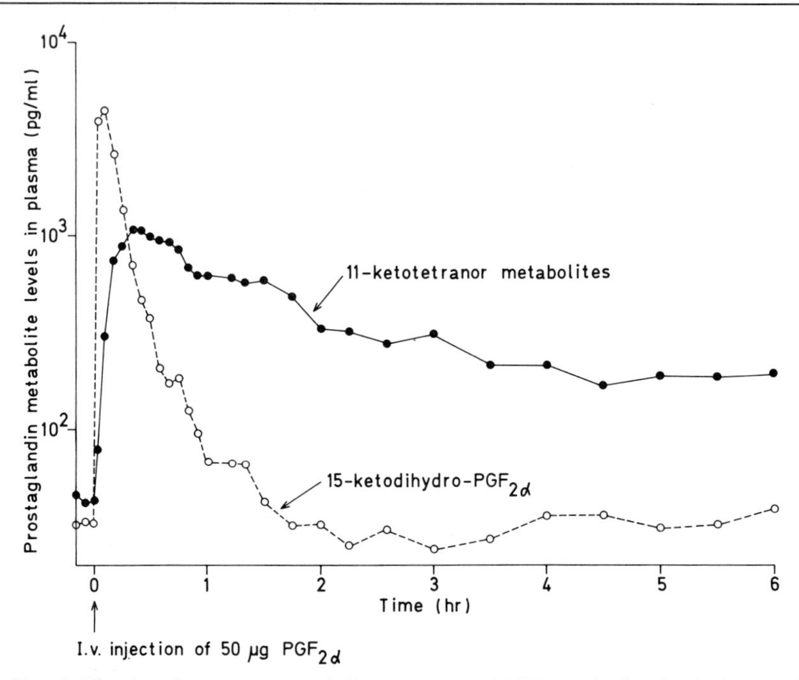

FIG. 5. Kinetics of appearance and disappearance of PGF metabolites in the human circulation after intravenous injection of 50 μg of $PGF_{2\alpha}$. The initially formed 15-ketodihydro-$PGF_{2\alpha}$ was measured by the radioimmunoassay described in this chapter; the assay for the late-appearing, long-lived 11-ketotetranor compounds is described in this volume [45].

$PGF_{2\alpha}$[18] indicate that the later formed, more degraded metabolites, such as 5α,7α-dihydroxy-11-ketotetranorprostane-1,16-dioic acid (Fig. 1) may be more reliable indicators of prostaglandin release, particularly if samples are collected with long intervals. Figure 5 shows the changing pattern of PGF products in human plasma after intravenous injection of 50 μg of $PGF_{2\alpha}$. The initially formed 15-ketodihydro metabolite as well as the later appearing 11-ketotetranor metabolites were assayed using radioimmunoassays described in this chapter and in this volume [45]. As Figure 5 shows, even after this comparatively high dose of the prostaglandin, the plasma concentration of 15-ketodihydro-$PGF_{2\alpha}$ had returned to normal after 1 hr 30 min, whereas the level of the 11-ketotetranor products was still fivefold the basal one after 6 hr. On the other hand, the concentration of these more degraded compounds never reached the same high level as 15-ketodihydro-$PGF_{2\alpha}$ did initially.

Even in a study where blood samples are taken very frequently, and

[18] E. Granström and H. Kindahl, submitted for publication.

15-ketodihydro-$PGF_{2\alpha}$ thus is considered the best parameter, it must be kept in mind that measured fluctuations in its concentration do not necessarily reflect changes in the $PGF_{2\alpha}$ production. First, using radioimmunoassay, it is likely that some 15-ketodihydro-$PGF_{1\alpha}$ is measured as well, for reasons discussed above. Second, and most important, PGF compounds may be formed in considerable amounts also from other prostaglandins, such as E compounds,[19,20] D compounds,[21] and also the endoperoxides,[22,23] and they may also be converted into at least prostaglandins of the E type[24] and perhaps into other compounds as well. As the biological activities of all these prostaglandins differ widely in nature, and the compounds probably play very different roles in the body, fluctuations in measured PGF metabolite levels become very difficult to interpret from a biological point of view.

Acknowledgments

This study was supported by grants from the Swedish Medical Research Council, project No. 03P-5804 and 03X-05915, the Swedish Council for Forestry and Agriculture Research, and the World Health Organization.

[19] M. Hamberg and B. Samuelsson, *J. Biol. Chem.* **246,** 1073 (1971).
[20] M. Hamberg and M. Wilson, *Adv. Biosci.* **9,** 39 (1973).
[21] C. K. Ellis, M. D. Smigel, J. A. Oates, O. Oelz, and B. J. Sweetman, *J. Biol. Chem.* **254,** 4152 (1979).
[22] C. Pace-Asciak and M. Nashat, *Biochim. Biophys. Acta* **388,** 243 (1975).
[23] E. Christ-Hazelhof, D. H. Nugteren, and D. A. van Dorp, *Biochim. Biophys. Acta* **450,** 450 (1976).
[24] C. Pace-Asciak and D. Miller, *Experientia* **30,** 590 (1974).

[45] Radioimmunoassay of 5α,7α-Dihydroxy-11-ketotetranorprostane-1,16-dioic Acid, a Major Prostaglandin F Metabolite in Blood and Urine

By ELISABETH GRANSTRÖM and HANS KINDAHL

The comparatively rapid disappearance of prostaglandin metabolites from the circulation makes the monitoring of prostaglandin production by plasma analyses alone somewhat unreliable, unless samples are taken very frequently.[1] Such a protocol may meet with difficulties, however: frequent blood sampling may not be possible in small laboratory animals

[1] E. Granström and B. Samuelsson, *Adv. Prostaglandin Thromboxane Res.* **5,** 1 (1978).

ISBN 0-12-181986-8

for practical reasons, and should be avoided in the human also for ethical reasons.

In such circumstances an alternative is measurements of metabolites in urine instead. Essentially all of a prostaglandin that reaches the blood stream will eventually be excreted into the urine as more degraded products, generally tetranor metabolites[2–5] (cf. Fig. 1 of this volume [44]). There is thus no risk of missing even a short surge of a prostaglandin release, if the total 24-hr portion of urine is collected. This approach also makes long-term studies possible.

A large number of successful studies have now been carried out on the total body production of various prostaglandins by measurements of their final tetranor degradation products in the urine.[2–5] Methods have been developed both for products of the E and F series.[6–12] Most of these studies were done with gas chromatographic–mass spectrometric methods.[6–9] However, a few radioimmunoassays also exist for one urinary prostaglandin metabolite,[10–12] viz. $5\alpha,7\alpha$-dihydroxy-11-ketotetranorprostane-1,16-dioic acid (see Fig. 1), a major urinary product of PGF metabolism in several species.[2]

As yet, no radioimmunoassay exists for the corresponding PGE metabolite, 7α-hydroxy-5,11-diketotetranorprostane-1,16-dioic acid,[2] although a few gas chromatography–mass spectrometry methods have been developed.[6,8] This metabolite is not likely to be stable,[cf. 13–15] how-

[2] B. Samuelsson, E. Granström, K. Gréen, M. Hamberg, and S. Hammarström, *Annu. Rev. Biochem.* **44,** 669 (1975).

[3] E. Granström, *in* "Prostaglandins and Thromboxanes" (F. Berti, B. Samuelsson, and G. P. Velo, eds.), p. 75. Plenum, New York, 1977.

[4] J. A. Oates, L. J. Roberts II, B. J. Sweetman, R. L. Maas, J. F. Gerkens, and D. F. Taber, *Adv. Prostaglandin Thromboxane Res.* **6,** 35 (1980).

[5] E. Granström, *in* "The Prostaglandin System" (F. Berti and G. P. Velo, eds.), p. 39. Plenum, New York, 1981.

[6] M. Hamberg, *Biochem. Biophys. Res. Commun.* **49,** 720 (1972).

[7] M. Hamberg, *Anal. Biochem.* **55,** 368 (1973).

[8] H. W. Seyberth, B. J. Sweetman, J. C. Frölich, and J. A. Oates, *Prostaglandins* **11,** 381 (1976).

[9] H. W. Seyberth, G. V. Segre, J. L. Morgan, B. J. Sweetman, J. T. Potts, Jr., and J. A. Oates, *N. Engl. J. Med.* **293,** 1278 (1975).

[10] S. Ohki, T. Hanyu, K. Imaki, N. Nakazawa, and F. Hirata, *Prostaglandins* **6,** 137 (1974).

[11] J. C. Cornette, K. T. Kirton, W. P. Schneider, F. S. Sun, R. A. Johnson, and E. G. Nidy, *Prostaglandins* **9,** 323 (1975).

[12] E. Granström and H. Kindahl, *Prostaglandins* **12,** 759 (1976).

[13] E. Granström and H. Kindahl, *Adv. Prostaglandin Thromboxane Res.* **6,** 181 (1980).

[14] F. A. Fitzpatrick, R. Aguirre, J. E. Pike, and F. H. Lincoln, *Prostaglandins* **19,** 917 (1980).

[15] E. Granström, M. Hamberg, G. Hansson, and H. Kindahl, *Prostaglandins* **19,** 933 (1980).

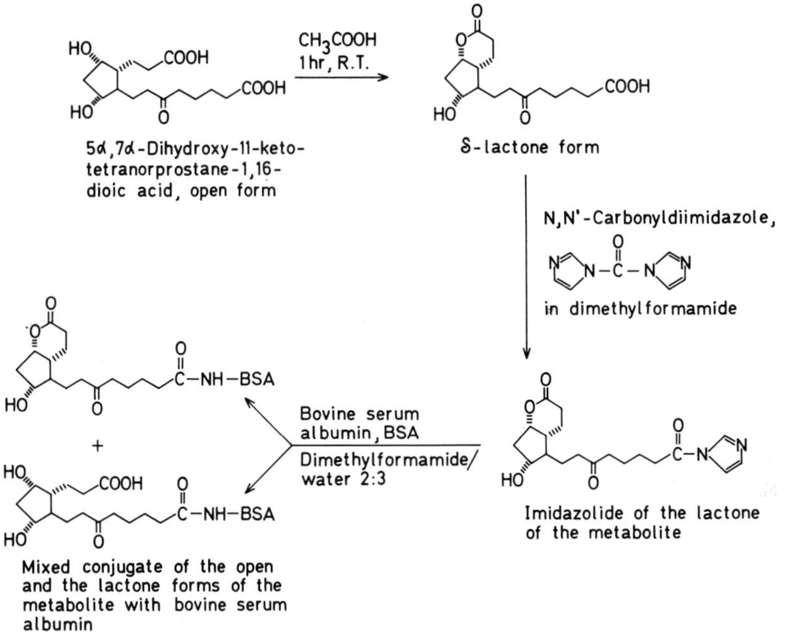

CH_3COOH
1 hr, R.T.

5α,7α-Dihydroxy-11-keto-tetranorprostane-1,16-dioic acid, open form

δ-lactone form

N,N'-Carbonyldiimidazole,
in dimethylformamide

Imidazolide of the lactone of the metabolite

Bovine serum albumin, BSA
Dimethylformamide/water 2:3

Mixed conjugate of the open and the lactone forms of the metabolite with bovine serum albumin

FIG. 1. Preparation of a bovine serum albumin conjugate of 5α,7α-dihydroxy-11-ketotetranorprostane-1,16-dioic acid for immunization. Coupling takes place exclusively at the ω carboxyl owing to prior blocking of the α carboxyl by induction of δ-lactonization.

ever, and may thus be unsuitable as a target, at least for radioimmunoassay measurements. Furthermore, no simple method exists for the protection of one of the two carboxyl groups, as is possible in the corresponding PGF compound (see below). Such protection is desirable when preparing an antigenic conjugate.

It has been demonstrated that the tetranor metabolites may also be possible to assay in plasma.[16] They may in fact be preferable as parameters of prostaglandin production, not only in urine, as mentioned above, but also in the circulation. One experiment upon which this conclusion was based was shown in this volume [44]. The tetranor products—generally 11-ketotetranor metabolites—were found to appear in the circulation shortly after the initially formed 15-ketodihydro metabolite, then reached a comparatively long-lasting peak concentration, and finally remained at elevated levels for many hours, long after the 15-ketodihydro metabolite had returned to its basal concentration.

This chapter describes a radioimmunoassay for a major PGF product,

[16] E. Granström, H. Kindahl, and M.-L. Swahn, submitted for publication.

5α,7α-dihydroxy-11-ketotetranorprostane-1,16-dioic acid, and its application to measurements in urine as well as in plasma. The assay is sufficiently nonspecific to allow measurements also of structurally related tetranor metabolites, for example, major PGF metabolites in other species than the human.

Development of the Radioimmunoassay

Preparation of the Metabolite

A few methods for the chemical synthesis of this compound or the related PGE metabolite have been published.[17,18] In our laboratory, however, we prefer to isolate the metabolite from biological material, generally from urine from patients receiving $PGF_{2\alpha}$ for termination of pregnancy.[12] Urine containing ^3H-labeled $PGF_{2\alpha}$ metabolites is pooled with this urine to provide a tracer for the purification of the compound, and methyl 5α,7α-dihydroxy-11-ketotetranorprostane-1,16-dioate is isolated according to previously published methods.[19,20]

Briefly, the acidified urine is percolated through a column of Amberlite XAD-2 (Rohm & Haas Co., Philadelphia, Pennsylvania), and the lipophilic products are eluted with methanol. This crude extract is then subjected to two consecutive reversed-phase partition chromatographies: in the first one, solvent system D is employed[19] (stationary phase, *n*-butanol, 4.0 ml per 4.5 g of support; moving phase, 0.6% acetic acid in water; support, hydrophobic Hyflo Super-Cel, 4.5 g per 50 mg of crude sample). The major peak from this chromatography contains mostly tetranor dioic acids. After esterification with diazomethane, these products are separated on a second reversed-phase partition column, solvent system F-50[19] (stationary phase, 10% heptane in chloroform; moving phase, 50% methanol in water). The two major peaks from this chromatography contain, respectively, the δ-lactone monomethyl ester and the open dimethyl ester of 5α,7α-dihydroxy-11-ketotetranorprostane-1,16-dioic acid. The less polar, dimethyl ester peak also contains small amounts of 7α,9α-dihydroxy-13-ketodinor- ω-dinorprostane-1,18-dioic acid, which can be removed using argentation thin-layer chromatography, employing the organic phase of ethyl acetate–methanol–water (160 : 10 : 100, v/v/v) as solvent system.[20]

After this minor metabolite has been removed, the material in the two

[17] C. Lin, *J. Org. Chem.* **41,** 4045 (1976).
[18] J. R. Boot, M. J. Foulis, N. J. A. Gutteridge, and C. W. Smith, *Prostaglandins* **8,** 439 (1974).
[19] E. Granström and B. Samuelsson, *J. Biol. Chem.* **246,** 5254 (1971).
[20] E. Granström and B. Samuelsson, *J. Biol. Chem.* **246,** 7470 (1971).

chromatographic peaks is combined and is subsequently subjected to mild alkaline hydrolysis (1 *M* NaOH in 50% methanol, 12 hr at room temperature), diluted with water, acidified to pH 3, and extracted using Sep-Pak (see this volume [58]).

This procedure results in pure 5α,7α-dihydroxy-11-ketotetranorprostane-1,16-dioic acid in an overall yield of approximately 80%, calculated from the amount present in the urine.

Coupling

The small prostaglandin molecules are normally rendered antigenic by coupling to the amino groups on proteins. In this particular case, the hapten is a dicarboxylic acid. In order to obtain a comparatively specific antibody it is desirable to perform the coupling exclusively at one of the two carboxyls. This requires protection of the other carboxyl, which in the case of a tetranor PGF_α compound can be conveniently done by induction of δ-lactonization at low pH between the carboxyl group at C-1 and the α-hydroxyl group at C-5[12,19] (cf. Fig. 1).

5α,7α-Dihydroxy-11-ketotetranorprostane-1,16-dioic acid, 10 mg, is kept in 1 ml of glacial acetic acid for 1 hr at room temperature. The acetic acid is then evaporated under a stream of nitrogen. The dry residue, i.e., the δ-lactone of the metabolite, is dissolved in 2 ml of dry dimethylformamide and kept under N_2 with stirring for 10 min at room temperature. *N,N'*-Carbonyldiimidazole,[21] 5 mg, is added, and stirring is continued for another 20 min. This solution is then added dropwise to a solution of 40 mg of bovine serum albumin in 3 ml of H_2O. Stirring is continued for 7 hr. The coupling procedure is summarized in Fig. 1.

The reaction mixture is subsequently dialyzed extensively, first against dimethylformamide–water (2:3 v/v) for 12 hr and then against several changes of water. The dialyzed conjugate is divided into eight equal portions, which is sufficient for the immunization of two rabbits, and lyophilized.

Immunization

Two rabbits are immunized after emulsification of the conjugate with Freund's adjuvant, according to the schedule outlined in this volume [44] (i.e., three doses of the conjugate per rabbit with weekly intervals, then a rest period and finally a booster dose). The animals are bled about 10 days after the boost, when the antibody titer has reached its maximum.

[21] U. Axen, *Prostaglandins* **5,** 45 (1974).

Properties of the Antibody

The metabolite was presumably coupled selectively with its ω-carboxyl group to the protein, since the α-carboxyl group was protected by ring closure to form a δ-lactone (Fig. 1). This δ-lactone is, however, easily hydrolyzed,[12,19] and thus the equilibrium between the open form and the δ-lactone form is probably reestablished in the aqueous solution employed during the final step of the coupling procedure and during the dialysis, and particularly during the subsequent long period at physiological pH in the animal. Thus, the produced antiserum is likely to contain antibodies to both forms. However, since the radioimmunoassay is also carried out in a similar environment, the same equilibrium is probably established also for the standard, the sample, and the tracer molecules of the metabolite. The heterogeneous population of antibodies is thus not likely to be a drawback for the assay.

The specificity of the antiserum is shown in Fig. 2.[22] The antibodies cross-react to a minor degree with C_{20} prostaglandins; somewhat more with F compounds and compounds with the 15-keto-13,14-dihydro structure. The central cluster of displacement curves shows that the antibodies cross-react to a somewhat greater extent with dinor metabolites, and that the structure at the ω end of these compounds is of minor importance. The group of displacement curves at the left indicates that the antibodies are nonselective with respect to tetranor metabolites and cross-react extensively with all these compounds. Again, the structure at the ω end does not influence the affinities of the compounds for the antibodies to any greater extent. These data taken together indicate that the coupling of the hapten to the protein carrier molecule really took place exclusively at the ω end (Fig. 1), since this part of the antigen would then not be recognized by the antibody.

Preparation of 5α,7α-Dihydroxy-11-ketotetranorprostane-1,16-dioic Acid for Use as Standard and as Labeled Ligand

The material to be used as standard and as tracer is prepared using the same methods as described above. However, for the preparation intended as standard, it is now necessary that the specific activity of the pure compound be known exactly, and also that it be sufficiently low not to interfere in the later assay. One possible method to accomplish has been described[12]: A human volunteer received 5 mg of [9β-^{3}H]PGF$_{2\alpha}$, specific activity 0.3 μCi/μmol, by intravenous infusion during 1 hr. This infusion was given after 4 days of pretreatment with indomethacin, 50 mg $\times$ 4, to

[22] E. Granström and H. Kindahl, *Adv. Prostaglandin Thromboxane Res.* **5,** 119 (1978).

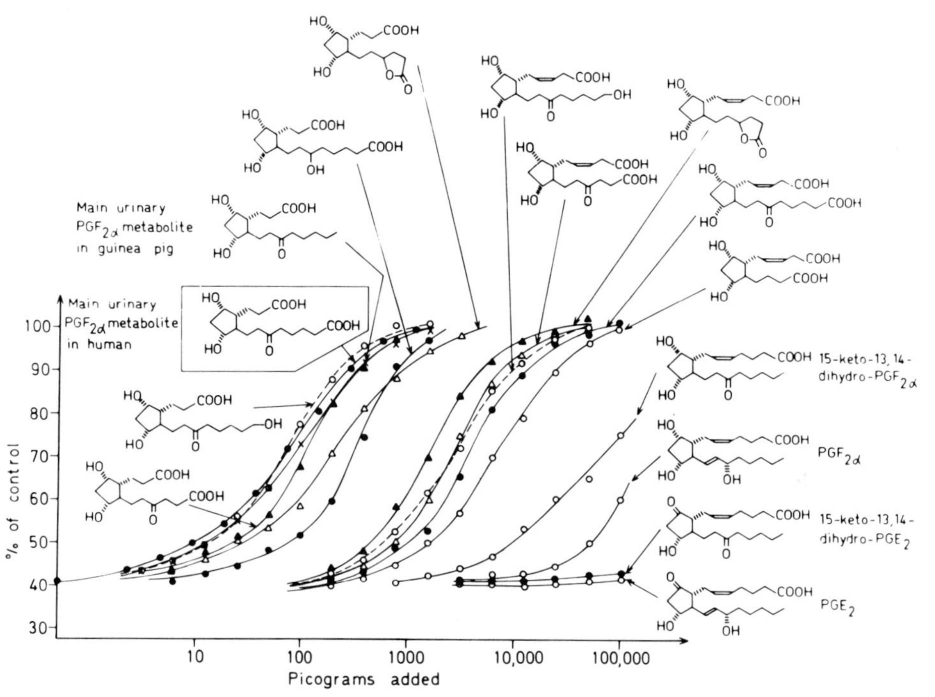

FIG. 2. Properties of an antiserum directed against the immunogen depicted in Fig. 1. For comments, see text. Reproduced from Granström and Kindahl,[22] with permission from Raven Press.

suppress the endogenous production. The compound was then isolated from urine as described above.

The labeled ligand is prepared accordingly; the infusion is done with [^{3}H]PGF$_{2\alpha}$ of the highest possible specific activity after indomethacin pretreatment.

However, there are other possible solutions to this problem. The metabolite can be prepared *in vivo* in an analogous way also from other species, e.g., the monkey, the rabbit, and the rat.[2] Another possible approach is based on the high cross-reactivity of the antibody with the corresponding monocarboxylic acid, 5α,7α-dihydroxy-11-ketotetranorprostanoic acid. It is in fact possible to use this heterologous compound both as standard and as tracer because of this similarity in structure. The monocarboxylic compound can be prepared in larger amounts and with exactly known specific activity by β-oxidation *in vitro* of 15-ketodihydro-PGF$_{2\alpha}$,[12] with either very high or very low specific activity for use as tracer and as standard, respectively. This approach essentially solves the problem with the possible endogenous contribution of unknown amounts, which can probably never be completely eliminated even with large doses of prostaglandin synthesis inhibitors. This is particularly important for the preparation of the labeled ligand, the higher specific activity of which will consequently improve the sensitivity of the assay.

However, the use of the two different compounds as labeled ligand and as tracer does not give exactly the same results (see below, under Problems Encountered).

Performance of the Radioimmunoassay

Reagents, Equipment, and Incubation Procedure

The practical details of this radioimmunoassay are essentially the same as those described for 15-keto-13,14-dihydro-$PGF_{2\alpha}$ in this volume [44] (see also Granström and Kindahl[22]). However, some modifications have been introduced.

Radioimmunoassay of Tetranor Metabolites in Urine. The amounts of the major $PGF_{2\alpha}$ metabolite in human urine are comparatively large: the daily excretion of $5\alpha,7\alpha$-dihydroxy-11-ketotetranorprostane-1,16-dioic acid is about 5–50 μg.[7] With an average daily urine volume of 1000–1500 ml, this means that the concentration of the metabolite is commonly about 3–50 pg/μl. These amounts fall reasonably well within the range of the standard curve (cf. Fig. 2), which means that assay of as little as 1–5 μl of undiluted urine normally gives quite acceptable results. The precision when pipetting such small amounts may be unacceptably low, however, and the preferred method for routine analyses in our laboratory is instead to work with urines after dilution between 1:20 and 1:100 and to analyze larger volumes instead. The high precision LKB diluter is normally used both for preparation and for the later analysis of these dilutions.

Also in other species—for example in the guinea pig, where the corresponding monocarboxylic acid is the major metabolite[2]—the endogenous total daily production is high enough to permit a considerable dilution prior to assay (see below).

The data obtained with radioimmunoassay of this metabolite in urine agree well with published gas chromatographic–mass spectrometric data.[7,12] It can thus be concluded that although urine contains large amounts of possibly interfering material,[22] this harmful influence seems to be largely eliminated by the high dilution permitted prior to assay.

Radioimmunoassay in Plasma. Assay of the 11-ketotetranor metabolites may be carried out in unextracted plasma. Basal plasma levels of these compounds in the human seem to be in the same order of magnitude as those of 15-keto-dihydro-$PGF_{2\alpha}$, i.e., about 10–50 pg/ml (see, however, below). Because of the higher sensitivity of this radioimmunoassay compared with the assay for 15-ketodihydro-$PGF_{2\alpha}$, the suitable sample volumes are lower in this assay, e.g., 0.1–0.4 ml of plasma. The assay is otherwise carried out in the same way as described above for 15-ketodihydro-$PGF_{2\alpha}$ in plasma. To improve separation of the free and antibody-bound fraction after the incubation, we normally add bovine γ-globulin, 0.2 ml of a 0.5% solution (see this volume [44]), to all sample tubes containing less than 0.3 ml of unextracted plasma. If this addition of globulin carrier is omitted, a slight opalescence of the final supernatant obtained after the precipitation step indicates that the γ-globulins of the sample were not completely precipitated. This would be deleterious for the interpretation of the results, since part of the antibody-bound ligand would then be counted together with the free fraction.

Application of the Radioimmunoassay

Assay of 11-Ketotetranor Metabolites in Urine

Also in this case we avoid the analysis of single samples and prefer the assay of series of samples during periods when changes in the prostaglandin production are expected (for discussion, see this volume [44]). To avoid the risk of nonspecific interference in the assay by changes in urinary salt concentration, all samples from one series should preferably be assayed at a comparable ionic strength. Consequently, the 24-hr portions of urine, or exact aliquots of them, are diluted to the same volume before assay. If urine is collected at irregular intervals, the dilution must be adjusted accordingly.

In a study on the role of $PGF_{2\alpha}$ as a luteolytic hormone in the guinea pig, the urine portions (collected once or twice a day) were diluted to give final volumes corresponding to 10 ml/hr. Small aliquots of these dilutions were then stored at −20° until assayed. Prior to assay, 0.1-ml aliquots of these diluted samples were further diluted 26-fold using the LKB diluter, and 50 and 100 μl of these final preparations were radioimmunoassayed in duplicate. The major PGF metabolite in this species is 5α,7α-dihydroxy-11-ketotetranorprostanoic acid,[2] and the homologous compound was thus employed also as standard and tracer.

Figure 3 shows the results from two consecutive estrous cycles in one guinea pig.[12] Day 1 of the cycle was identified by the appearance of estrous behavior and/or opening of the vaginal membrane. During the greater part of the cycle the PGF production was low and relatively constant. A few days before the next heat the prostaglandin synthesis rose

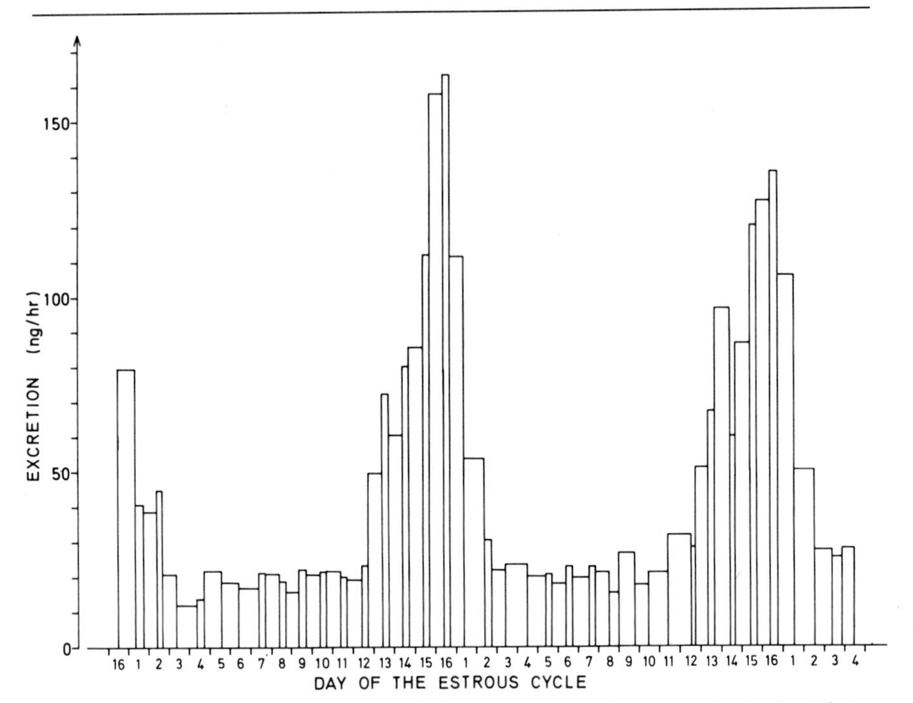

FIG. 3. Excretion of 11-ketotetranor PGF metabolites (mainly $5\alpha,7\alpha$-dihydroxy-11-keto-tetranorprostanoic acid) in the urine of a female guinea pig during two consecutive estrous cycles. Day 1 is the first day of heat. Reproduced from Granström and Kindahl,[12] with permission from the publisher.

markedly, and it reached a high peak during the last days of the cycle: the peak concentration was 7- to 8-fold the basal level.

$PGF_{2\alpha}$ production was also studied during human pregnancy and parturition using the same approach. In one subject urine was collected in 24-hr portions every second day from week 36. From 8 days before delivery all urine was collected; during the day of parturition in portions. All portions were diluted to give final volumes corresponding to 100 ml/hr. Of these preparations 50- and 100-μl aliquots were assayed in duplicate. This time the dioic acid was employed as standard; the monocarboxylic acid was still used, however, as tracer because of its considerably higher specific activity. Figure 4 shows the results from one such study. During the last month of pregnancy the urinary prostaglandin metabolite levels were elevated about threefold compared to the basal excretion; a very sharp increase occurred during the hours of labor and parturition. During the postpartal period the prostaglandin production slowly returned to normal.

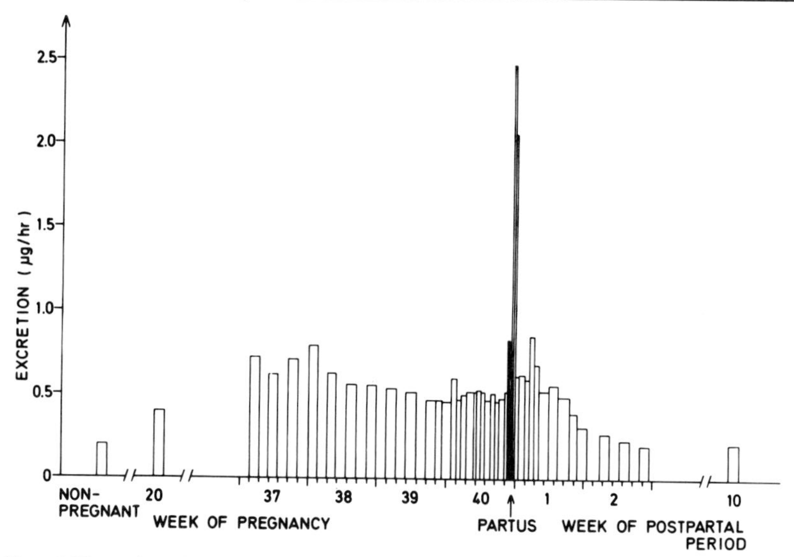

FIG. 4. Excretion of 11-ketotetranor PGF metabolites in one pregnant human female. Reproduced from Granström and Kindahl,[12] with permission from the publisher.

Assay of 11-Ketotetranor Metabolites in Plasma

In this volume [44] an experiment was described that dealt with the kinetics of appearance and disappearance of various PGF metabolites in the circulation (see Fig. 5 of this volume [44]). After an intravenous bolus injection of $PGF_{2\alpha}$ the initially formed 15-ketodihydro-$PGF_{2\alpha}$, which dominated the metabolite pattern during the first few minutes, was soon replaced in the circulation by shorter compounds. The rather complex irregular shape of the curve representing those more degraded products probably reflects the formation and disappearance of several structurally related metabolites.[2] As mentioned above, the antibody does not distinguish well between those compounds.

Judging from this sequence of events, it seems that a single sample or a few samples taken at any time during the first few hours after a prostaglandin release into the circulation might display elevated levels of the tetranor metabolites, in contrast to the 15-ketodihydro-$PGF_{2\alpha}$ concentration, which might be elevated only comparatively early after the release. To test this hypothesis we again studied the PGF production during late pregnancy, labor, and parturition in the human, this time with a series of blood samples instead of urine. This study was chosen because the roles of prostaglandins have been well established in human pregnancy, and also because the greatest prostaglandin production seems to occur as a relatively

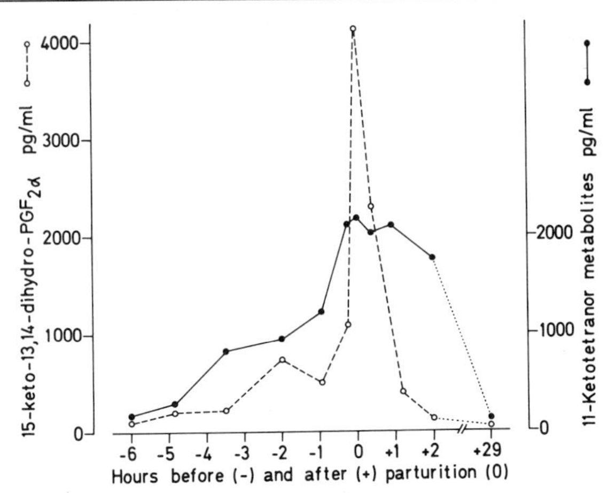

FIG. 5. Concentrations of 15-ketodihydro-$PGF_{2\alpha}$ (○---○) and 11-ketotetranor-PGF metabolites (●——●) in plasma during the peripartal period in one human female.

sharp peak almost confined to the hours of labor and expulsion of the fetus (cf. Fig. 4). This is in contrast to the events in several other species.[12,23,24]

Figure 5 shows the results in plasma samples from one subject. As expected,[25] levels of 15-ketodihydro-$PGF_{2\alpha}$ rose gradually during labor, reached a high peak shortly after the expulsion of the fetus, and had returned to normal about 2 hr later. The levels of the tetranor products also rose during labor to a similar degree. A more sustained elevation was seen around parturition and lasted for several hours, reflecting the considerably slower elimination of these compounds from the circulation.

Problems Encountered

Two problems were encountered in this particular assay. The first one appeared when the radioimmunoassay was applied to plasma samples: analyses of different amounts of the same plasma pool did not give completely proportional values; i.e., linearity was not perfect over the entire range of plasma volumes tried (25–500 μl). Blood levels calculated from analyses of plasma volumes below 100 μl consistently gave somewhat higher values than when larger plasma volumes were analyzed ($>$100 μl). Linearity in this higher range, however, was quite satisfactory. The rea-

[23] L.-E. Edqvist, H. Kindahl, and G. H. Stabenfeldt, *Prostaglandins* **16,** 111 (1978).

[24] R. L. Pashen and W. R. Allen, *J. Reprod. Fertil. Suppl.* **27,** 499 (1979).

[25] K. Gréen, M. Bygdeman, M. Toppozada, and N. Wiqvist, *Am. J. Obstet. Gynecol.* **120,** 25 (1974).

son for this discrepancy is not known; however, the same problem is frequently encountered in steroid as well as peptide radioimmunoassays.[26]

The most common solution to the problem, which also we found to be most satisfactory in this assay, is always to assay at least 100 μl of the plasma samples and to include 100 μl of a plasma pool also in the standard tubes. Several different pools of plasma from each species under study were tested for this purpose, and the ones that gave the lowest displacement of the bound tracer were chosen and employed in the respective studies. In all radioimmunoassay analyses two series of standard tubes were routinely included: one with and one without 100 μl of the selected plasma pool. For each sample, analyzed in different dilutions, comparison between results obtained from either of these standard curves maximally differed by 2–3 pg per tube.

The second problem applies to plasma as well as urinary analyses and was briefly mentioned earlier. The antibody cross-reacts extensively with the corresponding monocarboxylic metabolite, 5α,7α-dihydroxy-11-keto-tetranorprostanoic acid (Fig. 2), to an extent (87%) that allows the use of this heterologous compound as standard and tracer, instead of the less easily prepared and quantitated dioic acid metabolite. However, the antibody prefers the homologous dioic acid, which can readily be seen when standard curves of both compounds are included: the standard curve obtained with the monocarboxylic metabolite as the unlabeled ligand is found to the right of the homologous curve in the diagram. Depending on which of these standard curves is used for calculations, the obtained data will differ: values analyzed from the dioic standard curve will be somewhat lower. Since the biological material analyzed (plasma or urine) normally contains mixtures of a large number of structurally closely related tetranor metabolites, it is not self-evident which standard curve would give the most representative results. However, if the aim is not to determine the absolute levels of a metabolite, but rather to follow increases and decreases in metabolite levels, the choice of compound for the standard curve becomes less important.

11-Ketotetranor Metabolites as Parameters of Prostaglandin Production in Plasma and Urine

Obviously there may not be any particular advantage in measurements of the long-lived, late-appearing 11-ketotetranor metabolites in plasma during such dramatic events as parturition, where the time of highest

[26] K. Kirkham and W. Hunter (eds.) "Radioimmunoassay Methods." Churchill-Livingstone, Edinburgh, 1971.

prostaglandin production is easily defined by other means. However, in many other biological situations the peaks of prostaglandin production may go unaccompanied by simultaneous clinical signs. This is, for example, the case with the luteolytic period in many species, and the optimal sampling times may thus be difficult to identify in such a study. In those situations, when samples have to be taken randomly and perhaps also at long intervals, the monitoring of a long-lived plasma metabolite may give a safer indication of a preceding prostaglandin release.

In studies where repeated plasma samples cannot be taken (or trusted), an alternative is to assay urine for the presence of end products of prostaglandin metabolism. Two aspects have to be considered when this approach is chosen.

First, owing to the inevitable lag period between the release of a prostaglandin into the circulation and its appearance as metabolites in the urine,[12] no rapid fluctuations can be seen in a study of this design (cf. the luteolysis studies in Chapter 44 and in this chapter, Figs. 3 of both chapters). Temporal endocrine relationships may thus be difficult to study using this approach. Second, it is mandatory that all urine be collected during the period of interest. Diuresis may fluctuate considerably, and the prostaglandin metabolite concentration thus varies widely between different single urine portions for no other reason than a variable dilution by the kidneys. Collection of an entire 24-hr urine portion, for example, in clinical studies may be unexpectedly difficult, and estimates of the total daily prostaglandin production from such measurements must be viewed with this difficulty kept in mind.

Pronounced species differences are seen in prostaglandin metabolism[2–5]; however, the major part of the end products are tetranor compounds in most species. An antibody produced as described in this chapter, which is highly specific for tetranor compounds but nonselective regarding the structures in the ω side chain, may thus be employed in studies in many species. Absolute levels of any metabolite will obviously not be possible to establish because of the high cross-reaction with related compounds present in the sample. This may, however, not be necessary in many studies: normally the aim is to follow changes in the prostaglandin production, and measurement of several related end products simultaneously might give a more reliable picture of the events than if one single compound is monitored.

Finally, in interpreting changes in the plasma or urinary levels of these metabolites, the same limitations naturally apply that were discussed in Chapter 44: changes in measured PGF metabolite levels do not necessarily reflect changes in *PGF* production, since PGF metabolites may originate in many other compounds as well.

Acknowledgments

This study was supported by grants from the Swedish Medical Research Council, project No. 03P-5804 and 03X-05915, the Swedish Council for Forestry and Agriculture Research, and The World Health Organization.

Section III

Substrates, Reagents, and Standards

[46] Synthesis of Radiolabeled Fatty Acids

By HOWARD SPRECHER and SHANKAR K. SANKARAPPA

1-^{14}C-Labeled Fatty Acids

The procedure that we use for the synthesis of 1-^{14}C-labeled fatty acids is a modification of that originally described by Osbond *et al.*[1] for the synthesis of acetylenic acids. This pathway is illustrated below for the preparation of 7,10,13,16,19-[1-^{14}C] docosapentaenoic acid.

$$CH_3—CH_2—(C\equiv C—CH_2)_4Br + HC\equiv C—(CH_2)_5OH$$

$$\rightarrow CH_3—CH_2—(C\equiv C—CH_2)_5—(CH_2)_4—OH \rightarrow CH_3—CH_2(\overset{H}{C}=\overset{H}{C}—CH_2)_5—(CH_2)_4—Br$$

$$\rightarrow CH_3—CH_2(\overset{H}{C}=\overset{H}{C}—CH_2)_5—(CH_2)_4{}^{14}COOH$$

The synthesis can be divided into the five following steps: (1) preparation of 6-heptyn-1-ol; (2) synthesis of 1-bromo-2,5,8,11-tetradecatetrayne; (3) coupling of these two intermediates to give heneicosa-6,9,12,15,18-pentayn-1-ol; (4) reduction of this acetylenic alcohol to the ethylenic analog and subsequent purification of heneicosa-6,9,12,15,18-pentaen-1-ol; (5) conversion of this alcohol to 7,10,13,16,19-[1-^{14}C] docosapentaenoic acid.

Although the following presentation will be confined to the methods used for the preparation of this acid, this general procedure can be used for the synthesis of any all-cis acid that contains the skipped (methylene interrupted) pattern of unsaturation. The modifications of this procedure used for the preparation of intermediates required for the synthesis of other acids will be discussed in the appropriate sections. In addition, three reviews discuss in detail the synthesis of unsaturated fatty acids.[2-4]

Step 1. Synthesis of 6-Heptyn-1-ol

6-Heptyn-1-ol is prepared according to the following reaction sequence.

[1] J. M. Osbond, P. G. Philpott, and J. C. Wickens, *J. Chem. Soc.*, 2779 (1961).
[2] J. M. Osbond, *Prog. Chem. Fats Other Lipids* **9**, 121 (1966).
[3] W.-H. Kunau, *Angew. Chem., Int. Ed. Engl.* **15**, 61 (1976).
[4] H. Sprecher, *Prog. Chem. Fats Other Lipids* **15**, 219 (1978).

ISBN 0-12-181986-8

$$HO—(CH_2)_5—OH \xrightarrow{HBr} Br—(CH_2)_5—OH \xrightarrow[2.\ NH_3;NaC\equiv CH]{1.\ dihydropyran}$$

$$HC\equiv C—(CH_2)_5—O—\text{(tetrahydropyran-2-yl)} \xrightarrow{H+} HC\equiv C—(CH_2)_5—OH$$

1-Bromo-5-pentanol. To a liquid–liquid extractor mounted in an oil bath heated to 90–100° is added 208 g (2 mol) of 1,5-pentanediol and 500 g of 48% HBr. The reaction mixture is continuously extracted for 24 hr with boiling toluene to remove the bromohydrin as soon as it is produced. The toluene is then removed on a rotary evaporator to give the crude bromohydrin.

6-Heptyn-1-ol. The hydroxyl group of the bromohydrin is blocked with dihydropyran using the general procedure described by Jones and Mann.[5] To a three-neck flask fitted with a condensor, a dropping funnel and an overhead stirrer is added 1 mol of the crude bromohydrin and 0.2 ml of concentrated HCl. One mole of the bromohydrin is added dropwise over 30 min. The temperature of the reaction is kept below 60° by cooling in an ice bath. After stirring for 2 hr the reaction mixture is washed with a saturated solution of Na_2CO_3, and dried over anhydrous Na_2SO_4. The product is transferred to a round-bottom flask with ether, and the solvent is removed under reduced pressure.

To a 3-liter three-neck flask equipped with an overhead stirrer, dropping funnel, and Dewar condensor filled with acetone and Dry Ice is added approximately 1.5 liters of liquid ammonia. Acetylene is sequentially bubbled through a sulfuric acid trap, a Dry Ice trap, and then a soda trap into the liquid ammonia. Metallic sodium (a 1.1 molar excess) is then added in small pieces at such a rate that the solution does not turn dark blue.[6] After all the sodium is converted to sodium acetylide, the blocked derivative of the bromohydrin is added dropwise during 30 min. Stirring is continued for 3 hr, and then 10 g of solid NH_4Cl are added. The stirrer is stopped, and the liquid ammonia is allowed to evaporate overnight in the hood. When the flask comes to room temperature, the sides are cautiously rinsed down with small amounts of methanol and 200 ml of water are added. The product is recovered by extraction with ether, and the pooled ether extracts are washed with water until neutral. The extract is dried over Na_2SO_4, and the ether is removed under reduced pressure. To the residue is added 500 ml of methanol and 10 ml of sulfuric acid. The block-

[5] R. G. Jones and M. J. Mann, *J. Am. Chem. Soc.* **75,** 4048 (1953).

[6] K. N. Campbell and B. K. Campbell, *in* "Organic Synthesis" (N. Rabjohn, ed.), Vol. 4, p. 117. Wiley, New York, 1962.

ing group is removed by refluxing for 3 hr. The sulfuric acid is neutralized by addition of solid Na_2CO_3. Most of the methanol is removed under reduced pressure and the product is recovered by extraction with ether. Vacuum distillation yields the desired 6-heptyn-1-ol (bp 77–83°; 4 mm) in an overall yield of about 20–25% from the 1,5-pentanediol.

This procedure can be used for preparing ω-acetylenic alcohols containing 7 or more carbons. 3-Butyn-1-ol, 4-pentyn-1-ol, and 5-hexyn-1-ol are commercially available (Farchan Division, Story Chemical Corporation, Willoughby, Ohio). The methods used for the preparation of these short-chain ω-acetylenic alkyn-1-ols have been reviewed.[4]

Step 2. Synthesis of 1-Bromo-2,5,8,11-Tetradecatetrayne

1-Bromo-2,5,8,11-tetradecatetrayne is prepared according to the following reaction sequence.

$$CH_3—CH_2—C{\equiv}C—CH_2OH \xrightarrow{PBr_3} CH_3—CH_2—C{\equiv}C—CH_2Br$$

$$\xrightarrow[2.\ HC{\equiv}C—CH_2OH]{1.\ EtMgBr} CH_3—CH_2—(C{\equiv}C—CH_2)_2OH \xrightarrow{PBr_3} CH_3—CH_2—(C{\equiv}C—CH_2)_2Br$$

$$\xrightarrow[2.\ HC{\equiv}C—CH_2OH]{1.\ EtMgBr} CH_3—CH_2—(C{\equiv}C—CH_2)_3OH \xrightarrow{PBr_3} CH_3—CH_2—(C{\equiv}C—CH_2)_3Br$$

$$\xrightarrow[2.\ HC{\equiv}C—CH_2OH]{1.\ EtMgBr} CH_3—CH_2—(C{\equiv}C—CH_2)_4OH \xrightarrow{PBr_3} CH_3—CH_2—(C{\equiv}C—CH_2)_4Br$$

1-Bromo-2-Pentyne. A 3-liter three-neck round-bottom flask containing 147 g (1.75 mol) of 2-pentyn-1-ol (Farchan Division, Story Chemical Corporation, Willoughby, Ohio) in 1300 ml of ethyl ether is fitted with a condenser attached to a $CaCl_2$ drying tube and a dropping funnel. To the magnetically stirred solution is added 4.8 g of pyridine and 182 g (0.67 mol) of PBr_3 at a rate to maintain a gentle reflux. After addition is complete the reaction is refluxed for 3 hr and then poured on 500 ml of 2 N H_2SO_4 in ice. The water layer is extracted twice with ether and the pooled ether extracts are washed sequentially with water, 2 N Na_2CO_3, and water until neutral. Ether is removed under reduced pressure, and the desired 1-bromo-2-pentyne is recovered in a 75% yield by distillation at atmospheric pressure (bp 145–155°).

Although most 2-alkyn-1-ols are commercially available, they can also be prepared by coupling either a 1-bromoalkane or 1-iodoalkane with the sodium salt of the blocked derivative of propargyl alcohol in liquid ammonia. The boiling points of straight-chain 2-alkyn-1-ols and 1-bromo-2-alkynes have been summarized in tabular form in a review.[4]

Octa-2,5-diyn-1-ol. A 3-liter three-neck flask is equipped with a condenser, an overhead stirrer, and a dropping funnel. To the flask is added

92 g of Mg (3.8 mol) and 700 ml of dry tetrahydrofuran. A few crystals of iodine are added, and the Grignard reaction is initiated by the dropwise addition of ethyl bromide (381 g; 4.2 mol) in 300 ml of tetrahydrofuran. As soon as the reaction starts, as indicated by the loss of the yellow color, the flask is cooled to 0° by an ice bath and maintained at this temperature during the addition of the ethyl bromide (1–2 hr). The reaction is then stirred at room temperature for 2–3 hr, after which time the reaction is again cooled in an ice bath, and 108 g (1.9 mol) of propargyl alcohol in 100 ml of tetrahydofuran are added dropwise during the next 2 hr. Stirring is continued at room temperature for at least 4 hr or until ethane evolution ceases. The reaction mixture is again cooled in an ice bath, and 4.3 g of CuCN are added followed by 176 g (1.2 mol) of 1-bromo-2-pentyne in 100 ml of tetrahydofuran. The reaction is then stirred at room temperature for 48 hr, 2 g more of CuCN being added after 24 hr. The reaction mixture is poured into 2 *N* H_2SO_4 on ice and extracted three times with 300 ml of ether. The ether extract is washed with water and then at least six times with 2 *N* HCl to remove most of the CuCN. After a water wash, the organic phase is washed with 2 *N* Na_2CO_3 and finally with water until neutral. Ether is removed under reduced pressure and the octa-2,5-diyn-1-ol is recovered in a 65% yield by vacuum distillation (bp 77–82°/0.25 mm).

Acetylenic compounds containing two or more triple bonds are extremely susceptible to autoxidation and must be stored under nitrogen at −20°. If they are not to be used immediately it is advisable to seal them under argon and store at −70°.

1-Bromo-2,5-octadiyne (bp 79–85°/0.5 mm) is prepared in a yield of about 80% as described above for 1-bromo-2-pentyne. 2,5,8-Undecatriyn-1-ol (bp 130–135°/0.4 mm) is prepared in a yield of 45–50% as described for the synthesis of octa-2,5-diyn-1-ol.

1-Bromo-2,5,8-undecatriyne is synthesized in the usual way; however, we do not purify it by distillation, but immediately couple it with the di-Grignard complex of propargyl alcohol to give 2,5,8,11-tetradecatetrayn-1-ol in a yield of 28%. This alcohol is crystallized using the following protocol. A dark brown viscous oily residue is obtained after removal of the ether from the reaction workup. This residue is extracted repeatedly with aliquots of boiling hexane. Upon cooling to room temperature some oiling off will occur. Small amounts of ether are added until all the oil is dissolved. The solution is then heated and rapidly filtered through activated charcoal. If the solution is still dark yellow, the process is repeated. The filtrate is then rapidly cooled in an ice bath under nitrogen. The desired compound will crystallize out immediately. After an hour the crystals are recovered by vacuum filtration and immediately transferred to a standard-taper round-bottom flask. Trace amounts of solvent are removed under vacuum. The product is stored at $-70°$ under vacuum.

Step 3. Synthesis of Heneicosa-6,9,12,15,18-pentayn-1-ol

The di-Grignard complex of 6-heptyn-1-ol is prepared as described above for propargyl alcohol. This reaction mixture is then allowed to stand overnight at room temperature. The next morning 2,5,8,11-tetradecatetrayn-1-ol is converted to 1-bromo-2,5,8,11-tetradecatetrayne by addition of PBr_3 in the usual way except the reaction is stirred at room temperature for 2 hr. The product is rapidly isolated in the usual way and following addition of CuCN to the di-Grignard derivative of 6-heptyn-1-ol the 1-bromo-2,5,8,11-tetradecatetrayne is added in tetrahydrofuran and the reaction mixture is stirred at room temperature under nitrogen for 48 hr. Heneicosa-6,9,12,15,18-pentayn-1-ol (mp 74–75°) is recovered in a yield of 15% by crystallization from hexane–ether.

Step 4. Synthesis and Purification of Heneicosa-6,9,12,15,18-Pentaen-1-ol

To 25 ml of ethyl acetate is added 0.1 ml of synthetic quinoline, and 0.2 g of Lindlars catalyst[7] (Polysciences, Warrington, Pennsylvania). The reaction vessel is attached to a hydrogenator which is sequentially evacuated and flushed with hydrogen at least three times. To the reaction mixture is added 425 mg of heneicosa-6,9,12,15,18-pentayn-1-ol in 25 ml of ethyl acetate. The reaction mixture is stirred vigorously until hydrogen uptake ceases (5–10 min; with an uptake of 175 ml). The catalyst is removed by filtration and the ethyl acetate is removed under reduced pressure. The product is dissolved in ether and washed with 0.1 *N* HCl to remove quinoline and then with water until neutral. The heneicosa-6,9,12,15,18-pentaen-1-ol is purified by silicic acid column chromatography by eluting with 15% ether in hexane.

Even though reduction with Lindlar's catalyst is rapid and reasonably stoichiometric ($\pm$ 10%) some over reduction always occurs. The alcohol is thus converted to the acetate by stirring overnight at room temperature with 0.5 ml of pyridine in 2.5 ml of acetic anhydride. The acetate is then fractionated by argentation thin-layer chromatography (TLC) using 5% by weight of $AgNO_3$ in silica gel G by developing with ether–hexane (60:40, v/v). The desired acetate is recovered by extraction with $CHCl_3/CH_3OH/CH_3COOH/H_2O$ (50:39:1:10, v/v/v/v). To every 12 ml of extract is added 4 ml of 4 *N* NH_4OH. The bottom layer is washed once with

[7] H. Lindlar and R. Dubuis, *Org. Synth.* **46,** 89 (1966).

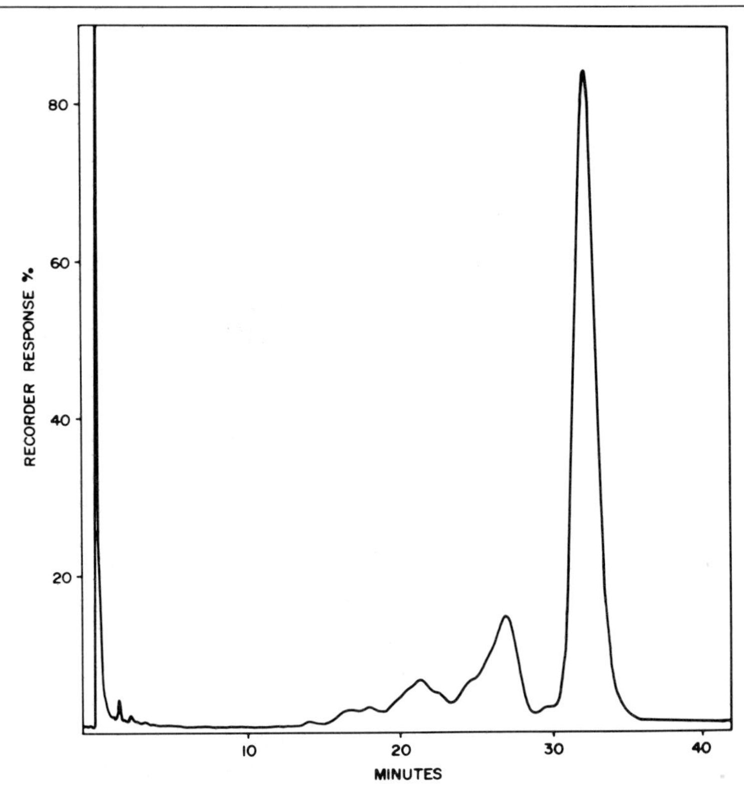

FIG. 1. Gas–liquid chromatogram of the acetate of 6,9,12,15,18-heneicosapentaen-1-ol prior to purification by argentation thin-layer chromatography. The glass columns (6 ft long; 2 mm internal diameter) were packed with 10% Altech CS-10 on 100/120-mesh Chromosorb W-AW. The column temperature was 180° with a nitrogen flow rate of 25 ml/min.

H_2O/CH_3OH (1 : 1, v/v) containing 5% NaCl using 4 ml of this solvent for every 12 ml of the original extract.[8]

Figures 1 and 2 show the purity of the acetate, as analyzed by gas–liquid chromatography (GLC), prior to and after purification by argentation TLC.

The acetate is then saponified by stirring overnight at room temperature with 4% KOH in EtOH–H_2O (9 : 1, v/v).

Step 5. Synthesis of 7,10,13,16,19-[1-^{14}C] *Docosapentaenoic Acid*

6,9,12,15,18-Heneicosapentaen-1-ol is converted to the mesylate by a modification of the procedure of Baumann and Mangold.[9] To 138 mg (0.46

[8] R. Sundler and B. Åkesson, *J. Biol. Chem.* **250,** 3359 (1975).
[9] W. J. Baumann and H. K. Mangold, *J. Org. Chem.* **29,** 3055 (1964).

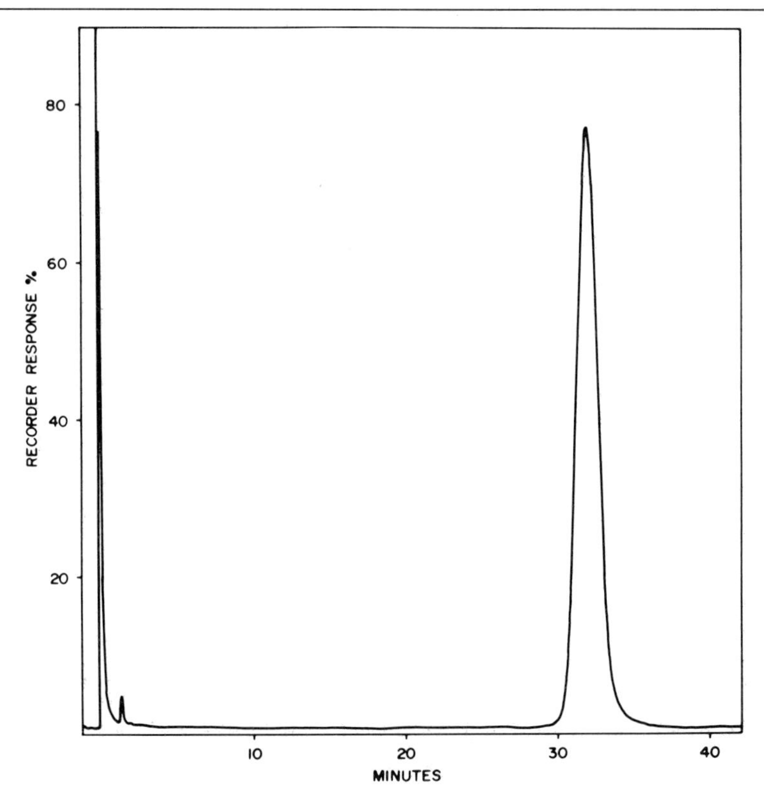

FIG. 2. Gas–liquid chromatogram of the acetate of 6,9,12,15,18-heneicosapentaen-1-ol after purification by argentation thin-layer chromatography. The glass columns (6 ft long; 2 mm internal diameter) were packed with 10% Altech CS-10 on 100/120-mesh Chromosorb W-AW. The column temperature was 180° with a nitrogen flow rate of 25 ml/min.

mmol) of alcohol in 1 ml of dry pyridine cooled in an ice bath is added a twofold molar excess of methane sulfonyl chloride. The reaction mixture is stirred at 0° for 4 hr, after which time 5 ml of water and 10 ml of ether are added. The aqueous layer is extracted 3 more times with 10 ml of ether. The ether is washed sequentially with 10 ml of water, 2 N H_2SO_4, water, 1% K_2CO_3, and finally with water until neutral. Water is removed by drying over anhydrous Na_2SO_4, and the ether is removed under nitrogen. The mesylate is dissolved in 2 ml of dry dimethyl sulfoxide which is added directly to a vial containing $Na^{14}CN$ (5 mCi; specific activity 49 Ci/mol). The reaction mixture is stirred for 2 hr at 80°, and the nitrile is recovered by extraction with ether.[10] Figure 3 shows a thin-layer radiochromatogram of this product. This crude nitrile is then added to 25 ml

[10] W. J. Baumann and H. K. Mangold, *J. Lipid Res.* **9**, 287 (1968).

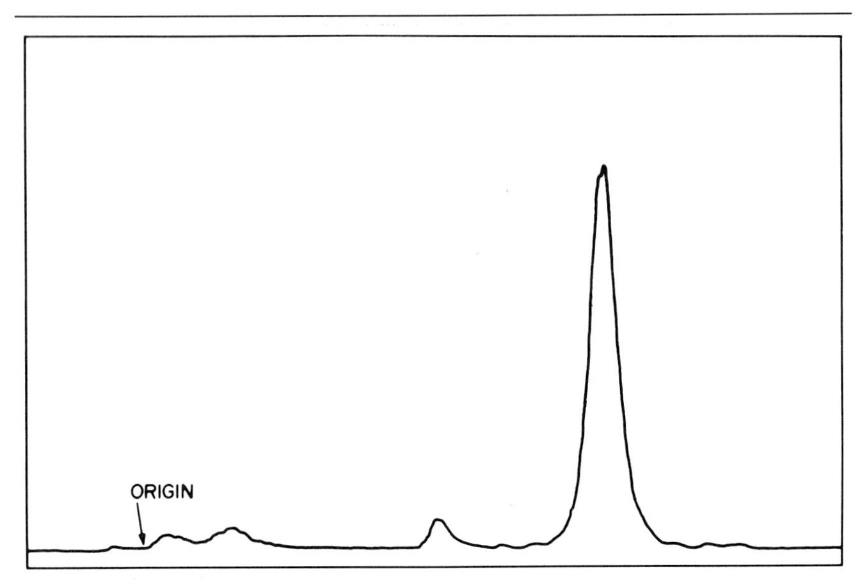

FIG. 3. Thin-layer radiochromatogram of the reaction products obtained after reaction of the mesylate of 6,9,12,15,18-heneicosapentaen-1-ol with $Na^{14}CN$. The solvent was petroleum ether–ether–acetic acid (80:20:2, v/v/v).

of anhydrous 25% HCl in methanol. The reaction is stirred for 2 hr at 0°, then 0.5 ml of water is added and stirring is continued overnight at room temperature. The reaction mixture is poured into 50 ml of ice water, and the methyl ester is recovered by extraction with ether. The methyl ester is purified by TLC by developing with hexane–ether–acetic acid (80:20:2; v/v/v). The methyl ester is recovered by extraction with ether and again passed through a column of silicic acid to remove silica gel particles.

Even though the alcohol used for the synthesis of a 1-^{14}C-labeled acids contains no detectable contaminants we frequently find small amounts of radioactive impurities when the methyl esters of radioactive fatty acids are analyzed by argentation TLC, GLC, or high-performance liquid chromatography (HPLC). The methyl ester is then again purified by argentation TLC. Figure 4 shows that the radiochemical purity of [methyl-1-^{14}C]7,10,13,16,19-docosapentaenoate is greater than 97% when analyzed by reversed-phase HPLC.[11]

After saponification the 7,10,13,16,19-[1-^{14}C]docosapentaenoic acid had a radiochemical purity greater than 97% when analyzed by either TLC or reversed-phase HPLC. The specific activity was 47.2 Ci/mol, and the overall radiochemical yield was 54%. Acids containing five or fewer

[11] M. VanRollins, M. I. Aveldaño, H. W. Sprecher, and L. A. Horrocks, this volume [62].

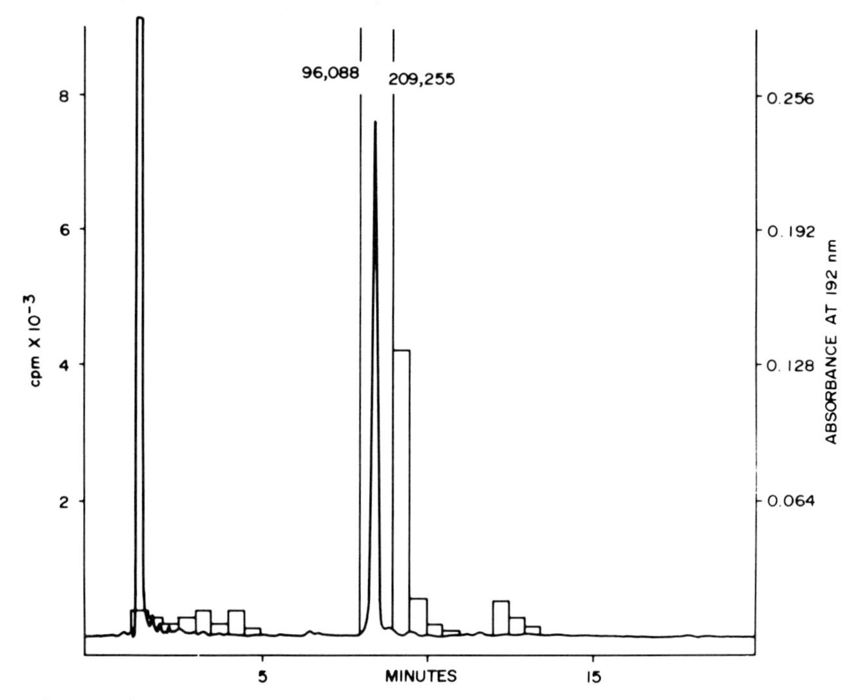

FIG. 4. High-performance liquid chromatogram of [methyl-1-^{14}C] 7,10,13,16,19-docosapentaenoate using a 25-cm Zorbax-ODS column. The solvent was acetonitrile–water (90:10, v/v) with a flow rate of 2 ml/min.

double bonds do not show any significant radiochemical decomposition for 1 year when sealed under argon in toluene–methanol (9:1, v/v) at −70°.

Tritiated Fatty Acids

Acids containing tritium are most frequently prepared by reducing acetylenic acids with tritium gas in the presence of Lindlar's catalyst. The desired acetylenic acids are prepared by coupling the appropriate substituted propargyl bromide with the di-Grignard complex of an ω-acetylenic acid. The ω-acetylenic acids are prepared according to the following reaction scheme.

$$HO—(CH_2)_x—OH \xrightarrow{SOCl_2} Cl—(CH_2)_x—Cl \xrightarrow{NaI} I—(CH_2)_x—Cl$$

$$\xrightarrow{NaC\equiv CH} HC\equiv C—(CH_2)_x—Cl \xrightarrow{NaCN} HC\equiv C—(CH_2)_x—CN$$

$$\xrightarrow{NaOH} HC\equiv C—(CH_2)_x—COOH$$

Many α-chloro-ω-iodoalkanes or α-chloro-ω-bromoalkanes are commercially available. If necessary they may be prepared by allowing α, -diols to react with thionyl chloride.[12] After purification by distillation, the dichloride is refluxed with NaI in acetone and the desired α-chloro-ω-iodoalkane is purified from residual starting material and the diiodoalkane by fractional distillation.[13] The iodine is then displaced by reaction with sodium acetylide in liquid ammonia.[6] The α-chloro-ω-alkynes are converted to nitriles by reaction with sodium cyanide in dimethyl sulfoxide.[14] Alkaline hydrolysis then converts the nitriles to ω-acetylenic acids.[15]

This procedure can readily be used to prepare ω-acetylenic acids when x is equal to or greater than 3. 4-Pentynoic acid is prepared by chromic acid oxidation of 4-pentyn-1-ol.[16]

The boiling and or melting points of the intermediates required for the synthesis of the most commonly required ω-acetylenic acids have been summarized in tabular form by Sprecher.[4]

Acknowledgments

The work described in this chapter was supported in part by U.S. Public Health Service NIH Contract N-01 HV82930 and NIH Grants AM 20387 and AM 18844.

[12] K. Ahmad, F. M. Bumpus, and F. M. Strong, *J. Am. Chem. Soc.* **70,** 3391 (1948).
[13] R. A. Raphael and F. Sondheimer, *J. Chem. Soc.*, p. 2100 (1953).
[14] R. A. Smiley and C. J. Arnold, *J. Org. Chem.* **25,** 257 (1960).
[15] W. Stoffel, *J. Am. Oil Chem. Soc.* **42,** 583 (1965).
[16] B. C. Holland and N. W. Gilman, *Synth. Commun.* **4,** 203 (1974).

[47] Preparation of Deuterated Arachidonic Acid

By DOUGLASS F. TABER, MARK A. PHILLIPS, and WALTER C. HUBBARD

The preparation of [$^2H_8$]arachidonic acid from the precursor tetra-acetylene eicosatetraynoic (ETYA) appears to be straightforward. Indeed, such an approach has been reported.[1,2] In our hands, however, the reported procedures gave rise to substantial double-bond isomerization and isotopic scrambling, leading to mixtures that yielded after purification only small quantities of the desired product. Our repeated need for multi-milligram quantities of [$^2H_8$]arachidonic acid led us to develop the method outlined below.

[1] U. H. Do, M. G. Sundaram, S. Ramachandran, and R. W. Bryant, *Lipids* **14,** 259 (1980).
[2] K. M. Green, M. Hamberg, B. Samuelsson, and J. C. Frölich, *Adv. Prostaglandin Thromboxane Res.* **5,** 44 (1978).

ISBN 0-12-181986-8

Procedure

General. ^{1}H NMR spectra were determined on a JEOLCO MH-100 spectrometer as solutions in $CDCl_3$. Chemical shifts are reported in δ units downfield from the internal reference tetramethylsilane. Couplings (J) are in hertz (Hz). The infrared spectra (IR) were recorded on a Perkin–Elmer 257 spectrometer as solutions in CCl_4 and are reported in reciprocal centimeters (cm^{-1}). Mass spectra (MS) were determined at 70 eV on an LKB 9000 gas chromatograph–mass spectrometer interfaced with a PDP-12 computer system and are reported as mass per unit charge (m/z), with intensities as a percentage of the peak of greatest ion current having $m/z \geq 100$. Organic chemicals were purchased from Aldrich Chemical Co. ETYA was a gift from Hoffmann-La Roche, Nutley, New Jersey, and was used as received. Solvent mixtures (e.g., 5% ethyl acetate–hexane) are by volume. R_f values indicated refer to thin-layer chromatography on Analtech 2.5 × 10 cm Uniplates coated with 250 μm silica gel GF. Column chromatography was carried out using the short-column technique,[3,4] modified by running the columns under air pressure (5–20 psig). We have found EM 7747 silica gel to be very effective in this application.

Preparation of Methyl [$^2H_8$]Arachidonate. A 100-ml 14/20 round-bottom flask equipped with a magnetic stirring bar and a water-cooled condensor was charged with ETYA (2.0 g, 6.76 mmol), methanol (50 ml), $BF_3 \cdot OEt_2$ (1 ml),[5] and methylene blue (25 mg). The mixture was flushed with N_2, stirred at reflux for 15 min under N_2, then allowed to cool. The resultant solution was diluted with 150 ml of 5% aqueous $NaHCO_3$ and extracted with CH_2Cl_2 (3 times, 100 ml). The organic extract was dried over K_2CO_3, filtered, and concentrated under vacuum to give a dark blue oil.

The reduction was carried out by a modification of the method of Brown.[6] Thus, a 250-ml flat-bottom hydrogenation vessel equipped with a sidearm inlet and a magnetic stirring bar was charged with $NiOAc \cdot 4H_2O$ (858 mg, 2.0 mmol) and 30 ml of methanol-OD (CH_3OD, Aldrich). The flask was attached to an atmospheric pressure hydrogenator and purged three times with deuterium gas. The stirrer was started. Then $NaBD_4$ (92 mg, 2.2 mmol; Stohler) in 1 ml of D_2O [to which previously had been added a small sliver of Na (ca. 5 mg)] was added in one portion, leading to instantaneous gas evolution and the formation of a thick black precipitate. After gas evolution had subsided, 1 ml of ethylenediamine was added,

[3] B. J. Hunt and W. Rigby, *Chem. Ind. (London)*, p. 1868 (1967).

[4] W. C. Still, M. Kahn, and A. Mitra, *J. Org. Chem.* **43,** 2923 (1978).

[5] P. K. Kadaba, *Synthesis* 1971, p. 316 (1971).

[6] C. A. Brown and V. K. Ahuja, *J. Chem. Soc. Chem. Commun.*, p. 553 (1973).

and stirring continued for 5 min. The stirrer was stopped, and blue oil from above was added, stirring was resumed, and gas uptake was monitored. After 120 min gas uptake had ceased, the total consumption being 615 ml at 26°. The resultant mixture was filtered through a pad of Celite with CH_2Cl_2. The CH_2Cl_2 solution was washed with 5% aqueous HC1, dried over K_2CO_3, filtered, and evaporated to give a residual blue oil that showed two main spots on TLC, R_f (10% EtOAc–hexane) = 0.41 and 0.46. The less polar spot cochromatographed with authentic nondeuterated methyl arachidonate.

The oil was redissolved in CH_2Cl_2 and 6 g of 60–100-mesh coarse-grade sililic acid was added. This sample was dried under vacuum and added to the top of a 50-g TLC mesh silica gel column equilibrated with 1.5% EtOAc–petroleum ether. The column was eluted with the same solvent and the first 400 ml were discarded. The next 200 ml were concentrated under vacuum to give the desired product as a colorless oil, 850 mg (2.6 mmol, 39%); R_f (10% EtOAc–hexane) = 0.46. ^{1}H NMR (δ): 0.90, t, J = 7, 3H; 1.3, m, 6 H; 1.69, quint, J = 7, 2 H; 2.05, m, 4H; 2.32, t, J = 7, 2 H; 2.82, s, 6 H; 3.67, s, 3 H. IR (cm^{-1}): 2920, 2850, 2240, 1730, 1615, 1428, 1150, 980. MS (m/z, %): 326 (17), 281 (25), 207 (64), 154 (81), 110 (100). This material gave a single peak on GC (2 m × 2 mm 3% OV-1, 215°, 20 ml/min, C = 19.7), identical with authentic, nondeuterated methyl arachidonate. The more polar material gave a molecular ion at m/z = 322, showing it to be partially deuterated material.

Preparation of $^2H_8$-Arachidonic Acid. A 50-ml round-bottom flask equipped with a magnetic stirring bar was charged with 278 mg (1.16 mmol) of methyl [$^2H_8$]arachidonate, 15 ml of dimethoxyethane, and 3 ml (9 mmol, 8 eq) of 3 N aqueous LiOH. The mixture was flushed with N_2, then stirred at 60° under an N_2 atmosphere for 2 hr. The reaction mixture was diluted with 5% aqueous hydrochloric acid and extracted with CH_2Cl_2. The organic extracts were combined, dried over Na_2SO_4, filtered, and concentrated under vaccuum.

The residue was taken up in methanol and chromatographed on a Waters μBondapak C_{18} 7.8 mm i.d. × 30 cm semipreparative column, eluting at 3 ml/min with (MeOH–H_2O–HOAc, 80:20:0.01, v/v/v). Crude acid corresponding to about 55 mg of starting ester was injected in each run, along with a tracer amount of tritiated arachidonic acid. Fractions of 3 ml were collected, and an aliquot of each was counted. The arachidonic acid usually appeared in fractions 52–64, with 95% recovery of applied radioactivity. The column was eluted with 90 ml of MeOH after each run, then reequilibrated with 100 ml of the MeOH–H_2O–HOAc eluent.

The pooled arachidonic acid-containing fractions were concentrated under vaccuum, diluted with 3% aqueous HCl, and extracted with CH_2Cl_2. The organic extracts were combined, dried over Na_2SO_4, filtered, and concentrated under vacuum to give the desired product as a colorless oil, 228 mg (0.73 mmol, 63%). Analysis of the methyl ester by GC–MS–SIM showed the molecular ion to be 326 (d_8), with a d_0 peak of 30 parts per thousand; an isotope distribution of 2% d_5, 7% d_6, 24% d_7, 58% d_8, 6% d_9, and 3% d_{10} was observed. Most of the d_0 peak was due to minor acetylenic impurities, as demonstrated by conversion[7] to 12-HETE, which showed a d_0 contamination (295/301) of 3.2 parts per thousand.

Discussion

We had felt that the extensive double-bond isomerization and isotopic scrambling observed in our previous preparations of deuterated arachidonic acid were probably due to the formation of intermediate π-allyl olefin complexes with adventitious acidic transition metal species. If the complex-forming metal ions were not the ones responsible for effecting catalytic hydrogenation, then inclusion of a strongly complexing Lewis base should allow cleaner conversion of the tetraacetylene to the desired product. To this end, ethylenediamine[6] was included in the reduction medium, and it proved to be of substantial benefit. The deuterated arachidonic acid prepared as outlined above appears to be of sufficient purity for chemical and biological transformation to the physiologically interesting oxygenated metabolites.

Acknowledgments

This investigation was supported by the National Institutes of Health grant GM 15431. We acknowledge support from Biomedical Research Support Grant RR-05424 for MS and NMR instrumentation. We thank Dr. James Hamilton of Hoffmann La Roche for a generous gift of ETYA.

[7] J. M. Boeynaems, A. R. Brash, J. A. Oates, and W. C. Hubbard, *Anal. Biochem.* **104,** 259 (1980).

[48] Preparation and Analysis of Radiolabeled Phosphatidylcholine and Phosphatidylethanolamine Containing ^{3}H- and ^{14}C-Labeled Polyunsaturated Fatty Acids

By HARUMI OKUYAMA and MAKOTO INOUE

Phosphatidylcholines with ^{14}C- or ^{3}H-labeled unsaturated fatty acids at the 2 position are commercially available. These include molecular species with oleic, elaidic, linoleic, eicosatrienoic, and arachidonic acids at the 2-position (Radiochemical Center and/or New England Nuclear). Since the difference in prices of these phosphatidylcholines and labeled precursor acids is relatively small, it is not worth preparing them in a laboratory scale from commercially available labeled fatty acids. When not available commercially, phosphatidylcholines with labeled fatty acids at either the 1 or the 2 position can be synthesized either by chemical acylation of monoacylglycerophosphocholine with fatty acid anhydrides or acyl chlorides[1,2] or by enzymic acylation with fatty acyl-CoAs.[3] The chemical methods have the advantage of yielding those with defined fatty acids, but involve synthesis on a larger scale and thus, require more labeled fatty acids. Furthermore, the yields in chemical syntheses of phosphatidylethanolamines with defined fatty acids have been relatively low in our hand. The enzymic method is suitable for biochemistry laboratories, but the products are contaminated with endogenous phosphatidylcholine or phosphatidylethanolamine from the enzyme preparation. This drawback of the latter method is minimized by using deoxycholate-treated microsomes as the enzyme source, as described below.

Principle. The reaction follows the following equations.

$$[^{14}\text{C}]\text{acyl-CoA} + \text{1-acyl-GPX} \rightarrow \text{1-acyl-2-}[^{14}\text{C}]\text{acyl-GPX} + \text{CoASH}$$
$$[^{14}\text{C}]\text{acyl-CoA} + \text{2-acyl-GPX} \rightarrow \text{1-}[^{14}\text{C}]\text{acyl-2-acyl-GPX} + \text{CoASH}$$

where GPX denotes *sn*-glycerol-3-phosphocholine or *sn*-glycerol-3-phosphoethanolamine. Reactions are followed continuously in the presence of DTNB by measuring the absorbance at 413 nm (ϵ_{413} 13,600).

[1] A. E. Brandt and W. E. M. Lands, *Biochim. Biophys. Acta* **144** 605 (1967).
[2] E. C. Robles and D. van den Berg, *Biochim. Biophys. Acta* **187,** 520 (1969).
[3] W. E. M. Lands and P. Hart, *J. Biol. Chem.* **240,** 1905 (1965).

ISBN 0-12-181986-8

Preparation of Labeled Unsaturated Acyl-CoAs According to a Modified[4,5] Seubert's Procedure[6]

Besides the labeled polyunsaturated fatty acids described above, [1-^{14}C]eicosapentaenoic acid (*n*-3) is available from Rosechem Products (2902 Gilroy St., Los Angeles, California 90039). In a typical reaction mixture, the molar ratio of fatty acid to CoA is ≥ 5. The yields are usually higher than 60% of the CoA used. To conserve the labeled fatty acids, the fatty acid:CoA molar ratio is decreased. A typical experiment with [^{14}C]arachidonic acid is described below.

Reagents

Labeled fatty acids; commercially available ^{14}C- or ^{3}H-labeled fatty acids are diluted with nonlabeled fatty acids to the desired specific radioactivities.

Santoquin (Ethoxyquin; 6-ethoxy-1,2-dihydro-2,2,4-trimethylquinoline, Monsanto Chemical Co.); purified by distillation under reduced pressure, bp 123–125°. It is a colorless or faint yellow liquid. Store in freezer.

Oxalyl chloride, distilled; stored in brown ampoules

Tetrahydrofuran, freshly distilled over excess $LiAlH_4$

CoA, commercially available

Procedure. Free fatty acid (1.5 mg of arachidonic acid, 3000 cpm/nmol) including Santoquin as antioxidant in an amount of 1% of the acid, is converted to acyl chloride by adding about 0.5 ml of oxalyl chloride and removing the excess oxalyl chloride completely with nitrogen. To avoid losing volatile arachidonoyl chloride by nitrogen, the temperature is kept below 25°. The removal of oxalyl chloride is judged by its odor. This step is repeated once. The oily residue is dissolved in 0.5 ml of freshly distilled tetrahydrofuran. The acid chloride solution is added dropwise to a solution of CoA (7 mg) in 5 ml of tetrahydrofuran–water (7:3), while the pH is maintained at about 8 (pH test paper) with 1 *N* NaOH and the solution is stirred vigorously with a magnetic bar. After adjusting the pH to 4–5 with a few drops of 10% $HClO_4$, the mixture is transferred to a 15-ml culture tube having a Teflon-coated sealer in the cap. The tetrahydrofuran is removed with nitrogen, and the volume of the water phase is adjusted to 2 ml by adding water. After adding 5 ml of light petroleum

[4] H. Okuyama, W. E. M. Lands, W. W. Christie, and F. D. Gunstone, *J. Biol. Chem.* **244,** 6514 (1969).

[5] R. C. Reitz, W. E. M. Lands, W. W. Christie, and R. T. Holman, *J. Biol. Chem.* **243,** 2241 (1968).

[6] W. Seubert, *Biochem. Prep.* **7,** 80 (1960).

ether and mixing vigorously, 0.2 ml of 10% $HClO_4$ is added while cooling in ice bath. The mixture is centrifuged at 3000 rpm for 3 min with the tubes capped, and the clear ether and water phases are removed carefully. Sometimes, bulky fluff is formed in the ether phase, but it can be broken by tapping the tube on the hand. The white precipitate in the interphase and on the wall is collected by centrifugation and washed twice each with 5 ml of petroleum ether. The remaining ether is removed by nitrogen, and the residue is dissolved in 1 ml of water. After adjusting the pH to 4–5 with 5% $NaHCO_3$, the solution is centrifuged at 4000 rpm for 5 min. The insoluble material floats at the top and the clear solution is taken carefully so as not to disturb the floating layer. If the final solutiion is not water-clear, the petroleum ether washings and the acid precipitation steps must be repeated. The yield is 33% of the arachidonic acid used. Santoquin in 5 μl of acetone is dispersed in the acyl-CoA solution in an amount of 1% of the acyl-CoA. Acyl-CoAs can be kept frozen at least for several months.

Determination of Acyl-CoAs. Since Santoquin interferes with the usual spectrophotometric assay of acyl-CoA (ϵ_{260} 15,400 liter M^{-1} cm^{-1} for the absorption of adenine and ϵ_{232} 9400 liter M^{-1} cm^{-1} for the thioester bond), the concentration of acyl-CoA is determined by following the CoA release using either the 1-acylglycerophosphocholine acyltransferase system (see below) or partially purified acyl-CoA hydrolase [precipitate of ammonium sulfate fractionation (40–80%) of rat brain high-speed supernatants]. Fatty acid has been confirmed by gas chromatography to show a single peak. Phosphorus is determined by the Eibl and Lands method.[7] Any of these assays can be used to determine the concentration of acyl-CoA, as long as the product is determined to contain the appropriate thioester linkage.

Preparation of Deoxycholate-Treated Microsomes[4,8]

Rat liver microsomes are prepared by collecting the particulate fraction that sediments between 20,000 *g* (20 min) and 100,000 *g* (90 min) from tissue homogenates in 10 volumes of 0.25 *M* sucrose–1 m*M* EDTA (pH 7.4). Usually 120 mg of microsomal protein are obtained from 10 g of wet tissue. Freshly prepared microsomes (60 mg of protein) suspended in 3 ml of 0.25 *M* sucrose containing 1 m*M* EDTA (pH 7.4), are mixed with 20 mg of sodium deoxycholate in 1 ml of 0.25 *M* sucrose–1 m*M* EDTA (pH 7.4). After being stirred for 10 min in an ice bath, 60 mg of bovine

[7] H. Eibl and W. E. M. Lands, *Anal. Biochem.* **30,** 51 (1969).

[8] H. Okuyama, H. Eibl, and W. E. M. Lands, *Biochim. Biophys. Acta* **248,** 263 (1971).

serum albumin in 2 ml of 0.25 *M* sucrose–1 m*M* EDTA (pH 7.4) are added, and the mixture is layered over a sucrose-density gradient made stepwise with 2 ml each of 1 *M* sucrose–1 m*M* EDTA (pH 7.4), 0.5 *M* sucrose–1 m*M* EDTA (pH 7.4), and 0.25 *M* sucrose–1 m*M* EDTA (pH 7.4). After centrifugation at 100,000 *g* for 90 min, the 0.5 *M* sucrose fraction and the cloudy part of 0.25 *M* sucrose fraction are diluted with 2 volumes of water and centrifuged at 100,000 *g* for 90 min. The pellets (11.8 mg and 5.4 mg of protein from 0.25 *M* sucrose and 0.5 *M* sucrose fractions, respectively) are resuspended in 2–3 ml of 0.25 *M* sucrose–1 m*M* EDTA (pH 7.4) and kept in liquid nitrogen. The oleoyl-CoA:1-acylglycerophosphocholine acyltransferase activities in these two fractions are at least 65 nmol/min per milligram of protein at 25°, and no measurable acyl-CoA hydrolase activities have been detected.

Preparation of Monoacylglycerophosphocholine and Monoacylglycerophosphoethanolamine, and Analysis of the Positional Distribution of Fatty Acids in Phospholipid Molecules

1-Acyl isomers are prepared by hydrolysis of phosphatidylcholine and phosphatidylethanolamine with snake venom phospholipase A_2. This reaction is also used to determine the positional distribution of fatty acids in phospholipids. Although the enzyme in snake venom is relatively stable and activity in crude venom is high, the incubation conditions must be examined carefully to avoid overhydrolysis.[8,9] Using excess enzyme with small amounts of phosphatidylcholine often yields free fatty acids and glycerophosphocholine, but very little 1-acylglycerophosphocholine.

Reagents

- Snake venom (*Crotalus adamanteus;* Sigma)
- Silica gel H
- Diethyl ether; when necessary, peroxide is removed by distillation over $LiAlH_4$
- Chloroform and methanol; distilled without pretreatment

Procedure. To 1 mg of phosphatidylcholine dissolved in 0.5 ml of diethyl ether is added 0.25 ml of 0.1 *M* Tris-HCl (pH 7.4), 0.25 ml of 0.1 *M* sodium borate (pH 7.4), and 50 μl of enzyme solution containing 0.25 mg of snake venom and 0.2 μmol of $CaCl_2$. The mixture in a culture tube is tightly sealed using a cap with a Teflon liner and is incubated for 0.5 hr at 37° while shaking vigorously with an evapomixer. Phosphatidylethanolamine is hydrolyzed completely in 1 hr with 0.5 ml of enzyme so-

[9] K. Yamada, H. Okuyama, Y. Endo, and H. Ikezawa, *Arch. Biochem. Biophys.* **183,** 281 (1977).

lution. The reaction is terminated by adding 20 μl of 0.1 *M* EDTA (pH 7.4). The lipid products are separated by silica gel H thin-layer chromatography with chloroform–methanol–acetic acid–water (25:15:4:2 v/v/v/v) as the developing (system A) solvent after extraction from the reaction mixture with chloroform–methanol or by solvent fractionation.

Liberated fatty acids are separated completely by extracting the reaction mixture 7 times each with 5 ml of diethyl ether; 1-acylglycerophosphocholine or 1-acylglycerophosphoethanolamine remaining in the water phase is extracted according to Bligh and Dyer's method[10] in a yield of at least 85%.

Preparation of 2-Acylglycerophosphocholine and 2-Acylglycerophosphoethanolamine

2-Acyl isomers are obtained either by I_2 hydrolysis of the corresponding plasmalogen[8,11] or by *Rhizopus* lipase hydrolysis of diacylglycerophospholipids.[12] Since the 2-acyl isomers easily isomerize to 1-acyl isomers, these must be prepared just before use. Choline plasmalogen and ethanolamine plasmalogen are purified from beef heart phospholipids by silicic acid column chromatography. Diacyl phospholipids are selectively removed by hydrolysis with cabbage phospholipase D.[8] The remaining plasmalogen-rich choline and ethanolamine phospholipids are treated as follows.

Reagents

I_2
$Na_2S_2O_3$
Boric acid

Procedure. A lipid sample containing 2.5 μmol of plasmalogen is suspended in 1.5 ml of petroleum ether and 1.5 ml of 0.1 *M* boric acid (pH is adjusted to 5.5). While shaking the mixture vigorously with a Vortex mixer, light petroleum ether saturated with I_2 is added dropwise until the iodine color remains for a few minutes. After adding 1.5 ml of methanol, the solution is left standing for 5 min with occasional shaking. Excess iodine, iodoaldehyde, and unhydrolyzed diacylglycerophospholipid are removed by 7 extractions with 10 ml of petroleum ether–diethyl ether (1:1, v/v). To the remaining lower layer containing 2-acylglycerophospholipid, 2 drops of 0.05 *M* $Na_2S_2O_3$ solution are added, and the solution is washed once with 10 ml of petroleum ether–diethyl ether (1:1, v/v). The 2-acyl

[10] E. G. Bligh and W. J. Dyer, *Can. J. Biochem. Physiol.* **37,** 911 (1959).

[11] H. Eibl and W. E. M. Lands, *Biochemistry* **9,** 423 (1970).

[12] A. J. Slotboom, G. H. de Haas, P. P. M. Bonsen, G. J. Burbach-Westerhuis, and L. L. M. van Deenen, *Chem. Phys. Lipids* **4,** 15 (1972).

isomer is then extracted from the aqueous solution with 8 ml of chloroform followed by 8 ml of chloroform–methanol (2:1), v/v). The combined chloroform extracts are washed with 2 ml of 0.1 *M* boric acid, and the chloroform solution is evaporated to dryness under nitrogen. The residue is suspended in 2 ml of 0.1 *M* boric acid or water. Recovery of 2-acylglycerophospholipids is about 70–80% of the starting plasmalogen. The product shows a single spot on silica gel H thin-layer chromatography with chloroform–methanol–acetic acid–water (25:15:4:2, v/v/v/v) as solvent.

Preparation of Phosphatidylcholine and Phosphatidylethanolamine with ^{3}H, ^{14}C-Labeled Polyunsaturated Fatty Acids

Both substrates of the acyltransferase system are detergents, and bulk addition of substrates to microsomes inactivates the enzyme. Hence, the substrates must be added little by little to obtain maximum yields.

Reagents

DTNB [5,5′-dithiobis-(2-nitrobenzoic acid)], 10 m*M* solution prepared by dissolving solid DTNB while adjusting the pH to 5.0–5.5

Procedure. For the preparation of phosphatidylcholine labeled at the 2-position, the incubation mixture in a cuvette consists of 150 μM 1-acylglycerophosphocholine, 1 m*M* DTNB, and deoxycholate-treated microsomes (0.1 mg of protein per milliliter) in 1 ml of 0.08 *M* Tris-HCl (pH 7.4). The reaction is started by adding 25 nmol of acyl-CoA, and the release of CoA is followed continuously by measuring the absorbance at 413 nm. When all the acyl-CoA is consumed (approximately after 4 min at 25°), a further 25 nmol of acyl-CoA are added and the incubation is continued. Then 150 nmol of 1-acylglycerophosphocholine and 25 nmol of acyl-CoA are added. This step is repeated until after 1 hr a total of 200 nmol of acyl-CoA and 375 nmol of 1-acylglycerophosphocholine have been added to the sample and the enzyme still retains one-third of its original activity. The radioactive product is purified by silica gel H thin-layer chromatography with the solvent system A. Labeled phosphatidylcholine is extracted from the silica gel once with 2 ml of chloroform–methanol–3% NH_4OH (6:5:1, v/v/v) and twice with 2 ml of chloroform–methanol (2:1, v/v). The combined extracts are centrifuged at 4000 rpm for 5 min to remove fine powders of silica gel, and the solvent is evaporated by nitrogen.

Typically phosphatidylcholine is obtained in a yield of 75% of the added acyl-CoA. The microsomal enzyme preparation used contains 12 nmol of phosphatidylcholine and 7 nmol of phosphatidylethanolamine, which comprises less than 7% of the product. Of the radioactivity found in phosphatidylcholine, 98% is determined to be at the 2 position.

For the preparation of phosphatidylcholine labeled at the 1 position, an incubation similar to that just described is carried out using freshly prepared 2-acyl-glycerophosphocholine at an initial concentration of 80 μM. When [^{14}C]stearoyl-CoA is used, 99% of the radioactivity is found at the 1 position; 83% of the radioactivity is found at the 1 position when [^{14}C]eicosatrienoyl-CoA is the substrate.

For the preparation of phosphatidylethanolamine labeled at the 1- or 2-positions, both 1-acyl- and 2-acylglycerophosphoethanolamine must be prepared immediately before use.[13] Incubations performed as described above but with 1-linoleoylglycerophosphoethanolamine yield phosphatidylethanolamine with 97 and 78% of radioactivities at the 2 position when using [^{14}C]eicosatrienoyl-CoA and [^{14}C]stearoyl-CoA, respectively. When 2-acylglycerophosphoethanolamine is the acceptor, radioactivities at the 1 position are 100% and 71% with [^{14}C]stearoyl-CoA and [^{14}C]eicosatrienoyl-CoA, respectively.[14]

[13] E. L. Gottfried and M. M. Rapport, *J. Biol. Chem.* **237,** 329 (1962).
[14] H. Okuyama, W. E. M. Lands, F. D. Gunstone, and J. A. Barve, *Biochemistry* **11,** 4392 (1972).

[49] Preparation of PGG_2 and PGH_2

By GUSTAV GRAFF

The current interest in studying the metabolism of prostaglandin endoperoxides and their pharmacological effects on vascular smooth muscle has created a demand for these substances. However, the instability of prostaglandin endoperoxides in aqueous solution[1,2] seems not only to have prevented its commercial production, but also has discouraged many laboratories from generating their own supply of these compounds.

The method described here is based on the short-time incubation of arachidonic acid with a suspension of lipid-depleted sheep vesicular microsomes and differs only slightly from previously published methods.[1-4] The key step in the present method is the rapid extraction of the microsomal suspension with precooled (−40°) solvents. This minimizes both fur-

[1] D. H. Nugteren and E. Hazelhof, *Biochim. Biophys. Acta* **326,** 448 (1973).
[2] M. Hamberg, J. Svensson, T. Wakabayashi, and B. Samuelsson, *Proc. Natl. Acad. Sci. U.S.A.* **71,** 345 (1974).
[3] F. B. Ubatuba and S. Moncada, *Prostaglandins* **13,** 1055 (1977).
[4] G. Graff, J. H. Stephenson, D. B. Glass, M. K. Haddox, and N. D. Goldberg, *J. Biol. Chem.* **253,** 7662 (1978).

ISBN 0-12-181986-8

ther biochemical metabolism and chemical decomposition of the endoperoxides. Under anhydrous conditions below −20°, the prostaglandin endoperoxides are stable for at least several years.

Materials

[1-^{14}C]5,8,11,14-Eicosatetraenoic (arachidonic) acid (Amersham)
5,8,11,14-Eicosatetraenoic acid (Nu Chek Prep)
12-Hydroxy-9-octadecenoic (ricinoleic) acid (Nu Chek Prep)
Prostaglandin A_2 (PGA_2; Sigma Chemical Co.)
Prostaglandin B_2 (PGB_2; Sigma Chemical Co.)
Prostaglandin E_2 (PGE_2; Sigma Chemical Co.)
Prostaglandin $F_{2\alpha}$ ($PGF_{2\alpha}$; Sigma Chemical Co.)
0.2 *M* Sucrose–1 m*M* EDTA, pH 7.4
Tris-HCl buffer, 0.1 *M*, pH 8.0
1 μM Hematin (Sigma Chemical Co.), 1 m*M* *p*-hydroxymercuribenzoate (Sigma Chemical Co.), 0.75% Tween 40, 10% glycerol in 100 m*M* Tris-HCl, pH 8.0
Petroleum ether (bp 30–60°), treated with concentrated sulfuric acid, washed with distilled water, and distilled over lithium aluminum hydride.
Ethyl acetate, spectral grade
Silicic acid, 100-mesh (Mallinckrodt); 500 g, washed 5 times each with 5 liters of distilled water. Fine particles not sedimenting over a period of 5 min are discarded. The washed silicic acid is activated in an oven for 12 hr at 150° and stored over phosphorus pentoxide in a desiccator.
Other chemicals used are of analytical quality and are obtained from Mallinckrodt, J. T. Baker, or Fisher Scientific Co.

Preparation of Sheep Vesicular Microsomes

Thaw 500 g of vesicular glands from noncastrated sheep in 2 liters of sucrose–EDTA solution (4°). Remove any fat from the glands and homogenize them with a Waring blender in 750 ml of ice cold sucrose–EDTA solution. Filter the homogenate through cheesecloth and centrifuge the filtrate for 20 min at 12,000 *g*. Reextract the pellet by homogenization with 500 ml of sucrose–EDTA solution and centrifuge at 12,000 g for 20 min. Combine the supernatant fractions and centrifuge at 100,000 *g* for 60 min. Resuspend the microsomal pellet in 80 ml of glass-distilled water, and homogenize the microsomal suspension with a Potter–Elvehjem homogenizer. Transfer the homogenate into a 1-liter round-bottom flask, freeze the homogenate in a Dry Ice–acetone bath, and lyophilize overnight. The residue (6.7 g) is powdered, and residual water is removed by drying over phosphorus pentoxide in a desiccator. Lipid is removed from the microsomal powder (6.7 g) at $-20°$ by four successive extractions with 150 ml each of *n*-butanol, acetone, and finally diethyl ether using a Brinkmann Polytron homogenizer. Separation of organic solvent from the microsomal suspension is accomplished by centrifugation at 2000 rpm in a Sorvall RC-2 table-top centrifuge in a 4° cold room. Remove any residual solvent from the lipid-depleted microsomes using high vacuum. Store the lipid-depleted microsomal powder in a tightly capped container at $-70°$. This enzyme preparation has been found to remain active for up to 4 years of storage.

Generation and Extraction of Prostaglandin Endoperoxides

Because of the lack of a specific inhibitor(s) of the hydroperoxidase,[5-7] generation of PGG_2 has thus far been possible only in mixture with PGH_2. However, if PGH_2 alone is desired, full advantage of the peroxidase activity can be taken by including a peroxidase cosubstrate such as phenol[7] in the incubation medium.

The incubation conditions described below are designed to generate both PGG_2 and PGH_2, or PGH_2 alone. Dilute 5 μCi of [1-^{14}C]arachidonic acid with 5 μmol of unlabeled fatty acid, evaporate the solvent, and add 4.0 ml of 100 m*M* Tris-HCl buffer, pH 8.0. Disperse the fatty acid by sonication for 2 min at 20° in a Branson bath-type sonicator. Homogenize 300 mg of lipid-depleted microsomal powder with a Potter–Elvehjem homogenizer in 4.0 ml of ice cold 100 m*M* Tris-HCl buffer, pH 8.0, containing 2 m*M* *p*-hydroxymercuribenzoate, 2 μM hematin, 0.75% Tween 40, and 10% glycerol. Initiate the cyclooxygenase reaction by adding 1.0 ml of the microsomal suspension to the arachidonic acid solution that is preequilibrated at 37°. After 15 sec of incubation, carried out with constant stirring and supply of oxygen gas (30 ml/min), add the remaining 3.0 ml of microsomal suspension. Continue the incubation for an additional 45 sec while bubbling oxygen gas through the sample. The supply of oxygen gas is necessary, since the initial oxygen concentration (230 μM)[8] in the incubation medium would be limiting.

Terminate the reaction by pouring the reaction mixture into a separatory funnel containing 100 ml of ethyl acetate–petroleum ether (1 : 1, v/v, precooled to approximately $-40°$) followed by the addition of 4.0 ml of

[5] F. J. van der Ouderaa, M. Buytenhek, D. H. Nugteren, and D. A. van Dorp, *Biochim. Biophys. Acta* **487,** 315 (1977).

[6] T. Miyamoto, N. Ogino, S. Yamamoto, and O. Hayaishi, *J. Biol. Chem.* **251,** 2629 (1976).

[7] M. E. Hemler, C. G. Crawford, and W. E. M. Lands, *Biochemistry* **17,** 1772 (1978).

[8] H. W. Cook, G. Ford, and W. E. M. Lands, *Anal. Biochem.* **96,** 341 (1979).

1 *M* citric acid. Shake the mixture vigorously, collect the organic phase, and keep the extract on Dry Ice. Reextract the aqueous phase once more with 100 ml of ethyl acetate–petroleum ether (−40°). Add 25 g of sodium sulfate to the combined organic extract and keep it on Dry Ice for 20 min. Remove sodium sulfate and ice crystals by vacuum filtration through a fritted-glass funnel. During filtration, the temperature of the extract is maintained between −10 and −20°. Transfer the filtrate into a 1-liter round-bottom flask and concentrate on a rotary evaporator under high vacuum to a final volume of about 1–2 ml. Add 50 ml of 10% ethyl acetate in petroleum ether (−40°) and reduce the volume to 5 ml under high vacuum. Repeat the latter step once more, and store the extract at −70°. Avoid heating the rotary flask during evaporation. This extraction procedure results in an overall isotope recovery of 90% ±9, with 17% ± 9 distributed in the aqueous phase and 74% ± 10 in the organic phase.

TLC Analysis of Reaction Products

TLC analysis of PGH_2 synthase reaction products is carried out in a cold room with TLC plates and eluting solvents equilibrated at 4°.

Apply an aliquot of the extract (~20,000 cpm) onto a silica gel G plate (0.25 mm thick; Mallinckrodt). In separate lanes spot 5 nmol each of PGA_2, arachidonic and ricinoleic acids as reference standards. Place the TLC plate into a TLC tank lined with filter paper (10 × 55 cm), and develop the plate to 15 cm above the sample origin with the solvent system consisting of ethyl acetate–2,2,4-trimethylpentane–petroleum ether–acetic acid (50:50:20:0.5, v/v/v/v). Remove the TLC plate from the tank, evaporate the solvent in a fume hood, and expose the plate to iodine vapor in order to visualize the separated components. Remove the iodine from the plate by heating it for 20 min in an 110° oven. Locate the radioactive components on the TLC plate with a radioisotope TLC scanner. An alternative route is to fractionate the sample into appropriate sections and locate the radioisotope-containing components by liquid scintillation counting.

A typical separation for PGH_2 synthase reaction products of incubations conducted in the absence of phenol is shown in Fig. 1. Unreacted arachidonic acid migrates with an R_f value of 0.63, and PGG_2 and PGH_2 migrate with R_f values of 0.41 and 0.27, respectively. A typical distribution of material is: 13% ± 12 unreacted arachidonic acid, 27% ± 7 PGG_2 (plus 15-keto-PGG_2), 37% ± 4 PGH_2, and 20% ± 3 of more polar products with R_f values of ≤ 0.20.

When incubations are conducted in the presence of 1 m*M* phenol, PGH_2 becomes the major reaction product (80–85%) (Fig. 2).

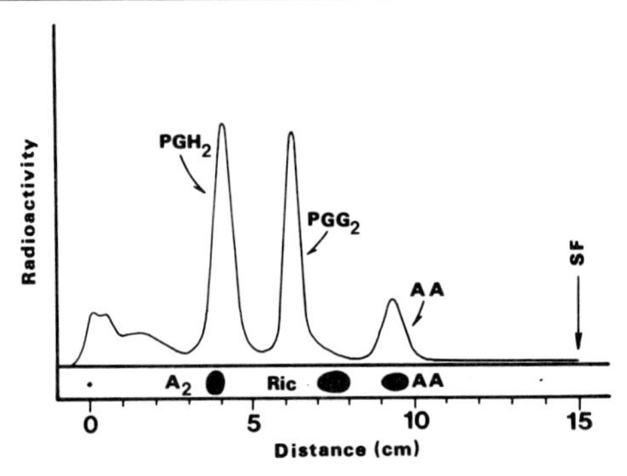

FIG. 1. Thin-layer chromatography of PGH_2 synthase products isolated after incubation of [1-^{14}C]arachidonic acid with lipid-depleted sheep vesicular gland microsomes in reaction medium devoid of phenol. Abbreviations used: AA, arachidonic acid, Ric, ricinoleic acid; A_2, prostaglandin A_2.

Purification of Prostaglandin Endoperoxides

Purification of prostaglandin endoperoxides is carried out with silicic acid columns and eluting solvents equilibrated in a $-20°$ cold room.

Prepare a slurry of 1.5 g of silicic acid in petroleum ether. Pour the slurry into a 1×10 cm column and allow the silicic acid to settle. Apply the sample (reduced to a volume of 1.0 ml) to the column, and elute stepwise at a flow rate of 0.5 ml/min with 120, 180, and 300 ml of 10, 15, and 20% ethyl acetate in petroleum ether, respectively. Fractions of 6 ml are collected. Continual monitoring of eluted radioactivity from the column can assure resolution of the components. A typical separation of reaction products isolated from incubations conducted in the absence of phenol is shown in Fig. 3. Minor differences in the elution profile frequently encountered are due to differences in column packing and flow rates. As shown in Fig. 3, unreacted arachidonic acid elutes with 10% ethyl acetate in fractions 1 to 20. The second peak, consisting of a mixture of PGG_2 and 15-keto-PGG_2,[9] is eluted with 15% ethyl acetate in fractions 21 to 50. The third peak, composed of pure PGH_2 (see Fig. 5B), elutes with 20% ethyl acetate in fractions 51 to 80. The overall isotope recovery from the column (fractions 1 through 80) is typically 75% $\pm$ 6.

Fractions containing PGH_2 are combined, concentrated by evaporation with a rotary evaporator, and stored in 10 ml of 10% ethyl acetate in petroleum ether at $-70°$. The final yield of PGH_2 from reaction mixtures lacking phenol ranges from 1.2 to 1.5 μmol (from 5 μmol of arachidonic acid).

[9] G. Graff, E. W. Dunham, T. P. Krick, and N. D. Goldberg, *Lipids* **14,** 334 (1979).

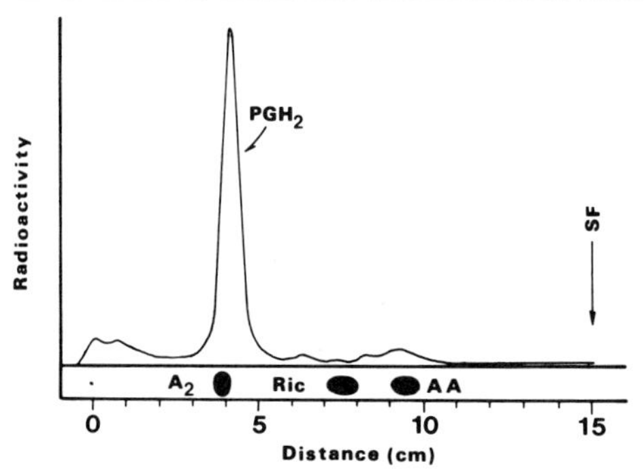

FIG. 2. Thin-layer chromatography of PGH_2 synthase products isolated after incubation of [1-^{14}C]arachidonic acid with sheep vesicular gland microsomes in reaction medium containing 1 mM phenol. Abbreviations are as in Fig. 1.

Fractions 21 to 50, containing both PGG_2 and its 15-keto analog, are combined and reduced to a final volume of about 1 ml by evaporation with a rotary evaporator. The sample is repurified on a silicic acid column of identical dimensions as described above. The column is eluted at $-20°$ with 275 ml of 10% ethyl acetate followed by 125 ml of 20% ethyl acetate in petroleum ether (Fig. 4). 15-Keto-PGG_2 elutes in fractions 21 to 41 with 10% ethyl acetate whereas PGG_2 elutes with 20% ethyl acetate in fractions 51 to 80. The fractions containing PGG_2 are combined, concen-

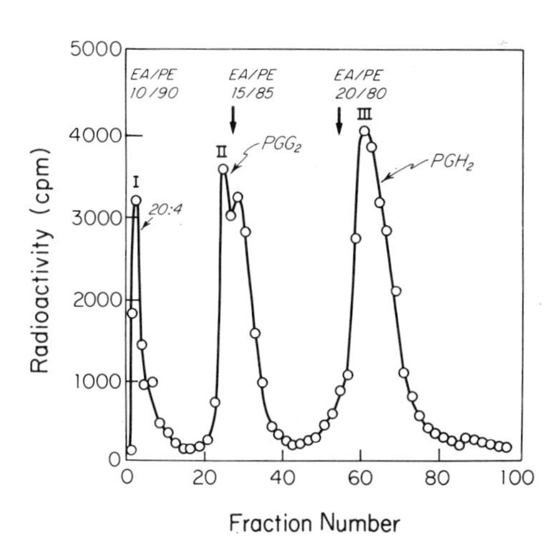

FIG. 3. Purification of PGH_2 synthase reaction products by silicic acid column chromatography at $-20°$ isolated from incubations conducted in the absence of phenol. The changing proportions of ethyl acetate (EA) to petroleum ether (PE) in the eluting solvent are indicated.

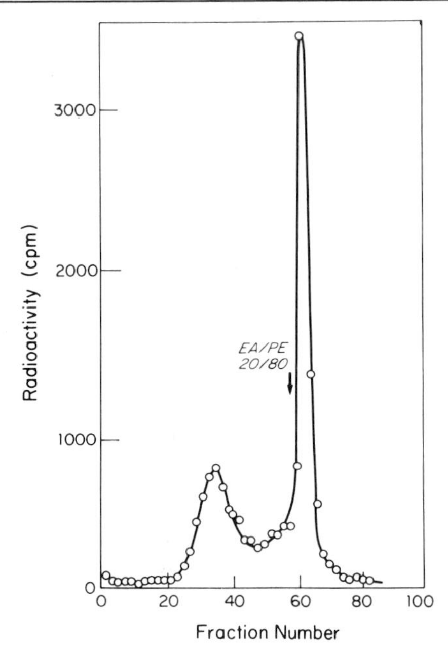

FIG. 4. Purification of PGG_2 (fractions 21–50, Fig. 3) by silicic acid column chromatography at $-20°$. Other conditions are the same as those described in Fig. 3.

trated, and stored in 10% ethyl acetate (10 ml) as described above. The yield of purified PGG_2 ranges between 0.4 to 0.5 μmol.

When PGH_2 is generated in incubations containing 1 m*M* phenol (Fig. 2), purification is accomplished in an identical manner as described in Fig. 3. The final yield of PGH_2 from these reactions ranges between 2.5 and 3 μmol.

Structural Identification

Place 50 nmol of the purified PGG_2 and/or PGH_2 into a screw-cap tube. Remove the solvent under a stream of nitrogen, and add 1 mg of triphenylphosphine in 1 ml of diethyl ether.[2] Allow to react for 60 min at room temperature. Evaporate the solvent under a stream of nitrogen. Dissolve the residue in 200 μl of ethyl acetate and apply the sample(s) to a silica gel H plate (E. Merck) (3.0 cm wide lane). In a separate lane apply 5 nmol of $PGF_{2\alpha}$, PGE_2, PGA_2, ricinoleic acid and arachidonic acid as reference standards. Place the TLC plate into a tank lined with filter paper (15 × 55 cm) and prequilibrated for 20 min with 100 ml of chloroform–methanol–acetic acid–water (90:8:1:0.8, v/v/v/v).[7] Develop the plate

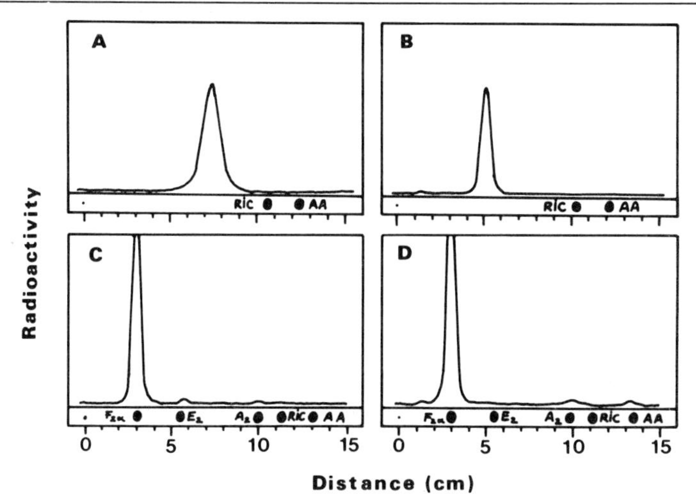

FIG. 5. Thin-layer chromatography (TLC) of purified PGG_2 ($R_f = 0.47$) (A) and PGH_2 ($R_f = 0.34$) (B) and their respective reaction products formed upon reduction with triphenylphosphine (C, D). TLC was conducted with the solvent systems ethyl acetate–petroleum ether–acetic acid (50:50:0.05, v/v/v) (A and B), and chloroform–methanol–acetic acid–water (90:8:1:0.8, v/v/v/v) (C and D). AA, arachidonic acid; RIC, ricinoleic acid; A_2, prostaglandin A_2; E_2, PGE_2; $F_{2\alpha}$, $PGF_{2\alpha}$.

to 15 cm above the sample origin. Remove the TLC plate from the tank, evaporate the solvent, and spray the lane containing the reference standards with a 0.1% ethanolic solution of 2,7-dichlorofluorescein. Visualize the separated components under UV light, and locate the radioisotope containing reaction products of PGG_2 and PGH_2 on the TLC plate as described above in the section on TLC analysis. Reduction of either PGG_2 or PGH_2 with triphenylphosphine results in the formation of $\geq 85\%$ $PGF_{2\alpha}$ (Fig. 5C,D). Scrape the area of the sample lane corresponding to the location of the $PGF_{2\alpha}$ standard into a test tube. Resuspend the silica gel scraping in 2 ml of chloroform–methanol (2:1, v/v), and apply the silica gel slurry to a Pasteur pipette previously filled with 0.5 cm of silicic acid in chloroform–methanol (2:1, v/v). Elute the component from silica gel with 10 ml of chloroform–methanol (2:1, v/v). Isotope recovery is determined by liquid scintillation counting.

Add 50 μmol of heptadecanoic and 12-hydroxy-9-octadecenoic acid to the sample. Evaporate the solvent and convert the free acids to methyl esters with diazomethane in ether. Add 50 nmol of methyl nonadecanoate to the sample and evaporate the solvent. Dissolve the residue in 500 μl of diethyl ether and transfer it into a 5 $\times$ 100 mm tube. Remove the solvent under a stream of nitrogen and add 50 μl of pyridine and 100 μl of bis(tri-

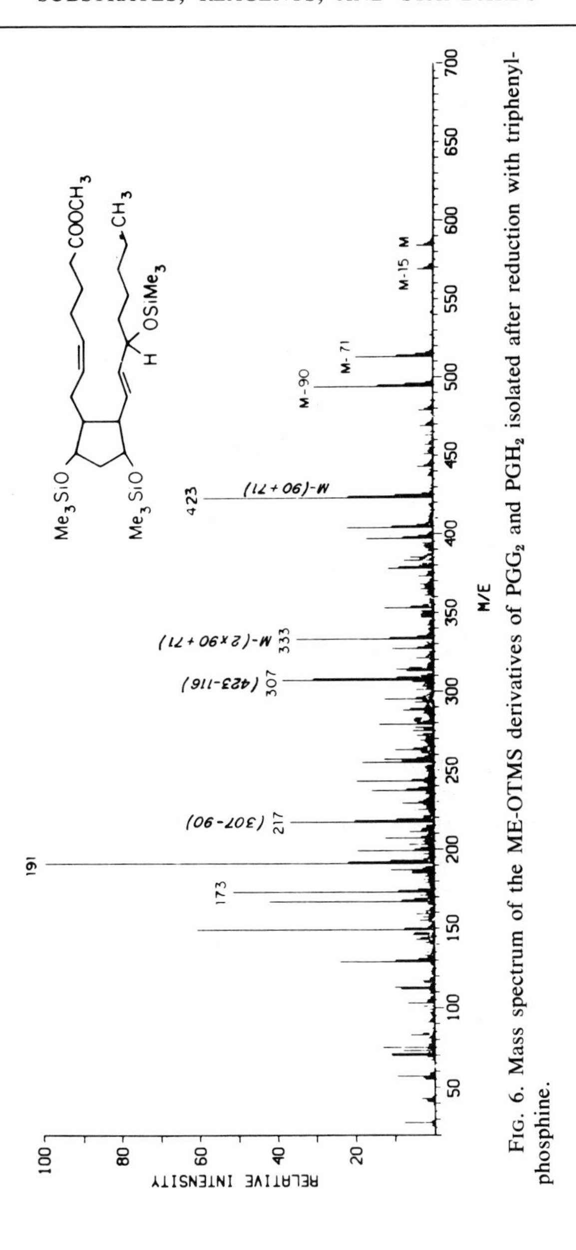

FIG. 6. Mass spectrum of the ME-OTMS derivatives of PGG_2 and PGH_2 isolated after reduction with triphenylphosphine.

methylsilyl)trifluoroacetamide. Place the tube on Dry Ice, and seal it with a torch. Heat the tube for 10 min at 130°, then cool and analyze an aliquot with a GLC equipped with a flame ionization detector and a 3-foot glass column $\frac{1}{8}$ inch in diameter packed with 5% OV-101 on Chromosorb W-AW. The yield of methylation and silylation of $PGF_{2\alpha}$ is determined with the aid of the internal standards. The ME-OTMS derivative of $PGF_{2\alpha}$, obtained upon reduction of PGG_2 and PGH_2 with triphenylphosphine, appears as a single component in the chromatogram with a ECL-value of 24.1.[10] The mass spectrum (Fig. 6) of the ME-OTMS derivative of $PGF_{2\alpha}$ derived from PGG_2 or PGH_2 is identical to that obtained from an authentic standard of $PGF_{2\alpha}$.

Comments

The use of lipid-depleted sheep vesicular gland microsomes has the advantage that preparations may be stored for long periods of time without significant loss in enzyme activity. Second, the potential for contamination with trace amounts of microsomal lipid is minimized. One minor disadvantage is that hematin added for reconstitution catalyzes the nonenzymic formation of the 15-keto analog from PGG_2. The formation of the keto derivative is considerably reduced (from 16 to 4%) when freshly prepared microsomes are used.[9] When phenol is added to enhance hydroperoxidase activity and PGH_2 formation, the interference from the nonenzymic reaction with hematin is not encountered.[9]

The use of diethyl ether for extraction and/or purification has been reported by several laboratories[1-3] but should probably be avoided to eliminate possible contamination of prostaglandin endoperoxides with ether-derived peroxides.[4]

Acknowledgments

This work was supported by Grants NS 05979, HL 06312, GM 28834 from the United States Public Health Service and by a grant from the March of Dimes Birth Defects Foundation 1-746.

[10] M. Hamberg and B. Samuelsson, *Proc. Natl. Acad. Sci. U.S.A.* **70,** 899 (1973).

[50] Preparation of 15-L-Hydroperoxy-5,8,11,13-eicosatetraenoic Acid (15-HPETE)

By GUSTAV GRAFF

Soybean lipoxygenase (EC 1.99.2.1) catalyzes the stereospecific oxygenation of numerous polyunsaturated fatty acids having both a *cis,cis*-1,4-pentadiene group and a methylene group in the n-8 position in the acyl chain.[1,2]

5,8,11,14-Eicosatetraenoic acid is among the fatty acids readily oxygenated by soybean lipoxygenase with formation of 15-L-hydroperoxy-5,8,11,13-eicosatetraenoic acid (15-HPETE).[1] This component, 15-HPETE, has become of interest to both enzymologists and pharmacologists, since it has been shown to alter the activity of several enzymes involved in generating vasoactive components. These include platelet lipoxygenase,[3] polymorphonuclear leukocyte lipoxygenase,[4] prostacyclin (PGI_2) synthase,[5] and PGH synthase.[6]

Since the initial report of 15-HPETE formation from 5,8,11,14-eicosatetraenoic acid by soybean lipoxygenase,[1] only one publication has appeared describing the isolation of 15-HPETE.[7] The method reported here represents a modification of that procedure and includes details not reported earlier. The present method is also designed to be conducted without sophisticated equipment.

Materials

- [1-^{14}C]5,8,11,14-Eicosatetraenoic acid (Amersham)
- 5,8,11,14-Eicosatetraenoic acid (Nu Chek Prep)
- Methylnonadecanoate, 12-hydroxyoctadecanoic, and 12-hydroxy-9-octadecenoic acid (Nu Chek Prep)
- Soybean lipoxygenase, type I, 170,000 units/mg protein (Sigma Chemical Co.)
- Tris-HCl buffer, 100 m*M*, pH 9.0
- Hydrochloric acid, 0.23 *N*

[1] M. Hamberg and B. Samuelsson, *J. Biol. Chem.* **242,** 5329 (1967).
[2] R. T. Holman, P. O. Egwim, and W. W. Christie, *J. Biol. Chem.* **244,** 1149 (1969).
[3] J. Y. Vanderhoek, R. W. Bryant, and J. M. Bailey, *J. Biol. Chem.* **255,** 5996 (1980).
[4] J. Y. Vanderhoek, R. W. Bryant, and J. M. Bailey, *J. Biol. Chem.* **255,** 10064 (1980).
[5] S. Moncada, R. T. Gryglewski, S. Bunting, and J. R. Vane, *Prostaglandins* **12,** 715 (1976).
[6] M. E. Hemler, H. W. Cook, and W. E. M. Lands, *Arch. Biochem. Biophys.* **193,** 340 (1979).
[7] C. G. Crawford, H. W. Cook, W. E. M. Lands, G. W. H. Van Alphen, *Life Sci.* **23,** 1255 (1978).

ISBN 0-12-181986-8

Ethyl acetate, spectral grade

Petroleum ether (bp 35–60°), treated with concentrated sulfuric acid, washed with distilled water, and distilled over lithium aluminum hydride

Silicic acid, 100-mesh; 500 g slurried, five times each with 5 liters of distilled water. Fine particles not sedimenting after 5 min are discarded. Washed silicic acid is activated in an oven for 12 hr at 150°.

Silica gel 60 H (MCB Manufacturing Chemists, Inc.)

Tris (trimethylsilyl)trifluoroacetamide (Aldrich Chemical Co.)

Other chemicals are of ACS grade

Soybean Lipoxygenase Reaction

Dilute [1-^{14}C]5,8,11,14-eicosatetraenoic acid (5 μCi) with 40 μmol of the unlabeled fatty acid to a final specific activity of 250 cpm/nmol. Remove the solvent under a stream of nitrogen, and disperse the eicosatetraenoic acid in 2 ml of 100 m*M* Tris-HCl buffer (pH 9.0) by sonication for 2 min at 20° in a Branson bath-type sonicator. Transfer the fatty acid suspension into a 250-ml Erlenmeyer flask containing 47 ml of 100 m*M* Tris-HCl buffer (pH 9.0) preequilibrated at 30°. Initiate the reaction by addition of 10 mg of soybean lipoxygenase dissolved in 1.0 ml of 100 m*M* Tris-HCl buffer (pH 9.0), and incubate for 5 min at 30° in an orbital shaker at 200 rpm, while supplying the reaction with oxygen gas (30 ml/min). This supply of oxygen gas is necessary because the oxygen concentration in the buffer (230 μM[8]) is rate limiting under the reaction conditions employed.

Extraction of Lipoxygenase Reaction Products

Terminate the reaction after 5 min by pouring the reaction mixture into a separatory funnel containing 100 ml of petroleum ether–ethyl acetate (1:1, v/v; precooled to −25°), and 20 ml of 0.23 *N* HCl. Shake the mixture vigorously, collect the organic phase, and place it on Dry Ice. The aqueous phase is extracted twice with 100 ml of cold petroleum ether–ethyl acetate (1:1, v/v). Add 50 g of anhydrous sodium sulfate to the combined extract and keep it on Dry Ice for 20 min with intermittent shaking. Sodium sulfate and ice crystals are removed by vacuum filtration through a fritted-glass funnel. During filtration, the temperature of the extract is maintained between −10° and −20°. Transfer the filtrate into a round-bottom flask. Concentrate the extract to about 1–2 ml by evapora-

[8] H. W. Cook, G. Ford, and W. E. M. Lands, *Anal. Biochem.* **96,** 341 (1979).

tion with a rotary evaporator. The temperature of the rotary flask during evaporation should not exceed 20°. Store the extract at −20° for further analysis and purification. The overall recovery of radioactivity in the extraction is 93% ± 10 (SEM), with 6% ± 3 of the radioisotope distributed in the aqueous phase and 87% ± 9 in the organic phase.

TLC Analysis of Reaction Products

Apply approximately 10,000 counts (40 nmol) of the lipoxygenase extract at room temperature to a 10 × 20 cm silica gel 60 H plate (0.25 mm thickness). In two separate lanes 1.5 cm wide, apply 10 nmol of arachidonic acid and 12-hydroxyoctadecanoic or 12-hydroxy-9-octadecenoic acid as reference standards. Place the TLC plate into a TLC tank lined with filter paper (15 × 55 cm) and preequilibrated at room temperature for 20 min with 100 ml of diethyl ether–petroleum ether–acetic acid (50:50:05, v/v/v). Develop the TLC plate to 13.5 cm above the sample origin. Remove the TLC plate from the tank, evaporate the solvent in a fume hood, and spray the TLC plate with a 0.1% ethanolic solution of 2,7-dichlorofluorescein. Visualize the separated components under UV light. Radioactive components on the TLC plate are located with a radioisotope TLC scanner (if available). An alternative method is to scrape the lane of the plate into appropriate sections and locate radioisotope-containing components by liquid scintillation counting.

A typical TLC separation of 1-^{14}C-labeled arachidonic acid-derived lipoxygenase products is shown in Fig. 1 (upper panel). Under the reaction conditions described (see section on soybean lipoxygenase reaction, above), 72% ± 7 (SEM) of the [1-^{14}C]arachidonic acid (R_f 0.62) is converted to 15-HPETE (R_f 0.36). Three additional minor products with R_f values 0.27, 0.16, and 0.00 are present. These latter components constitute a 5% ± 2, 14% ± 3 and 9% ± 7 of the radioactivity, respectively.

Purification

The steps described below for purification of 15-HPETE are carried out in a −20° cold room. Prepare a slurry of 3 g of silicic acid in 3% ethyl acetate in petroleum ether. Pour the slurry into a 1 × 10 cm glass column and allow the silicic acid to settle. Prior to application of the lipoxygenase extract to the column, the fraction of ethyl acetate in petroleum ether in the extract has to be reduced from 50% to 3%. This is accomplished by evaporation of the solvent under a stream of nitrogen to near dryness (~100–200 μl) followed by addition of 2 ml of 3% ethyl acetate. This procedure is repeated three more times before the sample is finally dissolved

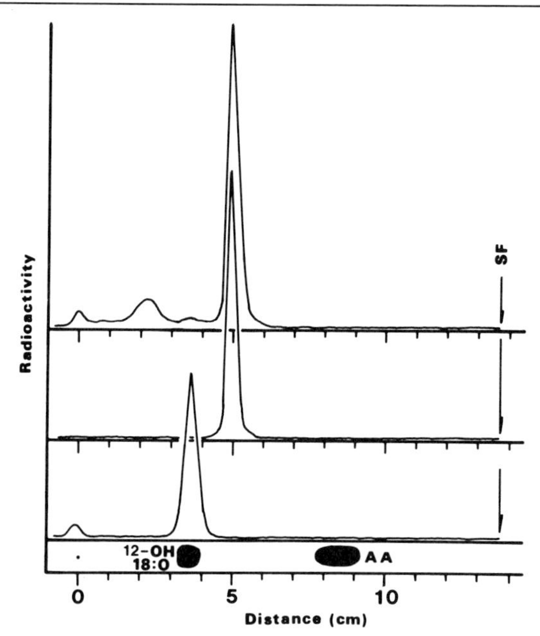

FIG. 1. Thin-layer chromatography of the reaction products isolated from incubation of [1-^{14}C]arachidonic acid with soybean lipoxygenase (upper panel), purified 15-HPETE (middle panel), and the reaction product of 15-HPETE after treatment with triphenylphosphine in diethyl ether (bottom panel).

in 1.0 ml of 3% ethyl acetate. Avoid evaporation of the sample to dryness. Apply the sample to the column and elute at 0.8 ml/min with 80 ml of 3% ethyl acetate then with 160 ml of 8% ethyl acetate in petroleum ether, and finally 50 ml of ethyl acetate. Fractions of 4 ml are collected. A typical column separation of a 1-^{14}C-labeled lipoxygenase reaction extract is shown in Fig. 2. A trace amount of [^{3}H]arachidonic acid, added as an internal standard to demonstrate the separation of fatty acid from its hydroperoxy counterpart, is eluted quantitatively with 3% ethyl acetate between fractions 1 and 20. 15-HPETE acid is eluted from the column between fractions 25 and 32 with 8% ethyl acetate in petroleum ether. A minor component, unresolved from 15-HPETE, emerges between fractions 33 and 60. TLC analysis of an aliquot of the combined fractions 25–32, representing 18 μmol ± 3 (SEM) (or 45% of the initial starting material), showed 15-HPETE as a single component with an R_f value of 0.36 (Fig. 1, middle panel). The components (9.4 ± 0.5 μmol) eluted after fraction 33 and those eluted with ethyl acetate (not shown) correspond to those found in the original extract with R_f values of 0.27, 0.16, and 0.00

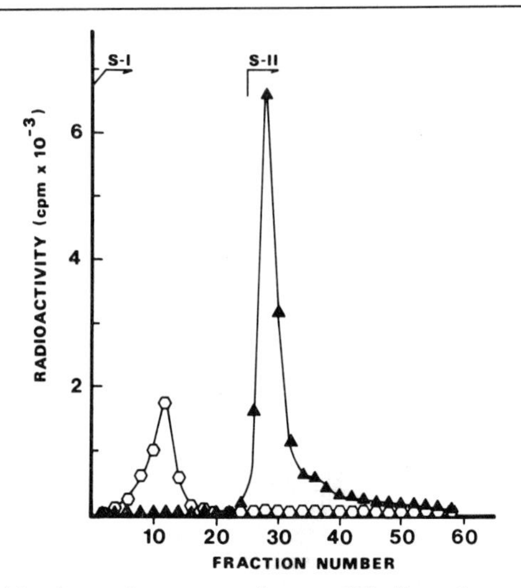

FIG. 2. Silicic acid column chromatography at −20° of reaction products isolated after incubation of [1-^{14}C]arachidonic acid with soybean lipoxygenase (▲—▲), containing [^{3}H]arachidonic acid as internal standard (○—○). The proportion of ethyl acetate in petroleum ether in the eluting solvent are indicated (S-I, 3% ethyl acetate; S-II, 8% ethyl acetate).

(Fig. 1, upper panel). The overall isotope recovery of radioactive material from the column is typically 77% ± 7 (SEM).

The only step reducing the yield of 15-HPETE appears to be the column chromatographic procedure. At room temperature the loss due to decomposition is at least 65%. The decomposition of 15-HPETE appears to be minimized to 28% when chromatography is carried out at −20°. Since the incubation conditions described fully convert arachidonic acid to its oxygenated products, the initial chromatographic elution step with 3% ethyl acetate may be omitted. This will reduce the time of exposure of 15-HPETE to silicic acid and perhaps reduce the extent of hydroperoxide decomposition.

Structural Identification

Place 150 nmol of purified [1-^{14}C]-15-HPETE into a screw-cap tube, containing a known amount (150 nmol) of 12-hydroxyoctadecanoic acid. Remove the solvent under a steam of nitrogen, and add 1 mg of triphenylphosphine in 1 ml of diethyl ether.[9] Allow to react for 60 min at room temperature. Evaporate the solvent to a final volume of ~200 μl, and apply

[9] M. Hamberg and B. Samuelsson, *Proc. Natl. Acad. Sci. U. S. A.* **71,** 3400 (1974).

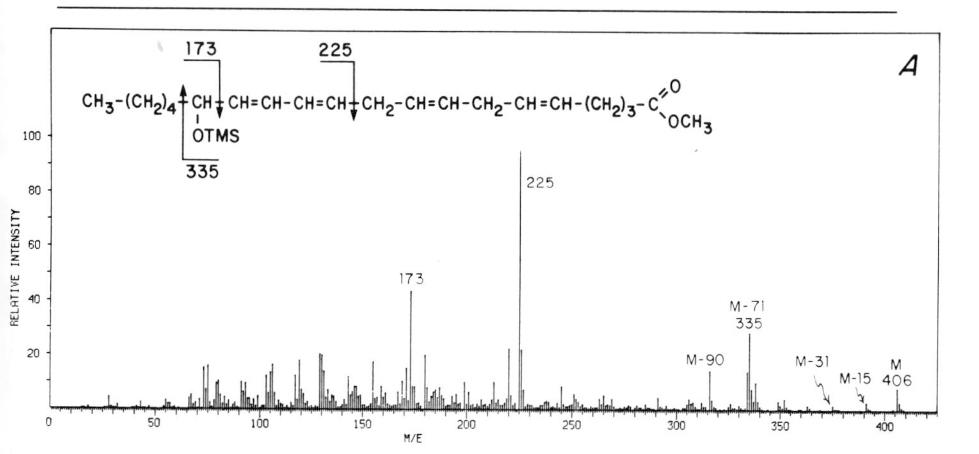

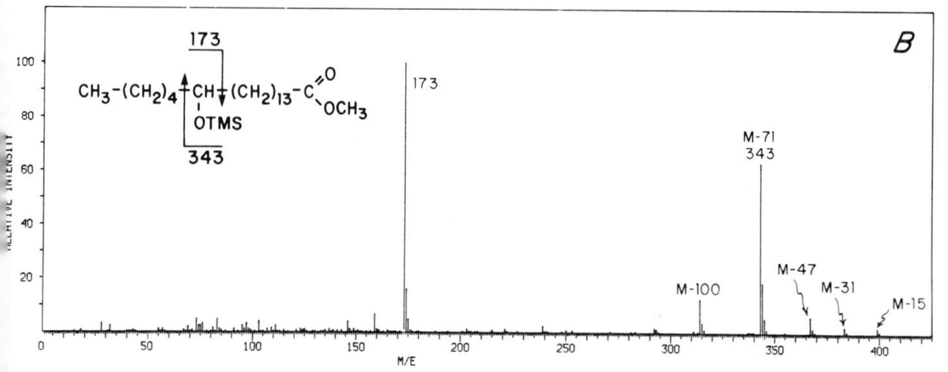

FIG. 3. Mass spectrum of the ME-OTMS derivative of (A) 15-HPETE after triphenylphosphine reduction; (B) 15-HPETE after triphenylphosphine reduction and catalytic hydrogenation.

the sample to a silica gel H plate (2.5 cm wide lanes). In separate lanes, apply 5 nmol [1-^{14}C]15-HPETE, arachidonic and 12-hydroxyoctadecanoic acids as reference standards.

Develop the TLC plate with diethyl ether–petroleum ether–acetic acid (50:50:0.5, v/v/v) and locate the reference standards as described above in the section on TLC analysis of reaction products. Scrape the area of the sample lane, corresponding to the location of the hydroxy fatty acid standard, into a test tube containing a known amount (i.e., 150 nmol) heptadecanoic acid. Resuspend the silica gel scraping in 1 ml of ethyl acetate, and transfer the silica gel slurry to a Pasteur pipette previously filled with 0.5 cm of silicic acid in ethyl acetate. Elute the components from silica gel with 5 ml of ethyl acetate. Evaporate the solvent and convert the same acids to methyl esters with ethereal diazomethane. Add a known amount (150 nmol) of methyl nonadecanoate, and evaporate the solvent. Dissolve the residue in 0.5 ml of diethyl ether, and transfer it into a 5 × 100 mm tube. Evaporate the solvent under a stream of nitrogen; add 50 μl of pyridine and 100 μl of Tris(trimethysilyl)trifluoroacetamide. Place the tube on Dry Ice and seal the tube with a torch. Heat the tube for 10 min at 130°, then cool, and analyze an aliquot by GLC. A 3-foot-long column packed with either OV 101 or SE-30 on 80/100 mesh Supelcoport (Supelco Inc.) are suitable for GLC. The yield of methylation and silylation is readily determined with the aid of the internal standards.

The methyl ester-trimethylsilyl (ME-OTMS) derivative of 15-HETE appears as a single component in the chromatogram with an ECL value of 21.3.[5] GLC–MS analysis exhibits the mass spectrum shown in Fig. 3A. The ions found at *m/e* 306, 355, and 225 indicate cleavages in the alkyl chain between C-15–C-16 and C-10–C-11, respectively, and locate the conjugated double-bond system in the 11 and 13 position and the OTMS group at C-15. Catalytic hydrogenation of the methyl ester of 15-HETE over platinum oxide in methanol, followed by conversion of 15-hydroxyeicosanoate to the ME-OTMS derivative, changes the retention from an ECL value of 21.3 to 22.3.[10] The mass spectrum of the ME-OTMS derivative of 15-hydroxyeicosanoic acid is shown in Fig. 3B. The ions at *m/e* 173 and 343, arising from α-cleavage at position C-14–C-15 and C-15–C-16, conclusively place the OTMS group into C-15 position.

Acknowledgments

This work was supported by grants NS 05979, HL 06312, GM 28834 from the United States Public Health Service and by a grant from the March of Dimes Birth Defects Foundation, 1-746.

[10] G. Graff, J. H. Stephenson, D. B. Glass, M. K. Haddox, and N. D. Goldberg, *J. Biol. Chem.* **253,** 7662 (1978).

[51] Preparation of [acetyl-^{3}H]Aspirin and Use in Quantitating PGH Synthase

By GERALD J. ROTH

Aspirin (acetylsalicylic acid) inhibits prostaglandin (PG) synthesis by acetylating a single serine residue of PGH synthase, thereby inactivating the enzyme (Fig. 1). The acetylated serine residue is related to an active-

ISBN 0-12-181986-8

site region of the enzyme. Spontaneous or catalytic inactivation of PGH synthase alters the aspirin-sensitive site of the enzyme and makes it unavailable to aspirin-mediated acetylation. The stoichiometry of the acetylation reaction is 1:1, one acetate transferred per enzyme monomer, M_r 70,000.[1–4] Aspirin can serve as a useful reagent in quantitating and characterizing PGH synthase.

Preparation of [acetyl-^{3}H]Aspirin

Principle. Acetylation of PGH synthase by [acetyl-^{3}H]aspirin results in site-specific, covalent modification and ^{3}H labeling of the protein (Fig. 1). [acetyl-^{3}H]Aspirin is synthesized by allowing [^{3}H]acetic anhydride to react with nonradioactive salicylic acid.[5]

Reagents

[^{3}H]Acetic anhydride, 400 Ci/mol (New England Nuclear), freshly distilled, 80% benzene
Salicylic acid
Pyridine

Procedure. The success of this synthesis depends to a large extent on the quality of the [^{3}H]acetic anhydride available, and three points should be made concerning the reagent. First, acetic anhydride is a symmetric molecule that donates an acetyl group during acetylation, leaving a molecule of acetate. Therefore, the specific radioactivity of a [^{3}H]acetyl product will be, at most, one half that of the original [^{3}H]acetic anhydride reagent. Second, when used as a quantitative reagent, [acetyl-^{3}H]aspirin, and hence the starting [^{3}H]acetic anhydride, must have a sufficiently high specific radioactivity to permit accurate measurement of the [acetyl-^{3}H]PGH synthase that is being assayed. We have found that [acetyl-^{3}H]aspirin of 100–200 Ci/mol gives satisfactory results in measuring the enzyme in human platelets and in ram seminal vesicles.[1,6] Third, [^{3}H]acetic anhydride frequently contains a nonvolatile, brownish contaminant that interferes with synthesis of [acetyl-^{3}H]aspirin and may inhibit aspirin-

[1] G. J. Roth, N. Stanford, and P. W. Majerus, *Proc. Natl. Acad. Sci. U.S.A.* **72,** 3073 (1975).
[2] G. J. Roth and C. J. Siok, *J. Biol. Chem.* **253,** 3782 (1978).
[3] G. J. Roth, C. J. Siok, and J. Ozols, *J. Biol. Chem.* **255,** 1301 (1980).
[4] F. J. van der Ouderaa, M. Buytenhek, D. H. Nugteren, and D. A. van Dorp, *Eur. J. Biochem.* **109,** 1 (1980).
[5] L. F. Fieser, "Organic Experiments," 2nd ed., p. 242. Raytheon Education Company, Lexington, Massachusetts, 1968.
[6] G. J. Roth and P. W. Majerus, *J. Clin. Invest.* **56,** 624 (1975).

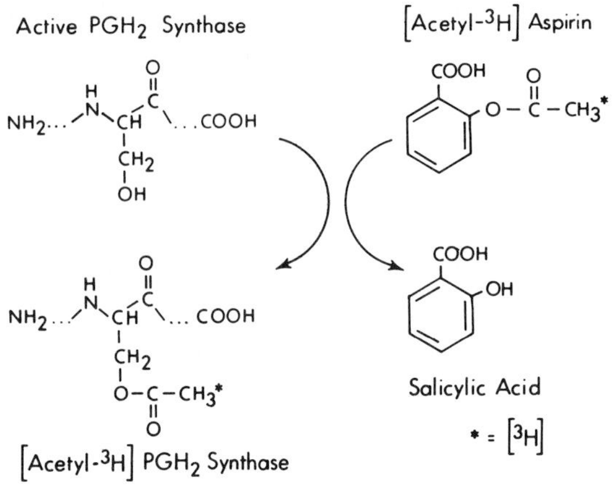

Active PGH_2 Synthase

$[Acetyl\text{-}^3H]$ Aspirin

Salicylic Acid

$[Acetyl\text{-}^3H]$ PGH_2 Synthase

$* = [^3H]$

FIG. 1. Interaction of aspirin with PGH_2 synthase. Aspirin selectively acetylates a single serine residue within the polypeptide chain of PGH_2 synthase, inactivating the enzyme. Use of [Acetyl-^{3}H]aspirin results in the transfer of [^{3}H]acetate to the enzyme and formation of [acetyl-^{3}H]PGH_2 synthase.

mediated acetylation of PGH synthase. The contaminant is removed by distillation, and therefore the [^{3}H]acetic anhydride must be distilled by the supplier just prior to shipment and used shortly thereafter.

[^{3}H]Acetic anhydride is commonly supplied as 20% acetic anhydride in 80% benzene by volume, and the solution can be used directly as supplied. An equimolar amount of salicylic acid (2.5 mmol) and a slight molar excess of the catalyst, pyridine (3.75 mmol) are added to [^{3}H]acetic anhydride (2.5 mmol, 400 Ci/mol, in 1 ml of benzene). After incubation at 37° for 2 hr, water (0.225 ml) is added to hydrolyze unreacted acetic anhydride, and the mixture is dried under vacuum for 1 hr. [^{3}H]Acetic acid, the by-product of aspirin synthesis and the later hydrolysis step, is volatile and is removed by drying under vacuum. To facilitate removal, three separate aliquots of ethanol (3.0 ml) and acetic acid (0.1 ml) are added, and the mixture is dried after each addition. The mixture should yield a whitish semisolid after the second or third drying step following addition of ethanol–acetic acid. Failure to form a solid may be due to the presence of the nonvolatile contaminant mentioned earlier. The dried material is dissolved in a minimal volume (3–6 ml) of anhydrous diethyl ether, followed by an equal volume of petroleum ether (boiling point, 38–47°). The solution is left undisturbed at $-20°$ for several hours, and the fluid phase is removed and discarded. Crystals of [acetyl-^{3}H]aspirin are dissolved in anhydrous ethanol at a concentration of about 50 m*M* and stored at −20°. Aspirin is stable for years under these conditions. The yield is about 50%.

The procedure is highly reproducible but depends on the quality of the [^{3}H]acetic anhydride available. A trial synthesis with nonradioactive reagents is advisable to make sure that the drying step will give a solid intermediate product which, in turn, will crystallize in ether–petroleum ether in good yield. Better results are obtained using relatively large amounts of [^{3}H]acetic anhydride, but satisfactory results have been obtained using 10-fold less of the various reagents in the same proportions as noted above. However, the yield and purity of the product are reduced when smaller quantities are used.

Properties. The final product contains about 90–95% aspirin and 5–10% salicylic acid, as determined by either gas–liquid or high-pressure liquid chromatography. The specific radioactivity is slightly less than half that of the starting [^{3}H]acetic anhydride and is determined by assaying for ^{3}H by means of scintillation spectrometry and for aspirin content (see below). All the ^{3}H present is found in the aspirin product as demonstrated by thin-layer chromatography on a Kodak Chromatogram sheet (Eastman) using the solvent system for salicylates[7] or by HPLC analysis and purification of the product.[8]

Comments. In contrast to [acetyl-^{3}H]aspirin, [aromatic ring-^{3}H]aspirin does not transfer a ^{3}H label during acetylation of PGH_2 synthase. Therefore, [aromatic ring-^{3}H]aspirin can be used as a control for enzyme acetylation experiments. The [aromatic ring-^{3}H]aspirin is synthesized by allowing nonradioactive acetic anhydride to react with [G-^{3}H]salicylic acid (New England Nuclear) using a procedure identical to that described above. The specific radioactivity of the [aromatic ring-^{3}H]aspirin will be the same as that of the [G-^{3}H]salicylic acid used for synthesis.

One may synthesize [^{14}C]aspirin, if desired, using the appropriate ^{14}C-labeled reagents. However, [^{14}C]acetic anhydride is much more expensive than the ^{3}H form.

For laboratory use, the appropriate amount of the radioactive aspirin preparation is dried under N_2 and simply dissolved in buffer. Aspirin is relatively stable in buffer at neutral pH, but freshly prepared solutions are used to avoid the potential problem of aspirin hydrolysis. Hydrolysis of aspirin stored in cold ethanol is minimal, but one may check for hydrolysis by simply redetermining the aspirin content of the stored material.

[7] K. Randerath, "Thin-Layer Chromatography," 2nd ed., p. 199. Academic Press, New York, 1966.

[8] S. L. Ali, *J. Chromatogr.* **126,** 651 (1976).

Assay of Aspirin Preparations

Hydroxamate Assay

Hydroxylamine reacts with acetylsalicylic acid to give acetyl hydroxamate which, in the presence of ferric salts, forms a colored complex that can be assayed by absorbance at 540 nm.[9] Aqueous solutions of aspirin (0–10 μmol) are mixed with 0.5 ml of 2 *M* neutral hydroxylamine (equal volumes of 4 *M* NH_2OH-HCl and 3.5 *M* NaOH), and water is added to give a final volume of 1.5 ml. After a 10 min incubation at room temperature, ferric chloride reagent (1.5 ml of equal volumes of 12% trichloroacetic acid, 3 *N* HCl, and 5% $FeCl_3$ in 0.1 *N* HCl) is added, and the absorbance at 540 nm is determined immediately. Known amounts of aspirin (purchased from BDH Chemicals Ltd.) serve as a standard for the assay. $A_{540\ nm} = 95\ cm^{-1}$ for 1 mmol of aspirin in the described assay.

Gas–Liquid Chromatography (GLC) Assay

Aspirin preparations may also be assayed by GLC after the method of Thomas *et al.*[10] Approximately 2 μmol of aspirin are converted to the trimethylsilyl form by incubation for a few minutes with 0.05 ml of *N,O*-bis(trimethysilyl)trifluoroacetamide (Pierce Chemical Co.), and an aliquot is chromatographed on OV-44 (Supelco) with component temperatures of 112° for the column, 180° for the injection port, and 220° for the detector. The content of both salicylic acid and aspirin in the preparation is easily determined by GLC, since the salicylic acid derivative elutes well before the aspirin derivative under the described conditions. Known amounts of aspirin and salicylic acid serve as standards for the assay.

High-Pressure Liquid Chromatography (HPLC) Assay

Aspirin preparations can be both assayed and purified by HPLC as described by Ali.[8] The sample is dried under N_2 and dissolved in methanol. An aliquot containing up to 1 μmol in 100 μl is injected on a reversed-phase column (Bondapak C_{18}, Waters Associates, Medford, Massachusetts). Asprin is eluted isocratically with water–methanol–acetic acid (52:48:1, v/v/v) at a flow rate of 1 ml/min and detected by absorbance at 240 nm. Aspirin elutes at about 4 min, and salicylic acid elutes at about 6 min under the described conditions. Therefore, both compounds are easily measured by the technique, using known amounts of each com-

[9] E. R. Stadtman, this series, Vol. 3, page 228.

[10] B. H. Thomas, G. Solomonraj, and B. B. Coldwell, *J. Pharm. Pharmacol.* **25,** 201 (1973).

pound as standards. Aspirin eluting from the column can be collected and thereby purified essentially free of salicylic acid and other contaminants.

Applications of [acetyl-^{3}H]Aspirin

Quantitation of PGH_2 Synthase

Aspirin acetylates PGH synthase rapidly (within minutes) at low concentrations (100 μM or less), acting as a site-specific, irreversible enzyme inhibitor.[1,6] At higher concentrations (mM) over longer time periods (hours or days), aspirin will "nonspecifically" acetylate a variety of proteins and nucleic acids.[11] In view of the distinctly different conditions for "specific" as contrasted with "nonspecific" acetylation, one can use low aspirin concentrations for brief time periods to give mainly "specific" acetylation of PGH synthase. For example, when platelets or fibroblasts (from human foreskin) are treated with 50 μM [acetyl-^{3}H]aspirin for 20 min at 37°, PGH synthase is the predominant acetylated protein in the cell.[6] This is seen more clearly by separating microsomal from soluble cytoplasmic proteins. Usually some degree of "nonspecific" acetylation of soluble proteins occurs, but PGH synthase is the sole acetylated microsomal protein.

By using [acetyl-^{3}H]aspirin of known specific radioactivity, one can measure active PGH synthase as [acetyl-^{3}H]protein, M_r 70,000; assuming a 1:1 stoichiometry and maximal acetylation of the available enzyme. The assay depends on gel electrophoresis to separate the [acetyl-^{3}H]PGH synthase from other acetylated proteins and from non-protein bound ^{3}H. The advantage of the assay lies in its ability to measure accurately the small amounts of active enzyme in intact cells. For example, by using this method with human platelets, one can assay 1 μg of PGH synthase per 1.5 mg of platelet protein (1×10^9 platelets).

The experimental procedure involves incubating cells (1 to 5×10^8 platelets or 1 to 5×10^6 human foreskin fibroblasts) in about 1 ml of Tris-saline, pH 7.4 with 50 μM [acetyl-^{3}H]aspirin for 20 min at 37°. These conditions give maximal acetylation of the enzyme in platelets.[6] If the enzyme preparation is taken from a tissue homogenate (ram seminal vesicle microsomes, for example), the acetylation reaction is performed in 0.02 M potassium phosphate buffer, pH 7.4, and a slightly higher aspirin concentration (100 μM) for a longer time period (30 min) is required. The extent of the acetylation reaction of the vesicular gland enzyme can be followed by assaying for the loss of enzyme activity by O_2 electrode

[11] R. N. Pinckard, D. Hawkins, and R. S. Farr, *Nature (London)* **219,** 68 (1968).

(assay performed at 30° in air-equilibrated 0.1 M potassium phosphate, pH 8.0, containing 0.7 mM phenol and 0.5 μM hemoglobin, using 100 μM arachidonate as substrate). Detergent-solubilized enzyme preparations are generally less susceptible to acetylation than the enzyme in whole cells or in microsomal preparations.

After incubation, cells are disrupted by sonication, and the mixture is centrifuged (100,000 g, 45 min, 4°) to give a microsomal pellet. This step removes soluble proteins and the bulk of non-protein-bound ^{3}H. The pellet is solubilized by sonication in 2% sodium dodecyl sulfate (SDS) containing 5% 2-mercaptoethanol and heated to 100° for 2 min. An aliquot of the sample (0.1 ml) containing the protein from 1 to 4 × 10^8 platelets (ca. 100–500 μg) is subjected to SDS–polyacrylamide gel electrophoresis in 7.5% gel according to the method of Weber and Osborn.[12] Gels are fixed and stained in 0.03% Coomassie Blue, 12% trichloroacetic acid, 50% methanol; destained in 10% methanol, 7% acetic acid; and sliced at 2-mm intervals. Gel slices are solubilized in scintillation fluid containing 3% tissue solubilizer (NCS, Amersham) for 18 hr at 50°.[6] The ^{3}H content in the gel slices is measured by scintillation spectrometry, and the background radioactivity found in gel slices on either side of the PGH synthase band, M_r 70,000, is subtracted. The recovery of [^{3}H]PGH synthase during gel electrophoresis is determined by adding a known amount of pure acetyl-^{3}H-labeled enzyme from sheep vesicular gland (see next section) to a gel run in an identical manner to a test gel run as described above. For example, pure [acetyl-^{3}H]PGH synthase containing 1000 cpm would be added to one gel and if the number of cpm in the band of interest (M_r 70,000) increased by 500 cpm, then recovery would be 50%. Using the recovery data, the specific radioactivity of the [acetyl-^{3}H]aspirin, and the ^{3}H content in the gel, M_r 70,000; one can determine the amount of active PGH synthase present in a given cell or enzyme preparation.

Preparation of [Acetyl-^{3}H]PGH Synthase

Pure [acetyl-^{3}H]PGH synthase is a useful reagent in immunological studies; as an antigen for immunization, and as a standard for radioimmune assay of the enzyme. Also, study of pure [acetyl-^{3}H]PGH synthase has provided structural informtion about the enzyme.[3]

[acetyl-^{3}H]PGH synthase has been prepared from ram seminal vesicles. Frozen vesicles (25 g) are homogenized in 100 ml of 0.1 M potas-

[12] K. Weber and M. Osborn, *in* "The Proteins" (H. Neurath and R. L. Hill, eds.), 3rd ed., Vol. 1, p. 180. Academic Press, New York, 1975.

sium phosphate, pH 8.0, 0.5 *M* KCl, 5 mg of bovine serum albumin per milliliter, 10 m*M* EDTA, 1 m*M* NADH, and 1 m*M* diethyldithiocarbamic acid (DDC); centrifuged (12,000 *g*, 10 min, 4°), and filtered through gauze. Microsomes are obtained from the supernatant by centrifugation (100,-000 *g*, 60 min, 4°) and suspended in 0.02 *M* potassium phosphate, pH 7.4, 1 m*M* EDTA, 1 m*M* NADH, and 1 m*M* DDC. [acetyl-^{3}H]Aspirin (100 μM, 100 Ci/mol) is added, and the mixture is incubated at 37° until no enzyme activity remains as measured by O_2 electrode (usually 20–30 min). The microsomes are recentrifuged (100,000 *g*, 60 min, 4°), suspended in 0.02 *M* potassium phosphate, pH 7.4, and solubilized by addition of 2% Tween 20. The mixture is recentrifuged (100,000 *g*, 60 min, 4°), and the solubilized enzyme is subjected to ion exchange column chromatography on DEAE-cellulose (20 × 2.5 cm) equilibrated with 0.02 *M* potassium phosphate, pH 7.4, containing 0.1% Tween 20. The enzyme does not adhere to DEAE-cellulose under these conditions. After the DEAE cellulose step, the enzyme is about 70% pure by SDS gel electrophoretic analysis, if the starting material had good activity (1200–1600 units of activity per 25 g of seminal vesicles; 1 unit of activity = 1 μmol fatty acid substrate consumed per minute at 30° by the initial rate of oxygen uptake using an O_2 electrode assay). The material from the DEAE step is dialyzed for 1 hr against water to decrease the ionic strength and then used as the 0% sucrose fraction in a preparative isoelectric focusing column. The column is poured with a 0 to 50% sucrose gradient, containing 0.1% Tween 20 and 2% Ampholines (80% of pH 5–8, and 20% of pH 3.5–10).

The focusing is performed at 0° for 36 hr at constant voltage (1000 V). The acetyl-^{3}H-labeled enzyme focuses at pH 6.2–6.5. Peak fractions are detected by ^{3}H content and pooled. Sodium dodecyl sulfate is added to give a concentration of 2%, and the material is dialyzed extensively against 0.001 *M* sodium phosphate, pH 7.0, 0.02% sodium azide, and 0.05% SDS, and lyophilized. The dry material is solubilized in 1% SDS, 0.01 *M* sodium phosphate, pH 7.0, and applied to a column of agarose (A-1.5m, Bio-Rad, Richmond, California) in 0.01 *M* sodium phosphate, pH 7.0, 0.02% sodium azide, and 0.05% SDS at room temperature. The acetyl-^{3}H-labeled enzyme elutes as a protein of M_r 70,000, detected by its ^{3}H content. Fractions containing the acetyl-^{3}H-labeled protein are pooled and constitute the preparation of purified [acetyl-^{3}H]PGH_2 synthase. The material gives a single band on SDS gel electrophoresis, M_r 70,000. The protein has a unique amino-terminal sequence (Ala-Asp-Pro-Gly-Ala-Pro-Ala-Pro-Val-Asn-Pro-Met-Gly-), with a defined amino acid composition[3] and absorption coefficient, $\epsilon_{280\,nm}$ = 130 mM^{-1} cm^{-1}. The purified enzyme contains a single [^{3}H]acetate per enzyme monomer.

Acknowledgments

This work was supported by NIH Grant HL 20974, and a Grant-in-Aid from the American Heart Association with funds contributed in part by the Connecticut Heart Association and the University of Connecticut Research Foundation. The author is an Established Investigator of the American Heart Association.

[52] Synthesis of Stable Thromboxane A_2 Analogs: Pinane Thromboxane A_2 (PTA_2) and Carbocyclic Thromboxane A_2 (CTA_2)

By K. C. NICOLAOU and RONALD L. MAGOLDA

Thromboxane A_2 (TxA_2) is an unstable substance (half-life ca 32 sec at pH 7.4 in aqueous solution at 37°) produced by blood platelets with potent vasoconstricting and thrombotic properties.[1] This important arachidonic acid cascade metabolite is generated from prostaglandin endoperoxide H_2 (PGH_2, Fig. 1) by thromboxane A_2 synthase. Although it has not yet been isolated or chemically synthesized, its structure was proposed in 1975 by Samuelsson *et al.* on the basis of its biosynthetic origin and chemical properties.[1] A great deal of biology surrounding this biomolecule has already been generated owing to its extremely interesting and unique biological profile, which is opposite to that of prostacyclin (PGI_2) (Fig. 1), a compound with antithrombotic and vasodilatory properties also generated from PGH_2.[2,3] Although both TxA_2 and PGI_2 are biologically very potent, they are relatively unstable chemically, being transformed rapidly to their stable metabolites, thromboxane B_2 (TxB_2) and 6-ketoprostaglandin $F_{1\alpha}$ (6-keto-$PGF_{1\alpha}$) (Fig. 1).

In view of the important physiological properties and chemical instability of TxA_2 we embarked on a program directed toward the synthesis of stable analogs of this molecule that might exhibit agonistic or antagonistic properties and/or inhibit thromboxane synthase. Such compounds would facilitate research in this area and may prove to be therapeutically useful. We describe here the synthesis of pinane thromboxane (PTA_2)[4] and car-

[1] M. Hamberg, J. Svensson, and B. Samuelsson, *Proc. Natl. Acad. Sci. U.S.A.* **72,** 2994 (1975).

[2] S. Moncada, R. Gryglewski, S. Bunting, and J. R. Vane, *Nature* (*London*) **263,** 663 (1976).

[3] R. A. Johnson, D. R. Morton, J. H. Kinner, R. R. Gorman, J. C. McGuire, F. F. Sun, N. Whittager, S. Bunting, J. Salmon, S. Moncada, and J. R. Vane, *Prostaglandins* **12,** 915 (1976).

[4] K. C. Nicolaou, R. L. Magolda, J. B. Smith, D. Aharony, E. F. Smith, and A. M. Lefer, *Proc. Natl. Acad. Sci. U.S.A.* **76,** 2566 (1979).

ISBN 0-12-181986-8

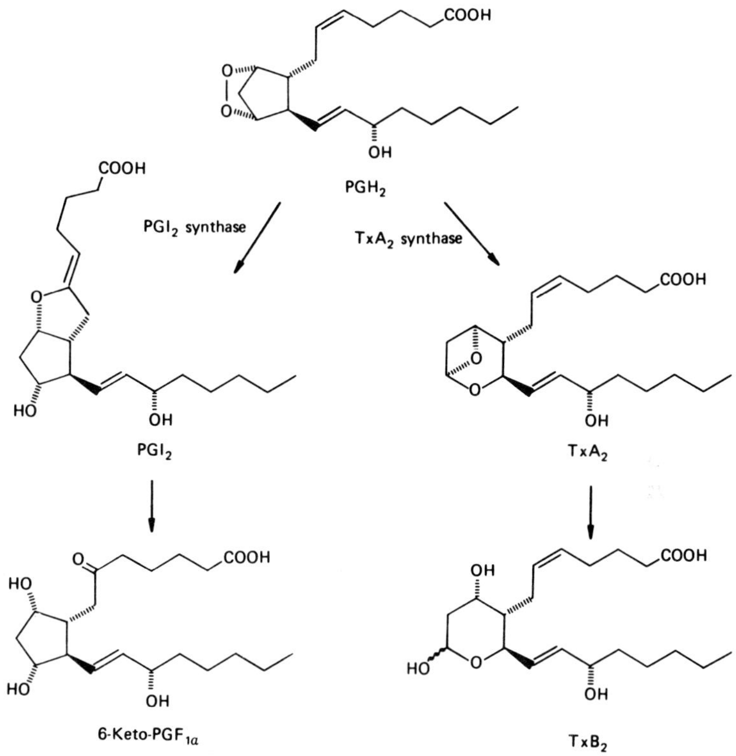

FIG. 1. Biosynthesis and degradation of thromboxane A_2 (TxA_2) and prostacyclin (PGI_2).

bocyclic thromboxane A_2 (CTA_2),[5,6] two stable and biologically active structural analogs of TxA_2 (TxA_2) (Fig. 1).

Synthesis of Pinane Thromboxane A_2 (PTA_2).[4]

The commercially available (−)-myrtenal (**1**)[7] underwent smooth 1,4-addition with the mixed cuprate derived from (±)-*trans*-lithio-1-octen-3-ol-*t*-butyldimethylsilyl ether and 1-pentynylcopper hexamethylphosphorus triamide complex to afford the adlehyde **3** in 80% yield. Reaction of **3** with methoxymethylenetriphenylphosphorane in toluene–tetrahydrofuran solution at 0° furnished the enol ether **4** (mixture of geometrical

[5] K. C. Nicolaou, R. L. Magolda, and D. A. Claremon, *J. Am. Chem. Soc.* **102,** 1404 (1980).

[6] A. M. Lefer, E. F. Smith III, H. Araki, J. B. Smith, D. Aharony, D. A. Claremon, R. L. Magolda, and K. C. Nicolaou, *Proc. Natl. Acad. Sci. U.S.A.* **77,** 1706 (1980).

[7] Aldrich Chemical Co., Milwaukee, Wisconsin 53233.

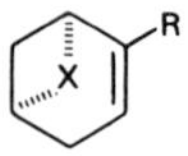

1, X = $C(CH_3)_2$, R = CHO
2, X = CH_2, R = CHO

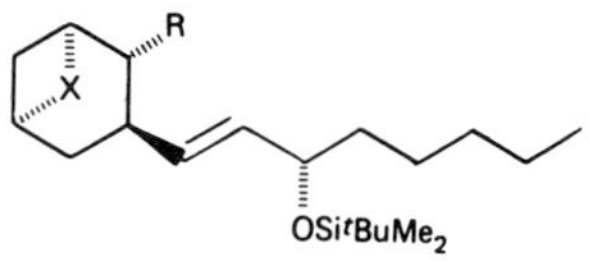

OSitBuMe$_2$

3, X = $C(CH_3)_2$, R = CHO
4, X = $C(CH_3)_2$, R = CH=CHOCH$_3$
5, X = $C(CH_3)_2$, R = CH_2CHO
6, X = $C(CH_3)_2$, R = $CH_2CH{=}CH(CH_2)_3COOCH_3$ (*cis*)
7, X = CH_2, R = CHO
8, X = CH_2, R = CH=CHOCH$_3$
9, X = CH_2, R = CH_2CHO
10, X = CH_2, R = $CH_2CH{=}CH(CH_2)_3COOCH_3$ (*cis*)

COOR
OH

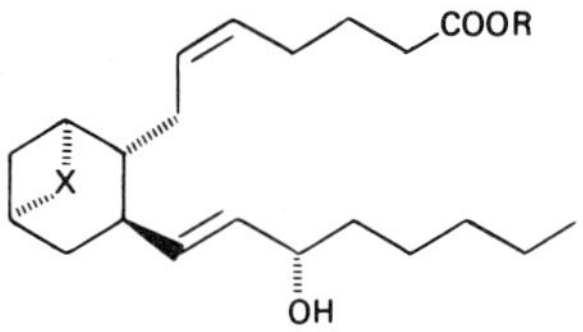

11, X = $C(CH_3)_2$, R = CH_3
12, X = $C(CH_3)_2$, R = H
13, X = CH_2, R = CH_3
14, X = CH_2, R = H

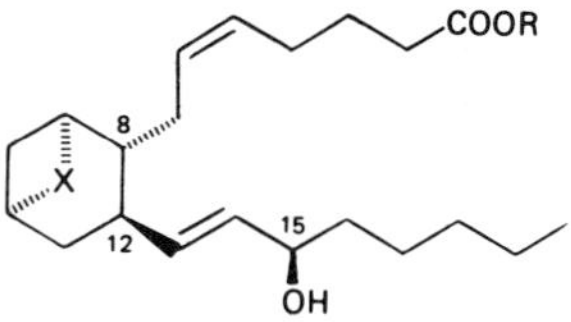

COOR
8
12
15
OH

15, X = $C(CH_3)_2$, R = CH_3
16, X = $C(CH_3)_2$, R = H
17, X = CH_2, R = CH_3 (8β, 12α, 15α)
18, X = CH_2, R = H (8β, 12α, 15α)

FIG. 2. Synthesis of pinane thromboxane A_2(PTA_2) and carbocyclic thromboxane A_2(CTA_2).

isomers, 94% yield) which was converted quantitatively to the aldehyde **5** by the action of $Hg(OAc)_2$–KI in aqueous tetrahydrofuran. The top side chain was completed by a Wittig reaction employing the sodium salt of 4-carboxybutylidenetriphenylphosphorane in dimethyl sulfoxide, leading after diazomethane treatment to the methyl ester **6** (mixture of C-15 epimers, PG numbering) in 80% yield. Deprotection of the hydroxy group by exposure to acetic acid–water–tetrahydrofuran (3:2:2, v/v/v) at 45° led to the methyl esters **11** and **15** (1:1, 100%), which were separated by preparative thin-layer chromatography (TLC) (silica, ether–petroleum ether 1:1, v/v) or flash column chromatography. Hydrolysis of the more polar isomer (**11**) with lithium hydroxide in aqueous tetrahydrofuran led to PTA_2 (**12**) in quantitative yield, whereas similar hydrolysis of the less polar compound (**15**) furnished the 15-epimer (PG numbering) **16** (100% yield). The stereochemistry of **12** and **16** was based on spectroscopic and chromatographic properties.[4] Both PTA_2 (**12**) and its 15-epimer are stable at 25° in solution or neat for prolonged periods of time.

Synthesis of Carbocyclic Thromboxane A_2[5]

The synthesis of carbocyclic thromboxane A_2 (CTA_2) starts with 2-formylbicyclo[3.1.1]hept-2-ene (**2**)[5] and proceeds along similar lines as for the preparation of PTA_2. Thus, 1,4-addition of the cuprate reagent obtained from (*t*)-*trans*-1-lithio-1-octen-3-ol-*t*-butyldimethylsilyl ether and 1-pentynylcopper–hexamethylphosphorus triamide (HMPT) complex to the aldehyde **2** produced, after exposure to anhydrous potassium carbonate in anhydrous methanol, the trans aldehyde **7** (mixture of 15-epimers, PG numbering) in 54% yield. The top chain was completed by (*a*) reaction of **7** with methoxymethylenetriphenylphosphorane in toluene–tetrahydrofuran furnishing the enol ether **8** (mixture of geometrical isomers, 81% yield; (*b*) generation of the aldehyde **9** (98% yield) with $Hg(OAc)_2$–KI in aqueous tetrahydrofuran; (*c*) Wittig reaction of **9** with the sodium salt of 4-carboxybutylidenetriphenylphosphorane in dimethyl sulfoxide leading after diazomethane treatment to the methyl ester **10** (mixture of C-15 epimers, PG numbering, 74% yield). The silyl ether was removed with acetic acid–water–tetrahydrofuran (3:2:2, v/v/v) at 45° to afford the two diastereoisomeric methyl esters **13** (65%) and **17** (33%) separated chromatographically. Hydrolysis of the more polar compound (**13**) with aqueous lithium hydroxide led to CTA_2 (**14**) in 95% yield, whereas similar hydrolysis of the less polar isomer (**17**) furnished its diastereoisomer **18** (reverse side chain stereochemistries 97% yield). The complete stereochemical structures of CTA_2 (**14**) and its diastereoisomer (**18**) were based on spectroscopic and chromatographic properties.[5] Both CTA_2 and its isomer were found to be stable at ambient temperature for prolonged periods of time either in solution or neat.

Biological Properties of PTA_2[4] and CTA_2[6]

PTA_2 at relatively low concentrations was found to inhibit cat coronary artery constriction induced by stable prostaglandin endoperoxide analogs. At slightly higher concentrations it inhibited platelet aggregation. At still higher concentrations it inhibited thromboxane synthase but had no effect on prostacyclin synthase. It was suggested that PTA_2 has a suitable biological profile as a potential antithrombotic agent.[4] CTA_2 is a potent cat coronary vasoconstrictor, acting at concentrations as low as 29 p*M*. At 1–5 p*M* CTA_2 inhibits arachidonic acid- and PG endoperoxide-induced aggregation of platelets. It also inhibits thromboxane synthase, but it does not inhibit prostacyclin synthesis. Thus CTA_2 dissociates coronary vasoconstriction from platelet-aggregating activity. *In vivo* CTA_2 induces myocardial ischemia and sudden death in rabbits by vasoconstriction, but in the absence of pulmonary or coronary thrombosis.[6]

Experimental Section

*Preparation of Aldehyde **3***

A solution of (±)-*trans*-1-iodo-3-*t*-butyldimethylsilyl-1-octen-3-ol (3.65 g, 9.92 mmol) in anhydrous ether (15 ml) was cooled to −78° under argon and treated dropwise, while magnetically stirred, with *t*-butyllithium (10.5 ml, 1.9 *M* in hexane, 19.95 mmol). After 3 hr stirring at −78° a solution of 1-pentynylcopper (1.4 g, 10.6 mmol) and HMPT (hexamethyl phosphorus triamide) (3.49 g ≡ 3.89 ml, 21.5 mmol) in anhydrous ether (2.5 ml) was added dropwise at −52.5° ± 2.5°, allowed to stir at that temperature for 15 min, and then cooled to −78° before aldehyde **1** (1.0 g, 6.6 mmol) in anhydrous ether (2 ml) was added dropwise. Stirring at −78° was continued for 4 hr (followed by TLC), then the reaction mixture was diluted with cold (0°) ether (200 ml) and quenched with saturated ammonium sulfate (75 ml). The organic layer was separated and washed successively with (*a*) saturated ammonium sulfate solution (75 ml), (*b*) water (75 ml); (*c*) 1% sulfuric acid (3 × 50 ml); (*d*) water (75 ml); (*e*) saturated sodium bicarbonate solution (75 ml); (*f*) brine (saturated aqueous solution of sodium chloride) (75 ml). The organic layer was dried with anhydrous magnesium sulfate and evaporated to give an oily residue that was purified by flash chromatography (230–400 mesh silica gel, 5% ether in petroleum ether) furnishing the aldehyde **3** (2.10 g, 80% yield) as an oil, $R_f = 0.42$.

*Preparation of Aldehyde **5** via Methoxyenol Ether **4***

Methoxymethyltriphenylphosphonium chloride (3.92 g, 11.4 mmol) was suspended in dry toluene (20.5 ml) in a flame-dried flask equipped with a magnetic stirrer. To this cold (0°) stirred suspension was added dropwise, under argon, a solution of lithium diisopropylamide (33.5 ml 0.33 *M*, 11.1 mmol, prepared from *n*-BuLi in pentane and diisopropylamine in THF). The resulting bright red solution was stirred at 0° for 15 min before the dropwise addition of the azeotropically dried (benzene) aldehyde **3** (1.49 g, 3.80 mmol). After allowing the reaction mixture to stir at 0° until complete by TLC (ca 30 min), the reaction was diluted with ice-cold ether (150 ml) and quenched with water (30 ml). The aqueous layer was reextracted with ether (50 ml), and the combined organic extracts were successively washed with water (2 × 50 ml) and brine (50 ml). Drying with magnesium sulfate and concentration under vacuum provided a residue that was subjected to flash column chromatography (silica, 2.5% ether in petroleum ether) furnishing the vinyl ether **4** (1.50 g, 94% yield) as a mixture of cis and trans (2:3) geometrical isomers, oil, $R_f = 0.75$.

The enol ether (450 mg, 1.07 mmol, cis/trans mixture) in water (2.7 ml)–tetrahydrofuran (27.3 ml) was stirred under argon at 25° and treated with $Hg(OAc)_2$ (1.02 g, 3.2 mmol). After stirring for 1 hr, the yellow mixture was poured into a 7% aqueous solution of potassium iodide (107 ml) and extracted with benzene (3 × 75 ml). The combined organic fractions were successively washed with (*a*) 7% aqueous potassium iodide (50 ml), (*b*) water (50 ml), and (*c*) brine (50 ml), dried with anhydrous magnesium sulfate, and evaporated to supply essentially pure aldehyde **5** (430 mg, 99% yield); oil; $R_f = 0.31$ (silica, 5% ether in petroleum ether).

*Preparation of Methyl Ester **6***

The crude aldehyde **5** (430 mg, 1.06 mmol) was azeotropically dried (benzene) and allowed to react with the ylid derived from (4-carboxybutyl)triphenylphosphonium bromide (1.41 g, 3.18 mmol) in dry dimethyl sulfoxide (7.40 ml) and dimsyl sodium (3.18 ml, 2 *M*, 6.36 mmol) at 25°. With the completion of the reaction (1 hr), the reaction mixture was diluted with ice-cold ether (100 ml) and carefully acidified with oxalic acid (1 *M*) to pH 3–4. The organic fractions resulting from repeated extractions of the aqueous phase with ether (4 × 50 ml) were combined, washed with water (50 ml) and brine (50 ml), dried with anhydrous magnesium sulfate, and concentrated under vacuum. The residue was diluted with ether (25 ml) and exposed to excess diazomethane (ether solution) at 0°. The resulting mixture was concentrated and purified by flash column chromatography (silica, 5% ether in petroleum ether) to supply the methyl ester **6** (420 mg, 80% yield) as an oil, $R_f = 0.65$.

*Removal of the Silyl Ether Protecting Group and Separation of Epimers **11** and **15***

Deprotection was achieved by stirring the silyl ether **6** (mixture of epimers, 175 mg, 0.35 mmol) at 45° for 10 hr with acetic acid–tetrahydrofuran–water (3:2:2, 10 ml). After cooling to room temperature, the reaction mixture was diluted with dichloromethane (100 ml) and water (10 ml). The combined organic fractions, after reextraction of the aqueous layer with dichloromethane (2 × 25 ml), were washed with (*a*) water (3 × 50 ml), (*b*) saturated sodium bicarbonate (2 × 25 ml), and (*c*) brine (25 ml). The solution was dried with anhydrous magnesium sulfate and evaporated to afford a residue that was subjected to preparative layer chromatography (PLC) (4 silica gel 60 F-254 plates, 20 cm × 20 cm × 0.50 mm; 50%

ether in petroleum ether), supplying the two hydroxymethyl esters **12** (67 mg, 51%, R_f = 53) and **16** (66 mg, 49%, R_f = 0.59) as oils.

*Preparation of PTA$_2$ (**12**) and 15-Epi-PTA$_2$ (**16**)*

The methyl ester **11** (56 mg, 0.14 mmol) in tetrahydrofuran (THF) and water (6.5 ml) was allowed to react with 1 *N* lithium hydroxide solution (0.70 mmol). After 12 hr stirring at 25° (complete by TLC), the THF was removed under vacuum and the aqueous solution mixed with ice-cold ether (50 ml) was acidified to pH 3–4 with 1 *N* oxalic acid at 0°. The aqueous phase was saturated with solid sodium chloride to aid extraction and extracted with ether (5 × 25 ml). The combined organic layer was extracted with water (50 ml) and brine (25 ml) before drying over anhydrous magnesium sulfate and concentrated under vacuum, affording essentially pure PTA$_2$ (**12**), which was repurified by preparative layer chromatography (2 silica plates, 20 cm × 20 cm × 0.25 mm, 50% ether in petroleum ether, 2 developments) to give pure PTA$_2$ (**12**) (52 mg, 98% yield) as an oil, R_f = 0.66 (silica, ether); IR (CCl_4) ν_{max} 3300 (OH), 1710 cm^{-1} (COOH); ^1H NMR ($CDCl_3$, 360 MHz) τ 2.95 (bs, 2H, OH, COOH), 4.55 (m, 4H, olefin), 5.88 (m, 1H, CHO), 7.60–8.78 (m, 24H), 8.80 (s, 3H, CH_3), 8.93 (s, 3H, CH_3), 9.12 (t, J = 7.5 Hz, 3H, CH_3); mass spectrum m/e (relative intensity) 376 (M^+, 1), 358 ($M^+ - H_2O$, 7), 71 (100%). The purity of this compound is estimated by ^1H NMR spectroscopy to be $\geqslant$98%.

The 15-epimer **16** of PTA$_2$ was prepared from the ester **15** in exactly the same way as described above for PTA$_2$.

*Preparation of Aldehyde **7***

A solution of (+)-*trans*-1-iodo-3-*t*-butyldimethylsilyl-1-octen-3-ol (4.74 g, 12.9 mmol) in anhydrous ether (8.5 ml) was cooled to −78° under argon and treated dropwise, while magnetically stirred, with *t*-butyllithium (21.5 ml, 1.2 *M* in pentane, 25.8 mmol). After 3 hr stirring at −78°, an ethereal solution of 1-pentynylcopper–HMPT complex prepared from 1-pentynylcopper (1.73 g, 13.2 mmol) and HMPT (4.80 ml, 4.32 g, 25.5 mmol) in ether (3 ml) was added dropwise at −52.5° ± 2.5°, allowed to stir at that temperature for 15 min, and then cooled to −78° before the aldehyde **2** (400 mg, 2.23 mmol) in ether (3 ml) was added dropwise. Stirring at −78° was continued for 4 hr (or until TLC indicated complete reaction), and then the reaction mixture was diluted with cold (0°) ether (250 ml) and quenched with saturated ammonium sulfate solution (100 ml). The organic phase was washed with saturated ammonium sulfate solution (100 ml). The organic phase was washed successively with (*a*) saturated ammonium sulfate (75 ml), (*b*) water (100 ml), (*c*) 1% sulfuric acid (3 × 75 ml), (*d*) water (100 ml), (*e*) saturated sodium bicarbonate solution (100 ml), and (*f*) brine (100 ml).

The organic layer was dried with anhydrous magnesium sulfate and evaporated to give an oily residue that was subjected to flash column chromatography (silica, 2.5% ether in petroleum ether) furnishing the aldehyde **7** (662 mg, 56%) as an oil, $R_f = 0.39$ (silica, 5% ether in petroleum ether). The ^{1}H NMR spectrum of this material indicated a trans/cis mixture of ca 2 : 1. Before proceeding, therefore, this material was completely epimerized to the desired trans aldehyde **7** by basic treatment as follows: The cis/trans mixture of aldehydes obtained as above (366 mg, 1 mmole) was dissolved in absolute methanol (20 ml) and treated under argon with anhydrous potassium carbonate (69 mg, 0.5 mmol) at room temperature for 12 hr. The mixture was then diluted with ether (100 ml) and brine (25 ml) and neutralized with 1 *N* oxalic acid (1.0 ml). The organic layer was separated, and the aqueous phase was extracted with ether (2 × 50 ml). The combined ether extract was washed with brine (25 ml), dried with anhydrous magnesium sulfate, and evaporated to afford essentially pure trans aldehyde **7** (348 mg, 95%).

*Preparation of Aldehyde **9** via Methoxyenol Ether **8***

Methoxymethyltriphenylphosphonium chloride (1.41 g, 4.1 mmol) was suspended in dry toluene (7.0 ml) in the flame-dried flask equipped with a magnetic stirrer. To this cold (0°), stirred suspension was added dropwise, under argon, a solution of lithium diisopropylamide (12.4 ml 0.33 *M*, 4.1 mmol; prepared from *n*-BuLi in pentane and diisopropylamine in THF). The bright red solution was stirred at 0° for 15 min before the aldehyde **7** (500 mg, 1.4 mmol) in toluene (1 ml) was added dropwise at the same temperature. After 30 min, TLC indicated complete reaction and the mixture was quenched with ice-water (30 ml) and ether (100 ml). The organic layer was separated, and the aqueous phase was reextracted with ether (50 ml). The combined organic solution was washed with brine (25 ml), dried with magnesium sulfate and evaporated to afford an oily residue that was subjected to flash column chromatography (silica, 5% ether in petroleum ether) furnishing pure methoxyenol ether **8** (436 mg, 81%), oil, $R_f = 0.61$ (mixture of cis and trans geometrical isomers ca 3 : 2).

The enol ether **8** (197 mg, 0.5 mmol, mixture of cis/trans isomers) in THF (13.6 ml) and water (1.5 ml) stirred under argon at room temperature was treated with $Hg(OAc)_2$ (477 mg, 1.5 mmol). After 1 hr stirring at 25°, the yellow mixture was poured onto a 7% aqueous solution of potassium iodide (100 ml) and extracted with benzene (2 × 75 ml). The combined organic layer was washed with (*a*) 7% aqueous potassium iodide solution (40 ml) and (*b*) brine (25 ml), dried with magnesium sulfate, and evaporated to afford essentially pure aldehyde **9** (186 mg, 98%), oil; $R_f = 0.29$ (silica, 5% ether in petroleum ether).

*Preparation of Methyl Ester **10***

The aldehyde **9** (186 mg, 0.49 mmol) obtained above was allowed to react without further purification with the ylid derived from (4-carboxybutyl)triphenylphosphonium bromide (1.33 g, 1.5 mmol) in dry dimethyl sulfoxide (1.5 ml) and dimsyl sodium (1.5 ml, 2 *M*, 3 mmol) at 25°. The reaction was complete in 1 hr and was then diluted with ice-cold ether (50 ml) and acidified carefully with 1 *N* oxalic acid to pH 3–4. The organic layer was separated and the aqueous phase was extracted with ether (3 × 50 ml); the combined organic solution was washed with water (50 ml) and brine (50 ml) and dried with magnesium sulfate. The solvents were removed, and the resulting mixture was taken up in ether (20 ml) and esterified with ethereal diazomethane at 0°. After complete reaction (indicated by TLC), the mixture was concentrated and flash-chromatographed (silica, 5% ether in petroleum ether) to afford the methyl ester silyl ether **10** (mixture of diastereoisomers, 178 mg, 74%), $R_f = 0.28$.

*Removal of Silyl Ether Protecting Group and Separation of Isomers **13** and **17***

The silyl ether **10** (mixture of diastereoisomers, 245 mg, 0.5 mmol) in acetic acid–tetrahydrofuran–water (3:2:2, v/v/v, 14 ml) was stirred under argon at 45° for 10 hr. After cooling to room temperature, the reaction mixture was diluted with dichloromethane (75 ml) and the organic phase was separated. The aqueous phase was reextracted with dichloromethane (75 ml), and the combined organic solution was washed with water (50 ml) and brine (50 ml), dried with magnesium sulfate, and concentrated to afford a mixture of hydroxy epimers. Separation of the two isomers was performed by preparative layer chromatography (8 silica plates; 20 × 20 cm × 0.25 mm; 10% ethyl acetate in petroleum ether, 3 elutions) furnishing the pure isomers, the more polar CTA_2 methyl ester (**13**), oil, $R_f = 0.29$ (123 mg, 65%) and the less polar isomer, methyl ester (**17**), oil, $R_f = 0.33$ (62 mg, 33%).

*Preparation of CTA_2 (**14**) and Its Diastereoisomer (**18**)*

The methyl ester **13** (50 mg, 0.14 mmol) in THF (6.5 ml) and water (6.5 ml) was treated at room temperature with stirring under argon with 1 *N* lithium hydroxide solution (0.70 ml, 0.70 mmol). After 12 hr stirring at ambient temperature (to complete reaction as indicated by TLC), the THF was removed under vacuum and the aqueous solution mixed with ether (50 ml) was acidified to pH 3–4 with 1 *N* oxalic acid at 0°. The aqueous phase was saturated with solid sodium chloride to aid extraction of the product into the organic phase and extracted a total of four times with ether (30 ml each). The combined ether solution was washed with water (25 ml) and brine (25 ml) before drying with magnesium sulfate; it was concentrated to afford essentially pure CTA_2 (**14**), which was repurified by preparative layer chromatography (2 silica plates; 20 × 20 cm × 0.25 mm; ether–petroleum ether, 1:1, v/v) (47 mg, 95% yield), oil, R_f = 0.21; IR (CCl_4) ν_{max} 3333 (OH), 1709 cm^{-1} (COOH); ^1H NMR ($CDCl_3$, 360 MHz) τ 2.90 (bs, 2H, OH, COOH), 4.48 (m, 2H, olefin), 4.62 (m, 2H, olefin), 5.85 (q, J = 6 Hz, 1H, C*H*O), 7.68 (m, 3H), 7.75 (m, 1H), 7.83 (m, 4H), 8.00 (m, 2H), 8.27 (m, 2H), 8.37 (m, 2H), 8.47 (m, 5H), 8.70 (m, 6H), 9.02 (t, J = 9.0 Hz, 1H), 9.12 (t, J = 7.5 Hz, 3H, CH_3); mass spectrum *m/e* (relative intensity) 330 ($m^+ - H_2O$, 3), 203 (33), 67 (100%). The purity of this material is estimated by ^1H NMR spectroscopy to be ≥98%.

The isomer **18** of CTA_2 was prepared from the ester **17** in exactly the same way as described above for CTA_2.

Acknowledgments

Financial support from the National Institutes of Health is gratefully acknowledged.

[53] Purification and Characterization of Leukotrienes from Mastocytoma Cells

By ROBERT C. MURPHY and W. RODNEY MATHEWS

The leukotriene pathway of arachidonic acid metabolism can be conveniently activated by stimulation of CXBGABMCT-1 cells.[1] This tumor is a transplantable mouse mastocytoma that has been described as one of several new mastocytomas induced by Abelson murine leukemia virus[2] in pristane-treated mice. When this is carried *in vivo* as an ascites tumor in the peritoneal cavity of certain strains of mice, it has been found to reproducibly yield leukotrienes after *in vitro* incubation with the calcium ionophore A23187. This cell line has also been successfully carried in tissue culture,[2] but the production of leukotrienes from tissue culture has not

[1] R. C. Murphy, S. Hammarström, and B. Samuelsson, *Proc. Natl. Acad. Sci. U.S.A.* **76,** 4275 (1979).

[2] G. R. Mendoza and H. Metzger, *J. Immunol.* **117,** 1573 (1976).

TABLE I
TYPICAL YIELDS OF ARACHIDONIC ACID LIPOXYGENASE PRODUCTS[a]

Product	Amount (μg/10^7 cells)
LTC_4	0.125
11-*trans*-LTC_4	0.02
LTD_4	0.002
LTB_4	0.1
5(*S*),12(*R*)-DHETE	0.07
5(*S*),12(*S*)-DHETE	0.06

[a] CXBGABMCT-1 cells were stimulated with the calcium ionophore A23187. LT, leukotriene; DHETE, dihydroxyeicosatetraenoic acid.

been fully evaluated. Advantages of the *in vivo* ascites tumor include a rather stable cell population for leukotriene production after numerous passages, and the large number of cells that can be harvested from each animal, typically 2 to 5 $\times$ 10^8 cells. Leukotriene C_4 and B_4 are the major leukotrienes produced (Table I) by these cells; smaller amounts of 11-*trans*-LTC_4 and LTD_4 are also produced. Several other lipoxygenase and cyclooxygenase products of arachidonic acid are also produced.

Reagents

Ca^{2+}-free buffer: The following reagents are made up to 1 liter with distilled water and adjusted to pH 7.0 with 30% NaOH: NaCl, 8.77 g; KCl, 0.28 g; Na_2HPO_4, 0.42 g; KH_2PO_4, 0.47 g; and glucose, 1.0 g.

Pristane (2,6,10,14-tetramethylpentadecane; Aldrich Chemical Co.)

A23187 (2 mg/ml ethanol, Calbiochem)

$CaCl_2$: 2.7 g of CaCl·2 H_2O dissolved in 100 ml of distilled water

NH_4Cl, 1% in distilled water

Silicic acid (Silicar cc-7, Mallinckrodt)

XAD-7 (Mallinckrodt)

Ascites Tumor Production

Murine mastocytoma cells (CXBGABMCT-1, gen. 12), obtained from Litton Bionetics (Bethesda, Maryland) are currently carried in $CB6F_1$ mice (Jackson Laboratory) having previously been carried successfully in CXBG mice.[2] Prior to the introduction of tumor cells, the mice must be primed. Priming refers to the injection of 0.5 ml of pristane into the peritoneal cavity of each mouse from 2 to 8 weeks prior to the introduction of the tumor. This procedure facilitates the growth of the tumor presumably by induction of a granuloma in the peritoneal cavity. An ampoule of frozen cells as supplied is immediately thawed and injected interperitoneally into one or two mice with a large (16 gauge) needle. Several (4–8) weeks are required before these mice have enough cells to be harvested, which can be detected by swelling of the abdomen. Once the cell line is established, primed mice are injected with 3 to 6 $\times$ 10^7 cells; approximately 2 weeks are required before harvesting. This amount of tumor will result in all mice dying within 3 weeks.

The tumor cells are harvested in Ca^{2+}-free incubation buffer. Mice are anesthetized in an ether chamber, decapitated, and placed on a tray ventral side up. The skin over the abdomen is nicked near the umbilicus without cutting the peritoneal musculature, and the skin is carefully pulled toward the head. Approximately 5 ml of incubation buffer is injected into the exposed peritoneal cavity and gently massaged for 30 sec. A small hole is cut in the peritoneum, and ascites fluid is removed with a Pasteur pipette. The cavity is washed with 5 ml of fresh buffer, which is then added to the ascites fluid stored in 50-ml plastic centrifuge tubes on ice. The cell suspensions are centrifuged at 400 *g* for 10 min, resuspended in fresh incubation buffer, counted, and diluted to a final concentration of 10^7 cells/ml with additional buffer. The cell line is maintained by injecting 3 to 6 $\times$ 10^7 cells each into previously primed mice. It is convenient to prime a desired number of new mice when the tumor cells are transferred.

Occasionally red blood cells can contaminate the CXBGABMCT-1 cells depending on the severity and tissue involvement of the ascites tumor. These can be removed by the method of Boyle[3] for the lysis of red blood cells using buffered ammonium chloride. After the red blood cell lysis, the cells are resuspended in fresh buffer at 10^7 cells/ml.

Incubation

Washed cells in Ca^{2+}-free buffer are warmed to 37° before stimulation, which proceeds in two steps. First, the calcium ionophore A23187 solution (0.05 ml per 10 ml of cell suspension) is added dropwise to a final concentration of 10 μg/ml. After 10 min, $CaCl_2$ (0.1 ml per 10 ml of cell suspension) is added slowly to a final concentration of 1.9m *M*. This initiates the leukotriene biosynthesis. Further, this reverse stimulation procedure has been found to be more reproducible than the addition of the A23187 ionophore to a Ca^{2+}-containing buffer. After 20 min, the reaction is stopped either by centrifugation at 1000 *g* for 10 minutes or by the addi-

[3] W. Boyle, *Transplantation* **6,** 761 (1968).

tion of ethanol (4 vol per volume of cell suspension). The former has the advantage of being a cleaner preparation for further purification of leukotrienes in the supernatant and is used in large-volume incubations. The latter is used to isolate all arachidonic acid metabolites and for time course studies. After centrifugation, the supernatant is made to 80% ethanol and stored overnight at 4°. Soluble proteins precipitate and are removed by filtration with Whatman No. 3 paper. The solution is evaporated to dryness with a rotary evaporator at 30°. The crude residue at this point has been found to be stable for long periods of time at 0°.

Purification Procedures

Two procedures have been used to purify leukotrienes from CXBGABMCT-1 cells, one for large-scale incubations (≥50 ml) and another for smaller incubations.

Method 1. The large-scale procedure involves an initial adsorption to XAD resin followed by silicic acid chromatography and two steps of reversed-phase high-pressure liquid chromatography (reversed-phase HPLC).

Amberlite XAD-7 and -8 have both been used and work equally well; however, XAD-8 is no longer available. Both of these resins require substantial preparation prior to use. This involves exhaustive extraction first with ethanol (1 liter per pound of resin) under vacuum several times followed by extraction of the resin with similar amounts of solvents in the following order: chloroform, methanol, and then 5 times with distilled water. The XAD resin is then stirred with 1 *N* HCl at 50° for 1 hr, washed neutral, and then stirred with 1 *N* NaOH at 50° for 1 hr. The resin is washed until neutral and further washed with distilled water five times before it can be stored at 4° in distilled water.

Clean XAD-7 suspended in distilled water is poured into a glass column at least 2.5 cm in diameter to a bed volume equal to the original cell incubation volume. The residue from the ethanolic supernatant is dissolved in 0.5 volume of 1% NH_4Cl and applied to the column. After elution of this solution the column is washed with 1 bed volume of 1% NH_4Cl. The leukotrienes are eluted with 1.5 bed volumes of absolute ethanol, which is then evaporated to near dryness at 30°. Activated silicic acid (0.5 g) is added to the residue along with 10 ml of ethanol, and the sample is taken to complete dryness as indicated by the silicic acid tumbling free in the round-bottom flask of the rotary evaporator.

Neutral silicic acid is activated by storing at 100° for at least 1 hr before use. A glass column (1 × 20 cm) is packed by gravity with a slurry of

3.5 g of silicic acid in 20 ml of ethyl acetate. The residue from XAD-7 that has been adsorbed onto silicic acid is poured onto the top of this silicic acid column, taking care to have a small volume of ethyl acetate always above the silicic acid. The column is then eluted successively with 50 ml each of ethyl acetate, 5% methanol, 10% methanol, 70% methanol in ethyl acetate (v/v), and 100% methanol. The ethyl acetate fraction contains LTB_4 and the 5,12-dihydroxyeicosatetraenoic acids, and the 70% and 100% methanol fractions contain LTC_4 and LTD_4. These silicic acid fractions are evaporated to dryness below 30° and can be stored at 0° prior to reversed-phase HPLC.

Preparative and analytical reversed-phase HPLC are used for the final purification of leukotrienes. The preparative column (10 × 250 mm) is packed with RSIL C_{18}, 10 μm particles (Alltech) and eluted with methanol–water–acetic acid, (69:31:0.02, v/v/v, adjusted to pH 5.7 with ammonium hyroxide) at 4 ml/min. The analytical column (4.6 × 250 mm) is packed with Nucleosil C_{18}, 5 μm particles (Macherey Nagel) and eluted with methanol–water–acetic acid (65:35:0.02, v/v/v) at a flow rate of 1 ml/min. The mobile phase is made with HPLC grade solvents as accurately as possible since the leukotriene retention times are greatly affected by polarity (methanol content) and pH.

The dried silicic acid fractions are dissolved in 1.0 ml of 30% methanol–water (v/v) and centrifuged at 10,000 *g* in a microcentrifuge if necessary to remove insoluble material. It has been found that large injection volumes can be used without noticeable HPLC peak broadening if the sample methanol content is substantially lower than the mobile phase. Using an HPLC ultraviolet detector set for absorption at 280 nm, the LTC_4 is seen to elute with a retention time approximately 15 min (Table II). The LTC_4 containing eluate is evaporated at 30° under reduced pressure, redissolved in 1 ml of 30% methanol and chromatographed on the analytical Nucleosil column. Pure LTC_4 and 11-*trans*-C_4 are obtained using the 280 nm detection to guide collection of the eluting components typically with retention times in Table II. The LTB_4 and the isomeric 5,12-dihydroxyeicosatetraenoic acids are purified from the silicic acid, ethyl acetate fractions in an identical manner.

Method 2. Small incubation volumes of CXBGABMCT-1 cells are amenable to a much simplified leukotriene isolation procedure. After the XAD-7 (prepared as described in method 1) adsorption step, the ethanol eluent is evaporated to dryness and dissolved in 2 ml of 30% methanol. It is often necessary to centrifuge (10,000 *g*) this sample to remove insoluble material. This is injected directly onto an analytical Nucleosil C_{18} column, and leukotrienes are eluted with methanol–water–acetic acid

TABLE II
CHROMATOGRAPHY OF LEUKOTRIENES ON REVERSED-PHASE HIGH-PRESSURE LIQUID CHROMATOGRAPHY COLUMNS[a]

	Retention time (min)		
Molecule	RSIL	Nucleosil I	Nucleosil II
LTC_4	14	31	10
11-*trans*-LTC_4	15	34	12
LTD_4	28	70	24
LTB_4	31.5	28	27
5(*S*),12(*R*)-DHETE	25	26	25
5(*S*),12(*S*)-DHETE	28	28	23
PGB_2	18.5	17	16.8

[a] Mobile phases are as described in the text; Nucleosil I used the pH 4 (unadjusted) system; nucleosil II refers to pH 5.7.

(65:35:0.02, v/v/v, pH 5.7) mobile phase as indicated by the UV monitor set at 280 nm. This procedure permits the isolation and quantitation of both LTC_4 and LTB_4.

Purified leukotrienes are relatively stable when stored under argon in the HPLC mobile phase at $-70°$. Samples have been stored for several months without appreciable degradation. LTC_4 from CXBGABMCT-1 cells has been found to be stable in saline at 1 μg/ml when frozen at $-20°$ for more than a month. These compounds, however, have been found to be unstable when dried and exposed to room temperature or above.

Characterization of Leukotrienes

Leukotrienes have a very characteristic UV absorption of a conjugated triene with a triplet profile centered at 280 nm for LTC_4 and LTD_4 and at 270 nm for LTB_4 and the isomeric 5,12-dihydroxyeicosatetraenoic acids. These metabolites obey the Beer–Lambert law, and molar extinction coefficients are listed in Table III. Reasonable quantities of leukotrienes purified by reversed-phase HPLC can be quantitated, knowing the volume of the solution and using the F280 factors listed in Table III to multiply the observed absorbance at 280 nm to obtain the concentration in micrograms per milliliter. Care must be exercised adequately to correct for solvent absorption in such measurements. The HPLC retention times along with UV absorption profile are necessary, but not sufficient evidence to establish the identity of these leukotrienes.

The guinea pig ilium bioassay as described in this volume [54] can be used to verify the presence of the myotropic leukotrienes LTC_4, LTD_4,

TABLE III
ULTRAVIOLET ABSORPTION CHARACTERISTICS OF VARIOUS LEUKOTRIENES

Compound	λ_{max} (molar extinction) (nm)			F280[a]	Reference[b]
LTC_4	270 (32,000)	280 (40,000)	290 (31,000)	15.6	1
11-*trans*-LTC_4	268 (32,000)	278 (40,000)	290 (31,000)	15.6	2
LTD_4	270 (32,000)	280 (40,000)	290 (31,000)	12.4	3
LTE_4	270 (31,000)	280 (40,000)	290 (31,000)	11.0	4
LTB_4	260 (38,000)	270 (50,000)	280 (39,000)	8.6	5
5(*S*),12(*R*)-DHETE	258 (42,000)	268 (56,000)	280 (44,000)	7.6	6
5(*S*),12(*S*)-DHETE	258 (42,000)	268 (56,000)	280 (44,000)	7.6	6

[a] Absorbance measured at 280 nm of leukotriene multiplied by F280 corresponds to concentration in micrograms per milliliter of solution.

[b] Key to references: (1) E. J. Corey, P. A. Clark, G. Goto, A. Marfat, C. Mioskowski, B. Samuelsson, and S. Hammarström, *J. Am. Chem. Soc.* **102,** 1436 (1980); (2) D. A. Clark, G. Goto, A. Marfat, E. J. Corey, S. Hammarström, and B. Samuelsson, *Biochem. Biophys. Res. Commun.* **94,** 1133 (1980); (3) R. A. Lewis, K. F. Austen, J. M. Drazen, D. A. Clark, A. Marfat, and E. J. Corey, *Proc. Natl. Acad. Sci. U.S.A.* **77,** 3710 (1980); (4) R. A. Lewis, J. M. Drazen, K. F. Austen, D. A. Clark, and E. J. Corey, *Biochem. Biophys. Res. Commun.* **96,** 271 (1980); (5) E. J. Corey, P. B. Hopkins, J. E. Munroe, A. Marfat, and S. Hashimoto, *J. Am. Chem. Soc.* **102,** 7984 (1980); (6) P. Borgeat and B. Samuelsson, *J. Biol. Chem.* **254,** 3643 (1979).

TABLE IV
GAS CHROMATOGRAPHIC RETENTION INDEX AND DIAGNOSTIC MASS SPECTRAL IONS OF VARIOUS LEUKOTRIENES[a]

Compound	Equivalent chain length[b]	$M^{\dagger}$	Selected mass spectral ions[c] *m/z* (relative intensity)
LTB_4	23.6	494 (0.1)	129(100), 203(23), 217(35), 293(17), 383(5), 404(0.5), 463(0.2), 479(0.2)
5(*S*),12(*R*)-DHETE	25.0	494 (0.0)	129(100), 203(15), 217(33), 293(26), 383(6), 404(0.4), 463(0.6), 479(0.1)
5(*S*),12(*S*)-DHETE	25.0	494 (0.0)	129(100), 203(19), 217(41), 293(27), 383(7), 404(0.7), 463(0.7), 479(0.2)

[a] Chromatographed as the methyl ester, bis(trimethylsilyl) ether derivative.

[b] Carbon number corresponding to saturated fatty acid (methyl ester).

[c] Electron impact ionization, 70 eV, Finnigan 3200 quadrupole mass spectrometer.

and LTE_4 eluting from the HPLC. LTB_4 and the isomeric 5,12-dihydroxy-eicostetraenoic acids can be further subjected to gas chromatography and gas chromatography–mass spectrometry analysis. Table IV lists the retention indices and diagnostic ions of the methyl ester trimethylsilyl ether derivative of these lipoxygenase products. Such gas-phase procedures are invaluable in the complete characterization of these lipoxygenase products available only in nanomolar amounts.

Acknowledgments

This work was supported by a grant from NIH (HL 25785).

[54] Production and Purification of Slow-Reacting Substance (SRS) from RBL-1 Cells

By CHARLES W. PARKER, SANDRA F. FALKENHEIN, and MARY M. HUBER

Considerable circumstantial and some direct evidence indicates that slow-reacting substance (SRS) is an important mediator of bronchoconstriction in human bronchial asthma. SRS was shown in 1940 to be released during immediate hypersensitivity reactions in lung, but elucidation of its structure proved to be unexpectedly difficult.[1,2] In 1976 and 1977 our laboratory reported that SRS was a metabolite of arachidonic acid (AA) and presented several lines of evidence that it was produced by a lipoxygenase acting selectively at the 5 position of AA.[1,3,4] We later reported that SRS was a mixture of several products that were closely related structurally, each of which had a sulfur-containing side chain bound in thioether linkage.[5] After degradation with sodium metal in liquid ammonia, a hydroxylated fatty acid comigrating chromatographically

[1] C. W. Parker, *in* "Immunopharmacology of the Lung" (H. H. Newball, ed.) in press. Dekker, New York.

[2] C. W. Parker, *J. Allergy Clin. Immunol.* **63,** 1 (1979).

[3] B. A. Jakschik, S. Falkenhein, and C. W. Parker, *Proc. Natl. Acad. Sci. U.S.A.* **74,** 4577 (1977).

[4] C. W. Parker, *in* "Asthma: Physiology, Immunopharmacology, and Treatment" (L. M. Lichtenstein, K. F. Austen, and A. S. Simon, eds.), Vol. 2, pp. 301–313. Academic Press, New York, 1977.

[5] C. W. Parker, B. A. Jakschik, M. M. Huber, and S. F. Falkenhein, *Biochem. Biophys. Res. Commun.* **89,** 1186 (1979).

ISBN 0-12-181986-8

with 5-hydroxyeicosatetraenoic acid (5-HETE) and with an absorbance maximum at 270–280 nm was obtained, providing further evidence that SRS was a 5-lipoxygenase product and that several conjugated double bonds were present in the fatty acid portion of the molecule.[1,6] These observations were confirmed and extended by Samuelsson and his colleagues, who demonstrated that SRS underwent a characteristic shift in its ultraviolet absorbance in the presence of soybean lipoxygenase and inferred that the double bonds were at the 7, 9, 11, and 14 positions of the fatty acid moiety.[7] Later that same year we provided evidence that most of the SRS had either a glutathionyl or a cysteinylglycyl side chain.[8] This was soon confirmed by Hammarström *et al.*[9,10] and by Morris *et al.*[11] and their colleagues. Corey and his colleagues[9] and Rokasch and his collaborators at Merck[12] independently prepared a number of possible SRS molecules synthetically and showed that one of the major species of SRS apparently was 6-glutathionyl-5-hydroxy-7-*trans*,9-*trans*, 11-*cis*, 14-*cis*-eicosatetraenoic acid (also termed glutathionyl SRS or leukotriene C_4). The other was later shown to be 6-*S*-cysteinyglycyl-5-hydroxy-7-*trans*,9-*trans*,11-*cis*,14-*cis*-eicosatetraenoic acid (cysteinylglycyl SRS or leukotriene D_4).[10] We further demonstrated that an SRS with a cysteinyl side chain (cysteinyl SRS or leukotriene E_4) also was demonstrable, particularly when long incubation times were used.[13,14]

The most useful source of cells for SRS biosynthesis in these initial studies of SRS structure was the RBL-1 cell. The RBL-1 cell line is a neoplastic cell with morphological and chemical characteristics of basophils that is cultivable *in vitro* as a pure cell population. This line was originally

[6] C. W. Parker, *in* "Advances in Inflammation Research" (G. Weissmann, ed.), Vol. II, p. 1. Academic Press, New York, 1981.

[7] R. C. Murphy, S. Hammarström, and B. Samuelsson, *Proc. Natl. Acad. Sci. U.S.A.* **76,** 4275 (1979).

[8] C. W. Parker, M. M. Huber, M. K. Hoffman, and S. F. Falkenhein, *Prostaglandins* **18,** 673 (1979).

[9] S. Hammarström, R. C. Murphy, B. Samuelsson, D. A. Clark, C. Mioskowski, and E. J. Corey, *Biochem. Biophys. Res. Commun.* **91,** 1266 (1979).

[10] L. Örning, S. Hammarström, and B. Samuelsson, *Proc. Natl. Acad. Sci. U.S.A.* **77,** 2014 (1980).

[11] H. R. Morris, G. W. Taylor, P. J. Piper, M. N. Samhoun, and J. R. Tippins, *Prostaglandins* **19,** 185 (1980).

[12] J. Rokasch, Y. Girard, Y. Guindon, J. P. Atkinson, M. Larue, R. N. Young, P. Masson, and G. Holme, *Tetrahedron Lett.* **21,** 1485 (1980).

[13] C. W. Parker, *Proc. Int. Symp. Biochem. Acute Allerg. React. 4th; Kroc Found. Ser.* **4,** 23 (1980).

[14] C. W. Parker, *Prostaglandins* **20,** 863 (1980).

identified in a strain of rats treated with a carcinogen in England in 1973.[15] Because SRS is released during IgE-mediated hypersensitivity responses and RBL-1 cells contain IgE receptors, our laboratory evaluated these cells as a possible biosynthetic source of SRS. While immunological stimuli failed to produce SRS, the divalent cation ionophore A23187 released large amounts of spasmogenic activity, which was indistinguishable from anaphylactically generated SRS by all the criteria then available.[1,6,16] With further analysis it became apparent that these cells made several different SRS species, which varied in their relative amounts depending on the time and conditions of stimulation.[3,5,13,14] During the first few minutes of the response to A23187 at 37°, the major product was glutathionyl SRS (LTC), which was subsequently converted to cysteinylglycyl SRS (LTD_4, and finally to cysteinyl SRS (LTE) (Fig. 1). These cells also made prostaglandin D_2 (PGD_2), 5-S-hydroxy-6,8,11,14-eicosatetraenoic acid (5-HETE), and 5-S-12-D-hydroxy-7,9,11,14-eicosatetraenoic acid (5,12-di-HETE) during exposure to the ionophore.[17] This chapter describes the production, purification, and characterization of unlabeled and radioactively labeled SRS from RBL-1 cells.

Production of SRS in RBL-1 Cells

RBL-1 cells can be grown in ordinary tissue culture medium in quantities of up to 24 liters. The medium used routinely in this laboratory is Eagle's Minimal Essential Medium containing 10% fetal or newborn calf serum and antibiotic–antimycotic mixture, 1:100 (Grand Island). The cells are normally maintained at a density of 1 to 3 $\times$ 10^6 cells/ml and subcultured every 2–3 days.

Prior to stimulation with A23187, the cells are washed and resuspended at a density of 1 to 1.5 $\times$ 10^7 cells/ml in medium MCM, a tissue culture medium containing 150 mM NaCl, 3.7 mM KCl, 3.0 mM Na_2HPO_4, 3.5 mM KH_2PO_4, 0.9 mM $CaCl_2$, and 5.6 mM dextrose, pH 7.0 (16). As a rule SRS is generated in medium containing 0.5 mg of bovine serum albumin per milliliter, although albumin-free media also are satisfactory (when the albumin is omitted the concentration of ionophore

[15] E. Eccleston, B. J. Leonard, J. S. Lowe, and H. J. Welford, *Nature (London), New Biol.* **244,** 73 (1973).

[16] B. A. Jakschik, A. Kulczycki, H. H. MacDonald, and C. W. Parker, *J. Immunol.* **199,** 618 (1977).

[17] S. F. Falkenhein, H. H. MacDonald, M. M. Huber, D. Koch, and C. W. Parker, *J. Immunol.* **125,** 163 (1980).

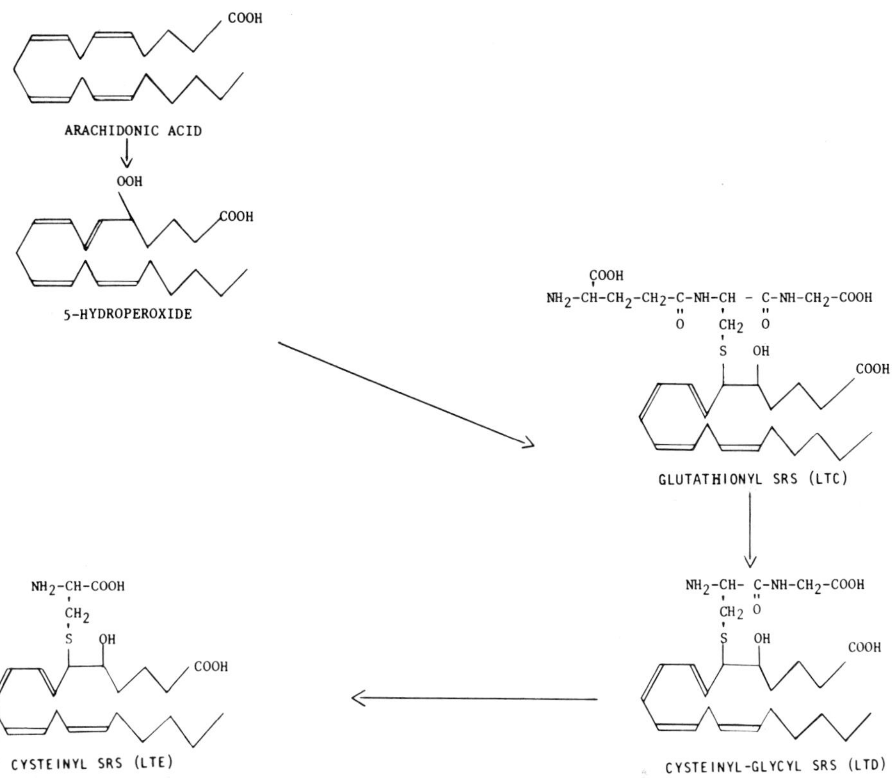

FIG. 1. Sequential conversion of arachidonic acid to various slow-reacting substance (SRS) species by RBL-1 cells.

is reduced by 60%). Just before stimulation with ionophore, unlabeled AA or a mixture of [1-C^{14}]AA[18] (usually 50 μCi at an original specific activity of 60 mCi/mmol), and unlabeled AA is added at a final total AA concentration of 21 μM and the cells are placed in a water bath at 37°. It may also be desirable to add a freshly prepared and neutralized solution of cysteine hydrochloride at a final concentration of 5 mM at the time of stimulation

[18] Although a single strain of RBL-1 cells has been used as a source of SRS over the entire course of our studies, some variation in SRS production has been experienced, so it is desirable to maintain a frozen stock of cells that can be thawed and used as a replacement if the biosynthetic capacity of the cultured cells appears to be changing. In addition, on two occasions the cells have been passed back through Lou-162 strain rats by injecting 6-week-old animals intraperitoneally, obtaining cells after one to several weeks and re-adapting them to grow in tissue culture.

with A23187.[19] SRS biosynthesis is initiated by the 0.01 volume of a concentrated solution of A23187 in dimethyl sulfoxide to a final concentration of 5 μg/ml. After mixing the incubation is then continued at 37° with occasional gentle shaking for various time periods (2–3 minutes for LTC_4, 10–15 min for LTD_4, and 60 min for LTE_4). Since LTD_4 is the most active of the different SRS species in the guinea pig ileal muscle bioassay system, the overall amount of SRS spasmogenic activity is greatest when a 10–15-min incubation time is used. With a 10–15-min incubation period, normally 300–1500 bioassay units of SRS (180–900 pmol) are generated per 1×10^7 RBL-1 cells. For a large batch of 2×10^{10} cells, this provides 600,000–3,000,000 SRS units for purification.[21]

After completion of the incubation, the cell suspension is centrifuged at 4° at 1125 *g* and the cells are discarded. The cell supernatant is then immediately purified on Amberlite XAD-7 columns as described below.

Various isotopes can be incorporated into SRS by variations in the above procedure. For incorporation of ^{35}S, 250 ml of cells at 1.5×10^6 cells/ml are cultured for 16 hr in cysteine-free tissue culture medium with [^{35}S]cysteine (approximately 80 mCi/mmol, 4μCi/ml).[5,7] The cells are then washed, and the SRS is generated in the usual way. For incorporation of radiolabeled glycine into SRS, the cells are cultured in a similar volume of Eagle's medium for 16 hr in the presence of [^{14}C]glycine (approximately 100 mCi/mmol, 4 μCi/ml).[14]

Purification of SRS

The representative purification procedure given below is suitable for amounts of SRS generated by 2×10^{10} cells. This purification procedure provides an average overall yield of SRS bioreactivity of 50%. SRS (2000 ml) containing cell supernatant is applied directly as an aqueous solution at ambient temperature to a 890-ml Amberlite XAD-7 column (Mal-

[19] Through most of the time in which we have studied SRS synthesis in RBL-1 cells the addition of cysteine has exerted little or no effect on the total amount of spasmogenic activity recovered, but the quantity of SRS activity has been somewhat improved when cysteine is present in the original incubation mixture. The role of cysteine may be to prevent the LTD_4 from being broken down to LTE_4.[20]

[20] D. Sok, J. Pai, V. Atrache, and C. J. Sih, *Proc. Natl. Acad. Sci. U.S.A.* **77,** 6481 (1980).

[21] Alternatively, 4 volumes of ethanol can be added to the supernatant, the precipitated protein and other insoluble material can be removed by centrifugation at 20,000 *g*, and the samples can be stored at −80° for 1–2 weeks before further SRS purification. Although an acceptable yield of SRS activity is obtained using this procedure, it is inconvenient in that most of the large volume of ethanol that has been added must be removed before the sample can be applied to the Amberlite XAD-7 resin column.

linckrodt).[16,22] The resin is prepared for a use in SRS purification by washing once with 2 liters of methylene chloride, twice with 1 liter of *n*-propanol, three times with 1 liter of water and 1 liter of acetone in succession, three times with 1 liter of 95% ethanol, and once with 20 liters of water. The SRS solution is recycled three times through the column at a slow column flow rate over a period of 90 min. The column is then drained and washed with 500 ml of water, which removes many of the water-soluble contaminants, including most of the histamine, serotonin, and inorganic salts. The SRS is eluted together with other fatty acids with 1000 ml of 80% ethanol. The ethanol–water eluate may be collected as several large fractions. Most of the SRS activity elutes in the 600–800 ml ethanol–water fraction. The 80% ethanol–water eluate is concentrated to 10 ml and diluted with an equal volume of methanol.

The resulting suspension is divided into two parts and applied to two 100-ml columns of silicic acid. The column is packed with equal amounts of Silicar CC7 (Mallinckrodt), and 100-mesh silica gel (Mallinckrodt). A rapidly setting fraction is prepared from the silica gel by suspending the gel three times in anhydrous methanol, allowing it to settle over a 30-min period, removing the slowly sedimenting gel each time, and drying. The two types of silicic acid are mixed in a 1:1 ratio (v/v), activated at 100° for 1 hr, equilibrated with methanol–chloroform, 3:7 (both reagent grade, redistilled), and placed at 4°. After application of the sample the column is eluted by gravity successively with 150 ml of methanol–chloroform (3:7), 250 ml of methanol–chloroform (1:1), and 250 ml of methanol.[23] Five fractions (labeled 1–5, in order of elution and containing 150, 175, 75, 175, 75 ml, respectively) are collected. The usual flow rate is about 40 ml/hr.

Fraction 1 contains more than 90% of the arachidonic acid and most of the 5,12-di-HETE, 5-HETE, and PGD_2, although these products are also present to a limited extent in later fractions. Fraction 2 contains 80–90% of the LTD_4 and LTE_4. Fraction 4 contains 80–90% of the LTC_4 and variable but significant amounts of LTD_4. Fractions 3 and 5 usually contain less than 10% of the total spasmogenic activity and ordinarily can be discarded. After completion of the silicic acid column, SRS-containing solutions may be stored under nitrogen at −20° as obtained off the column or concentrated first to 10–15 ml under a stream of nitrogen. In either case, the loss of SRS activity at −20° over a 3-month period is normally small (less than 25%.)

The SRS is further purified on two high-pressure liquid chromatogra-

[22] R. P. Orange, R. C. Murphy, M. L. Karnovsky, and K. F. Austen, *J. Immunol.* **110,** 760 (1973).

[23] K. Strandberg and B. Uvnäs, *Acta Physiol. Scand.* **82,** 358 (1971).

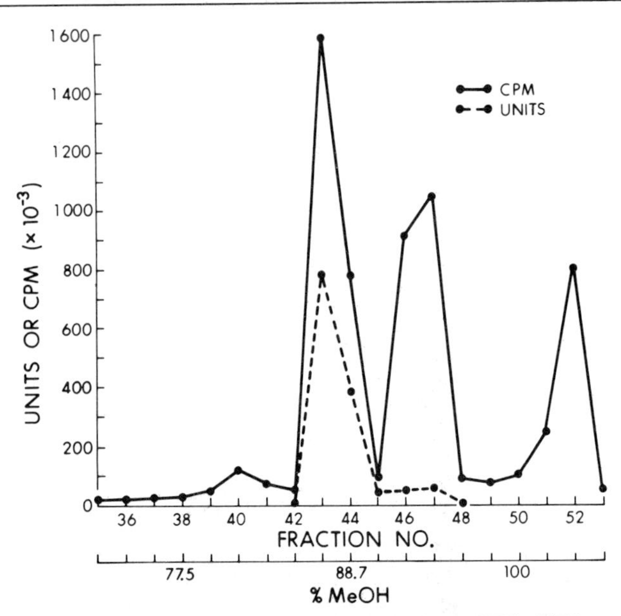

FIG. 2. Preparative C_{18} HPLC of slow-reacting substance (SRS) (LTD_4 and LTE_4)-rich silicic acid column fractions on a Unimetric RP18 column in a methanol–water neutral pH system. For precise chromatography conditions see text. LTD_4 and LTE_4 elute at 88% and 93% MEOH, reespectively, on this column.

phy (HPLC) columns.[24] The initial HPLC column is a preparative C_{18} column (Unimetric RP18, 10μm, 8 × 250 mm).[14] The column is operated at ambient temperature using a linear gradient from 6% methanol–0.6% *t*-pentanol–93.4% water buffered with 0.01 *M* sodium phosphate, pH 7.4, (solvent A) to 99.40% methanol–0.6% pentanol (solvent B) at a flow rate of 2 ml/min collecting 2-ml fractions and increasing the gradient by 2%/min. The *t*-pentanol is used because it dramatically improves the yield of SRS activity from the column. For loading either the fraction 2's or the fraction 4's from the 2 silicic acid columns are combined (2×10^{10} cell equivalents of SRS) and concentrated to 2.0 ml and 6.0 ml of solvent A (the initial solvent used on the column), are added. Eight separate runs are then made on the preparative Unimetric C_{18} column, loading 1 ml (approximately 2.5×10^9 RBL-1 cell SRS equivalents) each time. The column is monitored for radioactivity (if appropriate), ultraviolet absorbance at 280 nm (allowing for absorbance by the *t*-pentanol), and spas-

[24] Using the two HPLC columns, the preparative thin-layer chromatography (TLC) and DEAE-cellulose column chromatography steps used by our laboratory in the early stages of our SRS work[3] are no longer necessary.

mogenic activity. LTC_4 and LTD_4 elute sharply and reproducibly with UV 280 peaks at or very near 88% methanol (Fig. 2) whereas LTE_4 comes off at 92–94% methanol, PGD_2 at 81–82% methanol, 5-HETE usually at 97–98% methanol, 5,12-diHETE at 94% methanol, and AA at 100% methanol. Wherever possible, synthetic LTC_4, LTD_4, and LTE_4 should be used as reference standards to verify the elution characteristics of the column.

The SRS from the preparative Unimetric column is further purified on a smaller C_{18} column from another manufacturer (Waters, μBondapak) at pH 5.4[7,21,24] or, less desirably, on an analytical Varian Micropak MCH column in the same neutral phosphate–aqueous methanol–*t*-pentanol solvent system described above. The acidic μBondapak column is equilbrated at room temperature with methanol–water–acetic acid, 700:300:1 (v/v/v), adjusted to pH 5.4 with ammonium hydroxide, and eluted isocratically at a flow rate of 1 ml/min. The usual form of LTC_4 elutes at 9 min (Fig. 3), LTD_4 at 13 min, LTE_4 at 16 min, 5-HETE at 46 min, and AA at 75 min on this column. The 11-*trans*-SRSs elute slightly behind the 11-*cis*-SRSs. Their separation may be improved by reducing the proportion of methanol to water in the mobile phase or by changing the flow rate or buffer pH. At this stage of the purification the SRSs exhibit the characteristic ultraviolet absorption spectra of SRS (see below) and are essentially pure.

Properties

Storage and Solubility of Purified SRS. The final purified SRS preparations are usually stored in the HPLC solvents in which they eluted from the columns and appear to be stable more or less indefinitely provided they are maintained at −70°. However, loss of activity may occur when solutions are concentrated, particularly if a nitrogen or argon atmosphere is not used. SRS can also be maintained frozen in water for a period of at least several months with relatively little loss of activity. Isomerization of the 11-*cis* to an 11-*trans* double bond with loss of spasmogenic activity can be a problem during storage, particularly in aqueous solution without organic solvent. The purified SRSs have especially high solubility in 60% methanol–water, buffered at neutral pH. SRSs also have good solubility in aqueous solution without organic solvent at neutral pH. Their solubility in pure ethanol or methanol is limited. At pH 3.5 or below, highly purified preparations of SRS can be extracted into ether and other nonpolar organic solvents in good yield. However, a variety of reducing agents readily inactivate SRS in acidic aqueous solutions.

Ultraviolet Absorbance Characteristics. The usual forms of LTC,

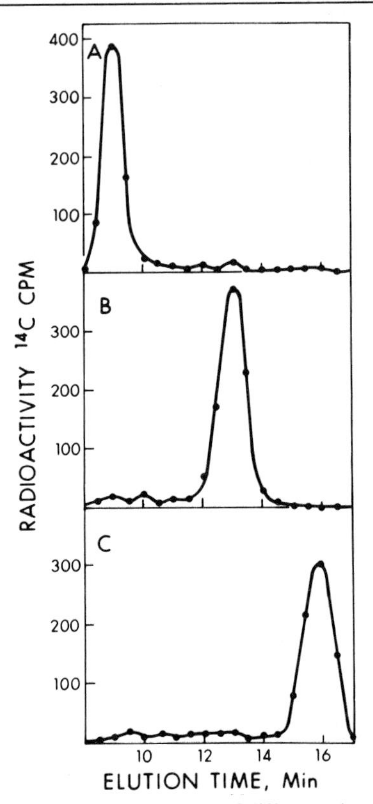

FIG. 3. Analytical C_{18} HPLC chromatography of different slow-reacting substance (SRS) species on a μBondapak (Waters) column in a MEOH water–acetic acid system adjusted to pH 5.4 with NH_4OH. For precise chromatography conditions, see text. (A) LTC_4; (B) LTD_4; (C) LTE_4. Redrawn from Parker *et al.*[25]

LTD, and LTE all have absorbance maxima at or very near 280 n*M*, and secondary peaks at 270 and 292 n*M*, (Fig. 4). The molar extinction coefficient at 280 n*M* is about 40,000.[7,8] The 11-*trans* forms of LTC_4, LTD_4, and LTE_4 have maxima at or near 278 n*M*. Most forms of SRS are susceptible to digestion by soybean lipoxygenase. The digestion is usually carried out at room temperature at 10 μg/ml concentrations of the enzyme as it is obtained commercially, in Tyrode's solution at pH 7.5 in an oxygen-containing atmosphere.[7] The enzyme produces a new absorbance maximum at 308–310 nm with shoulders at 295 and 323 and a decrease in 280 nm absorbance due to conversion of the origianl triene system to a

[25] C. W. Parker, D. Koch, M. M. Huber, and S. F. Falkenhein, *Biochem. Biophys. Res. Commun.* **97,** 1038 (1980).

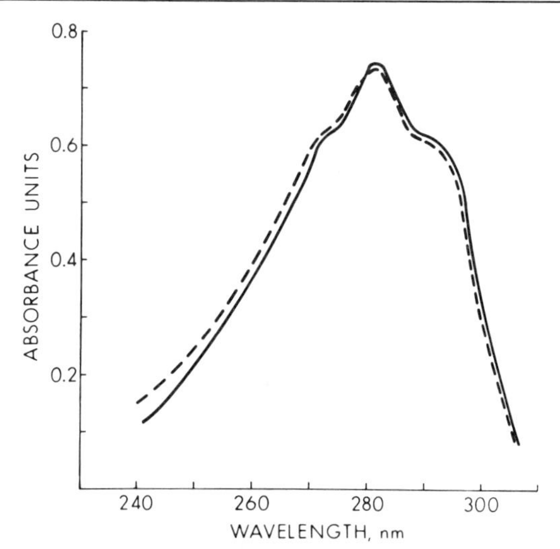

FIG. 4. Ultraviolet absorption spectra to LTD_4 (—) and LTE_4 (---) from RBL-1 cells in 50% methanol–water. Taken from Parker.[14]

tetraene. This absorbance shift is usually observable within 10–15 min, although a longer time may be required for LTC_4. The shift is not seen with the 11-*trans*-SRSs. (See also this volume [76].)

Thin-Layer Chromatography of Purified SRS. When [^{3}H] or [^{14}C]AA is used to label the SRS, the absence of contaminating AA metabolites and AA in the final purified SRS preparations can be verified by TLC on silica gel plates and radioautography. Samples are chromatographed both in a propanol–ammonia–water (90:45:15, v/v/v) system (R_fs of 0.5 for LTC_4, 0.6 for LTD_4, 0.62 for LTE_4, 0.8 for PGD_2, 0.8 for 5-HETE, 0.77 for 5,12-di-HETE, and 0.88 for AA) and a benzene–ether–ethanol–acetic acid (50:40:2:0.2, v/v/v/v) system (R_fs of less than 0.05 for all of the SRSs, 0.18 for 5,12-di-HETE, 0.46 for 5-HETE, 0.08 for PGD_2, and 0.77 for AA).[3,8,16,17]

Spasmogenic Activity of Purified SRS. Purified SRS preparations should produce well-defined contractile responses in terminal guinea pig ileal muscle strips in quantities of as little as 0.2–3.0 pmol, the exact amount depending on the species of SRS, the size of the bioassay chamber, and the sensitivity of the ileal smooth muscle preparation on that particular day. While elution positions on HPLC columns will ordinarily be the major basis for defining the type of SRS, the character of the contractile response does differ to some extent with different SRSs (see this volume [76]).

If further verification is needed in regard to the nature of the SRS as few as 2400 pmol of purified SRS suffice for amino acid analysis. The molar ratios of glutamic acid and glycine to SRS will easily distinguish LTC, LTD, and LTE from one another. All the SRSs give somewhat low values for cysteine[8] (recovered primarily as half-cysteine) because the sulfur group of cysteine is attached to the fatty acid chain, decreasing its recovery. Cysteine is usually recovered almost quantitatively in a standard amino acid analysis.

[55] Preparation, Purification, and Structure Elucidation of Slow-Reacting Substance of Anaphylaxis from Guinea Pig Lung

By PRISCILLA J. PIPER, J. R. TIPPINS, H. R. MORRIS, and G. W. TAYLOR

Slow-reacting substance of anaphylaxis (SRS-A) from the sensitized guinea pig lung was described in 1940 by Kellaway and Trethewie[1] and has, since that time, been the standard research material among a number of slow-reacting substances generated both immunologically and nonimmunologically. Therefore its structure elucidation has been of prime importance in the understanding of its role in immediate hypersensitivity reactions such as asthma.

Preparation

In our laboratory SRS-A is prepared from the perfused lung by the method of Engineer, *et al.*,[2] a modification of the original method described by Brocklehurst.[3] Male guinea pigs (Dunkin Hartley, 300 g) are sensitized by the injection of ovalbumin (Sigma Grade II, 100 mg ml^{-1}), 100 mg intraperitoneally and 100 mg subcutaneously. After 21 days, the animals are killed and the thorax is opened. The apex of the heart is removed, and the heart, lungs, and trachea are dissected free. A cannula is inserted via the cut right ventricle into the pulmonary artery and clipped in place, proximal to the bifurcation of the artery. The lungs are then suspended in a water jacket maintained at 37°. The lungs are inflated several times with a syringe, while being perfused with Tyrode's solution containing indomethacin (2.8×10^{-6} *M*) at 10 ml min^{-1} and 37°. The trachea is

[1] C. H. Kellaway and E. R. Trethewie, *Q. J. Exp. Physiol.* **30,** 121 (1940).
[2] D. M. Engineer, H. R. Morris, P. J. Piper, and P. Sirois, *Br. J. Pharmacol.* **64,** 211 (1978).
[3] W. E. Brocklehurst, *J. Physiol.* (*London*) **151,** 416 (1960).

ISBN 0-12-181986-8

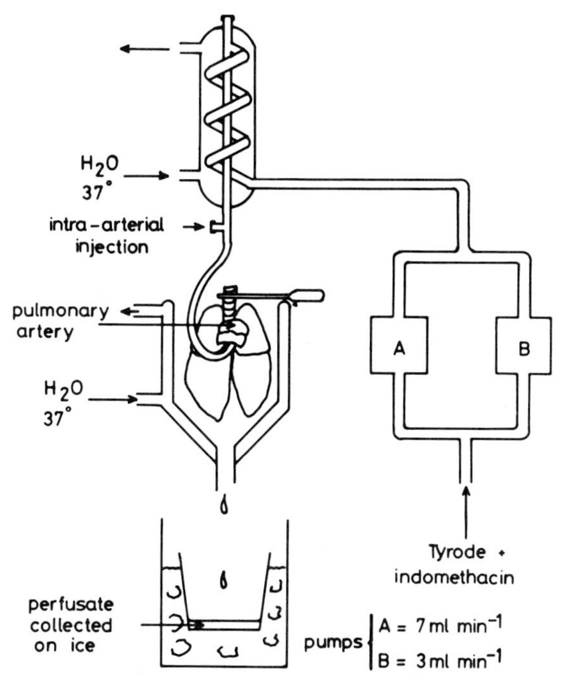

FIG. 1. Diagram of the apparatus used for the perfusion of guinea pig lungs and production of slow-reacting substance of anaphylaxis (SRS-A). The diagram shows the inflated lungs, with the trachea clamped, suspended in a water-jacketed bath. The lungs are perfused with warm Tyrode's solution via the pulmonary artery, and the perfusate is collected in an ice-cold polythene beaker.

then clamped with the lungs inflated. After 3 min the perfusion rate is reduced to 3 ml min^{-1} and the perfusion is continued for 15 min. The lungs are then challenged with 1 ml of ovalbumin (Sigma Grade III, 10 mg ml^{-1}). The perfusate is collected for 12 min in a polythene beaker on ice. The pooled perfusates from a number of lungs are collected in a flask, kept on ice, and bubbled with nitrogen. The apparatus is illustrated in Fig. 1.

Extraction

The SRS-A is extracted from the perfusate by adsorption onto charcoal and elution with 80% ethanol. The perfusate is initially centrifuged at 2000 *g* for 10 min to remove blood and the supernatant is then mixed with 2.5 mg of activated charcoal per milliliter (Sutcliff, Speakman & Co., 110 quality). The charcoal is deposited by a second centrifugation at 2000 *g* for 10 min, and the supernatant is decanted. The charcoal is washed in a small volume of distilled water (one-fifth volume of perfusate) and recentrifuged at 2000 *g* for 10 min. The SRS-A is then eluted from the charcoal by washing it twice with 30 ml of 80% ethanol per 100 ml of original perfusate volume, and separating the charcoal by centrifugation at 2000 *g* for 10 min. The ethanol washes are pooled and rotary evaporated under reduced pressure at 35°. The aqueous extract is then dried either by continuing the rotary evaporation or by freeze-drying. The lyophilizsed material is stored under nitrogen at −20° until used for purification. The charcoal is rinsed from the centrifuge bottles, washed, dried, and stored under vacuum. This charcoal, which is used for subsequent extractions, greatly increases the yield of SRS-A from the perfusate.

Purification

The procedure developed for purification of SRS-A was designed to minimize losses of SRS-A through adsorption, destruction, etc., while obtaining the maximum amount of physical data relevant to the structure. The purification method adopted introduced several significant improvements over existing methods. The first is the finding that material from the G-15 column could be extracted into diethyl ether at pH 3 with a yield of more than 95%. This finding is not consistent with the reported presence of a sulfate group in the molecule. The second is the introduction of high-pressure liquid chromatography (HPLC) (Fig. 2). This results in the production of very pure material, from which the first ultraviolet spectrum was obtained.[4] From this spectrum (the now familiar triplet, λ_{max} 280 nm), the presence of a nonaromatic conjugated triene[5] was postulated. The HPLC separation was later extended,[6] and this resulted in four closely related compounds eluting from the column. These are dealt with later in this chapter.

The lyophilized preparations of SRS-A eluted from charcoal are dissolved in methanol–water–0.880 ammonia (40:40:20, v/v/v), and material from a large number of guinea pigs is combined. This material is rotary evaporated to dryness at 35°. The dried material is then redissolved in 2 ml of the same solvent, centrifuged at 1000 *g* for 5 min to remove insoluble salts and excess charcoal, and then loaded onto the Sephadex G-15

[4] H. R. Morris, G. W. Taylor, P. J. Piper, P. Sirois, and J. R. Tippins, *FEBS Lett.* **87,** 203 (1978).

[5] H. R. Morris, P. J. Piper, G. W. Taylor, and J. R. Tippins, *Br. J. Pharmacol.* **67,** 179 (1979).

[6] H. R. Morris, G. W. Taylor, P. J. Piper, and J. R. Tippins, *in* "Prostaglandins and Inflammation," *Agents Actions Suppl.* **6,** 27 (1979).

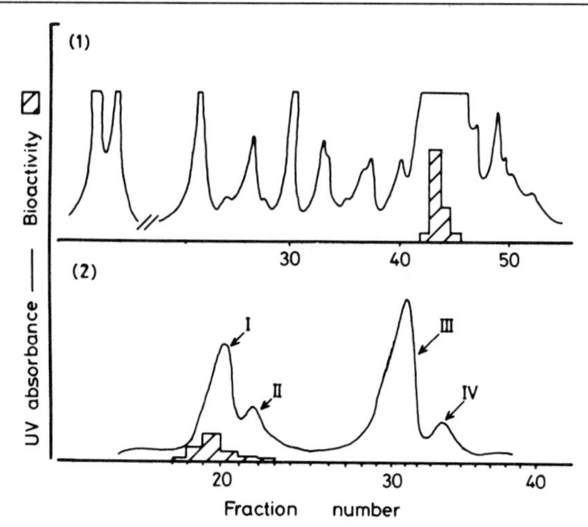

FIG. 2. High-pressure liquid chromatography (HPLC) elution profile of SRS-A. A batch of SRS-A was purified on two HPLC systems: (1) HPLC 1 using a methanol–water gradient, and (2) HPLC 2 using an *n*-propanol–acetic acid–water gradient. The characteristics of peaks I, II, III, and IV are explained in the text. Reproduced with permission of Birkhäuser Verlag, Basel, Switzerland, from Morris *et al.*[6]

column (180 × 1 cm, Pharmacia) under unit gravity. The column is eluted with the methanol–water–ammonia solvent at a rate of 12 ml/hr. The eluate is monitored at 275 nm with a Cecil spectrophotometer and by bioassay on the guinea pig ileum smooth muscle strip. The active samples are bulked and dried on a rotary evaporator. SRS-A is found to elute from the G-15 column with material absorbing strongly at 275 nm (found to contain tyrosine and phenylalanine) and the molecular weight could therefore be estimated to be 200–400. The elution solvent is found to enhance the stability of SRS-A during this step, with yields in excess of 90%.

The partially purified SRS-A is dissolved in cold pH 2.8 acetic acid. It is then rapidly extracted twice into peroxide-free diethyl ether at 0°, and the ether phase is dried under nitrogen and finally on a vacuum pump. Peroxide-free ether is prepared by shaking 250 ml of diethyl ether with ferrous sulfate (6 g of $FeSO_4 \cdot 7\ H_2O$, 0.5 ml of H_2SO_4 11 ml H_2O). The ether is then washed with water, 10% sodium carbonate, again with water, and then distilled. The first 10 ml of distillate is discarded and the remainder collected and used for the extraction.[7] The ether extraction is found to give a high yield (>95%) with no biological activity remaining in the aqueous phase. Most of the solids present in the sample after G-15 separation are removed by this step.

[7] A. I. Vogel, "Practical Organic Chemistry," p. 163. Longman Group, London, 1974.

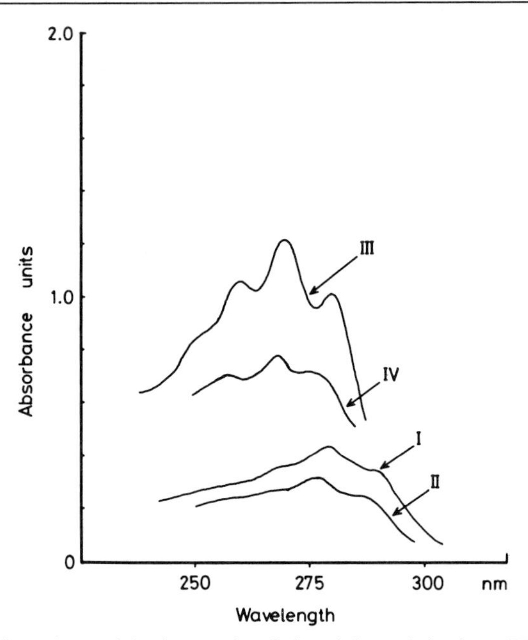

FIG. 3. Ultraviolet spectra of the four peaks of ultraviolet activity from HPLC 2. Peak I is the biologically active compound identified as SRS-A (leukotriene D_4). Peak III is leukotriene B_4. Reproduced with permission of Birkhäuser Verlag, Basel, Switzerland, from Morris *et al.*[6]

The ether-soluble material is then purified by two stages of HPLC on a Waters instrument using a reversed-phase μBondapak C_{18} column. The sample is dissolved in 200–300 μl of methanol and mixed with an equal volume of water in the loading syringe. The sample is then injected onto the column and eluted isocratically in 50% aqueous methanol (5 min) followed by a linear gradient (10 min) to 100% methanol. Samples are dried on a vacuum pump.

The active material from the first HPLC is loaded onto a second HPLC in 30% propanol in 5% aqueous acetic acid (200–500 μl) and eluted isocratically in this solvent for 10 min followed by a linear gradient (20 min) to 40% propanol in aqueous acetic acid. Samples are rapidly dried on a vacuum pump.

In both HPLC separations the columns are pumped at 2 ml min^{-1}, and the eluate is monitored at 254 and 280 nm. At all stages of purification the position and concentration of SRS-A eluting from the columns is determined by bioassay on guinea pig ileum smooth muscle, against a laboratory standard of partially purified, charcoal-extracted SRS-A, in the presence of mepyramine and hyoscine. Contractions of the ileum to SRS-A are tested for inhibition by FPL 55712, the specific SRS-A antagonist.

HPLC in the methanol–water system results in the elution of two peaks of biological activity and is interpreted as being due to the presence of a protonated and an unprotonated form of SRS-A. When the unprotonated material is acidified and rechromatographed, it elutes much later at a position corresponding to the protonated material. This material was used to obtain the first ultraviolet spectrum of SRS-A.[4]

The material eluting from the second HPLC, using the propanol:acetic acid:water system, appears as four peaks of UV absorbance (Fig. 3). Of these only peaks I and II are active on the guinea pig ileum. Compound I gives the same spectrum as that described for SRS-A (a broad triplet, λ_{max} 280 nm, with shoulders at 270 and 290 nm). Compound II (λ_{max} 278 nm) is weakly active on the guinea pig ileum and has a UV spectrum similar to that for SRS-A but is shifted hypsochromically by 2–3 nm. Compound III(λ_{max} 270 nm) and compound IV (λ_{max} 267 nm) both exhibit the triply conjugated spectra similar to compounds I and II and are similarly related by a cis-trans isomerization.[6]

Physicochemical Properties of SRS-A

Biological activity destroyed by
1. HCl, 0.1 *M*, RT,[a] 30 min
2. Acetylation
Methanol acetic anhydride, 4:1 v/v, RT, 1 min
Pyridine acetic anhydride 1:10 v/v, RT 10 min
Fluram treatment
3. Methylation
CH_2N_2, RT, 30 min
Methanol–HCl, RT, 30 min
Methanol–BF_3, 37°, 30 min
4. Catalytic hydrogenation. NiB, RT and 55° in MeOH
5. CNBr treatment
6. Arylsulfatase
Soybean lipoxygenase
Mammalian arachidonate lipoxygenase

[a] RT, room temperature.

Structure Determination

Evidence for the presence of various chemical groups present in the SRS-A molecule was gained by examining the stability of the biological activity of the pure material after brief exposure to various reagents. These are outlined in the table.

Short acetylation showed that an α-amino group was present in the molecule and essential for activity. Fluram destruction confirmed this primary amine function.

Methylation with CH_3OH/HCl or CH_2N_2 resulted in a loss of biological activity, indicating the presence of a free carboxyl group in the molecule. Inactivation by CNBr suggested the presence of a thioether linkage in SRS-A. Inactivation of SRS-A by soybean lipoxygenase was first shown in our laboratory[2] and indicated the presence of a *cis-cis*-1,4-pentadiene moiety in the molecule. The specificity of this enzyme and the putative precursor role of arachidonic acid[8] indicated that the $\Delta^{14,15}$ double bonds of the precursor were present in the molecule.

Amino acid analysis of tubes taken across the peaks of UV activity from the second HPLC was performed on a Beckman 121 MB autoanalyzer after hydrolysis under vacuum in 6 *M* HCl for 16 hr at 110°. To determine the sequence of the amino acids in the molecule, samples from these tubes were dansylated and hydrolyzed and the dansyl derivatives were examined by thin-layer chromatography.[9]

The results showed the presence of cysteine and glycine in compounds I and II, and profiles of amino acid, UV absorbance, and biological activity were coincident across these peaks. Compounds III and IV were devoid of amino acids. Dansylation and amino acid analysis showed that these amino acids were present as the dipeptide cysteinylglycine.

The trimethylsilyl ether of the *N*-acetylmethyl ester was chosen as a suitable derivative from which to determine a mass spectrum of SRS-A. To aid interpretation of the mass spectrum, the *N*-acetyl derivative was prepared using a 1:1 mixture of acetic anhydride and d_6 acetic anhydride. Ions containing the *N*-acetyl moiety therefore appeared as 1:1 doublets, three mass units apart.

SRS-A was converted to the trimethylsilyl ether of the *N*-acetyl-(1:1, d_3:h_3) methyl ester and analyzed mass spectrometrically (EI mode) by the mixture analysis method[10] over a temperature range of 120–350° on a Kratos MS 50 mass spectrometer using a mass range of 1200 scanned at an accelerating voltage of 8 kV and electron beam energy of 70 eV. A mass spectrum containing the 1:1 isotope label was observed at 200–240° (Fig. 4). From the fragmentation pattern and mass measurements, and by comparison with a previously obtained spectrum for rat basophil leukemia cell SRS, the structure of SRS-A was shown to be a dipeptidyl substituted C_{20} tetraunsaturated peptidolipid. Taken in conjunction with the data obtained from the physicochemical examination, this enabled the

[8] P. J. Piper, J. R. Tippins, H. R. Morris, and G. W. Taylor, *in* "Arachidonic Acid Metabolism in Inflammation and Thrombosis," Agents Actions Suppl. **4,** 37 (1979).

[9] B. S. Hartley, *Biochem. J.* **119,** 805 (1970).

[10] H. R. Morris, D. H. Williams, and R. P. Ambler, *Biochem. J.* **125,** 189 (1971).

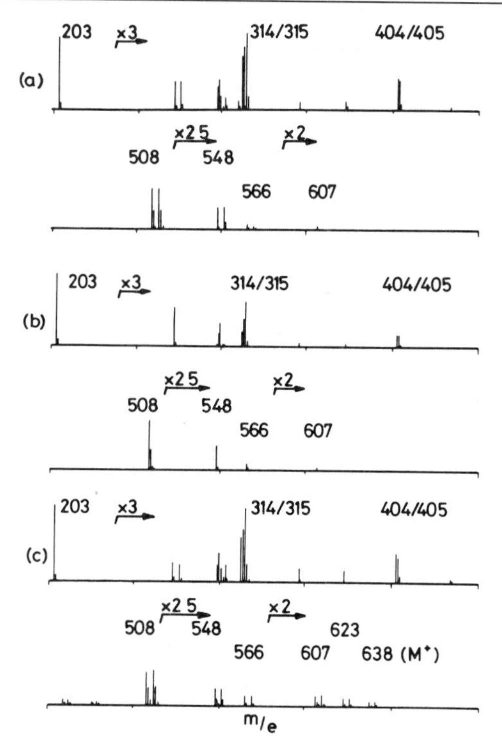

FIG. 4. Mass spectra of SRS-A and RBL-1 SRS. (a) Electron impact mass spectrum of the trimethylsilyl ether of the *N*-acetyl($CH_3CO:CD_3CO$, 1:1) methyl ester of SRS-A from guinea pig lung. (b) Mass spectrum of the same derivative (excluding isotopic labeling on the acetyl group) of the control sample of RBL-1 SRS. (c) Mass spectrum of the same derivative of RBL-1 SRS. Reproduced, with permission of Macmillan Journals, London, from Morris *et al.*[11]

molecule to be identified as 5-hydroxy-6-cysteinylglycinyl-7,9,11,14-eicosatetraenoic acid.[11]

The stereochemistry of the molecule was determined by extensive comparison with synthetic material (provided by Dr. J. Rokach, Merck Frosst Laboratories, Quebec, Canada). They were compared by (*a*) bioassay on guinea pig ileum smooth muscle and antagonism of the responses by FPL 55712; (*b*) HPLC analysis of the free materials and the acetylmethyl ester derivatives, using both the methanol–water system and the propanol–acetic acid–water system; (*c*) conversion by soybean lipoxygenase and examination of the products; and (*d*) UV and mass spectrometric analysis.

[11] H. R. Morris, G. W. Taylor, P. J. Piper, and J. R. Tippins, *Nature* (*London*) **285,** 104 (1980).

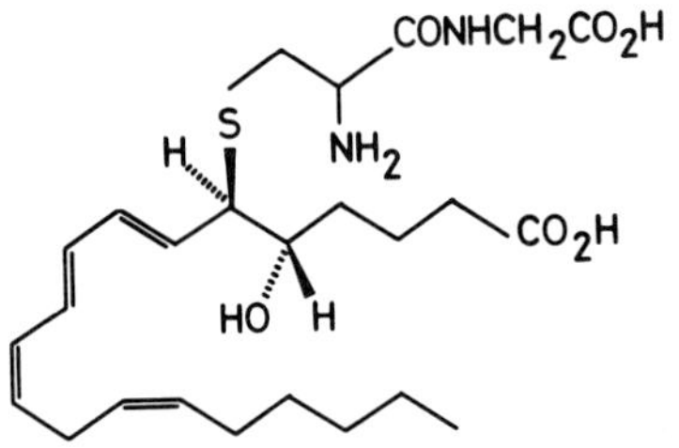

FIG. 5. Structure of SRS-A from guinea pig lung: 5(*S*)-hydroxy-6(*R*)-cysteinylglycinyl-7,9-*trans*-11,14-*cis*-eicosatetraenoic acid. Reproduced, with permission of Geron-X Inc., Los Altos, California, from Morris *et al.*[12]

Both synthetic and natural SRS-A were found to give identical dose–response curves on the guinea pig ileum, and both were inhibited by FPL 55712. They were found to have the same retention time on both HPLC systems. When equivalent amounts of each material were mixed and rechromatographed, a single UV peak was eluted, indicating a similarity in structure.

Treatment of both synthetic and natural SRS-A with soybean lipoxygenase under identical conditions led to production of identical UV spectra (λ_{max} 308 nm). The UV spectra of both materials were superimposable and identical to that originally described for SRS-A. The synthetic material, when derivatized by N-acetylation, esterification, and trimethylsilylation and analyzed by the direct probe mixture analysis method, gave an identical mass spectrum to that obtained for natural SRS-A. Thus, the structure of SRS-A was determined as 5(*S*)-hydroxy-6(*R*)-cysteinylglycinyl-7,9-*trans*-11,14-*cis*-eicosatetraenoic acid[12] (Fig. 5). This was subsequently given the trivial name leukotriene D_4.[13]

Compound III from the second HPLC was also analyzed by mass spectrometry. The compound was converted to the trimethylsilyl derivative of the carboxylic methyl ester and analyzed in the manner described for compound I. The mass spectrum was interpreted as arising from 5(*S*), 12(*R*)-dihydroxy-6,8,10,14-eicosatetraenoic acid.[6] This was subsequently given the trivial name LTB_4.[13] Compound IV is the all-trans triene isomer of LTB_4.

[12] H. R. Morris, G. W. Taylor, J. Rokach, Y. Girard, P. J. Piper, J. R. Tippins, and M. N. Samhoun, *Prostaglandins* **20,** 601 (1980).

[13] B. Samuelsson, P. Borgeat, S. Hammarström, and R. C. Murphy, *Prostaglandins* **17,** 785 (1979).

Summary

The work that we have described had originally three main aims: (*a*) to design a new purification system for SRS-A from which we would obtain pure material for the structural analysis; (*b*) to define the functional groups in the pure material by spectrophotometric, chemical, and enzymic inactivation methods; and (*c*) to deduce the complete covalent structure by an accepted spectroscopic method capable of defining structure in atomic detail.

These aims have been achieved. The structure of SRS-A, the physiologically more relevant example of the SRSs that were studied, because it was derived immunologically from an animal model of an acute hypersensitivity reaction, has been rigorously defined. Of paramount importance in the determination of this structure was the mass spectrometric analysis of the intact molecule. Degradative and comparative studies are not capable of unequivocally defining structure. For example, the mass spectrum clearly showed the absence of an amide or similar C-terminal blocking groups or, as has been suggested, a sulfone[14] in the molecule; such conclusions could not be drawn from comparative chromatographic data even on multiple systems. Mass spectrometric analysis of the intact molecule could overcome these problems by allowing the complete covalent structure to be collated from the information obtained from each fragmentation. The use of stable isotopes and accurate mass measurement removed possible ambiguities in the interpretation, and the sensitivity and specificity of mass spectrometry made it the method of choice for the structural analysis.

Acknowledgments

We thank the Medical Research Council and the Asthma Research Council for grants and Dr. J. Rokach, Merck Frosst Laboratories, for synthetic LTD_4.

[14] H. Ohnishi, H. Kosuzume, Y. Kitamura, K. Yamaguchi, M. Nobuhara, Y. Suzuki, S. Yoshida, H. Tomioka, and A. Kumagai, *Prostaglandins* **20,** 655 (1980).

[56] Physical Chemistry, Stability, and Handling of Prostaglandins E_2, $F_{2\alpha}$, D_2, and I_2: A Critical Summary

By RANDALL G. STEHLE

The sheer volume of prostaglandin literature is staggering. Surprisingly few publications, however, have appeared concerning the nonclinical aspects of prostaglandin research. Formulation and handling of such highly unstable molecules as prostaglandin E_2 (PGE_2) have proved to be a challenge to the investigator. To understand and circumvent such persistent stability problems, information from many areas of chemistry has been required. Because of the emphasis on clinical research, many areas have been insufficiently explored. For example, it has not been determined why some PGE_2 analog having similar structures—and anticipated similar stability—are significantly less stable than PGE_2 itself in ethanolic solution.

This chapter focuses mainly on the E family of prostaglandins, where more formal kinetics-oriented work has been done, and for which stability work has been done on a variety of dosage forms. Such prostaglandins as $PGF_{2\alpha}$ pose little in the way of stability problems, and PGD_2 and PGI_2 are in preliminary stages of investigation and little information is available on them.

Prostaglandin E_2

Physical Chemistry

Prostaglandin E_2 is a 20-carbon carboxylic acid with both polar and nonpolar properties. This "mixed-bag" nature assures low solubility in such nonpolar solvents as *n*-hexane and 2,2,4-trimethylpentane. The prostaglandin has its greatest affinity for solvents of intermediate polarity, such as *n*-alkanols, methylene chloride, ethyl acetate, and chloroform. The latter solvents, with limited (or no) water miscibility, can act as very efficient extractants. The free acid form of the molecule, similar to (the more stable) $PGF_{2\alpha}$, which has a 9-α-hydroxyl group in place of the keto group, has a pK_a of ca. 4.9–5.0.[1] PGE_2 as the free acid has water solubilities of 1.05 mg/ml at 25° and 1.23 mg/ml at 37° (the corresponding value for the slightly more polar $PGF_{2\alpha}$ is 1.50 mg/ml at 25°[1]). As with $PGF_{2\alpha}$, solubility increases with pH until, when enough anion is present to give a

[1] T. J. Roseman and S. H. Yalkowsky, *J. Pharm. Sci.* **62**, 1680 (1973).

ISBN 0-12-181986-8

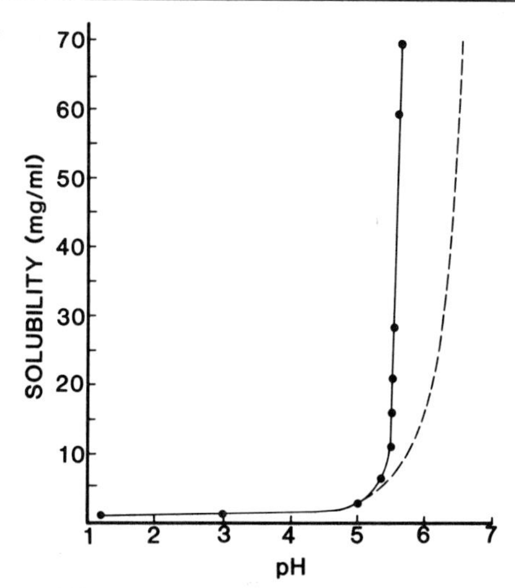

FIG. 1. Aqueous solubility of PGE_2 as a function of pH at 37°. The dashed line is the normally expected pH- and pK_a-determined solubility (S) profile, where S_0 = 1.23 mg/ml and pK_a = 4.9.

solubility of ca. 5 mg/ml, the solubility vs pH profile is almost vertical (Fig. 1).[2] The 5 mg/ml (ca. 0.014 *M*) level can be considered the critical micelle concentration (CMC); the solution becomes predominantly micellar at concentrations ≥5 mg/ml. The compound shows only mild surface activity at concentrations below 5 mg/ml.[3] PGE_1 (lacking the Δ^5 of PGE_2) has a 25° water solubility of only ca. 7.5 μg/ml. It shows similar solubility–pH behavior, with the curve of Fig. 1 displaced perhaps 1.5 pH units higher.

In the neat (solid) state, PGE_2 is a white-to-slightly yellow amorphous-appearing powder or is clearly crystalline in nature. The melting point is in the 60–69° range, with "purer," e.g., white and more obviously crystalline, material melting at the high end of that range. The heat of fusion of this compound appears also to be related to purity, with estimated values ranging from 5 to 9 kcal/mol[2] and purity has been measured, using differential scanning calorimetry (DSC).[4]

The molecular configuration of crystalline PGE_2 has been assessed by X-ray diffraction[5] and appears to have the α (C-1–C-7) chain and β (C-13–C-20) chain well separated. In solution, PGE_2 may or may not assume

[2] R. G. Stehle, unpublished data, 1973.
[3] E. L. Rowe, unpublished data, 1970.
[4] S. Nichols, unpublished data, 1974.
[5] J. W. Edmonds and W. L. Duax, *Prostaglandins* **5**, 275 (1974).

what is termed the "hairpin configuration," where the two side chains are closely and specifically aligned through hydrophobic interactions. This side-chain alignment is most prominent in protic media.[6,7] For example, in water, the chain-chain association is favored because there is *more* solvent (water) structure to disrupt if the chains should assume a different configuration(s). For $PGF_{2\alpha}$, a very similar prostaglandin, circular dichroism behavior (change in absorption between right and left circularly polarized light as a function of wavelength) shows progressive changes as one goes from pH 4 aqueous buffer to 95% ethanol to butanol and finally to acetonitrile.

Stability in Aqueous Systems

Reaction Products. The PGE_2 molecule, in any form, formula, or formulation, but particularly in water, readily dehydrates to PGA_2. Although the E→A reaction is theoretically reversible and PGE_2 has been synthesized from PGA analogs in the (*S*)-coral process,[8,9] given PGE_2 as the starting material, the equilibrium is far to the right. The reaction is effectively irreversible. This is primarily due to the β-hydroxyketo system in the 5-membered ring that distinguishes E prostaglandins. It is most probable that this dehydration is effected through an enol–enolate intermediate (Fig. 2).[10]

Other degradation products beyond the PGA_2 and PGB_2 molecules can appear in aqueous solution, particularly as a function of pH. The rates of formation of these products are very low to only comparable to that of the dehydration reaction. In acid solutions, reversible epimerization at C-15 can occur. This equilibrium reaction is probably independent of the main dehydration reaction. A typical reaction mixture would therefore contain amounts of 15-epi-PGA_2 and possibly 15-epi-PGB_2. In very mildly alkaline systems, the 8-isomer of PGE_2 can form,[11] this reaction also proceeding through the enol–enolate intermediate to an equilibrium mixture. With PGE_1, for example, the equilibrium position in ethanol–potassium acetate was estimated at ca. 9:1 PGE_1:8-iso-PGE_1. In strongly alkaline

[6] E. M. K. Leovey and N. H. Andersen, *J. Am. Chem. Soc.* **97,** 4148 (1975).

[7] N. H. Anderson, P. W. Ramwell, E. M. K. Leovey, and M. Johnson, *Adv. Prostaglandin Thromboxane Res.* **1,** 271 (1976).

[8] R. L. Spraggins, *Tetrahedron Lett.* **42,** 4343 (1972).

[9] W. P. Schneider, G. L. Bundy, and F. H. Lincoln, *J. Chem. Soc. Chem. Commun.*, p. 254 (1973).

[10] S. K. Perera and L. R. Fedor, *J. Am. Chem. Soc.* **101,** 7390 (1979).

[11] E. G. Daniels, W. C. Krueger, F. P. Kupieki, J. E. Pike, and W. P. Schneider, *J. Am. Chem. Soc.* **90,** 5894 (1968).

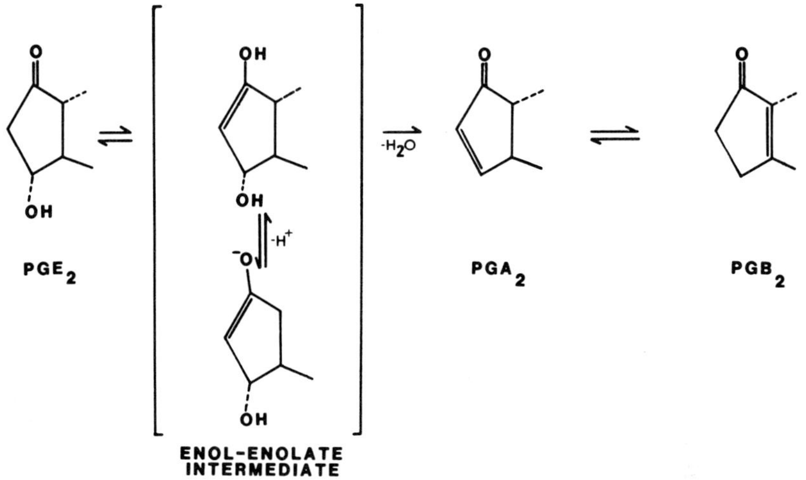

Fig. 2. The $PGE_2 \rightarrow PGA_2 \rightarrow PGB_2$ degradative scheme in aqueous solution, showing enol–enolate intermediate.

solutions the formation of PGB_2 is so rapid (within seconds with 0.5 *M* KOH in 9:1 methanol–water[12]) that no 8-iso-PGE_2 will be formed.

Mechanism of Reaction. Proceeding through the enol–enolate-like intermediate as shown in the preceding section, this reaction can be attributed to an E1cB mechanism.[10] The relative amount of enol–enolate present is dependent on the solvent. An aqueous medium apparently favors relatively large equilibrium amounts of the enol (as opposed to ketol form in which the molecule is usually shown).

The dehydration reaction at 25° is strongly enhanced[10,12] through specific acid and base catalyses, H_3O^+ and OH^-, respectively, at the low and high ends of the pH spectrum (Fig. 3). At very low pH (~1–2), the 11-hydroxyl ($pK_\alpha \sim -2$) can become directly protonated to form a good leaving group (H_2O), so that the subsequent E1 (sequential) or E2 (concerted) mechanism may act as an alternative pathway to the specific acid-catalyzed enolization pathway.

In media of sufficiently basic pH, direct abstraction of an acidic proton from C_{10} can compete with the enol–enolate pathway. This proton abstraction, however, can be interpreted as basic catalysis of enol formation.

The possibility of intramolecular general base catalysis of dehydration of PGE_2 has been proposed and demonstrated.[10,13] In the pH 5–7 range an

[12] R. G. Stehle and T. O. Oesterling, *J. Pharm. Sci.* **66,** 1590 (1977).

[13] R. G. Stehle and R. W. Smith, *J. Pharm. Sci.* **65,** 1844 (1976).

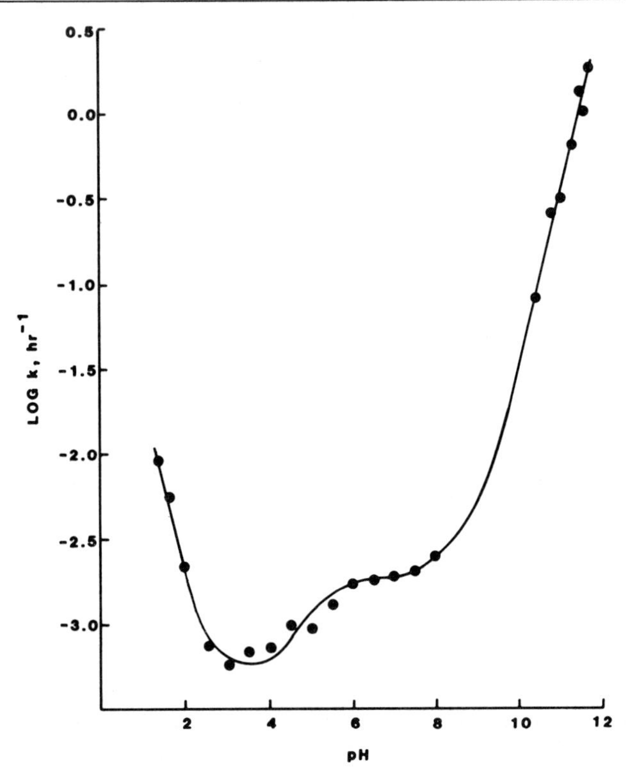

FIG. 3. Log rate–pH profile for PGE_2 degradation at 25°. From Thompson *et al.*[16]

increasing proportion of the PGE_2 is present as carboxylate anion. This charged part of the molecule can wrap around to the vicinity of the C-9–C-10 portion of the molecule to act as a general base catalyst of enol–enolate formation (Fig. 4). Because the carboxylate is on the *same* molecule, its *effective* concentration in this reaction can be up to several orders of magnitude greater than the bulk concentration of the prostaglandin itself. Studies with two E prostaglandins, PGE_1[14] and PGE_2 carbamoylmethyl ester[13] have illustrated that the effect is real.

Prostaglandin E_1 (alprostadil, PGE_1) showed greater stability than PGE_2 at pH 5–7.[14] This somewhat greater stability can be attributed to the relatively lower probability of PGE_1's carboxylate being in the vicinity of C-9–C-10 of that molecule. The many more possible configurations of the PGE_1 α (upper) side chain result from the increased number of bond rotations when the C-5–C-6 double bond of PGE_2 is eliminated. Another way

[14] D. C. Monkhouse, L. VanCampen, and A. J. Aguiar, *J. Pharm. Sci.* **62,** 576 (1973).

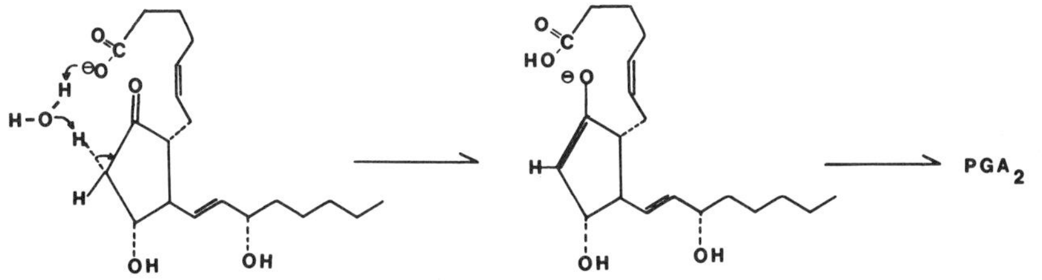

FIG. 4. Intramolecular general base catalysis of enol–enolate formation and dehydration of PGE_2.

of saying this is that a greater entropy decrease is required in PGE_1 relative to PGE_2 to fold the α side chain back to effect general base catalysis.

Figure 5 shows the total degradation rate–pH profile of PGE_2 at 60°.[15] Its inflection point is at pH $\sim$ 4.8. This is consistent with the pK_a 4.9 value found for $PGF_{2\alpha}$.[1] The curve's extremes represent specific acid and base catalyses, and there is probably an underlying spontaneous solvent (water) catalysis. Points on Fig. 5 are a compilation of data generated at Upjohn[12] and data from Thompson *et al.*[16] The curve in Fig. 5 was generated, assuming that first-order PGE_2 degradation can empirically be described by Eq. (1).

$$-\frac{d[PGE_2]}{dt} = k_{H_3O^+}[H^+] + k_{OH^-}[OH^-] + k_{H_2O}[H_2O] + k_{RCOO^+}[RCOO^-]_{effective}[PGE_2] \quad (1)$$

Estimates of these rate constants were obtainable through experiments at high and low pH values,[12] runs at intermediate pH with varying (phosphate and citrate) buffer concentration,[17] the experiments with PGE_2 carbamoylmethyl ester, and added acetate buffer.[13]

What stability can one expect at room temperature in aqueous solution? Looking back at Fig. 3, we see that maximum stability is obtained at ca. pH 3–4. Table I gives some estimated times for 10% loss from solution at 25°.

Stability in Nonaqueous Systems

Dipolar Aprotic Solvents. Dipolar aprotic solvents are solvents with a net dipole moment (i.e., hydrocarbons do not qualify) and have no reactive protons. Those dipolar aprotic solvents studied offered an environ-

[15] R. G. Stehle, unpublished data, 1977.

[16] G. F. Thompson, J. M. Collins, and L. M. Schmalzried, *J. Pharm. Sci.* **62,** 1738 (1973).

[17] R. G. Stehle, unpublished data, 1975.

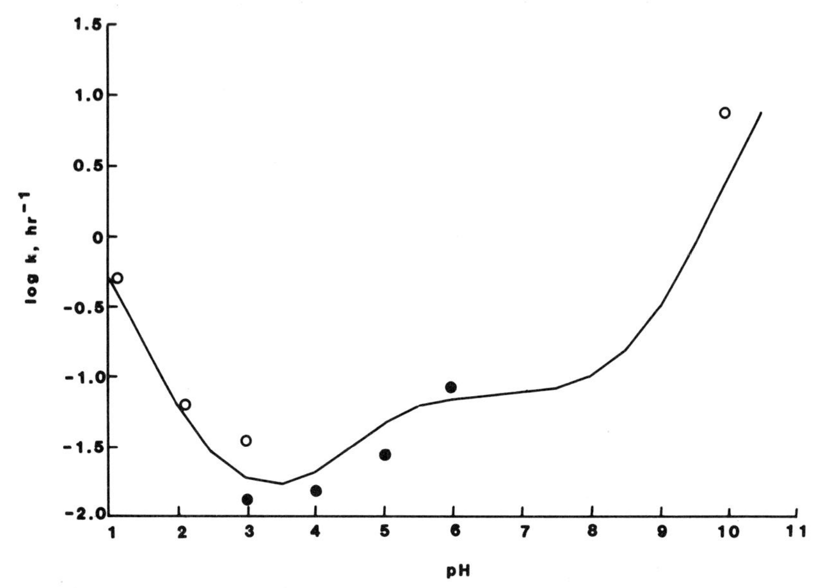

FIG. 5. (Log)rate constant–pH spectrum for PGE_2 degradation in aqueous solution at 60°. Curve is generated using Eq. (1).

ment in which PGE_2 has much improved stability over that in aqueous systems. The most studied solvent is *N*,*N*-dimethylacetamide (DMA). This solvent is of intermediate polarity and provides a medium in which the prostaglandin is very soluble (≥ 100 mg/ml), and it is completely miscible with water.

It may be postulated[18] that DMA and similar nonacidic or nonbasic solvents hydrogen-bond with the prostaglandin and thus stabilize the molecule through solvent–solute interactions. This is consistent with the finding that the stability of PGE_2 in DMA is very good at low (≤ 2.5 mg/ml) concentrations and drops off as initial concentration increases[19] (Fig. 6). As concentration of PGE_2 increases, solute–solute interactions may also increase, particularly if the solvent–solute interaction is not a particularly strong one. A 10% loss occurs only after ca. 36 months for 2–2.5 mg/ml solutions stored at room temperature in DMA. This compares to only a few *days* in water.

There are data available for many other similar solvents. Little and Yalkowsky[20] evaluated the stability of PGE_2 in triacetin at 5 mg/ml and

[18] C. A. Hampson and D. S. Hance, unpublished data, 1973.

[19] R. G. Stehle, unpublished data, 1972.

[20] C. L. Little and S. H. Yalkowsky, unpublished data, 1980.

TABLE I
AQUEOUS STABILITY OF PGE_2 AT 25°[a]

pH	k (hr^{-1})	Hours for 10% loss
3–4	0.00079	133
6	0.0020	53
8	0.0025	42
9	0.025	4.2
10	0.25	0.42 (25 min)

projected a 10% loss time of ca. 16 months at room temperature. At the 5 mg/ml level, the 10% degradation time for PGE_2 in ethyl acetate was estimated to be ca. 22 months at 25°.[21] Similarly, 10% loss was reported to occur in about 11 months in propylene carbonate. This latter solvent is apparently more reactive than the other aprotic solvents. The prostaglandin degrades in it by apparent second-order kinetics as opposed to first-order degradation in DMA, triacetin, ethyl acetate (and buffered water). In the earlier report[18] PGE_2 was found to degrade fairly rapidly in 1,2-dichloroethane. Repeat studies[22] in these laboratories at 60° showed a very slow initial decline followed by faster degradation, discoloration of the solution, and phase separation. This latter would cast doubt on any assays performed, since unchanged PGE_2 could partition into this separated phase. It was proposed that the solvent could be contaminated with, or could decompose to generate, HCl at high temperature. Futhermore, a symmetrical molecule like 1,2-dichloroethane may not be a true dipolar aprotic solvent.

A patent issued to American Cyanamid Co.[23] describes the use of triethyl citrate as a stabilizing solvent for PGE_2. This ester probably is quite similar to ethyl acetate as a medium for PGE_2. Triethyl citrate, however, has very low water solubility, placing a limit on its utility as a vehicle for parenteral formulas.

Excellent stability is also reported in methylene chloride, chloroform, and acetonitrile.[24]

Protic Solvents. Protic solvents have "available" protons. Such solvents as alcohols and glycols can be termed "amphoteric"; they can become protonated in very acidic media and lose a proton in strongly basic solutions. These solvents can therefore be considered acidic or basic

[21] C. A. Hampson and D. S. Hance, unpublished data, 1974.

[22] R. G. Stehle, unpublished data, 1973.

[23] American Cyanamid Co., U.S. Patent 4,211,793, "Stabilization of prostaglandin/E compounds by dissolving in triethyl citrate," 10 January 1980.

[24] W. Morozowich, personal communication.

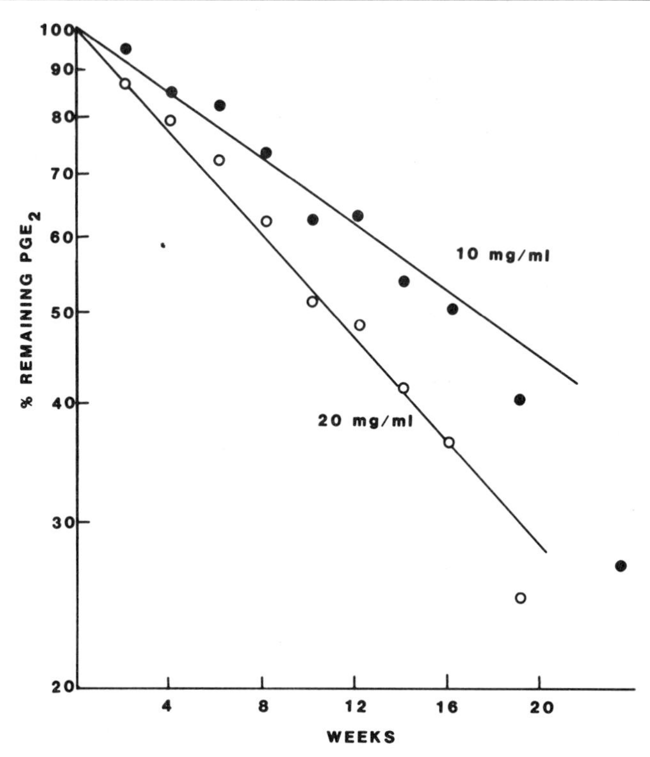

FIG. 6. Degradation of PGE_2 in *N*,*N*-dimethylacetamide at 60°.

$$ROH_2^{\oplus} \underset{}{\overset{pK_a < 0}{\rightleftarrows}} ROH \underset{}{\overset{pK_p > 15}{\rightleftarrows}} RO^{\ominus}$$

under proper conditions. Since the solvent is present in the prostaglandin solution at a high concentration (cf. 55.5 *M* for water), and even though only very weakly basic, proton transfer can occur. A relatively higher proportion of the PGE_2 will exist in such a solution in the enol–enolate form compared to the situation in aprotic solvents. One would therefore anticipate PGE_2 stability intermediate between that in water and the dipolar aprotic solvents. This is borne out with stability studies in the most representative of these solvents, ethanol.

The most noteworthy characteristic of the stability of PGE_2 in ethanol is the increasing degradation rates on a percentage basis per unit time as initial prostaglandin concentration, C_0, decreases below 5–10 mg/ml[25] (Fig. 7). Note the extremely rapid degradation at the 0.1 mg/ml level.

[25] R. G. Stehle, presentation at Academy of Pharmaceutical Science 29th Annual Meeting, San Antonio, Texas, November, 1980.

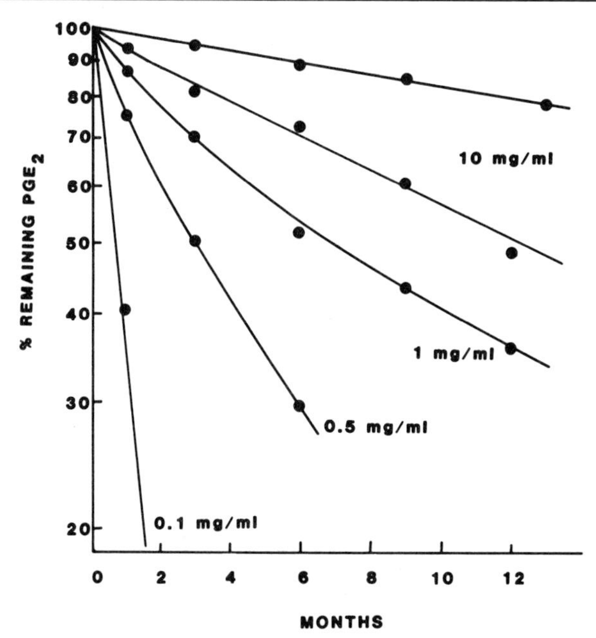

FIG. 7. PGE_2 stability at 25° absolute ethanol, showing effect of initial concentration and lot-to-lot variation. From R. G. Stehle.[25]

Solutions at concentrations as low as 10 μg/ml have shown significant rundown by the time initial assays were performed. Stability also varies between similarly prepared lots (Fig. 7).

At best one can expect only about a 6-month shelf life ($\leq$ 10% degradation) at 25°. The apparent reaction kinetics are second order (Fig. 8); reciprocal concentration vs time is linear. In absolute ethanolic solution PGE_2 typically loses about 10% potency in about 24–36 months at 4° for C_0 = 1–10 mg/ml.

A good portion of the lot-to-lot variation has been attributed[25] to contamination of the absolute alcohol by metal ions acquired in processing (manufacturer) or from their containers; the alcohol was originally supplied in metal tins with soldered seams. This was demonstrated by showing that concentrating the nonvolatile impurity in the alcohol by evaporation (before making the PGE_2 solution) increased the degradation rate of PGE_2 dissolved in the resulting solvent.

Experiments[25] with a variety of alcohol-soluble metal salts showed that PGE_2 is very sensitive to the presence of the metal ions. This is particularly true of di- and trivalent ions from groups 3a, 4a, and 8 of the periodic table. Particularly potent cations include Al^{3+}, Fe^{3+}, Sn^{2+}, Fe^{2+}, and

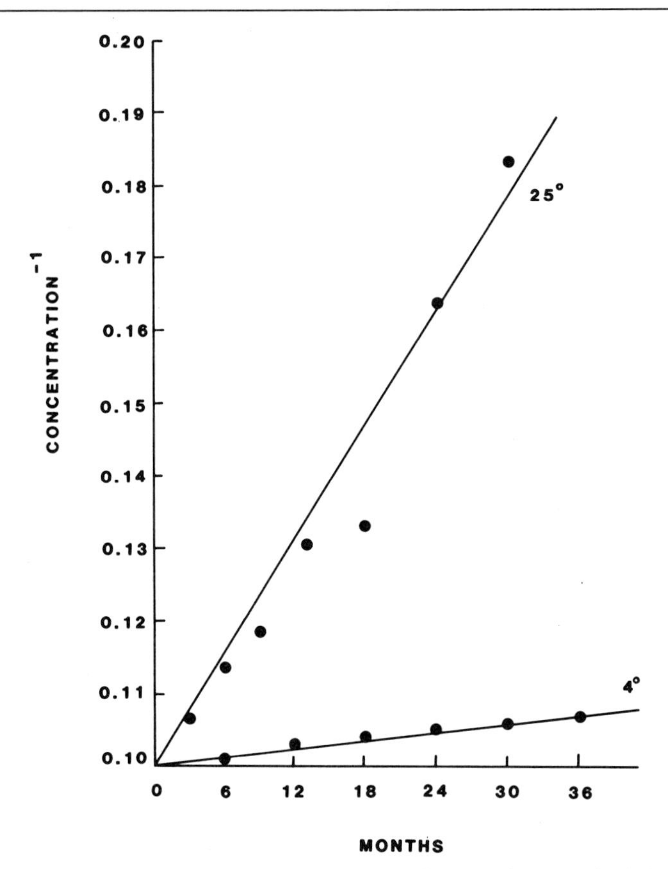

FIG. 8. Second-order kinetics for PGE_2 degradation in ethanol at 4° and 25°. From R. G. Stehle.[25]

Cu^{2+}. The prostaglandin degrades by first-order kinetics in these alcoholic solutions, whereas kinetics are second order with the weaker metal ions or metal ion-free solutions. The effectiveness of the metal ion catalysts is modulated by their accompanying anions. A salt such as $AlCl_3$ that dissociates more completely in ethanolic solution will have more catalytic effect than the triacetate salt of aluminum, which has less tendency to dissociate in solution. Similar results are seen for $FeCl_3$ vs ferric triacetate.

Finally, atomic absorption analysis of existing lots of ethanol have shown the presence of small amounts of Zn^{2+} and Sn^{2+}. The latter was present at ca. 10^{-5} *M* in one particularly bad lot of ethanol, enough to decrease significantly the stability of any E prostaglandin formulated in that lot of ethanol.

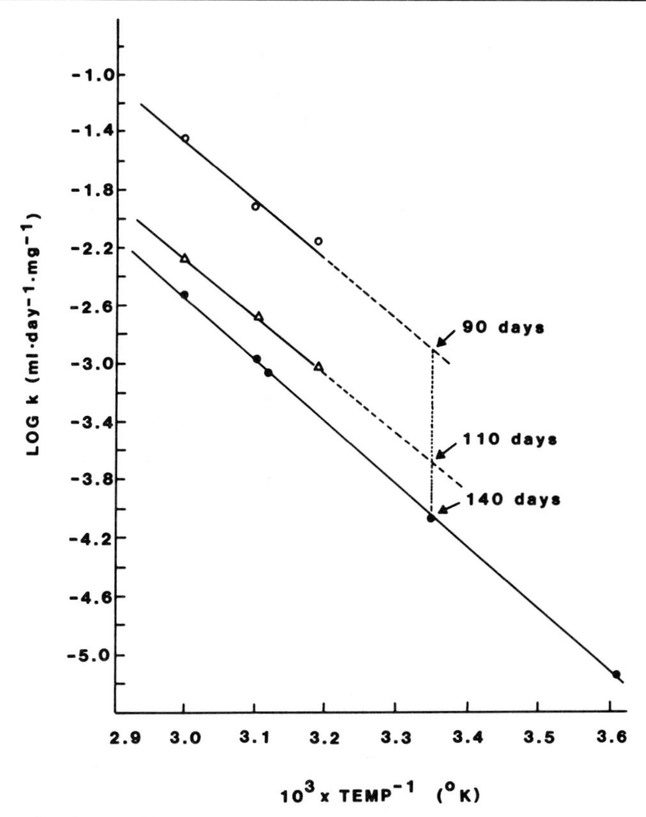

FIG. 9. Arrhenius plot for PGE_2 stability in redistilled-in-glass absolute ethanol, showing 10% loss times as a function of concentration at 25°: 1 mg/ml (○); 5 mg/ml (△); 10 mg/ml (●).

These metal ions act as "super-acid catalysts" in solution. They are highly efficient electron-withdrawing agents whose effect is magnified by their ability to exist in neutral solution at concentrations unattainable by hydrogen ions and by their polyvalent nature. They may act through acid catalysis of enol formation or by direct complexation with the 11-hydroxy group followed by concerted (E_2 mechanism) or stepwise (E_1 mechanism) elimination of the elements of water from the prostaglandin molecule. Redistillation of the ethanol in an all-glass apparatus will remove any contamination by metal ions.

Figure 9 compiles kinetic data for PGE_2 in redistilled-in-glass ethanol. Results are independent of container type (Teflon bottles vs glass ampoule). The slopes of the lines correspond to an activation energy of 19.4 kcal mol^{-1}. The best 10% loss time at 4° is about 3.8 years for the 10 mg/ml solution. That stability does decrease with initial concentration: 10% loss times at 25° are 140 days at 10 mg/ml, 110 days at 5 mg/ml, and 90 days at 1 mg/ml.

Stability of PGE_2 methyl ester[25] is superior to that of PGE_2 in ethanol. Other prostaglandins, however, have been found to have poorer stability. These include 17-phenyl-PGE_2 and PGE_1-alcohol. These latter two prostaglandins have significant changes in side-chain structure that may decrease chain–chain interaction. Exact correlation of chain structure with stability is not possible, however, at this time.

Glycols. PGE_2 has been dissolved in propylene glycol and propylene glycol–alcohol mixtures. Stability data for these solutions are not significantly different from data for pure ethanolic solutions.

Tertiary Alcohols—A Special Case? A patent issued to Monkhouse[26] claims extraordinary stability for PGE_2 in *t*-butanol. Stock solutions of PGE_2 at 1 mg/ml were studied at 37–80°. Although an Arrhenius plot submitted with the application predicted 25 years for 10% loss of PGE_2 from solution at 25°, examination of the graph revealed that this number is applicable to stability at ca. 8°. The actual 10% loss time at 25° appears to be ca. 1.3 years—still quite good. The *t*-butanol is probably approaching the behavior of a dipolar aprotic solvent. Sixteen months for 10% loss at 25° is fairly close to the value determined for PGE_1 in triacetin.[20] Moreover, in common with the aprotic solvents, degradation kinetics are first order. *t*-Butanol is miscible with water and may, because of steric hindrance by the methyl groups, have less ability to participate in acid–base reactions than primary or secondary alcohols.

Solid State Stability

Bulk Drug. This discussion now departs from the situation(s) where reaction kinetics are reasonably well known. Figure 10 shows typical profiles of room temperature stability of bulk PGE_2.[2] Ten percent loss times have ranged from less than 3 months to as high as 7–8 months. At 4°, in the refrigerator, the behavior is similar. The break in the curve occurs at around 24 months.

Onset of rapid degradation appears to coincide with the significant yellowing and "oiling out" of the solid mass. The composition of this oil has not been determined: it can be either a eutectic mixture of PGE_2 + PGA_2 + water or a different, "glass," form of PGE_2. Within the oil, geometric rearrangement of the PGE_2 molecules away from the rigid configuration in the crystals may allow some sort of intermolecular or intramo-

[26] D. C. Monkhouse, U.S. Patent 3,927,197, "Tertiary Alcohol Stabilized E-Series Prostaglandins," 16 December 1975.

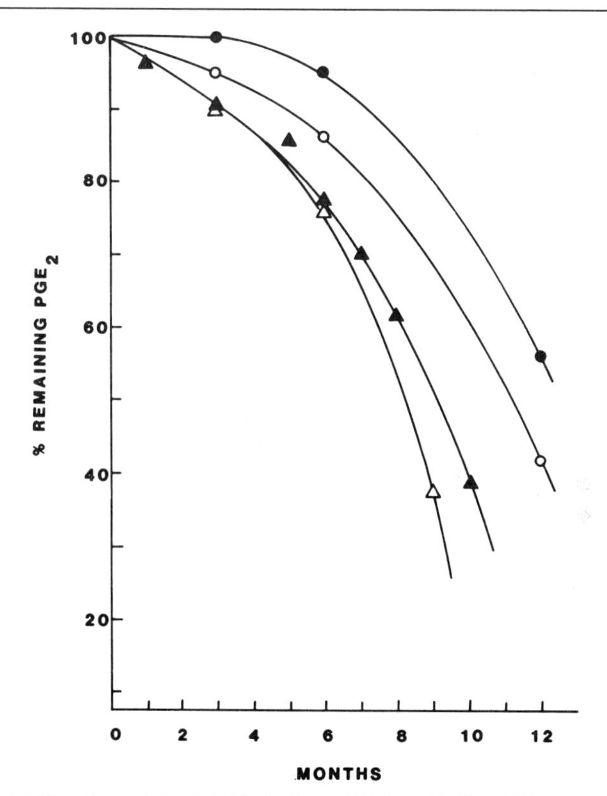

FIG. 10. Stability of four lots of PGE_2 in the solid state (R. G. Stehle, unpublished data, 1973).

lecular catalysis. The PGE_2 and/or the PGA_2 that is present may be responsible.

The lack of stability of PGE_2 in lyophilized powders can be attributed to two factors. First, the lyophilization process probably does not yield crystalline PGE_2. Dissolution of PGE_2 in a volatile solvent followed by evaporation of that solvent yields a glassy or oily mass. PGE_1, with a melting point of ca. 113°, 50° higher than that of PGE_2, has less tendency to become a glass under similar circumstances. Freeze drying PGE_1[27] or 17-phenyl PGE_2 from *t*-butanol, with PVP as a bulking agent gave "cakes" containing ca. 2% prostaglandin by weight, which showed ca. 10% loss after 2 years at 25°. The PGE_1 lyophilizate lost only 20% after 20 months at 47°. The second complicating factor was the residual water in the PGE_2

[27] R. W. Smith and R. G. Stehle, unpublished data, 1976.

lyophilizate. PGE_2 is extraordinarily sensitive to any kind of moisture when in the solid state.

Three crystalline esters of PGE_2 (*p*-acetamidophenyl-, *p*-benzamidophenyl-, and *p*-phenylphenacyl-) with melting points of approximately 100° were reported by Morozowich *et al.*[28] These showed virtually no degradation after 36 months at room temperature, supporting the theory that higher melting points lead to better solid-state stability. Low aqueous solubility of these and other compounds may cause low *in vivo* activity and restrict their usefulness as drugs.

The observation that the more crystalline-appearing lots of PGE_2 had higher heats of fusion[2] correlated with somewhat better solid-state stability. PGE_2 that was recrystallized repeatedly consisted of large white crystals, which melted at 67–69°, the high end of the melting point range observed for all lots of this prostaglandin. Once placed on a stability program, these larger crystals tended to turn yellow and become oils in an apparently random manner. Of two crystals in a vial, one might be white, as initially, whereas the other had yellowed and decomposed. The overall stability was only marginally better than the standard material at 4°. At 25°, however, 95% on the average remained after 8 months and about 75% after 12 months, a significant improvement.

PGE_2 as the amantidine salt showed extremely poor solid-state stability. Its 4° stability was comparable to that of PGE_2 at 25°. Poor stability was considered to be due to effect of the basic cationic part of the molecule.

Prostaglandin $F_{2\alpha}$

The free acid form of $PGF_{2\alpha}$ is a waxy solid that makes purification, handling, and weighing difficult. Therefore it has been synthesized and used as the tromethamine [tris(hydroxymethyl)methylamine (Tris)] salt, which is highly crystalline with a melting point of ca. 101°.[1] The aqueous solubility behavior of $PGF_{2\alpha}$ (Fig. 11) resembles that of PGE_2 (cf. Fig. 1). The intrinsic aqueous solubility, S_o, of the free acid is 1.5 mg/ml at 25°. This increases to ca. 3 mg/ml (2 S_0) at pH $\sim$ 4.9. Between pH 5.0 and 5.2, the solubility vs $-$pH profile is almost vertical due to micelle formation, so that a saturated solution contains ca. 140 mg/ml $PGF_{2\alpha}$ (in terms of the free acid) at pH 5.2. For detailed information on the interrelationships between pH, degree of neutralization, concentration, pK_a, and critical micelle concentration see Roseman and Yalkowsky.[1]

[28] W. Morozowich, T. O. Oesterling, W. L. Miller, C. F. Lawson, J. R. Weeks, R. G. Stehle, and S. L. Douglas, *J. Pharm. Sci.* **68,** 833 (1979).

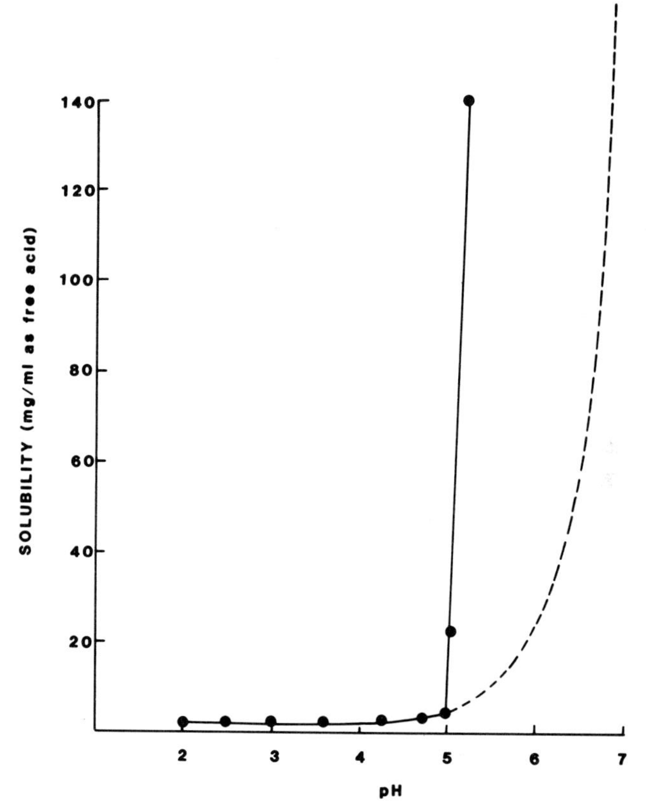

FIG. 11. Aqueous solubility of $PGF_{2\alpha}$ as a function of pH, at 25°. Dashed line is solubility based on $pK_a = 4.9$ and no micellization. From Roseman and Yalkowsky.[1]

$PGF_{2\alpha}$, like PGE_2, has very poor solubility in hydrocarbon solvents. Ethyl acetate and, in particular, chloroform are good solvents and are efficient extractants. Solubility in alcohols is also excellent. It was anticipated that, based on comparative *n*-octanol–water partition coefficients, $PGF_{2\alpha}$ would be as well absorbed buccally as heptanoic or octanoic acid (ca. 90% in 5 min).[29,30] $PGF_{2\alpha}$ is, in fact, poorly absorbed by the buccal route. The relatively high partition coefficient of $PGF_{2\alpha}$ in the octanol–water system may be attributed to solvent–solute interactions between the hydroxyl groups of $PGF_{2\alpha}$ and the octanol. The poor buccal absorbability of $PGF_{2\alpha}$ is better correlatable with its partition coefficient in the

[29] A. H. Beckett and A. C. Moffatt, *J. Pharm. Pharmacol.* **21,** Suppl., 144S (1969).
[30] R. G. Stehle, unpublished data, 1976.

n-heptane–water system. This partition coefficient is several orders of magnitude lower than those for heptanoic or octanoic acids.[29,30]

$PGF_{2\alpha}$ is the most stable of all the prostaglandins discussed in this chapter. This is reflected by the general lack of stability data in the literature. Its main degradation product in aqueous solution is the 15-epimer, with which it forms an equilibrium mixture. As with PGE_2, movement toward equilibrium is catalyzed by acid: the aqueous stability of $PGF_{2\alpha}$ increases with pH.[31] Analysis of $PGF_{2\alpha}$ and its 15-epimer is accomplished by forming the *p*-nitrophenacyl ester to generate UV absorptivity, followed by HPLC.[32] Using such a procedure, Roseman *et al.*[33] reported 40% conversion of $PGF_{2\alpha}$ to its epimer in a pH 3.0 solution after 30 days at 37°. At pH 8.0, no conversion was detectable after nearly a year at 25°.

Prostaglandin D_2

The literature on D prostaglandins is very limited. The most prominent feature governing their properties and stability behavior would be their structural similarity to E prostaglandins.

Prostaglandins of the D family are most readily prepared from suitable F prostaglandin analogs.[34]

Crystalline PGD_2 melts at 68°,[34] virtually the same as PGE_2. Water solubility as a function of pH and other properties would be expected to resemble those for PGE_2. Because PGD_2 can be considered an "upside-down PGE_2," one would not expect the stability behavior of PGD_2 in solution to differ greatly from that of PGE_2. The most rapid reaction would be a dehydration of PGD_2 to an inverted version of PGA_2 through an enolic intermediate. Ancillary but slower degradation routes would also be analogous to those for PGE_2. For instance, just as E prostaglandins form equilibrium amounts of their 8-isomers under mildly basic conditions, PGD_2 can yield an isomeric compound, Δ^{12}-13,14-dihydro-PGD_2.[35] In investigating the synthesis of C_1 esters of PGD_2, Morozowich[36] found that under mildly basic conditions (presence of a tertiary amine), fully 10–20% of the esterified products existed as the Δ^{12}-isomer. Under analogous conditions with PGE_2, less than 5% of the product had isomerized at the 8 position.

In summary, then, probably most of the comments regarding E prosta-

[31] T. J. Roseman, unpublished data, 1971.
[32] W. Morozowich and S. L. Douglas, *Prostaglandins* **10,** 19 (1975).
[33] T. J. Roseman, S. S. Butler, and S. L. Douglas, *J. Pharm. Sci.* **65,** 673 (1976).
[34] M. Hayashi and T. Tanouchi, *J. Org. Chem.* **38,** 2115 (1973).
[35] D. C. Peterson and G. L. Bundy, unpublished data, 1974.
[36] W. Morozowich, unpublished data, 1975.

glandins in various environments would also apply to D prostaglandins: they must be regarded as quite unstable, subject to catalytic degradation by water, H_3O^+, OH^-, and general bases.

Prostaglandin I_2

Physical Chemistry and Solution Stability

PGI_2 is known generically as prostacyclin, and in the older literature as PGX. It is technically a $PGF_{1\alpha}$ analog (9-deoxy-6,9α-epoxy-Δ^5-$PGF_{1\alpha}$) and is generally available as the sodium salt. Typical samples are amorphous in nature and have a fairly wide melting range (ca. 140–170°). Being essentially an aliphatic carboxylic acid, PGI_2's pK_a would be expected to be in the neighborhood of 5.0. An accurate value of this parameter does not appear in the literature. Rapid degradation occurs in aqueous solution to give 6-keto-$PGF_{1\alpha}$, which should have a similar pK_a.

The solubility behavior of PGI_2 in water should parallel that of E- and F-type prostaglandins as a function of pH; PGI_2's instability has, however, prevented measurement of solubility at acidic pH values. Both PGE_2 and $PGF_{2\alpha}$ form micellar solutions when the pH is high enough to allow a solubility of 5–10 mg/ml, and PGI_2 may do likewise. As the sodium salt, PGI_2 is also soluble in lower alcohols and dipolar aprotic solvents (DMA, DMF, DMSO), owing to the ability to hydrogen-bond with these solvents. As with PGE_2 and $PGF_{2\alpha}$, solubility in hydrocarbon solvents is very poor.

The most salient characteristic of PGI_2 is its extreme instability in aqueous solution, particularly at neutral and acidic pH.[37,38] For example, at pH 7.4, 50% of PGI_2 is degraded within less than 20 min at 4° and only 3–4 min at 25°. Degradation is so rapid that specialized techniques[37] are necessary when doing kinetic studies with this molecule.

The proposed reaction scheme for PGI_2 degradation appears in Fig. 12. The structural characteristic of PGI_2 responsible for its lability is the vinyl ether moiety. The specific acid catalytic rate constant, k_{H^+} (3.7×10^4 sec^{-1} M^{-1}), is extremely high for this type of compound. This can be explained through a release of ring strain when the PGI_2 molecule is protonated[37] or intramolecular general acid catalysis.

Figure 13[38] shows hydrolysis rate data for PGI_2 at zero buffer concentration. The inflection point of the curve in the pH 4.5–5.0 range (at the pK_a of the prostaglandin) is evidence of some type of intramolecular ef-

[37] M. J. Cho and M. A. Allen, *Prostaglandins* **15,** 943 (1978).

[38] Y. Chiang, A. J. Kresge, and M. J. Cho, *J. Chem. Soc. Chem. Commun.* p. 129 (1979).

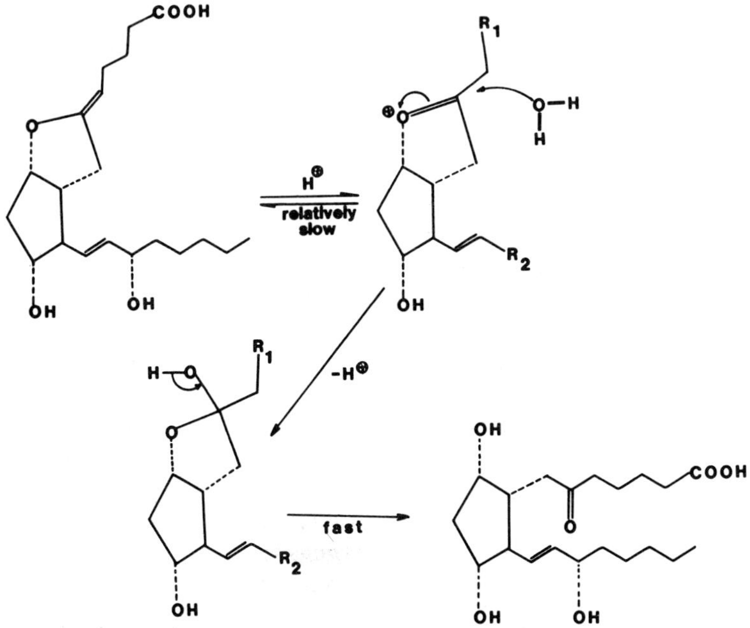

FIG. 12. Hydrolysis of PGI_2 to 6-keto-$PGF_{1\alpha}$.

fect. This is the reverse situation of the general intramolecular base catalysis seen in PGE_2 dehydration: for pH > 2, the hydrolysis rate is higher than it would be in the absence of intramolecular effect. This enhancement can be attributed[38] to either general acid catalysis by the nonionized carboxylic acid group or electrostatic stabilization of the positively charged transition state (Fig. 12) by carboxylate anion. The conjugate acid forms of buffer molecules also act as general acid catalysts in aqueous solution.[37] Below pH 2, the curve documents the effect of specific acid catalysis by ''external'' H_3O^+ in solution.

In summary, then, the observed rate constant may be expressed[37] as Eq. (2).

$$k_{obs} = k_{H^+}[H^+] + k_{HA}[HA] + k_{HA}[HA]_{eff} + k_{H_2O}[H_2O] \tag{2}$$

Specific acid catalysis	General acid catalysis by buffer components	Intramolecular general acid catalysis	Spontaneous solvent catalysis

where $[HA]_{eff}$ is the effective concentration of general acid in the vicinity of the reactive site, and will change with pH.

Actual 25° stability data at a constant ionic strength of 0.5 M are

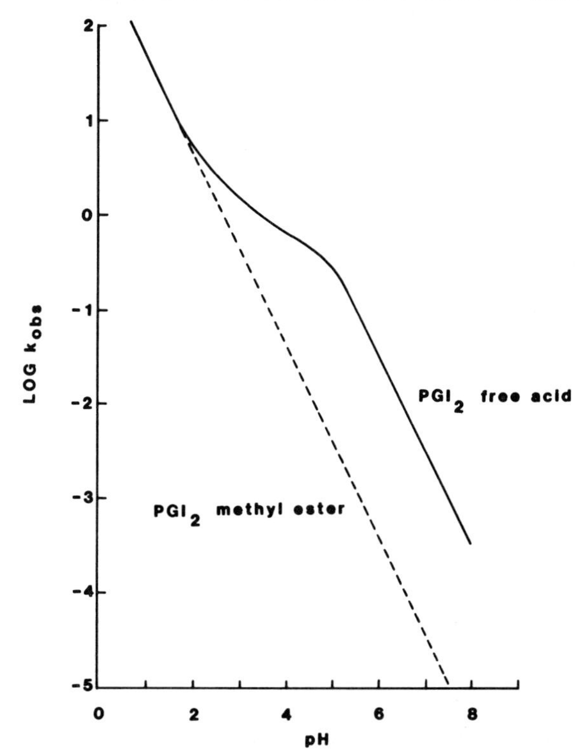

FIG. 13. Log (rate constant) vs pH profile for PGI_2 and PGI_2 methyl ester hydrolyses at 25°, showing evidence of general acid catalysis at pH > 3. μ = 0.1. Adapted from Chiang *et al.*[38]

shown in Fig. 14.[39] For example, at a pH of 10, one would expect 50% hydrolysis of PGI_2 after about 24 hr, with correspondingly higher loss times at high pH values.

Cooling the solutions below room temperature will slow the hydrolysis reaction. With the energy of activation for the reaction of 11.85 kcal/mol, the reaction rate will slow by a factor of two per 10° drop in temperature. Therefore the $t_{1/2}$ values in Fig. 14 could be increased by a factor of 4 by storing the solutions at about 0–4°. Further increase in storage time by freezing is inadvisable, since the solution components would be concentrated during the freezing process. Under these circumstances, catalysis by buffer components, etc., could be significant.

[39] M. J. Cho and M. A. Allen, *Int. J. Pharm.* **1,** 281 (1978).

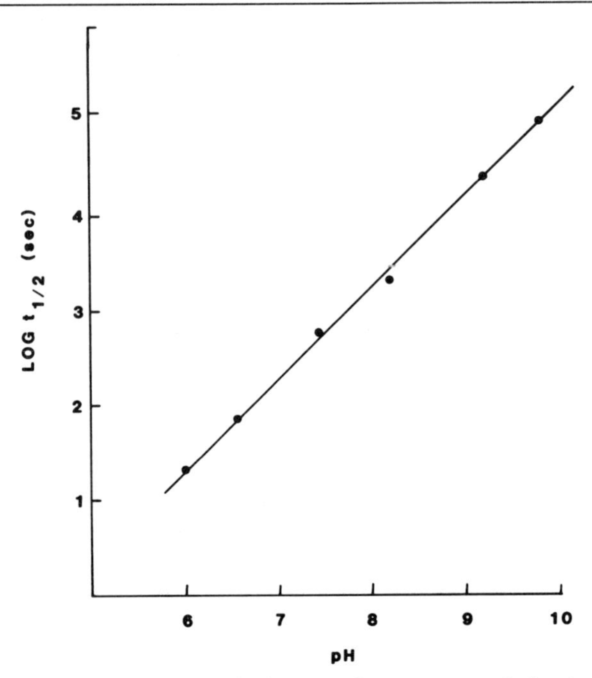

FIG. 14. Log of time for 50% loss of PGI_2 at 25° from aqueous solution (zero buffer concentration) as a function of pH. Adapted from Cho and Allen.[39]

Solid-State Stability

The critical factor in handling prostacyclin as the sodium salt in the solid state is total avoidance of moisture. Sufficient adsorbed moisture in the system can lead to solution formation on the microscopic level, and aqueous solution degradation to 6-keto-$PGF_{1\alpha}$ can readily occur.

15-Methyl Prostaglandins

Prostaglandins with a 15-methyl group can be effectively more potent due to inhibition of enzymic oxidation at C-15. However, the methyl group also confers a new type of chemical instability to the molecule, which is now a tertiary allylic alcohol. In addition to other degradative reactions associated with the basic type of prostaglandins, such alcohols can epimerize at C-15 through specific and general acid catalysis and formation of a stable carbonium ion. Either molecule can, furthermore, slowly undergo allylic arrangement to give 13-methyl-Δ^{14} products. The

TABLE II

STABILITY, ASSAY PROCEDURE, AND HANDLING INFORMATION FOR PROSTAGLANDINS

Prostaglandin	Relative stability		Assay procedure	Handling precautions
	Solid state	Aqueous solution		
E_2	Poor; store at $\leq 4°$; protect from moisture	Very poor; maximum stability at pH 3.5; see text	Quantitative TLC[a], HPLC[b]	"Dusty" compound; wear respirator when handling, weighing. Keep away from eyes and skin.
D_2	See PGE_2 above	See PGE_2 above	No known published assay; could be assayed by PGE_2 procedures.[a,b]	Unknown; should be similar to PGE_2
I_2 (as sodium salt)	Poor; store in freezer; must be protected from moisture	Extremely poor at all pH values; CO_2 from air can acidify solution and accelerate loss rate	UV end absorption[c]; extreme instability in solution precludes chromatographic analyses	Highly unstable nature precludes long action. Handle as other prostaglandins above. Degradation product is relatively inactive.
F_2 (as tromethamine (Tris) salt)	Excellent	Excellent for $pH \geq 5-6$	GLC[a] HPLC[b]	Very prone to "dusting"; irritant to the throat; wear respirator when handling

[a] T. J. Roseman, B. Sims, and R. G. Stehle, *Am. J. Hosp. Pharm.* **30,** 236 (1973).
[b] Morozowich and Douglas.[32]
[c] Cho and Allen.[37]

overall kinetics can be described as the sum of specific and general acid catalyses [Eq. (3)].[39]

$$k = k_{H^+} (H^+) + k_{H_2O} (H_2O) \tag{3}$$

Since the water concentration is independent of pH, the first term will predominate in strongly acid solutions. The epimerization can be slowed[39] from either direction [$(R) \rightarrow (S)$ or $(S) \rightarrow (R)$] by incorporation of a cationic surfactant in the aqueous solution. Such surfactants form positively charged micellar structures that incorporate (solubilize) the prostaglandin and inhibit catalysis by hydronium ion.

Handling of Prostaglandins: A General Note

Prostaglandins are extremely potent materials. For this reason precautions must be taken in handling them. Solid prostaglandins can disperse into the air very easily. Typical symptoms of exposure to airborne prostaglandins include coughing, "dry" and irritated throat; reddening, irritation, and swelling of the eyes and eyelids; and headache. The respirator mask, worn over the nose and mouth, prevents the irritation caused by such solid prostaglandins as PGE_1, PGE_2, and $PGF_{2\alpha}$-tromethamine salt. Handling of these substances should ideally be done in a confined area, such as a fume hood. Prostaglandins that are oils or that are in solution should not contact the skin. Protective gloves should be worn when handling these materials. This is especially true for prostaglandins with methyl groups at the 15 or 16 position, which confer resistance to enzymic deactivation and therefore more potency.

For cleanup of spills and disposal of waste, prostaglandins E_2, D_2, and $F_{2\alpha}$ can be deactivated with chlorine bleach,[40] which forms a chlorohydrin across the 5,6 double bond, along with less polar and essentially inactive products. E prostaglandins lacking the 5,6 double bond can be rapidly converted to essentially inactive B prostaglandins by treating with 0.1 *M* base. PGI_2 can be destroyed almost instantly by washing with mildly acid solution. The degradation product, 6-oxo-$PGF_{1\alpha}$, is essentially inactive.

Table II summarizes stability and handling precautions, and gives references for assay procedures for evaluating the potency of these prostaglandins.

[40] W. Morozowich, unpublished data, 1973.

[57] Synthesis of Prostacyclin Sodium Salt from $PGF_{2\alpha}$

By ROY A. JOHNSON

Prostacyclin [originally PGX, also PGI_2, prostaglandin I_2, or (5*Z*)-9-deoxy-5,9α-epoxy-$PGF_2\alpha$] was discovered[1] and characterized[2] in 1976. The chemical properties of prostacyclin are dominated by the reactivity of the vinyl (or enol) ether functionality involving carbons 5 and 6 and the oxygen at position 9. The vinyl ether functional group is subject to rapid

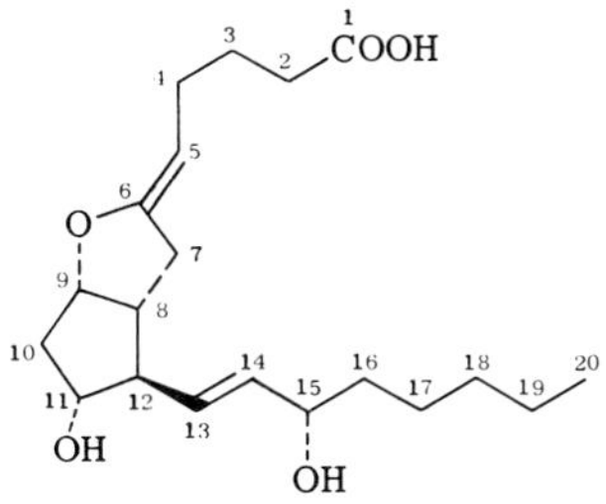

Prostacyclin (PGI_2)

hydrolysis and in prostacyclin is sensitive to hydrolytic conditions below a pH of about 9.[3] Hydrolysis results in the conversion of prostacyclin into 6-keto-$PGF_{1\alpha}$.[2] Because of this sensitivity to hydrolysis, preparation of prostacyclin in the free carboxylic acid form is exceedingly difficult and, to date, the isolation of a pure sample of the acid has not been described. Consequently, biological, pharmacological, and clinical studies of prostacyclin use the sodium salt of the carboxylic acid.[4] The synthesis presented here, therefore, is of prostacyclin sodium salt ($PGI_2 \cdot Na$). $PGI_2 \cdot Na$ is a white solid and is stable when stored so that it is protected from moisture and carbon dioxide. Aqueous solutions of $PGI_2 \cdot Na$ are also reasonably stable if buffered at a pH greater than 9.3.

A number of syntheses of prostacyclin salts have been reported; how-

[1] S. Moncada, R. Gryglewski, S. Bunting, and J. R. Vane, *Nature* (*London*) **263,** 663 (1976).
[2] R. A. Johnson, D. R. Morton, J. H. Kinner, R. R. Gorman, J. C. McGuire, F. F. Sun, N. Whittaker, S. Bunting, J. Salmon, S. Moncada, and J. R. Vane, *Prostaglandins* **12,** 915 (1976).
[3] Y. Chiang, A. J. Kresge, and M. J. Cho, *J. Chem. Soc. Chem. Commun.* **129** (1979).
[4] R. A. Johnson, F. H. Lincoln, J. L. Thompson, E. G. Nidy, S. A. Mizsak, and U. Axén, *J. Am. Chem. Soc.* **99,** 4182 (1977).

METHODS IN ENZYMOLOGY, VOL. 86

ISBN 0-12-181986-8

ever, they all employ the same synthetic strategy.[4–9] The present synthesis is derived from the original detailed descriptions of our PGI_2·Na synthesis[5] and includes several modifications that improve the procedure.[10] The general scheme is outlined in Fig. 1. A requirement of this synthesis is the necessity for either $PGF_{2\alpha}$ or $PGF_{2\alpha}$ methyl ester (**1**) as a starting material. Although two isomers (**2** and **3**) are formed in the iodocyclization reaction, this is of no important consequence since both give prostacyclin methyl ester (**4**) in the subsequent step. The scale of the present experiments may be increased or decreased by 10-fold without difficulty.

Procedure

General Experimental Procedures. Thin-layer chromatography (TLC) is done on 2.5 × 10-cm plates precoated with silica gel GF (Analtech). After development, the plates are visualized by first spraying with 1:1 methanol-sulfuric acid followed by charring on a hot plate. Solvents are distilled in glass (Burdick and Jackson). The term "brine" refers to a saturated aqueous solution of sodium chloride. "Concentrated under reduced pressure" refers to use of a rotary evaporating apparatus, such as the Büchi Rotavapor.

Prostaglandin $F_{2\alpha}$ Methyl Ester ($PGF_{2\alpha}$ Methyl Ester, **1**). Two methods for esterification of $PGF_{2\alpha}$ are outlined below in the event that the acid but not the methyl ester is available.

WITH DIAZOMETHANE. (CAUTION: Diazomethane must be used in an efficient hood.) Diazomethane is prepared by adding portions 1.5 g of *N*-methyl-*N'*-nitro-*N*-nitrosoguanidine (MUTAGEN!, Aldrich Chemical Co.) to a cooled (ice bath) and stirred mixture of 45% aqueous potassium hydroxide solution (5 ml, Mallinckrodt) and diethyl ether (20 ml). (Glassware used for diazomethane solutions should not be scratched, etched, or have ground surfaces.) After addition, the yellow ether layer is transferred to a second flask and dried over anhydrous sodium sulfate while being kept cool on ice. This ether solution of diazomethane may be used

[5] R. A. Johnson, F. H. Lincoln, E. G. Nidy, W. P. Schneider, J. L. Thompson, and U. Axén, *J. Am. Chem. Soc.* **100,** 7690 (1978).

[6] E. G. Corey, G. E. Keck, and I. Szekely, *J. Am. Chem. Soc.* **99,** 2006 (1977).

[7] N. Whittaker, *Tetrahedron Lett.*, p. 2805 (1977).

[8] I. Tömösközi, G. Galambos, V. Simonidesz, and G. Kovács, *Tetrahedron Lett.*, p. 2627 (1977).

[9] K. C. Nicolaou, W. E. Barnette, G. P. Gasic, R. L. Magolda, and W. J. Sipio, *J. Chem. Soc., Chem. Commun.*, p. 630 (1977).

[10] R. S. P. Hsi, W. T. Stolle, J. P. McGrath, and D. R. Morton, *J. Labelled Compd. Radiopharm.***18,** 1437 (1981). Several of the modifications incorporated in this manuscript were first introduced by Burris D. Tiffany.

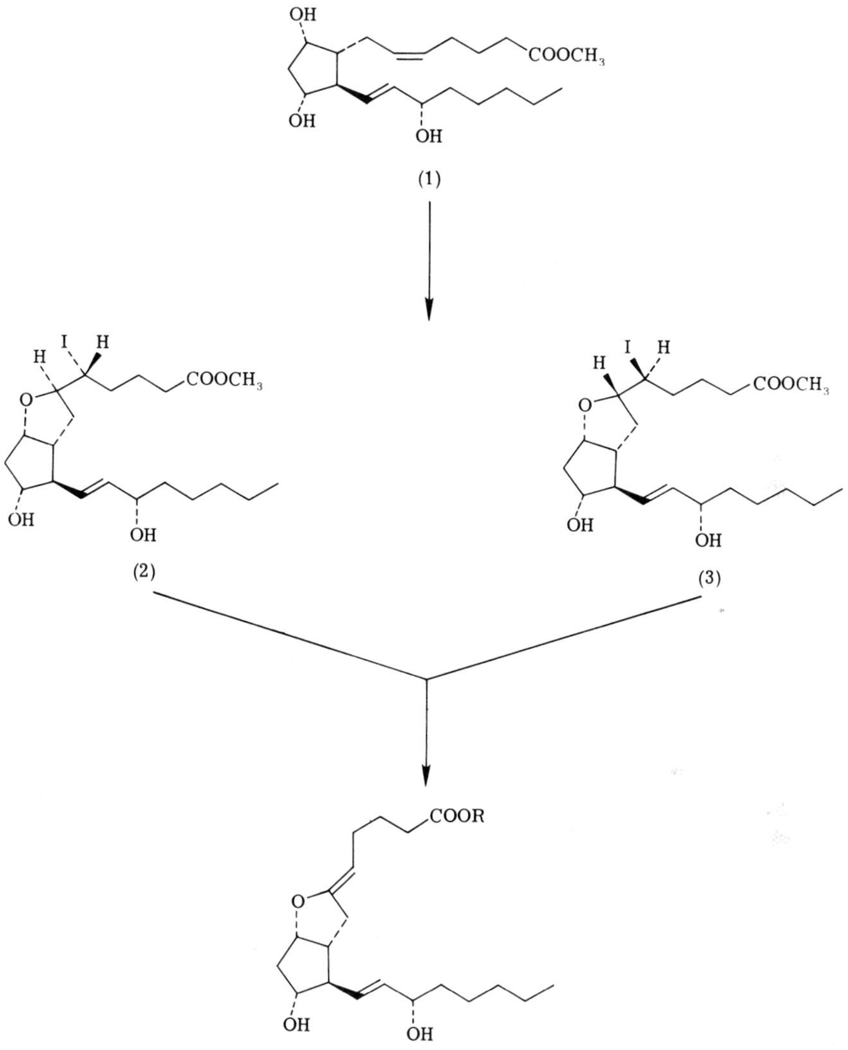

FIG. 1. Preparation of prostacyclin sodium salt ($PGI_2 \cdot Na$) (**5**).

to esterify the $PGF_{2\alpha}$. $PGF_{2\alpha}$ (1.0 g) may be dissolved in ether and the solution treated with the ethereal diazomethane solution. The reaction may be checked by silica gel TLC ($PGF_{2\alpha}$ methyl ester, *Rf* 0.48 in 5% methanol in ethyl acetate) and should be complete within 30 min. Excess diazomethane (yellow color) may be removed with a stream of nitrogen. Removal of solvent under reduced pressure (or under a nitrogen stream) will provide the oily $PGF_{2\alpha}$ methyl ester (**1**).

WITH DIISOPROPYLAMINE-METHYL IODIDE. To a solution of $PGF_{2\alpha}$ (1.0 g) in acetonitrile (20 ml, protected from light by aluminum foil wrapping the flask) are added *N,N*-diisopropylamine (1 ml) and methyl iodide (2 ml). The solution is stirred at room temperature and may be checked by TLC (as above) for completion. The reaction should be complete within 2 hr. The solution is concentrated under reduced pressure. The residue is taken up with ether (20 ml) and brine (20 ml). The layers are shaken thoroughly and the organic layer separated. The brine layer is extracted two more times with ether. The combined ether extracts are dried over sodium sulfate, filtered, and concentrated, giving the oily PGF_2 methyl ester (**1**).

Preparation of (5R,6R)-*5-Iodoprostaglandin* I_1 *Methyl Ester* (**2**) *and* (5S,6S)-*5-Iodoprostaglandin* I_1 *Methyl Ester* (**3**). A solution of $PGF_{2\alpha}$ methyl ester (1.0 g) in methylene chloride (20 ml) and 20 ml of a saturated aqueous solution of $NaHCO_3$ are placed together in a round-bottom flask cooled on an ice bath, and stirred with sufficient vigor to create an emulsion. Stirring and cooling is continued while a solution of iodine crystals (775 mg) in methylene chloride (50 ml) is added over a period of 10 min. The mixture is stirred at 0°C for an additional 50 min. Stirring is then stopped, and the layers are allowed to separate. After separating and collecting the lower methylene chloride layer, the aqueous layer is extracted with methylene chloride (30 ml). The combined methylene chloride layers are shaken with water (25 ml) containing sodium sulfite ($Na_2S_2O_3$, 6 g). The now colorless methylene layer is separated and further shaken with brine (25 ml) and again separated. The methylene chloride solution is dried over anhydrous sodium sulfate, filtered, and concentrated under reduced pressure, giving a pale yellow oil. The oil consists of an approximately 10:1 mixture of compounds **2** and **3**, but these need not be separated, as both yield the desired product in the subsequent chemical step. The mixture is not well separated by TLC (silica gel) in most systems and has R_f 0.47 in 1:1 acetone–hexane and R_f 0.48 in ethyl acetate. Physical properties for purified samples of **2** and **3** have been reported elsewhere.[5] For the purposes of this synthesis, the mixture is used without further purification in the next step.

Prostacyclin Methyl Ester (**4**). The mixture of **2** and **3** from above is dissolved in anhydrous diethyl ether (20 ml), and either 3 ml of 1,8-diazabicyclo[5.4.0]undec-7-ene (DBU, Aldrich Chemical Co.) or 3 ml of 1,5-diazabicyclo[4.3.0]non-5-ene (DBN) are added. The ether is removed under reduced pressure, and the residue is kept at room temperature for 1 hr. Water (20 ml), brine (20 ml), and ether–hexane–triethylamine

(1:1:0.02, v/v/v) (80 ml) are added. The mixture is shaken, and the lower aqueous layer is separated. The organic layer is again washed with brine, dried over anhydrous sodium sulfate, filtered, and concentrated at reduced pressure under 30°. The pale yellow oil (**4**) is immediately chromatographed over a column of neutral silica gel (60 g, 60–100 mesh, E. Merck Laboratories) packed as a slurry in acetone–hexane–triethylamine, (35:65:1). The column is eluted with acetone–hexane–triethylamine, (35:65:0.2). If 10-ml fractions are collected, the desired **4** will be found approximately between fractions 20 and 50 and may be detected by TLC on silica gel plates (1 × 4 in.); R_f 0.49 in ethyl acetate + 1% triethylamine, R_f 0.49 1:1 acetone-hexane + 1% triethylamine.

The fractions containing the product are pooled and concentrated under reduced pressure. The oily **4** may be used directly in the next step or may be crystallized by dissolving in ether (3 ml, containing 0.1% triethylamine) and adding hexane (3–5 ml, containing 0.1% triethylamine). As crystallization proceeds, additional hexane (20 ml total) is added in portions over a period of 2–3 hr. The fine, white crystals are collected by filtration, washed with a little hexane, and may be dried and stored in a vacuum desiccator. The yield is 600 mg (60%, based on **1**) of **4**, mp 56–58°.

Prostacyclin Sodium Salt (**5**). A solution of **4** (600 mg) in methanol (4 ml, degassed by sonication to remove dissolved CO_2 immediately before use) and 1.0 *N* aqueous sodium hydroxide (1.63 ml, freshly prepared and also degassed before use) is stirred under a nitrogen atmosphere at 40° (bath temperature) for 3 hr. Completeness of hydrolysis may be checked by silica gel TLC (1:1 acetone–hexane or ethyl acetate). If complete, only a heavy spot at the origin will be seen on the plate. Methanol (5 ml) is added to the solution, which is then concentrated under reduced pressure at 25° for 5–10 min. The residue is dissolved in water (1 ml), and acetonitrile (30 ml total) is added in small portions while stirring the solution vigorously. Fluffy white, semicrystalline solid precipitates and is collected by filtration and washed with a little dry acetonitrile. The product **5** (500 mg, 82%) is dried and stored in a vacuum desiccator.

High-Performance Liquid Chromatography (HPLC) Analysis of Prostacyclin Sodium Salt. The purity of prostacyclin sodium salt can be assayed if HPLC capabilities are available.[11] Required are an HPLC pump, injection valve, an ultraviolent spectrophotometer detector (capable of detecting in the range of 210–230 nm), a recorder, and a C_{18} reversed-phase analytical HPLC column. A solvent mixture consisting of 20% acetonitrile and 80% water that is buffered to pH 9.3 with boric acid (H_3BO_3) and sodium borate ($Na_2B_4O_7 \cdot x\ H_2O$) is used with this column. Five microliters of a solution of $PGI_2 \cdot Na$ (1 mg/ml concentration) are easily detected with a detector wavelength of 214 nm. Refer to Wynalda *et al.*[11] for an example of results using this system and illustration of separation from typical decomposition products (e.g., 6-keto-$PGF_{1\alpha}$) or contaminants found in PGI_2·Na.

Radiolabeled PGI_2·Na. For the preparation of radiolabeled PGI_2·Na, refer to Hsi *et al.*[10]

Acknowledgment

We are indebted to Frank H. Lincoln, Burris D. Tiffany, and Richard S. P. Hsi for providing many improvements in the procedures for synthesis of PGI_2·Na.

[11] M. A. Wynalda, F. H. Lincoln, and F. A. Fitzpatrick, *J. Chromatogr.* **176,** 413 (1979).

Section IV

General Separation Procedures

[58] Rapid Extraction of Arachidonic Acid Metabolites from Biological Samples Using Octadecylsilyl Silica

By WILLIAM S. POWELL[1]

Arachidonic acid (20:4),[2] monohydroxyeicosatetraenoic acids, prostaglandins (PGs), and thromboxanes (Txs) can be extracted from biological samples using a number of different procedures. In the past, the most common method for extracting 20:4 metabolites from tissue homogenates or subcellular fractions has been to extract acidified aqueous solutions with organic solvents, such as diethyl ether or ethyl acetate. PGs and PG metabolites can be extracted from body fluids, such as plasma and urine, by chromatography on columns of XAD-2 resin. Both of these methods are rather tedious and time consuming and often require evaporation of relatively large amounts of solvent. Moreover, they tend not to be very selective and often give fractions containing much extraneous material that must be removed by open column chromatography on silicic acid prior to analysis by high-pressure liquid chromatography (HPLC).

We have developed a simple chromatographic method to extract 20:4 and its metabolites from biological media using octadecylsilyl (ODS) silica.[3] ODS silica is a hydrophobic material formed by allowing silicic acid to react with octadecyltrichlorosilane. SEP-PAK C_{18} cartridges containing this material can be obtained from Waters Associates, Milford, Massachusetts. ODS silica has been used to extract various materials, including steroids,[4] peptides,[5] and drugs,[6] from biological samples. In general, aqueous solutions containing the substance to be extracted are passed through small columns of ODS silica. Solutes are retained on the basis of hydrophobic interactions with the stationary phase. Substances more polar than those which it is desired to extract can be eluted from the column with aqueous media containing relatively small amounts of miscible organic solvents (e.g., methanol, ethanol, or acetonitrile). The desired material is then eluted with a solvent containing a higher concentration of

[1] The author is a scholar of the conseil de la recherche en Santé du Québec.

[2] The abbreviations used are: 20:4, arachidonic acid; 15h-20:4, 15-hydroxy-5,8,11,13-eicosatetraenoic acid (15-HETE); PG, prostaglandin; Tx, thromboxane; ODS, octadecylsilyl; HPLC, high-pressure liquid chromatography.

[3] W. S. Powell, *Prostaglandins* **20,** 947 (1980).

[4] C. Shackleton and J. Whitney, *Clin. Chim. Acta* **107,** 231 (1980).

[5] H. P. J. Bennett, A. M. Hudson, C. McMartin, and G. E. Purdon, *Biochem. J.* **168,** 9 (1977).

[6] R. J. Allan, H. T. Goodman, and T. R. Watson, *J. Chromatogr.* **183,** 311 (1980).

ISBN 0-12-181986-8

the organic component. This method of extraction is therefore a simple form of reversed-phase liquid chromatography.

We developed a modification of this method that has proved to be very useful for the extraction of fatty acids and PGs.[3] We found that ODS silica could also act as a medium for normal-phase chromatography, possibly owing to the presence of underivatized silanol groups. When solutions of $PGF_{2\alpha}$ in benzene, for example, were passed through columns of ODS silica, the $PGF_{2\alpha}$ was completely adsorbed by the stationary phase. The $PGF_{2\alpha}$ could be eluted from the columns by using more polar organic solvents, such as diethyl ether, ethyl acetate, or methyl formate. There is therefore a second possible approach to the extraction of fatty acids and PGs from aqueous media using ODS silica, involving adsorption by reversed-phase chromatography as described above, followed by elution by normal-phase chromatography (using a series of increasingly polar organic solvents that are not miscible with water). We found the latter approach to have two main advantages. A relatively clean extract is obtained with a very small amount of residue, even from large samples (e.g., homogenates of 5–10 g of tissue). This enables samples to be chromatographed directly by HPLC without any further purification. Second, PG or hydroxy fatty acid fractions can be obtained in relatively small volumes of volatile organic solvents that can readily be evaporated under a stream of nitrogen.

A general scheme for the extraction of PGs and monohydroxy fatty acids using ODS silica is shown in Fig. 1. Depending on the type of sample and the compounds to be extracted, this scheme can be modified as discussed below; a Summary is presented at the end of this chapter. This chapter provides recommendations for nonpolar fatty acids, hydroxyeicosenoic acids, prostaglandins and thromboxanes, and polar prostaglandin metabolites.

Materials

Radioactive PGs, TxB_2, 20:4, and dipalmitoylphosphatidylcholine were purchased either from Amersham/Searle, Oakville, Ontario or New England Nuclear, Lachine, Quebec. 15-Hydroxy-5,8,11,13-[1-^{14}C]eicosatetraenoic acid (15h-20:4)[7] and [9β-^{3}H]$PGF_{2\alpha}$ were synthesized as described in the literature.[8] 19-Hydroxy[9β-^{3}H]$PGF_{2\alpha}$ was prepared as described elsewhere in this volume.[9]

[7] M. Hamberg and B. Samuelsson, *J. Biol. Chem.* **242,** 5329 (1967).
[8] E. Granström and B. Samuelsson, *Eur. J. Biochem.* **10,** 411 (1969).
[9] W. S. Powell, this volume [28].

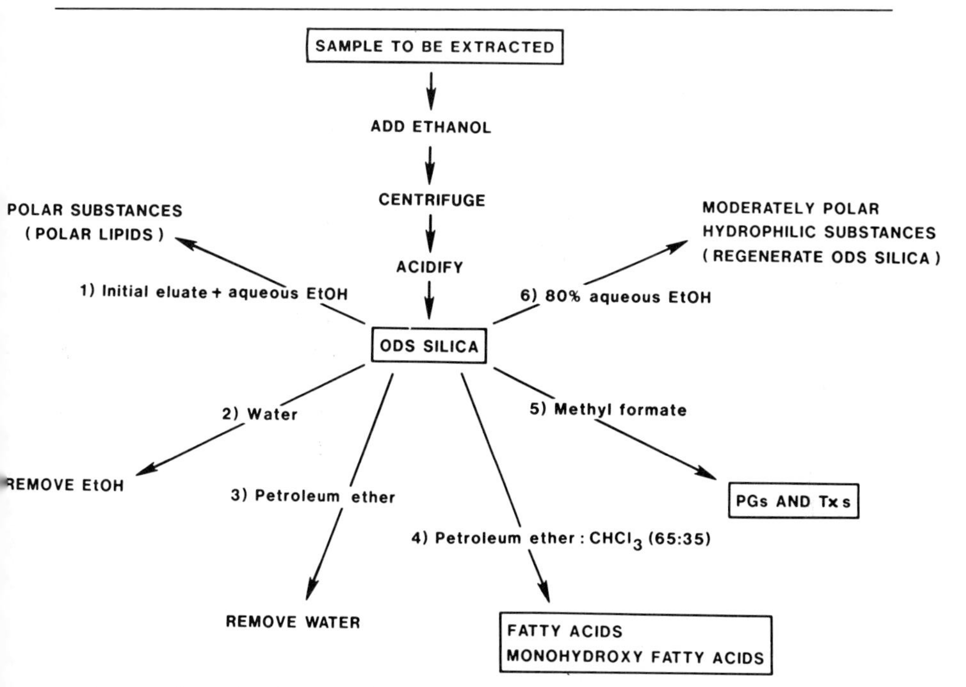

FIG. 1. Scheme for the extraction of arachidonic acid and some of its metabolites using octadecylsilyl (ODS) silica. The procedure may be modified as indicated in the text, depending on the nature of the substances to be extracted and the media. Body fluids can be passed through ODS silica after acidification without the addition of ethanol. The ODS silica is then washed with water, prior to elution with dilute aqueous ethanol (step 1).

Cartridges containing ODS silica (SEP-PAK C_{18} cartridges) were purchased from Waters Associates. If it is necessary (e.g., for extraction of large amounts of urine), ODS silica can be prepared by treating BioSil HA (Bio-Rad Laboratories, Richmond, California) with octadecyltrichlorosilane as described by Bennett *et al.*[5] The results obtained with this material appear to be the same as those with SEP-PAK C_{18} cartridges, although we have not tested it nearly as extensively. Methyl formate was purchased from Eastman Kodak, Rochester, New York.

Treatment of ODS Silica. Whether SEP-PAK C_{18} cartridges or ODS silica prepared as described above are used, it is necessary to wet the ODS silica with an organic solvent prior to use. This can be accomplished by passing ethanol (20 ml) through the ODS silica before it is used. The excess ethanol is then removed by passing water (20 ml) through the column.

Preparation of Samples

Homogenates or Subcellular Fractions. Incubations of various fractions with substrates are terminated by the addition of ethanol (2 volumes). For small volumes (up to ca. 5 ml incubation mixture) water is added to give a final concentration of 15% ethanol (the concentration of ethanol may vary, depending on the material to be extracted; see below). The mixtures are then centrifuged at 400 *g* for 10 min and the supernatants are acidified to a pH of ca. 3 and passed through an ODS silica cartridge using a syringe. The cartridge is then washed with aqueous ethanol, the concentration of ethanol being the same as that in the supernatant which was originally applied. For larger samples (incubation volumes up to 25 ml), the mixtures are centrifuged as described above, after the addition of ethanol. The supernatants are concentrated under vacuum to a volume identical to the original incubation volume and diluted with 2.3 volumes of water.[10] The mixtures are then acidified and passed through ODS silica as described above. For incubation volumes larger than 25 ml, it may be advisable to use more than one ODS silica cartridge or an open column of ODS silica.

Body Fluids. Body fluids such as plasma and urine can be applied directly to an ODS silica cartridge after acidification. The ODS silica is then washed with distilled water (20 ml) followed by aqueous ethanol. The percentage of ethanol is selected by criteria indicated in Fig. 2.

Elution of Polar Materials from ODS Silica

If the sample applied to the ODS silica is dissolved in 15% aqueous ethanol, many polar materials (e.g., phospholipids) will pass through the column without being retained. Alternatively, if solutes are applied in aqueous media, more polar materials will be removed by the subsequent elution with 15% aqueous ethanol. In order to elute the maximum amounts of polar contaminants, the highest concentration of ethanol that does not significantly reduce the recovery of the substances to be extracted should be used. The effects of ethanol concentration on the recoveries of 20:4, 15h-20:4, $PGF_{2\alpha}$, and 19-hydroxy-$PGF_{2\alpha}$ are shown in Fig. 2. Radioactive solutes were applied to ODS silica cartridges in acidified aqueous media (20 ml) containing various concentrations of ethanol. The cartridges were then eluted with the same concentration of aqueous ethanol (20 ml), followed by methyl formate (10 ml). The recovery values

[10] Assuming that the mixture contains a maximum of 50% ethanol after concentration under vacuum (probably an overestimate), the addition of 2.3 volumes of water will give a concentration of 15% ethanol.

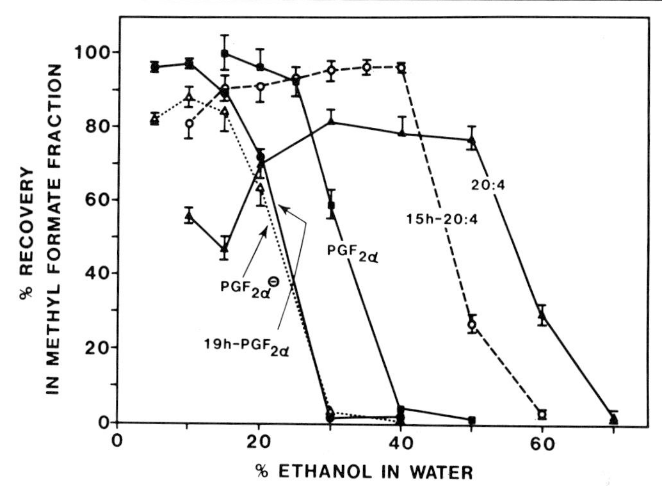

FIG. 2. Recoveries of 19-hydroxy[9β-^{3}H]$PGF_{2\alpha}$ (●), [9β-^{3}H]$PGF_{2\alpha}$ (■), 15h-[5,6,8,9,11,12,14,15-^{3}H]20:4 (○), and [5,6,8,9,11,12,14,15-^{3}H]20:4 (▲) in the methyl formate fraction as a function of the concentration of ethanol in the sample that was applied to the ODS silica. Standards dissolved in ethanol (10–20 μl) were added to various mixtures (20 ml) of ethanol in water. The solutions were acidified with 1 *N* HCl and passed through ODS silica cartridges, which were eluted with the same concentrations of ethanol in water (20 ml) and methyl formate (10 ml). The radioactivity in the methyl formate fractions was counted by liquid scintillation counting after evaporation of the solvent under a stream of nitrogen. A similar procedure was used for the extraction of $PGF_{2\alpha}$ from 0.05 *M* Tris-HCl, pH 7.5 ($PGF_{2\alpha}^{\ominus}$, △), except that the solution of $PGF_{2\alpha}$ was not acidified before being passed through the ODS silica cartridge. Determinations were carried out in quadruplicate, and values are mean ± SD.

refer to the percentages of solutes that were recovered in the methyl formate fractions.

It can be seen that the less polar the solute, the higher the concentration of ethanol that can be used without compromising recovery. With less polar solutes (20:4 and 15h-20:4) recovery in the methyl formate fraction is also reduced at low concentrations of ethanol. This effect is even more prominent for extraction of 20:4 from tissue homogenates.[3] In order to obtain optimal recovery of 20:4, the concentration of ethanol should be 30–40%. Prostaglandins such as 6-oxo-$PGF_{1\alpha}$, which have retention times lower than that of $PGF_{2\alpha}$ on reversed-phase chromatography, will be eluted from ODS silica with lower concentrations of ethanol than for $PGF_{2\alpha}$. In order to extract prostaglandins, TxB_2 and monohydroxy fatty acids from biological samples, we normally use 15% aqueous ethanol. A lower concentration of ethanol (10%) should be used for extraction of 19- or 20-hydroxyprostaglandins. Owing to a lack of standards we have not determined the optimal conditions for extraction of polar urinary metabolites of PGs. We did find, however, that 70–80% of the radioactive metabolites of $PGF_{2\alpha}$ in rabbit urine, including tetranor-$PGF_{1\alpha}$ and 13,14-dihydrotetranor-$PGF_{1\alpha}$, could be extracted from 5% aqueous ethanol by an open column of ODS silica.[3,11] We have only limited experience with the extraction of leukotrienes with ODS silica. Since leukotrienes have retention times intermediate between PGs and monohydroxy fatty acids on reversed-phase chromatography, one should be able to extract them from acidified aqueous media containing low concentrations of ethanol. We have successfully extracted leukotriene B_4 and related compounds from polymorphonuclear leukocytes using this procedure.

Effects of pH on the Extraction of PGs by ODS Silica

Most of the recovery studies described here were carried out by passing standards dissolved in acidified aqueous ethanol though ODS silica. This method can also be used for extracting PGs from media at neutral pH. Solutions of $PGF_{2\alpha}$ in 0.05 *M* Tris-HCl, pH 7.5, containing various amounts of ethanol were passed through cartridges of ODS silica ($PGF_{2\alpha}{}^{\ominus}$ in Fig. 2). Under these conditions $PGF_{2\alpha}$ should be present mainly as the carboxylate anion. This results in a shift to the left (toward lower ethanol concentrations) in the curve for recovery of $PGF_{2\alpha}$. Ionization of the carboxylic acid group of $PGF_{2\alpha}$ has about the same effect on recovery as the addition of a hydroxyl group (19-hydroxy-$PGF_{2\alpha}$, Fig. 2). Thus a lower concentration of ethanol (e.g., 10% for $PGF_{2\alpha}$) should be used for extraction of PGs from neutral or basic media. These results indicate that ODS silica could be used for extracting acid-labile PGs such as PGI_2, PGG_2, and PGH_2 directly from biological media without acidification.

Removal of Water from ODS Silica

After the sample is applied to the ODS silica, a certain amount of water will remain in the cartridge, even if air is forced through it. This can be annoying if the next solvent to be used is methyl formate, since it will dissolve most of the water, thus increasing the time required for evapora-

[11] Rabbits were injected with [9β-^{3}H]$PGF_{2\alpha}$, and urine was collected on ice for 36 hr. The urine was centrifuged, acidified, and passed through an open column of ODS silica (5 g). The column was eluted with water (100 ml), 5% ethanol in water (200 ml), benzene (100 ml), and methyl formate (50 ml). The methyl formate fraction was concentrated to dryness under vacuum, and the products were purified by reversed-phase and normal-phase HPLC.

tion of solvents. For extraction of PGs, most of the remaining water can be removed by passing petroleum ether (20 ml) through the ODS silica (see the table below). This is not successful for the extraction of less polar compounds, such as 15h-20:4 and 20:4, however, since petroleum ether dissolves some of the ethanol on the ODS silica, making the mobile phase more polar. Under these conditions the petroleum ether will elute ca. 50% of the 15h-20:4 and nearly all of the 20:4 extracted by the ODS silica (see the table). This problem can be circumvented if the ethanol in the ODS-silica is removed with 20 ml of water prior to elution with petroleum ether.

Elution of Monohydroxyeicosenoic Acids

Monohydroxyeicosenoic acids and 20:4 can be eluted from ODS silica with mixtures of petroleum ether (boiling range 30–60°) and chloroform. 20:4 and 15h-20:4 were dissolved in 30% and 15% aqueous ethanol, respectively, and passed through cartridges of ODS silica. The cartridges were eluted with the same concentration of ethanol, followed by water (20 ml), petroleum ether (20 ml for PGE_2 and 15h-20:4 and 10 ml for 20:4), and various concentrations of chloroform in petroleum ether (20 ml). The recoveries of 20:4 and 15h-20:4 in the chloroform petroleum ether fractions are shown in Fig. 3. A similar procedure was used for PGE_2, but in this case the cartridges were subsequently eluted with methyl formate. The recoveries of PGE_2 in the methyl formate fractions are also shown in Fig. 3. Elution of ODS silica with petroleum ether–chloroform (65 : 35) removes 82% of the applied 15h-20:4, but only 4% of the applied PGE_2. Subsequent elution with methyl formate results in 96% recovery of PGE_2 but only 9% recovery of 15h-20:4 in the methyl formate fraction.

Elution of Prostaglandins and Thromboxanes

Prostaglandins and thromboxanes can be eluted from ODS silica with more polar organic solvents such as diethyl ether, ethyl acetate, and methyl formate. We normally use methyl formate for this purpose. Because it is fairly polar, only 10 ml are required to elute PGs from the stationary phase. Larger volumes (e.g., 20 ml of diethyl ether) of less polar solvents are required for good recoveries. Methyl formate (boiling point 32°) is a very volatile solvent, enabling it to be removed rapidly under a stream of nitrogen.

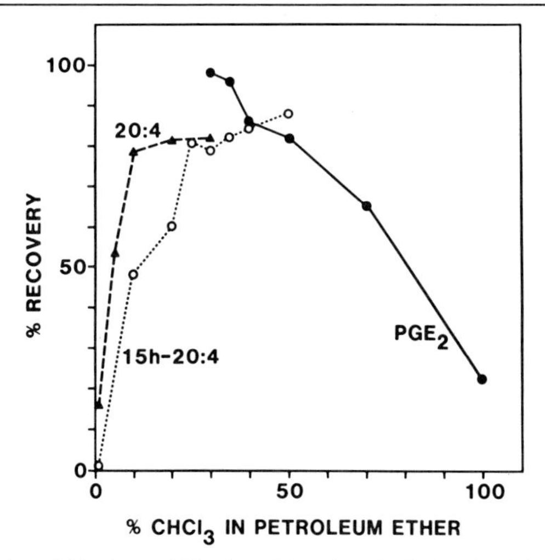

FIG. 3. Recoveries of 20:4 (arachidonic acid) (▲) and 15h-20:4 (15-hydroxy-5,8,11,13-eicosatetraenoic acid) (○) in the petroleum ether–$CHCl_3$ fraction and of PGE_2 (●) in the methyl formate fraction as a function of the percentage of $CHCl_3$ in petroleum ether. PGE_2 and 15h-20:4 were dissolved separately in 15% aqueous ethanol. After acidification, the solutions were applied to ODS silica cartridges, which were eluted with 15% aqueous ethanol (20 ml), water (20 ml), petroleum ether (boiling range 30–60°; 20 ml), various concentrations of $CHCl_3$ in petroleum ether (20 ml), and methyl formate (10 ml). The amounts of radioactivity in the petroleum ether–$CHCl_3$ (15h-20:4) and methyl formate (PGE_2) fractions were determined. The procedure for 20:4 was identical to that used for 15h-20:4 except that 30% ethanol was used instead of 15%, and 10 ml instead of 20 ml of petroleum ether was used. This amount of petroleum ether removed only 1–2% of the 20:4 on the cartridge. Subsequent elution of the ODS silica with a further 20 ml of petroleum ether (0% $CHCl_3$) removed 16% of the 20:4. Values are the means of triplicate determinations. We encountered some problems due to quenching when we counted the radioactivity in the petroleum ether–chloroform fractions after evaporation of the solvent. In order to determine recoveries, standards were therefore added to 20 ml of petroleum ether–chloroform, which was evaporated before counting.

Regeneration of ODS Silica

Cartridges of ODS silica can be reused many times, provided they are not used to extract very large samples or samples containing considerable amounts of particulate matter. The ODS silica should be washed with 80% aqueous ethanol (20 ml), followed by ethanol (20 ml) and water (20 ml) prior to reuse.

Recovery of Standards from Biological Samples

Tissue Homogenates. Bovine lung was homogenized in 5 volumes of 0.05 *M* Tris-HCl, pH 7.4, with a VirTis homogenizer. Aliquots (1 ml) of this homogenate were added to various standards dissolved in ethanol

(3 ml), and the mixtures were allowed to stand for 5 min. Water (16 ml) was added and the mixtures were centrifuged at 400 *g* for 10 min. The supernatants were acidified and passed through an ODS silica cartridge that was eluted with 15% aqueous ethanol (20 ml), petroleum ether (boiling range, 30–60°, 20 ml) and methyl formate (10 ml) as described above. Recoveries of radioactive standards in the methyl formate fractions were determined by liquid scintillation counting after evaporation of the solvent.

The recoveries of PGD_2, PGE_2, $PGF_{2\alpha}$, 6-oxo-$PGF_{1\alpha}$, TxB_2, and 13,14-dihydro-15-oxo-$PGF_{2\alpha}$ in methyl formate fractions were between 88% and 97%, whereas the recovery of 13,14-dihydro-15-oxo-PGE_2 (84%) was slightly lower (see the table). Only about 41% of 15h-20:4 appeared in the methyl formate fraction, since a considerable amount of this substance was eluted by the petroleum ether, which contained some ethanol, as discussed above. If it is desired to obtain a fraction containing both monohydroxyeicosenoic acids and PGs, the ODS silica can be washed with water prior to petroleum ether, as described above. Alternatively, the former compounds can be largely removed by elution with petroleum ether–chloroform (65:35). This solvent should not be used with 13,14-dihydro-15-oxo-PGE_2, however, since it will reduce recovery of this compound in the methyl formate fraction. Recoveries of 20:4 and dipalmitoylphosphatidylcholine in the methyl formate fraction are very low. The former compound is eluted by the petroleum ether, whereas the latter is not retained by the ODS silica.

Plasma and Urine. Radioactive standards dissolved in ethanol (50 μl) were added to human platelet-free plasma (5 ml) or urine (20 ml). The

RECOVERIES OF STANDARDS EXTRACTED FROM BOVINE LUNG HOMOGENATES, HUMAN PLASMA, AND HUMAN URINE WITH ODS SILICA [a]

Compound	Recovery (%)[b]		
	Homogenate	Plasma	Urine
PGD_2	88 ± 7	—	—
PGE_2	91 ± 3	96 ± 1	97 ± 2
$PGF_{2\alpha}$	97 ± 7	96 ± 2	102 ± 2
6-Oxo-$PGF_{1\alpha}$	89 ± 7	—	—
TxB_2	89 ± 6	—	—
13,14-Dihydro-15-oxo-PGE_2	84 ± 2	89 ± 6	—
13,14-Dihydro-15-oxo-$PGF_{2\alpha}$	93 ± 4	97 ± 5	—
15h-20:4	41 ± 5	—	—
Arachidonic acid	5 ± 1	—	—
Dipalmitoylphosphatidylcholine	3 ± 1	—	—

[a] Taken from Powell.[3]

[b] Samples were extracted using SEP-PAK C_{18} cartridges as described in the text. Values are the recoveries ±SD (n = 4) in the methyl formate fractions.

mixtures were acidified and passed through an ODS silica cartridge, which was eluted successively with water (20 ml), 15% aqueous ethanol (20 ml), petroleum ether (20 ml), and methyl formate (10 ml) as described above. The radioactivity in the methyl formate fractions was determined after evaporation of the solvent. Recoveries of PGE_2 and $PGF_{2\alpha}$ from urine were ca. 100%, whereas the recoveries of these compounds and their 13,14-dihydro-15-oxo metabolites from plasma were between 89% and 97% (see the table).

Summary

The following fractions can be obtained after extraction of biological samples with ODS silica.

1. Arachidonic acid or other nonpolar fatty acids. Apply sample to ODS silica in acidified 30% ethanol. Wash with 30% ethanol (20 ml), water (20 ml), and petroleum ether (10 ml). Elute 20:4 with petroleum ether–chloroform (1:1, 20 ml).

2. Monohydroxyeicosenoic acids. Apply sample in acidified 15–25% ethanol, depending on whether it is desired to obtain a subsequent fraction containing PGs. Wash ODS silica with the same concentration of ethanol (20 ml), water (20 ml), and petroleum ether (20 ml). Elute monohydroxyeicosenoic acids with petroleum ether–chloroform (65:35, 20 ml), chloroform (10 ml) or methyl formate (10 ml), depending on whether or not it is necessary to obtain a subsequent PG fraction.

3. Prostaglandins and thromboxanes. Apply sample to ODS silica in acidified 15% aqueous ethanol. Wash with 15% ethanol (20 ml) followed by petroleum ether (20 ml) or, if it is desired to extract monohydroxyeicosenoic acids, water (20 ml), and petroleum ether–chloroform (65:35, 20 ml). If nonpolar PGs such as 13,14-dihydro-15-oxo-PGE_2 are to be extracted, petroleum ether should be used. Elute PGs and Txs with methyl formate (10 ml).

4. Monohydroxyeicosenoic acids, PGs, and Txs. Apply sample in acidified 15% aqueous ethanol. Wash ODS silica with 15% aqueous ethanol (20 ml), water (20 ml) and petroleum ether (20 ml). Elute with methyl formate (10 ml).

5. Polar PG metabolites. Apply sample in acidified 0–10% aqueous ethanol (for ω-hydroxy-PGs 10% ethanol can be used). Wash ODS silica with the same concentration of ethanol (20 ml) followed by petroleum ether–chloroform (65:35). Elute PG metabolites with methyl formate (10 ml).

6. PGs extracted from neutral or basic media. Apply sample to the ODS silica in 10–15% aqueous ethanol at the desired pH. Wash with the

same concentration of ethanol (10 ml) (for greater speed, it should be possible to eliminate this step), followed by petroleum ether (20 ml). Elute with methyl formate (10 ml). For nonpolar products, better results might be obtained by washing the ODS silica with water prior to petroleum ether.

Acknowledgments

These studies were supported by grants from the Medical Research Council of Canada, the Quebec Heart Foundation, the Canadian Lung Association, and the Quebec Thoracic Society. The excellent technical assistance of Ms. C. Luttinger and Mr. G. Skinner is gratefully acknowledged.

[59] Extraction and Thin-Layer Chromatography of Arachidonic Acid Metabolites

By JOHN A. SALMON and RODERICK J. FLOWER

To review the many methods by which prostaglandins have been extracted and chromatographed would not be consistent with the aims of this series of monographs. Instead, we propose to concentrate on those techniques that we ourselves have found to be the most useful, indicating along the way the problems inherent in these procedures and suggesting ways in which these may be surmounted. The techniques used to extract and purify thromboxanes and leukotrienes are in most cases very similar to those used for prostaglandins, and so we will consider all these compounds together. Only one-dimensional thin-layer chromatography is described here: two-dimensional systems are considered by Granström (this volume [60]). For an exhaustive review of analytical methodology the reader is referred to Salmon and Flower.[1]

Extraction

Basic Methodological Philosophy

In some cases prostaglandins may be determined directly in aqueous samples (e.g., radioimmunoassay or bioassay of tissue perfusates and cell culture fluids) without recourse to extraction, but more usually, extraction into an organic solvent is a prerequisite to separation and/or

[1] J. A. Salmon and R. J. Flower, *in* "Hormones in Blood" (C. H. Gray and V. H. T. James, eds.), 3rd ed., Vol. 2, p. 237. Academic Press, New York, 1979.

ISBN 0-12-181986-8

quantitation. The advantages of extraction are that it eliminates protein, imparts some specificity to the assay and, by concentrating material, improves the sensitivity of the analysis. Disadvantages are that it is time-consuming and can itself be a major source of error: many solvents have a "blank" effect in subsequent assays, and the overall efficiency of the procedure is variable. It is essential to the analysis to include proper control samples enabling estimation of the "background," and to use a procedure that allows an estimation to be made of the recovery of the prostaglandins.

The choice of any particular extraction technique is governed by its efficiency, specificity, reproducibility, and practicability. As weakly acidic lipids, prostaglandins are readily extracted into organic solvents in their protonated form (thus separating them from many other compounds including some lipids, e.g., glycerides) and this forms the basis of most extraction procedures. Acidification must be performed cautiously, since decomposition of some prostaglandins occurs at low pH. For example, the β-hydroxyketone group present in PGE is susceptible to changes in pH; at acid pH, PGE is readily dehydrated to PGA, and at alkaline pH it is converted to PGB. In fact, overacidification of samples may well account for early reports of PGA in plasma samples whereas more recent, carefully controlled trials[2,3] have failed to detect this prostaglandin. Some investigators have used hydrochloric acid and other mineral acids, but we recommend that an organic acid with a relatively high pK_a be used, since this allows a more controlled adjustment of pH. Citric (0.5–2.0 *M*) and formic acids (1–3%, v/v) are most frequently employed; a final sample pH of 3.5–4.5 is suitable.

Chloroform, diethyl ether, and ethyl acetate are the solvents[4] most commonly used for extracting prostaglandins. After acidification the sample may be extracted directly into these solvents, by vigorous mixing of the phases. However, many samples (especially those containing phospholipids) when treated in this manner will produce emulsions making extraction difficult and inefficient. The formation of emulsions is a problem in most lipid work, and the cardinal rule is that they are easier to prevent than cure. The addition of a saturating amount of sodium chloride to the aqueous phase prior to extraction followed by centrifugation of the sample usually minimizes the problem. If an intractable emulsion should form

[2] J. C. Frölich, B. J. Sweetman, K. Carr, J. W. Hollifield, and J. A. Oates, *Prostaglandins* **10,** 185 (1975).

[3] K. Gréen and S. Steffenrud, *Anal. Biochem.* **76,** 606 (1976).

[4] Most authorities on lipid chemistry would insist on the inclusion of antioxidants, such as BHT, in any solvents used for extraction or chromatography, but we do not find this precaution essential when processing prostaglandins.

despite these precautions, it may sometimes be dispersed by the addition of a little ethanol at the emulsion–water interface. However, for anything other than the "cleanest" samples, we recommend the three-stage extraction procedure described below (based on the work of Frölich[5]), which entirely overcomes the emulsion problem.

Basic Extraction Technique

Reagents

ESSENTIAL REAGENTS

Acetone (Analar) cooled to 4°
Citric (0.5–2 *M*) or formic acid
n-Hexane or petroleum ether (40–60) (Analar)
Chloroform (Analar)

REAGENTS HIGHLY RECOMMENDED BUT NOT ESSENTIAL

Radioactive prostaglandins to act as internal markers. Tritiated compounds are the best because they are available at high specific activity (New England Nuclear, Boston, Mass. or Amersham International, Bucks, England). If analysis using gas chromatography–mass spectrometry (GC–MS) is employed, the prostaglandins should be labeled with stable isotopes (e.g., deuterium; see this volume [65–69]).

Method

A flow chart of the method is illustrated in Fig. 1. After the addition of a suitable amount of radioactive prostaglandin, the aqueous sample is shaken with cold acetone (2 volumes); this precipitates protein and some other macromolecules and, after centrifugation, enables the clear supernatant to be decanted into a clean extraction vessel. The precipitate is washed with a further volume of acetone, and the aqueous-acetone layers are combined. Neutral lipids in the aqueous-acetone extract are now extracted by shaking with petroleum ether or hexane (2 volumes), the upper phase being discarded. The remaining aqueous-acetone phase is acidified with citric or formic acid to pH 4.0–4.5 and then mixed with chloroform (2 volumes) and shaken. The separation of the aqueous and organic layers is vastly improved if the sample can be centrifuged, and then the organic phase (lower layer) is removed. The chloroform extraction should be repeated at least once more, and the combined organic layers can be evaporated under nitrogen or under vacuum. Using this procedure, recoveries

[5] J. C. Frölich, *in* "The Prostaglandins" (P. W. Ramwell, ed.), Vol. 3, p. 1. Plenum, New York, 1976.

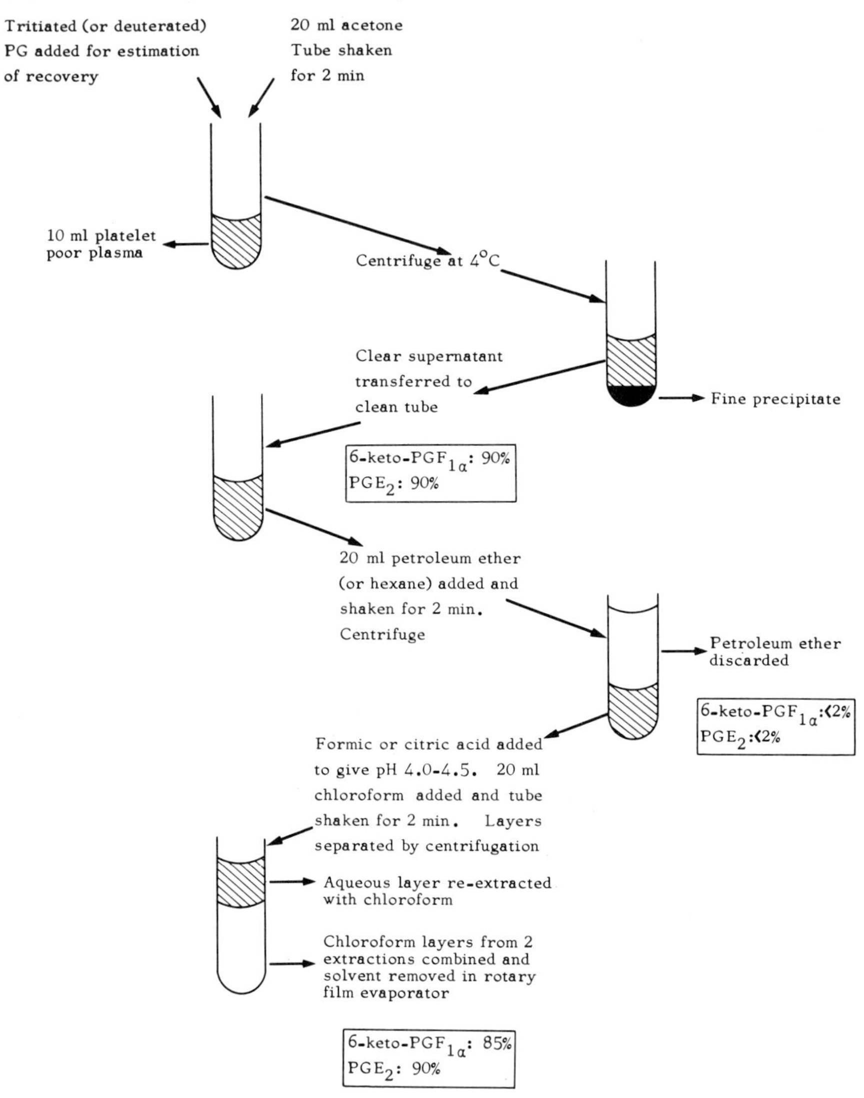

FIG. 1. Scheme for extracting prostaglandins from plasma. The recovery of [^{3}H]6-keto-$PGF_{1\alpha}$ and [^{3}H]PGE_2 at each stage is recorded in the boxes.

of PGE_2 are greater than 90% and the resulting extract is sufficiently "clean" to permit, for example, direct injection into a high-pressure liquid chromatograph. When processing large numbers of samples of small volumes (up to 5 ml), we have found a Multi-Tube Vortexer (SMI, Emeryville, California) to be particularly useful.

As mentioned above, this procedure is very efficient for extracting PGE_2 and most other prostaglandins and thromboxane B_2, but the recovery of 6-keto-$PGF_{1\alpha}$ is somewhat lower (approximately 80–85% in this laboratory). For this reason some investigators have preferred to replace chloroform with a more polar solvent, such as ethyl acetate. However, we have observed an apparent "decomposition" of 6-keto-$PGF_{1\alpha}$ when the compound was extracted directly into ethyl acetate from an aqueous sample adjusted to pH 3–4 with citric acid. The "decomposition" products are separated by thin-layer chromatography (TLC) (Fig. 2), but when subjected to GC–MS analysis they produce mass spectra identical to that for 6-keto-$PGF_{1\alpha}$ itself. This "decomposition" can be avoided if (*a*) the ethyl acetate extract is washed with water prior to evaporation; or (*b*) hydrochloric acid is used instead of citric acid to adjust the pH; or (*c*) chloroform is employed instead of ethyl acetate. These data suggest that citric acid can be extracted into the ethyl acetate (but not chloroform) and during concentration induces a modification of the 6-keto-$PGF_{1\alpha}$ structure (possibly tautomers). Investigators wishing to estimate 6-keto-$PGF_{1\alpha}$ should be aware of this possible source of artifacts.

Modifications of the Basic Method

The above procedure may be used for efficient extraction of prostaglandins from most types of sample, but some analysts prefer alternative schemes. For example, many investigators begin the extraction of prostaglandins from tissues by adding 4–5 volumes of ethanol, instead of acetone, as an initial protein denaturation step. The protein is then removed by filtration or centrifugation. Before processing the sample further, it is usually an advantage to remove most of the ethanol under reduced pressure using a rotary film evaporator. Some investigators have also included an additional step to improve the specificity of the extraction: the aqueous concentrate remaining after removal of the ethanol is adjusted to pH 8.0–8.5 (a more alkaline pH will induce dehydration of PGE to PGB) and the nonacidic lipids are extracted into organic solvents whereas the prostaglandins remain in the aqueous phase since they are fully ionized at pH 8. The aqueous phase is then adjusted to pH 3.5–4.5 and the sample is processed as described above. Alternatively, the aqueous concentrate could first be acidified and the prostaglandins extracted into organic solvents, then the organic extract is washed with phosphate or tris buffers (pH 8.0–8.5) and finally the aqueous layer is reacidified and the prostaglandins reextracted into organic solvents. The latter procedure is laborious and probably does not produce a superior extraction.

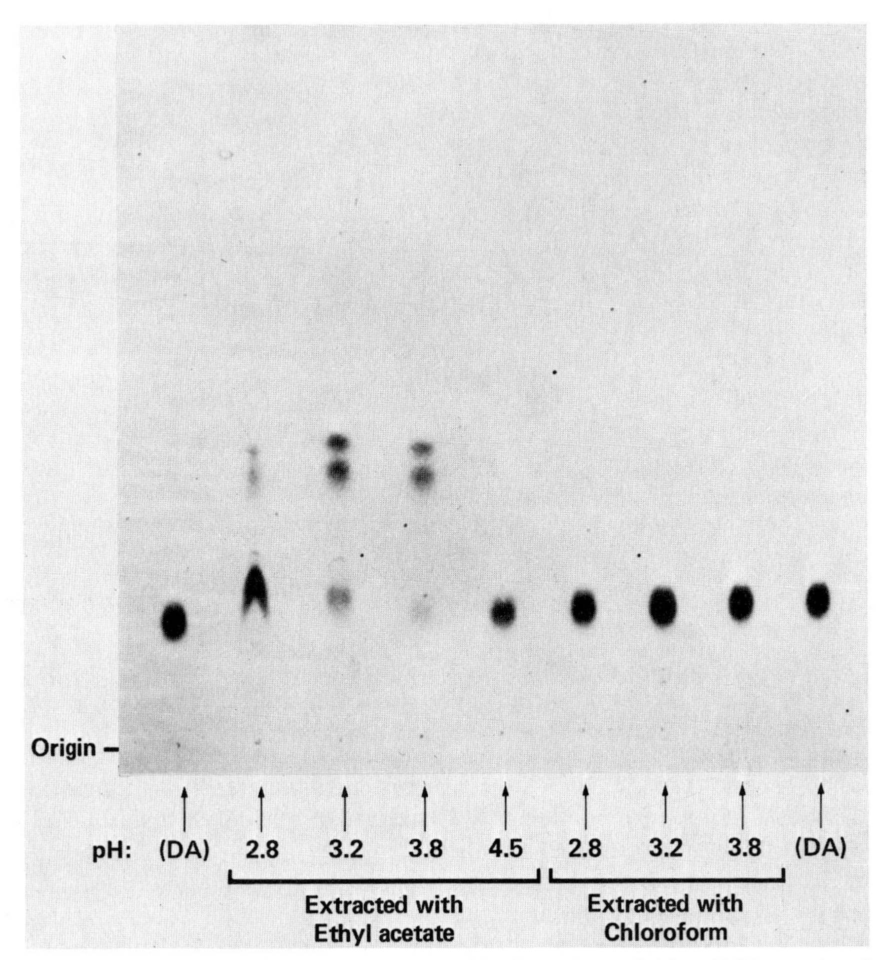

FIG. 2. Separation of apparent "decomposition" products of 6-keto-PGF$_{1\alpha}$ produced during extraction. 6-Keto-PGF$_{1\alpha}$ (20 μg) was either applied directly (DA) to the thin-layer chromatography plate (Uniplate; Analtech, Inc.) or extracted from an aqueous solution using either ethyl acetate or chloroform after adjusting the pH with citric acid as indicated. The developing solvent system was A (see Table I), and the products were visualized by spraying with phosphomolybdic acid.

Alternative Techniques for Extraction

Sometimes the sheer volume of the sample makes direct extraction into organic solvents by the "basic technique" impossible or at best very unwieldy. Fortunately, there are other techniques that can be applied in these cases.

One alternative method of extracting prostaglandins from aqueous solutions is the use of nonionic resins such as Amberlite XAD-2 (BDH, Poole, Dorset, England). A column of resin (e.g., 1 g per 100 ml of aqueous solution) is washed with water and methanol and then equilibrated in distilled water. The test solution is percolated slowly through the resin bed, followed by a wash with distilled water (2–3 bed volumes) to remove salts. Finally the prostaglandins are recovered by elution with methanol or ethanol (2–5 bed volumes). After evaporation of the alcohol, the residue can be subjected directly to further chromatographic purification, and after a thorough wash with the eluting solvent and reequilibration with buffer or water, the gel is ready to use again. Organic molecules bind most effectively to Amberlite when nonionized (i.e., most lipophilic), and therefore the binding of prostaglandins may be increased by lowering the pH of the aqueous solution, although in the authors' experience this is not usually necessary. An alternative method of recovery of the prostaglandins from the resin is by elution with 1 *N* NaOH; this elutes the compounds as the sodium salts. The latter method would not be suitable for PGE since it would be significantly dehydrated to form PGB.

Amberlite can be very useful; for example, a column can be placed under the outflow of a perfused organ, thereby trapping the prostaglandins immediately they are released. Another major use is in the removal of prostaglandins (or their metabolites) from fluids that contain low concentrations but can be obtained in large volumes, (e.g., urine[6,7] or amniotic fluid[8]). This technique reduces the practical problem of handling the large volumes of toxic and inflammable solvents that are required for other extraction procedures (see above). However, one serious disadvantage has been reported: Granström and Kindahl[9] demonstrated that a compound(s) was eluted from Amberlite that severly affected binding of prostaglandins to specific antibodies. Thus, if radioimmunoassay (RIA) is to be employed, extraction with Amberlite should probably be avoided. Although Amberlite XAD-2 has most frequently been employed, in this respect some other ion exchange resins may also be useful.[10]

A further method for separating prostaglandins from samples is by dialysis; this method does not concentrate nor specifically remove the prostaglandins, but if the final assay is particularly sensitive and specific and

[6] K. Gréen, *Biochim. Biophys. Acta* **231,** 419 (1971).

[7] B. Samuelsson, E. Granström, K. Gréen, and M. Hamberg, *Ann. N.Y. Acad. Sci.* **180,** 138 (1971).

[8] M. J. N. C. Keirse and A. C. Turnbull, *J. Obstet. Gynaecol. Br. Commonw.* **80,** 970 (1973).

[9] E. Granström and H. Kindahl, *Adv. Prostaglandin Thromboxane Res.* **5,** 199 (1978).

[10] D. J. Fretland, *Prostaglandins* **6,** 421 (1974).

can be performed directly on aqueous samples (e.g., RIA) it may be a useful procedure.[11]

Thin-Layer Chromatography

The extraction procedures described above efficiently remove all fatty acids, prostaglandins, and other hydroxy acids, but obviously if specific analysis of a particular compound is required some further purification is necessary. Many chromatographic procedures (e.g., silicic acid columns and high-pressure liquid; see this volume [61]) have been employed for analysis of prostaglandins, but TLC has much to commend it in terms of efficiency, simplicity, and economy of money and effort.

The major groups of prostaglandins (A, B, D, E, F, 6-keto-$PGF_{1\alpha}$ as well as thromboxane B_2) can be readily separated as free acids on silica gel G using various solvent systems.[1,12–16]

Choice of TLC Plate

In the past most investigators have had to prepare their own thin-layer plates; although with practice this can be performed satisfactorily, the silica layer may be of varying thickness and quality. Several commercial sources of precoated thin-layer plates are now available (e.g., E. Merck AG, Darmstadt, Germany; Anachem Ltd., Luton, Bedfordshire, England; Eastman Kodak Co., Rochester, New York; Whatman, Inc., Clifton, New Jersey), which are more expensive than "home-made" plates but do give very reproducible separations. The commercial plates are available coated onto glass, aluminum foil, or plastic; the latter two backing materials enable interesting zones to be cut out with scissors and then processed further. Our personal preference and strong recommendation are the LKD series of TLC plates (Whatman Inc.) since a 20 cm × 20 cm plate allows application of up to 19 samples on separate lanes (5-lane plates are also available) and the resolution of compounds is reproducible and excellent (see Fig. 3). Also, LKD plates have a preadsorbent zone that enables rapid application of samples; in most cases samples up to 50 μl can be spotted in one aliquot. Plates with a fluorescent indicator (F254) impreg-

[11] L. Levine and R. M. Gutierrez-Cernosek, *Prostaglandins* **3,** 785 (1973).

[12] K. Gréen and B. Samuelsson, *J. Lipid Res.* **5,** 117 (1964).

[13] M. Hamberg and B. Samuelsson, *J. Biol. Chem.* **241,** 257 (1966).

[14] N. H. Anderson, *J. Lipid Res.* **10,** 316 (1969).

[15] P. Ramwell and E. G. Daniels, *in* "Lipid Chromatographic Analysis" (G. V. Marinetti, ed.), Vol. 2, p. 313. Dekker, New York, 1969.

[16] K. Gréen, M. Hamberg, B. Samuelsson, and J. C. Frölich, *Adv. Prostaglandin Thromboxane Res.* **5,** 13 (1978).

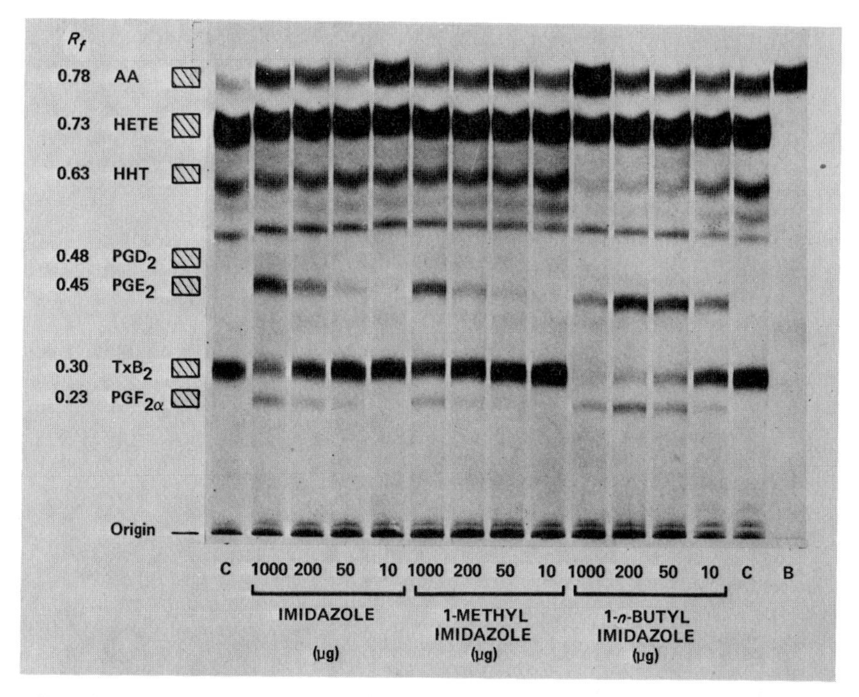

FIG. 3. An autoradiogram of thin-layer chromatography separation of the products obtained after incubation of [^{14}C]arachidonic acid with horse platelets in the presence of various concentrations of imidazole, 1-methylimidazole, and 1-*n*-butylimidazole (thromboxane synthase inhibitors). Lane C shows control incubation in the absence of inhibitors; B, incubation with boiled platelets. The developing solvent system was A (Table I); the hatched zones on the left of the autoradiogram together with the respective R_f values indicate the mobility of authentic standards.

nated in the silica are also available, and these may be of particular value for visualizing UV-absorbing compounds (e.g., hydroxyeicosatetraenoic acids, HETEs, formed from arachidonic acid by lipoxygenases). The LKD plates are individually sealed, and it is not essential to "activate" them; however, most other plates should be heated at 110° for 30 min and then allowed to cool in a desiccated box before use. Some investigators wash the plates in mixtures of solvents to remove contaminating impurities before use, but for most purposes this is not essential.

Spotting and Developing Procedure

Prostaglandins are most conveniently applied to the plates in chloroform–methanol (2:1, v/v) using microsyringes (e.g., Hamilton from Precision Sampling, Baton Rouge, Louisiana; or SGE, from Scientific Glass

Engineering Ltd., North Melbourne, Australia) or glass capillary tubes; in order to reduce spreading of the components the spot size should be kept to a minimum. However, if the LKD plates are used, the sample is simply streaked onto a preadsorbent zone allowing a much more rapid application. During development of the latter plates the components are concentrated at the interface between the preadsorbent zone and the silica gel before true chromatography commences (i.e., interface is equivalent to the origin). The plates are developed in tanks (e.g., from Shandon, Scientific Ltd., Willesden, London) containing approximately 100 ml of an appropriate solvent mixture for a 20 cm × 20 cm plate. It should be noted that Ramwell and Daniels[15] demonstrated that separation of the prostaglandins is impaired if the tank is equilibrated with solvent before development.

Choice of Solvent System

For most applications one of two solvent systems (A and B, Table I) is usually sufficient. Using them, most of the groups of prostaglandins and thromboxanes can be resolved; only the biologically less interesting PGA and PGB are not separated. The separation of arachidonate metabolites formed in blood platelets in the presence and in the absence of thromboxane synthase inhibitors using solvent mixture B is illustrated in Fig. 3. Some investigators use double developing systems to increase resolution of components, although we have not found this necessary when using LKD plates.

One point worth emphasizing is the need to use more than one developing system in order to confirm the presence of a particular prostaglandin. For example, several systems, including system B (Table I) do not permit separation of PGE from 6-keto-$PGF_{1\alpha}$, whereas some others do not separate 6-keto-$PGF_{1\alpha}$ from PGF. If metabolites of the prostaglandins are also present, these may have the same mobility as one of the parent prostaglandins (e.g., 13,14-dihydro-6,15-diketo-$PGF_{1\alpha}$ has a similar R_f to TxB_2 using system A[17]).

Although complete separation of many different prostaglandins and related compounds is achieved by TLC the possibility of contamination of zones with nonprostaglandin material in samples must be considered. For example, cholesterol cochromatographs with 12-HETE in several systems. Clearly, a combination of solvent systems is essential for a rigorous identification to be made.

The prostaglandin endoperoxides PGG_2 and PGH_2 may be separated

[17] P. M. Salzman, J. A. Salmon, and S. Moncada, *J. Pharmacol. Exp. Ther.* **215,** 240 (1980).

TABLE I

SEPARATION OF SOME ARACHIDONIC ACID METABOLITES AS FREE ACIDS BY THIN-LAYER CHROMATOGRAPHY

Solvent	Approximate R_f[b]											
system[a]	Arachidonic acid	12-HETE	HHT	PGE_1	PGE_2	$PGF_{1\alpha}$	$PGF_{2\alpha}$	PGD_2	PGA_2	PGB_2	6-Keto-$PGF_{1\alpha}$	TxB_2
A	0.85	0.79	0.74	0.36	0.36	0.25	0.25	0.48	0.61	0.61	0.18	0.32
B	0.78	0.73	0.63	0.45	0.45	0.23	0.23	0.48	0.53	0.53	0.45	0.30

[a] A, Organic (upper) phase only of ethyl acetate–2,2,4-trimethylpentane–acetic acid–water (110:50:20:100, v/v/v/v). B, chloroform–methanol–acetic acid–water (90:8:1:0.8, v/v/v/v).

[b] These data were obtained on Whatman LKD TLC plates. Mobilities on other plates will be in the same order, but the absolute R_f's may vary.

TABLE II
SEPARATION OF PROSTAGLANDINS AND THEIR PLASMA METABOLITES AS FREE ACIDS BY THIN-LAYER CHROMATOGRAPHY[a]

Compound	Approximate R_f
PGE_2	0.36
15-Keto-PGE_2	0.56
13,14-Dihydro-15-keto-PGE_2	0.59
$PGF_{2\alpha}$	0.25
15-Keto-$PGF_{2\alpha}$	0.40
13,14-Dihydro-15-keto-$PGF_{2\alpha}$	0.49
6-Keto-$PGF_{1\alpha}$	0.18
6,15-Diketo-$PGF_{1\alpha}$	0.30[b]
13,14-Dihydro-6-keto-$PGF_{1\alpha}$	0.23
13,14-Dihydro-6,15-diketo-$PGF_{1\alpha}$	0.32[b]
6-Keto-PGE_1	0.26
TxB_2	0.32

[a] Compounds were applied on LKD TLC plates (Whatman) and developed in solvent system A (see Table I).
[b] These compounds "streak" in this system.

from each other (and from their precursor, arachidonic acid) as free acids by TLC using a solvent system of ethyl acetate–trimethylpentane–acetic acid (50:50:0.5, v/v/v) (PGG_2, $R_f = 0.64$; PGH_2, $R_f = 0.46$).[18] Although some investigators[19] have purified endoperoxides using TLC, we do not recommend this technique (we prefer silicic acid column chromatography) because the endoperoxides rapidly isomerize on silica to PGD_2 and PGE_2 (approximate ratio 2:1). This decomposition may be limited by performing the TLC at low temperatures and by developing the plate *immediately* after the sample has been applied.

Since so many metabolites of the prostaglandins are formed, many with comparable polarities to the parent compounds, it is impossible to devise a system that separates arachidonic acid, hydroxy acids, prostaglandins, thromboxanes, and their metabolites. The solvent systems listed in Table II are useful for separation of the plasma metabolites (13,14-dihydro-; 15-keto-; and 13,14-dihydro-15-keto-) from the parent prostaglandin, but the metabolites may have a similar mobility to other

[18] M. Hamberg, J. Svensson, T. Wakabayashi, and B. Samuelsson, *Proc. Natl. Acad. Sci. U.S.A.* **71,** 345 (1974).
[19] D. H. Nugteren and E. Hazelhof, *Biochim. Biophys. Acta* **326,** 448 (1973).

prostaglandins (e.g., 13,14-dihydro-6,15-diketo-$PGF_{1\alpha}$ cochromatographs with TxB_2 using solvent system A).

More polar solvent systems than those described above are required to separate the urinary metabolites of the prostaglandins by TLC (e.g., water-saturated ethyl acetate containing 1% acetic acid). Because of the hydrophilic nature of these metabolites, it is usually advantageous to chromatograph the compounds as derivatives (e.g., methyl ester, methoxime; see Gréen *et al.*[20]) or to use reversed-phase column chromatography or paper chromatography.

The hydroperoxy- and hydroxyeicosatetraenoic acids, including the dihydroxy acid leukotriene B_4 (LTB_4) are best separated by TLC using a combination of diethyl ether, hexane, and acetic acid (60:40:1, v/v/v; see Table III).

The systems considered thus far have been designed to separate groups of prostaglandins. However, if resolution of individual prostaglandins with different numbers of double bonds is required, this can be achieved on plates of silica gel impregnated with 5% silver nitrate using similar solvent systems (see Table IV). The use of commercial precoated silver nitrate–silica gel TLC plates is not recommended as they deteriorate (darken) relatively quickly. Thin-layer plates of silica gel are easily impregnated with silver nitrate as follows: Dissolve 5 g of silver nitrate in 5 ml of distilled water; add to 95 ml of methanol, mix rapidly, and pour into a flat dish. The plate should be submerged in the solution for 1 min, then removed, air-dried and used immediately.

Plates impregnated with ferric chloride can also be used to separate prostaglandins on the basis of their unsaturation.[21]

Detection of Compounds on TLC Plates

After development of the plates, the prostaglandins may be detected by several techniques. We suggest the use of iodine vapor, as this is nondestructive (prostaglandins produce yellow-brown spots; but note that this is less sensitive for compounds containing only one double bond, e.g., 6-keto-$PGF_{1\alpha}$, and does not work on silver nitrate plates). When developed and dried, the plate can be put into a TLC tank containing a few iodine crystals for approximately 1 min, and this should easily reveal microgram amounts of prostaglandins. The brown color evaporates spontaneously after a few minutes; heavily stained areas may be cleared using

[20] K. Gréen, M. Hamberg, B. Samuelsson, M. Smigel, and J. C. Frölich, *Adv. Prostaglandin Thromboxane Res.* **5,** 39 (1978).

[21] J. A. F. Wickramasinghe and S. R. Shaw, *Prostaglandins* **4,** 903 (1974).

TABLE III
SEPARATION OF HYDROPEROXY AND HYDROXY ACIDS BY THIN-LAYER CHROMATOGRAPHY[a]

Compound	Approximate R_f
Arachidonic acid	0.70
12-HPETE	0.63
11-HPETE	0.63
15-HPETE	0.63
5-HPETE	0.60
12-HETE	0.61
11-HETE	0.59
15-HETE	0.61
5-HETE	0.40
5,12-DiHETE (LTB_4)	0.11
HHT	0.49
PGE_2	0.01
TxB_2	0.01

[a] Compounds applied on LKD TLC plates (Whatman) which were developed in solvent D (diethyl ether–hexane–acetic acid; 60:40:1, v/v/v).

warm air from a hair dryer. Spraying the plate with 10% phosphomolybdic acid in ethanol (prostaglandins give blue color on yellow background when the plate is heated) is another excellent method; it can be used for sensitive detection of compounds containing one double bond, and it can be used on silver nitrate plates, but it is destructive and not much more sensitive than iodine. Several other visualizing sprays are also avail-

TABLE IV
SEPARATION OF INDIVIDUAL PROSTAGLANDINS ON SILICA GEL G IMPREGNATED WITH 5% SILVER NITRATE

Solvent system[a]	Approximate R_f[b]							
	PGE_1	PGE_2	$PGF_{1\alpha}$	$PGF_{2\alpha}$	TxB_2	6-Keto-$PGF_{1\alpha}$	PGD_2	PGA_2
A	0.36	0.18	0.12	0.06	0.18	0.13	0.18	0.70
C	0.85	0.74	0.57	0.41	0.57	0.83	0.68	0.88

[a] Solvent system A (see Table I). Solvent system C: chloroform–methanol–acetic acid (80:10:10, v/v/v).

[b] Compounds applied to silica gel G thin-layer plate (Uniplate, Analtech Laboratories) impregnated with 5% silver nitrate (see text).

able[12,15,22] and are most conveniently applied using a pressurized spray gun (Shandon Scientific Co., Willesden, London, England).

Hydroperoxides (e.g., prostaglandin endoperoxides PGG_2 and PGH_2 and the hydroperoxy acids formed by lipoxygenase activity) are readily detected by spraying with freshly prepared ferrous thiocyanate (0.7 g of ferrous ammonium sulfate hexahydrate is dissolved in 10 ml of a 5%, w/v, solution of ammonium thiocyanate containing 1 ml of concentrated sulfuric acid). Hydroperoxides appear immediately as red spots but fade in a few minutes.

The above spray reagents permit detection of amounts of prostaglandins as low as 1 μg, and quantitation could be achieved using a microphotodensitometer, but this is rarely used for prostaglandin analyses.

Detection of [^{14}C]prostaglandins is most effectively accomplished by using autoradiography, radiochromatogram scanning (for example, using a Panax RTLS-IA scanner; Panax Ltd., Redhill, Surrey, England) or by scraping off zones and determining radioactivity by liquid scintillation counting. Although autoradiography may take several days, we strongly recommend this technique when time is not critical, since it provides the very best picture of the radioactivity on a plate (see Fig. 3). After autoradiography, we scrape off the important zones and determine the absolute radioactivity by liquid scintillation counting. Autoradiography of ^{14}C-labeled compounds is performed by leaving the TLC plate in contact with X-ray film (Kodak; from Eastman Kodak Inc., Rochester, New York) in the dark for 12 hr or more, depending upon the radioactivity on the plate. The film is developed with DX-80 (Kodak) and fixed in a solution of sodium thiosulfate, ammonium chloride and sodium metabisulfite (1 kg, 250 g, and 100 g, respectively, in 6 liters of water).

Tritiated prostaglandins may be detected by radiochromatogram scanning or scraping off the silica gel and determining the activity by liquid scintillation counting. Normal X-ray film is not suitable for autoradiography of tritiated compounds because of the low energy of the isotope; however, special nonplastic coated film is available (^{3}H-ultro-film; LKB, Stockholm, Sweden) or the plate may be sprayed with a scintillant (e.g., PPO in DMSO, which can be obtained from NEN, Boston, Massachusetts; or En^{3}Hance, also from NEN) prior to photographic development.

Recovery of Compounds from TLC plates

If the separated products are to be analyzed by another technique (e.g., radioimmunoassay, gas–liquid chromatography) they have to be

[22] J. E. Shaw and P. W. Ramwell, *Methods Biochem. Anal.* **17,** 325 (1969).

TABLE V

SEPARATION OF PROSTAGLANDINS AS METHYL ESTERS BY THIN-LAYER CHROMATOGRAPHY

Solvent system[a]	Approximate R_f[b]										
	Arachidonic acid	12-HETE	PGE_1	PGE_2	$PGF_{1\alpha}$	$PGF_{2\alpha}$	PGD_2	PGA_2	PGB_2	6-Keto-$PGF_{1\alpha}$[c]	TxB_2
E	0.91	0.89	0.33	0.34	0.17	0.18	0.59	0.77	0.78	0.42 (0.37) (0.12)	0.35
F	0.93	0.92	0.70	0.71	0.47	0.50	0.80	0.89	0.89	0.78 (0.53) (0.46)	0.60

[a] System E, organic (upper) phase only of ethyl acetate–2,2,4-trimethyl pentane–water (110:50:100, v/v/v). System F, Chloroform–methanol–water (90:8:0.8, v/v/v).

[b] Data obtained using Whatman LKD TLC plates.

[c] 6-Keto-$PGF_{1\alpha}$ methyl ester produced more than one spot in both solvent systems; the R_f of major spot is listed together with those of the minor spots in parentheses.

eluted from the silica gel. This step requires care; otherwise it may give low and inconsistent recoveries. If the product is present in a relatively large area we scrape off the silica and make a minicolumn with it; the compounds are then eluted with several column volumes of a suitable solvent (the prostaglandins are readily eluted with methanol or ethanol, although some analysts use acidified alcohols; diethyl ether may be preferred for elution of the monohydroxy acids, the HETEs). When using the LKD TLC plates we have found that zones of silica containing prostaglandins can be readily removed with a scalpel after slightly dampening the layer with water. The prostaglandins can then be extracted into organic solvents or buffers (e.g., 50 m*M* Tris buffer, pH 7.5–8.0); extraction with buffer produces good recoveries (approximate 70%) and is particularly useful if subsequent analyses are performed in buffer (e.g., bioassay, radioimmunoassay).

Elution of prostaglandins from TLC plates containing silver nitrate is aided by precipitation of the silver by adding hydrochloric acid and then extracting the prostaglandins into organic solvents.

Higher recoveries of prostaglandins from silica gel can be achieved if they are chromatographed as methyl esters. Thus, if the prostaglandins are separated on a preparative scale or if subsequent analysis is, or can be, performed using the methyl ester (e.g., gas chromatography), it is advantageous to methylate prior to TLC. The methyl esters of the various prostaglandin groups are separated in solvent systems similar to those employed for the free acids except that the acid is omitted (see Table V). Additionally, the solvent system described by Hamberg and Fredholm[23] (methanol–diethyl ether; 2:98, v/v) is useful for separating prostaglandins D_2 (R_f 0.75), E_2 (R_f 0.33), $F_{2\alpha}$ (R_f 0.17) and thromboxane B_2 (R_f 0.50) as their methyl esters. The methyl esters are readily eluted from the silica with ethyl acetate or methanol.

[23] M. Hamberg and B. B. Fredholm, *Biochim. Biophys. Acta* **431,** 189 (1976).

[60] Two-Dimensional Thin-Layer Chromatography of Prostaglandins and Related Compounds

By ELISABETH GRANSTRÖM

The growing complexity of the biochemistry of prostaglandins and related compounds gives rise to an ever increasing requirement for better separation methods. When analyzing a crude mixture of metabolites, it is often desirable to obtain an overall picture of all compounds present in the

ISBN 0-12-181986-8

sample. Provided resolution can be made high enough, thin-layer chromatography (TLC) may be suitable for this purpose, because it may simultaneously display compounds differing widely in polarity.

Because of its simplicity and rapidity, the common, one-dimensional TLC has been widely used since the first years of prostaglandin research.[1] Identical behavior of two compounds in TLC is commonly used as a criterion of identity.

However, it is often seen that compounds may migrate in a similar way even in different solvent systems. Thus, the comparatively late discovery of the prostacyclin pathway may to some extent have been caused by the cochromatography of its major end product, 6-keto-$PGF_{1\alpha}$, with either PGE or PGF compounds in the commonly employed thin-layer solvent systems.[2,3] For the same reason it is also likely that thromboxane formation may have been overlooked in the past, because TxB_2 might have been mistakenly identified as PGE_2 in many earlier biological experiments based solely on one-dimensional TLC analyses (cf. also Fig. 5).

Such mistakes can easily be avoided using two-dimensional TLC instead. A considerable increase in resolution can be achieved by combining two solvent systems with different chromatographic properties. Even minor differences in chromatographic behavior between two compounds can be clearly seen in a two-dimensional picture, particularly if the relative positions of the spots are reversed in the second system as compared with the first one. Furthermore, an assay can be obtained that can simultaneously provide information over the whole spectrum of metabolites, spanning over a wide range of polarity: from the unsubstituted precursor fatty acids, their mono-, di-, and trihydroxylated derivatives, prostaglandins, thromboxanes, and their highly polar β- and ω-oxidized end products of metabolism. If is seldom necessary to assay all these products in the same chromatography, and the solvent systems can then instead be selected to give increased separation of products within any of these categories.

Most workers in the field do not analyze such large amounts of material as to be able to visualize spots simply by spraying the plate with some TLC reagent. By far the most common type of study in this area is the use of radiolabeled substances in small amounts, and the problem of detection thus becomes one of radioactivity detection. Generally, radio-

[1] K. Gréen and B. Samuelsson, *J. Lipid Res.* **5,** 117 (1964).

[2] C. Pace-Asciak and L. S. Wolfe, *Biochemistry* **10,** 3657 (1971).

[3] F. Cottee, R. J. Flower, S. Moncada, J. A. Salmon, and J. R. Vane, *Prostaglandins* **14,** 413 (1977).

scanning is employed for this purpose; autoradiography is also commonly used.

Studies aimed at the conversion of precursor fatty acids into prostaglandins and related derivatives frequently employ ^{14}C-labeled substrate, most often [1-^{14}C]arachidonic acid. The sensitivity of either of the abovementioned detection methods is normally sufficient for detection of even rather small amounts of such products. However, in studies aiming at the further catabolism of the products, which may involve β-oxidation with the consequent loss of the label, such an approach is obviously not suitable. For this and other reasons as well, most scientists prefer to use ^{3}H-labeled compounds in such studies. Detection of products using either radioscanning, betacamera, or autoradiography is then considerably more difficult because of the weak energy of the ^{3}H radiation. This problem becomes particularly great in *in vivo* studies, when the total dose of radioactivity must be kept at a minimum, and also because the number of formed products may be very large.

This chapter describes two-dimensional TLC assay systems, suitable for several different types of analytical problems in the field of eicosanoid biochemistry. Detection of radioactive spots is achieved by autoradiography, employing either the traditional X-ray type of film or a more recently developed, highly sensitive film. This latter product, LKB Ultrofilm ^{3}H, renders the detection of even small amounts of tritium labeled products possible, and reduces the necessary exposure time also for ^{14}C-labeled substances.

Experimental Procedure

Thin-Layer Chromatography

Preparation of the TLC Plate. Precoated TLC plates (silica gel 60, thickness 0.2 mm, 20 × 20 cm, Merck, Darmstadt) were used in all experiments reported below. The plates were used after heat activation (1 hr, 120°).

The starts and ends of the two runs may be indicated by a system of penciled lines as shown in Fig. 1. Lines are drawn 2.5 cm from the bottom and the left edge, and 2 cm from the top and the right edge. Note: It is sometimes warned that irreversible adsorption of the sample may take place at the penciled lines. In our experience this has never happened; however, to avoid the possibility, the system of lines may naturally be added after the chromatography instead, and before autoradiography.

Application of the Sample. The sample should be sufficiently pure to allow its application as a small spot. If it is necessary to apply a larger aliquot than is judged possible for a spot a few millimeters in diameter, which gives the best results, a narrow band may be allowed, preferably not more than 1 cm wide. This technique will improve separation of components in very crude samples and is thus preferable to overloading at the starting point. The final spots after chromatography will then normally reflect this elongated shape only very close to the line of application; a few centimeters away the spots will display the normal round shape.

The starting point is in the lower left corner, 2.5 cm from each edge. This distance is chosen instead of the conventional 1.5–2 cm gap because it results in a somewhat slower passage of the solvent front over the starting place, thus providing a longer time for dissolution of the sample components with the consequent reduction of tailing phenomena. Such a modification may be necessary to compensate for the possibly higher sample load per unit area at the starting place in a technique like this.

It is recommended to include at least a few reference compounds in the sample: these should preferably be added to the sample before application to the TLC plate. The reference solutions must be made fresh at short intervals and regularly checked, since even a very small contamination with radioactive compounds will ruin the results. To avoid microchromatography at the starting point during application, the sample should be dissolved in a nonpolar solvent, preferably chloroform (if necessary with a small percentage of methanol).

Chromatography. The plate is run in the first solvent system until 2 cm from the top. The TLC plate is then air dried in a ventilation hood for not more than 30 min. Longer drying times will result in some irreversible adsorption along the line of chromatography, which is now the starting line for the second run. The TLC plate is turned 90° and run in the second solvent system, again until 2 cm from the top. The plate is left in the fume hood overnight: it must be absolutely free of solvent (particularly acetic acid) before application of the film.

Solvent Systems. The possible combinations of solvent systems are literally infinite. The ideal combination, however, is two solvent systems with similar polarity but giving different *relative* positions of the various compounds under study. A combination of a relatively polar first system with a nonpolar second one will result in spots spread along a hyperbolic curve, since only the least polar compounds will move in the second system. The opposite situation will obviously give the mirror image. Such combinations may be useful for particular separation problems.

Ideally, however, the two solvent systems should be of similar polarity, resulting in the compounds running more or less along a line with 45° slope from the lower left to the upper right corner (cf. Figs. 3–5), but that

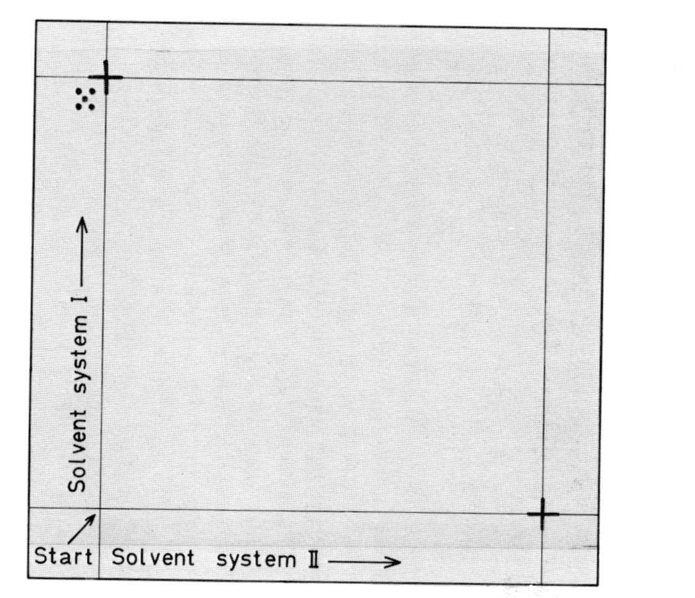

FIG. 1. Preparation of thin-layer plates for autoradiography. The thin lines denote the system of penciled lines; the thicker markings schematically illustrate suitable places for labeling with radioactivity prior to application of the two film sizes. Using the same radioactive solution, each TLC plate is also identified by some numbering code as suggested in the figure.

give widely differing relative positions of the sample components. Thus, as mentioned in the introduction, PGE_2 and TXB_2 have very similar R_f values in most TLC systems. Combinations of systems in which their R_f values have been reversed, however, give a distinct separation even though the actual differences in R_f values are minute. The same problem pertains to separations of PGE_2, $PGF_{2\alpha}$, and 6-keto-$PGF_{1\alpha}$, which also may have very similar R_f values[2,3] (see Fig. 5).

The following solvent system combinations have proved to be most useful in our laboratory. For compounds ranging in polarity from arachidonic acid or monohydroxy fatty acids to $PGF_{2\alpha}$ or 6-keto-$PGF_{1\alpha}$ (cf. Fig. 5), combination "23/15" or "10/15": first solvent system, "23" (ethyl acetate–hexane–acetic acid; 75:25:2, v/v/v) or "10" (organic phase of ethyl acetate–acetic acid–isooctane–water, 100:10:30:100, v/v/v/v); second solvent system, "15" (chloroform–ether–methanol–acetic acid, 45:45:5:2, v/v/v/v). For more polar compounds, e.g., ranging in polarity from nonpolar prostaglandin metabolites such as 15-keto-13,14-dihydro compounds to very polar, ω- and β-oxidized metabolites (cf. Figs. 3 and 4), combination "29/30": first solvent system, "29" (ethyl acetate–acetic acid–isooctane, 100:10:30, v/v/v); second solvent system, "30" (chloroform–*n*-butanol–acetic acid, 80:20:2, v/v/v). This solvent combination also gives a better separation between PGE_2 and TxB_2 than the above-mentioned alternatives.

The experiments reported in this chapter were all carried out using one of these solvent combinations. The plates were run in tanks without lining. Reproducibility was satisfactory: variations in R_f values seldom exceeded 5% between different runs.

Autoradiography

Preparation of the TLC Plate for Autoradiography. Before the film is applied, the TLC plate must be labeled with suitable information. This labeling serves two purposes: each chromatography can then be safely identified after the development of the film, and, most important, the exact positions of radioactive spots on the plate can be established, which is necessary for comparison with included reference compounds and for scraping out and quantification, if desired (see below).

The labeling can be conveniently done by careful application, using a microsyringe, of small amounts of a radioactive solution along some of the pencil lines, preferably as cross marks at the point of intersection of the lines. To avoid superimposition of radioactivity over any labeled metabolite, the marking can be done as suggested in Fig. 1. The two suggested alternatives for labeling the plates are suitable for the two different film sizes (see below), and both leave the two most interesting corners of the plate free.

To ensure sharp, well-defined lines, the radioactive compound used for this labeling must be relatively polar and added in a nonpolar solution; otherwise it may migrate on the plate. We normally employ aged, extensively autoxidized preparations of [^{14}C]arachidonic acid, dissolved in chloroform, for this purpose. Using the same solution, the plates are also labeled with some code to facilitate identification later (cf. Fig. 1).

Application of Film. The most sensitive film for autoradiography developed to date is LKB Ultrofilm ^{3}H (LKB Produkter, Bromma, Sweden). This film, originally called ^{3}H-Film, was developed for the purpose of autoradiographic analysis of tissues or chromatograms containing tritium.[4-6] However, it is also highly suitable for other β-emitters, such as ^{14}C, reducing the necessary exposure times considerably. It has been shown to be 12–64 times more sensitive than traditional X-ray films. The properties of this film have been described in detail.[5,6]

Ultrofilm ^{3}H differs from conventional X-ray films in several important respects. First, it lacks the protective "antiscratch" layer of gelatin, which normally absorbs more than 90% of the weak β-radiation of ^{3}H. Second, it is single coated with an extremely thin "monolayer" of rather coarsely grained X-ray type emulsion. Thus the background is reduced considerably and is also further improved by the film base lacking the traditional blue stain of most ordinary X-ray films. The emulsion is Singul-X RP with a high silver:gelatin ratio, which further increases the sensitivity to tritium.

LKB Ultrofilm ^{3}H is manufactured in two sizes: 35 × 43 cm and 24 × 30 cm. The bigger format is sufficient for the chromatographic area of four 20 × 20 cm TLC plates, although the plates will not be entirely covered. While still in its protective cover, the film is carefully cut in four pieces, 17.5 × 21.5 cm. Each piece is applied to one TLC plate as indicated in Fig. 1 (left panel), with the emulsion side facing the silica gel. The superfluous part to the left makes a convenient "handle" for the later development, fixation, and rinsing, and also for filing. Such a "handle" is particularly important for Ultrofilm: the lack of the anti-abrasion gelatin layer renders it extremely vulnerable to gross artifacts such as scratches

[4] S. Ullberg, *Sci. Tools,* Special Issue 1977, p. 2 (1977).

[5] B. Larsson and S. Ullberg, *Sci. Tools,* Special Issue 1977, p. 30 (1977).

[6] E. Elm and B. Larsson, *Sci. Tools* **26,** 24 (1979).

or fingerprints, which either remove the emulsion or produce blackening when the film is developed.

The smaller size film can either be used to cover one 20 × 20 cm TLC plate entirely or can be cut in three 10-cm wide strips (10 × 24 cm) and conveniently used for routine analyses of thin-layer chromatograms where the compounds are known to be found relatively close to the diagonal of the plate (Fig. 1, right panel). With the solvent system combinations recommended in this paper, most compounds are normally found in the area covered by this strip size. However, when unknown mixtures of compounds are analyzed, it is recommendable to use the bigger size.

When the 10 × 24 cm strips are applied as indicated in Fig. 1, the four corners protrude from the plate and can be used as "handles" during the later processing.

After application of the film, a glass plate (20 × 20 cm) is placed on top and is fixed in position with adhesive tape on at least three sides, or with bulldog clamps. Note: If adhesive tape is used, care must be taken that it is never removed while the film is in position. Pulling off tape creates little flashes of light that blacken the film. If necessary the tape must be cut open.

The sandwich of TLC plate, film, and glass plate is then placed in lighttight containers (black plastic bags, lighttight envelopes, boxes) and stored for the desired exposure time.

Exposure. The TLC plates can be stored at room temperature or in the cold. The film must be protected from moisture and heat. Undesired artifacts resulting from unsuitable storage are high background and latent image fading, i.e., the Ag atoms produced from the radiation are reoxidized to Ag^+ ions.

The necessary exposure time depends on many factors. When the same type of film and the same isotope is used in a number of experiments, the exposure time depends mainly on the total radioactivity on each TLC plate and the number and areas of the spots among which the radioactivity is distributed in each sample. Several other factors also influence, but to a lesser degree.

A general idea of the sensitivity of Ultrofilm ^{3}H can be obtained from Fig. 2. Mixtures of [9β-^{3}H]PGF$_{2\alpha}$ and |1-^{14}C|arachidonic acid were separated in a one-dimensional thin-layer chromatography [solvent system, ethyl acetate–n-heptane–acetic acid–methanol–water (45:40:4:2:1, by volume)]. The film was exposed for 2 weeks. After this time fully discernible spots could be seen with 2000 dpm of the ^{3}H-labeled compound and with 100 dpm of the ^{14}C-labeled substance. The areas of the PGF$_{2\alpha}$ spots were somewhat smaller than those of the arachidonic acid spots.

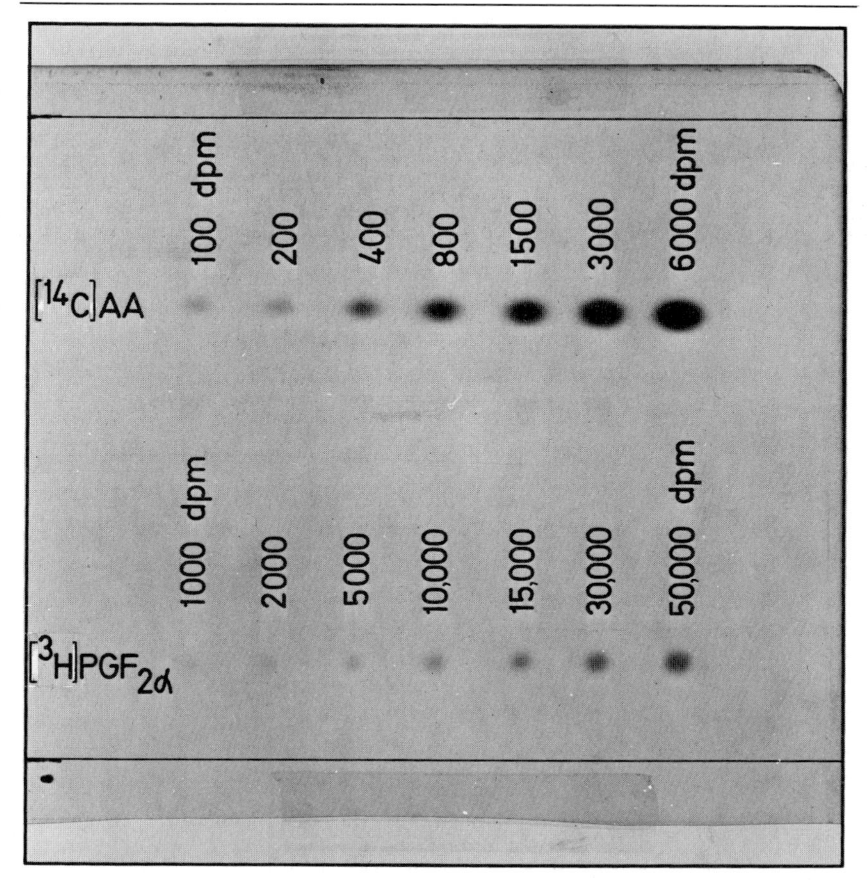

FIG. 2. Sensitivity of LKB Ultrofilm for ^{3}H and ^{14}C. Mixtures of [9β-^{3}H]$PGF_{2\alpha}$ and [1-^{14}C]arachidonic acid were separated in solvent system "8": ethyl acetate–*n*-heptane–methanol–acetic acid–water (45:40:2:4:1, by volume.) Although the sensitivity for the two isotopes is not directly comparable from this experiment because of the smaller spot size of $PGF_{2\alpha}$, a general idea can be obtained; after 2 weeks of exposure, spots containing 2000 dpm of ^{3}H or 100 dpm of ^{14}C were visible.

Detection of radioactivity can be further improved by prolonged exposure. However, the gradually increasing background counteracts this improvement, and it is doubtful that exposure times exceeding 1–2 months normally provide any advantage over shorter ones.

Development of the Film and Identification of Spots. For development, the manufacturer's procedure should be closely followed. If the results are deemed satisfactory and the autoradiography does not have to be repeated, the reference spots can now be visualized using traditional TLC spray reagents. Phosphomolybdic acid is the preferred reagent in our laboratory because the initial color of the spots developed with this reagent differs among different compounds and thus aids in their identification. Thus, PGE_2 is initially reddish-brown, PGD_2 greenish, $PGF_{2\alpha}$ blue-green, etc., with this reagent. The plate is carefully sprayed and put on a hot plate, and the gradual color development is observed.

Quantification. The relative amounts of each metabolite can easily be established after two-dimensional TLC by scraping out the corresponding silica gel, elution, and scintillation counting. Prior spraying and heating of the TLC plate to identify the reference spots should preferably be avoided, if the study is carried through this quantification step, or at least be done using very mild conditions.

The developed film is laid over the TLC plate, and the positions of the radioactive spots are indicated on the plate with pencil. After scraping the silica gel in these spots into scintillation vials, 10 ml of Instagel (Packard) are added directly: this polar scintillation cocktail gives sufficient elution of most substances from the silica gel. The vials are then counted in a scintillation counter. This procedure results in a higher efficiency than if prior elution with methanol or some other solvent is done before addition of the cocktail.

Using the appropriate equipment it is also possible to measure the relative amounts of the sample components directly from the autoradiogram, by estimation the darkness of the spots (a variation of densitometry).

Application of the Method

Figures 3–5 show some different experiments where crude mixtures of prostaglandins or related substances were analyzed by the described method.

Figure 3 demonstrates the highly complex pattern of $PGF_{2\alpha}$ metabolites appearing with time in the circulation of a rabbit. [9β-^{3}H]$PGF_{2\alpha}$ was injected into one ear vein of the animal, and blood samples were taken from the opposite ear. The first sample, taken about 1 min after the injection, showed the expected dominance of one single compound, which coincided with reference 15-keto-13,14-dihydro-$PGF_{2\alpha}$ (data not shown) (cf. Samuelsson *et al.*[7]). However, this initially formed metabolite was soon degraded into a multitude of other products. Twenty minutes after the injection, 15-ketodihydro-$PGF_{2\alpha}$ represented only 4% of the total sample radioactivity. The remainder was distributed more or less evenly over

[7] B. Samuelsson, E. Granström, K. Gréen, M. Hamberg, and S. Hammarström, *Annu. Rev. Biochem.* **44,** 669 (1975).

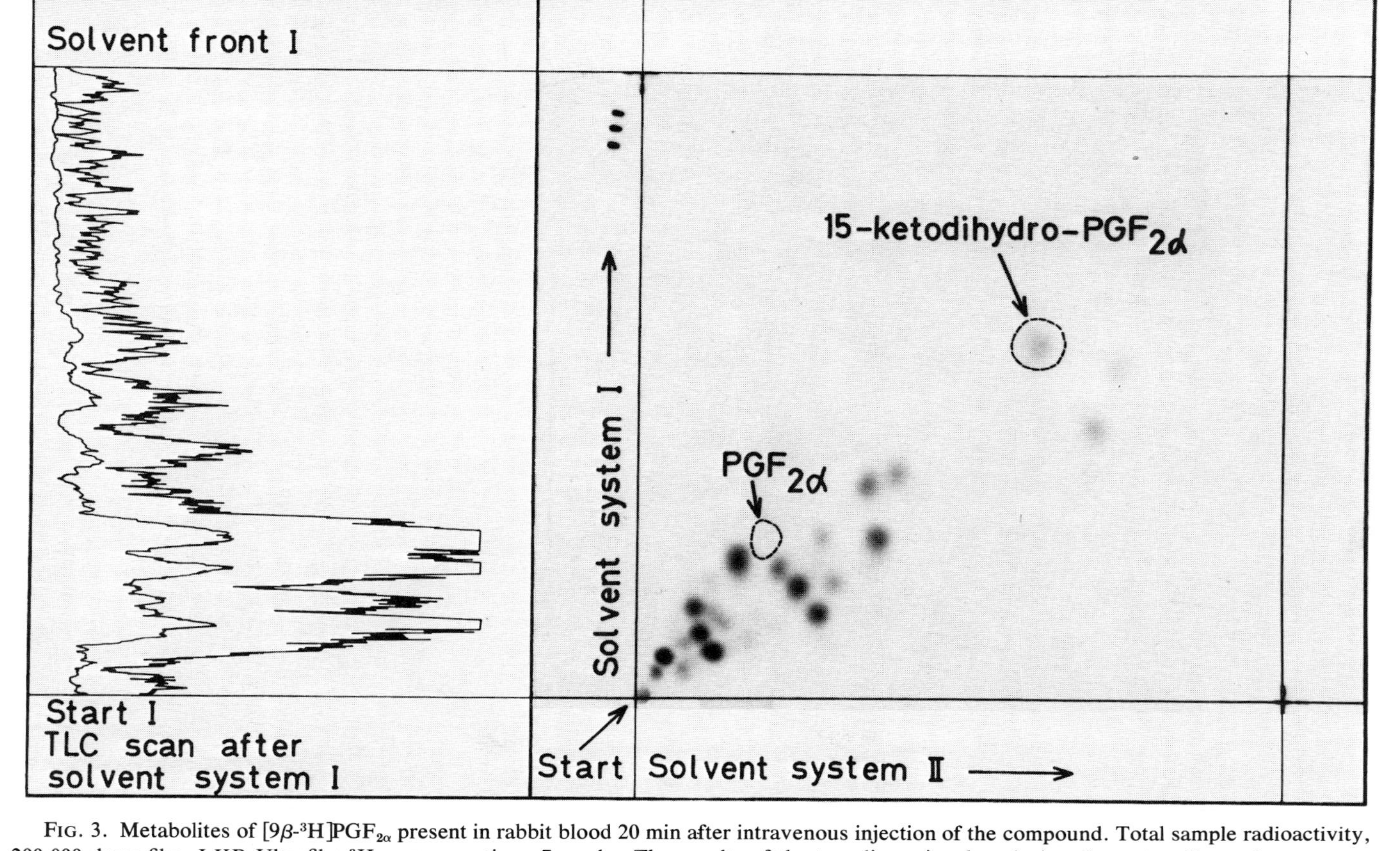

FIG. 3. Metabolites of [9β-^{3}H]PGF$_{2\alpha}$ present in rabbit blood 20 min after intravenous injection of the compound. Total sample radioactivity, 200,000 dpm; film, LKB Ultrofilm ^{3}H; exposure time, 7 weeks. The results of the two-dimensional analysis using autoradiography are compared with a radioscan obtained after one-dimensional TLC. Solvent system combination "29/30" (see Solvent Systems in Experimental Procedures section).

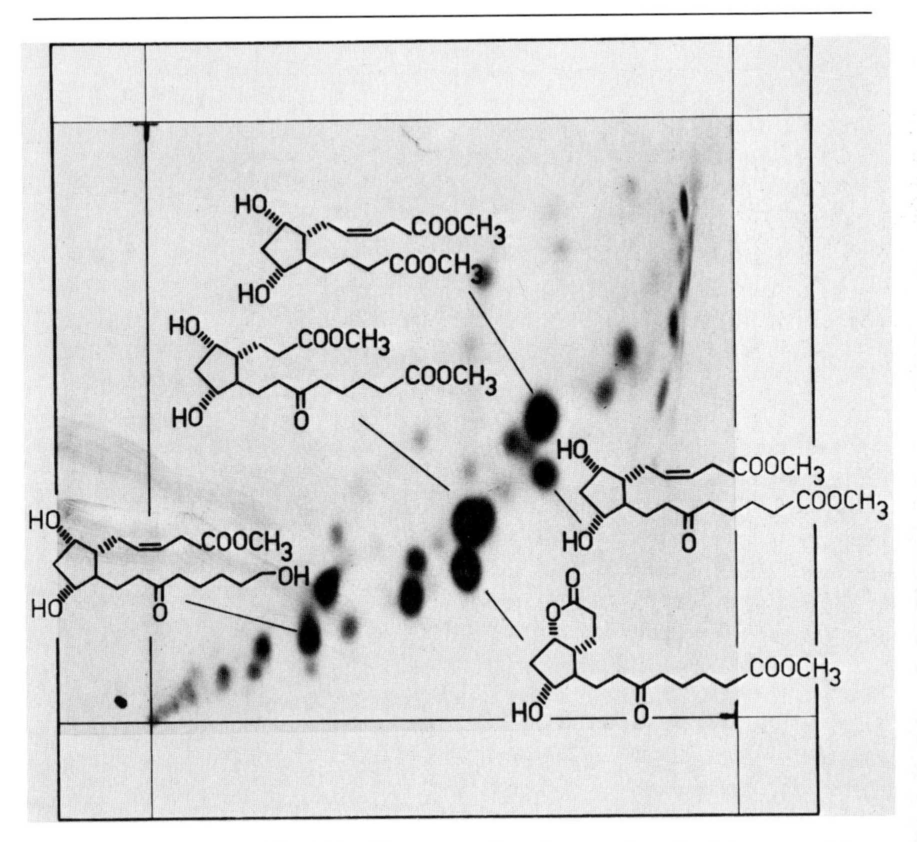

FIG. 4. Metabolites of [9β-^{3}H]PGF$_{2\alpha}$ excreted into human urine after intravenous injection of the compound. The extract (ethyl acetate phase from Sep-Pak) was esterified with diazomethane before TLC. Solvent system combination, 29/30 (see Experimental); total sample radioactivity, 300,000 dpm; film, LKB Ultrofilm ^{3}H; exposure time, 2 months.

some 25 different spots, none of them really dominating the picture (Fig. 3). Most of the polar compounds (found in the lower left part of the plate) have been identified, using among other methods cochromatography with reference compounds in the two-dimensional assay as criterion of identity. Conclusions about identity were supported by other analyses as well.[8] The majority of these products are tetranor metabolites, differing in structures in the ω side chain (11-keto or 11-hydroxy compounds, ω1 or ω2 hydroxylated compounds, dicarboxylic acids, etc. Of the dioic acids, some had been degraded by β-oxidation also from the ω end). All tetranor compounds occurred in two forms, well separated from each

[8] E. Granström and H. Kindahl, submitted for publication.

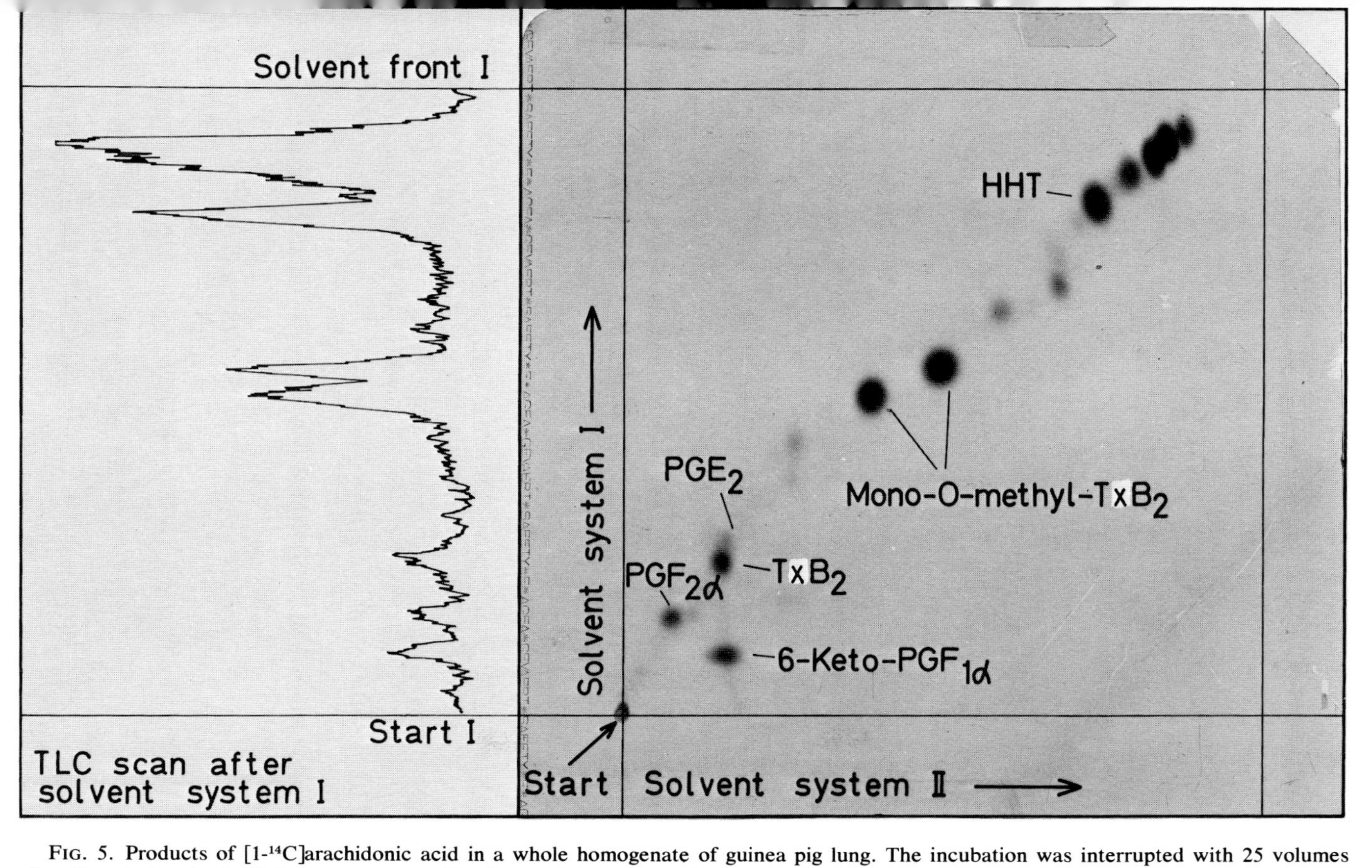

FIG. 5. Products of [1-^{14}C]arachidonic acid in a whole homogenate of guinea pig lung. The incubation was interrupted with 25 volumes of methanol after 1 min. Solvent system combination, 10/15 (see Experimental); total sample radioactivity, 30,000 dpm; film, Osray M-3; exposure time, 1 week.

other: in the open form and as the corresponding δ-lactone. A few dinor metabolites were also identified, and even C_{20} compounds were found in small amounts among these polar products.

The high resolving capacity of the two-dimensional TLC assay can be clearly seen by comparison with the traditional one-dimensional TLC analysis. The left part of Fig. 3 shows a radioscan (Berthold Dunnschicht-scanner II) obtained after chromatography in the first solvent system. Instead of about 25 different polar compounds, only 4 incompletely separated peaks are seen in the most polar part of the plate. These are preceded by three smaller peaks, one of which evidently represent 15-ketodihydro-$PGF_{2\alpha}$. Using the one-dimensional, traditional TLC assay only, it would not have been possible to recognize (*a*) the multitude of compounds appearing with time; (*b*) the total absence of the parent compound, $PGF_{2\alpha}$; (*c*) the identities of any metabolites with the possible exception of 15-ketodihydro-$PGF_{2\alpha}$. Many of the spots in the lower left cluster on the two-dimensional TLC plate were identified by cochromatography with reference compounds in the two-dimensional assay, as mentioned above.

Figure 4 shows a crude extract of human urine containing ^3H-labeled $PGF_{2\alpha}$ metabolites. This time the extract was esterified by treatment with diazomethane prior to TLC; nonetheless, the combination of solvent systems was the same as that in Fig. 3.

In earlier studies on $PGF_{2\alpha}$ metabolism in the human and other species,[7,9–11] a number of different chromatographic steps were required to disclose the multitude of metabolites as well as the properties of the various compounds. In those studies no reliable overall picture was ever obtained: the normal sequence of events was that peaks, often incompletely separated, from one complex chromatogram were further subdivided into new peaks in a subsequent chromatography using a different solvent system.[9–11] Since all steps furthermore lead to inevitable, and unfortunately variable, losses, the relative proportions of the different compounds were difficult to establish.

The two-dimensional TLC assay in Fig. 4 was run after one single extraction step, viz. Sep-Pak (this volume [58]). The lipophilic fraction from this extraction, obtained using either ethyl acetate or methyl formate, contained essentially all urinary radioactivity, and was directly subjected to TLC after methyl esterification. The autoradiogram thus displays all

[9] E. Granström and B. Samuelsson, *J. Biol. Chem.* **246,** 5254 (1971).
[10] E. Granström and B. Samuelsson, *J. Biol. Chem.* **246,** 7470 (1971).
[11] E. Granström, *Adv. BioSci.* **9,** 49 (1973).

the urinary products with only minor losses. An arbitrary idea of their relative proportions can be obtained from the blackness and areas of the spots; a more reliable picture is obtained after scraping out the respective compounds and counting, as described in the Experimental section.

The identities of a few of the compounds are indicated in Fig. 4. The two dark spots near the center of the plate represent the major metabolite, $5\alpha,7\alpha$-dihydroxy-11-ketotetranorprostane-1, 16-dioic acid, in its open and δ-lactone forms, respectively. The prominent spot above these two represents the second major urinary product, $7\alpha,9\alpha$-dihydroxydinor-ω-tetranor-prost-3-ene-1,14-dioic acid. A total of about 40 spots can, however, be seen; the identities of all these are not known, but a number of identified urinary PGF metabolites are given in Fig. 1 of this volume [44] for comparison with the present autoradiogram.

Figure 5 shows a somewhat different experiment. [1-^{14}C]Arachidonic acid was incubated with a homogenate of guinea pig lung. The incubation was interrupted after 1 min with a large excess of methanol to convert any TxA_2 present into mono-*O*-methyl TxB_2.[12,13] After ethyl acetate extraction, about 30,000 dpm of the crude sample were applied to the TLC plate, developed in system "10/15" with radioscanning after the first run, and then subjected to autoradiography. This time a conventional X-ray film was employed (Osray M-3, Agfa-Gevaert; exposure time, 1 week).

The expected products, 12-L-hydroxyheptadecatrienoic acid (HHT), PGE_2, $PGF_{2\alpha}$ and 6-keto-$PGF_{1\alpha}$, are clearly seen as distinct spots. Weaker spots with locations similar to those of PGD_2, 15-ketodihydro-$PGF_{2\alpha}$ and (nonenzymically formed) PGA_2–PGB_2 are also seen, although not indicated in Fig. 5, since those reference substances were not included in this run. The two dominating spots in the central part of the TLC plate are caused by the two epimers of mono-*O*-methyl TXB_2.[12,13] This time the radioscan, i.e., a one-dimensional analysis, gives about the same results (Fig. 5, left part), and judging from the major part of the plate no real advantage seems to be gained from a two-dimensional analysis. However, in the least polar part of the plate, immediately in front of the HHT peak, the radioscan reveals only one major peak, close to the solvent front and coinciding with reference arachidonic acid. Similar radioscans are very common in prostaglandin literature, and generally no attention is paid to this nonpolar peak: it is normally interpreted only as "unconverted arachidonic acid." The two-dimensional analysis, how-

[12] M. Hamberg, J. Svensson, and B. Samuelsson, *Proc. Natl. Acad. Sci. U.S.A.* **72,** 2994 (1975)

[13] E. Granström, H. Kindahl, and B. Samuelsson, *Prostaglandins,* **12,** 929 (1976).

ever, clearly demonstrates that no fewer than four different products are present in this area, even using these solvent systems that were not designed for separation of compounds of so low polarity. The compounds in these spots were not identified in this experiment; it is likely that some of them are monohydroxy fatty acids.[14] The least polar spot, which is the weakest one, is probably unconverted arachidonic acid. Thus the two-dimensional assay actually demonstrates an almost complete conversion of the precursor acid, a result definitely different from the one-dimensional radioscan. In our experience even highly unsuitable solvent systems generally give some resolution of compounds of similar polarity, which can be clearly seen using autoradiography; total confluence of spots with consequent misinterpretation of data is uncommon in this assay in contrast to the one-dimensional type.

We have occasionally seen chromatographic patterns deviating from the expected ones. A not uncommon phenomenon is cochromatography of relatively polar compounds, such as prostaglandins, with highly lipophilic substances, such as cholesterol, triglycerides, or long-chain fatty acids, in crude extracts of tissues rich in such compounds. The more polar substances probably dissolve in the steroid or, most often, the fatty acid component of the sample, and are never adsorbed to the silica gel. After the first chromatography the sample components are spread over a wider area on the plate; competition by the silica gel for the more polar compounds may now take place more successfully, and proper separation occurs during the second run. Figure 6 shows one example of this. In this experiment human endometrium was incubated with [^{14}C]arachidonic acid, and an ethyl acetate extract was subjected to two-dimensional TLC. Two parallel bands of spots are seen at the level of long-chain fatty acids and neutral steroids, respectively. The pattern of spots in these two bands indicate that they seem to contain the same components: obviously the silica gel failed to retain them to their proper R_f levels in the first run. This phenomenon may become quite extreme, and we have even seen chromatographies with a total lack of separation in the first run: all products cochromatographing with the long-chain fatty acid fraction and separating only during the second run. The reason why the prostaglandins and TxB_2 were found in their proper places in the experiment in Fig. 6 is likely to be the presence of comparatively large amounts of unlabeled material of all these compounds that were added as internal references.

This artifact may unfortunately be a rather common phenomenon and may naturally occur in other types of chromatographies as well. We never

[14] M. Hamberg and B. Samuelsson, *Biochem. Biophys. Res. Commun.* **61,** 942 (1974).

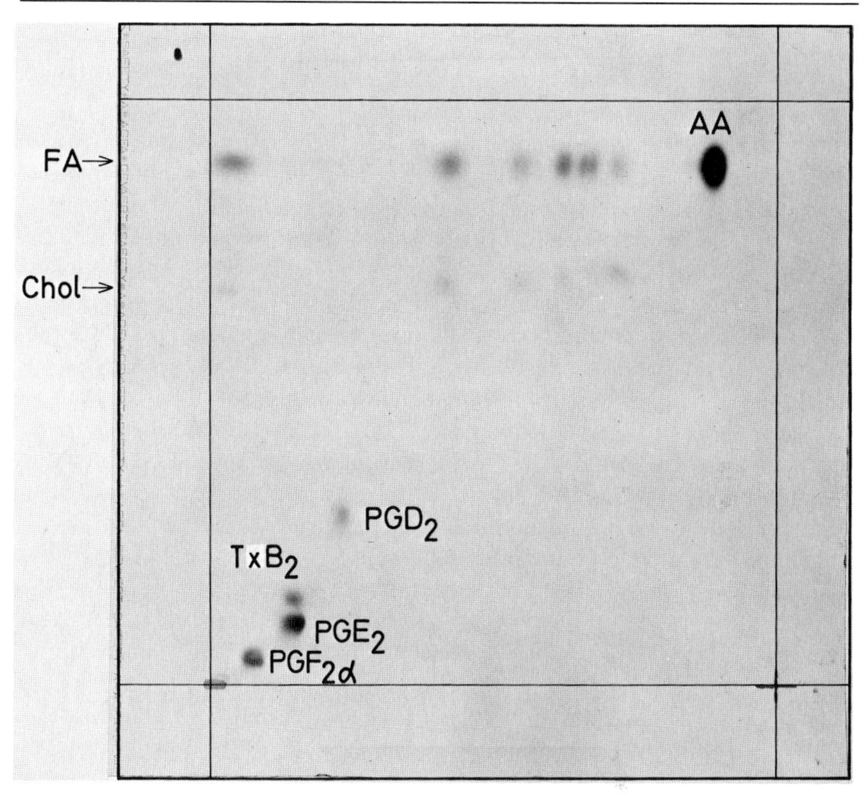

FIG. 6. Cochromatography of some ^{14}C-labeled arachidonate metabolites with endogenous lipophilic fractions of the sample. A biopsy of human endometrium was incubated with [1-^{14}C]arachidonic acid and the ethyl acetate extract subjected to two-dimensional TLC. Solvent system, 23/15; total sample radioactivity, 40,000 dpm; film, Osray M-3; exposure time, 1 week. Cochromatography of certain sample components in the first run with some constituents of the extract was revealed by the appearance of two bands of labeled products during the second run. The constituents causing this phenomenon were present in such large amounts that it was possible to visualize them with phosphomolybdic acid. Their identities were not conclusively established, however, the upper band of labeled metabolites appears at the R_f level of long-chain fatty acids (FA) and the lower band at the level of neutral steroids, such as cholesterol (Chol), as analyzed by a separate run in the first solvent system. The reason for the appearance of prostaglandins E_2, D_2, and $F_{2\alpha}$ and of thromboxane B_2 at their proper places was probably the presence of large amounts of unlabeled material of these substances added as internal references.

recognized its occurrence before we began to use the two-dimensional analysis. The implications of this serious source of error are naturally far-reaching for studies, qualitative as well as quantitative ones, relying only on one-dimensional TLC.

Discussion

Two-dimensional thin-layer chromatography or paper chromatography has been widely employed in other fields of biochemistry: for analyses of complex mixtures of steroids, amino acids, peptides, etc. In view of the tremendous increase in information obtained from a two-dimensional analysis, it is rather surprising that two-dimensional TLC has not gained widespread use in a field as complex as the prostaglandin area. Only a few laboratories have used the method (e.g., 15–17); however, in most cases either the employed solvent system combinations have been suboptimal, providing little improvement over the one-dimensional analysis, or detection methods with so low resolution have been used as to obviate the increase in resolving capacity obtained by the two-dimensional method.

Of the three possible detection methods—radioscanning, betacamera, and autoradiography—the first two are comparatively rapid methods and normally give results within hours. Even using the most modern instruments, however, we find resolution by these detection methods inferior to that obtained by the third method, autoradiography. Besides, the sensitivity for tritium is rather low for the first two methods.

Autoradiography, on the other hand, is unfortunately a rather slow process and may require days or weeks with low amounts of radioactivity. Using the conventional X-ray type of film, detection of tritium is very difficult: large doses and unacceptably long exposure times are necessary. To improve this situation various so-called enhancing or intensifying techniques have been developed over the years, by which the spots can be intensified, either during the autoradiography or after development of the film. However, these techniques are either very laborious or involve the handling of large amounts of radioactivity or of toxic chemicals, and a more attractive solution to the detection problem is the use of a more sensitive film. The LKB Ultrofilm ^{3}H represents a step forward in this respect, being 12–64 times more sensitive than the conventional X-ray films.

Employing this film for detection of autoradiography, a two-dimensional TLC assay was developed, which greatly facilitated the analysis of complex mixtures of labeled metabolites of arachidonic acid or prostaglandins. The assay was particularly suitable for *in vivo* studies, where the amount of isotope given must necessarily be kept at a minimum, and

[15] M. Lagarde, A. Gharib, and M. Dechavanne, *Clin. Chim. Acta* **79,** 255 (1977).

[16] J. Martyn Bailey, R. W. Bryant, S. J. Feinmark, and A. N. Makheja, *Prostaglandins* **13,** 479 (1977).

[17] J. E. Vincent and F. J. Zijlstra, *Prostaglandins* **14,** 1043 (1977).

where a large number of products are formed; however, it was also superior in *in vitro* studies.

The described method can be further developed and improved. For example, preliminary experiments using HPTLC plates have shown an even better resolution of components in complex mixtures. It must be kept in mind, however, that the risk of irreversible adsorption increases somewhat by the use of HPTLC plates with their smaller particle size.

The described method is mainly analytical. It can also be used preparatively for stable compounds, or even for more sensitive substances, if large amounts of radioactivity are used and the exposure times consequently can be reduced. To minimize destruction during the necessarily prolonged chromatography times, the TLC plates could be run in N_2 atmosphere and in tanks with lining.

For analytical purposes the two-dimensional assay offers further possibilities. Identification of compounds can be facilitated, not only by the increased resolution as discussed above, but also by the so-called "spot shift technique."[cf. 18] The sample components are then derivatized *in situ* after the first run: minor differences in properties between two compounds may thus be enhanced, or they may be separated according to a different principle in the second system.

Acknowledgments

This study was supported by grants from the Swedish Medical Research Council, project No. 03P-5804 and 03X-05915, and the World Health Organization.

[18] B. P. Lisboa, *in* "Lipid Chromatographic Analysis" (G. V. Marinetti, ed.), Vol 2, p.339. Dekker, New York, 1976.

[61] Separation of Arachidonic Acid Metabolites by High-Pressure Liquid Chromatography

By THOMAS ELING, BETH TAINER, ARIFF ALLY, and ROBERT WARNOCK

The most commonly used technique for the separation and quantitation of arachidonic acid (AA) metabolites involves the use of thin-layer chromatography (TLC) for separation, and radioisotopes for estimating the amounts of AA metabolites. The method suffers from poor resolution of chromatographic peaks on thin-layer chromatograms and the need to use several solvent systems to adequately separate the major AA metabo-

 ISBN 0-12-181986-8

lites. High-pressure liquid chromatography (HPLC) offers the advantage of high resolution and good reproducibility. High-pressure liquid chromatography using a silicic acid column followed by a reversed-phase column has been used to separate the cyclooxygenase metabolites of AA,[1,2] but separation of lipoxygenase metabolites, hydroxy fatty acids (HFA) was not reported. Other workers have successfully used silicic acid[3,4] and reversed-phase columns[4] for separation of HFA. No HPLC method previously described offers good resolution of both prostaglandins (PGs) and HFA. We describe here a method for rapid separation of PGs, thromboxanes (Txs), HFA, and AA with good resolution by reversed-phase HPLC.

Materials

The radiolabeled standards, [^{14}C]arachidonic acid, [^{3}H]arachidonic acid, [^{3}H]6-keto-$PGF_{1\alpha}$, [^{3}H]$PGF_{2\alpha}$, [^{3}H]TxB_2, [^{3}H]PGD_2, [^{3}H]PGA_2, [^{3}H]PGE_2, were obtained from New England Nuclear. HPLC grade solvents were used in all cases and were filtered through an organic (0.5 μm) Millipore filter before use. Water was deionized and filtered through a Hydro Ultra Pure water system. Acetonitrile, acetone, and reagent grade glacial acetic acid were obtained from Fisher Scientific. The [^{3}H]15-hydroxy-5,8,11,14-eicosatetraenoic acid (HETE) standard was prepared from [^{3}H]AA using soybean lipoxidase (Sigma) as described by Funk *et al.*[5] [^{3}H]12-L-Hydroxy-5,8,10-heptadecatrienoic acid (HHT) was prepared by incubating [^{3}H]AA and cold AA (NuChek Prep, Inc.) with human platelet microsomes (platelets from the Red Cross, Durham, North Carolina). 5-, 11- and 5-, 12- and 12-HETES were gifts from Frank Sun, Upjohn Co., Kalamazoo, Michigan. [^{3}H]12-HETE was also prepared by incubating [^{3}H]AA with platelet microsomes in the presence of 100 μM indomethacin.[6]

Methods

Separations of the PGs and other AA metabolites were done on a Waters Associates Model ALC/GPC 204 LC equipped with a Model

[1] A. R. Whorton, M. Smigel, J. A. Oates, and J. C. Frölich, *Biochim. Biophys. Acta* **529,** 176 (1978).

[2] A. R. Whorton, K. Carr, M. Smigel, L. Walker, K. Ellis, and J. A. Oates, *J. Chromatogr.* **163,** 64 (1979).

[3] N. A. Porter, R. A. Wolf, E. Yarbo, and H. Weenan, *Biochem. Biophys. Res. Commun.* **89,** 1058 (1979).

[4] P. Borgeat and B. Samuelsson, *J. Biol. Chem.* **254,** 7865 (1979).

[5] M. O. Funk, R. Isaac, and N. A. Porter, *Lipids* **11,** 113 (1976).

[6] R. W. Bryant, S. J. Feinnark, A. N. Makheja, and J. M. Baily, *J. Biol. Chem.* **253,** 8134 (1978).

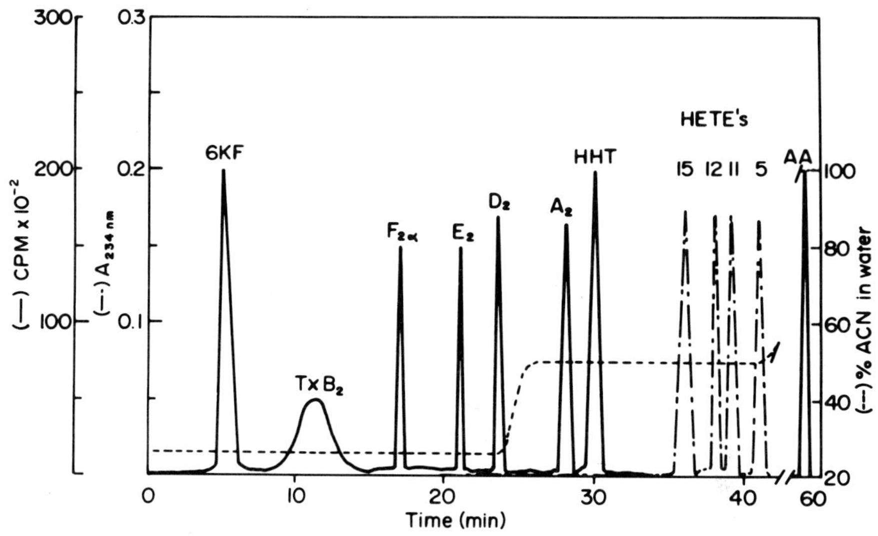

FIG. 1. Separation of prostaglandins and hydroxyeicosatetraenoic acids: use of gradient elution.

6000A high-pressure pumping system, a Model 660 solvent programer, a U6K injector, a 0.2 μm prefilter, and a RCM-100 radial compression module. The radial compression device was fitted with a 5 mm × 10 cm Radial Pak-A (C_{-18}) cartridge and eluted at 3 ml/min with varying ratios of water (pH 3.5) and acetonitrile. Elution of the HFA was followed by the measurement of the absorbance at 234 nm, and the eluate was collected in fractions every 30 sec with a Gilson fraction collector. These samples were mixed with 7 ml of Aquasol (New England Nuclear), and the radioactivity was measured by liquid scintillation spectrometry.

The column was equilibrated for at least 10 min under initial conditions: 26% acetonitrile in water (pH 3.5) at a flow rate of 3 ml/min. The sample and appropriate standards, usually dissolved in 50–200 μl of acetone, were injected, and the column was eluted isocratically for 23 min. As can be seen in Fig. 1, good separation of 6-keto-$PGF_{1\alpha}$, TxB_2, PGE_2, $PGF_{2\alpha}$, and PGD_2 was obtained. Each peak was sharp and symmetrical with little or no cross contamination. However, TxB_2 eluted as a symmetrical but broad peak.

After the initial isocratic period, a slightly convex gradient (No. 5 on the Waters 660 solvent programmer) was run from 26% to 80% acetonitrile in water over 12 min. This gradient was interrupted at 50% acetonitrile in water to allow the separate elution of PGA_2, HHT, and the various HETE isomers (Fig. 1). After 20 min, the gradient was resumed to allow

SUMMARY OF RETENTION TIMES FOR ARACHIDONIC ACID (AA) METABOLITES

AA metabolite	Retention[a,b] time (min)	Retention[a,c] time (min)
6-Keto-$PGF_{1\alpha}$	5	5
TxB_2	10.5	10.5
$PGF_{2\alpha}$	17	17
PGE_2	21	21
PGD_2	23.5	23.5
PGA_2	28	25
HHT	30	29
15-HETE	36	33
12-HETE	38	35
11-HETE	39	36
5-HETE	41	38
AA	57	43

[a] Retention time was obtained by elution of 5 mm C_{18} reversed-phase radial pack column under the following conditions at flow rate of 3 ml/min.

[b] Isocratic elution with 26% acetonitrile in water for 25 min followed by convex gradient to 80% ACN in water to 42 min.

[c] Isocratic elution with 26% acetonitrile in water for 25 min followed by isocratic elution of 50% acetonitrile in water for additional 20 min, followed by an isocratic elution in 100% acetonitrile for 10 min.

the elution of the relatively nonpolar AA. Upon completion of the run, the column was purged with 100% acetonitrile for 10 min to remove any residual material and then reequilibrated with 26% acetonitrile in water. The retention times from this HPLC system are summarized in the table.

The method as described above is capable of separating virtually all the AA metabolites (except leukotrienes) that are currently of major interest. However, many applications do not require the separation of some of these metabolites, and it is possible to modify the method to produce only the separations desired. The following are a few of the variations that have been useful. The separation of the PGs may be accomplished by using only the initial isocratic portion (26% acetonitrile in water) of the method. It is absolutely essential to purge the column with 100% acetonitrile after the 23-min isocratic run to remove the less polar material remaining on the column to prevent contamination of a following analysis.

Perhaps the most useful variation is one that separates 6-keto-$PGF_{1\alpha}$, TxB_2, other PGs, and HHT, but separates the HETEs as a single peak between HHT and AA. This is accomplished by omitting the 50% hold

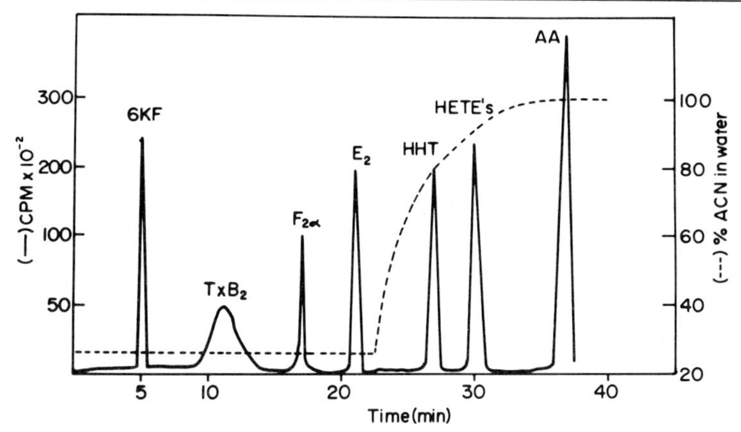

FIG. 2. Separation of prostaglandins from hydroxyeicosatetraenoic acids.

during the gradient portion of the run, allowing the gradient to proceed uninterrupted. Figure 2 demonstrates this variation, along with the corresponding effects on the elution of the compounds. This variation therefore permits separation of the individual metabolites of arachidonic acid produced only by PG synthase and significantly reduces the elution time.

The method described above requires the use of two pumps and a gradient programmer. It is possible, however, to use a variation of the method that does not require a second pump and gradient system. By using premixed water–acetonitrile solutions and a single multivalved inlet pump, one may switch to increasing concentrations of acetonitrile in a stepwise fashion. Figure 3 illustrates the separation of the arachidonic acid metabolites. The retention times were very similar to those obtained using the gradient method shown in the table. It should be noted that retention times vary somewhat among individual columns and between successive analyses on a single column. For these reasons, it is recommended that internal standards be added to unknown samples under analysis. For example, it is convenient to add a ^{3}H-labeled reference prostaglandin to a sample containing [^{14}C]prostaglandins and HFA formed from [^{14}C]arachidonic acid.

Remarks

Radiolabeled PGs and HFA were separated using a binary solvent system of acetonitrile and water at pH 3.5. The acidic environment suppresses the ionization of the fatty acid carboxyl group and allows migration based primarily on lipophilicity. This system permits complete

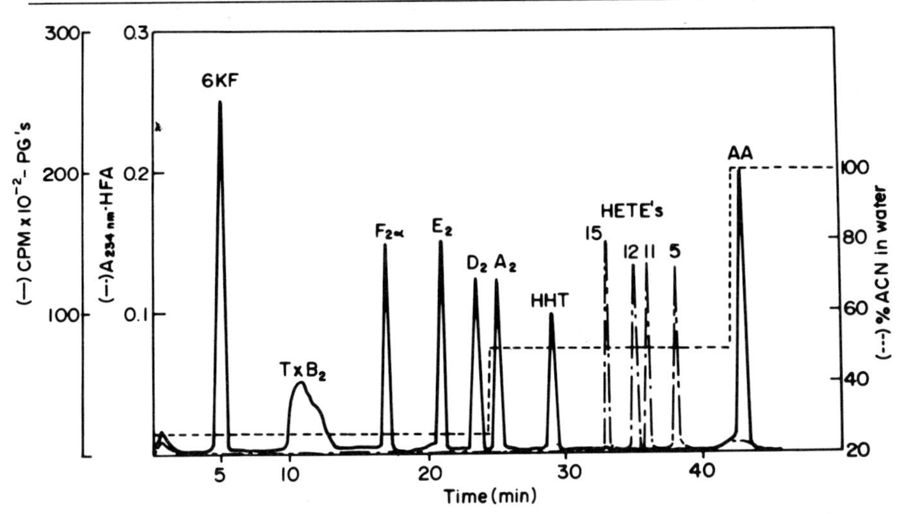

FIG. 3. Separation of prostaglandins and hydroxyeicosatetraenoic acids by step elution.

resolution of these lipids. The RCM 100 module allows for the use of fast flow rates, and thus shorter retention times, without sacrificing column efficiency as would occur with steel-jacketed columns. Under these conditions column efficiency for the individual primary prostaglandins ranged from 13,000 to 20,000 theoretical plates per meter, whereas thromboxane was eluted with an apparent efficiency of 1000 theoretical plates per meter. TxB_2 exists in equilibrium between an open and closed hemiacetal ring that could be responsible in part for these elution characteristics.

Quantitative determination of PG and HFA has presented a problem to many investigators. Biological assays are sensitive, but tedious, and generally a nonspecific biological response is observed. Gas-liquid chromotography–mass spectrometry (GC–MS) techniques are sensitive and reliable, but the cost is very high and therefore these techniques are restricted to a few research establishments. Radioimmunoassays are lower in cost, rapid, and possess the sensitivity and specificity to quantitate metabolites derivatized from endogenous arachidonic acid. However, antibodies to all the known arachidonic acid metabolites are not readily available, which thus limits the technique. Derivatized prostaglandins may be separated by HPLC and quantitated by spectroscopic methods.[7,8] How-

[7] F. A. Fitzpatrick, *Anal. Chem.* **48,** 499 (1976).

[8] M. V. Merritt and G. E. Bronson, *Anal. Biochem.* **80,** 392 (1977).

ever, derivatizations are laborious and may result in loss of PGs[9] and residual derivatization reagents can damage the column.

Many investigators, after using separation by TLC, quantitate arachidonic acid metabolites by the use of radioisotopes. These TLC methods suffer from low and variable recovery, which causes variable quantitation. Recovery of PGs or HFA from our HPLC system ranges from 75 to 85% and is very reproductible. This HPLC system is compatible with several types of detector systems. Some of the obvious advantages are that (*a*) sample preparation is simple; (*b*) the sample does not have to be derivatized or chemically altered; (*c*) the analysis is performed at ambient (25°) temperature, and thus thermally unstable compounds or intermediates of such compounds, e.g., hydroperoxy fatty acids, can be analyzed; (*d*) this HPLC system resolves closely related compounds, e.g., PGE_2 and PGD_2 or 12-HETE and 11-HETE; (*e*) fraction collection can be automated and later quantified by scintillation spectrometry, radioimmunoassay, bioassay, or mass spectrometry; (*f*) in experiments using radiolabeled arachidonic acid, the detection limit for prostaglandins by liquid scintillation spectrometry is better than 5×10^{-12} mol.

Summary

1. Total analysis time is short.
2. Reproducibility is ±0.5 min using the same column over a 3-month period for the primary prostaglandins.
3. The HPLC system is generally applicable to the analysis of all PG synthase and most lipoxygenase products and ensures optimization of the method for analysis of arachidonic acid metabolites.
4. The design, in the step isocratic mode, allows the utilization of a single pump with a three-way solvent selecting valve for the separation of these metabolites. This will significantly reduce the capital costs of acquiring such a system.

[9] J. Turk, S. J. Weiss, J. E. Davis, and P. Needleman, *Prostaglandins* **16,** 291 (1978).

[62] High-Pressure Liquid Chromatography of Underivatized Fatty Acids, Hydroxy Acids, and Prostanoids Having Different Chain Lengths and Double-Bond Positions

By MIKE VAN ROLLINS, MARTA I. AVELDAÑO, HOWARD W. SPRECHER, and LLOYD A. HORROCKS

Thin-layer chromatography (TLC) and high-pressure liquid chromatography (HPLC) methods are available that permit separation and direct visualization of submicrogram-to-milligram quantities of underivatized prostaglandins and lipoxygenase products.[1-4] The TLC procedure separates primarily on the basis of number of double bonds and hydrophilicity and is not affected by small differences in hydrophobicity. In contrast, the HPLC procedure separates fatty acids and metabolites that differ only in carbon number or in the number, position, or geometry of double bonds. HPLC may also be used to resolve and quantitate underivatized fatty acids obtained from natural mixtures, such as brain glycerophospholipids or hepatic triglycerides. HPLC is simple and highly reproducible, and the same column may be used for at least 400 runs with no evidence of deterioration.

HPLC of PGH Synthase and Lipoxygenase Products

The following separations are done using reversed-phase (C_{18}) columns and various mixtures of acetonitrile–aqueous phosphoric acid (pH 2). HPLC quality acetonitrile and phosphoric acid are purchased from MCB Manufacturing Chemists (Cincinnati, Ohio) or Burdick and Jackson Laboratories, Inc. (Muskegon, Michigan) and from Fisher Scientific Co. (Pittsburgh, Pennsylvania), respectively. Water is prepared by distillation followed by carbon adsorption–deionization–filtration using a Milli-Q system including an Organex-Q cartridge (Millipore Corporation, Beford, Massachusetts). This water is further purified with an Organic-pure water purifier (Barnstead, Boston). Just before use, all organic and aqueous solutions are filtered (0.5 μm FH and 0.22 μm GS, respectively;

[1] M. Van Rollins, S. H. K. Ho, J. E. Greenwald, M. S. Alexander, N. J. Dorman, L. K. Wong, and L. A. Horrocks, *Prog. Lipid. Res.* **20,** 783 (1981).

[2] M. Van Rollins, S. H. K. Ho, J. E. Greenwald, M. S. Alexander, N. J. Dorman, L. K. Wong, and L. A. Horrocks, *Prostaglandins* **20,** 571 (1980).

[3] A. Terragno, R. Rydzik, and N. A. Terragno, *Prostaglandins* **21,** 101 (1981).

[4] T. Eling, B. Tainer, A. Ally, and R. Warnock this volume [6].

ISBN 0-12-181986-8

Millipore). The columns used are either Ultrasphere-ODS (5 μm particles, 0.46 (i.d.) × 25 cm and 1 (i.d.) × 25 cm; Beckman, Irvine, California or Zorbax-ODS [6 μm particles, 0.46 (i.d.) × 25 and 0.46 (i.d.) × 15 cm; DuPont Co., Wilmington, Delaware]. Guard columns (0.46 (i.d.) × 4–5 cm or 1 (i.d.) × 5 cm) are prepared by slurry packing at 6000 psig the 5 μm particles of Spherisorb-ODS (Beckman) or Apex-ODS (Jones Chromatography, Columbus, Ohio) in acetone.

Chromatography is done using a Model 322MP system (Beckman), a column block heater (Jones Chromatography), and a variable-wavelength detector (LC-75, Perkin-Elmer, Norwalk, Connecticut) set at 192 nm. A microprocessor (Model 421, Beckman) is used to control the fraction collector (Instrumentation Specialties Co., Lincoln, Nebraska) and the gradients.

Arachidonic Acid as Precursor

Baseline separations of all major oxygenated metabolites (including $PGF_{2\alpha}$ and TxB_2) formed from arachidonic acid in platelets are readily achieved using a single ODS column (Figs. 1 and 2a). The superimposition of UV peaks from unlabeled standards with radioactive profiles facilitates identification of metabolites. Essentially all the radioactivity applied to the column is recovered[1] and, with baseline separations, accurate and precise radioactive distributions are determined.[1,2] Similar separations have been reported subsequently by others.[3] Thus the relationships between PGH synthase and lipoxygenase metabolites can be readily ascertained following one HPLC run. For example, the effects of PGH synthase inhibitors such as indomethacin,[1] purported lipoxygenase inhibitors (Figs. 2C, 2D), or combined PGH synthase–lipoxygenase inhibitors (Fig. 2B), can be assessed.[5] Alternatively, if only PGH synthase metabolites are being studied, a single isocratic, 16-min HPLC run will give baseline separations of all major prostaglandins including the positional isomers E_2 and D_2 (Fig. 3). The latter HPLC system is also advantageous in that radioimmunoassays may be applied directly to the aqueous eluent[6] for rapid, accurate quantitation.

Adrenic Acid (7,10,13,16-Docosatetraenoic Acid) as Precursor

Unlike TLC, HPLC can be used to separate prostanoids of varying chain lengths. Adrenic acid is a naturally occurring fatty acid produced by chain elongation of arachidonic acid. Thus adrenic acid is arachidonic

[5] T. E. Wilhelm, S. K. Sankarappa, M. VanRollins, and H. W. Sprecher, *Prostaglandins* **21**, 323 (1981).

[6] M. Van Rollins, unpublished results, 1981.

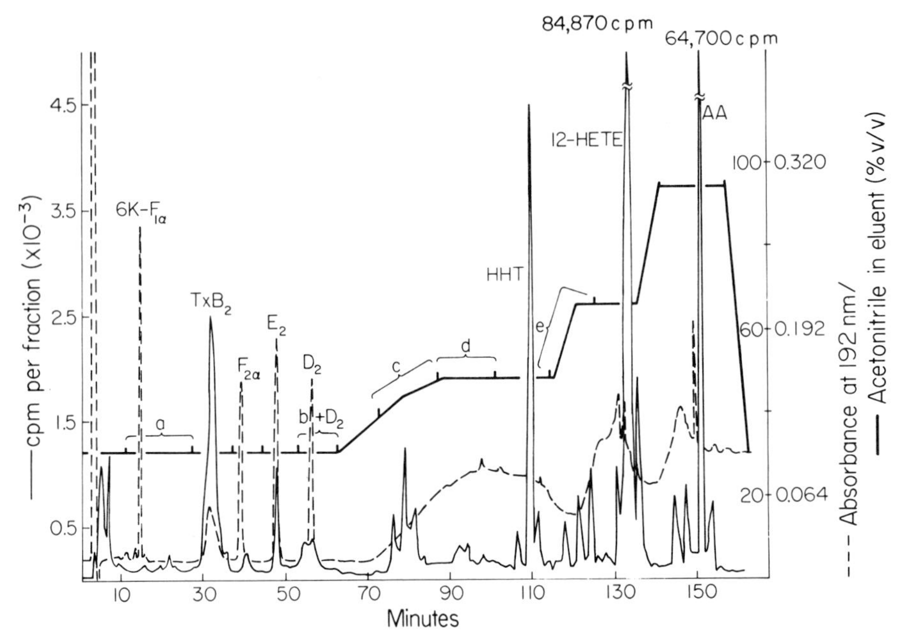

FIG. 1. HPLC of PGH synthase and lipoxygenase products from platelets. Metabolites from 410 ng of [1-^{14}C]arachidonic acid and 5 μg each of unlabeled prostaglandin standards were injected in 50 μl of methanol onto a Spherisorb-ODS guard column (0.46 × 4 cm) connected to an Ultrasphere-ODS column (0.46 × 25 cm). Elution was at 25° with a flow rate of 1 ml/min. For the initial 63 min after injection, the acetonitrile concentration (v/v) was 30.5%. Linear increases in acetonitrile concentration began at 63, 78, 114, and 134 min to 44% over 15 min, 48% over 10 min, 66% over 5 min, and to 95% over 5 min, respectively. Fractions were collected every 0.7 min (0.7 ml) and counted in minivials containing 4.4 ml of scintillation fluid. 6K-$F_{1\alpha}$, 6-ketoprostaglandin $F_{1\alpha}$; HHT, 12-hydroxyheptadecatrienoic acid; AA, arachidonic acid; HETE, 12-hydroxyeicosatetraenoic acid; TxB_2, thromboxane B_2. Reproduced from Van Rollins *et al.*[2]

acid with two methylene groups interposed at the carboxyl(α) end. Primary prostaglandins produced from adrenic acid can be separated from those produced from arachidonic acid with one exception: TxB_2 generated from arachidonic acid overlaps the prostacyclin metabolite obtained from adrenic acid (Fig. 4). This is not a problem, since the same tissue does not usually synthesize both prostacyclin and TxA_2 [it is believed that the same PGH synthase (cyclooxygenase) operates on adrenic acid as on arachidonic acid]. The simultaneous isolation of PGH synthase and lipoxygenase products from adrenic acid can also be done (Fig. 5). The structures of these prostanoids has been confirmed by GC–MS.[7] Thus

[7] H. W. Sprecher, M. Van Rollins, F. F. Sun, and P. Needleman, *J. Biol. Chem.*, in press.

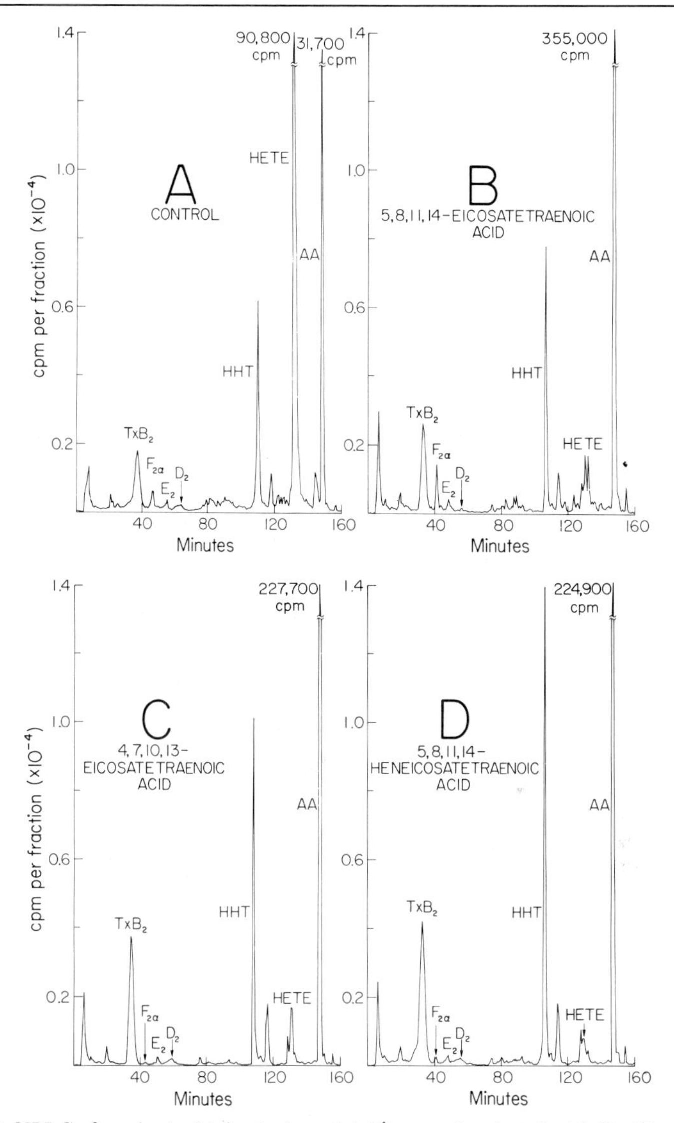

FIG. 2. HPLC of products obtained when platelets were incubated with (B–D) or without (A) acetylenic inhibitors. Washed human platelets (1.5×10^8) were incubated at 37° in the presence of [1-^{14}C]arachidonic acid (25 nmol). Some (2.5 nmol) or no acetylenic analogs of arachidonic acid were also present. Upon acidification and extraction with ethyl acetate, labeled metabolites were concentrated and resuspended in methanol. The HPLC conditions are the same as described in Fig. 1. Reproduced from Wilhelm *et al.*[5]

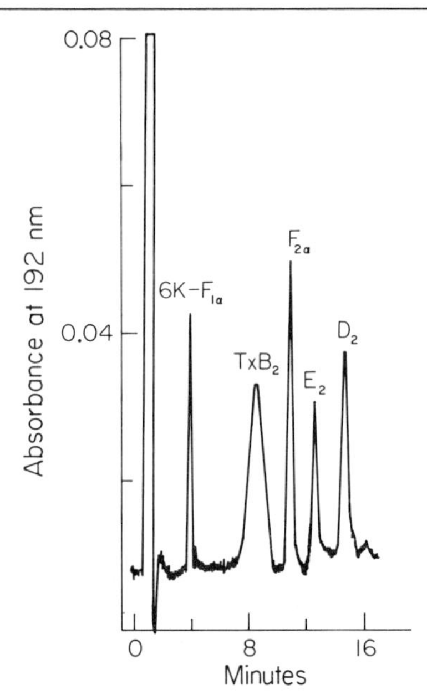

FIG. 3. Rapid isocratic separation of primary prostaglandins derived from arachidonic acid. Prostaglandin standards (less than 3 μg each) were injected in 20 μl of methanol onto a Zorbax-ODS column [0.46 (i.d.) × 25 cm] at 25°. Back pressure, 2000 psig; flow rate, 2 ml/min; acetonitrile concentration, 32%.

the present preparative procedure permits both the separation and the visualization of even-carbon isomers of prostanoids.

Odd-Carbon Chain Length Fatty Acids as Precursors

HPLC may also be used to resolve structures with very subtle differences in hydrophobicity, e.g., 19- or 21-carbon homologs. This is demonstrated in Fig. 6 for $PGF_{2\alpha}$ analogs, which are synthesized from 1-^{14}C-labeled 19:4, 20:4, or 21:4 by preparations of sheep vesicular gland in the presence of high amounts of ascorbic acid.[8] Thus, using ODS columns and various mixtures of acetonitrile–aqueous phosphoric acid, one may resolve either many metabolites with a wide range of polarities, e.g., from prostacyclin to free fatty acids, or a few metabolites with subtle differences in polarities (19-, 20-, 21-, or 22-carbon homologs).

[8] M. Van Rollins, L. LeDuc, M. I. Aveldaño, L. A. Horrocks, and H. W. Sprecher, in preparation.

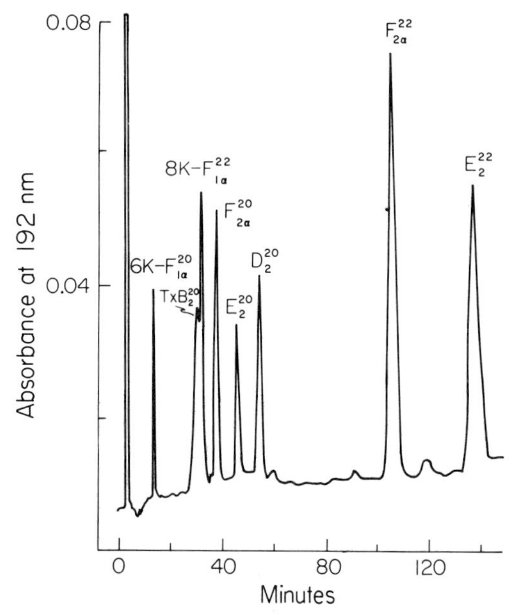

FIG. 4. Preparative HPLC of primary prostaglandins derived from arachidonic and adrenic acids. Prostaglandin standards were dissolved in 200 μl of methanol and injected onto an Apex-ODS guard column [0.46 (i.d.) × 4 cm] plus an Ultrasphere-ODS analytical column [1.0 (i.d.) × 25 cm] at 25°. Back pressure, 2310 psig; flow rate, 4.7 ml/min; acetonitrile concentration, 28%. Superscripts indicate 20- or 22-carbon homologs.

HPLC of Polyunsaturated Fatty Acids (PUFA)

The type and availability of PUFA substrates for PGH synthase and lipoxygenase enzymes are limiting factors in determining what oxygenated metabolites are found in different tissues. Positional isomers of arachidonic acid (20:4, *n*-6) are known to occur in mammalian tissues and are biosynthesized either by chain elongation of linolenic acid to form 20:4 (*n*-3) or by chain elongation plus desaturation of palmitoleic acid (16:1, *n*-7) to form 20:4 (*n*-7). Geometrical isomers of linoleic acid (18:2, *n*-6), the normal precursor of arachidonic acid, are also present in mammalian tissues and arise from the diet. Because of possible competition from other endogenous fatty acids, tracer studies are required to determine the extent to which such isomers are utilized by PGH synthase or lipoxygenases. Thus, it is desirable to be able to isolate and to quantitate the labeled precursors, substrates, and metabolites of PGH synthase and lipoxygenases.

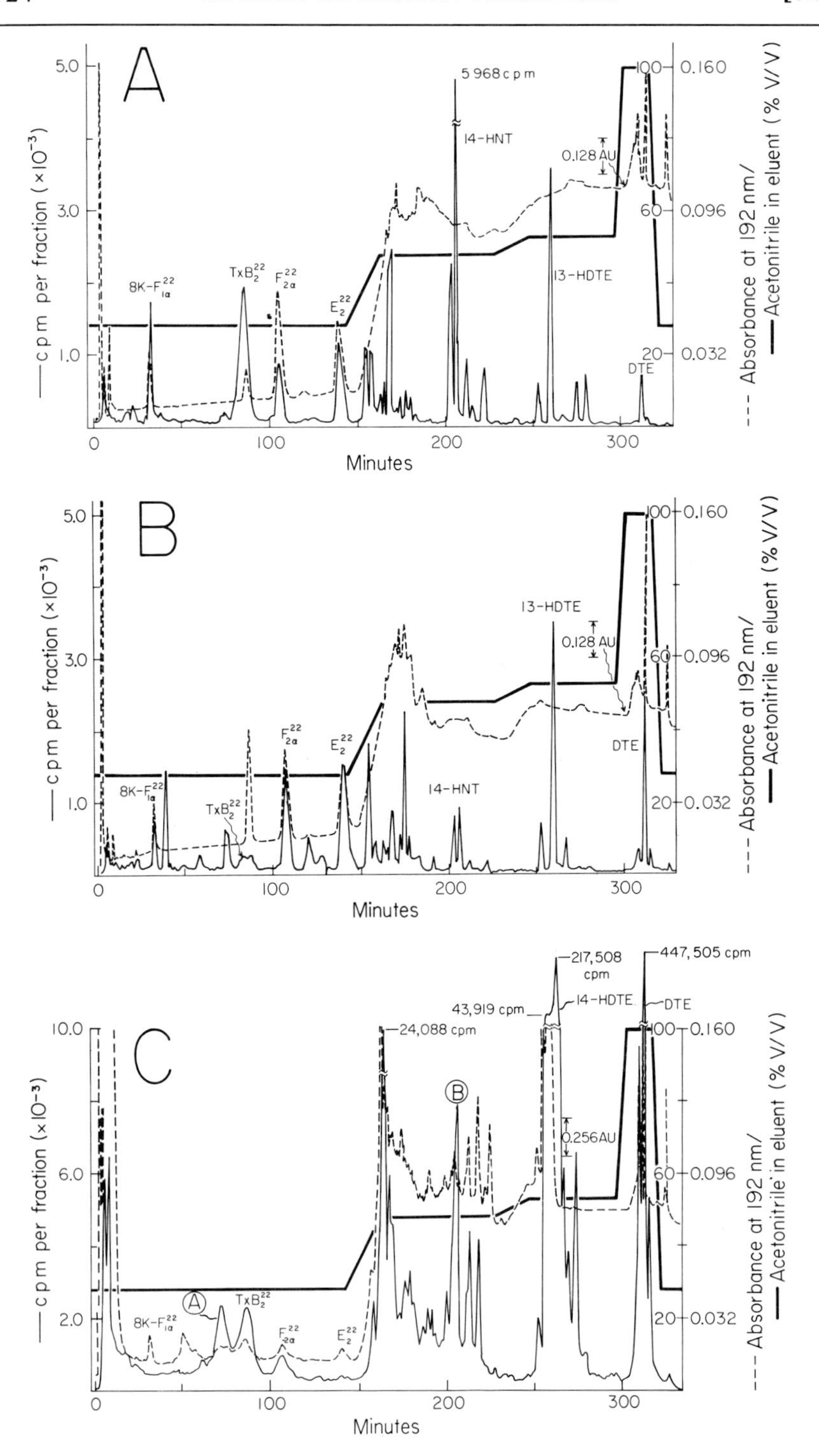
A
5 968 cpm
14-HNT
0.128 AU
13-HDTE
DTE
8K-F$_{1\alpha}^{22}$
TxB$_2^{22}$
F$_{2\alpha}^{22}$
E$_2^{22}$
— cpm per fraction (×10⁻³)
--- Absorbance at 192 nm/
— Acetonitrile in eluent (% V/V)
Minutes
B
13-HDTE
0.128 AU
DTE
14-HNT
C
217,508 cpm
447,505 cpm
43,919 cpm
14-HDTE
DTE
24,088 cpm
0.256AU

HPLC of Positional Isomers of Arachidonic Acid

Naturally occurring isomers (20:4, *n*-3 and 20:4, *n*-7) can be preparatively separated from arachidonic acid (20:4, *n*-6) (Fig. 7). Because these isomers are also directly visualized using UV light absorption, the mass may be calculated from their extinction coefficients, and their specific radioactivities may be directly determined within 60 min.[9]

HPLC of Geometrical Isomers of Arachidonic Acid Precursors

Naturally occurring geometrical isomers (9c,12t- or 9t,12c-18:2 and 9t,12t-18:2) can be separated from linoleic acid (9c,12c-18:2) (Fig. 8). The cis-trans and trans-cis isomers are not resolved from each other under these conditions.

HPLC of Natural Mixtures of Fatty Acids

After mild alkaline hydrolysis of glycerophospholipids or triacylglycerols, fatty acids may be extracted into hexane and resolved by HPLC (Fig. 9). These separations are achieved at greater acetonitrile concentrations (55–83%) than those used to resolve prostaglandins. At high acetonitrile concentrations (more than 80%), 22:6 (*n*-3) elutes before 18:3 (*n*-3), 20:4 (n-6) before 16:1, 22:5 (n-3) before 18:2, 22:4 (*n*-6) before

[9] M. I. Aveldaño de Caldironi, M. VanRollins, and L. A. Horrocks, *in* "Metabolism of Phospholipids in the Nervous System" (L. A. Horrocks, G. B. Ansell, and G. Porcellati, eds.), Raven, in press.

FIG. 5. Preparative HPLC of PGH synthase and lipoxygenase products of adrenic acid. [1-^{14}C]Adrenic acid was incubated with (A) medullary microsomes from hydronephrotic kidney; (B) the same as (A) but including the thromboxane synthase inhibitor imidazole; or (C) washed, human platelets. Products of labeled adrenic acid and at least 10 μg each of unlabeled standards were injected in 200 μl of methanol onto an Apex-ODS column [1.0 (i.d.) × 5 cm] connected to an Ultrasphere-ODS column [1 (i.d.) × 25 cm] at 23–25° The flow rate was 4.7 ml/min with an initial back pressure of 2300–2400 psig. The acetonitrile concentration was 28% for the first 145 min after injection. Linear increases in acetonitrile concentrations began at 145, 230, and 300 min to 48% over 20 min, 53% over 20 min, and 100% over 5 min, respectively. A fraction was collected every minute, and from each fraction a 1-ml aliquot was counted in 15 ml of scintillation fluid. Abbreviations: 8K-$F_{1\alpha}^{22}$, 1a,1b-dihomo-8-keto-$PGF_{1\alpha}$; TxB_2^{22}, 1a,1b-dihomothromboxane B_2; $F_{2\alpha}^{22}$, 1a,1b-dihomo-$PGF_{2\alpha}$; E_2^{22}, 1a,1b-dihomo-PGE_2; 14-HNT, 14-hydroxynonadecatrienoic acid; 13-HDTE, 13-hydroxydocosatetraenoic acid; 14-HDTE, 14-hydroxydocosatetraenoic acid; DTE, docosatetraenoic acid (adrenic acid); circled A, tentatively identified as trihydroxydocosatetraenoic acid; circled B, tentatively identified as 14-HNT.

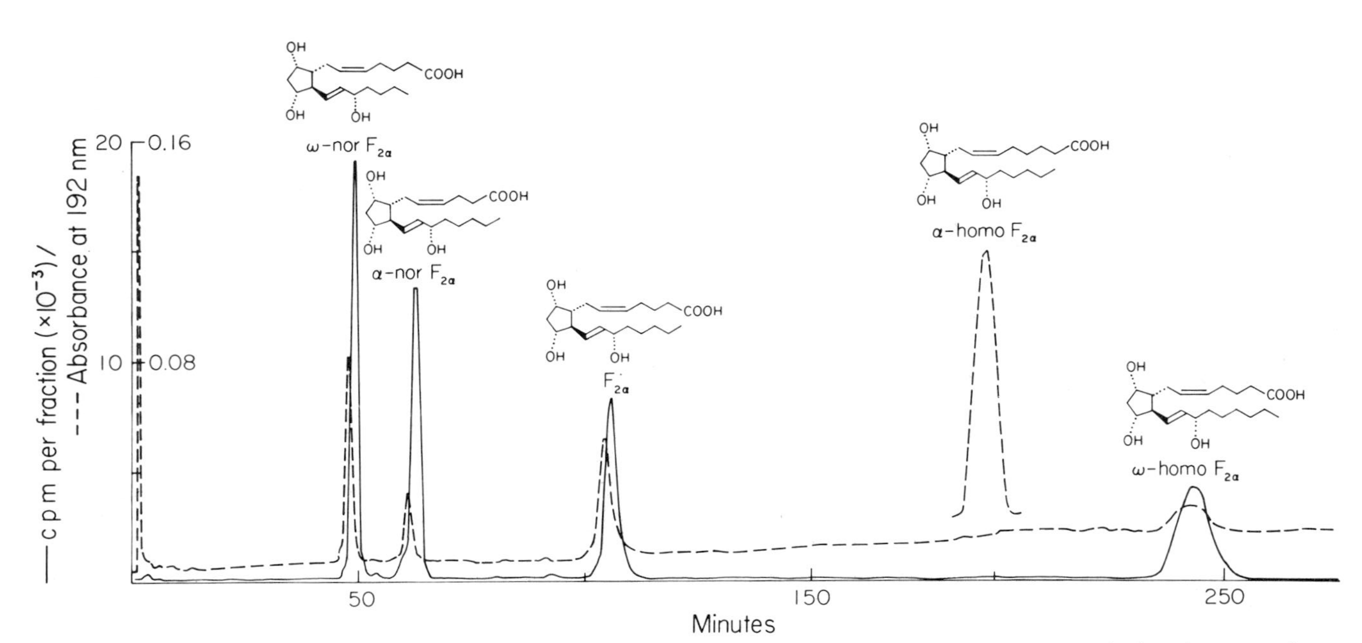

FIG. 6. HPLC of $F_{2\alpha}$ analogs of 19-, 20-, and 21-carbon chain lengths. $F_{2\alpha}$ analogs were synthesized by ram vesicular microsomes from [1-^{14}C]19:4 (*n*-6), 19:4 (*n*-5), 20:4 (*n*-6), 21:4 (*n*-7), or 21:4 (*n*-6). After being extracted into ethyl acetate and isolated by preparative HPLC, the $F_{2\alpha}$ analogs (5 μg each) were combined, dissolved in 20 μl of methanol, and injected onto two Zorbax ODS columns attached in series [0.46 (i.d.) × 25 cm and 0.46 (i.d.) × 15 cm]. Back pressure, 2980 psig at 35°; flow rate, 2 ml/min; acetonitrile concentration, 21%. Each fraction of 1 ml was collected and counted after addition of 5 ml of scintillation fluid. Note: The structure of α-homo $PGF_{2\alpha}$ (injected in a separate run) has not yet been confirmed by GC–MS.

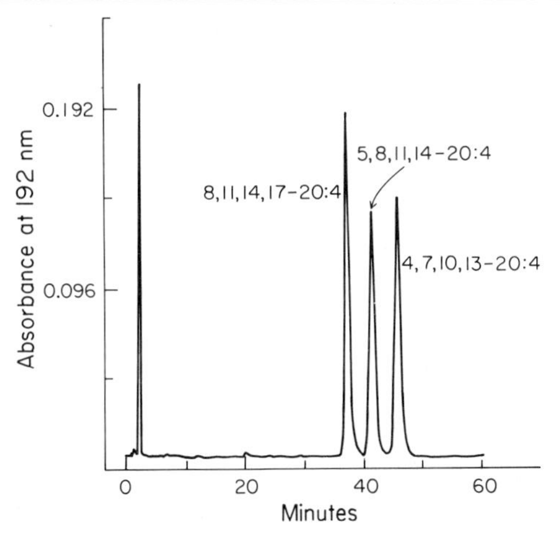

FIG. 7. HPLC of naturally occurring postitional isomers of arachidonic acid. Standards, dissolved in 20 μl of methanol, were injected onto Zorbax-ODS columns [0.46 (i.d.) × 25 cm plus 0.46 (i.d.) × 15 cm] at 35°. Back pressure, 2290 psig; flow rate, 2 ml/min; acetonitrile concentration, 58%.

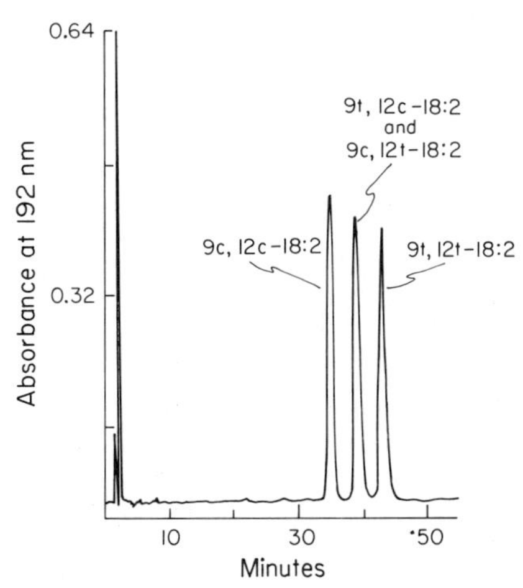

FIG. 8. HPLC of geometric isomers of linoleic acid. Standards (generously provided by Dr. E. A. Emken, Peoria, Illinois) dissolved in 20 μl of methanol were injected onto Zorbax-ODS columns [0.46 (i.d.) × 25 cm plus 0.46 (i.d.)] × 15 cm) at 35°. Back pressure, 2240 psig; flow rate, 2 ml/min; acetonitrile concentration, 60%.

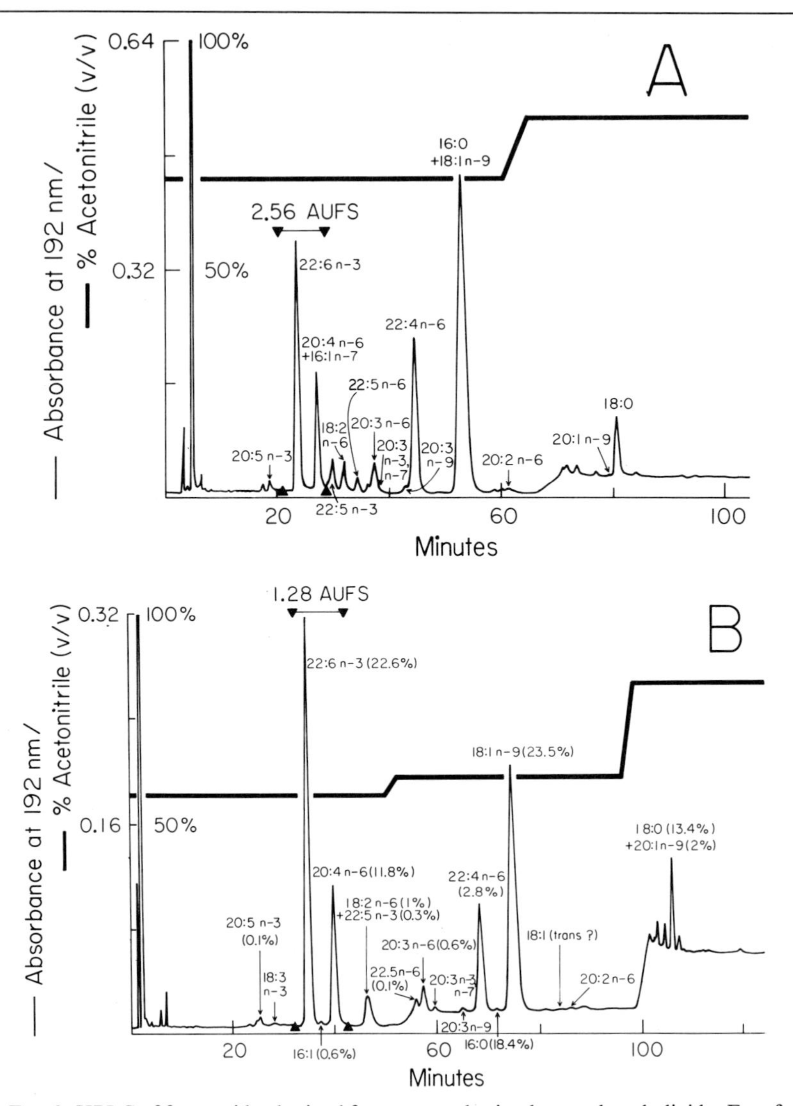

FIG. 9. HPLC of fatty acids obtained from mouse brain glycerophospholipids. Free fatty acids were prepared by saponification, extracted into hexane, concentrated under N_2, and suspended in methanol. Fatty acids (144 μg in 20 μl of methanol) were injected onto the column system described in Fig. 7. Two different programs were used: (A) 70% acetonitrile for 60 min, with an increase to 83% over the next 4 min; flow rate, 1 ml/min; (B) 57% acetonitrile for 50 min with an increase to 61% over the next 2 min and finally increasing at 96 min to 83% over 2 min; flow rate, 2 ml/min. Weight percentages were obtained by conversion of an aliquot of total free fatty acids to methyl esters and analysis by gas–liquid chromatography (GLC). Fatty acid identification was done by comparing retention times with those of reference standards and by GLC of the eluted peaks.

20:3 (*n*-6), 18:1 before 16:0, and 20:1 before 18:0. At lower concentrations of acetonitrile, the retention time for the initial component is increased more than that of the later member in each of the above pairs, and the order of elution is reversed. Because these selective shifts in PUFA may result in overlaps with other fatty acids, resolution is rigidly dependent upon exact acetonitrile concentrations. Optimal acetonitrile concentrations are different for Zorbax-ODS, Ultrasphere-ODS, and radial compression [1 (i.d.) × 10 cm, 5 μm particles, Waters] columns. However, the same elution pattern persists. Because of the above-mentioned changes in selectivity and the complexity of natural mixtures, it is necessary to establish separations at two different concentrations of acetonitrile. At 70% acetonitrile (Fig. 9A), the critical pairs 16:1–20:4 (*n*-6) and 16:0–18:1 are not resolved. Yet in the conditions shown in Fig. 9B, these fatty acids are resolved whereas 18:2–22:5 (*n*-3) and 18:0–20:1 are not.

Similar elution patterns are seen when the methyl ester derivatives, dissolved in acetonitrile, are injected onto ODS columns and eluted with acetonitrile–water mixtures (90–100%). This is useful, since the identification of fatty acids can be more firmly established by comparing retention times of unknowns with those of reference standards, both as free fatty acids and as methyl esters,[10] on the same ODS column.

A selective sensitivity in detecting endogenous PUFA at 192 nm is evident when weight percentages, determined after GLC, are compared for each peak, e.g., 22:6 (*n*-3) vs 18:1 in brain glycerophospholipids (Fig. 9). Most of this sensitivity is due to the high extinction coefficient (8000) for double bonds, which is additive per double bond. In contrast, 16:0 and 18:0 are visualized only at 0.02–0.04 AUFS settings, presumably because of the low extinction coefficient (less than 50) for carboxyl groups. It should also be noted that PUFA elute earlier than the less unsaturated homologs (in contrast to GLC), and therefore their peaks tend to be sharper and taller.

If standards are used to establish apparent extinction coefficients, the mass of each fatty acid may be determined at 192 nm. If fractions are collected for radioactivity measurements, the specific radioactivities, as well as the intermolecular distribution of radioactivity, can also be determined after the same HPLC run. In addition, the procedure may be used on a preparative scale. After elution, the fatty acids can be extracted from the phosphoric acid–acetonitrile solutions with hexane or dichloromethane, and subjected to further analysis, such as GC–MS, or to the Schmidt de-

[10] M. I. Aveldaño, M. Van Rollins, and L. A. Horrocks, in preparation.

carboxylation procedure to establish the intramolecular distribution of radioactivity.

Acknowledgments

This work was supported by Neuropathology Training Grant NS-07091 and Research Grants NS-08291, NS-10165, and NS-14381.

[63] Argentation–High-Pressure Liquid Chromatography of Prostaglandins and Monohydroxyeicosenoic Acids

By WILLIAM S. POWELL[1]

Arachidonic acid (20:4)[2] and 8,11,14-eicosatrienoic acid (20:3) are converted via lipoxygenases and prostaglandin endoperoxide synthase to a variety of prostaglandins and monohydroxy fatty acids in different tissues. Both normal-phase high-pressure liquid chromatography (HPLC) on silicic acid[3-5] and reversed-phase HPLC on octadecylsilyl silica[5-7] have proved to be very useful for the separation of these compounds. Each of these methods has some disadvantages, however. Normal-phase HPLC does not give very good separation of PGE_2 and 6-oxo-$PGF_{1\alpha}$. The latter compound gives rise to broad or multiple peaks owing to its existence in several isomeric forms.[3,8] Although reversed-phase HPLC is capable of resolving most prostaglandins, it necessitates the removal of aqueous solvents, which is time-consuming.

[1] The author is a scholar of the conseil de la recherche en santé du Québec.

[2] The abbreviations used are: 20:4, arachidonic acid; 20:3, 8,11,14-eisocatrienoic acid; 18:0, stearic acid; PG, prostaglandin; Tx, thromboxane; 5h-20:4, 5-hydroxy-6,8,11,14-eicosatetraenoic acid; 8h-20:4, 8-hydroxy-5,9,11,14-eicosatetraenoic acid; 9h-20:4, 9-hydroxy-5,7,11,14-eicosatetraenoic acid; 11h-20:4, 11-hydroxy-5,8,12,14-eicosatetraenoic acid; 12h-20:4, 12-hydroxy-5,8,10,14-eicosatetraenoic acid; 15h-20:4, 15-hydroxy-5,8,11,13-eicosatetraenoic acid; 15h-20:3, 15-hydroxy-8,11,13-eicosatrienoic acid; HPLC, high-pressure liquid chromatography; GC–MS, gas chromatography–mass spectrometry.

[3] W. S. Powell, *Prostaglandins* **20,** 947 (1980).

[4] J. M. Boeynaems, A. R. Brash, J. A. Oates, and W. C. Hubbard, *Anal. Biochem.* **104,** 259 (1980).

[5] A. R. Whorton, K. Carr, M. Smigel, L. Walker, K. Ellis, and J. A. Oates, *J. Chromatogr.* **163,** 64 (1979).

[6] M. Van Rollins, S. H. K. Ho, J. E. Greenwald, M. Alexander, N. J. Dorman, L. K. Wong, and L. A. Horrocks, *Prostaglandins* **20,** 571 (1980).

[7] I. Alam, K. Ohachi, and L. Levine, *Anal. Biochem.* **93,** 339 (1979).

[8] W. S. Powell and S. Solomon, *Adv. Prostaglandin Thromboxane Res.* **4,** 61 (1978).

ISBN 0-12-181986-8

Another method for the separation of prostaglandins is argentation chromatography. Silicic acid impregnated with silver nitrate has been used to separate prostaglandins of the 1, 2, and 3 series by thin-layer chromatography (TLC).[9] Argentation TLC can also be used to resolve metabolites of 20:4 produced by platelets.[10] A silver ion-loaded cation exchange column was reported to separate mixtures of the *p*-nitrophenacyl derivatives of prostaglandins A_2, B_2, E_1, E_2, $F_{1\alpha}$, and $F_{2\alpha}$ and some of their synthetic isomers and analogs.[11] We have used a similar column to separate arachidonic acid and 8,11,14-eicosatrienoic acid and some of their prostaglandin and monohydroxyeicosenoic acid metabolites.[12]

Standards for HPLC

[5,6,8,9,12,14,15-^{3}H]PGD_2 and 6-oxo[5,8,9,11,12,14,15-^{3}H]$PGF_{1\alpha}$ were purchased from New England Nuclear, Lachine, Quebec. [5,6-^{3}H]PGE_1, [5,6,8,11,12,14,15-^{3}H]PGE_2, [1-^{14}C]PGE_2, 13,14-dihydro-15-oxo-[5,6,8,11,12,14-^{3}H]$PGF_{2\alpha}$, [1-^{14}C]20:3, [1-^{14}C]20:4, [1-^{14}C]stearic acid ([1-^{14}C]18:0), and sodium boro[^{3}H]hydride were purchased from Amersham/Searle, Oakville, Ontario. [9β-^{3}H]$PGF_{1\alpha}$ and [9β-^{3}H]$PGF_{2\alpha}$ were synthesized by reduction of PGE_1 and PGE_2 with sodium boro[^{3}H]hydride as described in the literature.[13] 6-oxo[1-^{14}C]$PGF_{1\alpha}$,[8] [1-^{14}C]TxB_2[14] 5-hydroxy-6,8,11,14-[1-^{14}C]eicosatetraenoic acid (5h-[1-^{14}C]20:4),[4] 8-hydroxy-5,9,11,14-[1-^{14}C]eicosatetraenoic acid (8h-[1-^{14}C]20:4),[4] 9-hydroxy-5,7,11,14-[1-^{14}C]eicosatetraenoic acid (9h-[1-^{14}C]20:4),[4] 11-hydroxy-5,8,12,14-[1-^{14}C]eicosatetraenoic acid (11h-[1-^{14}C]20:4),[4] 12-hydroxy-5,8,10,14-[1-^{14}C]eicosatetraenoic acid (12h-[1-^{14}C]20:4)[14] and 15-hydroxy-5,8,11,13-[1-^{14}C]eicosatetraenoic acid (15h-[1-^{14}C]20:4)[15] were synthesized from [1-^{14}C]20:4, whereas 15-hydroxy-8,11,13-[1-^{14}C]eicosatrienoic acid (15h-[1-^{14}C]20:3)[15] was synthesized from [1-^{14}C]20:3 as described in the literature. The identities of the monohydroxyeisocatetraenoic acid derivatives were confirmed by gas chromatography–mass spectrometry (GC–MS).

[9] K. Gréen and B. Samuelsson, *J. Lipid Res.* **5,** 117 (1964).
[10] J. E. Greenwald, M. S. Alexander, M. VanRollins, L. K. Wong, and J. R. Bianchine, *Prostaglandins* **21,** 33 (1981).
[11] M. V. Merritt and G. E. Bronson, *Anal. Biochem.* **80,** 392 (1977).
[12] W. S. Powell, *Anal. Biochem.* **115,** 267 (1981).
[13] E. Granström and B. Samuelsson, *Eur. J. Biochem.* **10,** 411 (1969).
[14] M. Hamberg and B. Samuelsson, *Proc. Natl. Acad. Sci. U.S.A.* **71,** 3400 (1974).
[15] M. Hamberg and B. Samuelsson, *J. Biol. Chem.* **242,** 5329 (1967).

Equipment

Except as noted below, HPLC was carried out using a system from Waters Associates, Milford, Massachusetts. This consisted of two Model 6000 A solvent delivery systems, a Model 660 solvent programmer, and a Model U6K injector. The flow rate was 1.5 ml/min. In some cases (see Tables II and III) HPLC was carried out using a Milton Roy minipump equipped with a Valco injector. The flow rate was 2 ml/min. Radioactivity was detected in column eluates with a 2-channel radiocolumn monitor from Laboratorium Prof. Dr. Berthold, Wildbad, F.R.G. This enables detection of tritium and ^{14}C independently of one another.

Stationary Phase

The stationary phase for argentation HPLC was a cation exchange column (35×0.46 cm) of RSil CAT (particle size, 5 μm) which was obtained from Alltech Associates, Deerfield, Illinois. It consisted of silicic acid that had been treated to give a phenylsulfonic acid derivative. The column is supplied in the Na^+ form and must be treated with $AgNO_3$ to convert it to the Ag^+ form. It is first eluted with water (200 ml) to remove all the buffer. The following solvents are then pumped through the column: 1 *M* $AgNO_3$ (150 ml), water (300 ml or until the eluate is free of silver ions), methanol (150 ml), acetone (100 ml), ethyl acetate (100 ml), chloroform (100 ml), and hexane (100 ml). When not in use, the column is kept in hexane. One column gives fairly reproducible results over a long period of time. Although we do not use it continuously, our present column has been in use for the past year. Retention times tend to go down slightly with time, but this can be compensated for by using a slightly weaker solvent. At a flow rate of 2 ml/min, the back pressure is normally ca. 3000 psi. When the back pressure becomes much higher than this due to the accumulation of various impurities, the column can be washed with methanol–acetonitrile (1:1). Since neither methanol nor acetonitrile is miscible with hexane, when these solvents are used it is necessary to elute with an intermediate solvent, such as chloroform, before returning to hexane. The stationary phase can be reconditioned after washing with 100 ml of methanol and 100 ml of water and then 150 ml of 1 *M* silver nitrate as described above.

Mobile Phase

In argentation chromatography on cation exchange columns the main characteristics of the stationary phase that are responsible for the retention of solutes are the presence of (*a*) silver ions and (*b*) polar groups. Olefinic double bonds of the solute will interact with the silver ions of the

TABLE I
SOLVENT SYSTEMS FOR ARGENTATION HPLC

Solvent	Composition
A	Chloroform–methanol–acetic acid (79.5:20:0.5)
B	Chloroform–acetonitrile–acetic acid (79.5:20:0.5)
C	Chloroform–methanol–acetic acid (89.5:10:0.5)
D	Chloroform–acetonitrile–acetic acid (89.5:10:0.5)
E	Chloroform–methanol–acetonitrile–acetic acid (79.5:18:2:0.5)
F	Hexane–chloroform–acetonitrile–acetic acid (80:17.5:2:0.5)

stationary phase, whereas polar groups of the solute will interact with polar groups of the stationary phase, probably by hydrogen bonding to the sulfonate groups. The retention times of solutes depend largely on the summation of these two types of interactions. The relative contribution of each can be controlled by modifying the mobile phase in an appropriate manner. Acetonitrile interacts with the silver ions of the stationary phase,[11,16] and its addition to the mobile phase results in a reduction in the contribution of olefin-Ag^+ interactions to the retention time. Methanol, on the other hand, interacts with polar groups of the stationary phase and reduces the contribution of polar interactions to the retention of solutes. Most of the solvents described here contain various proportions of methanol and acetonitrile in chloroform : acetic acid or hexane : chloroform : acetic acid (Table I).

Figure 1 illustrates the effects on the retention times of 12h-20:4, PGE_2, $PGF_{1\alpha}$, and $PGF_{2\alpha}$ of altering the relative amounts of acetonitrile and methanol in the mobile phase. Various mixtures of solvents A and B containing 79.5% chloroform and 0.5% acetic acid were used to give different ratios of methanol to acetonitrile. When a low ratio of methanol to acetonitrile is used (Fig. 1A) the retention times of the solutes tested are proportional to their polarity (i.e., 12h-20:4 $<$ PGE_2 $<$ PGF_α). Prostaglandins of the 1 and 2 series are separated, but their retention times do not greatly differ. The major factor responsible for the retention of these solutes under these conditions thus appears to be polar–polar interactions between the solute and the stationary phase. At an intermediate ratio of acetonitrile to methanol (Fig. 1B), the retention times of all four solutes are very low owing to the suppression of both polar–polar and olefin–Ag^+ interactions by the high concentrations of the above two solvents. At a high ratio of methanol to acetonitrile (Fig. 1C), the concentration of acetonitrile is no longer sufficiently high to suppress olefin–Ag^+ interactions, which now make a large contribution to the retention times. Thus, 12h-

[16] M. Özcimder and W. E. Hammers, *J. Chromatogr.* **187,** 307 (1980).

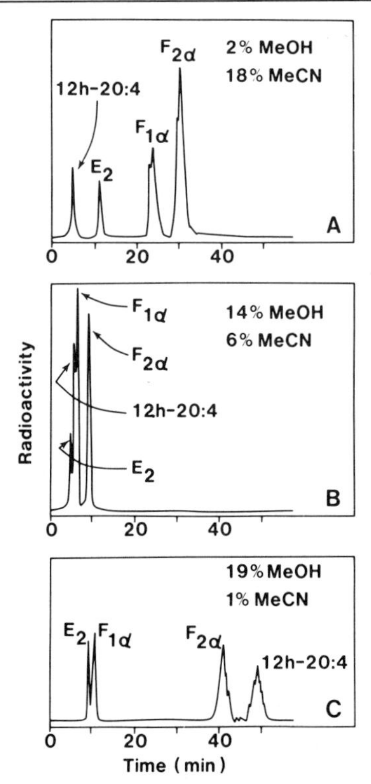

FIG. 1. High-pressure liquid radiochromatograms of 12h-[1-^{14}C]20:4, [1-^{14}C]PGE_2, [9β-^{3}H]$PGF_{1\alpha}$, and [9β-^{3}H]$PGF_{2\alpha}$ on a silver ion-loaded cation exchange column using various proportions of solvents A [$CHCl_3$:MeOH:HOAc (79.5:20:0.5)] and B [$CHCl_3$:MeCN:HOAc (79.5:20:0.5)] as mobile phases. The flow rate was 1.5 ml/min. (A) 10% solvent A, 90% solvent B; (B) 70% solvent A, 30% solvent B; (C) 95% solvent A, 5% solvent B.

20:4, by far the least polar of the four compounds in Fig. 1, now has the longest retention time because it has the greatest number of double bonds. $PGF_{1\alpha}$ and $PGF_{2\alpha}$, which differ from one another by the presence or absence of a 5,6-cis double bond, are very well separated in Fig. 1C. On the other hand, PGE_2 and $PGF_{1\alpha}$, which differ markedly in polarity, are not completely resolved using these conditions.

Figure 2 shows the retention times of PGs $F_{1\alpha}$, $F_{2\alpha}$, E_1, and E_2, 12h-20:4, and 18:0 as a function of the ratio of solvents A and B in the mobile phase. 18:0 is not retained by the stationary phase under these conditions, and its retention time is determined by the volume of mobile phase between the injector and the detector. At high concentrations of solvent B, the order of retention times is determined mainly by the polarity of the

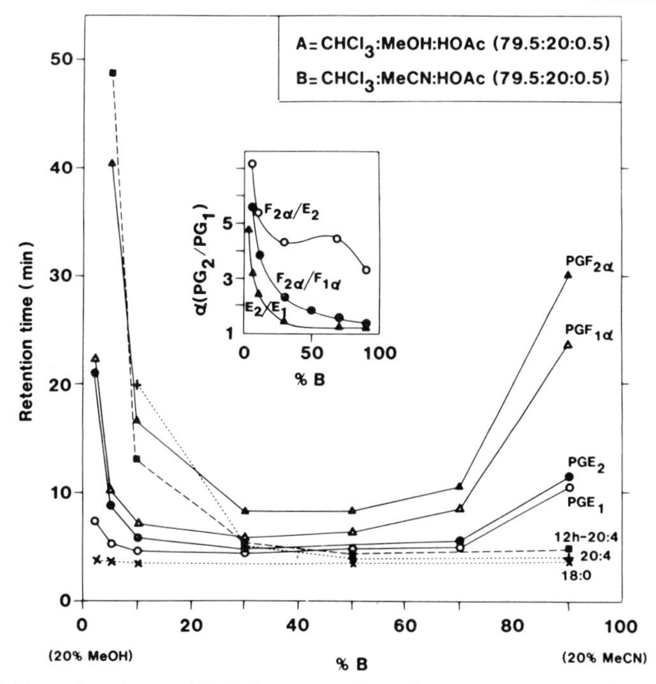

FIG. 2. Retention times of [1-^{14}C]18:0 (×), [1-^{14}C]20:4 (+), 12h-[1-^{14}C]20:4 (■), [5,6-^{3}H]PGE$_1$ (○), [1-^{14}C]PGE$_2$ (●), [9β-^{3}H]PGF$_1$α (△), and [9β-^{3}H]PGF$_2$α (▲) on a silver ion-loaded cation exchange column with various mixtures of solvents A and B (see Table I) as mobile phases. The flow rate was 1.5 ml/min. The inset shows the relative retentions ($\alpha = k_2'/k_1'$) for [5,6-^{3}H]PGE$_1$ and [1-^{14}C]PGE$_2$ (▲), [9β-^{3}H]PGF$_{1\alpha}$ and [9β-^{3}H]PGF$_{2\alpha}$ (●), and [1-^{14}C]PGE$_2$ and [9β-^{3}H]PGF$_{2\alpha}$ (○) as a function of the composition of the mobile phase. Taken from Powell.[12]

solute as discussed above (i.e., fatty acids < monohydroxy fatty acids < PGE < PGF).

As the concentration of solvent B is lowered, the number of olefinic double bonds in the solute become increasingly important in determining the retention time. This is illustrated by the inset to Fig. 2, which shows the relative retentions (α)[17] of various pairs of PGs as a function of the composition of the mobile phase. The values of α for the pairs PGF$_{2\alpha}$/PGF$_{1\alpha}$ and PGE$_2$/PGE$_1$ are maximal at low concentrations of solvent B in solvent A. Polar interactions also appear to be important under these conditions, however, since the maximum value of α for the pair

[17] $\alpha = k'PG_2/k'PG_1 = (t_{PG2} - t_M)/(t_{PG1} - t_M)$, where t_{PG2} and t_{PG1} are the retention times of different PGs and t_M is the retention time of 18:0 (i.e., the time taken by the mobile phase to pass through the system).

$PGF_{2\alpha}/PGE_2$ occurs at the lowest concentration of solvent B. Similar results are obtained for $PGF_{1\alpha}$ and PGE_1. It is possible that interaction between the silver ions of the stationary phase and the 13,14-double bonds of F prostaglandins could increase the proximity of their 9α-hydroxyl groups to the polar sulfonate groups of the stationary phase, resulting in increased retention times. Reversal of the stereochemistry about the 9-carbon atom (i.e., $PGF_{2\beta}$ derivatives) results in a considerable shortening of the retention time with solvents favoring olefin–Ag^+ interactions (see this volume [28]).

At low concentrations of solvent B, 20:4 and 12h-20:4 have retention times longer than those of PGs due to their greater numbers of double bonds. It is interesting that 12h-20:4, in spite of its greater polarity, has a shorter retention time than that of 20:4. This can be explained by the fact that 12h-20:4 has only 2 isolated cis double bonds as well as cis-trans pair of conjugated double bonds. The retention times of solutes with trans double bonds have been reported to be shorter than those of the corresponding cis isomers on argentation HPLC.[11,18] It is also possible that conjugated double bonds do not interact as strongly with the stationary phase as unconjugated double bonds. Low concentrations of solvent B in solvent A can also be used for the separation of other monohydroxy metabolites of 20:4 (Table II). A mixture of 5h-20:4, 11h-20:4, 12h-20:4, and 15h-20:4 can be completely resolved using this system; 8h-20:4 and 9h-20:4 are also separated, but the latter is not well resolved from 12h-20:4.

The retention times of a series of PGs using mixtures of solvents C and D, which contain a total of 10% of methanol and acetonitrile instead of 20% as in solvents A and B are shown in Fig. 4A (See later). This results in increased retention times for all the solutes tested with the exception of 18:0. Although the order of retention times of most PGs ($PGE_1 < $ 6-oxo-$PGF_{1\alpha} < PGE_2 < PGF_{2\alpha}$) is the same at low and high concentrations of solvent D in solvent C, it is apparent that separation of PGs of the 1 and 2 series is much better at low concentrations of solvent D. An exception is PGD_2, which, at high concentrations of solvent D, has the shortest retention time of all the PGs tested, as it does in normal phase chromatography with silicic acid. At low concentrations of solvent D, however, the retention time of PGD_2 is higher than those of PGE_1, PGE_2 and 6-oxo-$PGF_{1\alpha}$. This could be the result of a cooperative effect between the 13,14-double bond and the 9α-hydroxyl group of PGD_2 as discussed above for $PGF_{1\alpha}$ and $PGF_{2\alpha}$.

The interaction of solutes with the stationary phase in argentation

[18] S. Lam and E. Grushka, *J. Chromatogr. Sci.* **15**, 234 (1977).

TABLE II
RETENTION TIMES (t_R) OF SOME MONOHYDROXYEICOSATETRAENOIC ACIDS WITH A MOBILE PHASE CONSISTING OF CHLOROFORM–METHANOL–ACETONITRILE–ACETIC ACID (79.5:19.2:0.8:0.5)[a,b]

Compound	t_R (min)
5h-20:4	21.0
8h-20:4	35.3
9h-20:4	33.2
11h-20:4	25.6
12h-20:4	32.0
15h-20:4	14.4

[a] i.e., 4% solvent B in solvent A.

[b] HPLC was carried out with a Milton Roy minipump equipped with a Valco injector. The flow rate was 2 ml/min instead of 1.5 ml/min as used elsewhere. The silver ion-loaded cation exchange column used in this experiment was not the same as that used for the other experiments described in this chapter, with the exception of that described in Table III. The retention times were slightly shorter and the back pressure slightly lower with the new column, but otherwise the stationary phase behaved in the same manner as the previous one.

HPLC is very sensitive to the nature of the double bond. As noted above, *cis*-olefins appear to interact much more strongly with silver ions than *trans*-olefins. Substitution of tritium atoms for olefinic protium atoms also affects the retention of solutes, presumably because the carbon-tritium bond is shorter than the carbon-protium bond. This results in longer retention times for solutes with olefinic tritium atoms, especially with solvents that favor Ag^+-olefin interactions. This is illustrated in the inset to Fig. 3A, which shows the relative retentions of (*a*) [5,6,8,11,12,14,15-^{3}H]PGE$_2$ and [1-^{14}C]PGE$_2$ and (*b*) 6-oxo-[5,8,9,11,12,14,15-^{3}H]PGF$_{1\alpha}$ and 6-oxo-[1-^{14}C]PGF$_{1\alpha}$. This effect is more pronounced with [^{3}H]PGE$_2$, which has three olefinic tritium atoms, than with 6-oxo-[^{3}H]PGF$_{1\alpha}$, which has only one olefinic tritium atom. The greatest separation of [^{3}H]PGE$_2$ and [^{14}C]PGE$_2$ occurs at the lowest concentration of solvent D tested (i.e., 5%). This mobile phase can completely resolve mixtures of these two isomers. Similar results were obtained with [5,6,8,9,12,14,15-^{3}H]PGD$_2$ and

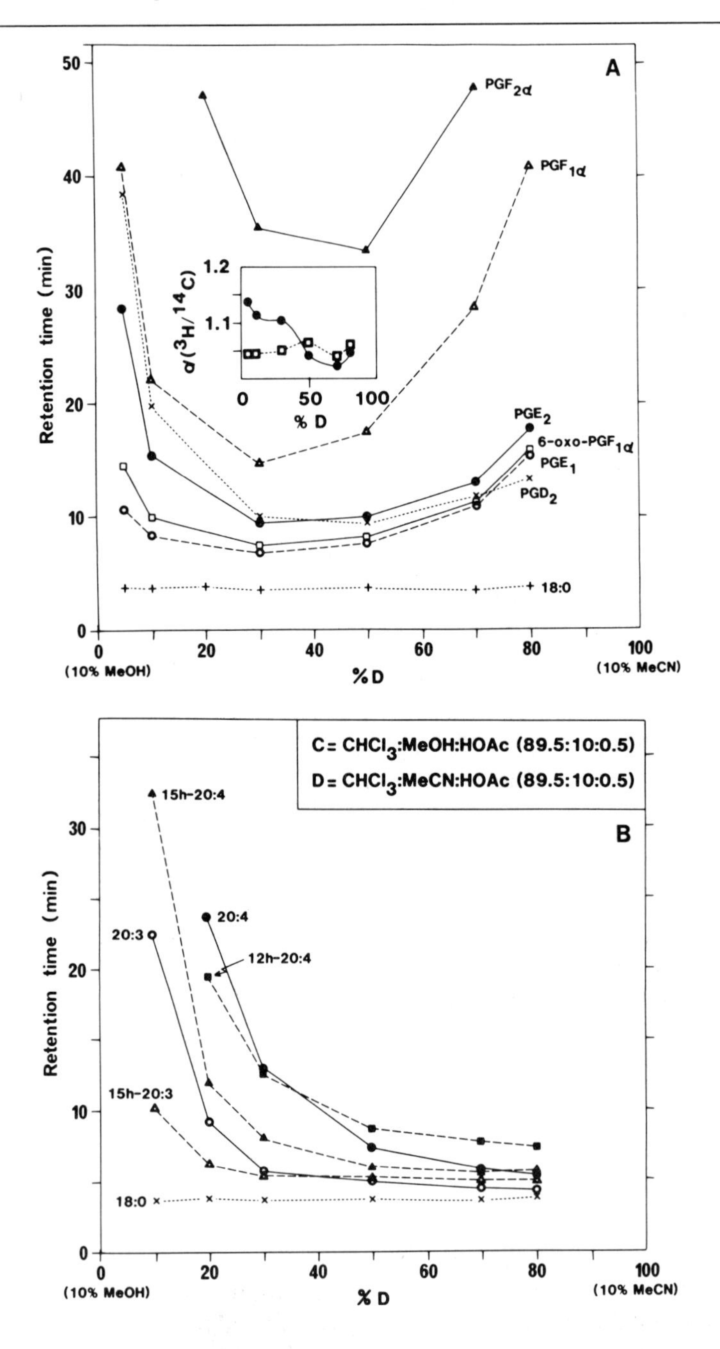
A
B
Retention time (min)
%D
(10% MeOH)
(10% MeCN)
$PGF_{2\alpha}$
$PGF_{1\alpha}$
PGE_2
6-oxo-$PGF_{1\alpha}$
PGE_1
PGD_2
18:0
$\alpha(^3H/^{14}C)$
% D
C= $CHCl_3$:MeOH:HOAc (89.5:10:0.5)
D= $CHCl_3$:MeCN:HOAc (89.5:10:0.5)
15h-20:4
20:4
20:3
12h-20:4
15h-20:3

[1-^{14}C]PGD$_2$. [5,6,8,9,11,12,14,15-^{3}H]20:4 and [1-^{14}C]20:4 can also be separated using argentation HPLC.

The ability of argentation HPLC to separate tritium- (and deuterium) labeled compounds from the corresponding natural isomers has both advantages and disadvantages. This method can be used to enhance the isotopic purity of compounds labeled with olefinic tritum or deuterium atoms (e.g., various metabolities of [5,6,8,9,11,12,14,15-^{2}H]20:4 which are used as internal standards in analysis by GC–MS). On the other hand, separation of labeled and unlabeled compounds could be a problem if compounds with olefinic tritium or deuterium atoms are used as internal standards to monitor recovery. This should not be a problem with compounds labeled on nonolefinic carbons (e.g., [9β-^{3}H]PGs, [5,6-^{3}H]PGE$_1$, and [3,3,4,4-^{2}H]PGs). The separation of labeled and unlabeled compounds can be minimized by increasing the amount of acetonitrile in the solvent (Fig. 3A, inset) or by reducing the polarity of the solvent (see below).

Mixtures of solvents C and D can also be used to separate 20:3 and 20:4 and some of their monohydroxyicosenoic acid metabolites (Fig. 3B). The longest retention times for these compounds occur with low concentrations of solvent D. The retention times of 20:3 and 20:4 under these conditions are longer than those of their monohydroxy metabolites, in agreement with the results obtained with mobile phases consisting of mixtures of solvents A and B (Fig. 2).

From the results described above, it is apparent that the degree of interaction between olefinic double bonds and the stationary phase can be increased by lowering the concentration of acetonitrile in the mobile phase. This can also be accomplished by increasing the polarity of the mobile phase while keeping the concentration of acetonitrile constant. This is illustrated in Fig. 4, which shows the retention times of a number of compounds using various mixtures of solvents E and F. The concentrations of acetonitrile (2%) and acetic acid (0.5%) are kept constant, while the polarity of the mobile phase is changed by going from methanol (18%) and chloroform (79.5%) to chloroform (17.5%) and hexane (80%). It should be noted that 0% F in Fig. 4 corresponds to 10% B in Fig. 2. In the presence of a less polar mobile phase, acetonitrile competes much more effectively

FIG. 3. Retention times of (A) [1-^{14}C]18:0 (+), [5,6,8,9,12,14,15-^{3}H]PGD$_2$ (×), [5,6-^{3}H]PGE$_1$ (○), [1-^{14}C]PGE$_2$ (●), [9β-^{3}H]PGF$_{1\alpha}$ (△), [9β-^{3}H]PGF$_{2\alpha}$ (▲) and 6-oxo-[1-^{14}C]PGF$_{1\alpha}$ (□); (B) [1-^{14}C]18:0 (×), [1-^{14}C]20:3 (○), [1-^{14}C]20:4 (●), 15h-[1-^{14}C]20:3 (△), 15h-[1-^{14}C]20:4 (▲), and 12h-[1-^{14}C]20:4 (■) on a silver ion-loaded cation exchange column with various mixtures of solvents C and D (see Table I) as mobile phases. The flow rate was 1.5 ml/min. The inset of Fig. 3A shows the relative retentions [$\alpha = k'(^3H)/k'(^{14}C)$] for [5,6,8,11,12,14,15-^{3}H]PGE$_2$ and [1-^{14}C]PGE$_2$ (●) and 6-oxo[5,8,9,11,12,14,15-^{3}H]PGF$_{1\alpha}$ and 6-oxo-[1-^{14}C]PGF$_{1\alpha}$ (□) as a function of the composition of the mobile phase. Taken from Powell.[12]

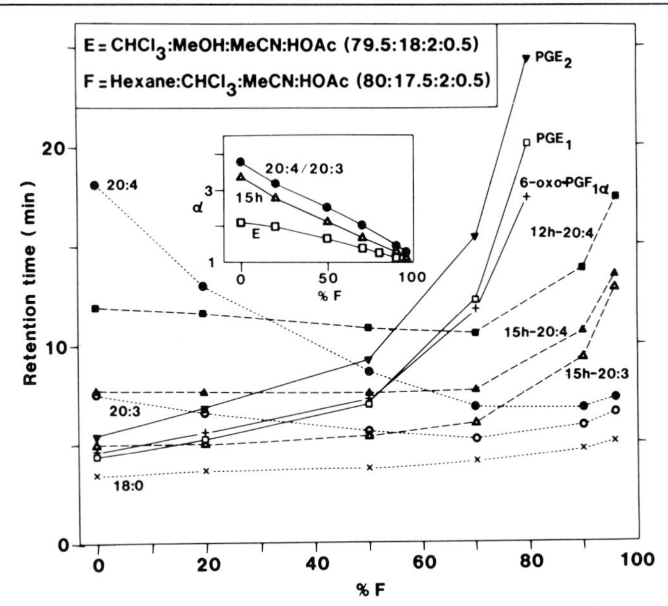

FIG. 4. Retention times of [1-^{14}C]18:0 (×), [1-^{14}C]20:3 (○), [1-^{14}C]20:4 (●), 15h-[1-^{14}C]20:3 (△), 15h-[1-^{14}C]20:4 (▲), 12h-[1-^{14}C]20:4 (■), 6-oxo-[1-^{14}C] PGF$_{1\alpha}$ (+), [5,6-^{3}H]PGE$_1$ (□), and [1-^{14}C]PGE$_2$ (▼) on a silver ion-loaded cation exchange column with various mixtures of solvents E and F (see Table I) as mobile phases. The flow rate was 1.5 ml/min. The inset shows the relative retentions (α) for [1-^{14}C]PGE$_2$ and [5,6-^{3}H]PGE$_1$ (□), 15h-[1-^{14}C]20:4 and 15h-[1-^{14}C]20:3 (△), and [1-^{14}C]20:4 and [1-^{14}C]20:3 (●). Taken from Powell.[12]

for the silver ions of the stationary phase, and solutes are consequently retained mainly on the basis of polar interactions. As the polarity of the mobile phase is increased, the retention of solutes becomes more and more dependent on the number of double bonds they possess. Thus the maximum degree of separation (α) for pairs of solutes differing from one another by the presence or the absence of one double bond is observed at 0% solvent F (i.e., the most polar mobile phase; Fig. 4, inset). The retention times of many PGs are rather short under these conditions, but they can be increased by reducing the concentration of acetonitrile in the mobile phase (e.g., to 1% or 0.4% as shown in Fig. 2).

Unsaturated fatty acids can also be separated by argentation HPLC in the absence of acetonitrile using polar solvents. Table III shows the retention times of a variety of compounds using methanol–acetic acid (99.8 : 0.2) as the mobile phase. With this solvent, the retention times of all PGs of the monoene series tested are shorter than those of PGs of the diene series. This is not true for the other mobile phases described above,

TABLE III
RETENTION TIMES (t_R) OF SOME FATTY ACIDS, MONOHYDROXY FATTY ACIDS, AND PROSTAGLANDINS WITH A MOBILE PHASE CONSISTING OF METHANOL–ACETIC ACID (99.8:0.2)[a]

Compound	t_R (min)	Compound	t_R (min)
18:0	2.4	15h-20:3	4.8
		20:3	19.0
PGE_1	3.1	15h-20:4	10.9
6-Oxo-$PGF_{1\alpha}$	3.4	5h-20:4	13.8
$PGF_{1\alpha}$	3.8	11h-20:4	21.4
		9h-20:4	22.4
PGE_2	5.1	8h-20:4	24.0
PGD_2	6.9	12h-10:4	25.1
$PGF_{2\alpha}$	10.8	20:4	65.0

[a] High-pressure liquid chromatography was carried out with a Milton Roy minipump equipped with a Valco injector. The flow rate was 2.0 ml/min. The silver ion-loaded cation exchange column was the same as that described in the footnote to Table II. All solutes were labeled with ^{14}C in the 1 position except for [5,6-^{3}H]PGE_1, [9β-^{3}H]$PGF_{1\alpha}$, [5,6,8, 9,12,14,15-^{3}H]PGD_2, and [9β-^{3}H]$PGF_{2\alpha}$.

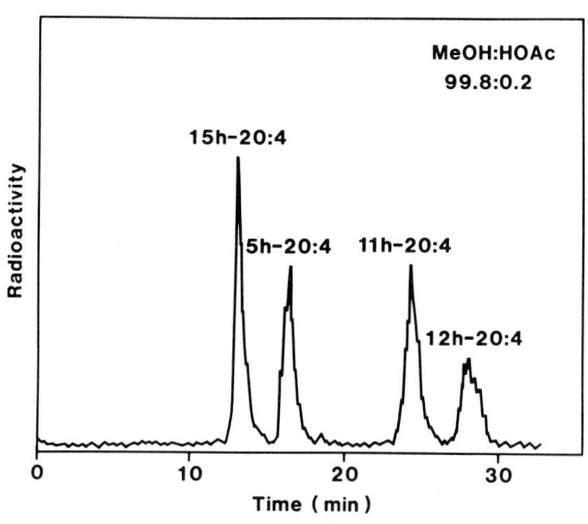

FIG. 5. High-pressure liquid radiochromatogram of 5h-[1-^{14}C]20:4, 11h-[1-^{14}C]20:4, 12h-[1-^{14}C]20:4, and 15h-[1-^{14}C]20:4 on a silver ion-loaded cation exchange column with methanol–acetic acid (99.8:0.2) as the mobile phase. HPLC was carried out using a Milton Roy mini-pump and a flow rate of 2 ml/min as described in the footnote to Table II.

with which $PGF_{1\alpha}$ has longer retention times than both PGE_2 and PGD_2. Thus a closer correlation is obtained between retention times and numbers of double bonds with methanol–acetic acid (99.8:0.2) as the mobile phase. Owing to the short retention times of most PGs with this solvent, it is not very useful for their purification. The retention times of monohydroxyeicosatetraenoic acids are longer, however, and mixtures of 5h-20:4, 11h-20:4, 12h-20:4, and 15h-20:4 can be completely resolved (Fig. 5). The retention times of 8h-20:4 and 9h-20:4, which are not well separated with this solvent, are intermediate between 11h-20:4 and 12h-20:4 (Table III). Methanol–acetic acid is also a very good solvent for the purification of PGs and other metabolites derived from 5,8,11,14,17-eicosapentaenoic acid.

Advantages and Disadvantages of Argentation HPLC

Argentation HPLC is a very effective technique because of the control that can be exerted over both the retention times of solutes and the selectivity of the stationary phase by changing the composition of the mobile phase. Argentation HPLC gives much better separation than reversed-phase HPLC of compounds differing from one another in their degree of unsaturation. Moreover, volatile organic solvents are used, which are much easier to remove than the aqueous solvents used in reversed-phase HPLC. Besides separating PGs and monohydroxyeicosenoic acids containing different numbers of double bonds, argentation HPLC also gives good resolution of compounds containing the same number of double bonds. It is very useful for separating mixtures containing 6-oxo-$PGF_{1\alpha}$, since a single sharp peak is obtained for this compound, in contrast to the broad or multiple peaks obtained after HPLC on silicic acid.[3] Argentation HPLC can also be used to enhance the isotopic purity of certain compounds labeled with deuterium or tritium. In our laboratory argentation HPLC is used mainly to separate a series of known compounds prior to analysis (e.g., PGE_1, PGE_2, and 6-oxo-$PGF_{1\alpha}$) or as a second step (after normal-phase HPLC) in the purification of metabolites of 20:3 or 20:4 prior to identification by GC–MS.

One possible problem with argentation HPLC is the fact that internal standards that have olefinic deuterium or tritium atoms can be separated from the corresponding unlabeled compounds. As noted above, this effect can be eliminated or minimized by (*a*) using compounds that are labeled on nonolefinic carbons or (*b*) increasing the amount of acetonitrile and/or reducing the polarity of the mobile phase. Another problem is that certain compounds do not chromatograph well using the column described here. Of the compounds tested, this includes only TxB_2 and 13,14-dihydro-15-

oxo-$PGF_{2\alpha}$, both of which gave broad or multiple peaks with relatively low retention times with the above solvent systems. Thus argentation HPLC cannot be recommended for the purification of these compounds, which are well separated from other PGs by normal-phase chromatography on silicic acid.[3] The capacity of silver ion-loaded cation exchange columns is somewhat lower than that of silicic acid columns. Although we have not tested this in much detail, it is possible to overload the column with several hundred micrograms of solute.

Acknowledgments

These studies were supported by grants from the Medical Research Council of Canada, the Quebec Heart Foundation, the Canadian Lung Association, and the Quebec Thoracic Society.

Section V

Gas Chromatography–Mass Spectrometry of Prostaglandin Derivatives

[64] Preparation of ^{18}O Derivatives of Eicosanoids for GC–MS Quantitative Analysis

By ROBERT C. MURPHY and KEITH L. CLAY

Quantitative mass spectrometry using stable isotope dilution techniques has emerged as a benchmark by which other quantitative methods can be evaluated. This is particularly true for the analysis of prostaglandins and other arachidonic acid metabolites. However, this type of analysis, which is unsurpassed in precision and accuracy, requires the availability of stable isotopically labeled species for each analyte. Deuterium-labeled isotopimers are available for only a few of the members of the large eicosanoid family.

One structural feature common to all arachidonic acid metabolites is the carboxylic acid moiety, which contains two oxygen atoms that can be exchanged in theory with oxygen-18 water to yield the $^{18}O_2$-labeled carboxylic acid [reactions (1) and (3)]. Oxygen-18 can also be incorporated upon ester hydrolysis [reaction (2)]. These exchange reactions can be catalyzed by both acid[1] and base[2] as well as enzymically by liver esterase.[3]

$$H_2{}^{18}O + R-\overset{\overset{O}{\|}}{C}-OH \longrightarrow R-\overset{\overset{^{18}O}{\|}}{C}-OH + H_2O \tag{1}$$

$$H_2{}^{18}O + R-\overset{\overset{O}{\|}}{C}-OCH_3 \longrightarrow R\overset{\overset{^{18}O}{\|}}{C}-OH + CH_3OH \tag{2}$$

$$H_2{}^{18}O + R-\overset{\overset{^{18}O}{\|}}{C}-OH \longrightarrow R-\overset{\overset{^{18}O}{\|}}{C}-{}^{18}OH + H_2O \tag{3}$$

There are two important considerations when using ^{18}O-labeled carboxylic acids: first, many eicosanoids are too unstable for direct chemical exchange reactions catalyzed by acid or base; and second, the loss of ^{18}O can occur through back exchange in aqueous solutions. Back exchange or loss of label is particularly rapid in plasma and tissue containing esterases and obviates the use of these labeled species in metabolic studies.[3]

Reagents

$H_2{}^{18}O$: 97–99 atom % ^{18}O (Yeda Chemicals, Israel)
Porcine esterase type II (Sigma, St. Louis, Missouri)
$KHPO_4$, 50 m*M*, pH 8.0
Li^{18}OH, 0.29 *N* in $H_2{}^{18}O$

[1] R. C. Murphy and K. L. Clay, *Biomed. Mass Spectrom.* **6,** 309 (1979).
[2] D. Samuel, *in* "Oxygenases" (O. Hayaishi, ed.), p. 32. Academic Press, New York, 1962.
[3] W. C. Pickett and R. C. Murphy, *Anal. Biochem.* **110,** 115 (1981).

ISBN 0-12-181986-8

Procedure

Method I. Acid-Catalyzed Exchange of Arachidonic Acid. Arachidonic acid labeled with two ^{18}O atoms in the carboxyl moiety can be obtained by heating arachidonic acid in $H_2{}^{18}O$ 1 *N* in HCl. HCl gas (Matheson) is bubbled into ice cold $H_2{}^{18}O$ until saturated. The resulting ^{18}O-labeled HCl is allowed to come to room temperature, at which point it is approximately 12 *N*. The actual normality of this solution can be obtained by titrating a small aliquot against a suitable base (e.g., Na_2CO_3). Dilution of the concentrated acid with $H_2{}^{18}O$ will then yield [^{18}O]HCl acid of the desired normality, usually 1 *N* [^{18}O]HCl is used.

In order to minimize the volume of expensive ^{18}O water used, reactions are carried out in small volumes, and cosolvents are used to increase the solubility of the lipid in the aqueous acid. Acetonitrile [^{18}O]HCl (50/50) has proved to be a satisfactory system for exchange of milligram quantities of arachidonic acid without the use of inordinately large water volumes. Other water-miscible solvents should also be suitable provided no dilution of the ^{18}O content is incurred. Exclusion of oxygen gas from the solution by bubbling with argon prior to heating is very important for prevention of oxidation of arachidonic acid.

Arachidonic acid (1 mg) in benzene is evaporated to dryness under argon in a screw-cap conical reaction tube of 0.3 ml total volume. The residue is dissolved in 50 μl of acetonitrile and then 50 μl of [^{18}O]HCl (1 *N*) is added. The vessel is purged with argon, sealed with a Teflon-lined cap, then heated 12–15 h at 60°. The free acid is recovered by removal of solvent by lyophilization.

Analysis of the resulting [^{18}O]arachidonic acid is best carried out by GC–MS of the *t*-butyldimethylsilyl (TBDMS) ester. The TBDMS derivative is prepared by addition of 50 μl of reaction solution, containing 150 mg of *t*-butyldimethylchlorosilane and 68 mg of imidazole in 1.0 ml of *N,N*-dimethylformamide, to a small amount of dried arachidonic acid. The mass spectrum of the resulting TBDMS ester of [^{18}O]arachidonic acid has a molecular ion at m/z 422 and an intense ion at m/z 365, which arises from the loss of a *t*-butyl group (M-57). This M-57 ion can be used to measure isotope abundance for the $^{18}O_2$, $^{18}O^{16}O$, and $^{16}O_2$ species (m/z 365, 363, and 361), which indicated the following mole atom percentage from the above exchange reaction: 97.1 atom % $^{18}O_2$, 2.4 atom % $^{18}O^{16}O$, and 0.5 atom % $^{16}O_2$.

Method II. Base-Catalyzed Hydrolysis of Methyl 5-Hydroxyeicosatetraenoate in $H_2{}^{18}O$. The methyl ester of 5-hydroxyeicosatetraenoic acid (5-HETE) is obtained by diazomethane methylation; the 5-HETE is synthesized from arachidonic acid.[4] One milligram is dissolved in 50 μl of meth-

[4] E. J. Corey, J. O. Albright, A. E. Barton, and S. Hashimoto, *J. Am. Chem. Soc.* **102,** 1435 (1980).

TABLE I
ISOTOPIC COMPOSITION OF 5-HETE AFTER ESTERIFICATION–HYDROLYSIS CYCLE IN $H_2{}^{18}O$[a]

	Experimental atom %[a]			Theoretical atom %[b]
Cycle	$^{16}O_2$	$^{18}O^{16}O$	$^{18}O_2$	$^{18}O_2$
1	8.4	82.2	9.3	0
2	5.8	50.1	44.0	40.5
3	0.8	38.8	60.4	58.7
4	0.6	31.1	68.3	68.7

[a] Atom percentage was calculated from the fragment ion abundances at m/z 203 ($^{16}O_2$), 205 ($^{18}O^{16}O$), and 207 ($^{18}O_2$) after correction for natural abundance. Mass spectra were recorded under electron impact, 70 eV, using a Finnigan 3200 quadrupole mass spectrometer.

[b] Based on 90 atom % $H_2{}^{18}O$ maintained for each cycle.

anol, transferred to a conical reaction vessel (Reactivial, Pierce Chemical Co.), and evaporated to dryness under vacuum (1×10^{-2} torr). Lithium [^{18}O]hydroxide is made by slowly adding 2 mg of lithium metal to 1 ml of $H_2{}^{18}O$. One hundred microliters of this 0.29 N $Li^{18}OH$ are added to the methyl 5-HETE; the reaction vessel is sealed with a Teflon cap under argon, and the mixture is shaken for 1 hr at room temperature. The 5-HETE free acid is isolated by careful acidification of the solution with 0.01 N formic acid to pH 4 and extraction with ethyl acetate (twice with 2 volumes). The [^{18}O]HETE is remethylated with diazomethane in ether. An aliquot is analyzed for ^{18}O content by GC–MS as indicated in Table I, cycle 1.

Incorporation of more than one ^{18}O in the carboxylate moiety is highly desirable, and this can be done simply by recycling this procedure of ester hydrolysis. Typical results of further ester hydrolysis are also indicated in Table I. It is important to exclude trace amounts of natural water, which will dilute the ^{18}O content of the $Li^{18}OH$ and thereby decrease the isotopic purity of the final product. The critical feature is to have a very low $^{16}O_2$ content, since in the quantitative analysis this form is the biological 5-HETE. The procedure described here can reduce $^{16}O_2$ content to 0.6 atom% while $^{18}O_2$ content of 5-HETE is 68.3% and $^{18}O^{16}O$ content is 31.1%, which is very close to a theoretical value of 90 atom % $H_2{}^{18}O$ maintained throughout the entire five esterification–hydrolysis cycles. After the last hydrolysis cycle the [$^{18}O_2$]5-HETE is stored as the free acid at $-20°$ in 50% aqueous methanol.

Method III. Esterase-Catalyzed Ester Hydrolysis for ^{18}O Incorporation. Porcine liver esterase (1 mg, 137 units) is dissolved in 0.15 ml of K buffer and lyophilized to dryness. The residue is then dissolved in 0.15 ml of $H_2{}^{18}O$ to reconstitute the buffer to pH 8.0 for optimum esterase activity. 6-Keto-$PGF_{1\alpha}$ methyl ester (1 mg) dissolved in 10 μl of methanol is added to the esterase in $H_2{}^{18}O$. The conical tube is sealed under argon and incubated at 37° for 1 hr. After this incubation, 0.5 ml of methanol, 1.5 ml of diethyl ether, and 1 ml of H_2O are added to extract unreacted 6-keto-$PGF_{1\alpha}$ methyl ester. The aqueous layer is acidified to pH 3 with 0.1 *N* formic acid and extracted with 2 ml of diethyl ether. The organic layer is dried with anhydrous sodium sulfate and diluted with 4 ml of hexane.

The labeled prostaglandin is purified by silicic acid chromatography. Silicar cc-4 (0.5 g, Mallinckrodt) previously activated at 120° for 4 hr is poured into a short column, and the hexane–ether solution of the prostaglandin is applied to the column. The column is washed with 15 ml of hexane–ether (2:1), then the prostaglandin is eluted with 15 ml of ethyl acetate. Table II lists the isotopic composition of the [$^{18}O_2$]6-keto-$PGF_{1\alpha}$ hydrolyzed at two different temperatures. Overall yields of prostaglandins are typically 90%.

The liver esterase catalyzes the hydrolysis of prostaglandin methyl esters with incorporation of ^{18}O [reaction (2)] and, furthermore, catalyzes the exchange of the oxygen atoms in free carboxylic acids with water [reactions (1) and (3)].[3] Table II lists the isotopic composition of several other prostaglandins and thromboxane B_2 exchanged with $H_2{}^{18}O$. The oxygen exchange and ester hydrolysis are time and temperature dependent, and for those prostaglandins, such as 6-keto-$PGF_{1\alpha}$, which do not readily

TABLE II

EXCHANGE OF CARBOXYL OXYGEN ATOMS OF EICOSANOIDS CATALYZED BY LIVER ESTERASE IN $H_2{}^{18}O$[a]

		Atom percent[b]		
Compound	Temperature (°C)	$^{18}O_2$	$^{18}O^{16}O$	^{16}O
$PGF_{2\alpha}$	25	92.3	7.3	0.5
PGD_2	25	88.5	9.2	2.3
PGE_2	25	70.5	25.5	4.0
TxB_2	25	62.0	13.7	24.3
PGB_2	25	44.3	13.6	42.3
6-Keto-$PGF_{1\alpha}$	25	31.3	45.7	23.0
6-Keto-$PGF_{1\alpha}$[c]	37	71.4	35.4	3.1

[a] Porcine liver esterase, 137 units, in 0.15 ml of $H_2{}^{18}O$ K buffer, pH 8.0 was used. Incubation was for 1 hr.

[b] Atom percentage was calculated from selected ion monitoring of diagnostic fragment ions under electron impact, 70 eV, GC–MS conditions.

[c] Incubation was for 30 min.

exchange at room temperature, incubations at 37° can be carried out in reasonable lengths of time. The [^{18}O]eicosanoids produced by these catalyzed exchange reactions can be purified by high-pressure liquid chromatography techniques described in Section IV of this volume.

Ideally, it is best to synthesize these standards at a scale where one can recrystalize the products to be able accurately to prepare standard solutions by gravimetric means. However, it is still possible to standardize a stock solution of an [^{18}O]eicosanoid for a GC–MS quantitative assay using the unlabeled eicosanoid as a reference standard. The methods for construction of a calibration curve as well as mass spectrometry and derivatization procedures for gas chromatography are described in this volume as well as in a previous volume.[5] [^{18}O]Leukotrienes can be further characterized by their unique ultraviolet absorption[6] and stock solutions quantitated.

In order to eliminate back exchange loss of the ^{18}O from the carboxylate moiety when analyses of plasma are performed, the ^{18}O internal standard is added to the plasma (1 ml) in 2 ml of absolute ethanol. This precipitates plasma proteins and inactivates the esterase activity. After centrifugation of the treated plasma, the supernatant is acidified to pH 3 with 0.1 *N* formic acid and extracted with 3 ml of chloroform. The organic layer dried with 0.5 g of anhydrous sodium sulfate and evaporated to dryness with a rotary evaporator at 30°. The residue is dissolved in hexane–ether (2:1, v/v) and purified by silicic acid chromatography as described above. For very small amounts of prostaglandins a further HPLC separation is often necessary before derivatization and GC–MS analysis to further reduce contaminating components.

Acknowledgments

This work was supported by NIH Grant HL 25785. The authors are grateful to Dr. John Pike (Upjohn Company) for the generous gift of prostagrandins.

[5] P. F. Crain, D. M. Desiderio, and J. A. McCloskey, this series, Vol. 35, p. 359.
[6] R. C. Murphy and W. R. Mathews, this volume [53].

[65] Preparation of Deuterium-Labeled Urinary Catabolites of $PGF_{2\alpha}$ as Standards for GC–MS

By C. R. PACE-ASCIAK[1] and N. S. EDWARDS

Deuterium labeled standards are required as both carriers and internal standards for quantitation of biological samples by GC–MS (mass fragmentography).[2–4] While deuterated primary prostaglandins (PGE_2, $PGF_{2\alpha}$, and 6-keto-$PGF_{1\alpha}$) with low protium background (approximately 0.5%) have been made available from the Upjohn Co., there is no readily available commercial source of either deuterated or undeuterated urinary catabolites of either of the prostaglandins. These products are needed in the complete profiling of these catabolites in biological samples by GC–MS.

We have discovered the suitability of commercially available deuterated (d_3) *O*-methoxylamine hydrochloride of very high deuterium : protium ratio for the introduction of three deuterium atoms into products containing a keto group. Since most of the urinary catabolites of the prostaglandins contain at least one keto group, this method can be applied for the preparation of deuterated standards of the urinary catabolites. We also describe a relatively simple biological method for the generation and purification of urinary catabolites of the prostaglandins. Although this chapter relates to catabolites of $PGF_{2\alpha}$ only, the method is quite general and can easily be applied to other prostaglandins as well.

Preparation of Unlabeled Urinary Catabolites

Animal Preparation. A procedure is described for the cannulation of the trachea, carotid artery, jugular vein, and urinary bladder. Male adult Wistar rats (200–400 g) are anesthetized with Inactin-BYK administered intraperitoneally. The initial dose is 120 mg/kg body weight. Supplemental doses of 5–10 mg are given as necessary to reach and maintain surgical anesthesia. After the rat is anesthetised, it is transferred to a rat table maintained at 37° for the rest of the experiment.

[1] Work reported in this paper was supported by a grant to C. P-A. (MT-4181) from the Medical Research Council of Canada.

[2] B. Samuelsson, M. Hamberg, and C. C. Sweeley, *Anal. Chem.* **38,** 301 (1970).

[3] U. Axen, L. Baczynskyj, D. J. Duchamp, K. T. Kirton, and J. F. Zieserl, Jr., *Adv. Biosci.* **9,** 15 (1972).

[4] K. Green, M. Hamberg, B. Samuelsson, and J. C. Frölich, *Adv. Prostaglandin Thromboxane Res.* **5,** 1 (1978).

ISBN 0-12-181986-8

A small (1–2 cm) incision is made in the throat and the submaxillary gland. Connective tissue and muscle layers are parted to expose the trachea. A ligature (000 silk, Ethicon A-54) is placed around the trachea. A cut is made above the ligature, and a beveled polyethylene cannula (Intramedic PE-260) is inserted. The cannula is secured by tightening of the ligature.

The carotid artery is located under the muscle layer beside the trachea and separated from the surrounding tissue. Two ligatures are placed around it. The distal ligature is tied securely, and a spring clamp is placed proximally to both ligatures. A beveled cannula (Intramedic PE-50) is attached to a needle (19 gauge) and syringe and filled with heparinized saline (50 IU of heparin per milliliter of normal saline). The cannula is inserted through a small cut in the artery, between the two ligatures. The proximal ligature is tightened, but not completely, to prevent slipping of the cannula. While the cannula is held in place, the spring clamp is carefully released. The cannula is pushed farther into the artery, then the proximal ligature is tightened completely. The distal ligature is also tied around the cannula to anchor it.

The jugular vein is located by extending the incision laterally 1–2 cm and clearing away fat and connective tissue. The portion of the vein immediately above the trapezoidal muscle is cleared from the surrounding tissue, and two ligatures are placed around it. The distal ligature is tightened completely. A small cut is made between the two ligatures, and the cannula (prepared as described above) is quickly inserted 1–2 cm into the vein, in the direction of the heart. The proximal ligature is tightened over the vein and cannula, and the distal ligature is tied around the cannula. After these cannulations, the throat wound is covered with a piece of moist gauze.

To locate the urinary bladder a 2 cm incision is made in the peritoneal wall, along the midline in the lower abdomen. The bladder is extruded, and two ligatures are placed around it. Gauze is placed under the bladder to absorb urine when the bladder is cut. A small cut is made in the distal end of the bladder, and a polyethylene cannula (PE-50) is inserted. The ligatures are tightened to secure the cannula and the gauze is removed. A clean moist piece of gauze is placed over the wound.

Infusion of $PGF_{2\alpha}$. The animals are infused with normal saline (2.4 ml/hr) for 0.5–1 hr during recovery from surgery, prior to infusion with $PGF_{2\alpha}$. Blood pressure is monitored continuously during the experiment by connecting the arterial cannula to a pressure transducer (Statham, Model P23 Db) filled with heparinized saline and connected to a recorder (Brush Model 2200). Mean blood pressure in normal Wistar rats after recovery is 131 $\pm$ 19 mm Hg (n = 14). Blood pressure tends to decrease slowly during the experiment, with an average change of 28 $\pm$ 22 mm Hg

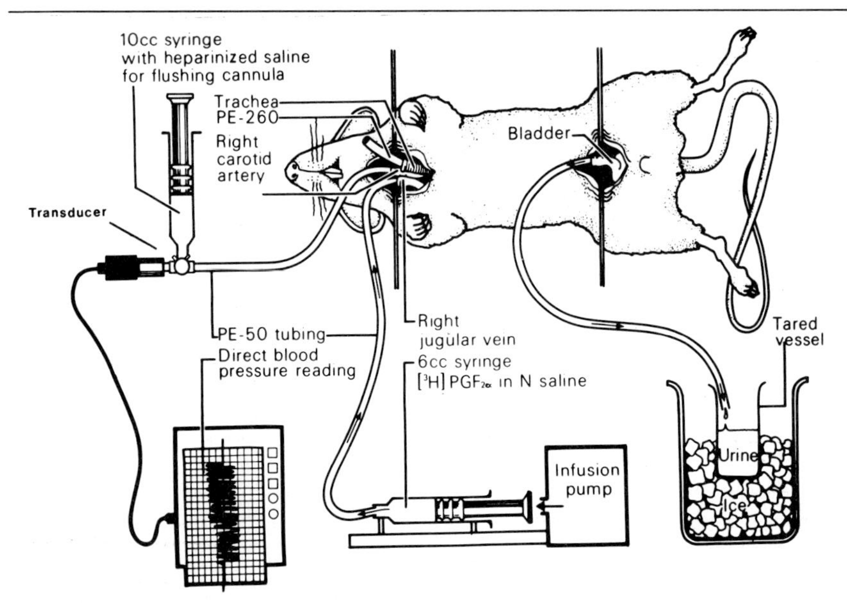

FIG. 1. General scheme showing the anesthetized rat preparation used for the infusion of $[^3H]PGF_{2\alpha}$ with simultaneous measurement of arterial blood pressure and collection of urine.

(n = 14) in male Wistar rats. Figure 1 shows setup of a typical experiment.

$[9\beta\text{-}^3H]PGF_{2\alpha}$ (New England Nuclear, 0.5 μCi) is diluted with unlabeled $PGF_{2\alpha}$THAM salt (0.86 mg/kg body weight) dissolved in normal saline (2.4 ml) and infused intravenously into the rat over 1 hr. This is followed by a further infusion of normal saline at the same rate. Urine is collected in tared vials on ice during 4 hr starting with the prostaglandin infusion. The average total volume of urine collected is 4.22 ± 2.68 ml (n = 14). Samples are stored at −20° until extracted as described below.

Isolation of Urinary Catabolites. Figure 2 outlines a general scheme used in the purification of the catabolites. Tritium-labeled prostaglandin catabolites are extracted from urine using XAD-2 chromatography. The urine samples are diluted with 3 volumes of distilled water, and an aliquot is taken to determine the amount of radioactive material present. The samples are acidified with 1 N HCl to pH 3 and applied to columns of XAD-2 resin (Rohm & Haas: 1 × 20 cm) previously equilibrated with water. The columns are eluted with water until the eluate is neutral. Tritium-labeled material is then eluted with 25 ml of acetone. The acetone-soluble fraction is dried under vacuum and redissolved in methanol. Aliquots of the methanol solution are counted to determine recovery of ma-

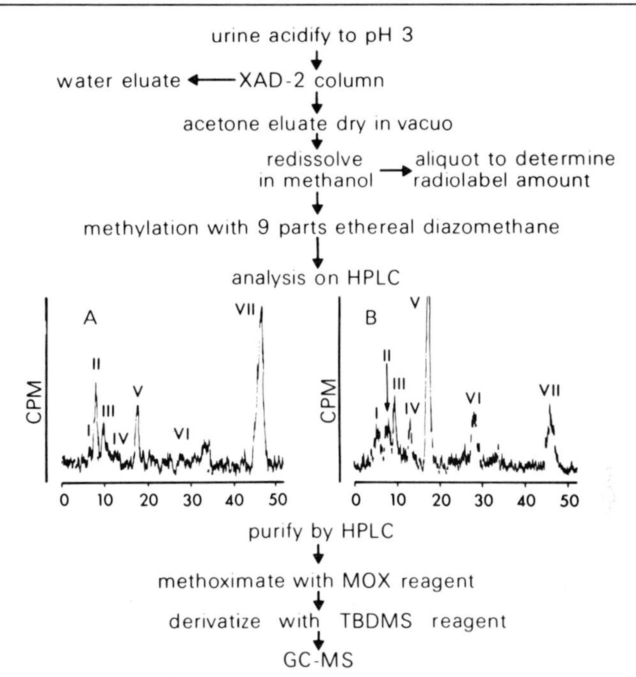

FIG. 2. Flow diagram outlining the extraction, purification, and derivatization of urinary catabolites of $PGF_{2\alpha}$. HPLC profiles represent urinary extracts from (A) a 22-day-old male rat (North American Wistar) and (B) an adult female rat (Japanese Wistar).

terial from the column. The average recovery of tritium-labeled products from these columns is 94.4 ± 3.5% (n = 14) of the amount applied.

Purification of Urinary Catabolites by HPLC. Urinary catabolites of $PGF_{2\alpha}$ are separated as their methyl esters using reversed-phase HPLC (Fig. 2). Urinary extracts are dissolved in methanol and allowed to react with a freshly prepared distilled solution of diazomethane in diethyl ether (9 volumes) in the dark for 10 min.[5] The samples are then dried with a fine stream of nitrogen in a well ventilated hood and redissolved in methanol. An aliqout of each sample is analyzed by reversed-phase HPLC (fatty acid analysis column, Waters Associates) using acetonitrile–water–benzene–acetic acid, 220:780:2:4, v/v/v/v (1.0 ml/min) as developing solvent.[5] $PGF_{2\alpha}$ catabolites containing tritium are monitored through an online radioactivity monitor (Berthold). Using this detector the effluent from the column is continuously mixed with PCS liquid scintillant (ratio approximately 1:4, respectively) before passing through a liquid scintillation detector equipped with a flow-through cell (Berthold) for tritium de-

[5] C. R. Pace-Asciak and N. S. Edwards, *J. Biol. Chem.* **255,** 6106 (1980).

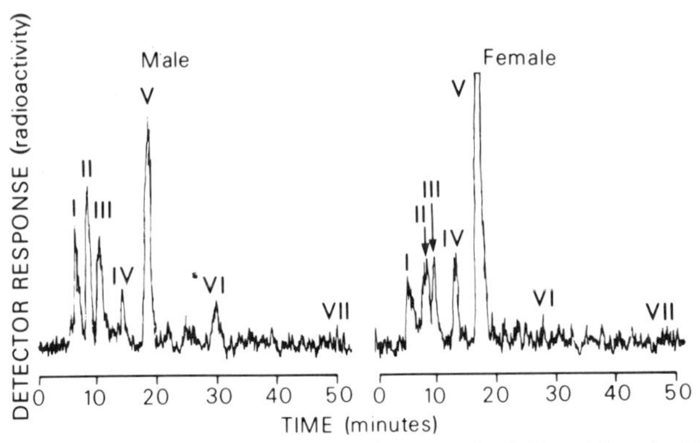

FIG. 3. HPLC profiles of urinary catabolites from a male (left) and female (right) adult rat (North American Wistar).

tection. The precise amount of each product is quantitated by collection of the effluent emerging from the radioactivity detector into scintillation vials followed by counting of these vials in a Beckman LS-255 liquid scintillation counter. When the retention volumes of the various peaks have been determined accurately, larger aliquots of the samples are injected for preparative runs and the column effluent is collected directly in fractions without mixing with liquid scintillant. These fractions are dried thoroughly *in vacuo* and redissolved in methanol. Aliquots are counted to estimate the amount of radiolabel present and hence the amount of unlabeled product. Purity of each product is verified by repeated analysis on HPLC.

Figure 3 shows a typical profile of $PGF_{2\alpha}$ urinary catabolites in an adult male rat. Four major peaks are clearly observed (I, II, III, V). Of these products II, III, and V contain a 15-keto group (see Scheme 1). The distribution and intensity of these products and two other products (VI and VII) varies with age and sex of the rat (see Figs. 2 and 3). Thus V is the principal urinary product in the adult female rat ($39 \pm 4\%$, $n = 6$; Fig. 3), and VII is the principal product in the 22-day-old rat $>50\%$, $n = 2$; Fig. 2)[5] of either sex. These observations permit choice of different types of rats for the somewhat "specific" generation of the required catabolite.

Preparation of (d_3) Deuterium-Labeled Catabolites

Each sample purified by HPLC as the methyl ester (1–100 μg) is dissolved in a glass-stoppered test tube in 100 μl of a 2% solution of d_3-methylhydroxylamine hydrochloride (Regis Chemical Co) in anhydrous pyridine (Pierce Chemical Co). After 16 hr at room temperature, the solvent is

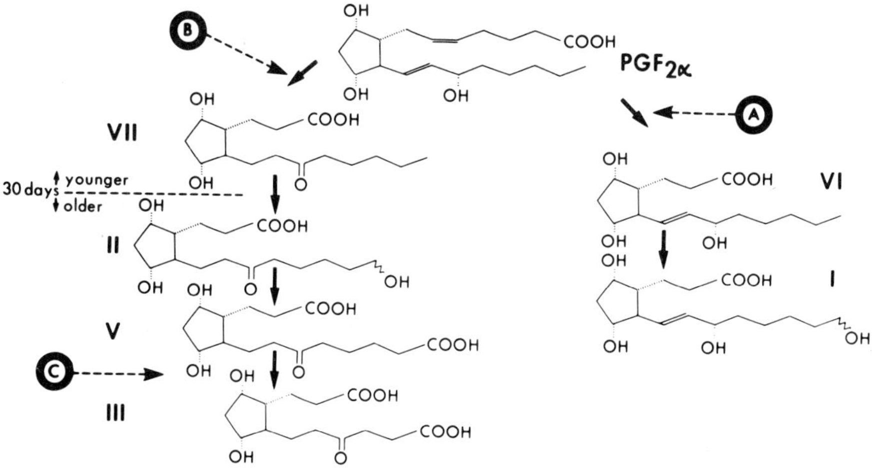

SCHEME 1. Catabolism of $PGF_{2\alpha}$ in the rat *in vivo* via two separate pathways. Pathway A appears to be stimulated in the Japanese strain of Wistar rat due to a decrease in the activity of pathway B. Pathway C is mostly absent in the female.

taken to dryness, and the sample is extracted with ethyl acetate and water. The methyl ester d_3-methoxime (MeMO) derivative can be stored in ethyl acetate at $-20°$ for future use in spiking of samples for GC–MS.

GC–MS Assay of Deuterium-Labeled Urinary Catabolites of $PGF_{2\alpha}$

The MeMO derivative is assayed as the *t*-butyldimethylsilyl ether derivative (TBDMS) because of the ease of preparation and subsequent handling and storage of this derivative (not easily hydrolyzed) and because of prominent fragment ions in the high region of the spectrum i.e., M-57 (loss of a *t*-butyl group). A reference standard of this derivative can be stored in hexane at $-20°$ for over 1 year without decomposition. This derivative is prepared by heating (1–10 μg of sample in 100 μl of *t*-butyldimethylchlorosilane–imidazole–dimethyl formamide reagent mix (Applied Science), at 60° for 20 min. After cooling, the solution is diluted with water (1 ml) and extracted once with hexane (2 ml). The hexane solution can be stored as such or after concentration.

Figure 4 shows prominent fragment ions monitored by mass fragmentography for each of the four keto-containing urinary catabolites of $PGF_{2\alpha}$, i.e., products II, III, V, and VII (see Scheme 1). The protium background of product VII is in the range of 0.03%, and slight protium interference is observed with other products, raising the background to 0.1–0.7%. Figure 4 also shows the feasibility of analyzing all four prod-

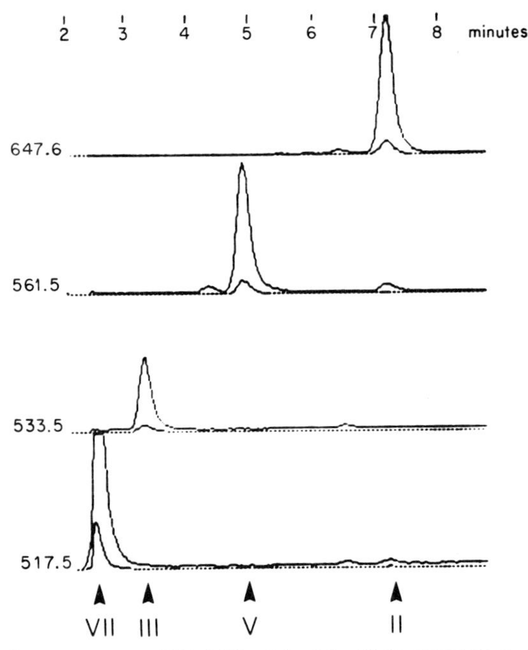

FIG. 4. Mass fragmentogram (GC–MS) of the Me-d_3MO-TBDMS derivatives of a mixture of catabolites II, III, V, and VII (see Scheme I). Gas chromatography was performed on 3% OV-1 on Gas Chrom Q (Applied Science) using a 2 ft. by $\frac{1}{8}$ inch column. Injection of the sample was at 220° with a 4° per minute increase in temperature up to 300°. Mass spectrometry was performed on a Hewlett-Packard quadrapole mass spectrometer.

ucts in the same sample, since complete resolution can be achieved under normal GC conditions.

An application of this method is shown in Fig. 5. Urine from a spontaneously hypertensive rat which had received an infusion of $PGF_{2\alpha}$, (as described above) was assayed. Together with deuterated (d_3) urinary catabolites II, III, V, and VII was added a mixture of d_4 and d_0 PGE_2, $PGF_{2\alpha}$, and 6-keto-$PGF_{1\alpha}$ to show the feasibility of measurement not only of a mixture of the catabolites alone, but also in the presence of the primary prostaglandins. As can be seen in Fig. 5, the mass spectrometer can act as a specific detector during the chromatographic run capable of completely resolving overlapping peaks (see catabolite V and PGE_2).

Summary

A procedure is described for the preparation and purification of six major urinary catabolites of $PGF_{2\alpha}$. Four of these products contain a 15-

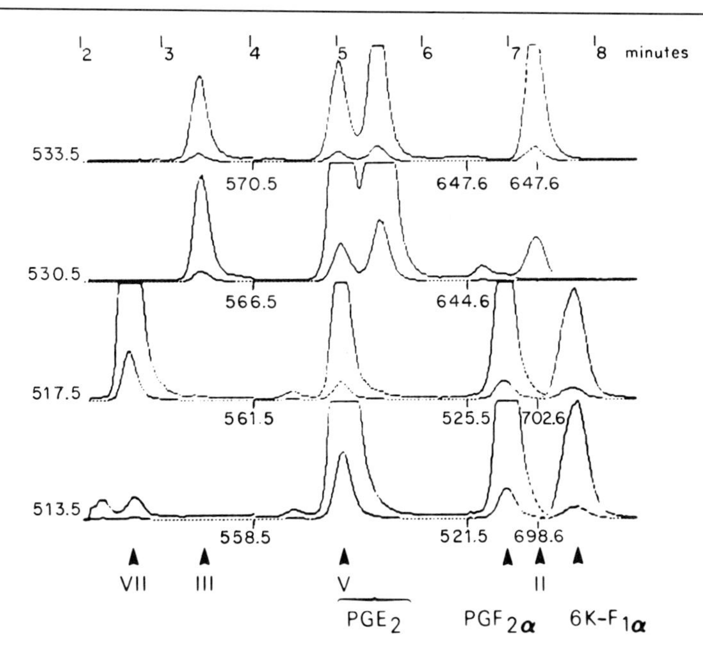

FIG. 5. Resolution of d_0 and d_3 catabolites of $PGF_{2\alpha}$ (II, III, V, and VII) from a mixture of d_0 and d_4 PGE_2, $PGF_{2\alpha}$, and 6-keto-$PGF_{1\alpha}$. Urine from a spontaneously hypertensive rat which received an infusion of $PGF_{2\alpha}$ served as sample for undeuterated $PGF_{2\alpha}$ catabolites; all other products were spiked.

keto group and consequently can be converted into the d_3-methyl oxime derivatives to serve as internal standards for mass fragmentography. Also described is the resolution of these products on GC allowing the simultaneous assay of all products in a mixture by GC–MS.

[66] Quantification of the PGD_2 Urinary Metabolite 9α-Hydroxy-11,15-dioxo-2,3,18,19-tetranorprost-5-ene-1,20-dioic Acid by Stable Isotope Dilution Mass Spectrometric Assay

By L. JACKSON ROBERTS II

It has become well established that the most accurate means to assess production of prostaglandins *in vivo* in man is quantification of circulating and urinary metabolites of prostaglandins, in contrast to quantification of the parent compounds.[1–3] Reliable assays for metabolites of PGE_2,

[1] B. Samuelsson, *Adv. Biosci.* **9,** 7 (1973).

ISBN 0-12-181986-8

$PGF_{2\alpha}$, TxB_2, and PGI_2 have been available; however, a method for quantification of a metabolite of PGD_2 has not been previously developed.

We reported the finding of the urinary excretion of markedly increased quantities of metabolites of PGD_2 in two patients with systemic mastocytosis associated with life-threatening attacks of flushing and hypotension and presented indirect evidence suggesting that PGD_2 is a previously unrecognized important mast cell-derived mediator of these hypotensive attacks.[4] The identification of these metabolites of PGD_2 in human urine was actually the first evidence for production of PGD_2 *in vivo* in man. Because of the implication that PGD_2 is an important mediator of the symptoms and signs of systemic mastocytosis, it seemed clear that a means accurately to assess the magnitude of production of PGD_2 *in vivo* would be essential in the diagnosis and treatment of individual patients with this disease.

The metabolic fate of PGD_2 was studied in the nonhuman primate.[5] In this study it was found that PGD_2 is converted both to metabolites retaining the PGD ring structure as well as to metabolites with a PGF ring. This conversion of PGD_2, in part, to PGF-ring metabolites also appears to occur in man, as in addition to PGD ring metabolites, the excretion of large quantities of several PGF ring metabolites was also found in the two patients described above with systemic mastocytosis.[6] Previous studies of the metabolic fate of $PGF_{2\alpha}$ in man, however, have provided no indication that $PGF_{2\alpha}$ is converted *in vivo* to PGD ring metabolites. Therefore, although quantification of a PGF ring metabolite could provide an indirect index of PGD_2 production, this would be associated with ambiguity regarding the source of its formation, PGD_2 versus $PGF_{2\alpha}$. Quantification of a metabolite of PGD_2 that retains the original PGD ring, therefore, seems to be the only reliable means to assess accurately the endogenous production of PGD_2 in man.

The method chosen for quantification of a PGD_2 urinary metabolite was a stable isotope dilution mass spectrometric assay because of the high degree of specificity, accuracy, and sensitivity associated with this assay methodology.[7,8] In this chapter is described the methods developed for

[2] B. Samuelsson and K. Gréen, *Biochem. Med.* **11,** 298 (1974).

[3] B. Samuelsson, E. Granström, K. Gréen, M. Hamberg, and S. Hammerström, *Annu. Rev. Biochem.* **254,** 4152 (1979).

[4] L. J. Roberts, II, B. J. Sweetman, R. A. Lewis, K. F. Austen, and J. A. Oates, *N. Engl. J. Med.* **303,** 1400 (1980).

[5] C. K. Ellis, M. D. Smigel, J. A. Oates, O. Oelz, and B. J. Sweetman, *J. Biol. Chem.* **254,** 4152 (1979).

[6] L. J. Roberts, II, unpublished data.

[7] E. Granström and B. Samuelsson, *Adv. Prostaglandin Thromboxane Res.* **5,** 1 (1978).

[8] K. Gréen, M. Hamberg, B. Samuelsson, M. Smigel, and J. C. Frölich, *Adv. Prostaglandin Thromboxane Res.* **5,** p. 39 (1978).

isolation and preparation of a deuterium- and tritium-labeled internal standard and specific assay procedures for quantification of the PGD_2 urinary metabolite 9α-hydroxy-11,15-dioxo-2,3,18,19-tetranorprost-5-ene-1,20-dioic acid (PGD-M).

Preparation of the Internal Standard

Required for a stable isotope dilution mass spectrometric assay is an internal standard of the compound being assayed that has been labeled with both stable and radioactive isotopes. Neither isotopically labeled nor unlabeled metabolites of PGD_2 are commercially available, nor have procedures for their chemical synthesis been developed. Infusion of labeled PGD_2 into an experimental animal could provide a means to generate biologically the labeled PGD_2 urinary metabolites. Such procedures usually are not only very expensive because of low yields of metabolites obtained, but can also be unsatisfactory because of excessive dilution of labeled metabolites with endogenous unlabeled metabolites. The excretion of large quantities of PGD_2 metabolites into urine of the two patients described above with mastocytosis, however, provided a source from which sufficient quantities of unlabeled PGD_2 metabolites could be isolated. After isolation and purification of PGD-M from the urine of these patients, the internal standard was prepared by a method developed for simultaneous incorporation of both deuterium and tritium labels into the unlabeled metabolite.

Isolation and Purification of Unlabeled PGD-M

Unlabeled PGD-M was isolated and purified from 5 liters of urine collected from the patients with mastocytosis. In the absence of a radiolabeled metabolite, column fractions containing PGD-M were identified during chromatographic purification procedures by analyzing aliquots of collected fractions with mass spectrometric-selected ion monitoring of major ions in the mass spectrum of PGD-M.[5]

Five liters of pooled urine were acidified to pH 3.2 with 6 *N* HCL and extracted with 1500 ml of Amberlite XAD-2 (Mallinckrodt, St. Louis, Missouri) as previously described.[9] Tritium-labeled 7α-hydroxy-5,11-diketotetranorprostane-1,16-dioic acid (a generous gift from Merck Sharp and Dohme, Rahway, New Jersey) was added to the urine prior to the extraction to calculate recovery, which was 95%. The residue obtained following solvent evaporation was then dissolved in water, acidified to pH 3.2 with 10% formic acid, and partitioned into ethyl acetate. The residue

[9] L. J. Roberts, II, B. J. Sweetman, N. A. Payne, and J. A. Oates, *J. Biol. Chem.* **252,** 7415 (1977).

obtained after solvent evaporation was then applied to a 50-g column of silicic acid (CC-4, Mallinckrodt, St. Louis, Missouri) and eluted with approximately 2000 ml of ethyl acetate.

Further purification of material obtained after silicic acid chromatography was accomplished by high-pressure liquid chromatography (HPLC). The material was divided into fourths, and each fourth was chromatographed separately on a 7.8 mm i.d. × 30 cm μPorasil column (Waters Associates, Milford, Massachusetts) using a linear gradient of 100% solvent A (chloroform–acetic acid, 500:0.5, v/v) to 50% solvent B (chloroform–methanol–acetic acid, 450:50:0.5, v/v/v) over 1 hr at 4 ml/min in 4-ml fractions. PGD-M was found to elute in each case with a retention volume of 116–136 ml. PGD-M was then further purified on a 3.9 mm i.d. × 30 cm fatty acid analysis column (Waters Associates, Milford, Massachusetts) under isocratic conditions with the solvent water: acetonitrile–acetic acid (500:40:0.5, v/v/v), 1 ml/min, 1 ml fractions. PGD-M eluted with a retention volume of 28–32 ml.

Incorporation of Deuterium and Tritium Labels into Unlabeled PGD-M

Available commercially are the necessary deuterium-labeled reagents for the preparation of deuterium-labeled diazomethane of high isotopic purity for methylation reactions (Deutero-Diazald Prep Set, Aldrich Chemical Co., Milwaukee, Wisconsin). This provides a simple method for preparation of a deuterium-labeled methyl ester of PGD-M. Formation of methyl esters using ethereal diazomethane is normally carried out in the presence of methanol, which greatly facilitates the reaction by serving as a proton donor (A, Fig. 1). Methylation using deuterium-labeled ethereal diazomethane results in the formation of a methyl ester with two deuterium atoms (B, Fig. 1). If, however, unlabeled methanol is replaced with a tritium donor during the methylation reaction, a tritium atom instead of a hydrogen atom will be incorporated into the methyl ester (C, Fig. 1). The specific activity of the methyl ester formed by this procedure depends on the specific activity of the tritium donor used. Satisfactory for this purpose is tritiated water, which is available with a very high specific activity (90 mCi/mmol; New England Nuclear, Boston, Massachusetts). Because a molecule of water contains two hydrogen atoms, a monocarboxylic acid methylated in the presence of 90 mCi/mmol tritiated water would theoretically yield an ester with one-half the molar specific activity of the water, whereas a dicarboxylic acid (such as PGD-M) would yield the bis-esterified compound with the same molar specific activity as the water.

The PGD-M obtained following the purification procedures outlined above was only partially pure, and visible impurities were present. Prior

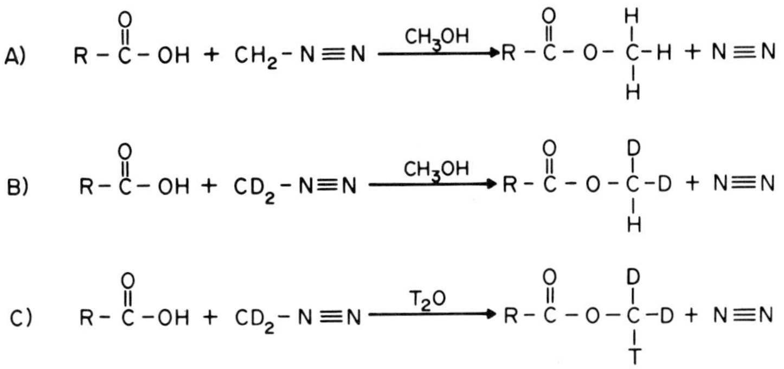

$$\text{A)}\quad R-\overset{O}{\overset{\|}{C}}-OH + CH_2-N\equiv N \xrightarrow{CH_3OH} R-\overset{O}{\overset{\|}{C}}-O-\underset{H}{\overset{H}{C}}-H + N\equiv N$$

$$\text{B)}\quad R-\overset{O}{\overset{\|}{C}}-OH + CD_2-N\equiv N \xrightarrow{CH_3OH} R-\overset{O}{\overset{\|}{C}}-O-\underset{H}{\overset{D}{C}}-D + N\equiv N$$

$$\text{C)}\quad R-\overset{O}{\overset{\|}{C}}-OH + CD_2-N\equiv N \xrightarrow{T_2O} R-\overset{O}{\overset{\|}{C}}-O-\underset{T}{\overset{D}{C}}-D + N\equiv N$$

FIG. 1. Formation of methyl esters of carboxylic acids using diazomethane. (A) Methyl ester formed using unlabeled diazomethane in the presence of unlabeled methanol. (B) Methyl ester formed using deuterium-labeled diazomethane in the presence of unlabeled methanol. (C) Methyl ester formed using deuterium-labeled diazomethane in the presence of tritiated water.

to methylation, the material obtained was washed twice with 20 μl of tritiated water (90 mCi/mmol) to exchange hydrogen atoms with tritium atoms in carboxylic acid compounds or other impurities in the sample that could serve as proton donors during the methylation reaction. The sample was then dissolved in 25 μl of tritiated water and 250 μl of acetone, and then excess deuterium-labeled ethereal diazomethane was added and the solution was allowed to stand at room temperature for 5 min. Acetone was added to effect miscibility between the tritiated water and the ethereal diazomethane.

After esterification, the PGD-M methyl ester was further purified by HPLC on a μPorasil column with the solvent chloroform under isocratic conditions, 2 ml/min, 2-ml fractions. The chromatogram obtained (Fig. 2) revealed that a large quantity of radiolabeled impurities were separated by this purification step. The peak containing PGD-M eluted at approximately 21–25 ml retention volume. When the polarity of the PGD-M-methyl ester was altered by conversion of an aliquot of the sample to an oxime derivative and rechromatographed, a substantial portion of the radioactivity did not coelute with PGD-M, indicating the presence of additional radiolabeled impurities. However, gas chromatographic and mass spectrometric-selected ion monitoring analysis of the PGD-M-methyl ester obtained after the isocratic chloroform HPLC step was free of interfering peaks.

The mass spectrum of PGD-M obtained prior to the treatment with deuterium-labeled diazomethane as an unlabeled methyl ester, trimethylsilyl oxime, trimethylsilyl ether derivative is shown at the top of Fig. 3. The mass spectrum of PGD-M obtained after treatment with the deute-

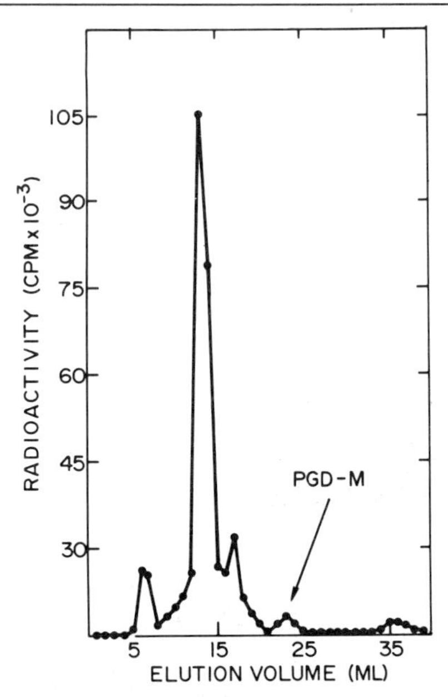

FIG. 2. High-pressure liquid chromatography on a μPorasil column of PGD-M methyl ester obtained after esterification with deuterium-labeled diazomethane in the presence of tritiated water. Solvent, chloroform; conditions, isocratic; flow rate, 2 ml/min; fraction volume, 2 ml.

rium-labeled diazomethane and conversion to a trimethylsilyl oxime, trimethylsilyl ether derivative is shown at the bottom of Fig. 3. Ions containing both ester groups, such as the intense ion at m/z 511 ($M^+ - 89$, loss of Me_3SiO), were shifted up four mass units in the deuterium-labeled PGD-M. Selected ion monitoring of m/z 511/515 in the labeled PGD-M revealed a $d_0:d_4$ ratio of 0.030.

Standardization of the PGD-M internal standard was accomplished by gas chromatography–mass spectrometry (GC–MS) against $PGF_{2\alpha}$. An aliquot of PGD-M and a known quantity of $PGF_{2\alpha}$ methyl ester were combined and coderivatized first by treatment with 3% hydroxylamine·HCL in pyridine at room temperature for 18 hr, converting the methyl ester of PGD-M to an oxime derivative. The pyridine was evaporated under a stream of nitrogen followed by the addition of water and extraction into ethyl acetate. The PGD-M was then converted to a trimethylsilyl oxime, trimethylsilyl ether derivative, and the $PGF_{2\alpha}$ to a trimethylsilyl ether derivative, by treatment with *N,O*-bis(trimethylsilyl)trifluoroacetamide in pyridine at 70° for 30 min. Quantification was accomplished by comparing the areas under the total ion current peaks of PGD-M and $PGF_{2\alpha}$ using a

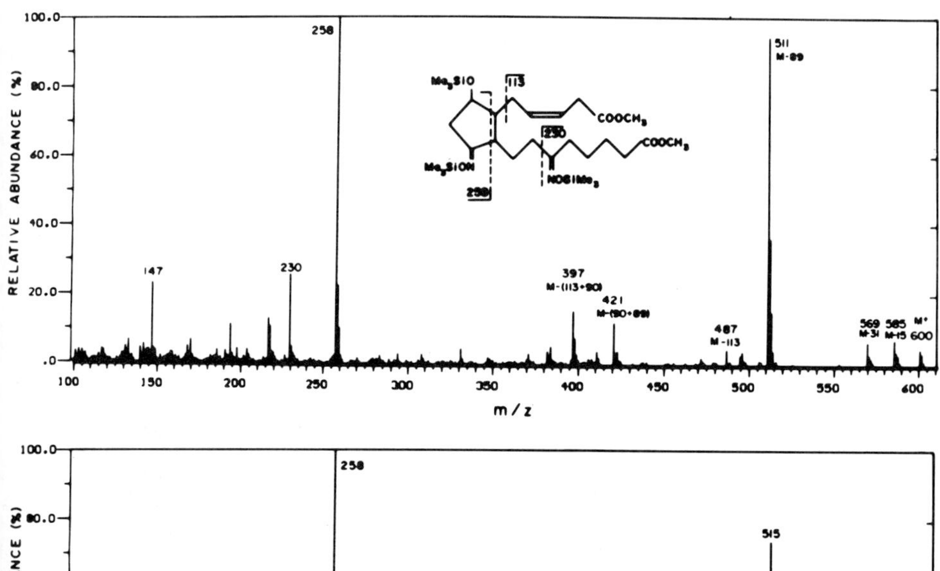

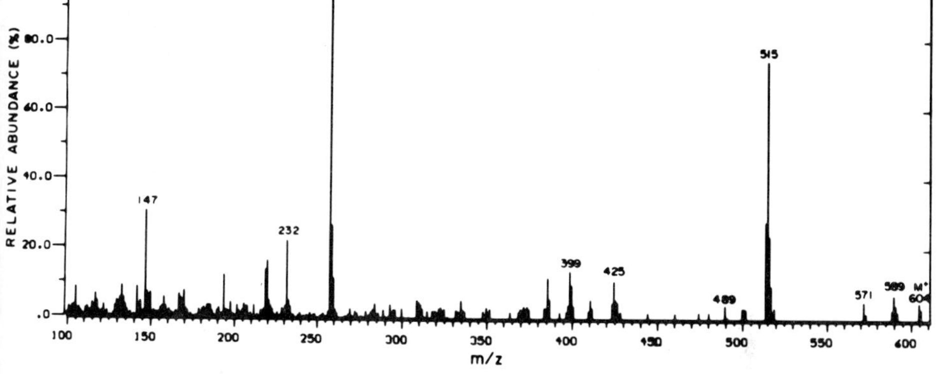

FIG. 3. *Top:* Mass spectrum of unlabeled PGD-M as a methyl ester, trimethylsilyl oxime, trimethylsilyl ether derivative. *Bottom:* Mass spectrum of PGD-M after esterification with deuterium-labeled diazomethane and conversion to a trimethylsilyl oxime, trimethylsilyl ether derivative.

6-foot Dexsil-300 gas chromatographic column operated isothermally at 230°.

Procedure for Quantification of Endogenous Urinary PGD-M

For quantification of urinary PGD-M, 5–10% of a 24-hr urine collection is assayed. Extraction of the urine is accomplished by Amberlite XAD-2 chromatography. The PGD-M internal standard is in the form of a methyl ester, whereas the endogenous PGD-M is in the free acid form. A comparison was made of the recovery and rates of elution from Amberlite XAD-2 between the free acid form of 11α-hydroxy-9,15-diketo-2,3,4,5-

tetranorprostane-1,20-dioic acid versus the methyl ester derivative, and these were found to be different. Therefore, the internal standard of PGD-M is not added to the urine until after the extraction is performed. To calculate recovery from the extraction, 50,000 cpm of the free acid of $PGF_{2\alpha}$ is initially added to the urine. The urine is acidified to pH 3.2 with 2 *N* HCl and percolated through approximately a 2.1 × 15 cm column of Amberlite XAD-2. The column is washed to neutrality with distilled water and subsequently eluted with approximately 100 ml of acetone. The recoveries are invariably 85–100%. Urines with lower recoveries are discarded, and another aliquot is extracted. One percent of the extracted urine is assayed for $PGF_{2\alpha}$ radioactivity to calculate recovery. The quantity of PGD-M quantified at the end of the assay is corrected for the extraction loss.

After extraction, solvents are evaporated under reduced pressure in a conical flask, and the residue is dissolved in 3.5 ml of methanol, 7 ml of ethereal diazomethane are added, and the mixture is allowed to stand at room temperature for 10 min. It is critical with this step, for accurate quantification, that excess diazomethane be added to ensure complete methylation of the urinary PGD-M. To assess whether excess diazomethane was added, the solvent mixture is evaporated under reduced pressure in a rotary evaporator while noting the color of the solvent distillate obtained. If the distillate is not yellow, a yellow color indicating excess diazomethane, the methylation procedure is repeated until complete methylation has been assured. After methylation, 35 ng of the PGD-M methyl ester internal standard is added.

The extracted residue is then transferred to a separatory funnel with methanol; 2 volumes of water are added and then extracted twice into an equal volume of diethyl ether. The diethyl ether is evaporated under reduced pressure, the residue is dissolved in methylene chloride–diethyl ether (95:5, v/v) and applied to approximately a 2-g column of silicic acid. The column is then washed with approximately 3 column volumes of methylene chloride: diethyl ether (95:5, v/v). This solvent mixture does not elute PGD-M, but it does elute substantial quantities of less polar impurities, which, if not removed at this stage, tend to overload the HPLC column employed at the next purification step, resulting in a substantial reduction in purification of PGD-M than is otherwise attainable. PGD-M is then subsequently eluted from the silicic acid column with approximately 3 column volumes of methylene chloride–diethyl ether (150:50, v/v).

Further purification of PGD-M is accomplished by HPLC. The residue obtained after silicic acid chromatography is dissolved in chloroform and subjected to chromatography on a 3.9 mm i.d. × 30 cm μPorasil column under isocratic conditions, flow rate 2 ml/min, 2-ml fractions. PGD-M

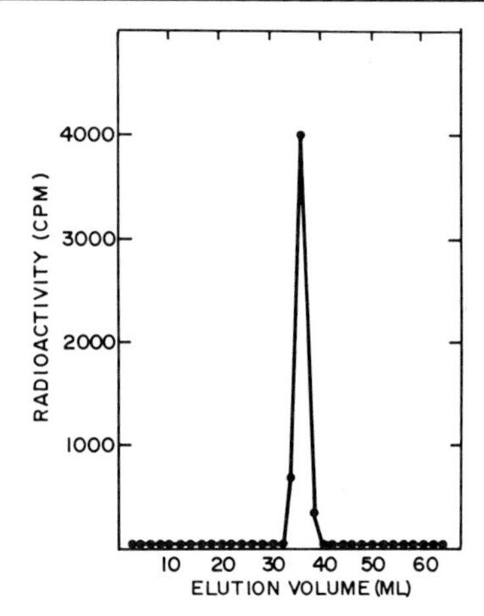

FIG. 4. High-pressure liquid chromatography on a μPorasil column of urinary PGD-M methyl ester and the PGD-M methyl ester internal standard. Solvent, chloroform; conditions, isocratic; flow rate, 2 ml/min; fraction volume, 2 ml.

characteristically elutes with a retention volume of approximately 34–38 ml (Fig. 4). The PGD-M peak is pooled and then subjected to reversed-phase chromatography on a 3.9 mm i.d. × 30 cm fatty acid analysis column under isocratic conditions with the solvent water–acetonitrile (75:25; (v/v), flow rate 2 ml/min, 2-ml fractions. PGD-M characteristically elutes from this column with a retention volume of approximately 18–22 ml (Fig. 5). PGD-M is extracted from the aqueous solvent with diethyl ether.

For quantification by GC–MS, the PGD-M is converted to a trimethylsilyl oxime, trimethylsilyl ether derivative as described above. This derivative was chosen for analysis because of the intense $M^+ - 89$ ion present at high mass in the mass spectrum of this derivative (*m/z* 511 for endogeneous PGD-M and *m/z* 515 for [$^2H_4$]PGD-M) (Fig. 2). This ion represents the loss of Me_3SiO from a trimethylsilyl oxime group. The mass spectrum of the methyl ester, *O*-methyloxime, trimethylsilyl ether derivative of PGD-M,[5] a derivative more customarily used with ketone containing prostaglandins, does not contain an intense ion of high mass and is therefore a less suitable derivative for selected ion monitoring. Quantification is accomplished by determining ion peak ratios of *m/z* 511 and *m/z* 515 using a Hewlett-Packard 5980A combined gas chromatograph–mass spectrometer with a 3-foot OV-1 column operated isothermally at 240°.

An outline of the procedures followed for this assay of PGD-M is summarized in Fig. 6.

A representative selected ion current chromatogram of *m/z* 511/515

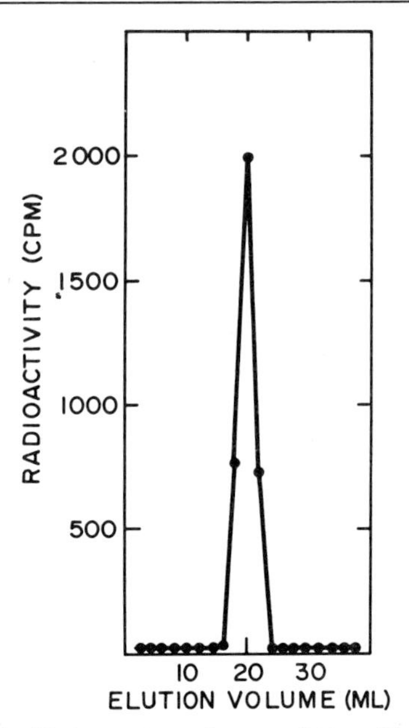

FIG. 5. High-pressure liquid chromatography on a fatty acid analysis column of urinary PGD-M methyl ester and PGD-M methyl ester internal standard. Solvent, water–acetonitrite (75:25, v/v); conditions, isocratic; flow rate, 2 ml; fraction volume, 2 ml.

obtained following purification of urinary PGD-M as outlined above is shown in Fig. 7. The level of PGD-M excreted by normal individuals in three males and one female has been found to range from 240 to 416 ng in 24 hr with a mean ± SD of 286 ± 75 ng in 24 hr. These levels expressed in picograms per milligram of urinary creatinine range from 132 to 353 with a mean ± SD of 193 ± 87.

Our previous study in which marked overproduction of PGD_2 was discovered in two patients with systemic mastocytosis provided the stimulus for the development of this stable isotope dilution assay for PGD-M. Initially, the finding of overproduction of PGD_2 in those two patients was semiquantitative. However, we have since quantified the level of excretion of urinary PGD-M in these two patients. The level of PGD-M found in our first patient was 59,676 ng/24 hr, and the level of excretion found in urine from the second patient was 8274 ng/24 hr. This use of this assay has now allowed us further to document overproduction of PGD_2 in many additional patients with systemic mastocytosis and has proved to be invaluable in making rational clinical decisions regarding whether or not to treat individual patients with prostaglandin biosynthesis inhibitors to control the manifestations of their disease.

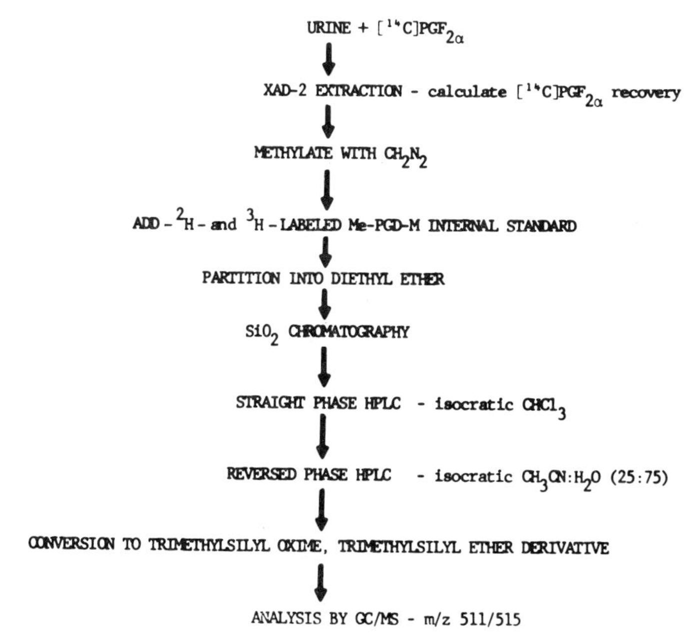

FIG. 6. Outline of procedures followed for the stable isotope dilution assay for urinary PGD-M.

This assay provides an accurate means to assess the endogenous production of PGD_2 in man. Future developments in the area of chemical synthesis of a labeled PGD-M internal standard should provide an accessible means to obtain the internal standard for stable isotope dilution assay of PGD-M. In addition, with the use of mass spectrometers with greater sensitivity and high resolution, combined with capillary gas chromatography, it should be possible substantially to reduce the number of purification procedures that are currently required prior to analysis of PGD-M by GC–MS. Such developments should greatly facilitate both the efficiency of analysis and sample capacity of the assay.

In addition to disorders of mast cell proliferation, this assay should now allow us to begin to explore the role of PGD_2 in other disease states. Disorders of immediate hypersensitivity involving abnormal mast cell activation are certainly an attractive area for investigation in this regard. As mentioned above, PGD_2 is also apparently metabolically transformed in part to metabolites with a PGF ring. It will, therefore, also be of interest to determine the production of PGD_2 in disorders in which the level of

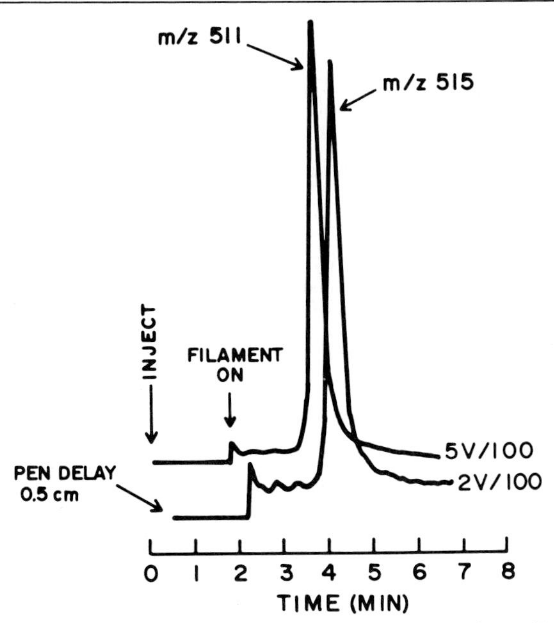

FIG. 7. Representative selected ion current chromatogram of *m/z* 511/515 obtained during quantification of PGD-M in human urine.

plasma or urinary metabolites of $PGF_{2\alpha}$ have been found to be elevated, such as pregnancy,[10] allergic asthma,[11] and medullary carcinoma of the thyroid.[12]

Acknowledgments

This work was supported in part by National Institutes of Health Grants GM 15431 and M01 RR0095. The expert technical assistance of Victor C. Folarin and Robin L. Fuchs was greatly appreciated.

[10] M. Hamberg, *Life Sci.* **14,** 247 (1974).

[11] K. Gréen, P. Hedqvist, and N. Svanborg, *Lancet* **1,** 1419 (1974).

[12] L. J. Roberts, II, W. C. Hubbard, Z. T. Bloomgarden, X. Y. Bertagna, T. J. McKenna, D. Rabinowitz, and J. A. Oates, *Trans. Assoc. Am. Physicians* **92,** 286 (1979).

[67] Quantitation of 15-Keto-13,14-dihydro-PGE_2 in Plasma by GC–MS[1]

By WALTER C. HUBBARD

Measurement of the 15-keto-13,14-dihydro metabolite of prostaglandin E_2 (PGE_2) in plasma is a more reliable index of the entry of PGE_2 into the circulation than direct quantitation of PGE_2.[1–3] Studies show that 15-keto-13,14-dihydroprostaglandin E_2 (15K-H_2-PGE_2) rapidly decomposes in plasma, physiological buffers, and volatile alcohols by first-order kinetics dependent upon pH, temperature, and albumin concentration.[4–7] The instability of 15K-H_2-PGE_2 suggests that entry of PGE_2 into the circulation as determined by measurement of 15K-H_2-PGE_2 in previous studies may have been underestimated.[8–16] The analytical problems associated with the instability of 15K-H_2-PGE_2 are further complicated by the necessity of frequent blood sampling due to the relatively short half-life of the PGE_2 metabolite in the circulation. In one study,[15] sampling of blood for quantitation of 15K-H_2-PGE_2 was as frequent as every 2–5 min, with 8–12 samples drawn within a period of 40–45 min. The acquisition of this large number of samples within such a short time period requires the use

[1] This work was supported by Grant GM 15431 from the National Institutes of Health.

[1] B. Samuelsson, E. Granström, K. Gréen, M. Hamberg, and S. Hammarström, *Annu. Rev. Biochem.* **44,** 695 (1975).

[2] B. Samuelsson, M. Goldyne, E. Granström, M. Hamberg, S. Hammarström, and C. Malmsten, *Annu. Rev. Biochem.* **47,** 997 (1978).

[3] E. Granström and B. Samuelsson, *Adv. Prostaglandin Thromboxane Res.* **5,** 1 (1978).

[4] M. D. Mitchell, H. Sors, and A. P. F. Flint, *Lancet* **2,** 558, (1977).

[5] E. Granström and H. Kindahl, *Adv. Prostaglandin Thromboxane Res.* **6,** 181 (1980).

[6] E. Granström, M. Hamberg, G. Hanson, and H. Kindahl, *Prostaglandins* **19,** 933 (1980).

[7] F. A. Fitzpatrick, R. Aguire, J. E. Pike, and F. H. Lincoln, *Prostaglandins* **19,** 917 (1980).

[8] B. Samuelsson and K. Gréen, *Biochem. Med.* **11,** 298 (1974).

[9] W. C. Hubbard and J. T. Watson, *Prostaglandins* **12,** 21 (1976).

[10] L. Levine, *Prostaglandins* **14,** 1125 (1977).

[11] D. Gordon, L. Myatt, A. Gordon-Wright, J. Hanson, and M. G. Elder, *Prostaglandins* **13,** 399 (1977).

[12] A. H. Tashjian, Jr., E. F. Voelkel, and L. Levine, *Biochem. Biophys. Res. Commun.* **74,** 199 (1977).

[13] A. H. Tashjian, Jr., E. F. Voelkel, and L. Levine, *Prostaglandins* **14,** 309 (1977).

[14] H. W. Seyberth, W. C. Hubbard, O. Oelz, B. J. Sweetman, J. T. Watson, and J. A. Oates, *Prostaglandins* **14,** 319 (1977).

[15] L. J. Roberts, II, W. C. Hubbard, Z. T. Bloomgarden, X. Y. Bertagna, T. J. McKenna, D. Rabinowitz, and J. A. Oates, *Trans. Assoc. Am. Physicians* **92,** 286 (1979).

[16] W. C. Hubbard, A. J. Hough, R. M. Johnson, and J. A. Oates, *Prostaglandins* **19,** 881 (1980).

ISBN 0-12-181986-8

of extraordinary procedures to minimize the loss of endogenous 15K-H_2-PGE_2, which can be appreciable within a period of 45 min. Moreover, it is essential that the PGE_2 metabolite be converted to a more stable derivative to minimize additional loss of 15K-H_2-PGE_2 during purification for analysis via combined gas-liquid chromatography–mass spectrometry (GC–MS).

The accuracy of quantitation of 15K-H_2-PGE_2 levels in the circulation via combined GC–MS analysis is not seriously affected if the internal standard of 15K-H_2-PGE_2 is added immediately upon collection of the samples. At this time, there are only two published GC–MS methods for quantitation of 15K-H_2-PGE_2 in plasma.[8,9] A serious impediment for development of additional GC–MS assays for quantitation of 15K-H_2-PGE_2 is the lack of availability of a suitable stable isotope analog of the PGE_2 metabolite as an internal standard and carrier.

This report details procedures for preparation of 3,3,4,4-tetradeutero-15K-H_2-PGE_2 for use as an internal standard for GC–MS assay of the PGE_2 metabolite, for minimization of loss of endogenous 15K-H_2-PGE_2 in blood samples prior to addition of the tetradeuterated internal standard, and for stabilization of 15K-H_2-PGE_2 via derivatization for prevention of additional loss during purification.

Biosynthesis, Purification, Evaluation, and Storage of 3,3,4,4-Tetradeutero-15K-H_2-PGE_2

3,3,4,4-Tetradeutero-15K-H_2-PGE_2 and 5,6,8,11,12,14-hexatritiated-15K-H_2-PGE_2 ([^{3}H]15K-H_2-PGE_2) are simultaneously prepared from 3,3,4,4-tetradeutero-PGE_2 (Merck, Sharp and Dohme Canada Limited) and 5,6,8,11,12,14,15-heptatritiated-PGE_2 ([^{3}H]PGE_2), respectively, via incubation with the 100,000 *g* subcellular fraction of either swine kidney or guinea pig liver employing modifications of described procedures.[17–19] Two milligrams 3,3,4,4-tetradeutero-PGE_2 and 250 μCi of [^{3}H]PGE_2 (100–200 Ci/mmol; Amersham) are incubated with the 100,000 *g* supernatant from 20 g of fresh swine kidney or 50 g of guinea pig liver for 20 min at 37°. EDTA (10^{-3} *M*), 5,8,11,14-eicosatetraynoic acid (10^{-4} *M*) and mercaptoethanol (0.06%) are added prior to homogenization, subcellular fractionation, and incubation procedures to prevent formation of 15K-H_2-PGE_2 from endogenous precursors. After termination of the reaction mixture via protein precipitation with cold acetone (3 × incubation volume), the reaction mixture is rapidly chilled in crushed ice. The follow-

[17] K. Gréen, E. Granström, B. Samuelsson, and U. Axén, *Anal. Biochem.* **54,** 434 (1973).

[18] M. Hamberg and V. Israelsson, *J. Biol. Chem.* **245,** 5107 (1972).

[19] K. Gréen, M. Hamberg, B. Samuelsson, M. Smigel, and J. C. Frölich, *Adv. Prostaglandin Thromboxane Res.* **5,** 39 (1978).

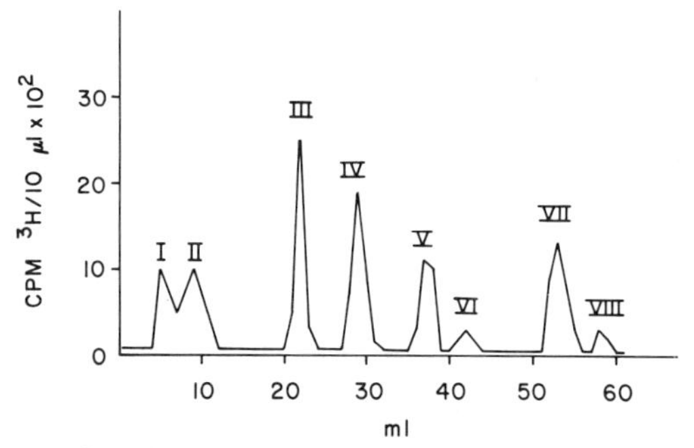

FIG. 1. HPLC separation of products of 3,3,4,4-tetradeutero-PGE_2 subsequent to incubation with the 100,000 *g* soluble fraction of swine kidney and guinea pig liver. Details of the methodology are presented in the text.

ing extraction procedures are performed at 0–4°. Neutral lipids are removed via extraction at neutral pH with cold (0–4°) petroleum ether. Acidic lipids including products derived from 3,3,4,4-tetradeutero-PGE_2 and [^{3}H]PGE_2 are extracted at pH 3 (titration with 6 *N* HCl) with cold (0–4°) chloroform (3 × incubation volume). Rapid separation of the aqueous–chloroform layers is achieved via centrifugation at 250 *g* for 5 min at 0–4°.

The aqueous (upper) layer is carefully aspirated and discarded. The chloroform (lower) layer is transferred to an evaporation flask and rotoevaporated to dryness under reduced pressure.

The residue in the rotoevaporation flask is transferred to a 1 cm × 18 cm silicic acid column (Porasil A; 37–75 μm particle size; Waters Associates) in 2–3 ml of chloroform–glacial acetic acid (500:1, v/v). The silicic acid column is eluted sequentially with 40 ml (10 × 4 ml fractions) of chloroform–glacial acetic acid (500:1, v/v) and 40 ml (10 × 4 ml) of chloroform–methanol–glacial acetic acid (450:50:1, v/v/v). The fractions with ^{3}H activity eluted from the column via chloroform–methanol–acetic acid are transferred to a rotary evaporation flask and rapidly rotoevaporated to dryness under reduced pressure. The residue is transferred to a 3.5-ml silanized glass vial with 3 ml (3 × 1 ml) of chloroform–methanol–glacial acetic acid (450:50:1, v/v/v), the solvent is removed under a nitrogen stream, and the sample is redissolved in 0.5 ml of chloroform–glacial acetic acid (500:1, v/v) for purification via high-performance liquid chromatography (HPLC).

Final purification of 3,3,4,4-tetradeutero-15K-H_2-PGE_2 is achieved via HPLC as described.[20] The 3,3,4,4-tetradeutero analog of 15K-H_2-PGE_2 is

[20] W. C. Hubbard, J. T. Watson, and B. J. Sweetman, *Biol. Biomed. Appl. Liquid Chromatogr., Chromatogr. Sci.* **10,** 31 (1979).

PRODUCTS FORMED FROM 3,3,4,4-TETRADEUTERO-PGE_2 BY 100,000 *g* SOLUBLE FRACTION OF SWINE KIDNEY AND GUINEA PIG LIVER

Peak(s)	Product(s)
I, II	Dehydration products of 15K-H_2-PGE_2
III	15-Keto-13,14-dihydro-PGE_2
IV	15-Keto-13,14-dihydro-$PGF_{2\alpha}$
V	13,14-Dihydro-PGE_2
VI	PGE_2 (unreacted)
VII	13,14-Dihydro-$PGF_{2\alpha}$
VIII	$PGF_{2\alpha}$

separated from a number of other products formed from 3,3,4,4-tetradeutero-PGE_2 formed in the 100,000 *g* supernatant fraction of swine kidney and guinea pig liver tissues as shown in Fig. 1. The identity of these products is listed in the table. Yields of 3,3,4,4-tetradeutero-15K-H_2-PGE_2 from 2 mg of 3,3,4,4-tetradeutero-PGE_2 are 8–16% (160–320 μg). The yield of 3,3,4,4-tetradeutero-15K-H_2-PGE_2 from 3,3,4,4-tetradeutero-PGE_2 varies inversely with the length of incubation time, 20 min incubation time being optimal.

The suitability of biosynthesized 3,3,4,4-tetradeutero-15K-H_2-PGE_2 as an internal standard and carrier for GC-MS quantitation of 15K-H_2-PGE_2 is evaluated via determination of the extent of dilution of deuterium-labeled compound with endogenous 15K-H_2-PGE_2 during the homogenization and incubation procedures. For this evaluation the ratio of unlabeled or nondeuterated ($^2H_0$) species to tetradeuterated ($^2H_4$) species of 3,3,4,4-tetradeutero-PGE_2 and biosynthesized 3,3,4,4-tetradeutero-15K-H_2-PGE_2 is determined. The $^2H_0 : {}^2H_4$ ratio of 3,3,4,4-tetradeutero-PGE_2 employed for the biosynthesis of 3,3,4,4-tetradeutero-15K-H_2-PGE_2 was 0.0020–0.0025 (2.0–2.5 parts per thousand). The $^2H_0 : {}^2H_4$ ratio of 3,3,4,4-tetradeutero-15K-H_2-PGE_2 biosynthesized with the crude enzyme preparation from swine kidney and guinea pig liver was 0.0025–0.0035 (2.5–3.5 parts per thousand). These data suggest that the dilution of 3,3,4,4-tetradeutero-15K-H_2-PGE_2 with endogenous 15K-H_2-PGE_2 is minimal and compares favorably with chemically synthesized 3,3,4,4-tetradeutero-15K-H_2-PGE_2 employed in earlier published studies.[8,9,20]

Purified 3,3,4,4-tetradeutero-15K-H_2-PGE_2 is stored at $-20°$ in ethanol : water (7 : 3) in a concentration of ~5 mg/ml. The deuterated analog of 15K-H_2-PGE_2 is diluted to the desired concentration with ethanol–water (7 : 3) immediately prior to its addition to a sample for analysis of 15K-H_2-PGE_2.

Addition of Internal Standard and Tracer

Blood samples (10–30 ml) are rapidly transferred to chilled 50-ml polypropylene centrifuge tubes (partially submerged in crushed ice) containing an anticoagulant and a mixture of deuterium-labeled and tritium-labeled 15K-H_2-PGE_2 (~0.5 μg and ~250,000 dpm, respectively) dissolved in ethanol–water (70:30). The blood sample should be added to the tube after removal of the needle to avoid extensive hemolysis of red cells. The blood is rapidly mixed with the anticoagulant and labeled analogs of 15K-H_2-PGE_2 by repeated gentle inversion of the centrifuge tube and chilled (replacement of centrifuge tube containing blood sample in crushed ice). Harsh shaking and rapid inversion of the centrifuge tube may increase hemolysis of erythrocytes. These procedures limit the loss of endogenous 15K-H_2-PGE_2 via degradation without proportionate degradative loss of the internal standard and carrier to the time required for sampling and equilibration of the blood sample with the internal standard and carrier present in the centrifuge tube (usually less than 1.5 min).

Isolation of Plasma; Protein Removal; Extraction; Solvent Removal

The plasma is isolated from the cellular components of whole blood by centrifugation at 2000 *g* for 15 min at 0–4°. The plasma is gently removed from the blood cells and transferred to a cold 110-ml glass centrifuge tube (partially submerged in crushed ice). The volume of plasma is recorded. Evidence of hemolysis of red cells (resulting from the addition of 70% ethanol to whole blood) may be seen. Usually, greater than 90% of the tritium label added to the blood sample is recovered in the plasma.

The plasma proteins are removed via precipitation with cold (0–4°) acetone (3 times plasma volume) followed by centrifugation at 750 *g* for 5 min at 0–4°. The acetone–aqueous layer is transferred to a second chilled 110-ml glass centrifuge tube (partially submerged in crushed ice) for extraction.

Neutral lipids are removed from the acetone–aqueous layer via extraction at neutral pH with cold (0–4°) petroleum ether (3 times plasma volume). After acidification of the acetone–aqueous layer to pH 3 with 6 *N* HCl, 15K-H_2-PGE_2 and other acidic lipids are extracted with cold (0–4°) chloroform (3 times plasma volume). Low speed centrifugation (250 *g* for 5 min at 0–4°) may be required for complete separation of the chloroform and aqueous layers. The aqueous (upper) layer is carefully removed via aspiration and discarded. The chloroform (lower) layer is transferred to a rotary evaporation flask and the solvent rapidly removed under reduced pressure. The plasma extract is transferred to a 3.5-ml silanized glass vial with chloroform for partial derivatization. Recovery of tritium label at this point of the procedure usually exceeds 60% of that added to whole blood.

Partial Derivatization (Oximation and Esterification)

The plasma extract containing 15K-H_2-PGE_2 is treated with 100 μl of a saturated solution of methoxyamine-HCl in pyridine (2 g/100 ml) at room temperature for 18 hr. After oximation, the pyridine is removed under a nitrogen stream, leaving a residue of plasma extract and crystalline methoxyamine HCl. The bismethyloxime derivative of 15K-H_2-PGE_2 and other lipids present in the residue are extracted with anhydrous diethyl ether (2 × 0.5 ml), leaving a residue consisting principally of crystalline methoxyamine-HCl. After transfer to a silanized 1.0-ml conical vial, the extract is dissolved in 100 μl of methanol and treated with excess ethereal diazomethane for 15 min at 0–4° (vials partially submerged in crushed ice) followed by removal of the excess ethereal diazomethane and methanol under a nitrogen stream. The bismethyloxime-methyl ester derivative of 15K-H_2-PGE_2 contained in the extract is then dissolved in 0.25–0.50 ml of chloroform for storage until purification by chromatography. Methoximation, extraction, and esterification is essentially quantitative resulting in no appreciable loss of tritium label.

Storage of Partially Derivatized Extracts

The bismethyloxime-methyl ester derivative of 15K-H_2-PGE_2 present in the plasma extract can be stored for a period of 10 days at −20° dissolved in chloroform without appreciable degradation of the partially derivatized PGE_2 metabolite. We have no experience with storage of samples under these conditions for longer periods of time.

Purification and Silylation

Initial purification of the bismethyloxime-methyl ester derivative of 15K-H_2-PGE_2 contained in the plasma extract is achieved via silicic acid chromatography. The plasma extract is applied to a 0.5 cm × 10 cm silicic acid column (Porasil A; 37–75 μm particle size; Waters Associates) in chloroform. The column is eluted sequentially with 10 ml (5 × 2 ml fractions) of chloroform and 20 ml (10 × 2 ml fractions) of chloroform–methanol (95:5, v/v). The portion of tritium label (usually less than 10%) added to the blood eluted by chloroform from this column cochromatographs with derivatized dehydration products of 15K-H_2-PGE_2 when analyzed via high-performance liquid chromatography (HPLC). This procedure removes most of pigment present in plasma and products of

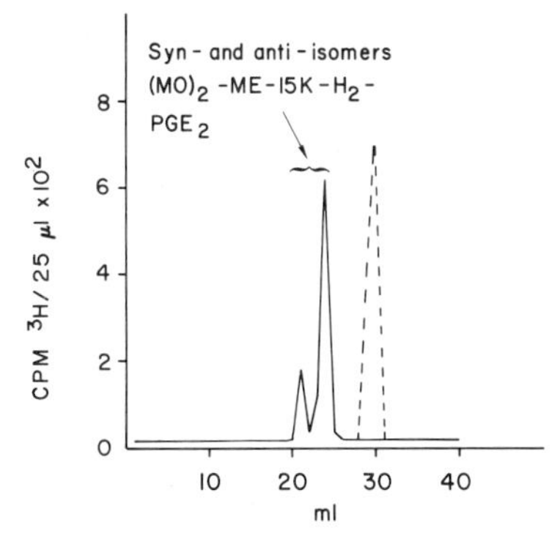

FIG. 2. HPLC separation of syn- and anti-isomers of the bismethyloxime-methyl ester derivative of 15K-H_2-PGE_2. A linear solvent program from chloroform to chloroform–methanol (96:4) over a period of 40 min was employed. The solvent flow rate was 1.0 ml/min. Fraction volume = 1.0 ml.

hemolyzed erythrocytes and greatly facilitates further purification via HPLC. The bismethyloxime-methyl ester derivative of 15K-H_2-PGE_2 is eluted by chloroform–methanol (95:5, v/v). The fractions containing ^3H activity are combined; the solvent is removed under reduced pressure and transferred to a 1.0-ml conical vial with 0.5 ml of chloroform. Recovery of tritium label added to blood after these procedures usually exceeds 45% of that added to the blood sample.

Further purification of the bismethyloxime-methyl ester derivative of 15K-H_2-PGE_2 is achieved via HPLC employing a straight-phase column via modification of described procedures.[9,18] Syn- and anti-isomers of the bismethyloxime-methyl ester derivative of 15K-H_2-PGE_2 are partially resolved during HPLC as shown in Fig. 2. The two peaks represented by the solid line in Fig. 2 depict elution of the syn- and anti-isomers of the bismethyloxime-methyl ester derivative of 15K-H_2-PGE_2 during straight-phase HPLC. The bismethyloxime-methyl ester derivative of 15K-H_2-PGE_2 contained in the larger peak is transferred to a 1-ml conical vial (either in one step or two steps if volume exceeds capacity of 1-ml vial) for silylation. The recovery of tritium added to blood after collection of the larger peak usually exceeds 25% of that added to the blood sample.

After complete removal of solvent under a nitrogen stream, the HPLC-purified bismethyloxime-methyl ester derivative of 15K-H_2-PGE_2 is converted to the trimethylsilyl ether derivative via reaction with excess silylating reagent. Ten microliters each of pyridine and either BSTFA or BSA (Pierce) are added to the 1.0-ml conical vial, and the vial is vortexted vigorously for 10–15 sec. The reaction is allowed to proceed for at least 3 hr (usually overnight).

Simultaneous Quantitation of 15-Keto-13,14-dihydroprostaglandin $F_{2\alpha}$

Simultaneous quantitation of 15-keto-13,14-dihydroprostaglandin $F_{2\alpha}$ (15K-H_2-$PGF_{2\alpha}$) and 15K-H_2-PGE_2 in plasma may be essential in certain studies.[15] The procedures described above can be used for simultaneous determination of levels of both prostaglandin metabolites in plasma or blood. The methyloxime-methyl ester derivative of 15K-H_2-$PGF_{2\alpha}$ is well resolved from the syn- and anti-isomers of the bismethyloxime-methyl ester derivative of 15K-H_2-PGF_2 as shown in Fig. 2 (dashed line in this figure depicts the retention volume of the methyloxime-methyl ester derivative of 15K-H_2-$PGF_{2\alpha}$). The syn- and anti-isomers of this derivative of 15K-H_2-$PGE_{2\alpha}$ are not resolved during HPLC purification.

GC–MS Analysis

Quantitation of 15K-H_2-PGE_2 in plasma employing GC–MS is performed by simultaneously monitoring characteristic fragment ions of the bismethyloxime-methyl ester–trimethylsilyl ether derivative of 15K-H_2-PGE_2 (m/z 375) and of 3,3,4,4-tetradeutero-15K-H_2-PGE_2 (m/z 379) as described.[9,19,20] Since the syn- and anti-isomers were partially resolved during HPLC, a single peak for 15K-H_2-PGE_2 and its internal standard may bc obtained. This is in contrast to analysis when the bismethyloxime derivative was synthesized subsequent to HPLC. GC–MS analysis of 15K-H_2-PGE_2 is facilitated by the removal of pyridine and excess silylating reagent under a nitrogen stream, then dissolving the derivatized PGE_2 metabolite and its internal standard in a hydrocarbon solvent such as heptane immediately prior to injection of a portion of the sample into the gas–liquid chromatograph interfaced with the mass spectrometer.

General Considerations

The use of high-purity reagents and solvents in the procedures described above cannot be overemphasized. Solvents recommended for plasma protein removal, extraction, and chromatography are "distilled in glass" quality and have very little residue after evaporation. The quality of reagents recommended are those specifically prepared for gas–liquid chromatography. The use of freshly opened silylating reagents and recently prepared diazomethane and oximation reagent is strongly suggested.

[68] Quantitation of the Major Urinary Metabolite of $PGF_{2\alpha}$ in the Human by GC–MS

By ALAN R. BRASH

The major urinary metabolite of $PGF_{1\alpha}$ and $PGF_{2\alpha}$ in man is $5\alpha,7\alpha$-dihydroxy-11-ketotetranorprostane-1,16-dioic acid (henceforth referred to as PGF-M).[1,2] PGF-M is also a significant urinary metabolite of PGE_2[3] and of PGD_2.[4] For this reason the compound cannot be regarded as a selective indicator of PGF biosynthesis. In fact PGF-M may be the ideal urinary metabolite to monitor in order to detect changes in biosynthesis of the primary prostaglandins as a group.

PGF-M exists as an equilibrium mixture of the dioic acid and the corresponding δ-lactone form (Fig. 1). The equilibrium is pH dependent. Quantitative conversion to the δ-lactone form was found to occur after 5 min treatment of the "open chain" acid at pH 1 and after 1 hr at pH 2, while the rate of lactonization was undetectable at pH 5 over the course of 2 hr at room temperature.[5] Conversely the δ-lactone was found to hydrolyze quantitatively to the acid after 2 hr at room temperature in pH 10 solution. At pH 8 the δ-lactone was stable for 2 hr at room temperature.[5] It is important to recognize that the objective in a quantitative analysis of PGF-M is measurement of the total amount of free acid plus δ-lactone in the sample; the proportions present as the free acid or δ-lactone are irrelevant, since this is merely dependent on sample pH.

Principle of the Method

The assay described below is based on stable isotope dilution with quantitative analysis by selected ion monitoring GC–MS.[6] The steps (shown in Fig. 2) are as follows:

1. Addition of $[^2H_3]$PGF-M to the urine sample as internal standard.
2. Equilibration of the internal standard with the endogenous metabolite by alkali treatment of the urine. The objective is not to ob-

[1] E. Granstöm and B. Samuelsson, *J. Biol. Chem.* **246,** 5254 (1971).
[2] M. Hamberg, *Anal. Biochem.* **55,** 368 (1973).
[3] M. Hamberg and M. Wilson, *Adv. Biosci.* **9,** 39, (1973).
[4] C. K. Ellis, M. D. Smigel, J. A. Oates, O. Oelz, and B. J. Sweetman, *J. Biol. Chem.* **254,** 4152 (1979).
[5] A. R. Brash, *in* "Prostaglandins, Prostacyclin and Thromboxanes Measurement" (J. M. Boeynaems and A. G. Herman, eds.), p. 123. Nijhoff, The Hague, 1980.
[6] A. R. Brash, T. A. Baillie, R. A. Clare, and G. H. Draffan, *Biochem. Med.* **16,** 77 (1976).

ISBN 0-12-181986-8

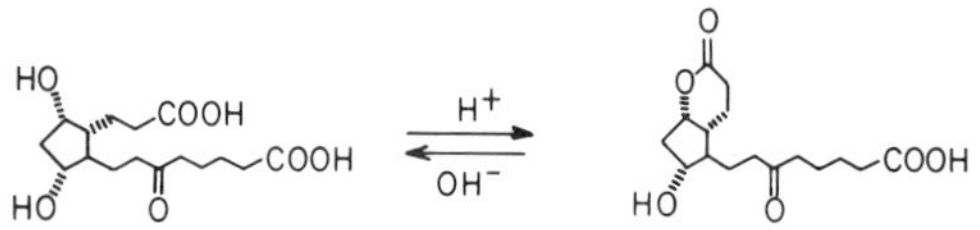

FIG. 1. The dioic acid and δ-lactone forms of 5α,7α-dihydroxy-11-ketotetranorprostane-1,16-dioic acid (PGF-M).

tain 100% formation of the dioic acid *per se,* but rather to convert both labeled and unlabeled metabolite to the same proportion of dioic acid prior to the extraction.

3. Extraction at acidic pH using XAD-2.
4. Conversion to the methyl ester derivative (unless specifically stated otherwise the term PGF-M methyl ester refers to the 1,16-dimethyl ester derivative).
5. Selective hydrolysis of the ester grouping at C-1 by treatment with mild alkali.
6. Interconversion of the resulting C-16 methyl ester derivative between the free acid form at C-1 and the corresponding δ-lactone. By use of appropriate pH manipulations and solvent extractions, PGF-M is isolated from neutral and basic compounds (when the metabolite is in the "open-chain" acid form at C-1) and from conventional urinary acids (when PGF-M is in the neutral δ lactone, C-16 methyl ester form).
7. Conversion back to the methyl ester derivative.
8. Silylation to the *t*-BDMS (*t*-butyldimethylsilyl) ether derivative.
9. Purification by thin-layer chromatography.
10. Quantitative analysis of the ratio of unlabeled to trideuterated molecules by selected ion monitoring GC–MS.

Preparation of Standards

Unlabeled PGF-M. Reference unlabeled PGF-M may be obtained by isolation from the urine of animals or man after the infusion of large quantities of $PGF_{2\alpha}$.[2,6] However the compound has been synthesized chemically and, in the past, reference samples have been available upon request from pharmaceutical companies.

[5 β-^{3}H; 4,6,6-$^2H_3$]PGF-M (Fig. 3)

Prostaglandin E_2 (60 mg) is dissolved in 10 ml of [0-^{2}H]carbitol (or [0-^{2}H]ethanol) containing 44 mg/ml of anhydrous potassium acetate. After standing for at least 3 days at room temperature (during which time

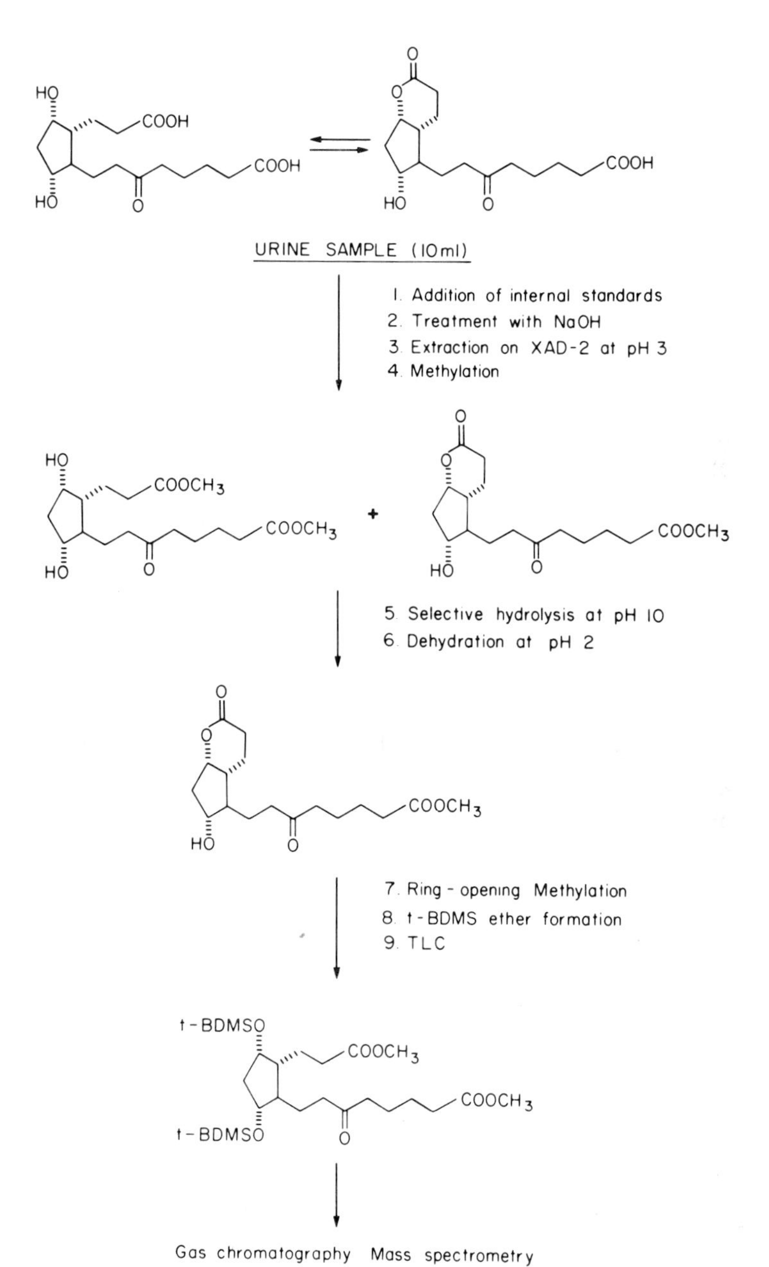

FIG. 2. Urinary assay procedure. Reprinted from Brash *et al.*,[6] with permission.

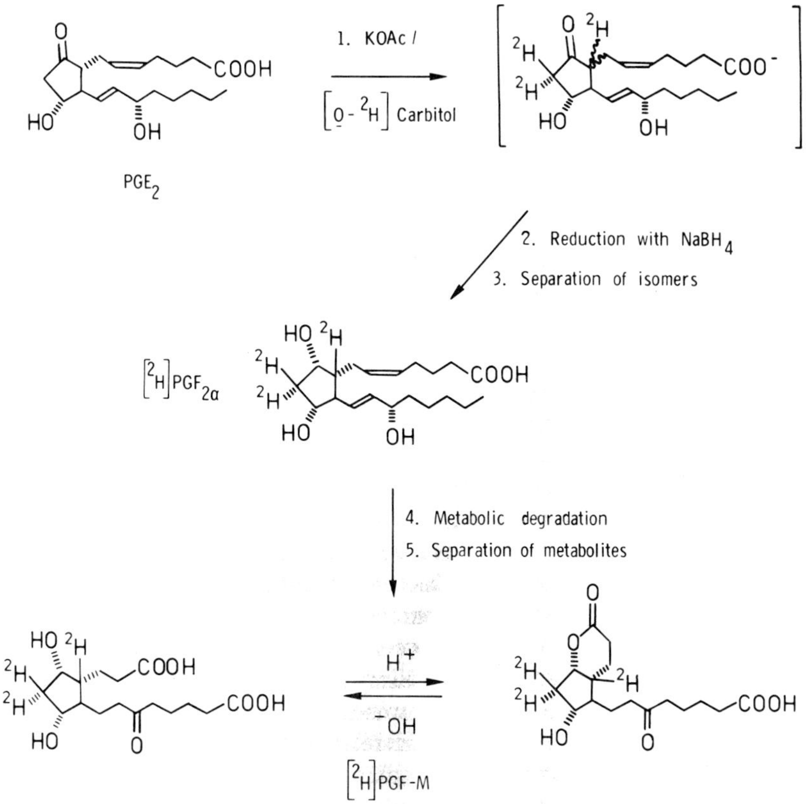

FIG. 3. Preparation of deuterium-labeled PGF-M. Reprinted from Brash *et al.*,[6] with permission.

alkali-catalyzed deuterium exchange occurs at C-8 and C-10), the solution is cooled to 0°, $NaBH_4$ (40 mg) is added, and reaction is allowed to proceed for 15 min at 0° and for a further 40 min at room temperature. After addition of water (100 ml) the solution is extracted with ether (2 × 100 ml), and the ethereal phases are evaporated to dryness. The mixture of deuterated PGF_2 isomers are subsequently separated by reversed-phase chromatography and the late-eluting "α" isomer is recovered. The reported yield is 14.5 mg of [8,10,10-$^2H_3$]$PGF_{2\alpha}$.[6] An improved procedure involving stereospecific reduction of the labeled PGE_2 exclusively to $PGF_{2\alpha}$ is available (Falardeau and Brash, this volume [69]). The deuterated $PGF_{2\alpha}$ is mixed with [9β-^3H]$PGF_{2\alpha}$ and administered in saline solution to an anesthetized rhesus monkey at a rate of 500 μg/kg per hour (total of 10 mg). Urine is collected until 2 hr after the infusion is ended. The required deuterated metabolite is isolated from the urine by reversed-phase and straight-phase chromatography, and the identification is confirmed by GC–MS.[6] The reported yield is 0.5 mg of [5β-^{3}H; 4,6,6-$^2H_3$]PGF-M. The ratio of unlabeled to trideuterated species was 0.02 when measured at m/z 397 and m/z 400 (M − t-BDMSOH) in the methyl ester t-BDMS ether derivative.

Assay of Urinary PGF-M (Fig. 2)

Steps 1 and 2. To an aliquot (10 ml) of a 24-hr urine collection is added [^{2}H]PGF-M methyl ester (200 ng) and 5 N NaOH (2 ml), and the mixture is allowed to stand at room temperature overnight.

Step 3. The solution is then acidified to pH 3 by the addition of 2 N HC1 and passed through a 2-g column of Amberlite XAD-2. The column is washed successively with water (15 ml) and n-heptane (5 ml), and the washings are discarded. The prostaglandins are then eluted with methanol (15 ml), and the solvent is evaporated under reduced pressure.

Step 4. The residue is dissolved in a mixture of 2% methanolic tetramethylammonium hydroxide (0.5 ml) and dimethylacetamide (2.0 ml), after which methyl iodide (50 μl) is added. Reaction is allowed to proceed at room temperature for 30 min and is quenched by the addition of water (5 ml). The mixture is then extracted with dichloroethane (2 × 5 ml), and the combined extracts are taken to dryness under a stream of nitrogen.

Steps 5 and 6. The methylated products are redissolved in methanol (0.2 ml), and 0.1 M sodium borate buffer (pH 10; 2.0 ml) is added. Hydrolysis is allowed to proceed at ambient temperature overnight, after which the reaction mixture is washed with dichloroethane (4 ml). The aqueous phase is acidified to pH 2 by addition of 0.5 N HC1 (0.5 ml), and the mixture is left to stand for 1 hr before extraction with dichloroethane (2 × 6 ml). The combined organic extracts are washed in turn with pH 8 borate buffer (2 ml) and water (0.5 ml) and are then taken to dryness under nitrogen.

Step 7. Remethylation of the hydrolysis product is carried out in a similar manner to the procedure described above. In this case, however, the solution of the substrate in methanolic tetramethylammonium hydroxide is allowed to stand at room temperature for 1 hr prior to the addition of dimethyl acetamide and methyl iodide. The reaction is allowed to proceed for 30 min, as before, quenched by the addition of water (5 ml), and extracted with dichloroethane (2 × 5 ml). The combined organic phases are washed with water (2 ml) and evaporated under nitrogen.

Step 8. The dry residue is then treated with the mixture of t-butyldimethylchlorosilane, imidazole, and dimethylformamide (200 μl; Applied Science) from a freshly opened ampoule, and the resulting solution is allowed to stand in a stoppered tube overnight. After the addition of water

(2 ml), the derivatized product is extracted into *n*-heptane (3 ml) and the extract is taken to dryness under nitrogen.

Step 9. Final purification is carried out by TLC (silica gel 60, Merck) using the solvent system ethyl acetate–*n*-heptane (2:3, v/v). Localization of the zone containing the PGF-M derivative is accomplished by the use of a yellow marker dye, 4-(dimethylamino)azobenzene, which has the same mobility ($R_f = 0.45$) in this solvent system. A 1.5 cm-wide zone, centered on the position of the marker dye spot, is scraped off the plate and extracted with ethyl acetate (3 ml). This extract is filtered, evaporated under nitrogen, and redissolved in ethyl acetate (20 μl) for mass spectrometric analysis.

Step 10. Aliquots (3 μl) of this solution are injected into the GC–MS instrument, which is focused to monitor the ion currents at m/z 397 and m/z 400. The gas chromatograph is equipped with a 6 foot × 2 mm i.d. column of 1% Dexsil-300 (or 1% OV-1) on Gas Chrom Q. Helium is used as carrier gas with a flow rate of 20 ml/min. Under these conditions the retention time of the PGF-M methyl ester *t*-BDMS ether derivative is about 5 min. The peak height ratio m/z 397/400 is measured, and the amount of PGF-M in the original urine sample is obtained by reference to a standard curve.

Preparation of Standard Curve

Known mixtures of unlabeled and deuterated PGF-M must be equilibrated in alkali prior to derivatization. If the standards can be quantified by radioactivity, then an alternative procedure may be more convenient. Formation of the methyl ester *t*-BDMS ether derivative serves to stop the δ-lactone ⇆ "open chain" interconversion (whereas formation of the methyl ester does not). The fully derivatized unlabeled and deuterated metabolites are purified separately by TLC to remove traces of the corresponding δ-lactones and then known mixtures are prepared.

Recovery through Assay Procedure

Recoveries were generally in the range 20–30%. In more concentrated urine samples the recovery tends to be lower due to the difficulty in dissolving the methylated urine extract in pH 10 buffer at step 5 of the assay procedure. This can be overcome by the use of five-fold larger volumes through steps 5 and 6.

Precision and Accuracy

These parameters were evaluated as ± 2% and 99.9%, respectively.[6]

Normal Levels in Urine

In male subjects the 24-hour urinary excretion of PGF-M was found to range from 10.6 to 66.9 μg (mean value 22.2 ± 13.2 μg/24 hr) and those for females from 6.5 to 18.8 μg/24 hr (mean value 11.6 ± 3.7 μg/24 hr.)[6] In 4 male subjects the metabolite levels were suppressed by 55%, 62%, 71%, and 76% on day 4 of treatment with 200 mg of indomethacin per day.[7]

[7] A. R. Brash and M. E. Conolly, *Prostaglandins* **15,** 983, (1978).

[69] Quantitation of Two Dinor Metabolites of Prostacyclin by GC–MS

By PIERRE FALARDEAU and ALAN R. BRASH

In order to monitor the biosynthesis of PGI_2 *in vivo,* a procedure was developed for measurement of 2,3-dinor-6-keto-$PGF_{1\alpha}$ and 6,15 diketo-13,14-dihydro-2,3-dinor-$PGF_{1\alpha}$.[1] These two compounds are significant urinary metabolites of systemically administered PGI_2 in all species studied so far (rat, monkey, man).[2–4] In common with other dinor analogs of 6-keto-$PGF_{1\alpha}$, these two metabolites can exist in several forms (Fig. 1) The γ-lactone form is favored under acidic conditions and this configuration is retained after extraction into organic solvent. The γ-lactone is sufficiently stable to be unaffected by a brief "back extraction" of the organic phase with mild aqueous base. However the γ-lactone ring is opened after 5–10 min of exposure to an aqueous solution of pH 8. Advantage was taken of these unusual properties in the design of a selective scheme for the purification of the metabolites from urine. With appropriate choice of conditions it is possible to purify the dinor analogs free of conventional acids, neutrals, and bases by simple extractions and manipulation of the pH.

The initial steps in the method development describe the preparation of the standards. A scheme originally used for the preparation of deuterium-labeled PGF metabolites from unlabeled PGE_2[5] is extended to in-

[1] P. Falardeau, J. A. Oates, and A. R. Brash, *Anal. Biochem.* **115,** 359–367 (1981).

[2] F. F. Sun and B. M. Taylor, *Biochemistry* **17,** 4096 (1978).

[3] F. F. Sun, B. M. Taylor, J. C. McGuire, P. T.-K. Wong, K. U. Malik, and J. C. McGiff, *in* "Prostacyclin" (J. R. Vane and S. Bergström, eds.), pp. 119–131. Raven, New York. 1979.

[4] B. Rosenkranz, C. Fischer, K. D. Weimer, and J. C. Frolich, *J. Biol. Chem.* **255,** 10194 (1980).

[5] A. R. Brash, T. A. Baillie, R. A. Clare, and G. H. Draffan, *Biochem. Med.* **16,** 77 (1976).

ISBN 0-12-181986-8

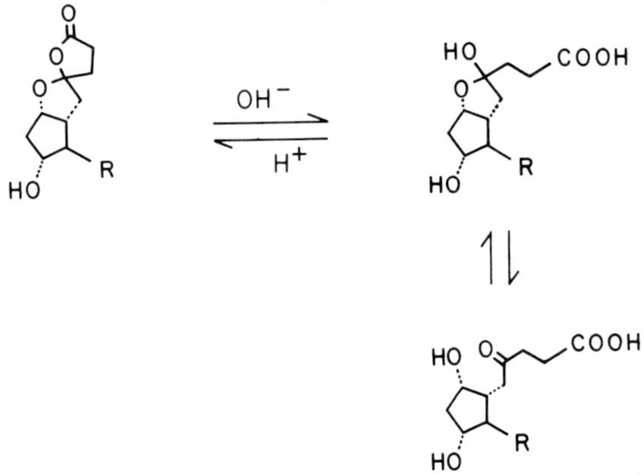

FIG. 1. Dinor metabolites of 6-keto-$PGF_{1\alpha}$ in the free acid and the lactone forms. For 2,3-dinor-6-keto-$PGF_{1\alpha}$ and 6,15-diketo-13,14-dihydro-2,3-dinor-$PGF_{1\alpha}$ the structure of the lower side chain is —CH=CHCH(OH)C_5H_{11} and —$CH_2CH_2COC_5H_{11}$, respectively. Reprinted from Falardeau *et al.*,[1] with permission.

clude preparation of labeled 6-keto-$PGF_{1\alpha}$ (Fig. 2). After labeling of the PGE_2 by deuterium exchange, [8,10,10-$^2H_3$]$PGF_{2\alpha}$ is obtained by stereoselective reduction.[6] A separate batch of deuterated PGE_2 is reduced with NaB^3H_4, affording a supply of tritium- and deuterium-labeled $PGF_{2\alpha}$. A procedure for small-scale conversion of $PGF_{2\alpha}$ to 6-keto-$PGF_{1\alpha}$ is utilized, and finally the labeled 6-keto-$PGF_{1\alpha}$ is enzymically transformed to the dinor compounds by the microorganism *Mycobacterium rhodochrous*. The 9β-^{3}H; 8,10,10-$^2H_3$-labeled compounds are used as internal standards in the assay of the corresponding unlabeled endogenous metabolites by stable isotope dilution and selected ion monitoring GC–MS.

Preparation of Standards

Synthesis of the Methyl Ester of [8,10,10-$^2H_3$]$PGF_{2\alpha}$ (Fig. 2)

PGE_2 (150 mg) is dissolved in 70 ml of [0^2H]carbitol ([0-^{2}H]ethanol is also satisfactory) in the presence of 3 g of potassium acetate (which is prepared anhydrous by gentle heating of the salt until molten). Over the course of several days at room temperature, a brown color develops in the solution although the main component is still PGE_2 as determined by thin-layer chromatography (TLC). The period of time allowed for reaction is a compromise between the improved deuterium exchange and the slow dehydration of the PGE_2. After 13 days the solution is diluted with 200 ml of

[6] E. J. Corey and R. K. Varma, *J. Am. Chem. Soc.* **93,** 7319 (1971).

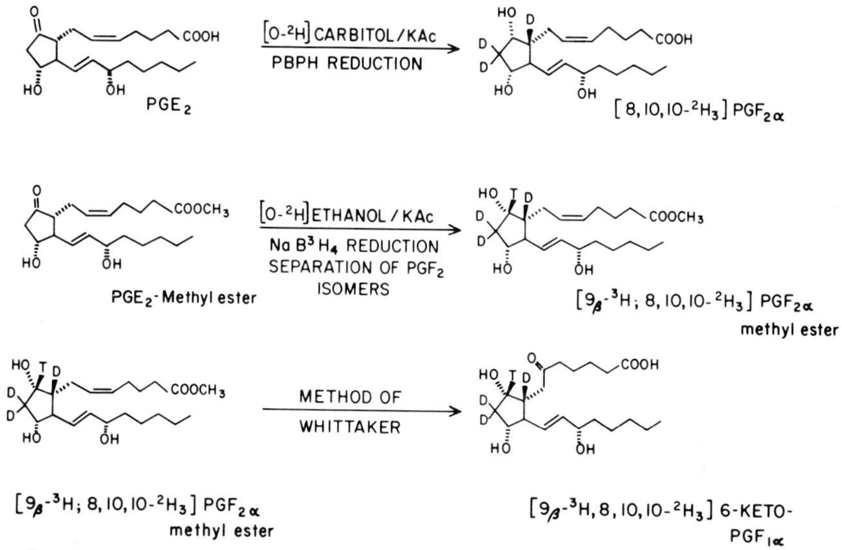

FIG. 2. Synthesis of the methyl esters of labeled prostaglandins E_2 and $F_{2\alpha}$ and of labeled 6-keto-$PGF_{1\alpha}$. Reprinted from Falardeau *et al.*,[1] with permission.

tetrahydrofuran and cooled to $-78°$ while stirring under an atmosphere of nitrogen. A 0.5 *M* solution of lithium perhydro-9b-boraphenalyl hydride (PBPH) in tetrahydrofuran (15 ml) is then added.[6] Care should be taken to avoid exposure of the solution of PBPH to air. After 20 min, the reaction is quenched by the addition of 20 ml of water, and the mixture is allowed to warm to room temperature. After dilution with 300 ml of water, the solution is extracted twice with 700 ml of ethyl acetate and the organic phases are discarded. The aqueous layer is then acidified to pH 3 with hydrochloric acid and extracted twice with 500 ml of ethyl acetate. After evaporation of the organic solvent and esterification with ethereal diazomethane, the mixture is purified on a 1.8 × 50 cm column of Sephadex LH-20, using a solvent system of heptane–chloroform–ethanol (10 : 10 : 1, v/v/v).[7] The reported yield is 70 mg of $[8,10,10\text{-}^2H_3]PGF_{2\alpha}$.[1]

Synthesis of the Methyl Ester of $[8,10,10\text{-}^2H_3]PGE_2$ (*Fig. 2*)

Prostaglandin E_2 methyl ester (30 mg) is allowed to equilibrate in 14 ml of deuterium-labeled ethanol in the presence of 630 mg of anhydrous potassium acetate at room temperature for 1 week. The volume of solvent is then reduced to 1 ml under vacuum; the solution is diluted with

[7] E. Änggård and H. Bergkvist, *J. Chromatogr.* **48**, 544 (1970).

10 ml of deuterium oxide and extracted twice with 40 ml of ethyl acetate. The combined organic phases are washed with 2 ml of deuterium oxide, and the ethyl acetate is evaporated under reduced pressure. The yellow residue is purified by reversed-phase HPLC using a semipreparative C_{18} μBondapak column (7.8 mm i.d. × 30 mm) and a solvent system of deuterium oxide–acetonitrile (60:40 v/v). A colorless oil is thus obtained, which, by TLC, is 90% the methyl ester of PGE_2 and 10% methyl ester of 8-iso-PGE_2 (yield; 15 mg).

Synthesis of the Methyl Ester of [9β-^3H; 8,10,10-$^2H_3$]$PGF_{2\alpha}$ *(Fig. 2)*

The methyl ester of [8,10,10-$^2H_3$]PGE_2 (15 mg, containing about 10% of the corresponding derivative of 8-iso-PGE_2) is dissolved in 1.2 ml of [0-^{2}H]ethanol, and the solution is cooled to 0°. One Curie of sodium borotritiide (5–15 Ci/mmol) is added, and the reaction is allowed to proceed, with stirring, at 0° for 30 min and then at room temperature for an additional 30 min. The excess borotritiide is then destroyed by the addition of 0.5 ml of 1 *N* acetic acid and the labile tritium removed under vacuum. The above procedure should be performed in laboratories suitably equipped for handling radioisotopes. The reported yield is 100 mCi of a mixture of the methyl esters of [9-^{3}H; 8,10,10-$^2H_3$]PGF_2 isomers.[1] The mixture is resolved on a column of Sephadex LH-20 (solvent system, heptane–chloroform–ethanol, 10:10:1, v/v/v)[7] to provide the methyl ester of [9β-^{3}H; 8,10,10-$^2H_3$]$PGF_{2\alpha}$(specific activity: 2.4 Ci/mmol).

Synthesis of [8,10,10-$^2H_3$]- *and* [9β-^3H; 8,10,10-$^2H_3$]6-*Keto*-$PGF_{1\alpha}$ *(Fig. 2)*

Separate batches of deuterium-labeled and tritium-plus-deuterium-labeled 6-keto-$PGF_{1\alpha}$ are prepared according to the technique of Whittaker,[8] starting with the appropriately labeled methyl ester of $PGF_{2\alpha}$. Comparatively small-scale synthesis of [9β-^{3}H; 8,10,10-$^2H_3$]6-keto-$PGF_{1\alpha}$ is required, and therefore the experimental details are given here.

The methyl ester of [9β-^{3}H; 8,10,10-$^2H_3$] (3.0 mg; specific activity: 2.4 Ci/mmol) is dissolved in 2.0 ml of diethyl ether and diluted with 10 mg of the methyl ester of [8,10,10-$^2H_3$]$PGF_{2\alpha}$. The solution is cooled to 0°, and 0.3 ml of a saturated aqueous solution of sodium bicarbonate is added, followed by 0.3 ml of a 2.5% solution of iodine in diethyl ether. Reaction is allowed to proceed overnight at 0°. The solution is then warmed to room temperature and reaction quenched by the addition of a few drops of an aqueous solution of sodium thiosulfate (until disappearance of the yellow color). The aqueous phase is diluted with 15 ml of water and extracted twice with 50 ml of ethyl acetate. The residue obtained after

[8] N. Whittaker, *Tetrahedron Lett.* **32,** 2805 (1977).

evaporation of the organic solvent is dissolved in 0.3 ml of 1,5-diazabicyclo[4.3.0.]non-5-ene. After 2 hr at room temperature the solution is diluted with 10 ml of water, quickly acidified to pH 3 with formic acid, and extracted with ethyl acetate. The residue obtained from the organic extract is then redissolved in 0.3 ml of cold methanol, to which is added 0.25 ml of 1 *N* aqueous potassium hydroxide. After 10 min at 0°, the solution is left at room temperature for 90 min. The solvent is then evaporated, and the residue is dissolved in 1 ml of a mixture of acetonitrile–water–acetic acid (20:80:0.1, v/v/v) and purified by HPLC on a preparative C_{18} μBondapak column (Waters Associates) using the same solvent mixture for elution. Tritium-plus-deuterium-labeled 6-keto-$PGF_{1\alpha}$ (specific activity: 140 mCi/mmol) is thus isolated in pure form, with a reported yield of 40%.[1]

Biosynthesis of [9β-^{3}H; 8,10,10-–^{2}H$_3$]-2,3-Dinor-6-keto-PGF$_{1\alpha}$ and [9β-^{3}H; 8,10,10-^{2}H$_3$]-6,15-Diketo-13,14-dihydro-2,3-dinor-PGF$_{1\alpha}$ (Fig. 2)

These compounds are prepared essentially as described by Sun *et al.*[9] Briefly, 600 ml of "growth medium" (2% glucose, 0.5% tryptone, and 0.3% yeast extract are inoculated with spores of *Mycobacterium rhodochrous*. The bacteria are allowed to grow at 28° for 72 hr, after which they are sedimented by gentle centrifugation and resuspended in 20 ml of 0.05 *M* Tris·HCl (pH 7.1). Twenty milligrams of [9β-^{3}H; 8,10,10-^{2}H$_3$]6-keto-$PGF_{1\alpha}$ (specific activity: 34 mCi/mmol) dissolved in 200 μl of dimethylformamide is added and incubated with the microbial suspension for 7 days at 28°. The suspension is subsequently centrifuged to remove the bacteria, and the solution is then acidified and extracted with ethyl acetate. The resulting mixture is purified by open-bed silicic acid chromatography followed by reversed-phase HPLC (semi-preparative C_{18} μBondapak column; isocratic elution with water–acetonitrile–acetic acid (76:24:0.1, v/v/v). The reported yield is 8 mg of [9β-^{3}H; 8,10,10-^{2}H$_3$]2,3-dinor-6-keto-$PGF_{1\alpha}$ and 3.5 mg of a less polar product identified as [9β-^{3}H; 8,10,10-^{2}H$_3$]6,15-diketo-13,14-dihydro-2,3-dinor-$PGF_{1\alpha}$.[1]

Unlabeled metabolites of 6-keto-$PGF_{1\alpha}$ are also prepared by this procedure.

Assay Procedure

Extraction and Purification

To a 20-ml aliquot of a 24-hr urine collection is added 75 ng each of 9β-^{3}H; 8,10,10-^{3}H$_3$-labeled 2,3-dinor-6-keto-$PGF_{1\alpha}$ and 6,15-diketo-13,14-

[9] F. F. Sun, B. M. Taylor, F. H. Lincoln, and O. K. Sebek, *Prostaglandins* **20,** 729 (1980).

dihydro-2,3-dinor-$PGF_{1\alpha}$. The pH of the urine is adjusted to pH 10–11 with 10 N sodium hydroxide, and the sample is left at room temperature for 15 min. The urine is then acidified to pH 3 with HCl, applied to a disposable 20-ml Clin Elut column (Analytichem International Inc.) and eluted with 2 × 20 ml of dichloromethane. The organic extract (25 ml) is back-extracted three times with 15 ml of 0.05 sodium borate buffer, pH 8, and the aqueous phases are discarded. The organic phase is evaporated to dryness under a stream of nitrogen. The residue is dissolved in 50 μl of pyridine, and then 1 ml of 0.05 M sodium borate pH 8, is added to the sample. After standing for 15 min at room temperature, the mixture is extracted twice with 10 ml of ethyl acetate, and the organic phases are discarded. The aqueous layer is acidified to approximately pH 2–3 by the addition of 10 μl of 4 N HCl and extracted with 4 ml of dichloromethane. The dichloromethane extract is washed with 1 ml of water and then evaporated under nitrogen and derivatized as described below.

For analysis of low-level samples (for example, from subjects receiving aspirin-like drugs), it is often necessary to include an additional step of purification using TLC. This procedure is performed prior to derivatization using plates of silica gel G (Analtech) and a solvent system of ethyl acetate plus 0.5% glacial acetic acid. Zones corresponding to 2,3-dinor-6-keto-$PGF_{1\alpha}$ (R_f = 0.32) and 6,15-diketo-13,14-dihydro-2,3-dinor-$PGF_{1\alpha}$ (R_f = 0.61) are eluted with 0.5 ml of 0.1% acetic acid in water and 2.5 ml of ethyl acetate. After phase separation the organic layer is transferred to a clean 5-ml Reactivial and evaporated under nitrogen. The sample is then ready for derivatization.

Derivatization

The sample is dissolved in 40 μl of a mixture of pyridine–water–triethylamine (10:10:1, v/v/v) and allowed to stand at room temperature for 1 hr. After evaporation of the solvent, the residue is dissolved in 30 μl of a saturated solution of methoxylamine hydrochloride in pyridine and kept overnight at room temperature. BSTFA (30 μl) is then added, and after 30 min at room temperature the solvent is evaporated under nitrogen and the *O*-methyloxime, trimethylsilyl ether, trimethylsilyl ester derivatives are dissolved in 15 μl of 5% BSTFA in hexane and analyzed by GC–MS.

When formation of the methyl ester is desired, esterification with ethereal diazomethane is performed immediately after the pyridine–water–triethylamine treatment.

Gas Chromatography–Mass Spectrometry

The gas chromatograph is equipped with a glass column (3 ft × 2 mm i.d.) packed with 3% SP 2250 on Supelcoport and operated at 240°, with

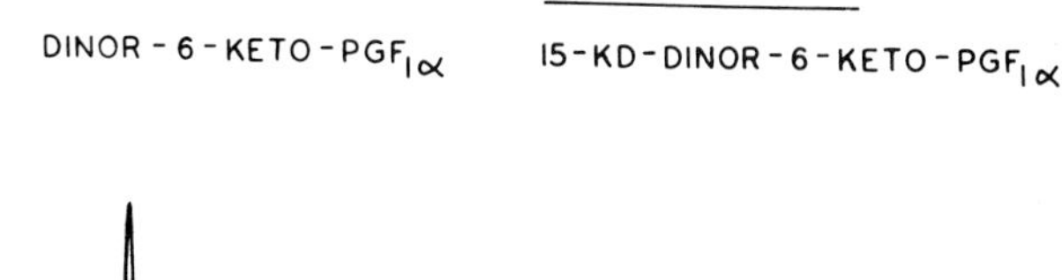

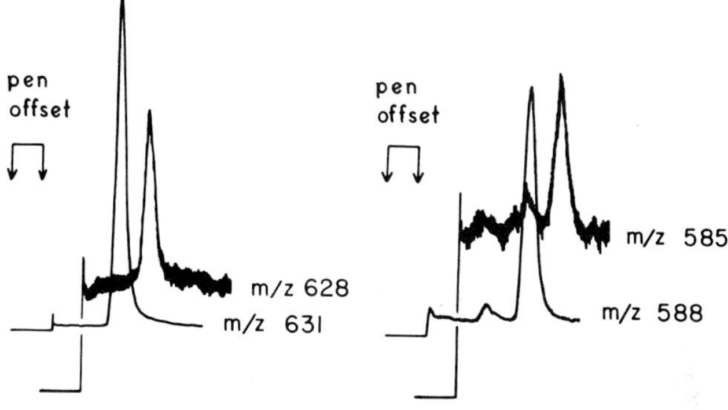

FIG. 3. Selected ion monitoring recordings of a urine extract analyzed as the trimethylsilyl ester, trimethylsilyl ether *O*-methyloxime derivatives. Reprinted from Falardeau *et al.*,[1] with permission.

helium as carrier gas (flow rate, 30 ml/min). Under these conditions the retention times of the trimethylsilyl ester, trimethylsilyl ether, *O*-methoxime derivatives of 2,3-dinor-6-keto-$PGF_{1\alpha}$ and 6,15-diketo-13,14-dihydro-2,3-dinor-$PGF_{1\alpha}$ are approximately 3 and 4 min, respectively. An ion source temperature of 260° and an electron energy of 70 eV are employed.

Selected ion monitoring recordings of the $M-31^+$ ions from both dinor metabolites isolated from a urine sample are shown in Fig. 3. In this example, the metabolites were analyzed together after injection of the derivatized urinary extract into the GC–MS instrument. The recordings represent the response obtained from the equivalent of a 10-ml urine extract injected on column.

Recovery through Assay Procedure

The final organic extracts contain approximately 80% of the radioactive internal standards added to the samples of urine. When further purification by TLC is employed, the recovery of 2,3-dinor-6-keto-$PGF_{1\alpha}$ is approximately 50% after the chromatography.

Precision and Accuracy

The reproducibility of the method was demonstrated by analysis of five 20-ml aliquots of a 24-hr urine collection. Standard deviations of 2% and 6% were obtained on the mean values for 2,3-dinor-6-keto-$PGF_{1\alpha}$ and

6,15-diketo-13,14-dihydro-2,3-dinor-$PGF_{1\alpha}$, respectively. Further aliquots of the same urine sample were analyzed after additions of known amounts of unlabeled metabolites. This gave values to 95% accuracy for both compounds.[1] By far the main factor affecting precision was variation in the sensitivity of the mass spectrometer. The low urinary levels of the 6,15-diketo-13,14-dihydro-2,3-dinor-$PGF_{1\alpha}$ metabolite demanded excellent sensitivity from the GC–MS instrument in order to make measurements from 20 ml of urine.

Metabolite Levels in Normal Subjects

In 24-hr urine collections from predominantly young adults the levels of 2,3-dinor-6-keto-$PGF_{1\alpha}$ were determined as 710 ± 264 ng (439 ± 201 ng/g creatinine) in males and 453 ± 124 ng (341 ± 93 ng/g creatinine) in females. The 24-hr urinary levels of 6,15-diketo-13,14-dihydro-2,3-dinor-$PGF_{1\alpha}$ were found to be 314 ng (179 ± 63 ng/g creatinine) in male subjects and 341 ± 93 ng/g creatinine in three nonpregnant healthy females.

Acknowledgments

This work was funded by NIH Grant GM 15431 and USPHS Grant HL 14192. P. F. received a Centennial Fellowship from the Medical Research Council of Canada.

[70] Quantitative Assay of Urinary 2,3-Dinor Thromboxane B_2 by GC–MS

By RICHARD L. MAAS, DOUGLASS F. TABER, and L. JACKSON ROBERTS II

Thromboxane A_2 (TxA_2) is a highly unstable platelet-aggregating factor biosynthesized from arachidonic acid by human platelets.[1] By virtue of its potent bioactivity, TxA_2 may be important in the genesis of cardiovascular and thrombotic disease. Thus, quantification of endogenous TxA_2 production in man may provide a reliable index of platelet activation *in vivo* and have considerable relevance to understanding the role of platelets in these conditions. Both *in vitro* and *in vivo*, TxA_2 is rapidly converted, at least in part, by hydrolysis of the oxetane ring to a stable product, TxB_2, the metabolic fate of which has been determined.[2–7] In man,

[1] M. Hamberg, J. Svensson, and B. Samuelsson, *Proc. Natl. Acad. Sci. U.S.A.* **72,** 2994 (1975).

[2] For evidence that TxA_2 can also react *in vitro* to form covalent derivatives with albumin,

ISBN 0-12-181986-8

the major urinary metabolite of TxB_2 is 2,3-dinor-TxB_2, a C_{18} homolog lacking two carbons on the upper side chain due to a single cycle of β-oxidation.[6]

The method described here for quantification of urinary dinor-TxB_2 is based on stable isotope dilution and GC–MS analysis with selected ion monitoring. This method offers the usual advantages of such metabolite assays, including avoidance of artifactual elevations in parent compound levels upon sampling, measurement of a compound that is concentrated in the blood or urine, and high specificity.[8,9] On the other hand, the method requires (*a*) preparation of a suitable low blank internal standard; and (*b*) extensive sample purification to ensure accurate GC–MS analysis, particularly when relatively low endogenous levels of dinor-TxB_2 are being quantified in patients taking inhibitors of the cyclooxygenase enzyme.

Preparation of Internal Standards

There are a number of possible methods that can be utilized to obtain the appropriate deuterium and tritium-labeled dinor-TxB_2 internal standard, including biosynthetic procedures, chemical synthesis, and combinations of the two. We have obtained tritium-labeled dinor-TxB_2 by bacterial conversion of biosynthetically prepared [$^3H_8$]TxB_2. Although we prepare [$^3H_8$]TxB_2 biosynthetically from [$^3H_8$]arachidonic acid, [$^3H_8$]TxB_2 is also available commercially (New England Nuclear, Boston, Massachusetts). Alternatively, [$^3H_8$]dinor-TxB_2 can also be obtained from urine after intraperitoneal injection of [$^3H_8$]TxB_2 into a guinea pig,[10] although the bacterial method of conversion of TxB_2 to dinor-TxB_2 is more efficient and convenient. The bacterial method of conversion of TxB_2 to dinor-TxB_2 can also be used to obtain deuterium-labeled dinor-TxB_2 from deuterium-labeled TxB_2. [$^2H_8$]TxB_2 is also available commercially (Ran Biochemicals). In addition, the preparation of [$^2H_4$]$PGF_{2\alpha}$ has been de-

see: J. Maclouf, H. Kindahl, E. Granström, and B. Samuelsson, *Prostaglandin Thromboxane Res.* **6,** 283 (1980).

[3] L. J. Roberts, II, B. J. Sweetman, J. L. Morgan, N. A. Payne, and J. A. Oates, *Prostaglandins* **13,** 631 (1977).

[4] L. J. Roberts, II, B. J. Sweetman, and J. A. Oates, *J. Biol. Chem.* **253,** 5305 (1978).

[5] H. Kindahl, *Prostaglandins* **13,** 619 (1977).

[6] L. J. Roberts, II, B. J. Sweetman, N. A. Payne, and J. A. Oates. *J. Biol. Chem.* **252,** 7415 (1977).

[7] L. J. Roberts, II, B. J. Sweetman, and J. A. Oates, *J. Biol. Chem.* **256,** 8384 (1981).

[8] B. Samuelsson, *Adv. Biosci.* **9,** 7 (1973).

[9] A. R. Brash, *in* "Prostaglandins, Prostacyclin, and Thromboxanes Measurement" (J. M. Boynaems and A. G. Herman, eds.), p. 123. Nijhoff, The Hague, 1980.

[10] J. Svensson, *Prostaglandins* **17,** 351 (1979).

scribed,[11] and synthetic methods exist for conversion of $PGF_{2\alpha}$ to TxB_2.[12] We also describe below a method for the chemical synthesis of [20,20,20-$^2H_3$]TxB_2. More recently we have developed a method for the direct chemical synthesis of [20,20,20-$^2H_3$]dinor-TxB_2, which is the means currently employed by us to obtain deuterium-labeled dinor-TxB_2.

Biosynthetic Preparation of [$^3H_8$]TxB_2. [$^3H_8$]TxB_2 is routinely prepared using a modification of procedure described elsewhere,[6] by incubation of [$^3H_8$]arachidonic acid (New England Nuclear, 60 Ci/mmol), 1 mCi, with 2.0 ml of an albumin-washed (2% lipid-free BSA in phosphate buffer) 1:1 suspension of sheep seminal vesicle and guinea pig lung microsomes, rich sources of cyclooxygenase and thromboxane synthase, respectively. After incubation for 10 min in 0.1 *M* KH_2PO_4–K_2HPO_4 buffer, pH 7.4 containing 1 m*M* EDTA, 2 m*M* phenol, and 0.2 mg of bovine hemoglobin per milliliter, reaction is quenched with 3 volumes of acetone. The precipitated protein is removed by centrifugation, and the supernatant is diluted, acidified, and extracted three times with ethyl acetate. The organic phase is evaporated to dryness and purified by reversed-phase HPLC as described under Assay Procedure. [$^3H_8$]TxB_2 is obtained in ca. 25% yield.

Bacterial Conversion of TxB_2 *to Dinor-*TxB_2

The procedure described by Sebek and colleagues is used.[13,14] Sterile medium (150 ml, pH 7.2) containing 2.0% glucose, 1.0% tryptone, and 0.5% yeast extract is inoculated with *Mycobacterium rhodochrous* (The Upjohn Company, Kalamazoo) from an agar slant stored at 4°. The bacterial suspension is allowed to grow under sterile conditions for 3–4 days at 25° with gentle shaking. The bacteria are then concentrated by centrifugation at 1200 *g* for 15 min, and resuspended in 10 ml of 0.25 *M* phosphate buffer, pH 7.0, without nutrients. Then 1.5 mg or less of the substrate TxB_2 is added in 25 μl of dimethyl acetamide, and the incubation is allowed to proceed for 1–3 days. The length of incubation is critical, and each day 10–25-μl aliquots should be removed from the incubation, the bacteria spun down, and the products analyzed as the Me-TMS or Me-MO-TMS derivatives by GC or GC–MS. If the incubation time is overextended, only small amounts of dinor-TxB_2 can be identified and several other metabolites are formed.[15] After the formation of a substantial quan-

[11] A. R. Brash, T. A. Baille, R. A. Clare, and G. H. Draffan, *Biochem. Med.* **16,** 77 (1979).

[12] W. P. Schneider and R. A. Morge, *Tetrahedron Lett.* **37,** 3283 (1976).

[13] N. A. Nelson, R. W. Jackson, and O. K. Sebek, *Prostaglandins* **16,** 85 (1978).

[14] F. F. Sun, B. M. Taylor, F. H. Lincoln, and O. K. Sebek, *Prostaglandins* **20,** 729 (1980).

[15] We have also recently noted considerable variation in the biosynthetic capacity of different batches of *Mycobacterium rhodochrous;* with two out of six preparations, little conversion has been observed.

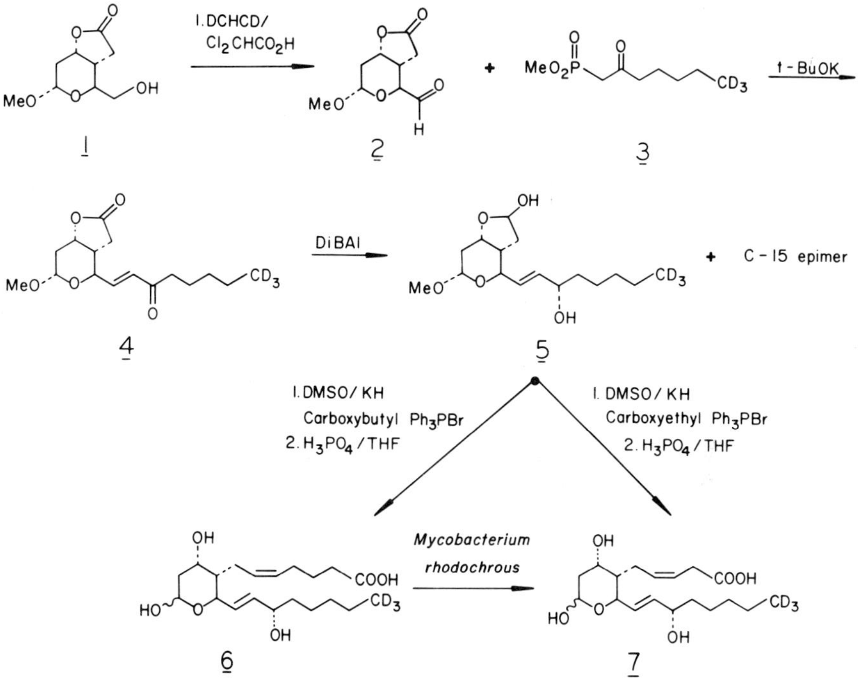

FIG. 1. Synthesis scheme showing reactions used to prepare [20,20,20-$^2H_3$]TxB_2 and dinor-TxB_2. The starting material, (**1**) was a gift from the Upjohn Company, synthesized according to Nelson and Jackson[16] and Kelly *et al.*[17] Compound 3 was prepared as described by Taber and Lee.[19]

tity of dinor-TxB_2, the incubation is terminated with 1.5 volumes of acetone, followed by centrifugation, dilution, and extraction of the supernatant with ethyl acetate, and evaporation under vacuum. The products are purified by reversed-phase HPLC, as described below, and then by normal-phase HPLC or TLC, and their identity is confirmed by GC–MS. The final yield from the starting material is 25–50%.

Chemical Synthesis of [20,20,20-$^2H_3$]dinor-TxB_2 and [20,20,20-$^2H_3$]TxB_2. The synthetic procedures outlined in Fig. 1 are similar to those described by Nelson, Kelly, and colleagues for the synthesis of TxB_2.[16,17] The starting material **1** was obtained from the Upjohn Company, although synthetic routes to **1** exist.[16–18] The procedures outlined incorporate three deuterium atoms at C-20 of TxB_2 and dinor-TxB_2. The incorporation of deuterium at the C-20 position is ideally suited for selected ion monitoring

[16] N. A. Nelson and R. W. Jackson, *Tetrahedron Lett.* **37,** 3275 (1976).

[17] R. C. Kelly, I. Schletter, and S. J. Stein, *Tetrahedron Lett.* **37,** 3279 (1976).

[18] S. Hannessian and P. Lavallee, *Can. J. Chem.* **55,** 562 (1977).

of either the Me-MO-TMS or the Me-*t*BDMS derivatives of dinor-TxB_2. The base ion in the mass spectrum of the Me-MO-TMS derivative of dinor-TxB_2 is m/z 301, the structure of which consists of the entire lower side chain and is therefore shifted upward to m/z 304 in the mass spectrum of [20,20,20-$^2H_3$]dinor-TxB_2. In the mass spectrum of the Me-tBDMS derivative of dinor-TxB_2 is an ion of high intensity at m/z 641, formed by the loss of $\cdot CMe_3$ from the molecular ion, and is shifted upward three mass units to m/z 644 in the mass spectrum of [20,20,20-$^2H_3$]dinor-TxB_2. The preparation of the deuterium-labeled phosphonate (**3**) has been previously described.[19]

Enone (**4**): Dicyclohexylcarbodiimide (600 mg) was added to a solution of **1** (202 mg, 1.0 mmol) in 5 ml of benzene and 0.5 ml of DMSO. The mixture was magnetically stirred, then 60 μl of dichloroacetic acid were added to give a thick precipitate. After 25 min, the reaction was quenched by dropwise addition of oxalic acid (254 mg) in 600 μl of methanol. The reaction mixture was diluted with water and extracted with ethyl acetate. The organic layer was dried over K_2CO_3, concentrated under vacuum, and chromatographed on 10 g of TLC mesh silica gel (EM7747, Scientific Products) with 8% acetone–methylene chloride to give the crude aldehyde **2** as a colorless oil, R_f (20% acetone–methylene chloride) = 0.41.

A solution of phosphonate (**3**) (122 mg, 0.6 mmol) in 5 ml of THF was stirred magnetically under a nitrogen atmosphere. Potassium *t*-butoxide (56 mg, 0.5 mmol) was added, followed by the freshly prepared aldehyde **2**. After 1 hr, the reaction mixture was diluted with half-saturated aqueous NaCl and extracted with ethyl acetate. The organic layer was dried over K_2CO_3, concentrated under vacuum, and chromatographed on 5 g of TLC mesh silica gel with 40% ethyl acetate–petroleum ether to give **4** (73 mg, 27% based on **1**) as an oil, R_f (1 : 1, ethyl acetate–hexane) = 0.33. ^{1}H-NMR: 1.30, m, 6H; 1.5–3.0, m, 7H; 3.46, s, 3H; 4.24, m, 1 H; 4.64, q, J = 5,1H; 4.88, m, 1H; 6.38, d, J = 16, 1H; 6.72, dd, J = 5, 16, 1H.

[20,20,20-$^2H_3$]TxB_2 (**6**): To a stirred solution of lactone (**4**) (35 mg, 0.12 mmol) in 5 ml of toluene in a dry ice–acetone bath was added diisobutyl aluminum hydride (1.0 ml of 1.0 *M* in toluene) in one portion. After 20 min, the mixture was diluted with aqueous NaH_2PO_4 and extracted with ethyl acetate. The organic layer was dried over Na_2SO_4 and concentrated to give crude **5** as an oil; 5 ml of DMSO were carefully added (foams) to KH (175 mg, 4.3 mmol) in a 25-ml round-bottom flask under N_2. The reaction mixture was stirred magnetically for 20 min to give a clear pale yellow solution. Carboxybutyltriphenylphosphonium bromide (930 mg, 2.1 mmol) (Aldrich) was added, to give a deep orange solution. The crude **5**

[19] D. F. Taber and C. H. Lee, *J. Labelled Compd. Radiopharm.* **14,** 599 (1978).

from above in 1 ml THF was added, and stirring continued for 1.5 hr. The reaction mixture was worked up and purified as described for [20,20,20-$^2H_3$]dinor-TxB_2 below, to give 6 mg of [20,20,20-$^2H_3$]TxB_2.

[20,20,20-$^2H_3$]dinor-TxB_2 (**7**): DMSO 5 ml, was carefully added to KH (206 mg, 515 mmol) in a 25-ml round-bottom flask under N_2. The mixture was stirred magnetically for 15 min to give a clear pale yellow solution. The flask was transferred to an ice-water bath (freezes), and carboxyethyltriphenylphosphonium chloride[20] (963 mg, 2.6 mmol) was added. The flask was removed from the cooling bath; as it warmed to room temperature, the DMSO melted to give a deep red solution. Crude **5** (prepared from 18 mg of (**4**) in 1 ml of THF was immediately added, and stirring continued for 2 hr. The reaction mixture was then diluted with 100 ml of 5% NaOH and extracted with 3 × 50 ml portions of ethyl acetate. The aqueous layer was then acidified with concentrated aqueous HCl and extracted with 3 × 50 ml of ethyl acetate, dried over Na_2SO_4, and evaporated to dryness under vacuum to give the mixed acetal as an oil.

Hydrolysis of the mixed acetal was performed essentially as described.[13] The oil from above was dissolved in 10.5 ml of freshly distilled THF in a 50-ml round-bottom flask, and an equal volume of 8.5% H_3PO_4 was added. The flask was stoppered and heated to 55° for 2 hr with magnetic stirring. The reaction mixture was diluted with 100 ml of saturated NaCl, to which was added ca. 5 μCi of [$^3H_8$]dinor-TxB_2, and then extracted with 3 × 50 ml of ethyl acetate. The organic layer was dried over Na_2SO_4 and concentrated under vacuum to give crude **7** as a mixture of C-15 epimers. Final purification of **7** is achieved using a semipreparative Waters μ-Porasil (7.6 mm × 30 cm, 10 μm silicic acid particles) column with a linear program from $CHCl_3$ to $CHCl_3$–MeOH–HOAc (500:50:11, v/v/v) over 1 hr at 4 ml/min, 4-ml fractions (elution time, 30 min) and reversed-phase HPLC (using a semipreparative column) as described under Assay Procedure, to give 6 mg of [20,20,20-$^2H_3$]dinor-TxB_2. The standard is added to the [$^3H_8$]dinor-TxB_2 prepared above to give a final specific activity of 94 Ci/mol, and is stored in ethanol containing 0.1% pyridine to prevent acetal formation. Under these conditions, the standard has been stable for over 2 years.

Assay Procedure for Urinary 2,3-Dinor-TxB_2

Step 1. Ether Extraction. To a 50–100 ml (100 ml if low levels are anticipated) aliquot of a 24 hr urine is added 330 ng of internal standard, containing 200,000 dpm of ^{3}H. The sample may then be frozen at −20° for periods of several weeks; after about 6 months, however, sample deterioration may be noticed. The urine sample is acidified to pH 3.5 by dropwise

[20] D. B. Denny and L. C. Smith, *J. Org. Chem.* **27,** 3404 (1962).

addition of 5 *N* HCl, and extracted with 3 volumes of diethyl ether. After drying over Na_2SO_4, the organic phase is decanted and evaporated to dryness on the rotary evaporator. Recovery of ^3H at this stage is typically 70–75%.

Alternative Extraction Procedures. Twenty-milliliter portions of the sample may be extracted by acidification to pH 3.5 and applied to 20-ml disposable Clin-Elut columns (Analytichem); after 3 min, the sample is eluted with two 20-ml portions of ethyl acetate. Amberlite XAD-2 is well suited to the extraction of large volumes of urine; procedures for its use have been described elsewhere.[21,22] Although both of these latter methods give very good recoveries (~90%), a higher degree of purification is achieved with ether extraction and is preferred.

Step 2. Silicic Acid Chromatography. The dry residue from step 1 is resuspended in 4 ml of ethyl acetate by vigorous sonication and use of a plastic rod, and the coarse suspension is applied to a 3-g column of Mallinckrodt CC-4 silicic acid, previously activated at 110°. The sample is eluted with 40 ml of ethyl acetate and evaporated to dryness on a rotary evaporator, giving 40–60% overall recovery of ^3H. Samples may be stored indefinitely at this stage in ethyl acetate at −60°.

Step 3. High-Pressure Liquid Chromatography (HPLC) (a) After evaporation to dryness, samples are dissolved in 1.0 ml of reversed-phase HPLC solvent composed of 18% acetonitrile, 82% water, 0.1% acetic acid (all HPLC organic solvents are Burdick and Jackson, distilled in glass grade). Any precipitate is removed by centrifugation or by Millipore filtration using filters 1.3 cm in diameter (pore size, 0.5 μm) in a 2.0-ml syringe fitted with a Luer tip. Samples are then injected onto a Waters Model 5000 HPLC system equipped with a U6K injector and a Waters fatty acid analysis column, 3.9 mm × 30 cm. Elution is accomplished under isocratic conditions using the above solvent at 2.0 ml/min, with collection of 2.0-ml fractions. Under these conditions, dinor-TxB_2 typically elutes at 18 min, TxB_2 at 30 min. Aliquots (1%) of alternate fractions bracketing the correct retention time are counted by liquid scintillation to establish the chromatographic profile, and the entire ^3H-labeled peak is pooled (isotopic separation of ^3H-,^2H-, and ^1H-labeled species is not significant under these conditions). The eluates are extracted twice with equal volumes of ethyl acetate and the extract evaporated to dryness under a stream of nitrogen with absolute ethanol used azeotropically to remove traces of water.

(b) The sample is stored in 250–500 μl of $CHCl_3$ for direct injection onto a normal-phase HPLC column (Waters μPorasil; 10 μm silicic acid particles) with a linear program from $CHCl_3$ to $CHCl_3$–MeOH–HOAc

[21] H. L. Bradlow, *Steroids* **11,** 265 (1968).

[22] M. Hamberg and B. Samuelsson, *J. Biol. Chem.* **246,** 6713 (1971).

(500:50:11, v/v/v), over 1 hr, 1 ml/min, 1 ml fractions. Using the above program, TxB_2 typically elutes at 34 min, dinor-TxB_2 at 38 min. Alternate fractions from fraction 30–40 are taken for liquid scintillation counting (1–2% aliquots), and the ^3H peak is pooled and evaporated to dryness in a 1-ml Reactivial. Alternatively, purification at this stage can be accomplished using a μPorasil column under isocratic conditions with the solvent $CHCl_3$–MeOH–HOAC (98.5:1.5:0.1, v/v/v), 2 ml/min, 2-ml fractions. With this program, dinor-TxB_2 elutes with a retention time of approximately 15 min.

Step 4. Derivatization and Final Purification (Optional). Samples are converted to methyl esters by dissolution in 25 μl of methanol and addition of 200 μl of ethereal diazomethane, followed by evaporation to dryness under nitrogen after 2–3 min. Conversion to the *t*BDMS derivative is achieved by addition of 100 μl of dimethylformamide containing 5 *M* *t*-butyldimethylchlorosilane from an ampoule (Applied Science). Care must be taken to remove the last vestiges of any moisture before reaction with *t*BDMS reagent. More economically, 5 *M* solutions of imidazole and *t*-butyldimethylchlorosilane (Petrarch Systems, Inc., Levittown, Pennsylvania) in dimethylformamide can be prepared (taking care to keep them well desiccated), and equal volumes of each can be added to the sample. The sample is heated to 70° for 12–18 hr, followed by addition of 300 μl of distilled water and extraction into 500 μl of hexane. The hexane extract is transferred to a clean 1-ml Reactivial with a syringe, and the sample is extracted with a second 500 μl of hexane; the hexane extracts are combined and evaporated to dryness, using xylene to remove any traces of dimethylformamide that may have extracted. Because of large variations in the potency of different batches of *t*BDMS-imidazole reagent, it is essential that the conditions required for complete derivatization be carefully checked with standards; otherwise partial derivatization may result. After quantitative conversion of dinor-TxB_2 to the Me-tBDMS derivative, there should be less than 1% tritium left in the aqueous phase after two hexane extractions, and the product should give a single peak on GC (*C* value 28.5, 3% SP2250 or OV-17).

Although the Me-tBDMS derivatives can now be analyzed by GC–MS, often with excellent results, the low levels of endogenous dinor-TxB_2 are frequently partially obscured by interfering peaks on selected ion monitoring analysis. Therefore, we have generally employed a final chromatographic step for rigorous sample purification. This can be efficiently performed by TLC using 10% ethyl acetate in hexane (5 × 20 cm plates, unactivated, Analtech; 2 samples per plate), $R_f = 0.52$. Unlabeled TxB_2 Me-*t*BDMS ($R_f = 0.55$), or any other convenient marker can be used as a standard; the appropriate zone is scraped off the plate into a 5-ml Reac-

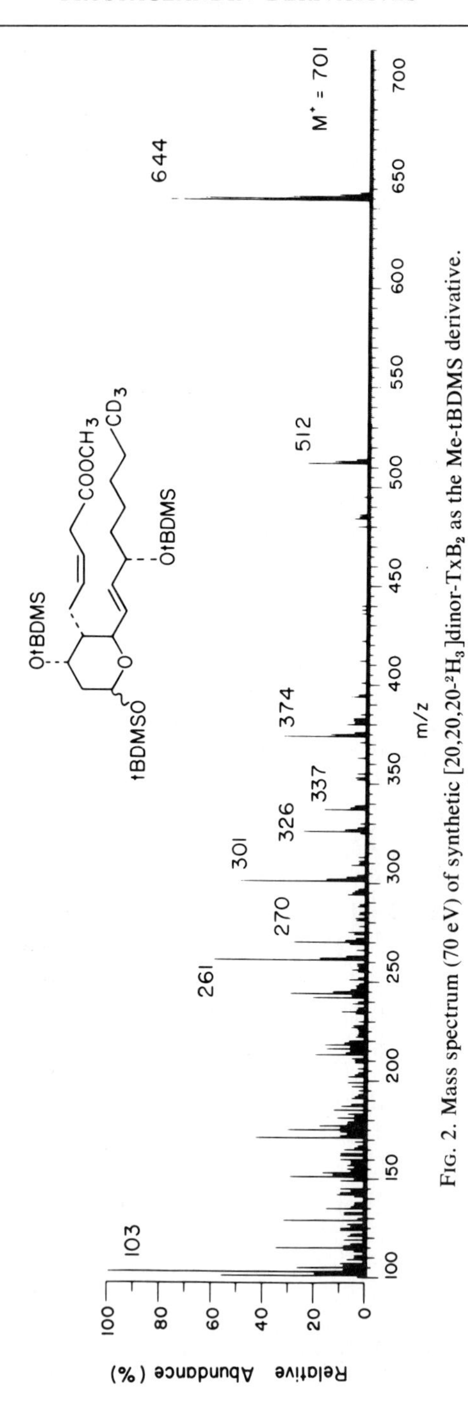

FIG. 2. Mass spectrum (70 eV) of synthetic [20,20,20-$^2H_3$]dinor-TxB_2 as the Me-tBDMS derivative.

tivial, diluted with 2 ml of ethyl acetate; after centrifugation, the supernatant is transferred to a clean vial. An alternative, lengthier procedure that gives excellent results involves a final step of normal-phase HPLC under isocratic conditions using hexane–$CHCl_3$ (10:90) (v/v), 1 ml/min, 1-ml fractions (retention time, 15 min). The Me-tBDMS derivatives are stable for many months at $-60°$ in heptane or ethyl acetate. The final recovery of dinor-TxB_2 through the entire procedure is typically 10–20%.

Step 5. Gas Chromatography–Mass Spectrometry. Samples are injected into a GC–MS system (Hewlett-Packard 5980A quadrapole) equipped with a 3-foot column of 1% SP2100 or OV-1 at 260–270°, with selective ion monitoring of M-57 ions (loss of $\cdot CMe_3$) at m/z 641 (endogenous dinor-TxB_2) and m/z 644 ([$^2H_3$]dinor-TxB_2 internal standard). Ion ratios are measured, and sample levels are calculated from a standard curve prepared at the time of sample analysis by injecting into the mass spectrometer known mixtures of [^1H]- and [$^2H_3$]dinor-TxB_2 coderivatized as the Me-*t*BDMS derivative. With good instrumental sensitivity each sample can be injected 2–3 times.

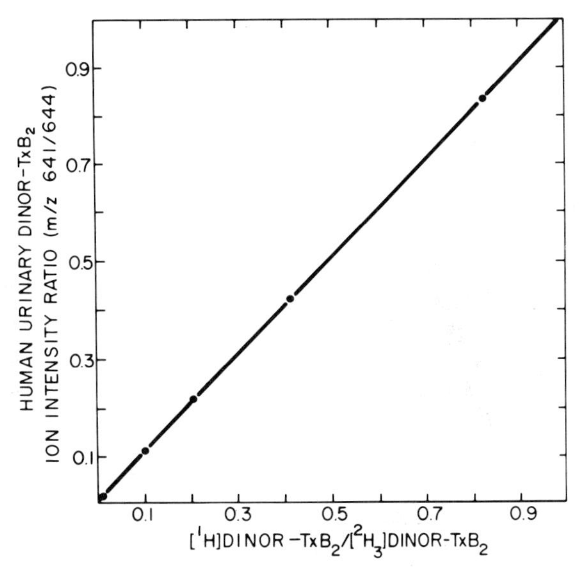

FIG. 3. Standard curve showing ion ratios obtained from selected ion monitoring of the M-57 ion for known mixtures of [^1H]- and [$^2H_3$]dinor-TxB_2 Me-tBDMS, at m/z 641 and 644. [$^2H_3$]Dinor-TxB_2 was prepared as described; [^1H]TxB_2 was prepared by a similar procedure and converted to dinor-TxB_2 with *Mycobacterium rhodochrous*.[13] Ratios approaching 1.0 were corrected by the method of J. R. Chapman and E. Bailey, *J. Chromatogr.* **89,** 215 (1974).

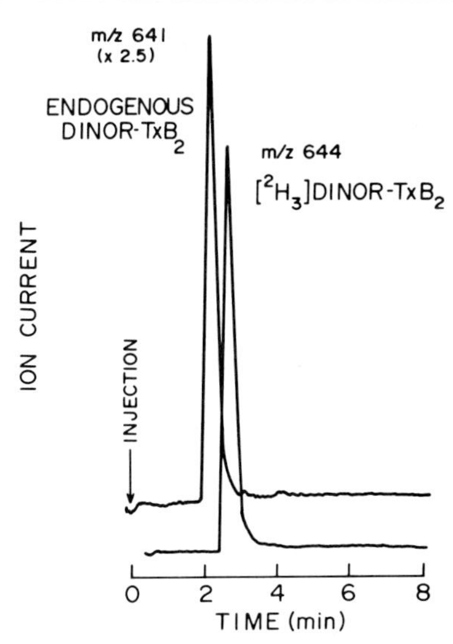

FIG. 4. Representative selected ion-monitoring chromatograms of ions m/z 641 (endogenous dinor-TxB_2) and m/z 644 (dinor-TxB_2 internal standard) obtained using the purification scheme for urinary dinor-TxB_2 described in the text.

Results and Discussion

A standard curve and representative selected ion-monitoring chromatograms obtained using the procedure described above are shown in Figs. 3 and 4. The precision of the assay is within 10%. Using this procedure, urinary dinor-TxB_2 excretion has been measured in six normal healthy males with results of 206± 58 pg/mg creatinine and 379± 166 ng/24 hr (mean ± SD)[23] (Fig. 5a). The relatively low levels of dinor-TxB_2 that are excreted into human urine present a rather formidable analytical challenge, thus necessitating the arduous purification scheme outlined here combined with a sensitive mass spectrometer.

Evidence for the validity of the assay as in index of *in vivo* thromboxane production was obtained from a patient with a thrombocytosis (platelet count, $2 \times 10^6/mm^3$), who demonstrated substantially increased urinary excretion of dinor-TxB_2 (Fig. 5b), suggesting a platelet origin for the production of thromboxane.[23] Moreover, two patients with variant angina, treated once with a 150–300 mg dose of aspirin, a dose known to result in substantial inhibition of the platelet cyclooxygenase,[24] showed marked reductions in urinary dinor-TxB_2 excretion (Fig. 5c).[25] Measure-

[23] R. L. Maas, L. J. Roberts, II, and J. A. Oates, *Clin. Res.* **28,** 319A (1980).

[24] J. W. Burch, N. Stanford, and P. W. Majerus, *J. Clin. Invest.* **61,** 314 (1978).

[25] R. M. Robertson, D. Robertson, L. J. Roberts, R. L. Maas, G. A. FitzGerald, G. C. Friesinger, and J. A. Oates, *N. Engl. J. Med.* **304,** 998 (1981).

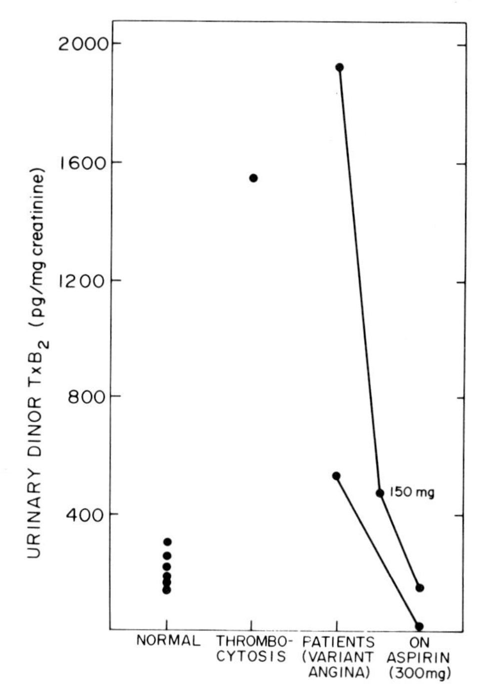

FIG. 5. Histogram showing (a) levels of dinor-TxB_2 in normal healthy males (206 ± 58 pg/mg creatinine); (b) elevated levels in a patient with a thrombocytosis of 2×10^6 platelets/mm^3; and (c) marked reduction of elevated urinary dinor-TxB_2 levels in two patients with variant angina receiving in one case 150 mg of aspirin (1 dose) then 300 mg and in the other one 300 mg dose.

ment of urinary dinor-TxB_2 excretion may thus represent a useful tool for investigating the role of platelets and thromboxane in human disease. Further improvements in the assay, utilizing capillary GC–MS or UV-absorbing derivatives to facilitate HPLC purification, will, it is hoped, permit more efficient use of the analytical methodology described here.

Acknowledgments

This work was supported by a grant from the National Institutes of Health, GM 15431. R. L. M. is supported by the Vivian Allen Fund, Vanderbilt Medical School. The generous gift of the lactone starting material, (compound **1**), from the Upjohn Company is gratefully acknowledged.

[71] Preparation of 2,3,4,5-Tetranor-Thromboxane B_2, the Major Urinary Catabolite of Thromboxane B_2 in the Rat

By C. R. PACE-ASCIAK[1] and N. S. EDWARDS

We have reported that 2,3,4,5-tetranor-TxB_2 is the principal urinary catabolite excreted by rats that had received a slow infusion of TxB_2 (Fig. 1).[2] This is in sharp contrast to data reported in two other species (human and primate)[3-5] in which the principal urinary catabolite was found to be 2,3-dinor-TxB_2. The excretion of a tetranor product by the rat was not entirely unexpected in view of our previous studies with $PGF_{2\alpha}$ in which we had shown that all urinary products of $PGF_{2\alpha}$ had been subjected to a double β-oxidation, i.e., tetranor products, prior to their elimination.[6] In this chapter we describe a procedure for the preparation of tetranor-TxB_2. We also describe its conversion into the d_3-methoxime derivative and some GC–MS properties that permit the simultaneous detection of TxB_2 and its dinor and tetranor catabolites in urine.

Preparation of Urinary Catabolite

TxB_2 (New England Nuclear, Boston, Massachusetts; specific activity 125 Ci/mmol) was diluted with unlabeled material (gift from Drs. U. Axen and J. E. Pike, The Upjohn Co.) to a specific activity of 100,000 dpm/μg in normal saline. It was infused intravenously at a rate of 0.016 μg per gram of body weight per minute for 60 min. into the rat prepared according to the procedure described in this volume [65]. Urine was collected during the infusion of TxB_2 and for a subsequent 3-hr period.

The urine was combined, diluted with water (10 volumes), acidified to pH 3 with 1 *N* HCl and passed through a short column of XAD-2 Amberlite resin (Rohm & Haas). After washing with 50 ml of water, the column was eluted with 50 ml of acetone in which >95% of the applied radioactiv-

[1] The work reported in this chapter was supported by a grant to C. P.-A. (MT-4181) from the Medical Research Council of Canada.

[2] C. R. Pace-Asciak, and N. S. Edwards, *Biochem. Biophys. Res. Commun.* **97,** 81 (1980).

[3] L. J. Roberts, B. J., Sweetman, N. A. Payne, and J. A. Oates, *J. Biol. Chem.* **252,** 7415 (1977).

[4] L. J. Roberts, B. J. Sweetman, L. J. Morgan, N. A. Payne, and J. A. Oates, *Prostaglandins* **13,** 631 (1977).

[5] H. Kindhal, *Prostaglandins* **13,** 619 (1977).

[6] C. R. Pace-Asciak, and N. S. Edwards, *J. Biol. Chem.* **255,** 6106 (1980).

ISBN 0-12-181986-8

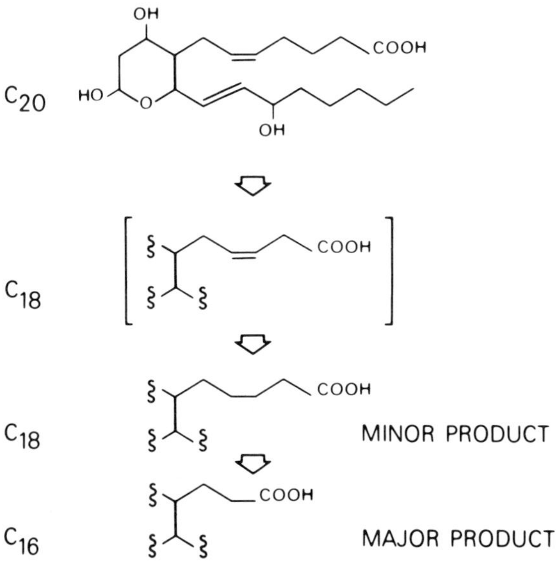

FIG. 1. Scheme outlining the pathway in the formation of the tetranor catabolite by the rat.

ity was obtained. The acetone fraction was evaporated to complete dryness, dissolved in 0.1 ml of methanol, and converted to the methyl ester derivative with 0.9 ml of a freshly prepared distilled solution of diazomethane in diethyl ether (10 min, 23°).

The tetranor catabolite could be purified by preparative thin-layer chromatography (silica gel G, Brinkmann) using chloroform–methanol–acetic acid–water (90:9:1:0.65, v/v/v/v) as developing solvent (R_f = 0.36 compared to 0.40 for TxB_2) or by HPLC using a Waters fatty acid analysis column with an acetonitrile–water solvent mixture (33:67, v/v; R_t = 8 min).

Derivatives for GC–MS

a. Unlabeled Catabolite. The purified catabolite (methyl ester) was converted into the methoxime (MO) and *t*-butyldimethylsilyl ether (tBDMS) derivative by consecutive reaction with MOX reagent (Pierce, 100 μl overnight, 23°) and *t*-butyldimethylchlorosilane reagent kit (Applied Sciences, 100 μl, 20 min, 60°). The sample was then diluted with water and extracted with hexane as described in this volume [65].

b. d_3-Labeled Catabolite. Since TxB_2 and its catabolites contain a keto group, they can easily be labeled with d_3-methoxylamine hydrochloride to produce internal standards and carriers for mass fragmentographic assays

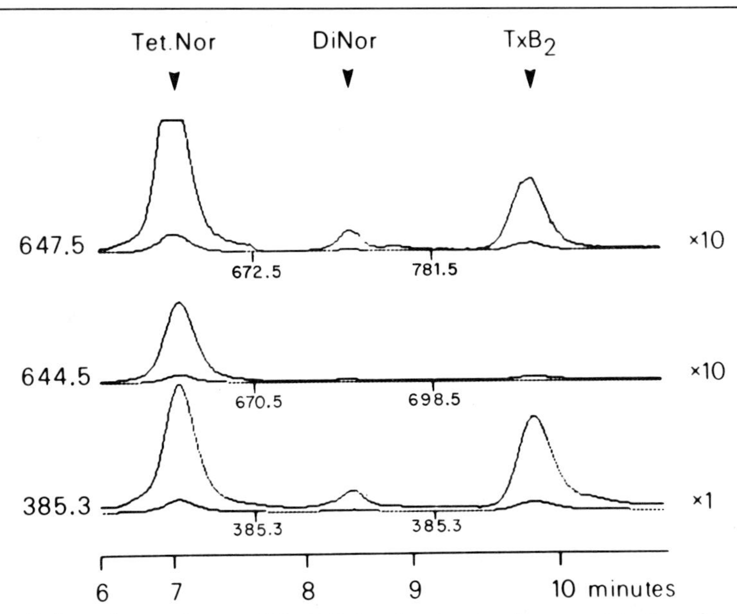

FIG. 2. Mass fragmentogram (GC–MS) showing the simultaneous detection of tetranor, dinor, and TxB_2 products in the same sample of urine from a rat that had received a slow infusion of TxB_2. d_3 products were added to this sample prior to analysis by GC–MS. The record shown is in real time collected during the chromatographic run. Gas chromatography was performed on 3% OV-1 on Gas-Chrom Q (Applied Science) using a 2-foot × $\frac{1}{8}$ inch column. Injection of the sample was at 220° with a 4°/min increase in temperature up to 300°. Mass spectrometry was performed using a Hewlett-Packard quadrapole mass spectrometer.

of these products in biological fluids. The samples were derivatized into d_3-MO and tBDMS as described in this volume [65].

Figure 2 shows a mass fragmentogram of the simultaneous assay of TxB_2 (C value = 29.2) and its dinor (C value = 28.1) and tetranor (C value = 27.0) urinary catabolites in the urine of a rat that had received an infusion of TxB_2. This figure is intended to illustrate the following points.

1. The three compounds can easily be resolved from each other under normal GC operating conditions (3%OV-1).

2. The unique mass spectral fragmentation pattern of TxB_2 and its catabolites can be put to use for the monitoring of TxB_2 catabolites (*a*) *as a class* using the common fragment ion at m/z 385 (base peak) due to the fragment containing C-12 to C-20 resulting from cleavage of the C-8–C-12 bond (consequently this fragment lacks the 12-oxime group); and (*b*) *individually,* making use of fragments containing the 12-oxime group, i.e., m/z 644 (d_0) vs 647 (d_3) for the tetranor catabolite and 698 (d_0) vs 701 (d_3)

for TxB_2 itself. These fragments arise out of loss of 57 (*t*-butyl) from the molecular ion of each compound.

Making use of the mass spectrometer as a selective and specific detector, it is possible to select appropriate fragment ions for each product and analyze and quantitate primary prostaglandins and their urinary catabolites, *all in the same sample*.

Summary

The preparation of 2,3,4,5-tetranor-TxB_2 is described. Its conversion into d_3-MO derivative allows the preparation of a deuterium-labeled product for use in mass fragmentography. An example is given of the potential application of this standard for the simultaneous measurement of this catabolite from the dinor catabolite and TxB_2 by GC–MS.

[72] Measurement of 5-Hydroxyeicosatetraenoic Acid (5-HETE) in Biological Fluids by GC–MS

By MARTIN L. OGLETREE, KENNETH SCHLESINGER, MARY NETTLEMAN, and WALTER C. HUBBARD

Lipoxygenation of arachidonic acid at C-5 produces 5(*S*)-hydroperoxy-6-*trans*-8,11,14-*cis*-eicosatetraenoic acid (5-HPETE). Further metabolism of 5-HPETE proceeds via two known pathways. The major pathway of 5-HPETE metabolism is conversion to 5(*S*)-hydroxy-6-*trans*-8,11,14-*cis*-eicosatetraenoic acid (5-HETE).[1] 5-HPETE is also converted to leukotriene A_4 (LTA_4), a reactive precursor of other leukotrienes.[2] LTA_4 is converted to leukotriene B_4 (LTB_4) a potent chemotactic agent,[3] and leukotrienes C_4 and D_4, bioassayable as slow-reacting substances.[4]

Measurement of 5-HETE is employed as an index of 5-lipoxygenation of arachidonic acid. An assay for 5-HETE employing high-performance liquid chromatography (HPLC) has been described and is useful for measurement of multinanogram quantities of 5-HETE.[5] This chapter details a highly sensitive and selective assay for 5-HETE employing racemic octa-

[1] P. Borgeat, M. Hamberg, and B. Samuelsson, *J. Biol. Chem.* **251,** 7816 (1976).

[2] O. Radmark, C. Malmsten, B. Samuelsson, G. Goto, A. Marfat, and E. J. Corey, *J. Biol. Chem.* **255,** 11828 (1980).

[3] A. W. Ford-Hutchison, M. A. Bray, M. V. Doig, M. E. Shipley, and M. J. H. Smith, *Nature (London)* **286,** 264 (1980).

[4] P. Hedqvist, S.-E. Dahlen, L. Gustafsson, S. Hammarström, and B. Samuelsson, *Acta Physiol. Scand.* **110,** 331 (1980).

[5] P. Borgeat and B. Samuelsson, *Proc. Natl. Acad. Sci. U.S.A.* **76,** 2148 (1979).

ISBN 0-12-181986-8

deuterated 5-HETE for stable isotope dilution measurements with quantitation via combined gas-liquid chromatography–mass spectrometry (GC–MS). The assay method is useful for samples containing as little as 500 picograms of 5-HETE.

Preparation, Identification, Standardization, and Storage of Internal Standard

Octadeuterated arachidonic acid is prepared from 5,8,11,14-eicosatetraynoic acid as described by Taber in this volume.[6] A mixture of octadeuterated (d_8) 5-HETE and tritiated 5-HETE ($\sim 2 \times 10^5$ dpm ^{3}H per microgram of octadeuterated compound) is prepared chemically from d_8-arachidonic acid and [5,6,8,9,11,12,14,15-^{3}H]arachidonic acid (New England Nuclear, 60–100 Ci/mmol) by reaction with H_2O_2 in the presence of Cu^{2+}, as described by Boeynaems *et al.*[7] Alternatively, 5-HETE can be synthesized chemically by iodolactonization of arachidonic acid and dehydroiodination of the iodolactone.[8] For use as internal standard, the ratio of nondeuterated 5-HETE to d_8-5-HETE should be less than 0.6%.

Two independent methods are employed for standardization of d_8-5-HETE, due to the tendency of 5-HETE to form a δ-lactone. Ultraviolet spectrophotometry at $\lambda_{max}^{MeOH} = 235$ nm ($\epsilon = 27{,}000$)[9] and gas-liquid chromatography with tricosanoic acid as internal standard should give agreement within ±5%.

Stock solutions of d_8-5-HETE in phosphate buffer (pH 9.5) containing ethanol (3:1) are stored under nitrogen at $-70°$. Purity of d_8-5-HETE is checked periodically by straight phase HPLC (silica, μPorasil 3.9 × 300 mm column) of the methyl ester employing a solvent system of hexane with isopropanol (99.7:0.3).

Biological Fluids

Blood and Lymph Samples. Blood and lymph should be collected in tubes containing anticoagulant and a lipoxygenase inhibitor, such as nordihydroguaiaretic acid (10 μ*M*).[10] 5,8,11,14-Eicosatetraynoic acid is not recommended as an inhibitor of *ex vivo* 5-HETE biosynthesis.[1,11] Samples

[6] D. F. Taber, M. A. Phillips, and W. C. Hubbard, this volume [47].

[7] J. M. Boeynaems, A. R. Brash, J. A. Oates, and W. C. Hubbard, *Anal. Biochem.* **104,** 259 (1980).

[8] W. C. Hubbard, M. A. Phillips, and D. F. Taber, *Prostaglandins* **23,** 61 (1982).

[9] M. Hamberg and B. Samuelsson, *J. Biol. Chem.* **242,** 5329 (1967).

[10] M. Hamberg, *Biochim. Biophys. Acta* **431,** 651 (1976).

[11] G. M. Bokoch and P. W. Reed, *Fed. Proc. Fed. Am. Soc. Exp. Biol.* **40,** 658 (1981).

are centrifuged at 2000 *g* for 15 min at 0–4° to remove cellular components. Measured volumes of cell-free supernatant (5–to 15 ml) are transferred to 110-ml glass centrifuge tubes followed by addition of internal standard and tracer (0.50 μg d_8-5-HETE and 1 × 10^5 dpm [$^3H_8$]5-HETE).

Leukocyte Incubations in Vitro. In vitro, 0.25-ml aliquots of polymorphonuclear leukocytes in plasma (5 × 10^7 cells/ml) are prepared by the method of Boyum[12] and are incubated in the presence and in the absence of stimulus. For leukocyte incubations in Krebs-Ringer bicarbonate medium, procedures described by Bokoch and Reed[13] are employed. Incubations are terminated by addition of 4 volumes of methanol, after which octadeuterated and tritiated 5-HETE internal standard (0.50 μg) is added.

Extraction of 5-HETE from Biological Fluids. Plasma proteins are removed by precipitation with acetone (3 times plasma volume) and centrifugation at 750 *g* for 5 min. The supernatant is adjusted to pH 11.0–11.5 with 0.1 *N* NaOH for extraction of neutral lipids into hexane (3 times plasma volume). After removal of the hexane (upper) layer, the pH is readjusted to 11.0–11.5, and extraction into hexane (3 times plasma volume) is repeated. The aqueous (lower) phase is then titrated to pH 7.0 with 0.1 *N* HCl and extracted with ethyl acetate or chloroform (3 times plasma volume). The organic phase is transferred to a round-bottom flask, and the solvent is removed under reduced pressure. The extract is transferred to a 1.0-ml silanized conical vial with chloroform–methanol (3:1, v/v) and stored under nitrogen at 0° until purification by HPLC. Recovery of 5-HETE from plasma employing these procedures usually exceeds 75%.

The procedures described above have been modified in studies employing leukocyte suspensions and other conditions in which small volumes (≤2 ml) of aqueous media are present. After termination of the reaction by addition of methanol or acetone or after cell removal by filtration or centrifugation, an equal volume of distilled water or buffer solution is added. For solutions containing plasma, neutral lipids are removed after titration to pH 11.0–11.5 as described above. Samples are adjusted to pH 7.0 by addition of 0.05 *M* phosphate buffer (pH 7.0, 10% sample volume). If necessary, the pH is lowered to 7.0 with glacial acetic acid for extraction into either ethyl acetate or chloroform. The organic solvent is transferred to a rotary evaporation flask and removed under reduced pressure. The extract is transferred to a 1.0-ml silanized conical vial in chloroform–methanol (3:1, v/v) and stored under nitrogen at 0° until purification via HPLC. Recovery of 5-HETE usually exceeds 75%.

[12] A. Boyum, *Scand J. Immunol. Suppl.* **5,** 9 (1976).
[13] G. M. Bokoch and P. W. Reed, *J. Biol. Chem.* **255,** 10223 1980).

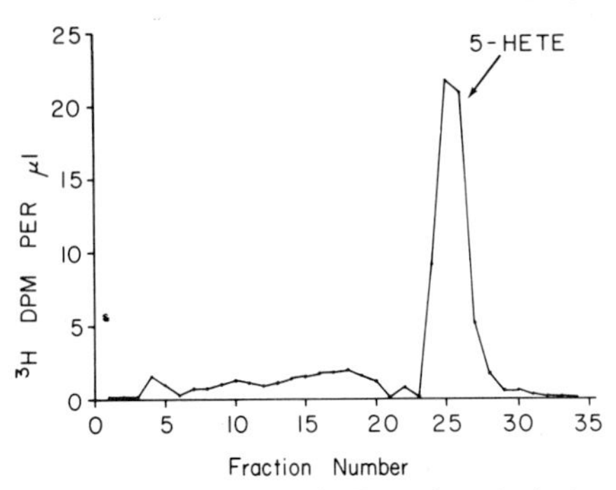

FIG. 1. Reversed-phase HPLC (μBondapak C_{18} column) elution pattern for 5-HETE added to and extracted from human blood plasma. Solvent, methanol–water–acetic acid (75:25:0.01); flow rate, 1 ml/min; fraction volume, 1 ml. Of the added radioactivity, 58% was recovered in fractions 24–27.

Purification by Reversed-Phase HPLC

Solvent is evaporated from the sample extract under a stream of nitrogen, and the residue is dissolved in a small volume of methanol (≤ 50 μl) for purification by reversed-phase HPLC. 5-HETE is eluted from a μBondapak C_{18} column (Waters Associates) with a solvent of methanol–water–glacial acetic acid (75:25:0.01, v/v/v) delivered at a flow rate of 1–2 ml/min. One milliliter fractions are collected. Identification of fractions containing 5-HETE is achieved by detection of radioactivity (50-μl aliquots) with or without simultaneous measurement of optical density at 235 nm. Figure 1 shows the reversed-phase HPLC elution profile of radiolabeled 5-HETE following extraction from a sample of human blood plasma. Recovery of [^{3}H]5-HETE usually exceeds 50% of that added to the sample. Fractions with radioactivity or the peak absorption at 235 nm containing ^{3}H label are pooled, and the solvent is removed under reduced pressure. The residue is transferred to a 1.0-ml silanized conical vial with chloroform–methanol (3:1, v/v).

Derivatization

After removal of solvent under a nitrogen stream, the residue, usually yellow in color, is dissolved in methanol (100 μl), and the methyl ester is prepared by reaction with excess ethereal diazomethane for 20 min at room temperature. After removal of methanol and excess ethereal diazomethane under a nitrogen stream, the trimethylsilyl (TMS) ether derivative is prepared by adding to the residue excess *N*,*O*-bis(trimethylsilyl)trifluoracetamide (BSTFA, 10 μl) and pyridine (10 μl). The reaction is allowed to proceed for at least 2 hr at room temperature. Usually the reaction is allowed to proceed overnight.

Boeynaems and Hubbard[14] have also prepared the methyl ester-*t*-butyldimethylsilyl ether derivative of 5-HETE, which produces a relatively intense fragment ion (electron ionization) at m/z 391 (M^+-57, loss of $C(CH_3)_3$ of the *t*-butyldimethylsilyl ether moiety).

Analysis by Gas Chromatography–Mass Spectrometry (GC–MS)

Pyridine and excess BSTFA are removed under a nitrogen stream, and the residue is dissolved in heptane (10 μl) immediately prior to injection into the gas chromatograph interfaced with the mass spectrometer. A glass column of 3% OV-1 on Gas Chrom Q (1 m × 2 mm) at 210° is employed with helium as a carrier gas at a flow rate of ~25 ml/min. Because of the thermal lability of the ME-TMS derivative of 5-HETE, the temperatures of the injection port and the interface line between the gas–liquid chromatograph and the mass spectrometer are maintained at 220°.

Endogenous 5-HETE in purified sample extracts is quantitated by simultaneously monitoring the currents of characteristic fragment ions of the ME-TMS derivative of 5-HETE (m/z 305) and of 5,6,8,9,11,12,14,15-octadeutero-5-HETE (m/z 313). The methyl ester-*t*-butyldimethylsilyl ether derivative of 5-HETE has also been employed for quantitation of 5-HETE. Characteristic fragment ions monitored for this derivative of native 5-HETE and d_8-5-HETE are m/z 391 and 399, respectively. The use of this derivative of 5-HETE requires higher operational temperatures of the gas chromatograph, injection port, and transfer line, resulting in considerable loss of both the internal standard and endogenous 5-HETE due to thermal degradation.

General Considerations

Care is necessary in handling of samples, especially before addition of internal standard, and in the use and storage of internal standard because of the lability of 5-HETE. Under acidic conditions, 5-HETE forms a 1,5-δ-lactone[7] with chromatographic properties distinctly different from those of 5-HETE. The purification procedures are designed for elution of the free acid form of 5-HETE. Except during reversed-phase HPLC, samples should be maintained at neutral to alkaline pH to prevent lactonization of 5-HETE.

[14] J. M. Boeynaems and W. C. Hubbard, *in* "Prostaglandins, Prostacyclin and Thromboxane Measurement" (J. M. Boeynaems and A. G. Herman, eds.), pp. 167–181. Nijhoff, Amsterdam, 1980.

Reagents and solvents used in this assay should be of very high purity. Solvents recommended are "distilled in glass" quality specifically prepared for HPLC. Reagents for derivatization should be those specifically recommended for gas–liquid chromatography.

Acknowledgments

The authors thank Drs. J. M. Boeynaems and G. M. Bokoch for their advice in developing this assay and Dr. D. Taber for the generous gift of octadeuterated arachidonic acid. This work was supported by Grants GM 15431, HL 19153, and HL 26198 from the National Institutes of Health.

[73] Open Tubular Glass Capillary Gas Chromatography for Separating Eicosanoids

By JACQUES MACLOUF and MICHEL RIGAUD

Gas–liquid chromatography involves a partition of the molecules of the sample components to be analyzed between the stationary liquid phase and the carrier gas within the column. In traditional packed columns, the low speed of diffusion of the sample components between the two phases is the limiting factor of the efficiency (i.e., separation power) and of the time of analysis; open tubular columns shorten both, and provide a gain in sensitivity of $\times 100$. In the petrochemical field, metal open tubular capillary columns have been used with chromatographic efficiencies exceeding 10.000 plates/meter. Unfortunately, for biochemical problems, heated metal is incompatible with many substances, and glass capillary columns are required as for packed columns. Technological problems inherent to the preparation of thermostable glass open tubular capillary columns were overcome, partially, in 1968 with the development of a stable, uniform film of liquid partitioning phase on the inner surface of the glass capillary.[1]

Instrumental Aspects

Preparation of the Columns

We have used various procedures for the preparation of capillary columns,[2] but we will limit ourselves to the most satisfactory ones. Mainly,

[1] K. Grob, *Helv. Chim. Acta* **51,** 718 (1968).

[2] J. Maclouf, M. Rigaud, J. Durand, and P. Chebroux, *Prostaglandins* **11,** 999 (1976).

ISBN 0-12-181986-8

we will outline some of the practical details or modifications of the original methods that we have found to improve the column preparation. Capillary tubes, either soda-lime for polyglycol phases, such as Carbowax, or borosilicate glass for other phases, such as polysiloxanes, were drawn by a machine similar to that described by Desty *et al.*[3] Any commercial glass-drawing machine is sufficient for this purpose provided the final internal diameter of the capillary is about 0.2–0.25 mm. Uncoated glass capillaries (e.g., Supelco Inc., Bellefonte, Pennsylvania) may also be purchased. All commercial ready-made columns that we have tested were less efficient, relatively short-lived, and catalytically active as compared to our home-made ones. This last point is especially critical when sensitive solutes, particularly unsaturated eicosanoids, are to be analyzed quantitatively in the low nanogram range by this technique (see Selection of Optimal Chromatographic Conditions). We caution that 6–12 months are necessary to handle this technique in a reproducible way, and that 1–4 months are required for full optimization even of commercial columns.

Carbowax and Carbowax–Terephthalic Acid Columns. These columns were prepared exactly following the procedure of Alexander *et al.*[4] The inner-glass soda-lime surface, etched by a uniform layer of microcrystalline sodium chloride, was nearly totally inactivated, and the technique was quite reproducible as indicated in Table I. The number of plates per meter always exceeded 2500, and the coating efficiency always exceeded 55%. The partition ratio, k, for stearic acid methyl ester at the indicated temperatures was quite low. For example, it would be approximately 5 to 10-fold more with packed columns. The temperature limit of these columns is about 240°. Such columns are invaluable to check the purity of the polyunsaturated fatty acid precursors that are used as substrates for *in vitro* transformations by cyclooxygenase or lipoxygenase enzymes; however, they are unsuitable for prostaglandin analysis. The latter must be silylated prior to chromatography; consequently, there may be silyl exchange between the hydroxyl group of the polyglycol columns and the hydroxyl group of the prostaglandins. This chemical reaction alters the properties of the columns.

Low Polarity Phase Columns. The surface silanol groups of borosilicate capillaries were totally silylated by filling one-fourth of the column with the clear supernatant of hexamethyldisilazane–dimethylchlorosilane (5:1, v/v). A vacuum was applied at both ends of the column, which were then fused with a flame. After heating the column at 210° for 24–48 hr, the excess reagent was flushed with dry nitrogen, and the column was washed with 2 ml of methanol and 2 ml of toluene, and then dried with nitrogen. A

[3] D. H. Desty, J. N. Haresnape, and B. H. F. Whyman, *Anal. Chem.* **32,** 302 (1960).
[4] G. Alexander, G. Garzo, and G. Palyi, *J. Chromatogr.* **91,** 25 (1974).

TABLE I
CHROMATOGRAPHIC CHARACTERISTICS OF SOME CARBOWAX GLASS CAPILLARY COLUMNS

Column No.	Length (m)	I.d.[a] (mm)	Theoretical plates (n)	n/m	$HETP_{meas}$[b] (mm)	$HETP_{calc}$[c] (mm)	Coating efficiency (%)	β[d]	df[e] (μm)	$\bar{u}$[f] (cm/sec)	Partition ratio ($k_{°C}$)	Stationary phase
1	21	0.25	60.200	2870	0.348	0.195	56	468	0.13	34.9	2.90_{200}	CW 20 MAT
2	29	0.34	73.300	2520	0.395	0.256	65	500	0.18	34	1.95_{200}	CW 20 MAT
3	42	0.35	120.000	2950	0.349	0.255	73	426	0.21	17	3.30_{200}	FFAP
4	31	0.36	100.800	3190	0.313	0.266	85	624	0.15	31	1.92_{200}	FFAP
5	35	0.30	105.000	3000	0.333	0.224	67	499	0.15	27	2.31	FFAP
6	41	0.36	111.000	2714	0.368	0.300	82	457	0.21	18	3.38_{190}	FFAP
7	42	0.30	128.000	3000	0.331	0.219	69	525	0.14	26	2.15_{190}	CW 20 M
8	40	0.30	98.000	2450	0.410	0.237	58	573	0.13	24	3.21_{190}	CW 20 M
9	23	0.30	63.000	2700	0.364	0.228	63	451	0.15	23	5.61_{180}	CW 20 M
10	29	0.28	82.587	2872	0.348	0.230	66	451	0.16	20	3.98_{190}	CW 20 M

[a] Internal diameter.
[b] Height equivalent to one theoretical plate (measured).
[c] HETP calculated.
[d] Phase ratio of the column.
[e] Phase thickness.
[f] Linear carrier gas velocity.

thermostable ionic surface agent, benzyltriphenylphosphonium chloride, was added according to Rutten and Luyten.[5] The preliminary silylation stabilizes the tensioactive agent on the inner glass surface. This modification, intended initially for steroid analysis,[6] reduced the background significantly compared to the original method. A low background is a prerequisite when working at high sensitivities. Table II illustrates the characteristics of seven columns (OV-1, SE-52, SE-54 phases) prepared according to this procedure. The number of plates per meter exceeded 3000, the coating efficiency exceeded 65%, and the partition ratio was quite homogeneous. On one column, more than 2000 injections were performed without deterioration of its chromatographic properties. The temperature limit of these columns is about 280°, which makes them suitable for analysis of eicosanoids.

Thermostable Glass Capillary (GC) Columns. Silicone prepolymer can be bound *in situ* to the inner glass surface of a capillary by a modification of our original method.[7] Methyl and phenyl silicone prepolymers, commercially available from Petrarch Systems Inc, Levittown, Pennsylvania, are now used. After reaction,[7] residual column catalytic activity is reduced by silanization (see above). The thermostability and low bleeding of these columns permit their use at extreme sensitivities (Table III).

Injection System and Connections

Most chromatographs are not adapted to capillaries or the manufacturers provide apparatus and equipment only complementary to their columns. Therefore, the equipment has to be adjusted to the columns, very often with some difficulty. Gas leaks and dead volumes at the proximal or distal end of the column are the main problems. Additionally, the quantitative analysis of compounds ranging from molecular weight 300 to 600 requires that all the sample is transferred onto the column; this requirement is incompatible with commercial split/splitness injectors. We use two types of injectors satisfactorily adapted to capillary columns.

Solid Injector. Figure 1 shows a scheme of a Ros injector[8] and its adaptation to a Packard Model 428 chromatograph that was transformed for glass capillary column chromatography. The proximal part of the column is mounted so that it is as close as possible (2–3 mm) to the injection needle in its low position. This reduces the dead volume, for compatibility with the internal volume of the capillary column. Leaks will be avoided if

[5] G. A. F. M. Rutten and J. A. Luyten, *J. Chromatogr.* **73,** 177 (1972).

[6] F. Berthous, *in* Thèse es-sciences, Brest, France 42, 208 (1977).

[7] M. Rigaud, P. Chebroux, J. Durand, J. Maclouf, and C. Madani, *Tetrahedron Lett.* **44,** 3935 (1976).

[8] A. Ros, *J. Chromatogr.* **3,** 252 (1965).

TABLE II
CHROMATOGRAPHIC CHARACTERISTICS OF SOME LOW POLARITY-PHASE GLASS CAPILLARY COLUMNS

Column No.	Length (m)	I.d.[a] (mm)	Theoretical plates (n)	n/m	$HETP_{meas}$[b] (mm)	$HETP_{calc}$[c] (mm)	Efficiency (%)	β[d]	df[e] (μm)	$\bar{u}$[f] (cm/sec)	Partition ratio ($k_{°C}$)	Stationary phase
1	22	0.25	72.000	3256	0.307	0.197	64	499	0.12	28	4.16_{220}	SE-54
2	22	0.28	67.000	2982	0.335	0.229	68	499	0.14	21	4.09_{220}	OV-1
3	20	0.26	67.500	3300	0.303	0.214	65	499	0.13	28	4.20_{220}	OV-1
4	21	0.30	80.000	3814	0.262	0.246	94	499	0.15	16	4.32_{220}	SE-52
5	24	0.26	84.000	3493	0.286	0.216	76	499	0.13	21	4.31_{220}	OV-1
6	30	0.26	95.000	3100	0.322	0.205	64	499		24	3.93_{220}	OV-1
7	23	0.27	95.000	4100	0.243	0.228	94	499	0.13	23	4.29	OV-1

[a] Internal diameter.
[b] Height equivalent to one theoretical plate (measured).
[c] HETP calculated.
[d] Phase ratio of the column.
[e] Phase thickness.
[f] Linear carrier gas velocity.

TABLE III
CHROMATOGRAPHIC CHARACTERISTICS OF SOME THERMOSTABLE GLASS CAPILLARY COLUMNS

Column No.	Length (m)	I.d.[a] (mm)	Theoretical plates (n)	n/m[b]	n′/m[c]	df[d] (μ)	$\bar{u}$[e] (cm/sec)	Partition ratio ($k_{°C}$)	Initial monomer[f] (v/v)	Prepolymer concentration (%)
1	23	0.31	58.000	2500	1600	0.25	21	3.82_{240}	DMCS Fluka	10
2	30	0.32	78.000	2600	1900	0.41	32	6.02_{230}	DMCS DCDØS 19/1	10
3	13	0.36	30.000	2300	1600	0.40	25	5.25_{230}	DMCS DCDØS 19/1	10
4	13	0.36	34.000	2700	1600	0.24	22	3.18_{230}	DMCS DCDØS 19/1	10
6	30	0.31	89.000	3000	2000	0.30	21	4.60_{230}	DMCS DCDØS 19/1	10
7	52	0.31	133.000	2600	1700	0.28	18	4.27_{230}	DMCS DCDØS 19/1	10
8	29	0.31	67.000	2300	1400	0.22	30	3.42_{230}	DMCS DCDØS 19/1	10
9	36	0.31	72.123	2000	1600	0.50		7.55_{230}	DMCS/DCDØS 19/1	15
10	30	0.31	90.000	3000	1800	0.23	14	3.50_{230}	DMCS DCDØS 19/1	10
11	25	0.28	70.000	2800	1800	0.24	31	4.17_{230}	DMCS	20
12	26	0.28	79.000	3000	2000	0.27	26	4.57_{230}	DMCS	20

[a] Internal diameter.
[b] Number (n) of plates per meter (m), calculated.
[c] Number of plates per meter, measured.
[d] Phase thickness.
[e] Gas velocity.
[f] DMCS, dimethyldichlorosilane; DCDØs, dichlorodiphenylsilane.

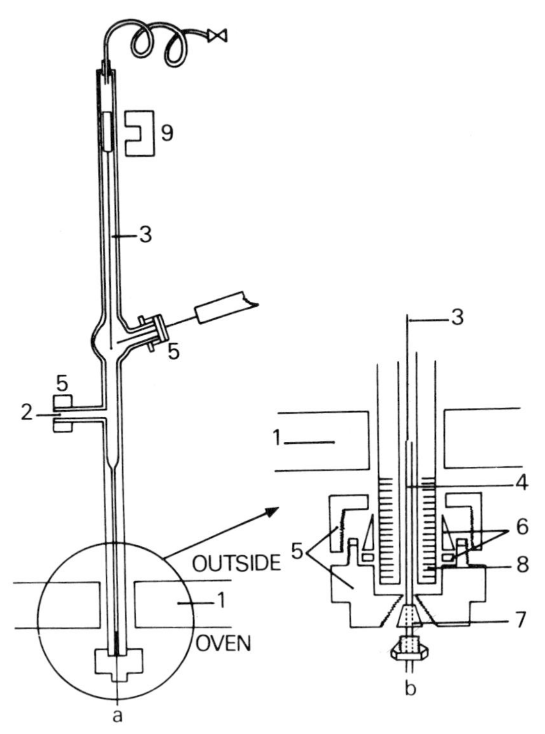

FIG. 1. Schematic representation of a whole glass solid injector to a Packard Model 428 gas chromatograph: (a) general scheme; (b) connection between capillary column and injector. 1, Injector heated block; 2, carrier gas arrival; 3, glass injection needle; 4, capillary column; 5, metal swagelocks ($\frac{1}{4}$ inch); 6, Teflon or Vespel ferrules (Alltech) + back ferrules stainless steel ($\frac{1}{4}$ inch); 7, "guarnizzione" (Carlo Erba) or Vespel ferrules ($\frac{1}{16}$ inch); 8, ground glass; 9, magnet and metallic part of the needle. The inside capillary part of the injector is 0.5 mm, and it is about 7 mm elsewhere. The outside diameter should be adapted to $\frac{1}{4}$-inch swagelocks. In our model the glass needle length is approximately 150 mm.

the bottom part of the injector has been ground. Thermostable Teflon or Vespel ferrules (Alltech Assoc., Illinois) are recommended. Pieces Nos. 5 and 6 should be mounted first and then heated for 2–3 hr at 250°, tightened again after cooling at 100°, then the column should be mounted. This allows better adherence of the swagelocks to the glass via the Vespel ferrules. With frequent use, the glass needle should be cleaned every 2 weeks with concentrated chromosulfuric acid solution. After immersion for 2–3 hr, the needle tip should be carefully rinsed with distilled water and wiped with a Kleenex. After insertion, it should be silanized by 1–2 μl of *N,O*-bis(trimethylsilyl)trifluoroacetamide–trimethylchlorosilane (1%) before use. The reproducibility of the injector was ± 4.2% or better.

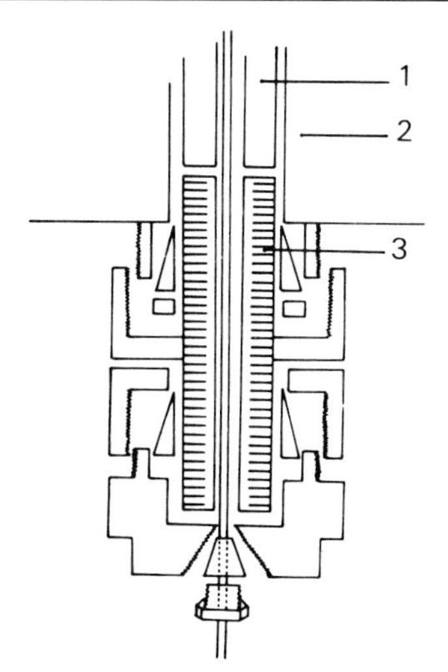

FIG. 2. Schematic representation of glass capillary mounting in the flame ionization detector of a Packard Model 428 gas chromatograph: 1, quartz jet; 2, detector block; 3, additional glass connection, ground on the outside. Other details as in Fig. 1.

One dominant advantage of this injector is that it allows a concentration of the sample on the needle. This is extremely useful when minute amounts of compounds must be analyzed.

On-Column Injector (*Carlo-Erba*). This injector allows the direct transfer of the whole sample onto the proximal part of the column. Therefore, all the possibilities of catalytic destructions due to contact of the biological material with metallic or glass connections are excluded. Further, the very low temperature in the initial length of the column at the time of injection, avoids any possible destruction of labile compounds in the syringe needle. One drawback of this injector, as compared to the Ros type, is that maximally 1–2 μl of liquid can be injected in order to avoid overloading of the column.

Detector Coupling

The connection of the column to the detector must be as short as possible for the same reasons as for the injector.

Flame Ionization Detection.[2,9] In the Packard Model 428 chromatograph, it is possible to mount the distal end of the column inside the quartz jet of the flame that leads to a theoretical "zero" dead volume detector (Fig. 2).

[9] J. Maclouf, H. de la Baume, J. P. Caen, H. Rabinovitch, and M. Rigaud, *Anal. Biochem.* **109,** 147 (1980).

Electron-Capture Detection.[10] The large "dead" volumes of electron-capture (EC) detectors neutralize the superiority and optimal performance of capillary columns. When capillaries are mounted on EC–GC instruments, a scavenger gas must be added at the distal end of the column to maximize the column and detector performances.[10] The Brechbühler detector[11] with a low dead volume may avoid these problems although it is not yet adaptable to most commercial instruments.

Mass Spectrometer Used as a Detector: Multiple Ion Detection. To adjust capillary columns to a mass spectrometer, a scavenger gas must be added at the distal end of the column. We have used an LKB 2091 mass spectrometer equipped with a double jet separator. A scavenger gas supplied at 30 ml/min compensated for the low flow rate of the column (2 ml/min). The scavenger gas is usually the carrier gas, but it may also be the ionization gas in case of chemical ionization use.

Selection of Optimal Chromatographic Conditions

The optimization of the column performance is essential to profit from the advantage of open tubular glass capillary columns. Two features should be tested before any use of the column for biological samples: (*a*) efficiency, i.e., separation power;[12] (*b*)the absence of activity, i.e., no adsorption or "tailing" and the absence of destructive catalytic properties of the inner glass wall.

Efficiency

When temperature programming is used, which is often the case in order to separate substances varying from a molecular weight ranging from 300 to more than 600 in the same analysis, we suggest that the flow rate be adjusted from an empirical formula obtained from our experimental data and supported by theoretical calculations of others.[13]

$$t_R(\text{hexane}) = L(4.25 \pm 0.43)$$

where the retention time for hexane (t_R) is expressed in seconds and L, the column length, in meters. The gas flow rate should be adjusted in agreement with the calculated value of t_R by injecting hexane at 200°, isothermal. One should also keep in mind that the flow rate of the carrier gas decreases as the temperature increases. When using capillary columns,

[10] F. A. Fitzpatrick, D. A. Stringfellow, J. Maclouf, and M. Rigaud, *J. Chromatogr.* **177,** 51 (1979).

[11] B. Brechbüller, L. Gay, and H. Jaeger, *Chromatographia* **10,** 478 (1977).

[12] L. S. Ettre, *Chromatographia* **8,** 291 (1975).

[13] D. W. Grant and M. G. Hollis, *J. Chromatogr.* **158,** 3 (1978).

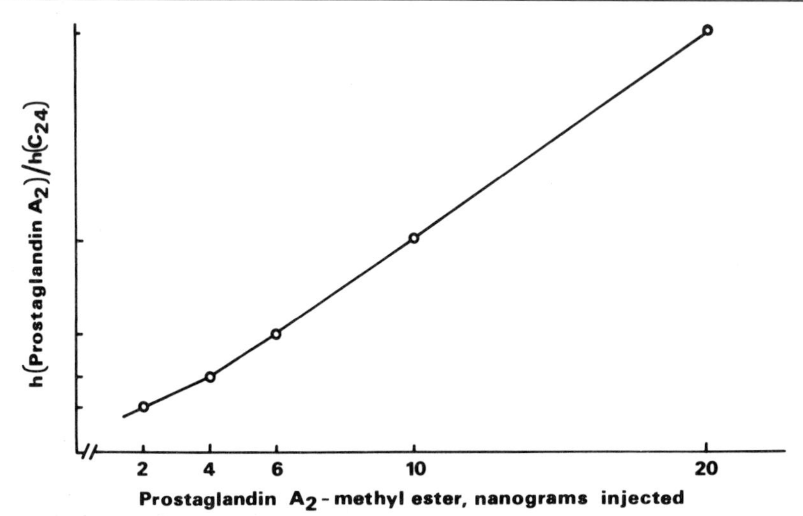

FIG. 3. Test of the column activity. Ratio of peak height (h) (prostaglandin A_2 methyl ester, various concentrations/C_{24}, 20 ng) as a function of injected prostaglandin. Note the absence of linearity at 2 ng.

the rate of temperature rise is inversely proportional to the column length. Inappropriate programmation, e.g., too fast, may bring a dramatic loss in efficiency.

Control of Column Deactivation

Even after coating with a liquid phase, the inner glass surface of the column may still retain some catalytic activity. The estimation of this property is critical when the column is intended for quantitative analysis. A good test is the injection of a partially derivated prostaglandin. We inject prostaglandin A_2 (PGA_2) methyl ester, which has an underivatized 9-keto group as well as a 15-hydroxyl function. Various concentrations (2–20 ng) of prostaglandin A_2 methyl ester mixed with a constant amount of C_{24} (20 ng) are injected at 240°, isothermal. The results are plotted as the ratio of height of peaks of PGA_2/C_{24} as a function of the PGA_2 concentration. The curve obtained provides an idea of the linearity of the response. Usually when the column is active, low concentrations of PGA_2 are excluded from the linear portion of the curve. Such an example is given in Fig. 3. After some months of intensive use, it is possible to observe a gradual activation of the columns as reflected by a progressive increase in the tailing of the peaks. In that case, we recommend cutting about 20 cm at each end of the column and mounting it again. This is generally sufficient to get rid of the column parts where either the phase has totally disappeared or has accumulated in discontinuous droplets, thus provoking adsorption phenomena.

Assay Method

The high efficiency of capillary columns allows, simultaneously, total separation and sensitive measurements of all the compounds from a mixture, with a minimum handling of the biological samples. This avoids some preliminary purification steps used occasionally with packed columns. However, it is obvious that the capillary column technique has limits that should be carefully tested before any involvement in complex kinetic, pharmacological, or biochemical problems. Often these need an adjustment of the incubation conditions for compatibility with the technique, as illustrated below from arachidonic acid metabolism studies on human blood platelets. For all the analytical steps described, chemicals and reagents of the highest purity are recommended (e.g., Mallinckrodt nanograde, Merck Lichrosolv, Burdick and Jackson distilled in glass).

Preparation of the Biological Samples[9]

Washed human platelets (2.5 to 3 × 10^8 platelets per milliliter) were incubated with the appropriate aggregation inducer. In order to get enough products for the analysis, 0.5–2.0 ml of the incubate were transferred into 12-ml glass extraction tubes containing sufficient ice-cold 0.01 *N* HCl to acidify the sample to pH 3. Three internal standards were added: 12-hydroxy stearic acid, synthetic $PGF_{2\beta}$, and 2a,2b-dihomoprostaglandin $F_{1\alpha}$, 500 ng each. The addition of several internal standards is a critical factor when quantitative analysis are performed in the low-nanogram range. The amount added should correspond to the expected mass of the biological products. Thereby the attenuation can remain constant in the part of the profile where eicosanoids and internal standards are eluted. This judgment requires preliminary experiments. The sample is extracted twice (1 min each) using twice the volume of diethyl ether (Merck Darmstadt, Germany, stabilized with 7 ppm of 2,6-di-*t*-butyl-4-methylphenol in order to avoid autoxidation of the unsaturated products by the solvent). The ether phases are pooled in a Teflon-capped tube and evaporated at 40° under nitrogen. Ethyl acetate is also acceptable, but its evaporation is more tedious. Since derivatization needs anhydrous conditions, we add 1–2 ml of dichloromethane or toluene after ether extraction, and evaporate again to obtain an azeotropic effect that will eliminate traces of water. If water remains, the process should be repeated. The biological extracts or the standards are subsequently derivatized unless further purifications of the sample are made at this stage. These derivatizations should be performed immediately (i.e., less than 48 hr) before the analysis.

Esterification

Methylation. The dry sample is dissolved in 20 μl of methanol, and an ethereal diazomethane solution is added in excess (400 μl). The reaction proceeds for 5 min at room temperature in the dark, and the excess reagent is removed by a stream of nitrogen. Various procedures have been described to prepare diazomethane, but we use the procedure of Fales *et al.*[14] with a reaction vial (Kontes Glass Co., Vineland, New Jersey). The method involves the action of an alkali (sodium hydroxide, 5 *N*) on *N*-methyl-*N'*-nitro-*N*-nitrosoguanidine (Aldrich Chemical Co.). All details for preparation are sent with the glass device. Care should be taken to cool the apparatus before initiating the reaction and to avoid glass particles in the tube, which may cause explosions. Add sodium hydroxide slowly in two aliquots. This apparatus is useful in providing a reproductible and safe method to get diazomethane in solution. Diazomethane is kept in Teflon-capped tubes at $-30°$ until use for 2–3 weeks (provided it is still yellow).

Pentafluorobenzyl ester.[10] This procedure was originally described by Wickramasinghe.[15] The sample or standard is dissolved in acetonitrile (10 μl) containing 0.022 nmol of pentafluorobenzylbromide (Pierce Chemical Co, Rockford, Illinois) and 10 μl of acetonitrile containing 0.077 mmol of diisopropylethylamine (Pierce Chemical Co.). It was heated at 40° for 5 min, and the excess reagent was evaporated.[16]

Both esterification methods have the advantage of removing excess reagent by evaporation.

Derivatization of the Carbonyl Group

Only anhydrous pyridine stored under potassium hydroxide pellets or molecular sieve should be used.

Methyloxime.[2,9] Eicosanoid esters are dissolved in 100 μl of a stock solution of saturated *O*-methylhydroxylamine hydrochloride (Applied Science Laboratories) in anhydrous pyridine. The reaction proceeds for 1 hr at 40°, and the pyridine in excess is evaporated until a white-dry precipitate appears. Other alkyl or aryl oximes can be prepared analogously.

Derivatization of Hydroxyl Groups[2,9]

The favored method is silylation. The dried ester–oxime–eicosanoid derivatives are dissolved in 200 μl of a hexane/*N*,*O*-bis(trimethylsilyl)trifluoroacetamide, 1% trimethylchlorosilane (BSTFA/TMCS, Applied Sci-

[14] H. M. Fales, J. M. Jaouni, and J. F. Babashak, *Anal. Chem.* **45,** 2302 (1973).

[15] J. Wickramasinghe, W. Morozowich, W. Hamlin, and R. Shaw, *J. Pharm. Sci.* **62,** 1428 (1973).

[16] F. Fitzpatrick, M. Wynalda, and D. Kaiser, *Anal. Chem.* **49,** 1032 (1977).

ence Laboratories), 100 : 100, v/v. After 2 hr at 40°, the samples are evaporated to dryness and redissolved in 100 μl of a hexane solution containing 10% BSTFA. Great care should be taken during all these steps in order to avoid the artifactual introduction of plastifiers and phthalates, which have a dramatic effect on the profiles. Avoid any direct contact of the rubber material with organic solvents. Since silyl derivatives can be hydrolyzed, the samples should be anhydrous before injection. In case of hydrolysis, dry the sample and treat again with the silylating agent.

From these derivatized products we have also found that catalytic hydrogenation could be performed on the sample without any additional derivatization and injected afterward. For such a procedure, i.e., reduction of double bonds and analysis, glass capillary column gas chromatography provides the best analytical system to detect the separation of unsaturated eicosanoids derived from arachidonic acid or any other polyunsaturated fatty acid precursor.

Application of the Technique to the Separation of Eicosanoids

Flame Ionization Detector

Separation of Standards. Complete resolution of fatty acids and eicosanoids was possible on the same profile (Fig. 4). The separation of related unsaturated components was complete in most cases. It was possible to resolve C16 : 1 from C16 : 0; or C18 : 2 from C18 : 0; or C20 : 4 from C20 : 3; that is successively peaks 1 and 2, 3 and 4, 6, and 8. The resolution between C20 : 5 and C20 : 4 (peaks 5 and 6), although incomplete, is clearly visible, and complete resolution can be achieved easily with Carbowax columns. Prostaglandin E_2 was totally separated from PGE_1, i.e., peaks 18 and 19; however, the formation of methyloximes is generally accompanied by the appearance of minor and major peaks that correspond to syn- and anti-isomers. These isomers can be totally separated, (PGE_2 corresponds to peak 18′ and 18 and PGE_1 corresponds to peak 19′ and 19) or partially separated (6-keto-$PGF_{1\alpha}$, 21′ and 21, or 6,15-diketo-$PGF_{1\alpha}$, 22-grouped peaks). Certain components such as thromboxane B_2 produced only a single peak. Hydroxy fatty acids can be separated as their position isomers, such as 5- and 12-hydroxyeicosatrienoic acids (peaks 11 and 10); the same holds true for trihydroxyeicosatrienoic acids isomers (peaks 15 and 15′). Obviously, the three nonphysiological compounds introduced as internal standards, 12-hydroxystearic acid, $PGF_{2\beta}$ and 2a2b-dihomo-$PGF_{1\alpha}$ (peaks 9, 14, and 23) have been selected so that they elute at different times or temperatures of the program. Quantitation is done using the ratio of the area of the peak to be analyzed : area of the internal standard peak, calculating the ratios corresponding to each internal standard.

Although these columns present little adsorption of the compounds, it

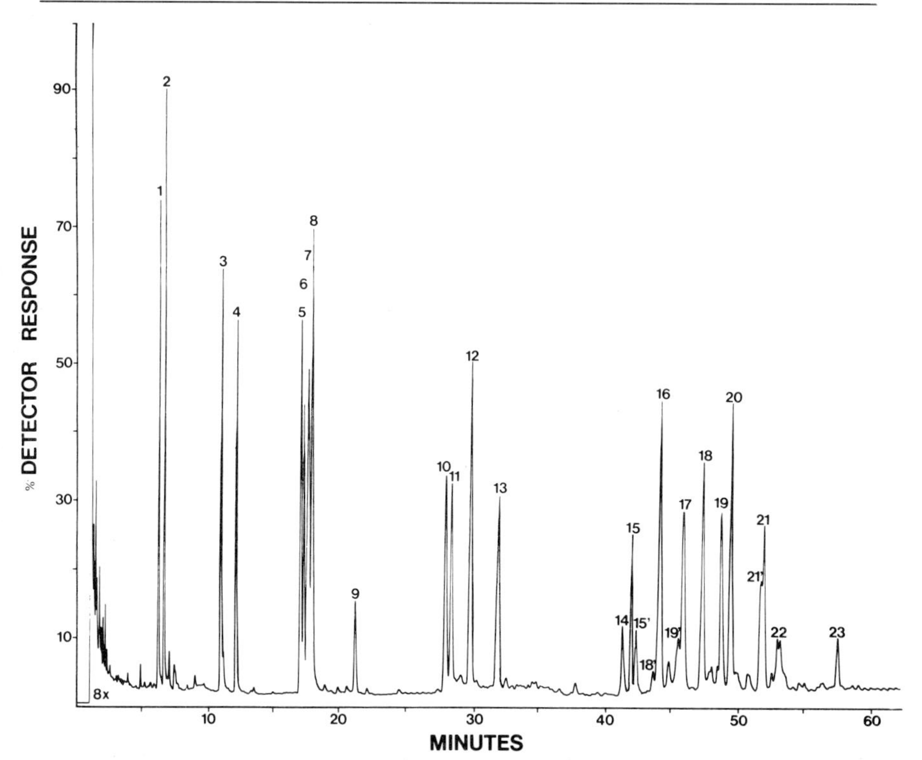

FIG. 4. Separation of fatty acids and eicosanoids with glass capillary gas chromatography flame ionization detection (methyl ester methyloxime trimethylsilyl ethers) (PG, prostaglandin, Tx, thromboxane). Peaks 1, palmitoleic acid (C16 : 1); 2, Palmitic acid (C16 : 0); 3, linoleic acid (C18 : 2ω6); 4, stearic acid (C18 : 0); 5, eicosapentaenoic acid (C20 : 5ω3); 6, arachidonic acid (C20 : 4ω6); 7, HHT (12-hydroxy-5,8,10,heptadecatrienoic acid); 8, dihomo-γ-linolenic acid (C20 : 3ω6); 9, 12-hydroxystearic acid; 10, 12-HETE (12-hydroxy-5,-8,10,14-eicosatetraenoic acid); 11, 5-HETE (5-hydroxy-6,8,11,14-eicosatetraenoic acid); 12, erucic acid (C22 : 1); 13, EPHETA (10-hydroxy-11,12-epoxy-5,8,14-eicosatrienoic acid); 14, $PGF_{2\beta}$; 15, THETE (8,9,12-trihydroxy-5,10,14-eicosatrienoic acid + 8,11,12-trihydroxy-5,9,-14-eicosatrienoic acid); 16, $PGF_{2\alpha}$; 17, PGD_2; 18, PGE_2; 19, PGE_1; 20, TxB_2; 21, 6-keto-$PGF_{1\alpha}$; 22, 6,15-diketo-$PGF_{1\alpha}$; 23, 2a,2b-dihomo-$PGF_{1\alpha}$. Conditions: column, Se-52 type (thermostable) 28 m, 0.2 mm internal diameter; carrier gas, helium, 1.5 ml/min; temperature programming, 175–250°, 1°/min; attenuation, 8 ×; approximately 15 ng on column; chromatograph, Packard Model 428, all glass injector, Ros type.

is better to use an integrator to measure peak area than peak height. We use a 500-ng internal standard for 2 μg of each compound and perform calculations on the biological sample to which 500 ng of each internal standard have been added. The quantification is done according to

$$(x') = \left(\frac{S'\mathrm{x}}{S' \text{ int } 1}\right)_{\text{biological}} \times \left(\frac{S \text{ int } 1}{S\mathrm{x}}\right)_{\text{standard}} \times (x)$$

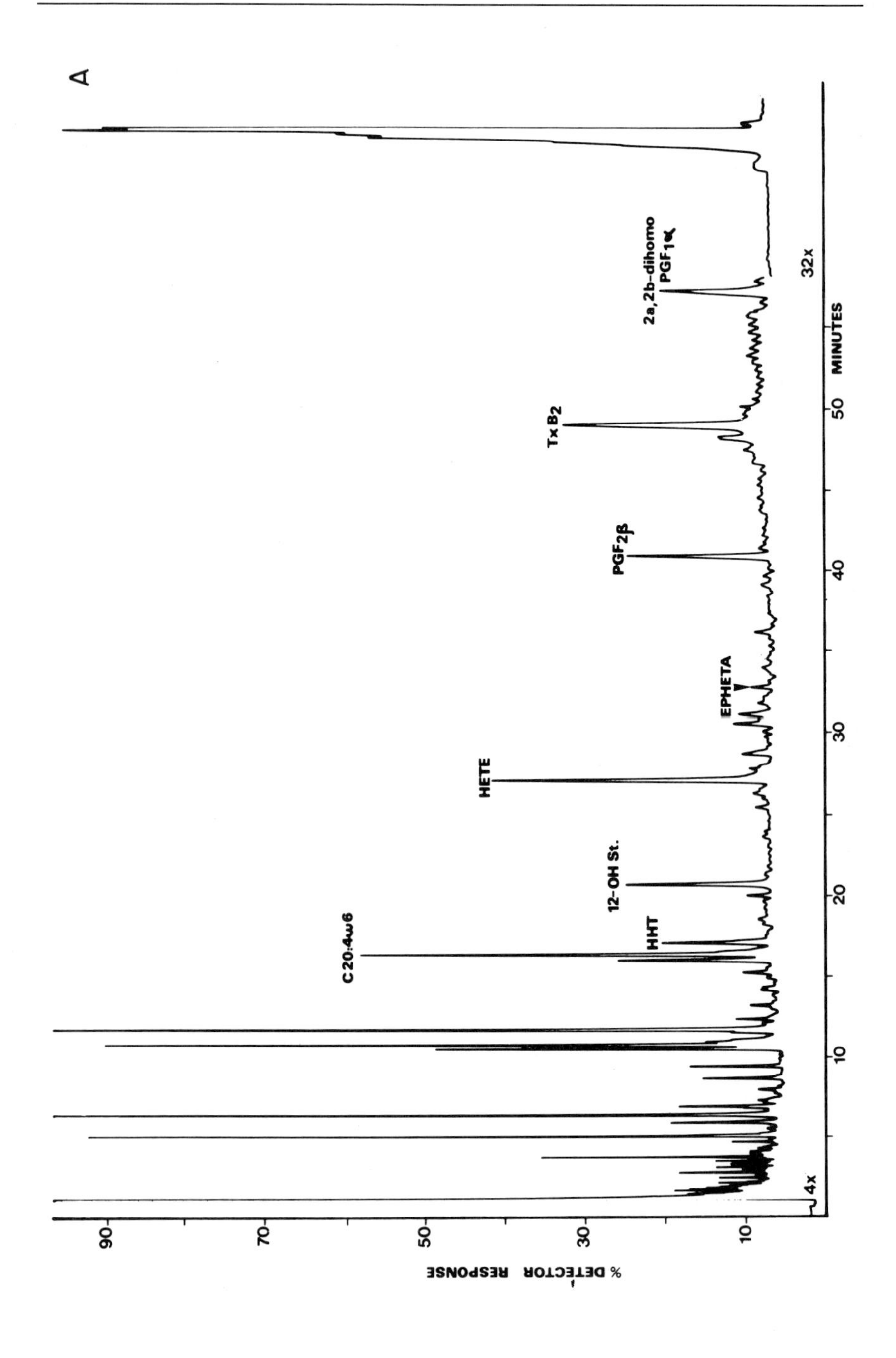
A
C20:4ω6
HHT
12-OH St.
HETE
EPHETA
PGF$_{2β}$
TxB$_2$
2a,2b-dihomo PGF$_{1α}$
4x
32x
10
20
30
40
50
MINUTES
90
70
50
30
10
% DETECTOR RESPONSE

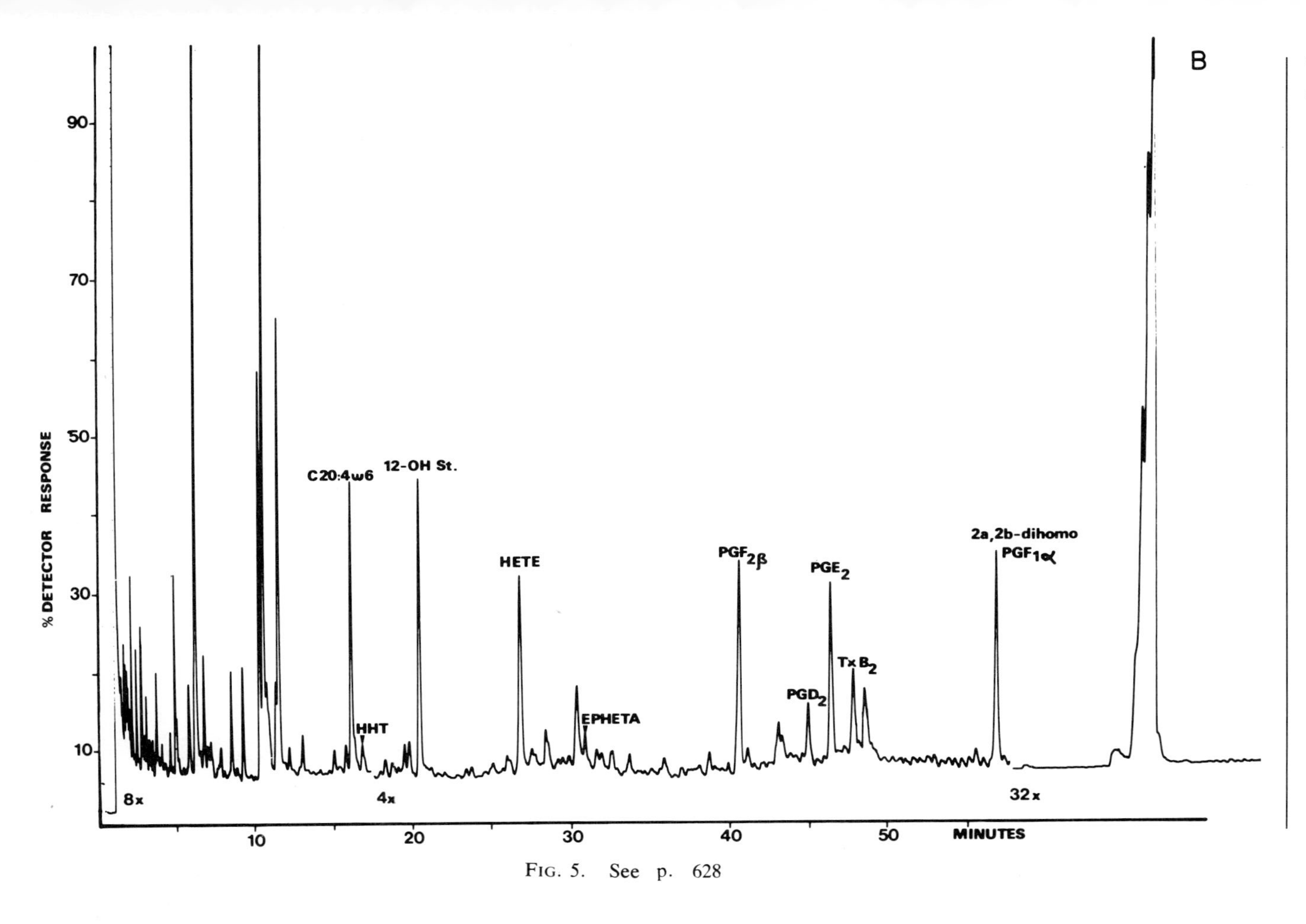

FIG. 5. See p. 628

where x' is the quantity of eicosanoid x in the biological extract; S'x and S′int 1 are the peak area of eicosanoid x and of internal standard 1 in the sample; Sx and S int are the peak area of eicosanoid x and of internal standard 1 in the standards reference test tube; and x is the known quantity of eicosanoid in the reference tube. This operation is repeated with all internal standards. The reference test tube is injected every day.

Arachidonic Acid Metabolism of Washed Human Platelets. Arachidonic acid and products from its oxidative metabolism can be monitored after stimulation of washed human platelets by the ionophore A 23187 (Fig. 5A). The amount of platelets necessary for this profile is quite low as compared to other techniques, and the concentration of inducer is also reasonable. Not so long ago[17] this simultaneous monitoring could be done only by using a gas chromatography–mass spectrometric technique using packed columns. From such a control metabolic profile, the influence of pharmacological agents on arachidonic acid metabolism can be investigated. Figure 5B represents the same stimulation after preincubation of the platelets with a thromboxane synthase inhibitor, imidazole. Comparison of the two profiles indicates a reduction of thromboxane formation with the concomitant appearance of PGE_2 produced by the diversion of PGH_2. It can be noted that (*a*) the metabolic profile from one analysis provides information concerning the regulation of the several different enzymatic oxidative pathways; (*b*) the summing of all released arachidonic acid and its oxygenated metabolites provides an indirect approach for the evaluation of phospholipases activities after external stimulation of a cell.

Electron Capture Detector

Thermostable methylphenylsiloxane glass capillary columns have been adapted to an electron-capture detector.[10] Although the dead volume of the detector reduced the potential efficiency of the column, it was possible to separate structurally similar F prostaglandins (Fig. 6). The low amount of injected prostaglandin (here approximately 500 pg on column, but it is possible to detect as little as 20 pg[10]) allows to reach sensitivities matched only by radioimmunoassay. Unfortunately, electron-capture detection suffers from limitations instrinsic to the isolation and manipulation of picogram amounts of material. Further, the temperature of the column

[17] M. Hamberg, J. Svensson, and B. Samuelsson, *Proc. Natl. Acad. Sci. U. S. A.* **71,** 3824 (1974).

FIG. 5. Analysis of arachidonic acid and oxygenated metabolites after stimulation of washed human platelets (1 ml) by the Ca^{2+} ionophore A 23187 (2 μM) in the absence (A) or in the presence (B) of the thromboxane synthase inhibitor, imidazole (300 μM). Conditions as in Fig. 4.

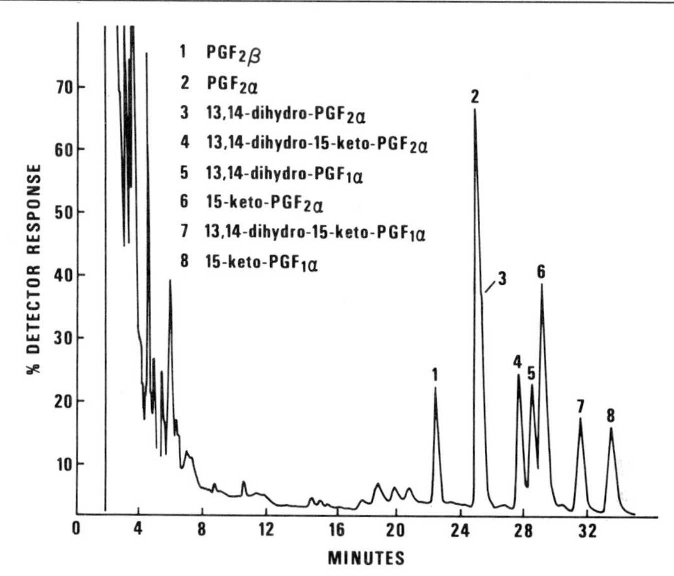

FIG. 6. Separation of pentafluorobenzyl ester trimethylsilyl ethers of F prostaglandins with glass capillary gas chromatography electron capture detection. Conditions: Column OV-1 type (thermostable) 25 m, 0.3 mm internal diameter; carrier gas, helium 22.5 cm/sec; make-up gas, argon–methane (90 : 10), 15 ml/min; isothermal, 250°; detector, 300°, injector, 250°; ^{63}Ni, 15 mCi; chromatograph, Hewlett-Packard Model 5713; Vandenberg all-glass solventless injector (approximately 500 pg on column).

necessary to elute high molecular weight pentafluorobenzyl ester derivatives is currently incompatible for the additional separation of fatty acids and lipoxygenase products. Temperature programming necessary to achieve this purpose is not suited for electron-capture detection. Additional modifications of the detection system may improve greatly the use of this technique.

Mass Fragmentometry Detection

Using mass fragmentometry and glass capillary columns, detection limits exceed those obtained with other detectors. "Cyclooxygenase profiles" from mice peritoneal macrophages were effected using this system. When 3×10^6 cells were incubated with exogenous arachidonic acid, the distribution of products arising from the cyclooxygenase pathway was as follows: 6-keto-$PGF_{1\alpha}$, 42.24% ± 6.08% ($n = 4$); TxB_2, 5.44% ± 4.79% ($n = 4$); PGE_2, 31.76% ± 4.52% ($n = 4$); PGD_2, 10.75% ± 3.90% ($n = 4$); $PGF_{2\alpha}$, 9.81% ± 2.83% ($n = 4$). Figure 7 shows the recording of a synthetic mixture of prostaglandins and thromboxane using the mass fragmentometry detection.

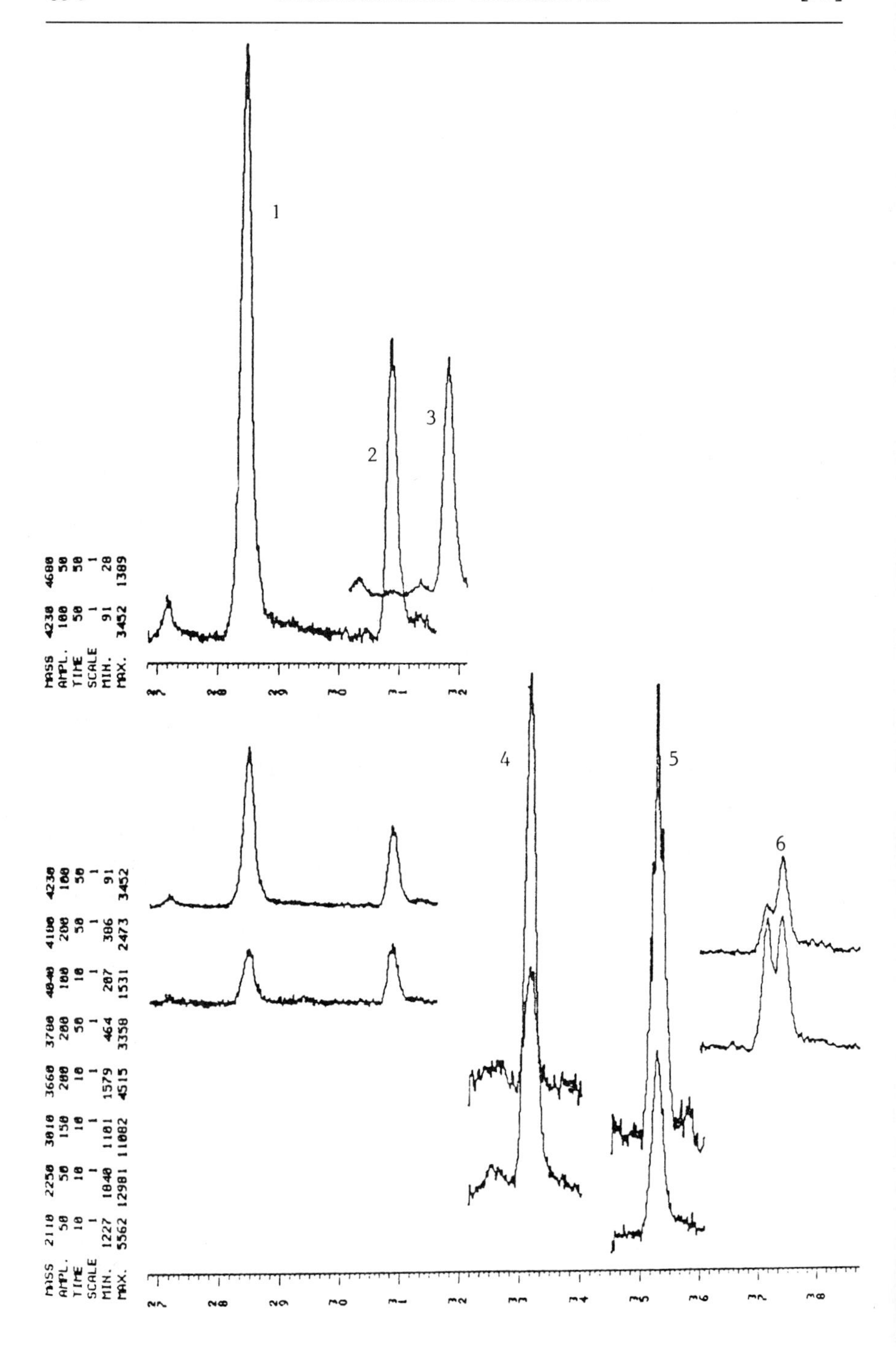
1
2
3
4
5
6
MASS 4230 4600
AMPL. 100 50
TIME 50 50
SCALE 1 1
MIN. 91 20
MAX. 3452 1309
MASS 2110 2250 3010 3660 3700 4040 4100 4230
AMPL. 50 50 150 200 200 100 200 100
TIME 10 10 10 10 50 10 50 50
SCALE 1 1 1 1 1 1 1 1
MIN. 1227 1040 1101 1579 464 207 306 91
MAX. 5562 12981 11082 4515 3358 1531 2473 3452
27 28 29 30 31 32
27 28 29 30 31 32 33 34 35 36 37 38

Concluding Remarks

Only two analytical methods can compete to achieve total chromatographic separation of all known eicosanoids: open tubular glass capillary column gas chromatography and high-performance liquid chromatography. Surprisingly, both have been used sparingly. The optical detection systems of HPLC are, ordinarily, not sensitive enough to detect prostaglandins present in most biological samples, which explains its limited use. Although detection limits with HPLC can be extended by monitoring enzymic conversions of radiolabeled precursor fatty acids, accurate quantitation is impossible without correcting for isotope dilution effects and heterogeneous distribution of labeled precursor into accessible lipid pools. In contrast, glass capillary gas chromatography is fully adapted to study the complex metabolism of eicosanoids because (*a*) it can separate mixtures of all eicosanoid metabolites into their distinct component parts; (*b*) its high resolution eliminates the need for purification prior to gas chromatography; (*c*) its high sensitivity can detect eicosanoids in minimal amounts of biological material from incubations that rely exclusively on endogenous fatty acid substrate; (*d*) it facilitates studies on two distinct oxidative metabolic pathways, the cyclooxygenase and the lipoxygenase, within one comprehensive, yet simple, analytical technique. In this context it also reflects the activity of the premier enzyme of eicosanoid metabolism, the phospholipase enzyme.

The diversity of eicosanoids, their structural similarity, and the existence of different levels of unsaturation of eicosanoids create the need for analytical methods with extraordinary resolution. Glass capillary gas chromatography is paramount among such methods. A more widespread use of this technique will improve our biochemical, physiological, and pharmacological understanding of eicosanoid metabolism.

Acknowledgments

The authors are indebted to Dr. J. E. Pike (The Upjohn Company, Kalamazoo, Michigan) for furnishing the prostanoid and thromboxane standards as well as to Dr. F. A. Fitzpatrick (Upjohn) for his electron capture profile.

FIG. 7. Multiple ion detection of 1, $PGF_{2\beta}$, m/z = 404 (M^+-2x90); 2, $PGF_{2\alpha}$, m/z = 404 (M^+-2x90); 3, PGD_2, m/z = 468 (M^+-71); 4, PGE_2, m/z = 225 and m/z = 366; 5, TxB_2, m/z = 211 (301-TMSOH); 6, 6-keto-$PGF_{1\alpha}$, m/z = 378 (M^+-2x90-71) and m/z = 418 (M^+-31-2x90). Column, OV-1 type (20 m × 0.26 mm i.d.); standard mixture of derivatized PGS and TxB_2 (500 pg of each injected onto the column) as ME.MO-TMS.

Section VI

Biological Methods

[74] Use of Microwave Techniques to Inactivate Brain Enzymes Rapidly

By CLAUDIO GALLI and GIORGIO RACAGNI

This chapter describes a method (microwave MW irradiation) generally used to sacrifice small laboratory animals (rats, mice), when levels of endogenous compounds undergoing rapid postmortem changes, are to be measured. In fact, brain concentrations of various neuromodulators, metabolic products, and intermediary metabolites are rapidly modified after decapitation, following activation of various enzyme systems.

Levels of oxygenated metabolites of eicosapolyenoic fatty acids (the eicosanoid system), and mainly those deriving from arachidonate, in tissues, are the result of a balance between their formation—which can be stimulated even by unspecific factors such as mechanical trauma[1,2] and tissue hypoxia[3–5]—and their metabolic transformation[6,7] or removal.[8]

The sequence of events involved in the eicosanoid system may be summarized as follows:

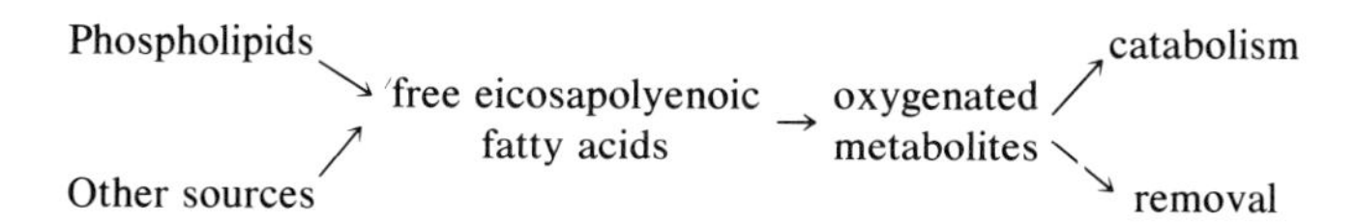

Availability of precursor fatty acid(s) as free substrate(s) for conversion to prostaglandin(s) (PG) and related compounds, following stimulation of lipolytic enzymes, is considered one of the limiting steps of PG synthesis. Thus, the endogenous basal levels of free eicosapolyenoic fatty acids and of their oxygenated metabolites are difficult to determine in tissues, such as the central nervous system (CNS), where stimulation of free fatty acid, and especially of free arachidonic acid, release[9] and of PG for-

[1] S. H. Ferreira and J. R. Vane, *Nature* (*London*) **16,** 868 (1967).
[2] R. J. Flower and G. J. Blackwell, *Biochem. Pharmacol.* **25,** 285 (1976).
[3] J. C. McGiff, K. Crowshaw, N. A. Terragno, and A. J. Lonigro, *Circ. Res. Suppl.* **26,** 121 (1970).
[4] B. M. Jaffe, C. W. Parker, G. R. Marshall, and P. Needleman, *Biochem. Biophys. Res. Commun.* **49,** 799 (1970).
[5] A. Wennmalm, Phan-Chank, and M. Junstad, *Acta Physiol. Scand.* **9,** 133 (1974).
[6] E. Änggård and B. Samuelsson, *J. Biol. Chem.* **239,** 4087 (1964).
[7] E. Änggård, C. Larsson, and B. Samuelsson, *Acta Physiol. Scand.* **81,** 396 (1971).
[8] L. Z. Bito and H. Davson, *J. Physiol.* (*London*) **236,** 39 (1974).
[9] N. G. Bazan, *Biochim. Biophys. Acta* **218,** 1 (1970).

ISBN 0-12-181986-8

mation[10] have been observed after decapitation. It is thus necessary to inactivate metabolic enzymes in brain as rapidly as possible during the sacrifice procedure in order to avoid postmortem artifactual changes and to assess accurately the levels of the above metabolites *in vivo*. Conventional methods are generally based upon immersion or decapitation into liquid nitrogen or Freon-12, or both, to stop brain enzyme activity. However, since enzyme inactivation by freezing methods is dependent upon the thermal conductivity of the tissue, up to 90 sec may be required to completely block enzyme activities in all brain areas.[11] The time of inactivation may be considerably reduced (tenths of a second) by the use of more rapid freezing methods, such as the freeze-blowing technique.[12] However, the freeze-blowing method does not allow dissection of brain regions and nuclei. Furthermore, enzyme systems may recover during thawing and additional artifactual changes may thus be produced.

Methodology of Microwave (MW) Irradiation. The procedure is based upon the rapid inactivation of all brain enzymes by rapid exposure of the head of small experimental animals to focused microwave (MW) irradiation, followed by the animal's death.[13] Methods are also available for total body MW irradiation, and applications have been made to the determination of compounds such as cyclic nucleotides in peripheral tissues. However, methods based on total body irradiation have never been applied to the study of products of the PGH synthase and lipooxygenase pathways in tissues.

Principle of Method.[14] The use of high-intensity microwave irradiation has the advantage of sacrificing an animal while simultaneously inactivating brain enzymes in milliseconds. The principle of inactivation is the heat lability of most enzymes, which are irreversibly denatured by exposure to temperatures in the range of 55–90°. This procedure, thus, greatly reduces the artifacts occurring with the use of other methods of sacrifice and also allows one to dissect brain regions accurately at room temperature.[13,15,16] The minimum energy required for enzyme inactivation by MW irradiation depends upon the heat lability of the enzyme system and its postmortem activity. The enzymes involved in the oxidative metabolism of eicosapolyenoic fatty acids, especially PGH synthase, appear to be very labile to heat denaturation.

[10] E. Bosisio, C. Galli, G. Galli, S. Nicosia, C. Spagnuolo, and L. Tosi, *Prostaglandins* **11,** 773 (1976).

[11] D.F. Swaab, *J. Neurochem.* **18,** 2085 (1971).

[12] R. L. Veech, R. L. Harris, D. Veloso, and E. H. Veech, *J. Neurochem.* **20,** 183 (1973).

[13] W. B. Stavinoha, B. Pepelko, and P. Smith, *Pharmacologist* **12,** 257 (1970).

[14] R. H. Lenox, P. V. Brown, and J. L. Meyerhoff, *Trends In Neuro Sciences* **2,** 106 (1979).

[15] G. J. Balcom, R. H. Lenox, and J. L. Meyerhoff, *J. Neurochem.* **24,** 609 (1975).

[16] M. J. Schmidt, D. E. Schmidt, and G. A. Robinson, *Science* **173,** 1142 (1971).

General Factors Affecting the Reliability of the Methodology. The presently available MW techniques present some general limitations that should be considered also in the study of brain PG. It was observed, in fact, that inactivation of enzymes throughout the brain was not uniform[17,18] and that different brain regions require different durations of microwave exposure in order to stabilize the contents, e.g., cyclic AMP levels.[19,20] The patterns of regional heating and enzyme inactivation in brain depend upon the patterns of energy absorption and the thermal diffusion properties of the tissue.

Other factors that influence the effectiveness and regional patterns of brain enzyme inactivation by MW techniques are the orientation and mobility of the animal relative to the MW field. Significant problems with regard to reproducibility from animal to animal may, thus, arise from movements of an animal within the MW field. From a theoretical point of view, freezing methods, in contrast, give a more reproducible enzyme inactivation, since the freezing pattern is dependent only upon the type and amount of materials composing the object (tissue) and their distribution.[11]

The dependence of the inactivation pattern by MW irradiation upon the position of the animal's head during exposure leads to the requirement to immobilize the animal during sacrifice. Forced immobilization, used however in most sacrifice methods, may introduce artifacts, since it is a potent stressor. This effect can be minimized by introducing the animal into the irradiation chamber several times before the final MW irradiation in order to reduce the stress due to the animal's constraint. The optimal conditions for MW irradiation, in the absence of stress artifacts, is obtained with total body irradiation, a procedure that allows the sacrifice of freely moving animals.

Quality of the Equipment. Power source parameters must be controlled in order to achieve reproducible enzyme inactivation. The net power absorbed by a given load depends upon the difference between the incident and the reflected power. The available instrumentation has been generally designed to control the intensity and duration of the incident power and to minimize the reflected power.

General Specifications of the MW Oven Used. The instruments used in our laboratory have the following technical characteristics: cooling system, forced air cooled magnetron, water-cooled energy absorbing load

[17] L. L. Butcher and S. G. Butcher, *Life Sci.* **19,** 1079 (1976).

[18] R. H. Lenox, O. P. Gandhi, J. L. Meyerhoff, and H. M. Grove, *IEEE Trans. Microwave Theory Tech.* **24,** 58 (1976).

[19] R. H. Lenox, J. L. Meyerhoff, O. P. Gandhi, and H. L. Wray, *J. Clin. Nucleotide Res.* **3,** 367 (1977).

[20] W. B. Stavinoha, *in* "Cholinergic Mechanisms and Psychopharmacology" (D. Jenden, ed.), p. 169. Plenum, New York, 1978.

termination; current, 15 Amp; power consumption, idle 275 W, operating at 3500 W; power output, 1300 W; frequency, 2450 Mhz; control timer, 0.25–15 sec; dial setting, 0.1 sec/div; reset, automatic reset after each operating cycle.

Adequate Enzyme Inactivation and Stability of Products To Be Analyzed. As a consequence of the nonuniformity of the MW field generated within the brain, it is essential that MW irradiation be of sufficient duration to inactivate enzymes in the least heated regions. For optimal MW exposure of the brain, however, not only an irreversible inactivation of the enzyme system, but also thermal stability of the substrate measured and preservation of the structural integrity of the tissue are required.

Temperatures reached within the tissue during irradiation are difficult to measure owing to the limited technology available and to the distortion of the field due to the presence of *in situ* thermistors. Temperatures of the order of 70–100° have been measured a few seconds after exposure.[14,17,18] It should be considered also that the stability of products to heat exposure in tissues is generally greater than that of the purified products *in vitro,* possibly owing to the presence of stabilizing factors or conditions *in vivo.* The stability of PG in brain to MW irradiation has been checked by evaluating the recovery, as a single product, of radiolabeled PG injected intracerebrally immediately before MW irradiation of the animal's head. Around 95% of the recovered radioactivity had the same chromatographic characteristics of the injected PG. Indirect evidence of the stability of endogenous PG in brain to MW irradiation of the animal's head derives from the observation that, e.g., levels of $PGF_{2\alpha}$ in brains removed 90 sec after the onset of experimentally induced convulsions in the rat sacrificed by MW irradiation were virtually identical to those measured, under the same experimental conditions, when brains were frozen immediately after

TABLE I
LEVELS OF $PGF_{2\alpha}$ (PG/MG PROTEIN) IN BRAIN CORTEX OF RATS[a,b]

	After convulsions	
Basal	Microwave irradiation	Decapitation and freezing
15 ± 3	672 ± 89	694 ± 118

[a] Rats were sacrificed by focused microwave (MW) irradiation or by decapitation followed by immediate freezing of the tissue, 90 sec after the onset of experimentally induced convulsions, in respect to basal levels.

[b] Values are the average ± SEM of determinations carried out in 6 animals. Basal values were measured in brain cortex obtained from nontreated animals after sacrifice by MW irradiation. Samples of brain cortical tissue of convulsant animals were obtained after MW irradiation or decapitation and immediate freezing of the tissue. $PGF_{2\alpha}$ was measured by radioimmunoassay.

decapitation (Table I). Under the last conditions, however, the tissue cannot be adequately dissected. Also, MW irradiation does not affect the composition of brain structural phospholipids and their fatty acids.[21]

Operative Procedure

Immediately before sacrifice, groups of animals to be irradiated with focused MW are brought to the room where the MW oven is located. One animal is placed in a closely fitting cylindrical Plexiglas container of a size selected according to the species (mice, rats) and size (body weight) of the animal used. The animal's head protrudes into a separate compartment of the container, on which MW are to be focused. With the most commonly used MW ovens for brain irradiation, maximum body weight of the irradiated animal (rat) is around 300 g. Plastic adaptors for animals of the size of newborn rats and mice (10–20 g of body weight) can be easily made.

The loaded container is then placed in the irradiation chamber, which is shielded to avoid dispersion of MW energy outside the instrument, with the compartment containing the animal's head placed in the cavity where MW are focused. The time of irradiation is selected, through the dial on the instrument, depending upon the energy characteristics of the oven and the thickness of the layers (especially the skull), interposed between the surface of the brain and that of the animal's head. The last parameter depends, obviously, mainly upon the age of the animal. The following irradiation times are recommended, using instruments with the previously described general specifications: for adult rats (body weight 150–300 g), periods of 2–4 sec are adequate, whereas for adult mice (body weight 20–40 g), the values should be 0.7–1.5 sec. For newborn animals the irradiation times should be of the order of 0.5–1 sec, due to the thinness of the skull and the high water content of the brain. Immediately after irradiation, the heated heads are cooled in crushed ice and then decapitated by a guillotine. After complete cooling, skulls are opened and the whole brain is removed and used for further analysis. Tissue can be used as such or frozen for coronal and sagittal dissection in order to isolate specific brain nuclei (punching techniques).

Inactivation of the Enzymes Involved in Neurotrasmitter and Arachidonic Acid Metabolism

The use of focused MW irradiation for animal sacrifice prevents postmortem changes of various neurochemical parameters, e.g., levels of neuromediators, giving essentially the same results as those obtained with the freeze-blowing technique, which is the fastest method for inactivation

[21] C. Galli and C. Spagnuolo, *J. Neurochem.* **26,** 401 (1976).

TABLE II
LEVELS OF DIFFERENT COMPOUNDS IN RAT BRAIN AFTER SACRIFICE BY VARIOUS PROCEDURES

Compound	Microwave irradiation	Freeze blowing	Decapitation
Cyclic AMP (pmol/mg protein)	0.80 ± 0.09	1.09 ± 0.08	7.81 ± 0.50[a]
Cyclic GMP (pmol/mg protein)	6.10 ± 0.54	5.12 ± 0.06	7.02 ± 0.82
GABA (nmol/mg protein)	12.03 ± 0.41	11.15 ± 0.25	24.15 ± 2.31[a]
DOPA (pmol/mg protein)	2.60 ± 0.20	2.73 ± 0.26	1.41 ± 0.13[a]
3-Methoxytyramine (pmol/g tissue)	140 ± 10	—	1540 ± 113[a]
Normetanephrine (pmol/mg protein)	1.12 ± 0.08	—	18.10 ± 2.24[a]
Choline (pmol/mg protein)	300 ± 34	280 ± 54	1450 ± 130[a]

[a] $p < 0.01$ versus microwave-irradiated brains. Values represent the mean ± SEM. Cyclic AMP and 3-methoxytyramine were measured in striatum; cyclic GMP and GABA in cerebellum; DOPA and choline in whole brain; normetanephrine in hypothalamus. Brain tissues were left at room temperature for 3–5 min after either microwave irradiation or decapitation.

of brain enzymes. Table II presents data obtained in our laboratory[22–25] on levels of various neuromediators and metabolites in different brain areas obtained from animals sacrificed by MW irradiation in comparison with those measured in the same areas after decapitation. Data are also compared with those reported in the same areas after freeze-blowing sacrifice.[26] Levels of the various products measured after MW irradiation or freeze-blowing are virtually identical, whereas in the samples obtained after decapitation they vary, either increasing (cyclic AMP, 3-methoxytyramine, normetanephrine, GABA, and choline), or decreasing (DOPA).

Inactivation of the Enzymes Involved in Arachidonic Acid Metabolism

The blockade of the reactions involved in release of free arachidonic acid (FAA) from brain lipids and formation of metabolites of PGH synthase and lipoxygenase by the use of the MW technique is supported by the observations that the rise of FAA and of its oxygenated metabolites observed in rat brain after decapitation is prevented by this method. Levels of FAA in brain areas of animals sacrificed by focused MW irra-

[22] I. Mocchetti, L. De Angelis, and G. Racagni, *J. Neurochem.* **37,** 1607 (1981).

[23] G. Racagni, D. L. Cheney, G. Zsilla, and E. Costa, *Neuropharmacology* **15,** 723 (1976).

[24] G. Racagni, A. Groppetti, F. Cattabeni, C. L. Galli, and M. Parenti, *Life Sci.* **23,** 1763 (1978).

[25] A. Maggi, F. Cattabeni, F. Bruno, and G. Racagni, *Brain Res.* **133,** 382 (1977).

[26] A. Guidotti, D. L. Cheney, M. Trabucchi, M. Doteuchi, C. Wang, and R. A. Hawking, *Neuropharmacology* **13,** 115 (1974).

TABLE III
LEVELS OF FREE ARACHIDONIC ACID (FAA) AND OF $PGF_{2\alpha}$ AND THROMBOXANE B_2 (TxB_2) IN BRAIN CORTEX OF RATS SACRIFICED BY MICROWAVE IRRADIATION OR DECAPITATION FOLLOWED BY HOMOGENIZATION IN ORGANIC SOLVENTS (FAA) OR ACIDIC BUFFER ($PGF_{2\alpha}$ AND TxB_2)[a]

Compound	Microwave irradiation	Decapitation	
		0.5 Min	5 Min
FAA (μg/mg protein)	0.54 ± 0.19	0.90 ± 0.16	2.54 ± 0.64[b]
$PGF_{2\alpha}$ (pg/mg protein)	82 ± 14	621 ± 95[b]	565 ± 80[b]
TxB_2 (pg/mg protein)	48 ± 6	243 ± 19[b]	267 ± 28[b]

[a] Values are the average ± SEM of duplicate determinations carried out in 6 animals for each experimental condition.

[b] Statistically different (Student's t test) from microwave-irradiated samples ($p < 0.005$). Levels of FAA were measured by gas chromatography of the free fatty acid fraction isolated from brain lipids, made quantitative by the use of an internal standard; $PGF_{2\alpha}$ and TxB_2 were measured by radioimmunoassay.

diation were much lower than those measured by decapitation followed by a time interval before analysis[21,27] and were not statistically different from those measured after decapitation and immediate analysis (less than 30 sec between decapitation and homogenization in organic solvents[21]), or analysis after decapitation and immediate freezing.[28] Furthermore, levels of $PGF_{2\alpha}$,[10] and thromboxane B_2 (TxB_2),[29] derived through PGH synthase, and of HETE,[30] derived through the lipoxygenase, were much lower in the brain of animals sacrificed by MW irradiation than in those sacrificed by decapitation followed by a time interval before analysis and were of the same order of those measured in brain tissue frozen in liquid nitrogen immediately after sacrifice.[28] In synthesis, the use of focused MW irradiation prevents both the release of FAA occurring after decapitation followed by a time interval before analysis and the accumulation of PG and Tx observed in brain even after the shortest possible interval (less than 30 sec) between decapitation and analysis (Table III).

Since catabolism of PG is considered to be nonexistent in brain,[31-33] increments of PG levels observed in MW-irradiated brains from animals

[27] R. J. Cenedella, C. Galli, and R. Paoletti, *Lipids* **10,** 290 (1975).

[28] J. Marion and L. S. Wolfe, *Prostaglandins* **16,** 99 (1978).

[29] C. Spagnuolo, L. Sautebin, G. Galli, G. Racagni, C. Galli, S. Mazzari, and M. Finesso, *Prostaglandins* **18,** 53 (1979).

[30] L. Sautebin, C. Spagnuolo, C. Galli, and G. Galli, *Prostaglandins* **16,** 985 (1978).

[31] E. Änggård, C. Larsson, and B. Samuelsson, *Acta Physiol. Scand.* **81,** 396 (1971).

[32] J. Nakano, A. V. Prancan, and S. E. Moore, *Brain Res.* **39,** 397 (1972).

[33] C. R. Pace-Asciak, *Adv. Prostaglandins Thromboxane Res.* **4,** 45 (1978).

subjected to various experimental treatments reflect stimulation of synthetic processes.

In conclusion, the use of focused MW irradiation of the head of small laboratory animals, such as mice and rats, allows one to determine correctly endogenous levels of free PG-precursor fatty acids (e.g., FAA) and their oxygenated metabolites in brain regions and discrete nuclei, both in basal conditions (when they are very low) or after experimental treatments such as administration of centrally stimulating drugs, and hypoxia (when they are markedly increased).

[75] Platelet Aggregation and the Influence of Prostaglandins

By JON M. GERRARD

Platelets are blood cells with an important role in hemostasis. Aggregation is a quick and simple method for assessing their function *in vitro*. Since many prostaglandins and related compounds have important effects on platelet function, aggregation can be a useful tool in prostaglandin research. Its uses include (*a*) a bioassay of certain prostaglandins and thromboxanes; (*b*) the assay of platelet enzymes needed for synthesis of prostaglandins and thromboxanes; (*c*) the study of drugs that influence these processes; and (*d*) the evaluation of patient groups for alterations in platelet prostaglandin and thromboxane production. The use of aggregometry for these studies and its limitations will be discussed.

Preparation of Platelet-Rich and Platelet-Poor Plasma

Plasticware (polypropylene or polycarbonate) or siliconized glassware is used throughout. Platelet-rich plasma (PRP) is prepared following anticoagulation of blood using citrate or heparin. While several different formulations of citrate can be used, we use 93 m*M* sodium citrate, 7 m*M* citric acid, 0.14*M* dextrose, pH 6.5, and mix 1 volume of this anticoagulant with 9 volumes of blood. In rare instances where it is desired to compare the function of platelets in PRP from individuals with very different hematocrits, the citrate concentration must be adjusted to account for changes in hematocrit.[1] The anticoagulated blood is centrifuged at 100 *g* for 20 min at room temperature to separate PRP. PRP can also be prepared by using shorter centrifugation times with higher *g* forces, by keep-

[1] J. G. Kelton, P. Powers, J. Julian, V. Boland, C. J. Carter, M. Gent, and J. Hirsh, *Blood* **56,** 38 (1980).

ISBN 0-12-181986-8

ing the product of the *g* force in the middle of the tube times the time in minutes between 1750 and 2400 (i.e., 200 $g \times 10$ min = 2000). In some individuals a second shorter centrifugation (120 *g* for 5 min) may be necessary to eliminate a small degree of red blood cell contamination. In individuals with a platelet count of less than 50,000/μl a slower spin is desirable (70 *g* for 20 min). For animal blood samples, the conditions of centrifugation required may vary with the species. Platelet-poor plasma (PPP) is prepared by centrifugation of anticoagulated blood or PRP at 5000 *g* for 10 min at 4°. For comparison of different patient groups, it is wise to standardize platelet counts at 200,000/μl or 300,000/μl by diluting PRP with PPP.

Preparation of Washed Platelets

For many purposes (i.e., biochemical studies, evaluation of plasma factors, alteration or standardization of albumin concentration) washed platelets are desirable. The novice should be aware that there is "art" as well as science involved in washing platelets, and success depends on careful handling of these cells at all steps. Akkerman *et al.*[2] have evaluated several washing procedures and found gel filtration to give the most metabolically and functionally intact platelets. However, they found considerable variation from one sample to another within a given technique. The citrate wash technique described below was not evaluated by Akkerman, but has been found in my laboratory regularly to give superb platelets as evaluated morphologically and functionally. In some hands, the EDTA wash, a procedure widely used in prostaglandin research, may be as good, though my experience coincides with that of Akkerman that such platelets tend to have more pseudopods and be functionally less intact. EDTA has a much higher affinity for calcium and magnesium than citrate and may leach these important cations from the membrane in addition to chelating them in the medium, leading to a mild to severe platelet injury that may be irreversible in some circumstances (note that even short exposure of platelets to EDTA at 37° produces irreversible changes[3]). Nevertheless, EDTA-washed platelets may be adequate for and useful in the study of arachidonic acid metabolism and the preparation of platelets for certain enzyme assays. The albumin wash technique,[4] has been used primarily to study platelet procoagulant activity. As it has not been used much in prostaglandin research, it will not be discussed here.

[2] J. W. N. Akkerman, M. H. M. Doucet-de-Bruine, G. Gorter, S. de Graaf, S. Hefne, J. P. M. Lips, A. Numeuer, and J. Over, *Thromb. Haemostasis* **39,** 146 (1978).

[3] M. B. Zucker and R. A. Grant, *Blood* **52,** 505 (1978).

[4] P. N. Walsh, D. C. B. Mills, and J. G. White, *Br. J. Haematol.* **36,** 281 (1977).

Centrifugation Methods

For all centrifugation methods, the centrifugation speed should be the lowest that will pellet the platelets. This speed will vary depending on the density of the medium from 1000 *g* for 10 min in plasma to 300 *g* for 10 min in buffers with no albumin or plasma. The use of higher *g* forces for longer times will make the platelets harder to resuspend and may cause platelet injury.

Citrate Wash. Add 1 ml of citrate anticoagulant (93 m*M* sodium citrate, 7 m*M* citric acid, 0.14 *M* dextrose, pH 6.5) per milliliter of PRP, chill to 4° for 5 min, then centrifuge at 1000 *g* for 10 min. Resuspend in one-tenth the original volume of the desired buffer by sucking the cells gently in and out through the tip of a plastic or siliconized glass Pasteur pipette and then make up to the desired volume. We routinely resuspend in Hank's balanced salt solution (HBSS), pH 7.4 (136 m*M* NaCl, 5.4 m*M* KCl, 0.44 m*M* KH_2PO_4, 0.34 m*M* $Na_2HPO_4 \cdot 7\ H_2O$, 5.6 m*M* D-glucose) either with or without albumin depending on the desired use of the platelets. For studies using ADP, epinephrine, or arachidonyl monoglyceride aggregation, we resuspend in HBSS containing 5–10% platelet-poor plasma. Where more than one wash in necessary (i.e., to remove the remaining plasma for thrombin aggregation) or desired, the once-washed platelet suspension is diluted 1:0.5 with citrate anticoagulant, centrifuged, and resuspended as above. Platelets should be incubated for 5–15 min at 37° to restore discoid shape and functional properties before performing aggregometry studies.

An alternative citrate wash procedure has been described by Mustard.[5] PRP is prepared from blood drawn into acid citrate dextrose (85 m*M* trisodium citrate, 65 m*M* citric acid, 110 m*M* D-glucose) (1 volume of anticoagulant to 6 volumes of blood) and then pelleted directly at 37°. Details of the resuspending buffers can be found in the original description of the method.

EDTA Wash. Chill PRP to 4°, add EDTA to a final concentration of 7.7 m*M*, and centrifuge at 1000 *g* for 10 min at 4°. Resuspend in the desired buffer. This washing procedure can be repeated several times as with the citrate wash. Baenziger and Majerus[6] advise using room temperature instead of 4°. Experience in my laboratory suggests that 4° is usually more satisfactory. Baenziger and Majerus also advise against initial resuspension in Tris-saline buffer as spontaneous aggregation may occur, though such buffers can be used after a second wash. The washed

[5] J. F. Mustard, D. W. Perry, N. G. Ardlie, and M. A. Packham, *Br. J. Haematol.* **22,** 193 (1972).

[6] N. L. Baenziger and P. W. Majerus, this series, Vol. 31, p. 149.

platelets can be kept at 4° until just before the aggregation studies, when they should be rewarmed to 37°.

Gel-filtered platelets. Gel-filtered platelets are prepared by applying citrated PRP to a Sepharose 2B column (Lages *et al.*[7] recommend first thoroughly washing the Sepharose 2B with acetone and 0.9% NaCl) prepared after deaeration of the gel. The column is equilibrated with buffer, then the platelets are applied. The size of the column depends on the amount of PRP to be applied and the degree of separation of platelets from the plasma proteins. (Usually the gel volume should be 5–10 times the plasma volume for good separation.) The platelets, which are eluted at the void volume, are collected by visual observation of the change in opacity of the column effluent. Data of Akkerman *et al.*[2] suggest that for most purposes Ca^{2+}-free Tyrode solution (137 m*M* NaCl, 2.68 m*M* KCl, 0.42 m*M* NaH_2PO_4, 1.7 m*M* $MgCl_2$, 11.9 m*M* $NaHCO_3$, 5.6 m*M* D-glucose, 2 g of bovine albumin per liter, pH 7.3) is an excellent buffer to use although other buffers also may be satisfactory.

Choice of Resuspending Buffer

Resuspend platelets at 200,000 to 500,000/μl for optimum studies of platelet aggregation. In choosing a resuspending buffer, be aware of several factors. Once separated from plasma, washed platelets tend to become rapidly depleted in ATP unless glucose (usually 5.6 m*M*) is added to the resuspending medium. Some potassium should also be included in washing and resuspension buffers since platelet levels may become depleted during washing. Apyrase is frequently added to degrade any ADP released into the plasma during washing procedures, since such ADP can make the platelets refractory to further ADP stimulation. Heparin is sometimes added to the first resuspension buffer to prevent effects of small amounts of thrombin that may be generated. For investigators interested in prostaglandins, one of the most important considerations in choosing a resuspending buffer relates to the albumin concentration. Albumin binds arachidonic acid, and the concentration of this fatty acid required to produce aggregation is thus critically dependent on the albumin level of the medium. In plasma, the dose of arachidonic acid needed to produce aggregation varies from 0.2 to 1.5 m*M* depending on the intrinsic sensitivity of the platelets used. In contrast, platelets washed three times with the EDTA wash to remove plasma and then resuspended in HBSS, pH 7.4, aggregate using 0.3–5 μM arachidonic acid, a concentration approximately 1000 times lower.[8] If too high a concentration of arachidonic

[7] B. Lages, M. C. Scrutton, and H. Holmsen, *J. Lab. Clin. Med.* **85,** 811 (1975).
[8] J. M. Gerrard, J. C. Peller, T. P. Krick, and J. G. White, *Prostaglandins* **14,** 39 (1977).

acid is used relative to the concentration of the albumin, platelet lysis results. Such lysis may occur with as little as 20 μM arachidonic acid in the absence of any albumin. Lysis has been mistaken for aggregation by some investigators. Where it occurs it is not inhibitable by aspirin. If aspirin pretreatment of the platelets (100 μM for 15 min) does not completely block arachidonic acid aggregation, the fatty acid is probably causing cell lysis. The concentration of albumin also has an effect on the dose of prostaglandin endoperoxides and A23187 (a calcium ionophore), though to a lesser extent than with arachidonic acid. Since albumin can enhance the conversion of PGH_2 to PGD_2 this consideration is important to studies of prostaglandin endoperoxides and their metabolism by platelets.

For most aggregating agents, fibrinogen may be necessary for the aggregation. Since platelet alpha granules contain fibrinogen, it is not essential to add fibrinogen with an agent that causes granule secretion, but added fibrinogen is needed for ADP aggregation under circumstances where no secretion is produced. For ADP or epinephrine aggregation of citrate-washed platelets, these cells can be resuspended in HBSS containing 5–10% PPP, or in some circumstances in buffers containing albumin and fibrinogen.[5] These washed platelets aggregate at similar concentrations of these agents to those used in native plasma. Fibrinogen used for this purpose may need to be treated with diisopropylfluorophosphate to avoid any contribution of procoagulant material contaminating the protein preparation.[9]

External calcium or magnesium are also essential for platelet aggregation. Since albumin binds calcium, the concentration needed may vary with the concentration of albumin. Changing the calcium level has important effects on the second wave of platelet aggregation and on the release of arachidonic acid from platelet phospholipids, the second-wave aggregation and arachidonic acid release being suppressed at higher calcium concentrations.[9,10]

For arachidonyl monoglyceride, we routinely use citrate-washed platelets resuspended in 10% plasma, as we have found these conditions to be optimal, perhaps relating in part to the optimum conditions for delivering arachidonyl monoglyceride in liposomes to platelets.[11]

Platelet Aggregation

The platelet aggregometer measures the transmission of light through a stirred cuvette maintained at a constant temperature, usually 37° (Fig.

[9] J. F. Mustard, D. W. Perry, R. L. Kinlough-Rathbone, and M. A. Packham, *Am. J. Physiol.* **228,** 175 (1975).

[10] M. J. Stuart, J. M. Gerrard, and J. G. White, *Blood* **55,** 418 (1980).

[11] J. M. Gerrard and G. Graff, *Prostaglandins Med.* **4,** 419 (1980).

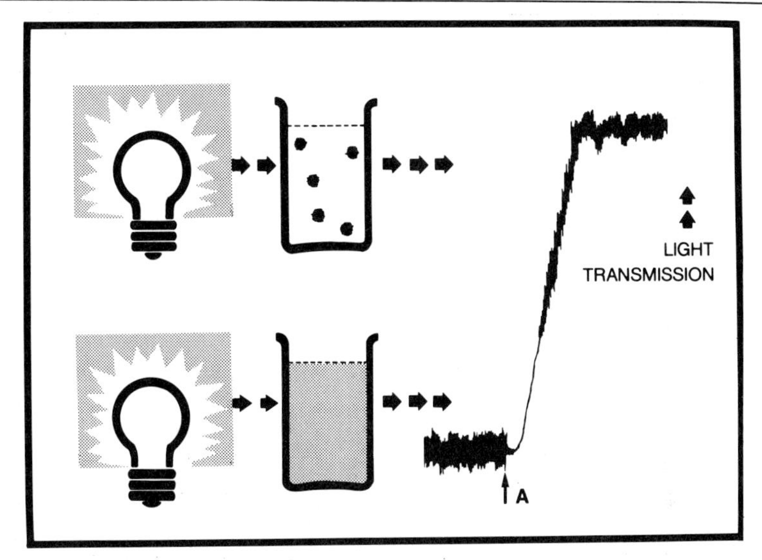

FIG. 1. The platelet aggregometer evaluated platelet function by its ability to measure light transmission through a suspension of stirred platelets. As shown in the tracing on the right, the initial suspension is opaque and the resting discoid platelets give baseline oscillations. As aggregation proceeds, platelets form larger and larger clumps and the solution clears until maximum aggregation is achieved.

1). The upper and lower limits of the recording pen are first set using the suspending medium without the platelets (PPP or washed platelet resuspending buffer) as 100% aggregation, and the platelet suspension (PRP or washed platelets) as 0% aggregation. In Fig. 1 a sample aggregation tracing is shown to facilitate understanding of the test and its uses. At the bottom of Fig. 1, light is transmitted through a stirred relatively opaque solution of platelets (0% aggregation). The birefringence of stirred resting platelets that are discoid in shape produces oscillations in the light transmission as recorded at the right. In the example shown, following addition of an agonist at A, there is first a change in the baseline oscillations to give a straight line. The initial loss of birefringence results from a change in platelet shape to a more spherical form with pseudopods. Subsequent to this, the extent of light transmission increases to the extent that platelets clump together into aggregates. When all platelets are clumped into a few large aggregates, the remainder of the solution is clear (100% aggregation). Since the assay conditions depend critically on measurement of light transmission, it is very important to be aware of and to allow for changes in light transmission resulting from addition of the agonist solution if careful quantitation is desired.

Since platelet aggregation responses may vary with time, and since spontaneous aggregation may occur under some circumstances, it is always essential to add a buffer control, and in evaluating the effects of inhibitors, such controls should be run at the beginning and the end of an experiment.

Buffers and Solvents for Use with Specific Agonists

In general, water-soluble reagents can be made up satisfactorily in HBSS or Tris-buffered saline. For compounds that are not readily soluble in water, two organic solvents that are water miscible are frequently used —ethanol and dimethyl sulfoxide. Both of these are inhibitory and/or toxic to platelets in high concentrations. The final concentration of either after addition to the platelet suspension should be kept less than 0.5% whenever possible and must never be more than 1%. Liposomes have been found useful to deliver arachidonyl monoglyceride to platelets and may be useful for other agents in the future.

Epinephrine (as the chloride salt), *ADP*, and *thrombin* are all water soluble and pose no particular problem in reconstituting in HBSS or Tris-buffered saline. *Collagen* can be prepared as a suspension or, as acid-soluble collagen, can be solubilized in a dilute solution of acetic acid. *A23187* is soluble in dimethyl sulfoxide or ethanol and can be added in a few microliters to the platelet suspension. *Lysophosphatidic acid* is soluble in 50% ethanol, or it can be used as a suspension in an aqueous buffer such as HBSS. *Arachidonic acid* can be used as the sodium, potassium, or ammonium salt in an alkaline buffer (i.e., 0.1 *M* Tris, pH 8.5) in an ethanol–carbonate solution, or can be added dispersed in a plasma or albumin solution. For PGG_2, and PGH_2, we have routinely evaporated the solvent under N_2 and then added the platelet suspension immediately. *Thromboxane* A_2 can be prepared on addition of prostaglandin endoperoxides to a preparation containing thromboxane synthase. *Arachidonyl monoglyceride* needs to be made up in liposomes. Liposomes made from beef heart phosphatidylcholine or from phosphatidylinositol from yeast or soybeans are satisfactory. Liposomes can be made satisfactorily by sonication of 20 m*M* phospholipid with 8 m*M* arachidonyl monoglyceride in distilled water using either a Bronson probe sonicator (3×10 sec at 4°) or a bath-type sonicator (15 min at 37°).[11]

Analysis of Aggregation Patterns

Understanding patterns of platelet aggregation with various agents and conditions is essential to use of the aggregometer. In this section, emphasis will be placed on changes in patterns with the type or dose of agonist. Use of platelet aggregometery in a quantitative or semiquantitative fashion requires attention to changes with dose in order to choose a parameter to measure to evaluate greater or lesser degrees of aggregation. Patterns of platelet aggregation of particular relevance to prostaglandin research can be grouped into three categories as shown in Fig. 2. Epinephrine, the agonist shown in pattern 1 of this figure is unique in that it causes a first wave of aggregation in which platelets clump together with little or no change in shape. No granule secretion or synthesis of prostaglandins occurs during the first wave of epinephrine aggregation. During the second wave of epinephrine-induced aggregation, shape change does occur together with the synthesis of prostaglandins and thromboxanes and the secretion of platelet granule contents. Studies with inhibitors and congenitally deficient platelets suggest that in most circumstances some degree of both thromboxane synthesis and secretion are necessary for this second wave.

A number of agonists, including ADP, thrombin, lysophosphatidic acid, prostaglandin endoperoxides, and thromboxane A_2, when added to platelets in PRP produce pattern 2 of platelet aggregation as shown in Fig. 2. Low doses of agonist cause shape change only, whereas slightly higher

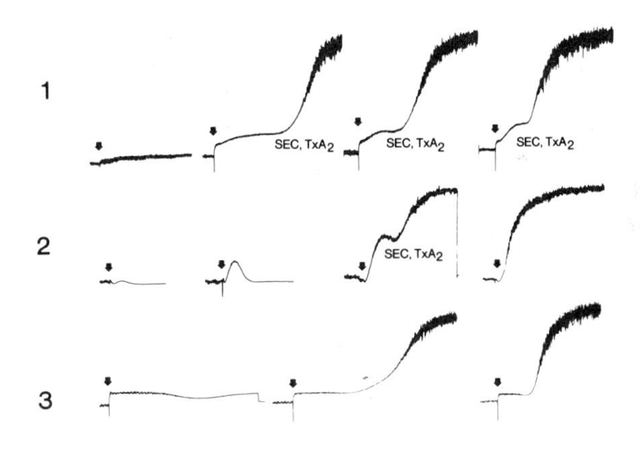

FIG. 2. Three types of aggregation patterns important for investigators studying prostaglandins are shown (see text). For each type the change in pattern with increasing dose of agonist is shown so that the reader can appreciate how the curves change with dose. Quantitation of aggregation depends on the use of a parameter that will vary with dose. The choice of parameter varies with the pattern of aggregation seen, as discussed in the text. In some circumstances, there are two waves of aggregation. Where this occurs, endogenous thromboxane A_2 (TxA_2) production and secretion (SEC) are associated with the second wave of aggregation. Inhibitors of the prostaglandin endoperoxide synthase, such as aspirin and indomethacin, will prevent the development of the second wave of aggregation under such circumstances.

doses of agonist produce a reversible first wave. Still higher doses produce a biphasic response with two waves of aggregation, and at the highest doses only a single complete wave of aggregation is seen. The intermediate form in which there are two waves is less prominent with platelets from certain individuals and with some agents, though it may occur with all these agents in PRP or in washed platelet preparations in which there is substantial plasma or albumin. In general, the extent to which a second wave is obtainable with ADP and with the other agents depends on the concentration of calcium with a greater likelihood of a second wave response at a lower calcium concentration (i.e., Mustard *et al.*[9] showed a second wave and secretion with ADP stirred with washed platelets in 1 m*M* Mg and no Ca, but no second wave or secretion with 1 m*M* Mg and 2 m*M* Ca).

The third pattern of aggregation shown occurs with collagen, arachidonic acid, and A23187 in platelet-rich plasma, with arachidonyl monoglyceride (AMG) added to platelets in 10% plasma, and with collagen, arachidonic acid, prostaglandin endoperoxides, and thromboxane A_2 added to platelets washed and resuspended in the absence of plasma. Rarely A23187 or AMG have been seen or reported to produce a double wave when added to PRP as shown in pattern 2. In many circumstances, aggregation tends to be an all-or-none phenomenon as shown in pattern 3 of Fig. 2, with the major change with increasing agonist concentration above threshold levels being in the time to onset of aggregation or the slope of aggregation. In some circumstances, particularly with washed platelets, aggregation may progress in incremented steps (i.e., 20% aggregation at a lower dose, 40% aggregation at a higher dose, and 100% aggregation at the highest dose (each having a single irreversible wave).

Quantification of Platelet Aggregation

Platelet aggregation can be used in a quantitative or semiquantitative fashion to analyze various questions. For most purposes, it is desirable to choose a parameter of the aggregation response that changes in a linear fashion proportional to the dose or the log of the dose of the agonist. When this does occur it is usually only in a relatively limited dose range. Use of platelet aggregation in a quantitative fashion relies on establishing this dose range and the response. A problem is that each individual's platelets may respond a little differently, so that a dose response in one individual may be quite different from that in another. For this reason, comparisons between different individuals is often done using the dose to achieve a specific effect (i.e., 50% aggregation). The variation in response also means each platelet preparation must be tested for its dose response before using an inhibitor (i.e., PGI_2) if it is desired to quantitate the amount of agonist or inhibitor.

In studying inhibitors, there is a danger in using too high or too low a dose of agonist (i.e., if the initial dose response shows that 2 μM ADP gives 20% aggregation, 4 μM ADP gives 50% aggregation, 6 μM ADP gives 80% aggregation, 8 μM ADP gives 90% aggregation, 10 μM ADP gives 95% aggregation, and 30 μM gives 96% aggregation, then the appropriate dose to check the effect of an inhibitor is 4–6 μM ADP. If 30 μM ADP is used, considerable inhibition could be achieved with very little change in the aggregation pattern.

The parameter to be used in assessing changes in aggregation with agonist concentration varies with the type of agonist. For agonists in pattern 1 (Fig. 2), the time to 50% aggregation, or the area under the aggregation curve up to 3 or 5 min tends to be most useful. For agonists with pattern 2, the height (percentage aggregation) of the first wave or the area under the aggregation curve is useful. For agonists with pattern 3, the time to 50% aggregation, the slope of the aggregation curve, or the area under the aggregation curve is usually most rewarding; however, in some circumstances the percentage of aggregation is useful. Measurements of granule secretion may be more accurate than evaluation of the aggregation tracing at high agonist doses. In general our experience is that except for aggregating agents in pattern 3, slope is not reliable, although some investigators have managed to achieve conditions where slope can be used with agonists of pattern 2. Some investigators have used the time to onset of aggregation or the time to 100% aggregation. Our experience is that a higher degree of accuracy can be achieved using the time to an intermediate degree of aggregation (usually 50%, but sometimes 10% or 90%), since the start and completion of aggregation as determined from an aggregation curve is somewhat arbitrary.

With all quantitative uses of platelet aggregation, attention needs to be given to keeping the preparation of platelets stable in their response for some length of time. In PRP, platelets usually produce a rapid rise in pH that will alter aggregation patterns. The rise in pH can be avoided by replacing the air in the tube with a CO_2/O_2 mixture and stoppering the tube. For additional steps to reduce variability in platelet aggregation, the investigator should consult Newhouse and Clark.[12]

Platelet Aggregation as a Bioassay for Prostaglandin Endoperoxides and Thromboxane A_2

Platelet aggregation can be used quantitatively as described above to detect prostaglandin endoperoxides and thromboxane A_2. As with other bioassays, it is usual to reduce interference with the assay from other

[12] P. Newhouse and C. Clark, *in* "Platelet Function" (D. A. Triplett, ed.), p. 63. American Society of Clinical Pathologists, Chicago, 1980.

agonists. As a minimum, most investigators have pretreated the platelets with aspirin or indomethacin to inhibit any platelet response to arachidonic acid. In some circumstances, it may be important to add heparin or hirudin to inhibit effects due to thrombin. Platelets can be made refractory to ADP, yet maintain their responsiveness to endoperoxides and thromboxane A_2, by preincubating the platelet sample with ADP.[13] Some EDTA-washed platelet preparations are also refractory to ADP but respond well to endoperoxides. Some investigators have succeeded in using a combination of creatine phosphate and creatine phosphokinase to convert ADP to ATP to prepare a platelet suspension that responds nicely to endoperoxides and thromboxane A_2 but does not respond to ADP. However, others have found that such preparations do not respond well to endoperoxides or thromboxane A_2 when ADP is inhibited. It is possible that some preparations of creatine phosphate or creatine phosphokinase have contaminants that inhibit aggregation to endoperoxides. However, given the current controversy, this methodology cannot be recommended.

Platelet Aggregation as a Bioassay for PGI_2, PGD_2, and PGE_1

These three prostaglandins can be assayed owing to their inhibitory effects on platelet aggregation. Bioassays have been devised in which a standard dose of agonist (usually ADP) is added to PRP (as explained above this dose should be one that gives 50–80% aggregation). The dose of PG needed to give various degrees of inhibition is then established by adding the PG in varying concentrations at a fixed time (usually 30 sec) before addition of ADP on the aggregometer. The solution containing an unknown concentration of PG can then be added in a similar fashion and its concentration determined by comparison with the effect of the calibrating doses. In these experiments it is important to recalibrate the platelet response at intervals during a long experiment and both before and after a short experiment to check that the platelet response to ADP, and its inhibition by PG, has not changed substantially during the assay. The assay is not selective and will detect any inhibitory PG or indeed other inhibitor. An approach to identify a selective effect of PGI_2 has been to use the short half-life of PGI_2 in neutral or acid aqueous solutions to show that the effect of PGI_2 in the test sample behaves like PGI_2 in this respect.

Assay of Platelet Enzymes

In general, aggregation can be used only as a quick, crude evaluation for the presence, absence, or substantial change in activity of an enzyme. Further biochemical quantitation is an essential subsequent step.

[13] J. M. Gerrard and J. G. White, *Prog. Hemostasis Thromb.* **4,** 87 (1978).

Prostaglandin Endoperoxide Synthetase (*Cyclooxygenase*). Add arachidonic acid to PRP or washed platelets and follow aggregation. A response to arachidonic acid in plasma or in washed platelets indicates that the enzyme is present. With washed platelets, the aggregation should always be retested with aspirin-treated samples to make sure this abolishes the response (see earlier discussion regarding lysis). The platelets should also be tested with an endoperoxide or endoperoxide analog to be sure that the cells respond to these compounds and do not have another defect. Variation in the inherent responsiveness of platelets unrelated to the activity of this enzyme may make comparisons of enzyme present in different individuals difficult, although an enhanced responsiveness of diabetics to arachidonic acid has beed used to indicate an enhanced enzyme activity.[14] In view of the major effect of albumin on the platelet response to arachidonic acid, it may be desireable to standardize the amount of albumin in the platelet suspension.

Thromboxane Synthase. Gorman *et al.*[15] have used platelets in a fashion to detect abnormal thromboxane synthase activity (no response to PGH_2, but normal response to a thromboxane analog,) but their interpretation has been questioned.[16] Since thromboxane A_2 is more active than PGH_2, platelet microsomes containing thromboxane synthase when mixed with PGG_2 or PGH_2 produce considerably more aggregation than PGG_2 or PGH_2 alone, and this difference could be used to evaluate this enzyme.[17] At present it seems wise to perform another assay of this enzyme in addition to platelet aggregation.

Diglyceride Lipase. Add arachidonyl monoglyceride to washed platelets in 10% plasma. Aggregation by this agent is believed to require this lipase, and the presence of aggregation to arachidonyl monoglyceride implies that this enzyme is present. If there is no aggregation, aggregation to arachidonic acid should be evaluated to check whether subsequent conversion and the response to arachidonic acid are intact.

Phospholipase C and A_2. There is currently still debate as to the precise events involved in release of arachidonic acid from platelet phosholipids, and until this is resolved only a few comments can be made. Second wave aggregation to epinephrine or ADP which can be blocked by pretreating the platelets with aspirin, implies that the platelets are capable of releasing arachidonic acid from their phospholipids, converting it to thromboxane A_2, and responding to this agent. Absence of such a second

[14] P. V. Halushka, D. Lurie, and J. A. Colwell, *N. Engl. J. Med.* **297,** 1306 (1977).

[15] R. R. Gorman, G. L. Bundy, D. C. Peterson, F. F. Sun, O. V. Miller, and F. A. Fitzpatrick, *Proc. Natl. Acad. Sci. U. S. A.* **74,** 4007 (1977).

[16] D. E. MacIntyre, *in* "Platelets in Biology and Pathology" (J. L. Gordon, ed.), Vol. 2, p. 211. Elsevier/North-Holland, Amsterdam, 1981.

[17] S. Moncada, P. Needleman, S. Bunting, and J. R. Vane, *Prostaglandins* **12,** 323 (1976).

wave in the presence of a normal response to arachidonic acid could be associated with defective release of arachidonic acid, but may also reflect defective platelet granule content or decreased metabolic ATP.

The Evaluation of Drugs

Two basic questions may be asked: What is the mechanism of action of an inhibitory drug; and What is its potency? To answer the first question, it is important to compare the pattern of inhibitory effects seen with that produced by known inhibitors. Agents that raise cyclic AMP, including PGI_2, have a general depressant action on aggregation by all agents (i.e., first-wave aggregation as much as second-wave aggregation), although at low concentrations there may be selective inhibition of arachidonic acid aggregation compared to inhibition of thrombin, collagen, or ADP.[18] Agents that inhibit the prostaglandin endoperoxide synthase enzyme inhibit second-wave epinephrine or ADP aggregation and inhibit aggregation by arachidonic acid, but not aggregation due to endoperoxides. A selective diglyceride lipase inhibitor should inhibit aggregation due to arachidonyl monoglyceride, but not aggregation produced by arachidonic acid. A selective thromboxane receptor antagonist inhibits aggregation produced by thromboxane, but not aggregation due to ADP.

To assess the potency of a drug, the conditions are similar to those used to quantitate inhibitory PGs described earlier. However, for some circumstances, it is desirable to use an aggregating agent other than ADP. For an agent that inhibits the prostaglandin endoperoxide synthase, arachidonic acid aggregation or second-wave epinephrine aggregation are more useful than ADP aggregation, since ADP is less affected by endoperoxide synthase inhibitors.

The Study of Patient Groups

In comparing samples from different inidividuals, it becomes important to standardize platelet counts, to be particularly careful about good venipuncture technique and red cell contamination of samples, and to use a standard time interval after drawing the blood for the test. Variation in any of the above can influence results.

Acknowledgments

This work was supported in part by MRC Grant 7396, by a grant from the National Cancer Institute of Canada, and by an MRC Scholarship.

[18] G. H. R. Rao, K. R. Reddy, and J. G. White, *Prostaglandins Med.* **6,** 75 (1981).

[76] Pharmacologic Characterization of Slow-Reacting Substances

By CHARLES W. PARKER, MARY M. HUBER, and SANDRA F. FALKENHEIN

The slow-reacting substances (SRSs) have been defined as a family of unsaturated C_{20} fatty acids hydroxylated at the 5 position and substituted with one of several sulfur-containing side chains at the 6 position.[1,2] A variety of normal cell types or neoplastic cell lines have been shown to produce SRS, including RBL-1 cells (a rat basophilic cell line), human basophils, rat mast cells, rat monocytes or macrophages, and human neutrophils as well as more complex tissues such as the lung. Until their structural characterization, the SRSs were detected largely or entirely by their smooth muscle contracting activity. Guinea pig ileal muscle is the smooth muscle preparation that has been most frequently used for SRS measurements.[3] SRSs also produce contraction of tracheal–bronchial smooth muscle in several species including man, guinea pig and rat colon, and rat stomach. Peripheral lung airway smooth muscle is also apparently sensitive to SRS,[4] although because of the presence of other smooth muscle elements in peripheral lung preparations, this has been more difficult to prove conclusively. In contrast to some of the other spasmogenic agents, the SRSs show little or no activity in smooth muscle preparations such as the rat duodenum, chick rectum, and cat jejunum. In the skin they increase vascular permeability and contract (LTC_4) or dilate (LTD_4) cutaneous vessels.[4] The SRSs also lower systemic blood pressure in anesthetized guinea pigs, decrease heart rate and cardiac muscle contractility, and alter Purkinjie nerve cell function.

Now that the structure of SRS is known, studies of SRS metabolism would preferably involve the use of quantitative and specific assay systems such as radioimmunoassay and isotope dilution mass spectroscopy analysis or, at the very least, purification to the point that the behavior of the active material on high-pressure liquid chromatography (HPLC) columns and its ultraviolet absorbance characteristics can be evaluated. On the other hand, the absolute amounts of SRS produced by most tissues

[1] C. W. Parker, *Proc. Int. Symp. Biochem. Acute Allergic Reaction, 4th, Kroc Found. Ser.* (1978).

[2] C. W. Parker, *Adv. Inflammation Res.* **4,** 1 (1982).

[3] W. E. Brocklehurst, *Allergy* **6,** 539 (1962).

[4] J. M. Drazen, K. F. Austen, R. B. Lewis, D. A. Clark, G. Goto, A. Marfat, and E. J. Corey, *Proc. Natl. Acad. Sci. U.S.A.* **77,** 4354 (1980).

ISBN 0-12-181986-8

are extremely small, and measurements of spasmogenic activity require much less material than most forms of direct structural analysis. Therefore, measurements of SRS spasmogenic activity will continue to be useful for biological studies, particularly in laboratories that lack the experience and facilities for mass spectroscopic analysis. This review will describe how SRS smooth muscle contracting activity is measured, including some of the pitfalls and problems in interpretation that need to be considered. While a detailed description will be given only for the guinea pig ileal smooth muscle bioassay system, the general approach with regard to verification of specificity is readily applicable to other SRS activities.

Characteristics of the Guinea Pig Ileal Smooth Muscle Bioassay System

Longitudinal strips of intestine are prepared from the terminal ileum of 200–300 g Hartley guinea pigs.[3] Where desired, strips may be taken from higher in the ileum or even the jejunum. As a rule analyses are made on the entire intestinal wall, although with some care and practice isolated strips of longitudinal smooth muscle largely free of other intestinal elements can be prepared. Whole intestinal ileal strips and isolated ileal smooth muscle preparations have a similar sensitivity to SRS. After washing, the intestinal strips are suspended between two strings, in small (usually 2–5 ml) chambers containing continuously oxygenated Tyrode's solution. One of the strings is attached to a transducer permitting measurements to be made of changes in gut tension. As a rule, at least four strips are placed in chambers to allow for simultaneous measurements of several samples and the possibility that some of the strips may prove to be unsuitable for SRS measurements. Indeed, not infrequently all the strips from a given animal are unsatisfactory for bioassay and a new animal has to be used.

Once the gut preparations have been hung, it takes at least several hours of washing and standardization before they are ready for SRS measurements. The strips are washed repeatedly (every 5 min) with Tyrode's solution over a period of several hours, adding 40 ng of histamine 15–20 sec before each wash; then each strip is standardized twice with graded amounts of histamine ranging from 5 to 40 ng. Our laboratory uses a semiautomatic tissue adapter that regularly washes the muscle strips, and a multiple channel recorder to permit continuous measurements. Tyrode's buffer containing 1 μM pyrilamine maleate and 1 μM atropine sulfate is then added to the chamber and used throughout the remainder of the experiment, in order to block any histamine or acetylcholine that might be present in unknown samples as well as to decrease spontaneous contractile activity. These blocking agents are not required for purified SRS samples, but are usually used even here, since they do not affect the SRS response.

The strips are then pulsed three times for 1 min with 2–4 units of SRS. [By definition, 1 unit of SRS produces a contractile response equivalent to 5 ng of histamine base; for LTC_4 this represents about 0.6 pmol of SRS.[5]] Graded amounts of SRS standard covering a range from 0.5 to 5 units (0.12 in the case of LTD_4 1.2 n*M* solutions in the 2.5-ml chamber) used in our laboratory are then added to determine the sensitivity of each strip to SRS. Once suitable strips (adequate sensitivity with low spontaneous activity) have been identified, they are usually satisfactory for analysis for at least 8–12 hr, and measurements may be undertaken on SRS-containing samples. However, the responsiveness of the strips to the SRS standard must be redetermined regularly (every 15–30 min) during the bioassay. When critical comparisons are being made of SRS samples that may not differ markedly in their SRS content, it is highly desirable that they be analyzed on the same smooth muscle strip. Contractions in response to SRS are usually observed over a period of 1–2 min, and then the strips are carefully washed and observed for return of muscle tension to the original baseline before additional analyses are made. The addition of large amounts of SRS activity to the chambers should be avoided, since very extensive washing may then be required before the strip is ready for further additions. Glutathionyl SRS (LTC_4) is particularly difficult to re-

[5] Although the SRS unit is usually defined as being equivalent to 5 ng of histamine, the quantity of SRS in a unit may vary quite considerably from laboratory to laboratory. While standardization with histamine has been very useful it makes more sense to standardize primarily with SRS itself particularly, if pure standard is available. The purified SRS standard may be either chemically synthesized or highly purified biosynthetically derived SRS. In either case LTC_4, LTD_4, and LTE_4 are evaluated individually as pure solutions. Small (5–50 μCi) volumes of stock 0.2–10 μM solutions of SRS standard were added to the chambers, the exact concentrations depending on the potency of the SRS and to a certain extent the sensitivity of other gut preparations. The standard is usually added in the HPLC buffer in which it has been purified. Alternatively (and less satisfactorily), crude SRS preparations prepared from a large batch of RBL-1 cells[6] and standardized repeatedly on a minimum of 12 different intestinal strips versus histamine may be used as a standard. Such preparations are usually stored in the form of a lyophilized powder in small aliquots, making up a fresh solution of the standard on each day a SRS bioassay is performed. Even with highly purified SRS standards it is important that the storage conditions be optimal. The most satisfactory method of storage is probably in the HPLC buffer used for purification or 50% methanol–water at −80°. SRS breakdown during storage can be detected by losses in spasmogenic activity, a shift in the absorption maximum from 280 to 278 nm (isomerization of the 11-*cis* double bond), loss of overall 270–290 nm absorbancy, or a change in elution behavior on HPLC columns.

[6] B. A. Jakschik, A. Kulczycki, H. H. Macdonald, and Parker, C. W., *J. Immunol.* **119,** 618 (1977).

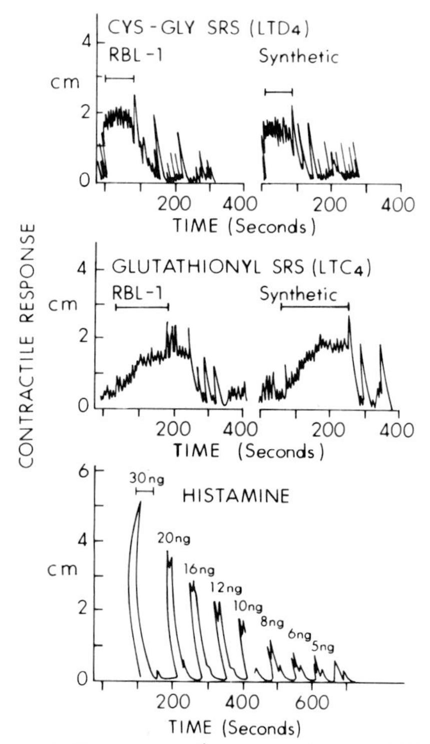

FIG. 1. Typical guinea pig ileal contractions to synthetic and RBL-1 slow-reacting substance (SRS) (LTC_4 and LTD_4) and histamine. The brackets indicate the period in which samples were exposed to SRS before washing. The synthetic SRS was generously provided by Dr. J. Rokasch (Merck).

move, and washing periods of many minutes may be required, even with relatively small amounts of this material. Each unknown sample should be analyzed repetitively using at least 2, and preferably 3, different-sized aliquots to determine the reproducibility and linearity of the response. Absolute amounts of SRS spasmogenic activity are calculated by reference to a standard dose-response curve for authentic SRS (or histamine, see above) plotted on a semilog scale.

Comparative Responses of Different Purified SRSs

Typical SRS contractions for LTC_4 and LTD_4 are shown in Fig. 1. All of the SRSs produce responses that are sustained in their presence over a period of many minutes without a diminution in intensity.[7–11] While the

[7] C. W. Parker, S. F. Falkenhein, and M. M. Huber, *Prostaglandins* **20,** 863 (1980).

[8] R. P. Örning, S. Hammarström, and B. Samuelsson, *Proc. Natl. Acad. Sci. U.S.A.* **77,** 2014 (1980).

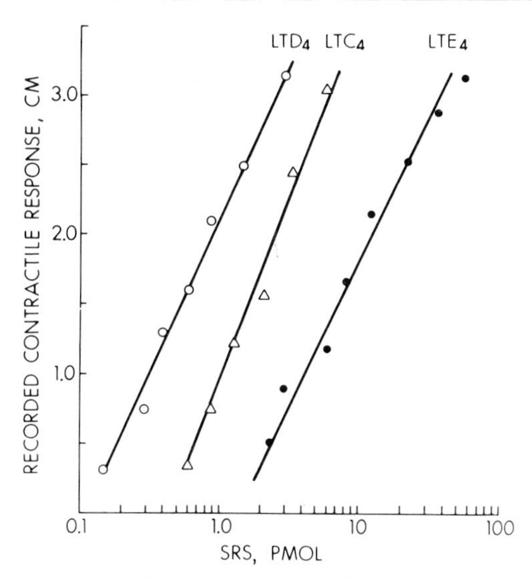

FIG. 2. Comparative contractile response on the guinea pig ileum to the three major SRS species produced biosynthetically in RBL-1 cells.

response is characteristically slower in onset than that for histamine, the slow onset of contraction is much more evident at lower than at higher SRS concentrations. Semilog plots of dose-response curves for the different SRSs as determined in our laboratory are shown in Fig. 2. The most potent is LTD_4 followed by LTC_4 and then LTE_4[7]; LTD_4 and LTE_4 produce parallel responses. The response to LTC_4 is often not parallel to the others, although this may vary with the smooth muscle preparation. The threshold for demonstration of an ileal contraction for LTD_4 is of the order of 0.1 pmol (a 40 p*M* concentration in our 2.5-ml chamber bioassay system). Depending on the strip used, LTD_4 is about 30–100 times more potent than histamine on a molar basis in the ileal smooth muscle system. In our hands, 11-*trans*-LTC_4 is nearly equivalent to 11-*cis*-LTC_4 although others have reported that the 11-*trans*-LTC_4 is considerably less active. 11-*trans*-LTD_4 and LTE_4 are less active than the corresponding 11-*cis* isomers although the precise ratios must await further study, since the chemically synthesized SRS standards themselves may exist as isomeric mixtures. LTC_4 and LTD_4 differ in the character of their contractions in the ileal muscle system in that the response to LTC_4 appears more slowly

[9] D. Sok, J. Pai, V. Atrache, and C. J. Sih, *Proc. Natl. Acad. Sci. U.S.A.* **77,** 6481 (1980).

[10] C. W. Parker, M. M. Huber, M. K. Hoffman, and S. F. Falkenhein, *Prostaglandins* **18,** 673 (1979).

[11] C. W. Parker, B. A. Jakschik, M. M. Huber, and S. F. Falkenhein, *Biochem. Biophys. Res. Commun.* **89,** 1186 (1979).

(Fig. 1). As already indicated, LTC_4 is difficult to remove by washing, making precise determination of its potency more difficult than for LTD_4. This difference may help explain some of the variation that has been reported by different laboratories in the sensitivity of ileal muscle strips to LTC_4.

Measurements of SRS Activity in Crude Extracts

SRS activity is frequently measured in crude tissue extracts by bioassay. Since substantial SRS degrading activity may be present, systemic experimentation may be needed to determine the optimal time for generation of maximal SRS activity. The possibility of losses of SRS activity during storage also must be considered. In studies to date A23187 has usually been the most effective stimulus to SRS formation, although it remains to be demonstrated that this will be true for all tissues. In general, the tissue is incubated with the SRS generating stimulus for 5–15 min at 37°. The suspension is then chilled, and the tissue or cells are removed by sedimentation or centrifugation. For SRS measurements made the same day, it is usually satisfactory to store the supernatant at 0° until bioassay. If measurements are going to be delayed for several days, ethanol is added to supernatants to a final concentration of 80% (v/v), and after centrifugation at 20,000 *g* the soluble fraction is obtained by aspiration and stored at −20°. The ethanol removes most of the protein and some of the salt present in the original reaction mixture.

On the day of the bioassay, as close to the time of actual assay as possible, the ethanol and water are removed from the samples by evaporation under nitrogen or argon. The samples are then reconstituted by the addition of water or Tyrode's solution, diluted if necessary with Tyrode's solution and maintained at 0° until analysis. If the activity in the sample is exceptionally high, it may be possible to analyze small aliquots of the sample directly without removal of the organic solvent, but the final ethanol concentration in the bioassay chamber should be 1% or less (some strips may not tolerate even this amount of ethanol). The possibility of interference or nonspecific contraction in the bioassay from salts, particularly those containing K^+, or any reagent used during the SRS generating step that has not been thoroughly evaluated previously, needs to be considered. One approach to verifying the absence of interfering activity is to add known amounts of SRS as an internal standard to the sample while it is still in the bioassay chamber (after completing the analysis for endogenous SRS activity) and seeing if the expected increment in smooth muscle tension is obtained. In addition, buffer solutions containing the reagent in question can be evaluated in the presence and in the absence of known amounts of SRS standard.[12]

Products of the SRS generating cells themselves may cause nonspecific effects in the bioassay such as spontaneous muscle contraction and irritability, particularly when highly concentrated cellular suspensions are being studied and the cells are not removed before the extraction step with ethanol. Interference in the assay can often be avoided by use of small aliquots of sample, but if this procedure is followed, small amounts of SRS activity may easily be missed. Other criteria apart from the character of the contraction itself that may be helpful in verifying that SRS is being produced include showing that the amount of activity does indeed increase with increasing amounts of tissue and that little or no activity is present in the absence of a stimulus to SRS formation. Since SRS is not preformed, unstimulated samples should contain little or no activity.

Verification of the Specificity of the SRS Response

Selective Blocking Agents. Now that the structure of SRS is known, it can be anticipated that structural analogs of SRS that competitively inhibit the contractile response by interfering with SRS binding will soon be available. In the meantime FPL-55712, sodium 7-(3,4-acetyl-3-hydroxy-2-propylphenoxy)-2-hydroxypropoxyl- 4-oxo-8-propyl-4*H*-1-benzopyran-2-carboxylate, and related chromone-2-carboxylic acids that were identified empiricially as SRS blockers, appear to be of considerable value in verifying the specificity of SRS responses. While FPL-55712 has not been shown to be a structural analog of SRS and is not entirely specific for SRS, its KI_{50} for inhibition of contraction in the guinea pig ileal bioassay system is 1000-fold or more lower for SRS than for PGE_2, $PGF_{2\alpha}$, 5-hydroxytryptamine, histamine, bradykinin, or acetylcholine.[13] The FPL-55712 is prepared as a stock 1 mg/ml solution in dimethyl sulfoxide, diluted further in buffer if necessary, and added in a small volume (10–50 μl) to the bioassay chamber either 30 sec before the SRS or during an SRS-induced contraction (when the response is maximal), keeping the final concentration of dimethyl sulfoxide in the chamber below 0.25% (v/v). Normally 10 ng/ml concentrations of FPL-55712 markedly reduce SRS responses in the guinea pig ileal system provided submaximal amounts of SRS are used (for example, an amount of SRS producing 50% of a full-scale response on the recorder) whereas 50–100 ng/ml concentrations of the inhibitor are required to reverse, largely or entirely, an existing contraction (Fig. 3). At lower concentrations of FPL-55712 the

[12] S. F. Falkenhein, H. MacDonald, M. M. Huber, D. Koch, and C. W. Parker, *J. Immunol.* **125,** 163 (1980).

[13] J. Augstein, J. B. Farmer, T. B. Lee, P. Sheard, and M. L. Tattersall, *Nature* (*London*) *New Biol.* **245,** 215 (1973).

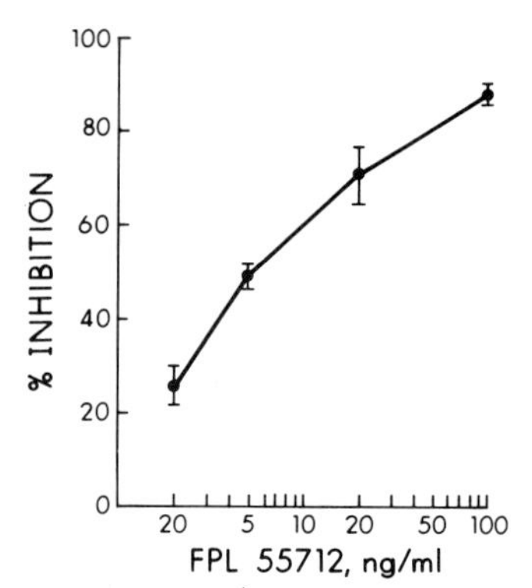

FIG. 3. Dose-dependent FPL-55712 inhibition of the guinea ileal response to constant amounts of slow-reacting substance standard from RBL-1 cells. Taken from Jakschik *et al.*[6]

dose-response curve is shifted in a parallel fashion to the right (higher concentrations of SRS). Responses to LTC_4 are less readily inhibited than responses to LTD_4 and LTE_4. Blocking studies with FPL-55712 are best delayed until the end of the bioassay day, since the muscle strip is frequently not usable for subsequent analyses.

Degradation Studies. Another way of determining whether or not a spasmogenic response is due to SRS is to see whether the activity is stable to selected enzymes or extremes of pH or temperature. Enzymes that may inactivate SRS include arylsulfatase and lipoxygenase.

Most of the SRSs that have been studied to date are inactivated by type B arylsulfatases,[14] at least as these enzymes are presently obtained commercially. The inactivation is seen with both crude and purified SRS preparations. Based on several lines of evidence the decrease in SRS activity appears to be due mainly to a contaminating protease.

1. Studies with 1-^{14}C- and ^{35}S-labeled SRS indicate that the sulfur containing side chain is not split from the SRS molecule, as would be expected if a sulfate ester were being cleaved.[15]
2. The enzyme does remove the glycine moiety from [^{14}C] glycine-labeled LTC_4 and LTD_4.
3. Arylsulfatases are not known to degrade sulfide linkages.
4. In contrast to an earlier paper from the same laboratory,[14] it has been reported that highly purified preparations of type B arylsulfatase do not inactivate the SRS.[16] Even though the inactivation ap-

[14] S. I. Wasserman, E. J. Goetzl, and K. F. Austen, *J. Immunol.* **114,** 645 (1975).

[15] C. W. Parker, D. A. Koch, M. M. Huber, and S. F. Falkenhein, *Prostaglandins* **20,** 887 (1980).

[16] S. Holgate, C. Winslow, R. Lewis, and K. F. Austen, *Fed. Proc., Fed. Am. Soc. Exp. Biol.* **40,** 1069 (1981).

parently does not involve the cleavage of an aryl sulfate ester linkage and the possibility must be considered that future type B arylsulfatase enzyme preparations might not contain inactivating activity, up to now the effect has been sufficiently reproducible that it has been useful as a criterion for SRS activity.

The enzyme [usually Sigma type V (limpet) arylsulfatase, although the patellar enzyme is also suitable] is suspended in 0.15 *M* NaCl, 0.01 *M* phosphate, pH 7.4, at a concentration of 50–80 units of enzyme (arylsulfatase) activity per milliliter, and dialyzed for 1 h versus 0.2 *M* acetate, pH 5.7.[15] Aliquots (0.2 ml) of the dialyzed enzyme solution are then added to dried samples of SRS (100–1500 SRS units). The pH is adjusted if necessary to 5.7, and samples are incubated for 2–3 hr at 37°. At the completion of the incubation, the pH is adjusted to less than 1.5 with 1 *N* HCl. Samples are then heated for 10 min in a boiling water bath, and centrifuged to remove the enzyme. After adjustment of the supernatant to neutral pH with dilute NaOH, the SRS is bioassayed in the usual way. As a control SRS samples are incubated at 37° in 0.2 *M* acetate at pH 5.7 over the same time period in the absence of enzyme, the enzyme is then added and samples are immediately acidified and boiled. LTC_4 and LTD_4 normally show 65–80% inactivation under the above conditions as compared with 15–20% inactivation in buffer without enzyme. This varies to some extent with the enzyme preparation. With LTD_4 the loss of activity is accomplished by a change in chromatographic behavior since the digested LTD_4 now coelutes with LTE_4. Much less, if any, inactivation is observed with LTE_4, because the glycine portion of the sulfur containing side chain has already been removed.

For inactivation of SRS with lipoxygenase, a 4 n*M* solution of purified or partially purified SRS in Tyrode's solution at pH 7.5 is incubated at room temperature with 10 μg/ml of commercial soybean lipoxygenase.[17] The inactivation is facilitated by bubbling O_2 into the reaction mixture during the incubation. As discussed in the section on preparation of SRS from RBL-1 cells, the reaction may be followed spectrophotometrically at 280 and 310 n*M* as well as in the guinea pig ileal bioassay system. Ordinarily LTD_4 and LTE_4 are at least 75% inactivated within 45–60 min; 11-*cis*-LTC_4 is inactivated more slowly by the enzyme. 11-*trans*-LTC_4 and LTD_4 are not susceptible to inactivation because the enzyme requires a 1,5-pentadiene with both double bonds in the cis configuration.

Other enzymes that have been reported to produce SRS inactivation include γ-glutamyltranspeptidase[7,8] and leucine aminopeptidase.[9] Again it appears that degradation of the sulfur-containing side chain is involved.[7]

SRS is also inactivated by heating at acid pH, the precise conditions

[17] R. C. Murphy, S. Hammarström, and B. Samuelsson, *Proc. Natl. Acad. Sci. U.S.A.* **76,** 4275 (1979).

varying rather widely with its state of purity. This can be combined with HPLC if the SRS is thought to be radiochemically pure, since the inactivation product migrates very differently from the original SRS.[11] With crude SRS, heating at 37° at pH 3 for 1–2 hr will destroy most of the SRS activity. With highly purified SRS, however, the SRS must be boiled in 0.05 *N* HCl for 90–120 min to produce 70–80% inactivation.[11] On the other hand, even crude SRS is ordinarily stable to heating in dilute (0.1 *N*) NaOH at 37° for 30 min.

Analysis for Other Spasmogens. Other spasmogenic substances that are frequently present in biological samples include histamine, bradykinin, the prostaglandins, serotonin, platelet-activating factor, and acetylcholine. Since the slow sustained contractions seen with the SRSs differ markedly from the more rapid and transient responses seen with most of these substances, the characteristics of the contraction itself may be of considerable value. Not infrequently, combined responses are seen with features both of SRS and an agent producing a more rapid contraction. In addition to the use of specific antagonists for agents such as histamine, acetylcholine, and serotonin, samples may be analyzed directly for some of these spasmogens. For example, histamine can be measured by enzymatic or fluorometric methods. If large amounts of histamine are present, then it is important to verify that sufficient antihistamine is present in the bioassay buffer to block the histamine in the sample. Similarly, the prostaglandins can be determined by radioimmunoassay, and the use of other smooth muscle preparations may be helpful. Many of the smooth muscle preparations that are prominently affected by prostaglandins or histamine are relatively or totally unresponsive to SRS. For example, the chicken rectum responds well to PGE_2 and poorly or not at all to SRS, whereas the reverse is true for the guinea pig ileum. Trypsin and other proteolytic enzymes that normally cause marked losses of bradykinin activity can be used to evaluate the role of bradykinin in the response.

Use of Metabolic Inhibitors. Since SRS is a lipoxygenase product formed de novo at the time of the stimulus by a combination of oxidized arachidonic acid with glutathione, lipoxygenase inhibitors and agents that lower intracellular glutathione concentrations interfere with its biosynthesis.[12,18] In addition, the role of prostaglandins in the response can be evaluated by the use of a pharmacologic inhibitor of their biosynthesis such as indomethacin. SRS biosynthesis is unaffected or even increased by indomethacin under conditions in which prostaglandin biosynthesis is largely or completely inhibited. Therefore, if the spasmogenic activity is unaffected or increased by concentrations of indomethacin of 1 μM or higher, major contributions of prostaglandins in the spasmogenic response are

[18] C. W. Parker, C. M. Fischman, and H. J. Wedner, *Proc. Natl. Acad. Sci. U.S.A.* **77,** 6876 (1980).

unlikely. Nonetheless, direct documentation that the PGH synthase (e.g., indomethacin) inhibitor is indeed blocking prostaglandin biosynthesis is obviously highly desirable.

Bioassay of Partially Purified SRS

Even though most studies will not require that individual samples be routinely purified before analysis for SRS activity, it is important to show initially that the activity under study does indeed fractionate as expected for SRS. Therefore, in addition to the criteria described above, partial purification of the spasmogenic activity before bioassay is an important aspect of any SRS study.

For example, Sephadex LH-20 absorption chromatography can be used to separate almost completely SRS from prostaglandins in crude lung extracts.[19] Two milliliters of SRS-containing medium that has been freed of lung fragments by centrifugation (or extracted by the addition of 8 ml of 95% ethanol, centrifuged to remove precipitated proteins and concentrated back to 2 ml) are added to a tube containing 6 ml of *n*-butanol and 50 mg of Sephadex LH-20. The mixture is evaporated under reduced pressure at 30–55° on a rotatory evaporator. When the volume has been reduced to 1 ml, 2 ml of ethyl acetate and 1 ml of *n*-butanol are added. The resulting suspension is transferred to a 0.9 × 10 cm silanized glass column that gives a column of resin 0.8 × 8 cm. The gel is successively washed with 3 ml of ethyl acetate–*n*-butanol (1 : 1), 6 ml of *n*-butanol, and 4 ml of 80% ethanol. Water is added to the individual eluates, and the organic solvent is removed in a vacuum at 50°. The SRS is eluted in the 80% ethanol fraction, whereas all the prostaglandins elute in earlier fractions (Fig. 4).

Another useful procedure for initial fractionation of samples from heterogeneous tissues, is to apply the SRS in aqueous solution to small (2 ml) Amberlite XAD-7 columns, wash with water, and recover the SRS-rich fraction by elution with 80% (v/v) ethanol. As described in this volume [54] in connection with SRS purification of RBL-1 cells, this column removes inorganic salts and water-soluble spasmogens, although the prostaglandins and other fatty acids are eluted with SRS.

Regardless of the initial fractionation procedure, HPLC is highly desirable for further analysis of the sample because of its very high resolving power. For relatively crude SRS samples, we prefer the methanol–water and neutral pH HPLC buffer system in conjunction with Varian Micropak monomeric 300 × 4 *mm* C_{18} analytical column as described for RBL-1 cells (this volume [54]). Indeed we have been able to load unpurified RBL-1 cell supernatants directly on HPLC columns in the neutral pH HPLC system, recover most or all of the SRS activity, and verify that the

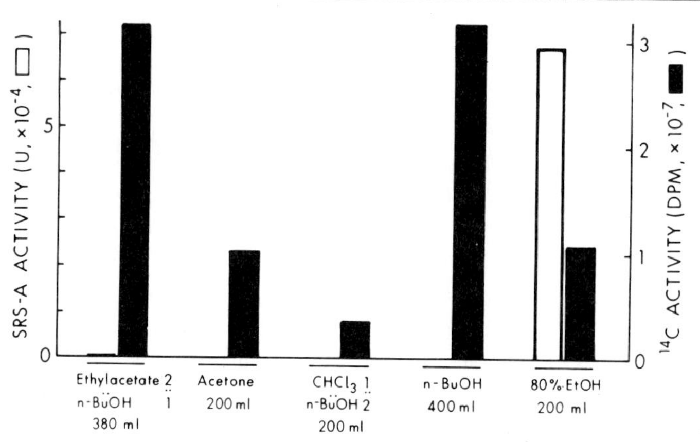

FIG. 4. Absorption chromatography of slow-reacting substance (SRS) from antigenically stimulated guinea pig lung fragments prelabeled with [1-^{14}C]arachidonic acid on a Sephadex LH-20 column. The prostaglandins and unmetabolized arachidonic acid are eluted almost exclusively in the first four column fractions. For further details see text. Taken from Watanabe-Kohno *et al.*[19]

spasmogenic activity elutes at exactly the same position as purified SRS. Direct HPLC of crude cell supernatants in acidic solvent systems is less desirable, since unpurified SRS solutions frequently lose substantial amounts of activity when they are maintained at acid pH. The possibility that an activity is something other than SRS can be largely eliminated using a suitably selected HPLC column and is almost completely excluded with two HPLC columns used in combination.

The major drawback to partial purification of SRS before analysis is the possible loss of activity during purification. However, even in a heterogeneous tissue such as the chopped guinea pig lung, with precautions to minimize oxidation we have been able to purify SRS to apparent homogeneity and still recover at least 50–60% of the original SRS activity.[19] Obviously, at this stage of our understanding of SRS metabolism, it must be kept in mind that even relatively minor structural variations in SRS molecules may affect their HPLC behavior, and it cannot necessarily be assumed that all naturally occurring SRSs will correspond to the major ones that have already been defined (LTC_4, LTD_4, LTE_4, and their 11-*trans* isomers). Nonetheless, there is already reason to believe that these will be the predominant, if not the sole, SRS species. While SRSs with different double-bond configurations or sets of substitution by hydroxy groups or sulfur may quite conceivably arise either naturally or artificially in general, these products are likely to have considerably reduced biological ac-

[19] S. Watanabe-Kohno and C. W. Parker, *J. Immunol.* **125,** 946 (1980).

tivities and will not contribute importantly to overall spasmogenic activity. If an activity elutes differently from known SRSs, the burden of proof is on the investigator to establish that a member of the SRS family is indeed involved. Of course, if an unnatural substrate such as either 5,8,11- or 5,8,11,14,17-eicosanoic acid rather than arachidonic acid is used during SRS generation, a difference in chromatographic behavior is to be expected.

Other Pharmacologic Effects of SRS

Other pharmacologic effects of SRS, such as those on vascular permeability, tracheal–bronchial muscle reactivity, or systemic blood pressure, may also be studied and may prove to be of value in the pharmacologic characterization of SRS. However, at the present time the standard bioassay system for SRS is the guinea pig ileal system, and it remains to be established that these other systems would be as reproducible or as relatively free from interference with other spasmogenic agents. Even the guinea pig ileal system is subject to considerable variation, and careful standardization and repetition of results is necessary for the verification of experimental observations.

Acknowledgments

This work was supported by U.S. Public Health Service Research Grant (CIRID) 1 P50 AI 15322, National Research Service Awards 5 T32 AI 00112 07, and the Howard Hughes Medical Institute at Washington University.

Author Index

Numbers in parentheses are reference numbers and indicate that an author's work is referred to although the name is not cited in the text.

A

B

C

D

E

F

I

J

K

L

M

N

O

P

Q

R

S

T

Y

Z

Subject Index

A

B

C

D

I

N

O

P

R

S

T

U

V

W